Classical and Relativistic Rational Extended Thermodynamics of Gases

Prof. Masaru Sugiyama (*left*) and Prof. Tommaso Ruggeri (*right*). Oberwolfach, March 2014 (@MFO; source http://owpdb.mfo.de/detail?photo_id=18572)

Tommaso Ruggeri • Masaru Sugiyama

Classical and Relativistic Rational Extended Thermodynamics of Gases

Springer

Tommaso Ruggeri
Department of Mathematics and Research
Center on Applied Mathematics
University of Bologna
Bologna, Italy

Masaru Sugiyama
Nagoya Institute of Technology
Nagoya, Japan

ISBN 978-3-030-59143-4 ISBN 978-3-030-59144-1 (eBook)
https://doi.org/10.1007/978-3-030-59144-1

This Springer imprint is published by the registered company Springer Nature Switzerland AG.
The registered company address is: Gewerbestrasse 11, 6330 Cham, Switzerland

To Ester and Tomoko
Francesca, Federico and Yutaka,
Kyeongjae, Azusa
Sofia and Humiya

Preface

The book aims to present the recent developments in Rational Extended Thermodynamics (RET) of gases in both classical and relativistic cases. The RET theory is a phenomenological field theory being capable of describing thermally nonequilibrium phenomena with steep gradients and rapid changes in space-time out of local equilibrium.

This book is the natural continuation of our previous book "Rational Extended Thermodynamics beyond the Monatomic Gas" published in 2015. However, the present book cannot be regarded as a second edition of the previous one because more than half part of this book deals with the new topics that are not covered by the previous title, and moreover the other half was revisited and streamlined to a considerable degree. We have reviewed some recent results also in the monatomic-gas case both in classical and relativistic frameworks. And, the relativistic RET with and without internal structure of a molecule is also explained extensively. Furthermore, we show some attempts to extend the RET theory from rarefied gas to dense gas. Another important point discussed here is the classical limit of the relativistic theory that gives us an optimal choice of moment variables. Also, the mixture theory with multi-temperature was largely revisited. We have added a relativistic model and the recent new results concerning the analogy between the model of mixture of gases and the Cucker-Smale model for flocking problem. The methodology of RET has been consistently adopted throughout both the previous and present books, but now the validity range of RET becomes much wider than that of the previous one. For these reasons, we have decided to adopt the new title for the present book.

The features of this book are summarized as follows:

- We firstly present a large introductory chapter that summarizes the problems in nonequilibrium thermodynamics and gives an overview of the book.
- One chapter is devoted to the mathematical structure of RET. The qualitative analysis of the differential system is made by taking into account the fact that, due to the convexity of the entropy, there exists a privileged field (*main field*)

by which the system becomes symmetric hyperbolic. The existence of the global smooth solution and the convergence to equilibrium are also studied.

- We summarize the mathematical aspects of nonlinear wave propagation phenomena. In particular, we discuss the theory of shock-wave phenomena.
- We discuss the old and new results in classical and relativistic RET of monatomic gases, which are necessary to understand the new progress in RET.
- We present a hyperbolic theory, that is, the RET theory of polyatomic rarefied gases with 14 fields (hereafter referred to as ET_{14}), which, in the parabolic limit, reduces to the Navier-Stokes and Fourier theory. The singular monatomic-gas limit of the theory is also considered.
- The ET_{14} theory gives us a complete phenomenological model, but its differential system is rather complex. For this reason, we have constructed a simplified theory with 6 fields (ET_6). This simplified theory preserves the main physical properties of the more complex ET_{14} theory. This is useful, in particular, when the bulk viscosity plays a more important role than the shear viscosity and the heat conductivity. This situation is observed in many gases such as rarefied hydrogen gas and carbon dioxide gas in some temperature ranges. This model is interesting because it is also valid in a situation far from equilibrium.
- In many experimental data of sound wave, the relaxation times of the rotational mode and of the vibrational mode of a molecule are quite different from each other. Therefore, in a case with high temperature where both rotational and vibrational modes exist, more than one molecular relaxation processes should be taken into account to make the RET theory more realistic. We construct new RET models (ET_7 and ET_{15}).
- We present a theory of *molecular* ET with an arbitrary number of field variables by using the method of closure based on both the maximum entropy principle and the entropy principle. We prove that the two closures are equivalent to each other.
- We present some typical applications of the RET theory: sound wave, shock wave, nozzle flow, acceleration wave, light scattering, heat conduction, and fluctuation. We compare the theoretical predictions with experimental data.
- Some recent attempts to construct a RET theory of dense gases are presented.
- A relativistic theory of polyatomic gases is presented. Its classical and ultra-relativistic limits are also discussed. In particular, the classical limit gives us the precise structure of hierarchies of moments.
- A theory of mixtures of gases with multi-temperature is presented together with a natural definition of the average temperature in both classical and relativistic regimes.
- An analogy between behaviors of mixture of gases in spatially homogeneous case and collective behaviors of many-body systems, e.g., aggregation of bacteria, flocking of birds, swarming of fish, herding of sheep, synchronous flashing of fireflies, and synchronization of coupled cells, is presented. In addition, we show the extension of the Cucker-Smale flocking model, that is, thermodynamical Cucker-Smale model and relativistic Cucker-Smale model.
- We summarize open problems and provide an outlook on future studies.

This book is designed for applied mathematicians, physicists, and engineers. We hope that the methodology presented can offer powerful models for possible applications to, for example, re-entry of a satellite into the atmosphere of a planet, semiconductors, microflows, and nanoscale phenomena.

Bologna, Italy
Nagoya, Japan
July 2020

Tommaso Ruggeri
Masaru Sugiyama

Acknowledgments

We are particularly indebted to our coauthors, Takashi Arima, Francesca Brini, Andrea Mentrelli, Sebastiano Pennisi, and Shigeru Taniguchi, who read and criticized the manuscript and continuously helped us improving almost all parts of the book.

Our perspectives on the field of Rational Extended Thermodynamics have been strongly influenced by the pioneering works of Ingo Müller, who introduced both of us to this beautiful research field.

The topic of this book belongs to three different research fields: thermomechanics of continuous media, kinetic theory of gases, and hyperbolic systems of balance laws. Therefore many colleagues, coauthors, and collaborators have affected our research works directly by common papers and discussions or indirectly through their papers. We sincerely thank all of them, although the list of their names is too long to write down here. But, we would like to mention Guy Boillat who made fundamental works on the mathematical problems, which are intimately related to the book.

We are also grateful to the Mathematical Research Institute of Oberwolfach that gave us the opportunity to work together on this book under the *Research in Pair* program in 2014 and 2018.

We want to thank Springer, in particular, the Executive Editor Dr. Francesca Bonadei and Dr. Michela Castrica for their help.

Contents

Symbols

$A^{\alpha\alpha_1\cdots\alpha_n}$	Moment tensor in relativistic RET
$A^{\alpha\beta\gamma}$	Triple tensor in relativistic RET
$A_{(ij)}$	Symmetric part of a tensor
$A_{[ij]}$	Antisymmetric part of a tensor
$A_{\langle ij\rangle}$	Deviatoric tensor
$D = 3 + f^i$	Degrees of freedom of a molecule
$F^{\alpha_1\alpha_2\ldots\alpha_N}$	Relativistic moment tensor of order N
F_A, F_{iA}, P_A	Densities, Fluxes, and Productions with a multi-index
$F_{k_1k_2\ldots k_N}$	Momentum-like field (or moment) of order N
G	Newton's gravitational constant
$G(t)$	Acceleration amplitude in an acceleration wave
$G_{kkk_1k_2\ldots k_M}$	Energy-like field (or moment) of order M
I	Extra variable in the distribution function of a polyatomic gas
I^R	Extra variable due to molecular rotational mode
I^V	Extra variable due to molecular vibrational mode
$I^{\alpha_1\cdots\alpha_n}$	Production tensor in relativistic RET
$I^{\beta\gamma}$	Production term in relativistic RET with 14 fields
$J_{\alpha\beta}$, $J_{i\alpha\beta}$	Symmetric matrices appearing in the MEP procedure
$K_m(x)$	Modified Bessel functions
K_n	Knudsen number
M_0	Mach number
$P^{\alpha_1\alpha_2\ldots\alpha_N}$	Relativistic production tensor of order N
$P_{k_1k_2\ldots k_N}$	Momentum-like production of order N
Q	Collision term or quadratic form for convexity of entropy
Q_i	Extra flux vector in a mixture of ET_6 gases
$Q_{kkk_1k_2\ldots k_M}$	Energy-like production term of order M
R	Scalar curvature
$R_{\mu\nu}$	Ricci curvature tensor
S	Dynamic structure factor or entropy density in a mixture
T	Absolute temperature
$T^{\alpha\beta}$	Energy-momentum tensor

$T_A^{\alpha\beta}$	Energy-momentum tensor in a relativistic mixture
U^α	Four-velocity
V^α	Particle flux
V_A^α	Particle flux in a relativistic mixture
Z, Ω	Partition functions
$Z = \Pi/p$	Ratio of dynamic pressure and equilibrium pressure
Δ, δ	Energy differences in a molecular mode
$\Delta^K, \Delta^R, \Delta^V$	Energy differences between nonequilibrium and local equilibrium energies
Γ	Gamma function or Lorentz factor
Γ_A	Lorentz factor of the component A in a relativistic mixture
Ω	Region occupied by a continuous medium in R^3
Π	Dynamic pressure
Π_θ	Dynamic pressure due to the difference of temperatures
$\Pi_{k_1 k_2 \ldots k_M}$	Dynamical pressure tensor in many moments
$[\![\cdot]\!]$	Jump across wave front
Σ	Entropy production or surface of a domain
Θ_α	Diffusion temperature
Θ_α	Diffusion temperature flux
α	Attenuation factor in linear waves or index of the component of mixture
$\alpha = (D-5)/2$	Power exponent related to the degrees of freedom of a molecule
α_λ	Attenuation per wavelength
$\bar{\Sigma}$	Entropy production of a principal subsystem
$\bar{h}^\alpha$	Entropy four-vector of a principal subsystem
$\boldsymbol{\Phi}^i$	Non-convective flux vector
$\bullet = d/dt$	Material time derivative
χ	Variable of the distribution function in the MEP procedure
δ	Shock thickness
δ_{ij}	Kronecker's delta
γ	Ratio of specific heats or ratio $(mc^2)/(k_B T)$
$\hat{P}_{ll}^{\mathfrak{a}}, \hat{P}_{ll}^{\mathfrak{b}}, \hat{P}_{ll}^{\mathfrak{c}}$	Production terms for $(\mathfrak{a}, \mathfrak{b}, \mathfrak{c})$ = (V,K,R), (R,K,V), (K,R,V)
$\hat{\boldsymbol{\Phi}}^i$	Non-convective flux at zero velocity
$\hat{\mathbf{F}}^0$	Density vector at zero velocity
$\hat{\mathbf{F}}^i$	Flux vector at zero velocity
$\hat{\mathbf{f}}$	Production at zero velocity
$\hat{c}_v$	Dimensionless specific heat at constant volume
$\hat{h}^\alpha$	Entropy four-vector at zero velocity
$\hat{h}'^\alpha$	Potential four-vector at zero velocity
κ	Thermal conductivity (heat conductivity)
λ	Characteristic velocity or one of the Lagrange multipliers
$\lambda, \lambda_i, \lambda_{ij}, \mu_i$	Lagrange multipliers
$\lambda^{(k)}$	k-characteristic velocity
λ_{max}	Maximum characteristic velocity

$\mathbf{A}^D$	Deviatoric part of a tensor
$\mathbf{A}^T$	Transpose of a tensor
$\mathbf{A}^r$	Matrices that appear in the Galilean invariance
$\mathbf{C} \equiv (C_i)$	Peculiar velocity
$\mathbf{D}$	Strain velocity tensor
$\mathbf{F}^0$	Density vector
$\mathbf{F}^i$	Flux vector
$\mathbf{J}_\alpha$	Diffusion flux
$\mathbf{X}$	Matrix that dictates the velocity dependence in the Galilean invariance
$\mathbf{c} \equiv (c_i)$	Molecular velocity
$\mathbf{d}$	Right characteristic eigenvector
$\mathbf{f}$	Production vector
$\mathbf{l}$	Left characteristic eigenvector
$\mathbf{m}_\alpha$	Production due to interchange of momentum in a mixture
$\mathbf{n}$	Unit normal vector
$\mathbf{q} \equiv (q_i)$	Heat flux
$\mathbf{t} \equiv (t_{ij})$	Stress tensor
$\mathbf{u}'$	Main field
$\mathbf{v} \equiv (v_i)$	Velocity
$\mathbf{v}'$	First block of the main field vector corresponding to conservation laws
$\mathbf{w}'$	Second block of the main field vector corresponding to balance laws
$\mathbf{x} \equiv (x_i)$	Position vector
$\mathscr{A}$	Amplitude of an acceleration wave or affinity in the Meixner theory
$\mathscr{D}$	Thermal diffusion
$\mathscr{E}$	Specific internal energy in the Meixner theory
$\mathscr{G}$	Chemical potential in nonequilibrium
$\mathscr{H}$	Hamiltonian of a gas system
$\mathscr{H}^I$	Hamiltonian of molecular internal mode
$\mathscr{H}^K$	Hamiltonian of molecular translational motion
$\mathscr{H}^R$	Hamiltonian of molecular rotational mode
$\mathscr{H}^V$	Hamiltonian of molecular vibrational mode
$\mathscr{H}^{K+U}$	Hamiltonian of molecular translational motion and inter molecular potential
$\mathscr{P}$	Nonequilibrium pressure or pressure in the Meixner theory
$\mathscr{P}(\lambda)$	Characteristic polynomial
$\mathscr{S}$	Specific entropy density in the Meixner theory
$\mathscr{T}$	Temperature in nonequilibrium
$\mathfrak{t}, \mathfrak{t}_{\langle ij \rangle}, \mathfrak{s}_i$	Random forces
μ	Shear viscosity, Lagrange multiplier, or ratio of molecular masses
$\mu^{(\bar{k})}$	$\bar{k}$-characteristic velocity of the equilibrium subsystem

∇	Nabla vector
ν	Bulk viscosity
ω	Frequency
ω^{α}	Interchange of dynamic pressure in a mixture of ET_6 gases
$\omega_{cs}(\mathbf{x}, \mathbf{y})$	Interaction weight function in Cucker-Smale model
$\partial_\alpha \equiv \partial/\partial x^\alpha$	Relativistic notation of space-time derivative
$\partial_i \equiv \partial/\partial x_i$	Spatial derivative
$\partial_t \equiv \partial/\partial t$	Time derivative
$\mathscr{F}$	Total density vector in a mixture
$\mathscr{F}_i$	Total flux vector in a mixture
$\mathscr{G}_i$	Intrinsic flux vector in a mixture
$\mathscr{P}$	Total production vector in a mixture
ψ	Weighting function
ρ	Mass density
σ^{rr}	Radial viscous stress component
$\sigma_{\langle ij\rangle}$	Shear stress (deviatoric part of the viscous stress tensor)
σ_{ij}	Viscous stress tensor
τ_α	Production due to interchange of mass in a mixture
τ, τ_δ	Relaxation times
τ_Π	Relaxation time for the dynamical pressure
τ_σ	Relaxation time for the shear (deviatoric viscous) stress tensor
τ_q	Relaxation time for the heat flux
θ^I	Nonequilibrium temperature for molecular internal mode
θ^K	Nonequilibrium temperature for molecular translational mode
θ^R	Nonequilibrium temperature for molecular rotational mode
θ^V	Nonequilibrium temperature for molecular vibrational mode
θ^{K+U}	Nonequilibrium temperature for molecular kinetic and potential energies
ε	Specific internal energy density
ε^I	Specific internal energy density due to molecular internal mode
ε^K	Specific internal energy density due to molecular translational mode
ε^R	Specific internal density due to molecular rotational mode
ε^V	Specific internal density due to molecular vibrational mode
ε^{K+U}	Specific internal energy density due to molecular kinetic and potential energies
φ	Weighting function or ratio of bulk and shear viscosities
φ_i or $\hat{h}^i$	Entropy flux at zero velocity
$\lVert\cdot\rVert_2$	L^2 norm
$\xi, \xi^{(1)}, \xi^{(2)}$	Internal variables in the Mexiner theory
c	Light speed or concentration
c_0, c_s, a_0	Sound velocity in an equilibrium state
c_v	Specific heat at constant volume
c_v^I	Specific heat at constant volume due to molecular internal mode

c_v^K	Specific heat at constant volume due to molecular translational mode
c_v^R	Specific heat at constant volume due to molecular rotational mode
c_v^V	Specific heat at constant volume due to molecular vibrational mode
c_v^{K+U}	Specific heat at constant volume due to molecular kinetic and potential energies
e	Relativistic energy density
e_{Synge}	Relativistic Synge energy
e_α	Production due to interchange of energy in a mixture
f	Distribution function
f^i	Internal degrees of freedom of a molecule
$f^{(G)}$	Grad distribution function
$f^{(I)}$	Distribution function for the internal mode I
$f^{(M)}$	Maxwellian distribution function
$f_N^{(\alpha)}$	Truncated distribution function with truncation index N and order α
f_E, $f^{(E)}$	Equilibrium distribution function for a polyatomic gas
f_J	Jüttner distribution function
f_N	Truncated distribution function with truncation index N
$f_{KR}^{(E)}$	Distribution function defined in (8.33)
$f_{KV}^{(E)}$	Distribution function defined in (8.36)
$f_K^{(E)}$	Distribution function defined in (7.39) or (8.31)
$f_{RV}^{(E)}$	Distribution function defined in (8.38)
g	Chemical potential
$g^{\alpha\beta}$	Inverse metric tensor
$g_{\alpha\beta}$	Metric tensor
h^0 or h	Entropy density
h^α	Entropy four-vector
h^i	Entropy flux
$h^{\alpha\beta}$	Projector
h'^α	Potential four-vector
k	Nonequilibrium entropy density or complex wave number
k_B	Boltzmann constant
m	Atomic mass
n	Particle number density
p	Equilibrium pressure
p^K	Pressure due to molecular translational mode
p^{K+U}	Pressure due to molecular kinetic and potential energies
p^α	Contravariant four-momentum
p_α	Covariant four-momentum
q^α	Heat flux four-vector
q^r	Radial heat flux component
s	Shock velocity or specific entropy density in equilibrium

s^I	Specific entropy density due to molecular internal mode
s^K	Specific entropy density due to molecular translational mode
s^R	Specific entropy density due to molecular rotational mode
s^V	Specific entropy density due to molecular vibrational mode
s^{K+U}	Specific entropy density due to molecular kinetic and potential energies
t	Time
$t^{\langle\alpha\beta\rangle}$	Relativistic viscous deviatoric stress
t_{cr}	Critical time
u'_A	Main field multi-index components
$u', u'_\alpha, u'_{\alpha\beta}$	Relativistic main field
u_α	Diffusion velocity
v_{ph}	Phase velocity
$\boldsymbol{\xi} \equiv (\xi_i)$	Molecular velocity

Chapter 1
Introduction and Overview

> Quelli che s'innamorano di pratica senza scienza son come il nocchiere, che entra in naviglio senza timone o bussola, che mai ha certezza dove si vada.
>
> *Leonardo Da Vinci*

Abstract In this chapter, before going into details, we give an overview of the present book, starting with a short history of nonequilibrium thermodynamics and the upbringing of Rational Extended Thermodynamics (RET) of rarefied monatomic gases. The new version of RET includes the 14-field theory of rarefied polyatomic gases that reduces to the classical Navier–Stokes and Fourier theory in the parabolic limit (*Maxwellian iteration*), to the 13-field RET theory of monatomic gases in a singular monatomic-gas limit, and to the RET theory with 6 fields as a subsystem. The 6-field theory is the minimal dissipative system, where the dissipation is only due to the dynamic pressure, after the Euler system of perfect fluids. For rarefied polyatomic gases, we discuss a theory of *molecular* ET with an arbitrary number of field variables by using the method of closure based on either the maximum entropy principle or the entropy principle. It can be proved that the two methods are equivalent to each other. Several applications of the RET theory of polyatomic gases are reviewed as well.

In the case of high temperature where both molecular rotational and vibrational modes exist, these modes should be taken into account individually to make the RET theory more realistic. A relativistic theory of polyatomic gas is presented, and its classical and ultra-relativistic limits are also discussed. In particular, the classical limit gives a precise structure of hierarchies of moments. We discuss also some recent tentatives for constructing a phenomenological RET theory of dense gases.

Moreover, in both classical and relativistic frameworks, we discuss the theory of a mixture of gases with multi-temperature, i.e., a mixture in which each component has its own temperature. We also show an analogy between the behavior of mixture

T. Ruggeri, M. Sugiyama, *Classical and Relativistic Rational Extended Thermodynamics of Gases*, https://doi.org/10.1007/978-3-030-59144-1_1

of gases and the collective behavior of many-body system, and extend the Cucker-Smale flocking model to the model with the temperature field and also to the model in the relativistic framework.

The qualitative analysis of the differential system is also done by taking into account the fact that, due to the convexity of the entropy, there exists a privileged field (*main field*) such that the system becomes symmetric hyperbolic. The existence of global smooth solutions and the convergence to equilibrium are also discussed.

1.1 Dawn of Thermodynamics

The nineteenth century witnessed the birth of thermodynamics and its great developments. Thermodynamics was expected to address the demands of the new industrial mode of production with revolutionary new technology such as a steam engine that transforms heat into work. Many scientists made fundamental contributions to this new discipline: Carnot, Mayer, Joule, Helmholtz, Clausius, Kelvin (Thomson), Maxwell, Gibbs, Duhem, Nernst, Carathéodory and many others.

Three fundamental laws of thermodynamics have been established. The first is the law of conservation of energy for thermodynamic systems. The second law selects appropriate evolution of a system by introducing a physical quantity called entropy. The third law is concerned with the value of the entropy at zero absolute temperature. In this way, we recognize the existence of a universal competition between energy and entropy in nature [1, 2]. On the basis of these laws, theory of thermodynamics has been successfully constructed, and has been utilized in various theoretical fields and diverse practical applications. See, for examples, the text books [3–6].

Thermodynamics, in particular the second law of thermodynamics, raised the following question: Can thermodynamics be traced back, in some way, to mechanics? Thermodynamics predicts the so-called arrow of time (irreversibility), but the equations of motion for molecules in mechanics have a strict symmetry with respect to the time reversal (reversibility). If thermodynamics is a branch of mechanics, how does the irreversibility emerge? Boltzmann tried to answer to this fundamental problem by using the newly developed H-theorem in the kinetic theory. However, it was revealed that Boltzmann had introduced implicitly an assumption, sometimes called the assumption of molecular chaos, into the collision term in his equation (Boltzmann equation) and had broken the time reversal symmetry.

Thermodynamic studies focusing especially on irreversible processes were commenced also in the nineteenth century. For example, Fourier's law of heat conduction, Navier–Stokes' law of viscous flow, Fick's law of mass diffusion, Ohm's law of electrical conduction, and thermo-electrical coupling effects such as the Seebeck effect and the Peltier effect were found empirically.

From the pioneering works of Onsager, Eckart, Meixner, Prigogine, and others, *thermodynamics of irreversible processes* (TIP) was established as a systematic nonequilibrium thermodynamic theory in the middle of the twentieth century. For its details, see the monumental book by de Groot and Mazur [7]. The Navier–Stokes and Fourier (NSF) theory for viscous and heat-conducting fluids [7, 8], for example, can be regarded as one of the typical TIP theories. TIP has been useful in various practical situations involving nonequilibrium processes such as mass diffusion, viscous flow, heat conduction, chemical reaction, electrical conduction. If we want to analyze nonequilibrium phenomena keeping its validity range in mind, TIP continues to be useful.

1.2 Physics of Macroscopic Systems

It is well known that there exist three levels of description for macroscopic physical systems, i.e., systems composed of huge number of particles such as molecules, atoms, ions, electrons:

- **Macroscopic level:** We use the *thermodynamic description* for systems evolving in *irreversible* processes. Examples are descriptions of viscous fluid flow by using the Navier–Stokes equation, and of heat conduction by using the energy conservation law with Fourier's law.
- **Mesoscopic level:** For rarefied gases, for example, we have the *kinetic-theoretical description* based on the Boltzmann equation that governs the time-evolution of the one-body distribution function. Another example is the description of the Brownian motion of a small particle immersed in a liquid by using the stochastic differential equations (for example, Langevin equation) or the Fokker-Planck equation.
- **Microscopic level:** We have the *statistical-mechanical description* of Hamiltonian systems, which have the *time-reversal symmetry*. The distribution function (or density matrix) is governed by the Liouville(-von Neumann) equation.

To establish the links among these three levels in a rigorous way has been a notoriously hard task. It still remains an open problem, and is related to the sixth Hilbert problem posed by David Hilbert in his famous speech at the Mathematical Congress in Paris (1900). His problem concerns the role of mathematics in physics: *Mathematical Treatment of the Axioms of Physics*, in which he wrote:

> .....Thus Boltzmann's work on the principles of mechanics suggests the problem of developing mathematically the limiting processes, there merely indicated, which lead from the atomistic view to the laws of the motion of continua.

We quote here only a few of the contributions presented recently on the sixth Hilbert problem: Saint-Raymond [9], Slemrod [10], and Gorban and Karlin [11].

1.2.1 Links Among the Three Levels of Description

We discuss some possible links among the three levels of description by considering specific examples. Figure 1.1 summarizes the contents of this subsection (see for more details [12]).

- **From discrete particle model to continuum model:** No complete answer has been given until now to this question [13]. In reality, Morrey in 1955 [14] tried to show the logical steps necessary to prove that the Euler equations for a compressible fluid can be derived from Newton's law of a particle-system under a suitable scaling procedure. The main difficulty in the mathematical problem in Morrey's program is that we do not know the behavior of the Hamiltonian system with a huge number of degrees of freedom for a long time span and no global regular solution exists for the Euler system.
- **From particle model to Boltzmann equation:** Grad in 1949 [15] posed a question: In what kind of limiting process the description given by the Boltzmann equation is expected to hold? In modern terminology, the Boltzmann—Grad limit is the one among the so-called large-scale limits applied to the derivation of kinetic equations from microscopic dynamics of particles. The asymptotic solution of the Cauchy problem for the BBGKY hierarchy for many-particle systems interacting via short-range potentials in the Boltzmann-Grad limit can be described by a certain hierarchy of equations called Boltzmann hierarchy.

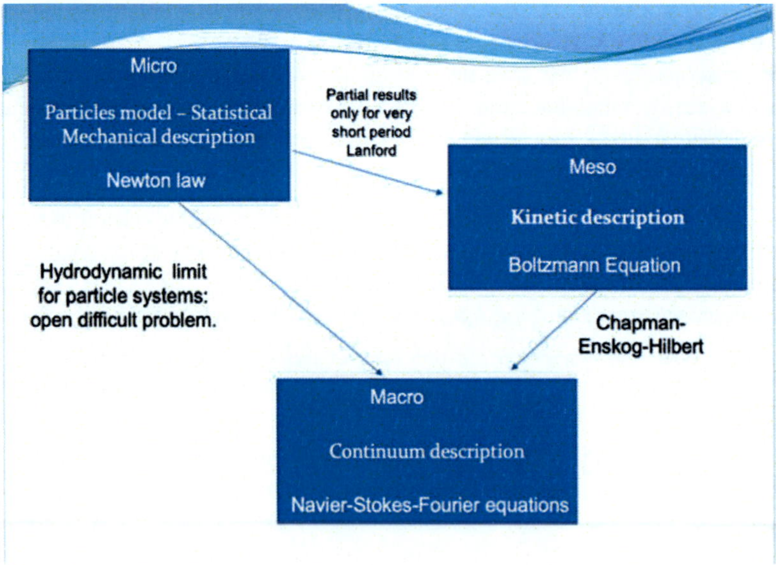

Fig. 1.1 Three different levels of description for macroscopic physical systems: micro-, meso-, and macro-scopic levels, and possible links among the three

Lanford, in 1983, wrote a Lecture Note "*On a derivation of the Boltzmann equation*" [16] and he proved a theorem which states that the Boltzmann equation gives an approximate description of the behavior of a system of classical particles interacting via short-range interactions. The description becomes exact in the limit: the number of particles becomes infinite and the range of the inter-particle forces goes to zero. It is pointed out, however, as follows: the above-mentioned theorem says only that the Boltzmann equation holds for a very short time span, no larger than one-fifth of the mean free time!

The Boltzmann equation, despite its enormous success, has been beset by a number of problems. Its mathematical properties still remain obscure. In particular, no satisfactory global existence theorem has been proved so far.

- **From Boltzmann equation to continuum theory:** On a formal level, within the context of the Boltzmann equation, the behavior in the hydrodynamic scale was analyzed first by Hilbert himself who developed an important technical tool, so-called Hilbert expansion, to approach the problem. Later Chapman and Enskog deduced the NSF theory [17]. Nishida [18] and Caflisch [19] derived the Euler equation, which is valid for a short period (before the occurrence of shocks).

Rational Extended Thermodynamics (RET) is, of course, concerned with the macroscopic level of description. However, it must be emphasized that, by using the Molecular Extended Thermodynamic (molecular ET) theory of rarefied gases, we can study also the link between the mesoscopic kinetic level and the macroscopic thermodynamic level. Indeed this extended thermodynamic theory is, within rarefied gases, a generalization of the NSF theory of a viscous and heat-conducting gas, and it has an intimate relationship with the moment theory of the Boltzmann equation, which is, as mentioned above, a typical equation in the mesoscopic level.

Before going into the detailed discussion of RET, we describe a brief history on nonequilibrium thermodynamics starting with TIP. Then we proceed to the first tentative of an extended thermodynamic theory, and lastly to the modern approach of RET. We give also the introductory perspective on the main topics of the present book without going into their details. The motivation of the present book, that is, the construction of the RET theory valid for more large class of materials such as polyatomic gases, dense gases, and mixtures of gases with multi-temperature is emphasized.

1.3 Thermodynamics of Irreversible Processes

In this section, we review briefly the theoretical structure of TIP through studying one-component viscous and heat-conducting fluids [20]. See also its historical explanation in [4]. The empirical Navier–Stokes and Fourier laws are naturally derived from TIP. We also point out its validity range.

In thermodynamics of continuous media, time-evolution of relevant densities is expressed in *balance form*: Let $\mathbf{F}^0(\mathbf{x}, t)$ be an $\mathbb{R}^N$-vector of the densities depending on the space variable $\mathbf{x} \equiv (x_i) \in \Omega \subset \mathbb{R}^3$ and the time $t \in \mathbb{R}^+$, then we have

$$\frac{d}{dt}\int_{\Omega} \mathbf{F}^0 \, d\Omega = -\int_{\Sigma} \boldsymbol{\Phi}^i n_i d\Sigma + \int_{\Omega} \mathbf{f} d\Omega, \tag{1.1}$$

where the first term on the right-hand side represents the inflows with the fluxes $\boldsymbol{\Phi}^i \in \mathbb{R}^N$ $(i = 1, 2, 3)$ through the surface Σ of Ω with unit outward normal vector $\mathbf{n} \equiv (n_i)$,[1] while the second term represents the productions with the production densities $\mathbf{f} \in \mathbb{R}^N$ in Ω. Under suitable regularity assumption, the system (1.1) can be put in local form:

$$\frac{\partial \mathbf{F}^0}{\partial t} + \frac{\partial \mathbf{F}^i}{\partial x^i} = \mathbf{f}, \qquad \mathbf{F}^i = \mathbf{F}^0 v_i + \boldsymbol{\Phi}^i, \tag{1.2}$$

where $\mathbf{v} \equiv (v_i)$ is the velocity.

Hereafter, let us study a one-component fluid as a typical example. These fields should satisfy the conservation law of mass, and the balance laws of momentum and energy:

$$\begin{aligned}
&\frac{\partial \rho}{\partial t} + \frac{\partial \rho v_i}{\partial x_i} = 0,\\
&\frac{\partial \rho v_j}{\partial t} + \frac{\partial}{\partial x_i}\left(\rho v_i v_j - t_{ij}\right) = \rho b_j,\\
&\frac{\partial}{\partial t}\left(\frac{\rho v^2}{2} + \rho\varepsilon\right) + \frac{\partial}{\partial x_i}\left\{\left(\frac{\rho v^2}{2} + \rho\varepsilon\right) v_i - t_{ik} v_k + q_i\right\} = \rho b_j v_j + r,
\end{aligned} \tag{1.3}$$

where $\rho, t_{ij}, \varepsilon$, and q_i $(i, j = 1, 2, 3)$ are mass density, stress tensor, specific internal energy, and heat flux, respectively, and $v^2 = v_i v_i$. The production terms are specific external body force b_j and heat supply r. If b_j and r are null, the system represents the conservation laws of mass, momentum, and energy. The system (1.3) is a particular case of (1.2) with

$$\mathbf{F}^0 = \begin{pmatrix} \rho \\ \rho v_j \\ \frac{1}{2}\rho v^2 + \rho\varepsilon \end{pmatrix}, \quad \mathbf{F}^i = \begin{pmatrix} \rho v_i \\ \rho v_i v_j - t_{ij} \\ \left(\frac{1}{2}\rho v^2 + \rho\varepsilon\right) v_i - t_{ik} v_k + q_i \end{pmatrix}, \quad \mathbf{f} = \begin{pmatrix} 0 \\ \rho b_j \\ \rho b_j v_j + r \end{pmatrix}. \tag{1.4}$$

As usual the stress tensor $t_{ij} (= -p\delta_{ij} + \sigma_{ij})$ can be decomposed into an isotropic part and a deviatoric part (symmetric traceless part) denoted by the brackets $\langle\ \rangle$:

$$t_{ij} = -(p + \Pi)\delta_{ij} + \sigma_{\langle ij \rangle}, \tag{1.5}$$

[1] We adopt the summation convention, i.e., we take summation over repeated indexes: $i, j = 1, 2, 3$.

where p is the equilibrium pressure, δ_{ij} is Kronecker's delta, σ_{ij} is the viscous stress tensor, and $\Pi(= -\sigma_{ll}/3)$ is the so-called dynamic pressure. The deviatoric tensor $\sigma_{\langle ij \rangle}$ is called shear stress tensor. As is well known the previous system (1.3) can be rewritten, for classical solutions, in the form (we omit both the body force and the heat supply from now on):

$$\begin{aligned} \dot{\rho} + \rho \frac{\partial v_j}{\partial x_j} &= 0, \\ \rho \dot{v}_i - \frac{\partial t_{ij}}{\partial x_j} &= 0, \\ \rho \dot{\varepsilon} - t_{ij} \frac{\partial v_i}{\partial x_j} + \frac{\partial q_i}{\partial x_i} &= 0, \end{aligned} \tag{1.6}$$

where a dot on a quantity, say $w(x_i, t)$, stands for the material time derivative operator:

$$\dot{w} = \frac{dw}{dt} = \frac{\partial w}{\partial t} + v_i \frac{\partial w}{\partial x_i}. \tag{1.7}$$

The system (1.3) or (1.6) has 5 equations with 14 field variables. Therefore we need appropriate constitutive relations so as to close the system. Let us study this problem as follows:

First of all, we recall the important work made by the mathematician Constantin Carathéodory with his *axiomatic approach to thermodynamics*. He postulated the so-called *principle of inaccessibility* that states [21] (see also [22, 23]): *In the neighborhood of any equilibrium state of a system (of any number of thermodynamic coordinates), there exist states that are inaccessible by reversible adiabatic processes*. Starting with this axiom and using some properties of Pfaffian forms, Carathéodory introduced the absolute temperature T as an integrating factor and justified rigorously the Gibbs equation in equilibrium thermodynamics:

$$T\,ds = d\varepsilon - \frac{p}{\rho^2}\,d\rho, \tag{1.8}$$

where s is the specific entropy density .

In a nonequilibrium case, TIP adopts a crucial assumption that, even though a system is globally out of equilibrium, the relation (1.8) is still valid in a small volume element. This is sometimes called *assumption of local equilibrium*. As for the validity criterion for the assumption, see, for example, the reference [24]. Then we have the relation:

$$\dot{s} = \frac{1}{T}\left(\dot{\varepsilon} - \frac{p}{\rho^2}\dot{\rho}\right). \tag{1.9}$$

Elimination of $\dot{\varepsilon}$ and $\dot{\rho}$ by using (1.6) gives, after some calculations, the relation:

$$\rho\dot{s} + \frac{\partial}{\partial x_i}\left(\frac{q_i}{T}\right) = \frac{1}{T}\sigma_{\langle ij\rangle}\frac{\partial v_{\langle i}}{\partial x_{j\rangle}} - \frac{1}{T}\Pi\frac{\partial v_i}{\partial x_i} - \frac{1}{T^2}q_i\frac{\partial T}{\partial x_i},$$

or equivalently:

$$\frac{\partial \rho s}{\partial t} + \frac{\partial}{\partial x_i}\left(\rho s v_i + \frac{q_i}{T}\right) = \frac{1}{T}\sigma_{\langle ij\rangle}\frac{\partial v_{\langle i}}{\partial x_{j\rangle}} - \frac{1}{T}\Pi\frac{\partial v_i}{\partial x_i} - \frac{1}{T^2}q_i\frac{\partial T}{\partial x_i}, \tag{1.10}$$

which can be seen as a balance equation of the entropy. Then we may have the following interpretation:

(intrinsic) entropy flux; $\varphi_i = \frac{q_i}{T}$,
entropy production : $\Sigma = \frac{1}{T}\sigma_{\langle ij\rangle}\frac{\partial v_{\langle i}}{\partial x_{j\rangle}} - \frac{1}{T}\Pi\frac{\partial v_i}{\partial x_i} - \frac{1}{T^2}q_i\frac{\partial T}{\partial x_i}$.

The entropy production is decomposed into the three products of the dissipative flux and the thermodynamic force:

Dissipative flux	Thermodynamic force	
shear stress $\sigma_{\langle ij\rangle}$	deviatoric velocity gradient	$\frac{1}{T}\frac{\partial v_{\langle i}}{\partial x_{j\rangle}}$,
dynamic pressure Π	divergence of velocity	$-\frac{1}{T}\frac{\partial v_i}{\partial x_i}$,
heat flux q_i	temperature gradient	$-\frac{1}{T^2}\frac{\partial T}{\partial x_i}$.

1.3.1 Laws of Navier–Stokes and of Fourier

From the second law of thermodynamics, the entropy production Σ must be non-negative. Assuming linear relations between the dissipative fluxes and the thermodynamic forces, we have the constitutive equations (phenomenological equations):

$$\begin{aligned}
\sigma_{\langle ij\rangle} &= 2\mu\frac{\partial v_{\langle i}}{\partial x_{j\rangle}} \qquad && \mu \geq 0,\\
\Pi &= -\nu\frac{\partial v_i}{\partial x_i} && \nu \geq 0,\\
q_i &= -\kappa\frac{\partial T}{\partial x_i} && \kappa \geq 0.
\end{aligned} \tag{1.11}$$

These are known as the laws of Navier–Stokes and of Fourier with μ and ν being the shear and bulk viscosities and κ the thermal conductivity. All of these coefficients may be functions of ρ and T.

Along with the thermal and caloric equations of state: $p = p(\rho, T), \varepsilon = \varepsilon(\rho, T)$, relations (1.11) are adopted as the constitutive equations of TIP. And the differential system (1.3) or (1.6) is closed, i.e., 5 equations for 5 unknowns.

1.3.2 Parabolic Structure and the Prediction of Infinite Wave-Speed in TIP

TIP raises the problem of infinite speed of waves. Historically, in order to avoid this unphysical prediction of TIP, extended thermodynamics (ET) [25] was conceived.

Let us, therefore, discuss firstly this problem through studying heat conduction in a fluid at rest with a constant mass density ρ (heat conduction in a rigid heat conductor). Inserting the Fourier law $(1.11)_3$ into the energy balance equation $(1.6)_3$, we obtain the equation of heat conduction:

$$\frac{\partial T}{\partial t} = \mathscr{D} \Delta T, \tag{1.12}$$

where

$$\mathscr{D} = \frac{\kappa}{\rho c_v} \quad \text{and} \quad c_v = \left(\frac{\partial \varepsilon}{\partial T}\right)_\rho$$

are, respectively, the thermal diffusion coefficient and the specific heat at constant volume. We assume, for simplicity, that $\mathscr{D}$ is constant.

The solution of (1.12) in an unbounded domain with an initial value $T(\mathbf{x}, 0)$ is given by

$$T(\mathbf{x}, t) = \frac{1}{(4\pi \mathscr{D} t)^{3/2}} \int_{-\infty}^{\infty} T(\mathbf{y}, 0) e^{-\frac{(\mathbf{y}-\mathbf{x})^2}{4\mathscr{D} t}} d\mathbf{y}. \tag{1.13}$$

We notice that $T(\mathbf{x}, t)$ is nonzero for any $\mathbf{x}$ if $t > 0$ even though the initial value $T(\mathbf{x}, 0)$ is nonzero only in a bounded domain. This phenomenon has been sometimes called paradox, because the temperature propagates with infinitely large speed. From a mathematical point of view, this is due to the parabolic character of the basic equation (1.12).

The assertion of infinite speed is, of course, beyond the validity range of TIP. It cannot describe properly such a rapid change because it is based on the local equilibrium assumption. On the other hand, it is well known that equations (1.12) and (1.13) have been utilized quite successfully in various practical situations.

Indeed, if we take carefully its validity range into account, and if we do not care about its unphysical predictions, we would have useful results from a practical point of view.

There is, however, a situation where the infinite speed should be avoided strictly. It is found in a relativistic thermodynamic case where the propagation speed of a wave should be less or equal to the light speed.

How can we construct a new thermodynamic theory that predicts only finite speeds of waves by generalizing TIP?

1.4 Cattaneo Equation

A heuristic argument made by Cattaneo [26] gives us an interesting suggestion for the question above.

The heat flux vector $\mathbf{q}$ is assumed, by Cattaneo, to be proportional not only to the gradient of temperature ∇T but also to the gradient of the time derivative of the temperature, $\nabla \dot{T}$, in such a way that

$$\mathbf{q} = -\kappa \left(\nabla T - \tau \nabla \dot{T} \right), \tag{1.14}$$

where the dot indicates now the time derivative. When τ is small, the operator in (1.14) seems to be approximated as follows:

$$\left(1 - \tau \frac{d}{dt} \right)^{-1} \approx 1 + \tau \frac{d}{dt}. \tag{1.15}$$

Then (1.14) is rewritten as

$$\tau \dot{\mathbf{q}} + \mathbf{q} = -\kappa \nabla T, \tag{1.16}$$

which is known as (classical) Cattaneo equation. It reduces to the Fourier law when $\tau \to 0$. Combining (1.16) with the energy equation $(1.6)_3$ in the present case:

$$\rho c_v \dot{T} + \operatorname{div} \mathbf{q} = 0, \tag{1.17}$$

we obtain

$$\tau \ddot{T} + \dot{T} = \mathscr{D} \Delta T. \tag{1.18}$$

This equation, called *telegraph equation*, is hyperbolic provided that $\tau > 0$. It predicts the propagation of heat pulses with finite speed:

$$V = \sqrt{\frac{\mathscr{D}}{\tau}}.$$

If the relaxation time τ is negligibly small, Eq. (1.18) reduces to the heat conduction equation (1.12). However, we should be careful about this because it is a singular limit process from hyperbolic to parabolic equation.

What are the essential points in Cattaneo's heuristic argument for constructing a new thermodynamic theory RET? These may be the following two:

1. Dissipative fluxes, such as the heat flux, should be introduced as independent variables into the theory in addition to the usual thermodynamic quantities in TIP.
2. Relaxation processes of these dissipative fluxes should be properly taken into consideration in order to make the theory hyperbolic.

1.4.1 Weak Points of Cattaneo Equation, Generalized Cattaneo Equation, and Second Sound

Although the Cattaneo equation was obtained by a simple heuristic argument, it has become a popular model in applications concerning rigid heat conductors. Huge literature on this subject exists. See, for example, the review papers of Joseph and Preziosi [27, 28], and the recent book of Straughan [29].

In reality, however, Cattaneo's argument has several weak points. Firstly, there is no justification of the assumption (1.14) in mesoscopic physics. Secondly, the approximation (1.15) is not well justified mathematically. In fact, the system composed of (1.17) and (1.14) is not hyperbolic and its solutions are unstable, while the system composed of (1.17) and (1.16) is hyperbolic! Thirdly, it is simple to verify that the classical Cattaneo equation with internal energy depending only on the temperature T is compatible with the entropy principle if and only if the ratio between the relaxation time and the heat conductivity is proportional to T^2. In fact, Coleman, Fabrizio, and Owen [30] noticed that, in order to be compatible with the entropy principle for a generic relaxation time and the heat conductivity, the internal energy should depend not only on T but also on the square of $\mathbf{q}$.

The assertion of Coleman, Fabrizio, and Owen was, however, criticized by Morro and Ruggeri [31], who revisited a previous paper of Ruggeri [32]. And they proposed a more general model than the Cattaneo model by adopting a system of equations of balance type according with the philosophy of RET. In particular, they proposed the following generalized Cattaneo equation:

$$\frac{\partial \alpha(T)\mathbf{q}}{\partial t} + \nabla\gamma(T) = -\nu\mathbf{q}, \tag{1.19}$$

where scalar functions $\alpha(T), \gamma(T)$, and $\nu(T)$ are uniquely determined by the entropy principle and the knowledge of second sound velocity as a function of the temperature. Equation (1.19) reduces to the Cattaneo equation in the case that the *thermal inertia* α is constant and we put $\tau = \alpha/\nu$ and $\kappa = \gamma\,'/\nu$ (the prime indicates the derivative with respect to the temperature). The generalized Cattaneo equation overcomes the above-mentioned weak points and is compatible with the internal energy depending only on the temperature T. The justification of the thermal inertia α was given in [33] in which a unified theory between phonon and superfluid helium was proved using a mixture theory of two fluids.

Therefore the Cattaneo equation is physically meaningful near equilibrium where only linear terms in nonequilibrium variables are present in the field equations. A recent paper of Spigler [34] gives a connection between the original Cattaneo proposal (1.14) and the Morro–Ruggeri equation (1.19). Also Sellitto, Zampoli, and Jordan recently compared a non-local model [35] with the one given in [31].

The importance of hyperbolic equations has been backed up by the experimental evidence in the so-called second sound, i.e., a heat wave. The second sound was observed first in liquid helium at a temperature below the λ-point [36], and then its existence in crystals was predicted [37]. Recently sophisticated experiments observing the propagation of a heat wave in an ultra-cold quantum gas were made [38]. In the generalized Cattaneo model (1.19), we can discover interesting new phenomena in crystals such as the passage from hot to cold shocks at a critical temperature [39, 40], the behavior of a simple wave, and shock formation [41].

Moreover from a mathematical point of view, the system is no more genuine nonlinear and is linearly degenerate only in a hyper-surface of the configuration space. This requires a different approach for what concerns the qualitative analysis [42].

On the basis of RET, we can show later that the Cattaneo equation is valid only for rigid heat conductors. Moreover it is not a constitutive equation (see Remark 1.2 in Sect. 1.8.3 and also Sect. 33.2) although many authors have thought it in this way. But, in the case of gas, it should be replaced by a balance equation with more complex structure.

Finally we recall that, starting from the pioneering works of Guyer and Krumhansl [43, 44], an interesting approach to this problem on the basis of the phonon-gas theory was taken by Dreyer and Struchtrup [45], by Larecki and Banach [46, 47], and recently by Cao et al. [48].

1.5 First Tentative of Extended Thermodynamics and Its Limitations

The first tentative approach to Extended Thermodynamics (ET) of a viscous and heat-conducting rarefied gas, made by Müller [49] in a classical context and by Müller (Ph.D. thesis) and independently by Israel [50] in a relativistic context, is based on the modification of the Gibbs equation (1.8) where the effects of dissipative fluxes are also taken into account. This point of view has been adopted by several authors and is a starting point of Extended Irreversible Thermodynamics (EIT), which has gained popularity through the book of Jou et al. [51]. See also the book [52].

The introduction of dissipative fluxes as independent variables together with the usual thermodynamic quantities into the theory has a deep implication. It is evident that ET has gone beyond the local equilibrium assumption because dissipative fluxes play an essential role to characterize a nonequilibrium state in the theory.

The applicability range of ET becomes, in this way, wider than that of TIP. In other words, ET is applicable to highly nonequilibrium phenomena where the local equilibrium assumption is no longer valid. Examples of such phenomena are shock waves, micro- and nano-flows, second sounds, light scattering, and so on. Thus ET is expected to be useful not only for relativistic cases but also for non-relativistic cases far from equilibrium.

Such an approach was, however, criticized by Ruggeri [53] because the entropy production depends strongly on the choice of the entropy flux, and there appear different field equations for a different entropy flux. Moreover the differential system is not a priori in the form of balance type. This implies, from a mathematical point of view, that it is not possible to define weak solutions and therefore it is impossible to study, in particular, shock waves. Then Ruggeri, by using the theory of symmetrization, proposed a model prescribed by a system of balance type. This model is symmetric hyperbolic for any fields. It satisfies the entropy principle and the convexity condition of the entropy. And the model reduces to the Navier–Stokes and Fourier model when relaxation times are small. Granted that Müller appreciated the innovative ideas of Ruggeri about the structure of balance equations and a symmetric hyperbolic system, he criticized that the system obtained by Ruggeri is not justified by the kinetic theory. Therefore Müller proposed a revision of ET [54] together with Liu, where the structure of the balance laws dictated by the moment theory associated with the Boltzmann equation was adopted as a starting point of their theory. Then, in a relativistic framework, a similar theory was constructed by Liu et al. [55]. This new approach was named Rational Extended Thermodynamics (RET). And the main results obtained at that time were summarized in the book by Müller and Ruggeri [25, 56].

As will be explained below, the mathematical framework of RET of rarefied monatomic gases consists of a hierarchy of balance laws. The same hierarchical structure can be seen in the system of moment equations in the kinetic theory with the truncation at some arbitrary order of moments. The RET theory, in this respect, resembles to the theory of moment equations. However, the closure of RET is achieved by means of the universal principles of continuum theories: *objectivity principle*, *entropy principle*, and *principle of causality and stability*.

1.6 Rational Thermodynamics and the Entropy Principle

Before we present the details of Rational Extended Thermodynamics, we need to recall the approach of the so-called *Rational Thermodynamics* (RT). RT is mainly due to the Truesdell school [57] and has the starting point in the fundamental paper of Coleman and Noll [58] where they reinterpreted the second law of thermodynamics as a selection rule for constitutive equations. We have seen that the basic system of equations of continuous media is in the form of balance laws (1.1) or (1.2). To close the system, constitutive equations are necessary. The idea of Coleman and Noll is to regard the entropy law:

$$\frac{\partial \rho s}{\partial t} + \frac{\partial}{\partial x_i}\left(\rho s v_i + \frac{q_i}{T}\right) \geqslant 0$$

as a constraint for the admissible physical constitutive equations. In this way, the prescription of the arrow of time that characterizes the irreversibility is made indirectly. In fact, if we pick up some constitutive equations, we are not sure whether the second law is certainly satisfied or not. Therefore, according with this new idea, we select at the beginning the class of admissible constitutive equations that satisfy the inequality for any initial and boundary data. This was a very important observation and is also very useful as we will see in Sect. 2.2 where we will characterize, in more precise way, the entropy principle. In passing, we have seen before that, from (1.10), how the Navier–Stokes and Fourier constitutive equations (1.11) are compatible with the entropy principle.

The limitation of the idea of Coleman and Noll is, however, in the postulation that the entropy flux is in the form of Clausius, i.e., the ratio of heat flux to temperature, q_i/T. Müller noticed with the help of kinetic-theoretical considerations that this requirement is too much restrictive, and he proposed to extend the entropy principle in such a way that both entropy density and entropy flux are regarded as constitutive quantities [59].

An interesting book based on the framework of mathematical methods of Rational Mechanics and Thermodynamics is the one by Šilhavý [60].

1.7 Other Approaches

We observe that, in addition to *Classical Irreversible Thermodynamics* (CIT or TIP), *Rational Thermodynamics* (RT), *Extended Irreversible Thermodynamics* (EIT) that we briefly summarized before, and *Rational Extended Thermodynamics* (RET), there are other approaches in nonequilibrium thermodynamics with exotic acronyms.

The most popular and interesting one may be GENERIC, an acronym for *General Equation for Non-Equilibrium Reversible-Irreversible Coupling* proposed by Grmela and Öttinger [61–63]. The general form of dynamic equation for a system with both reversible and irreversible dynamics is explored. See also the recent paper [64].

The relation between the Hamiltonian structure and the symmetric hyperbolic structure of the equations in nonequilibrium system was investigated recently by Peshkov et al. in [65]. This approach follows the idea of Godunov and Romenski based on symmetric hyperbolic systems that are derived from a variational principle [66]. Although, from a mathematical point of view, this approach seems to be similar to RET, the physical starting points are quite different from each other.

Another approach proposed by Müller, Reitebuch, and Weiss is the so-called *Consistent-Order Extended Thermodynamics* (COET) [67]. In this approach every nonequilibrium variable has a certain order of magnitude, and the set of variables adopted in the theory contains only variables up to a chosen order.

A different point of view of nonequilibrium thermodynamics based on a revisit of the entropy principle and on the introduction of the so-called *calortropy* is the subject of the books of Eu [68, 69].

Connection between nonequilibrium thermodynamics and variational principles is the subject of the book of Gyarmati [70].

A tentative list of some different approaches in nonequilibrium thermodynamics was presented by Cimmelli et al. (see [71] and references cited therein), and in the recent papers by Jou [72] and by Öttinger et al. [73]. An interesting review of nonequilibrium thermodynamics, elucidating different pathways to macroscopic equations, was published by Müller and Weiss [74].

Finally we recall that there exist parabolic-type approaches to nonequilibrium processes by using kinetic-theoretical considerations: the works by Torrilhon and Struchtrup [75, 76] via the regularization of moment-equations, and the works by Bobylev and Windfäll [77] through revisiting the Chapman–Enskog method.

1.8 Rational Extended Thermodynamics of Rarefied Monatomic Gas and the Kinetic Theory

The study of nonequilibrium phenomena in gases is particularly important. As explained in Sect. 1.2, we have two complementary approaches to rarefied gases, namely the *continuum approach* and the *kinetic approach*.

The continuum model consists in the description of the system by means of macroscopic equations (e.g., fluid-dynamic equations) obtained on the basis of conservation laws and appropriate constitutive equations. A typical example is the Navier–Stokes and Fourier model in TIP. The applicability of this classical macroscopic theory is, however, inherently restricted to a nonequilibrium state characterized by a small Knudsen number K_n, which is a measure to what extent the gas is rarefied:

$$K_n = \frac{\text{mean free path of molecule}}{\text{macroscopic characteristic length}}.$$

The transport coefficients associated to dissipation processes are not provided by the theory except for the sign. Usually we require experimental data on the coefficients.

The approach based on the kinetic theory [78–81] postulates that the state of a gas can be described by the velocity distribution function. The evolution of the distribution function is governed by the Boltzmann equation. The kinetic theory is applicable to a nonequilibrium state characterized by a large K_n, and the transport coefficients naturally emerge from the theory itself. The applicability range of the Boltzmann equation is limited to rarefied gases.

The RET theory, a generalization of the TIP theory, also belongs to the continuum approach but is applicable to a nonequilibrium state with larger K_n. In a sense, RET is a sort of bridge between TIP and the kinetic theory. An interesting point to be noticed is that, in the case of rarefied gas, there exists a common applicability range of the RET theory and the kinetic theory. Therefore, in such a range, the results from the two theories should be consistent with each other. Because of this, we can expect that the kinetic-theoretical considerations can motivate us to establish the mathematical structure of the RET theory.

1.8.1 Boltzmann Equation and the Moments

The kinetic theory describes a state of a rarefied gas by using the phase density (velocity distribution function) $f(\mathbf{x}, t, \mathbf{c})$, where $f(\mathbf{x}, t, \mathbf{c})d\mathbf{c}$ is the number density of (monatomic) molecules at the point $\mathbf{x}$ and time t that have velocities between $\mathbf{c}$ and $\mathbf{c} + d\mathbf{c}$. Time-evolution of the phase density is governed by the Boltzmann equation:

$$\partial_t f + c_i\, \partial_i f = Q, \tag{1.20}$$

where the right-hand side, the collision term, describes the effect of collisions between molecules. Here $\partial_t \equiv \frac{\partial}{\partial t}$ and $\partial_i \equiv \frac{\partial}{\partial x_i}$.

As is well known, almost all macroscopic thermodynamic quantities are identified as moments of the phase density:

$$F_{k_1 k_2 \cdots k_j} = \int_{\mathbb{R}^3} m f c_{k_1} c_{k_2} \cdots c_{k_j} d\mathbf{c}, \tag{1.21}$$

where m is the mass of a molecule. The moments satisfy a hierarchy of the balance laws in which the *flux* in one equation becomes the *density* in the next one:

$$
\begin{aligned}
&\partial_t F + \partial_i F_i = 0, \\
&\swarrow \\
&\partial_t F_{k_1} + \partial_i F_{ik_1} = 0, \\
&\swarrow \\
&\partial_t F_{k_1k_2} + \partial_i F_{ik_1k_2} = P_{\langle k_1k_2\rangle}, \\
&\swarrow \\
&\partial_t F_{k_1k_2k_3} + \partial_i F_{ik_1k_2k_3} = P_{k_1k_2k_3}, \\
&\vdots \\
&\partial_t F_{k_1k_2\cdots k_N} + \partial_i F_{ik_1k_2\cdots k_N} = P_{k_1k_2\cdots k_N}, \\
&\vdots
\end{aligned}
\tag{1.22}
$$

where $P_{\langle k_1k_2\rangle} = P_{k_1k_2} - \frac{1}{3}P_{ll}\delta_{k_1k_2}$ is the deviatoric part of the tensor $P_{k_1k_2}$ and

$$P_{k_1k_2\cdots k_j} = \int_{\mathbb{R}^3} mQc_{k_1}c_{k_2}\cdots c_{k_j}\,d\mathbf{c}.$$

Taking $P_{\langle kk\rangle} = 0$ into account, we notice that the first five equations for F, F_i, F_{kk} are exactly conservation laws. These equations correspond to the conservation laws (1.3) of mass, momentum, and energy.

Due to the structure of the hierarchy (1.22), we obtain, from $(1.22)_2$ and the trace of $(1.22)_3$, the relation $3(p+\Pi) = 2\rho\varepsilon$. As Π is a nonequilibrium variable vanishing in equilibrium, we have

$$p = \frac{2}{3}\rho\varepsilon \quad \text{and} \quad \Pi \equiv 0.$$

Then the gas under consideration is indeed monatomic, and the dynamic pressure vanishes identically.

For the quantities defined by

$$h^0 = -k_B \int_{\mathbb{R}^3} f \log f \, d\boldsymbol{c}, \quad h^i = -k_B \int_{\mathbb{R}^3} f \log f \, c^i \, d\boldsymbol{c} \tag{1.23}$$

with k_B being the Boltzmann constant, it is possible to prove the famous H-theorem as a consequence of the Boltzmann equation (1.20):

$$\partial_t h^0 + \partial_i h^i = \Sigma \geqslant 0. \tag{1.24}$$

This represents the balance law of entropy if we identify h^0, h^i, and Σ as the entropy density, the entropy flux, and the entropy production, respectively.

Remark 1.1 In reality, the expressions of the entropy density and the entropy flux (1.23) should be given by (see for example [25, 82])

$$h^0 = -k_B \int_{\mathbb{R}^3} f \log\left(\frac{f}{y}\right) d\boldsymbol{c}, \quad h^i = -k_B \int_{\mathbb{R}^3} f \log\left(\frac{f}{y}\right) c^i \, d\boldsymbol{c}, \tag{1.25}$$

where y is the constant having the same dimension of f:

$$y = e(2s+1)\frac{m^3}{h^3} \equiv eY$$

with h and s being the Planck constant and the spin quantum number of a particle, respectively. The constant y is due to the quantum effects: indistinguishability of identical particles, quantization of the phase-space, and the spin. We may rewrite (1.25) as follows:

$$h^0 = -k_B \int_{\mathbb{R}^3} f \left(\log\left(\frac{f}{Y}\right) - 1\right) d\boldsymbol{c}, \quad h^i = -k_B \int_{\mathbb{R}^3} f \left(\log\left(\frac{f}{Y}\right) - 1\right) c^i \, d\boldsymbol{c}. \tag{1.26}$$

We can see clearly that $(1.26)_1$ is the Boltzmann entropy [83]. And, in an equilibrium case, this gives exactly the same result (Sackur–Tetrode formula) derived from statistical mechanics. However, in many analyses below, the constant y has no essential role. This affects only the entropy constant. Therefore, throughout this book, keeping this Remark in mind, we will use the expressions (1.23) by putting formally "$y = 1$" unless otherwise mentioned.

1.8.2 *Closure of RET*

If we truncate the hierarchy (1.22) at the density with the tensorial order N:

$$\partial_t F + \partial_i F_i = 0,$$

$$\partial_t F_{k_1} + \partial_i F_{ik_1} = 0,$$

$$\partial_t F_{k_1k_2} + \partial_i F_{ik_1k_2} = P_{\langle k_1k_2\rangle},$$

$$\partial_t F_{k_1k_2k_3} + \partial_i F_{ik_1k_2k_3} = P_{k_1k_2k_3},$$

$$\vdots$$

$$\partial_t F_{k_1k_2\cdots k_N} + \partial_i F_{ik_1k_2\cdots k_N} = P_{k_1k_2\cdots k_N},$$

we encounter the *problem of closure* because the last flux and the productions are not in the list of the densities. The first idea of RET [25] was to consider the truncated system as a phenomenological system of continuum mechanics and then to regard the quantities out of the list as constitutive functions:

$$F_{k_1k_2\ldots k_Nk_{N+1}} \equiv F_{k_1k_2\ldots k_Nk_{N+1}}\left(F, F_{k_1}, F_{k_1k_2}, \ldots F_{k_1k_2\ldots k_N}\right),$$

$$P_{k_1k_2\ldots k_j} \equiv P_{k_1k_2\ldots k_j}\left(F, F_{k_1}, F_{k_1k_2}, \ldots F_{k_1k_2\ldots k_N}\right), \quad 2 \leqslant j \leqslant N.$$

The constitutive quantities at one point and time depend on the independent fields at the same point and time, i.e., they are local and instantaneous. As mentioned above, according to the continuum theory, the restrictions on the constitutive equations come from the *universal principles* (see Chap. 2 for details).

1.8.3 *Macroscopic Approach of RET of Monatomic Gas with 13 Fields*

The first attempt of RET was made in the 13-field case. Thirteen is a special number because, by comparison between the first 5 moments and conservation laws (1.28) below, it is possible to relate, in a unique way, the 13 moments $\{F, F_i, F_{ij}, F_{lli}\}$

to the 13 fields $\{\rho, v_i, t_{ij}, q_i\}$. Therefore, in the case of rarefied monatomic gas, the first thirteen moments have concrete physical meanings. The balance laws in this case are given by

$$\begin{aligned}
&\partial_t F + \partial_i F_i = 0, \\
&\partial_t F_{k_1} + \partial_i F_{ik_1} = 0, \\
&\partial_t F_{k_1k_2} + \partial_i F_{ik_1k_2} = P_{k_1k_2}, \\
&\partial_t F_{kkj} \;\; + \partial_i F_{kkij} \;\; = P_{kkj}.
\end{aligned} \tag{1.27}$$

The constitutive quantities (except for the velocity dependence) are $F_{i\langle k_1k_2\rangle}$, F_{lk_1kk}, $P_{\langle k_1k_2\rangle}$, and P_{kkj}.

The restrictions imposed by the universal principles—in particular the Galilean invariance and the entropy principle—are so strong that, at least for processes not too far from equilibrium, the system can be completely closed. In this case, the closed system is given, by employing the usual symbols, as [25, 54]:

$$\begin{aligned}
&\frac{\partial \rho}{\partial t} + \frac{\partial}{\partial x_k}(\rho v_k) = 0, \\
&\frac{\partial \rho v_i}{\partial t} + \frac{\partial}{\partial x_k}\left\{\rho v_i v_k + p\delta_{ik} - \sigma_{\langle ik\rangle}\right\} = 0, \\
&\frac{\partial}{\partial t}\left\{\rho v_i v_j + p\delta_{ij} - \sigma_{\langle ij\rangle}\right\} + \\
&\quad + \frac{\partial}{\partial x_k}\Big\{\rho v_i v_j v_k + p(v_i\delta_{jk} + v_j\delta_{ki} + v_k\delta_{ij}) - \sigma_{\langle ij\rangle}v_k - \sigma_{\langle jk\rangle}v_i - \sigma_{\langle ki\rangle}v_j \\
&\qquad + \frac{2}{5}(q_i\delta_{jk} + q_j\delta_{ki} + q_k\delta_{ij})\Big\} = \frac{\sigma_{\langle ij\rangle}}{\tau_\sigma}, \\
&\frac{\partial}{\partial t}\left\{\rho v^2 v_i + 2(\rho\varepsilon + p)\, v_i - 2\sigma_{\langle li\rangle}v_l + 2q_i\right\} + \\
&\quad + \frac{\partial}{\partial x_k}\Big\{\rho v^2 v_i v_k + p(v^2\delta_{ik} + 7v_i v_k) - \sigma_{\langle ik\rangle}v^2 - 2\sigma_{\langle li\rangle}v_l v_k - 2\sigma_{\langle lk\rangle}v_l v_i \\
&\qquad + \frac{4}{5}q_l v_l\delta_{ik} + \frac{14}{5}q_i v_k + \frac{14}{5}q_k v_i + 5\frac{p^2}{\rho}\delta_{ik} - 7\frac{p}{\rho}\sigma_{\langle ik\rangle}\Big\} \\
&= -2\frac{q_i}{\tau_q} + 2\frac{\sigma_{\langle ij\rangle}}{\tau_\sigma}v_j,
\end{aligned} \tag{1.28}$$

where τ_σ and τ_q are the relaxation times, which are related to the shear viscosity and the heat conductivity, respectively. Notice that the trace of Eq. (1.28)$_3$ is the energy conservation law, while the deviatoric part is the balance law for $\sigma_{\langle ij\rangle}$. Therefore the first five equations in (1.28) are the usual conservation laws of mass, momentum, and energy, while the remaining block represents the balance laws for the viscous stress σ_{ij} (in the present case of monatomic gas, $\sigma_{ij} = \sigma_{\langle ij \rangle}$) and for the heat flux q_i, respectively. They reduce to the Navier–Stokes and Fourier constitutive equations when the relaxation times are small [25].

We emphasize that we have assumed the system (1.27) that is motivated by the kinetic theory but, after that, the procedure in the construction of the theory has been carried out completely in a macroscopic way. In particular, we have not relied on the fact that the F's are related to the distribution function of the kinetic theory.

Remark 1.2 In the last equation of (1.28), we recognize the Cattaneo structure if we put $v_i = \sigma_{ij} = 0$. But as remarked in Sect. 1.4.1, the equation is more complex than the Cattaneo equation due to the interaction with the other fields. This is also evident in the general case of non-polytropic polyatomic gas (last equation of (6.33)). This is the proof that the Cattaneo equation is not a universal constitutive equation. The same is true for the Navier–Stokes and Fourier equations as we will see in Sect. 6.3.11 and in Chap. 33 in more general cases.

1.8.4 Grad Distribution

It is interesting to observe that the macroscopic universal principles give us the above result that is in perfect agreement with the kinetic-theoretical result derived from the Grad distribution function [84]. The idea of Grad is to use a perturbation method based on the expansion of the distribution function in terms of the Hermite polynomials around the Maxwellian distribution $f^{(M)}$:

$$f^{(M)} = \frac{\rho}{m} \sqrt{\frac{m}{2\pi k_B T}}^{\,3} e^{-\frac{m(C_1^2+C_2^2+C_3^2)}{2k_B T}}, \tag{1.29}$$

where $C_i = c_i - v_i$ is the peculiar velocity. He obtained the distribution function:

$$f^{(G)} = f^{(M)} \left\{ 1 - \frac{m}{2pk_B T} \sigma_{\langle ij \rangle} C_i C_j - \frac{1}{p} \frac{m}{k_B T} q_i C_i \left(1 - \frac{m}{5k_B T} C^2 \right) \right\}. \tag{1.30}$$

Inserting $f^{(G)}$ into the definition of the moments (1.21), we can accomplish the closure, and have the same result as that derived from the method of RET (1.28). This is the first success of RET because it is proved clearly that both the macroscopic approach and the kinetic Grad approach give the same result!

1.8.5 Closure via the Maximum Entropy Principle and Molecular ET of Monatomic Gases

The 13-moment theory has extensively shown its superiority over the Navier–Stokes and Fourier theory. However, in some situations, for example, high-frequency sound waves, light scattering with large scattering angle, shock waves of large Mach number, even the 13-moment theory cannot provide satisfactory results. In order to remedy this difficulty a larger number of moments are required. RET can provide a coherent theoretical framework for building theories with larger number of moments that give results in excellent agreement with experiments. In this sense, *RET is the theory of theories*. But, usually, it is too difficult to adopt the pure continuum approach for a system with such a large number of field variables.

Therefore it is necessary to recall that the field variables are the moments of a distribution function truncated at some order. And then the closure of the balance equations of the moments, which is known as the *maximum entropy principle* (MEP), should be introduced. This is the procedure of the so-called *molecular extended thermodynamics* (molecular ET). The principle of maximum entropy has its root in statistical mechanics. It is developed by Jaynes in the context of the theory of information based on the Shannon entropy [85, 86].

Nowadays the importance of MEP is recognized fully due to the numerous applications in many fields, for example, in the field of computer graphics.

MEP states that the probability distribution that represents the current state of knowledge in the best way is the one with the largest entropy. Another way of stating this is as follows: Take precisely stated prior data or testable information about a probability distribution function. Then consider the set of all trial probability distributions that would encode the prior data. Of those, one with maximal information entropy is the proper distribution, according to this principle. Concerning the applicability of MEP in nonequilibrium thermodynamics, this was originally given by Kogan in 1967 [87]. The same author noticed that Grad's distribution function maximizes the entropy. A precise equivalence between MEP and RET with 13 moments was proved for the first time by Dreyer in 1987 [88]. In this way the 13-moment theory can be obtained in three different ways: RET, Grad, and MEP. A remarkable point is that all closures are equivalent to each other!

The MEP procedure was applied, by Müller and Ruggeri in 1993 [56], also to the general case of any number of moments, where it was proved for the first time that the closed system is symmetric hyperbolic if one chooses the Lagrange multipliers as field variables. The MEP was proposed again and popularized three years later by Levermore [89]. The complete equivalence between the entropy principle and the MEP was finally proved by Boillat and Ruggeri [90]. Later MEP was formulated also in a quantum-mechanical context [91].

A comprehensive review of the state-of-the-art of the maximum entropy principle in both classical (MEP) and quantal (QMEP) formulations was presented by Trovato and Reggiani [92]. Interesting approach of MEP to radiation was given by Larecki and Banach [93, 94]. Some mathematical delicate questions concerning the domain of validity of the MEP was given in a series of papers by Junk and co-workers [95–97]. The validity of MEP was revisited by studying two extreme cases in the kinetic theory by Dreyer and Kunik [98]. The details on MEP in RET will be given in Chap. 4.

1.9 Rational Extended Thermodynamics of Rarefied Polyatomic Gas

Unfortunately the previous RET theory, being strictly connected with the kinetic theory, suffers from nearly the same limitations as the Boltzmann equation. Indeed, the previous RET is valid only for rarefied monatomic gases, where the specific internal energy ε and the pressure p are connected by the relation $2\rho\varepsilon = 3p$, and the dynamic pressure Π vanishes identically.

In the case of polyatomic gases, on the other hand, the rotational and vibrational degrees of freedom of a molecule, which are not present in monatomic gases, come into play in macroscopic phenomena [99]. From a mathematical standpoint, this effect is responsible for an intrinsic change in the structure of the system of field equations. A simple hierarchy of field equations as in the case of monatomic gas is no longer valid. In particular, the internal specific energy is no longer related to the pressure in a simple way.

1.9.1 Macroscopic Approach with 14 *Fields*

After some pioneering works [100–102], a *14-field* RET theory for rarefied polyatomic gases has been developed by Arima, Taniguchi, Ruggeri, and Sugiyama [103]. This theory adopts two parallel hierarchies (binary hierarchy) for the independent fields: the mass density, the velocity, the internal energy, the shear stress, the dynamic pressure, and the heat flux. One hierarchy consists of balance equations for the mass density, the momentum density and the momentum flux (*momentum-like* hierarchy), and the other one consists of balance equations for the energy density and the energy flux (*energy-like* hierarchy):

$$\begin{aligned}
&\partial_t F + \partial_i F_i = 0, \\
&\partial_t F_{k_1} + \partial_i F_{ik_1} = 0, \\
&\partial_t F_{k_1k_2} + \partial_i F_{ik_1k_2} = P_{k_1k_2}, \qquad \partial_t G_{ll} + \partial_k G_{llk} = 0, \\
&\qquad\qquad\qquad\qquad\qquad\qquad\quad\;\; \partial_t G_{lli} + \partial_k G_{llik} = Q_{lli},
\end{aligned} \tag{1.31}$$

where G_{ll} is the energy density, G_{llk} is the energy flux, G_{llik} is the flux of G_{lli}, and Q_{lli} is the production with respect to G_{lli}. These hierarchies cannot merge with each other in contrast to the case of rarefied monatomic gases because the specific internal energy (the intrinsic part of the energy density) is no longer related to the pressure (one of the intrinsic parts of the momentum flux) in a simple way.

By means of the closure procedure of the RET theory, the constitutive equations are determined explicitly by the thermal and caloric equations of state. Then, as will be explained in Chap. 6, we obtain the closed system for any non-polytropic gas. In particular, in the case of polytropic gas with the thermal and caloric equations of state given by

$$p = \frac{k_B}{m}\rho T \quad \text{and} \quad \varepsilon = \frac{D}{2}\frac{k_B}{m}T, \qquad (D = 3 + f^i) \tag{1.32}$$

where D is related to the degrees of freedom of a molecule given by the sum of the space dimension 3 for the translational motion and the contribution from the internal degrees of freedom $f^i\,(\geqslant 0)$, we have the following differential system [103]:

$$\begin{aligned}
&\frac{\partial \rho}{\partial t} + \frac{\partial}{\partial x_k}(\rho v_k) = 0,\\
&\frac{\partial \rho v_i}{\partial t} + \frac{\partial}{\partial x_k}\left\{\rho v_i v_k + (p+\Pi)\delta_{ik} - \sigma_{\langle ik\rangle}\right\} = 0,\\
&\frac{\partial}{\partial t}(\rho v^2 + 2\rho\varepsilon) + \frac{\partial}{\partial x_k}\left\{\rho v^2 v_k + 2(\rho\varepsilon + p + \Pi)v_k - 2\sigma_{\langle kl\rangle}v_l + 2q_k\right\} = 0,\\
&\frac{\partial}{\partial t}\left\{\rho v_i v_j + (p+\Pi)\delta_{ij} - \sigma_{\langle ij\rangle}\right\} +\\
&\quad + \frac{\partial}{\partial x_k}\left\{\rho v_i v_j v_k + (p+\Pi)(v_i\delta_{jk} + v_j\delta_{ki} + v_k\delta_{ij}) - \sigma_{\langle ij\rangle}v_k - \sigma_{\langle jk\rangle}v_i - \sigma_{\langle ki\rangle}v_j + \right.\\
&\qquad \left. + \frac{2}{D+2}(q_i\delta_{jk} + q_j\delta_{ki} + q_k\delta_{ij})\right\} = -\frac{\Pi\delta_{ij}}{\tau_\Pi} + \frac{\sigma_{\langle ij\rangle}}{\tau_\sigma},\\
&\frac{\partial}{\partial t}\left\{\rho v^2 v_i + 2\,(\rho\varepsilon + p + \Pi)\,v_i - 2\sigma_{\langle li\rangle}v_l + 2q_i\right\} +\\
&\quad + \frac{\partial}{\partial x_k}\left\{\rho v^2 v_i v_k + 2\rho\varepsilon v_i v_k + (p+\Pi)(v^2\delta_{ik} + 4v_i v_k) - \sigma_{\langle ik\rangle}v^2 - 2\sigma_{\langle li\rangle}v_l v_k - 2\sigma_{\langle lk\rangle}v_l v_i\right.\\
&\qquad + \frac{4}{D+2}q_l v_l\delta_{ik} + \frac{2D+8}{D+2}q_i v_k + \frac{2D+8}{D+2}q_k v_i + \frac{k_B}{m}T\left[(D+2)p + (D+4)\Pi\right]\delta_{ik} -\\
&\qquad \left. - \frac{k_B}{m}T(D+4)\sigma_{\langle ik\rangle}\right\} = -2\frac{q_i}{\tau_q} - 2\left(\frac{\Pi\delta_{ij}}{\tau_\Pi} - \frac{\sigma_{\langle ij\rangle}}{\tau_\sigma}\right)v_j,
\end{aligned} \tag{1.33}$$

where τ_σ, τ_Π, and τ_q are the relaxation times. The Navier–Stokes and Fourier theory is contained in the present theory as a limit of small relaxation times (the Maxwellian iteration [25, 104]) as shown in [103] (see also [105]). Details will be explained in Chap. 6, Sect. 6.3.11.

The validity of the theory has been confirmed by comparing its predictions to experimental data in the cases of ultrasonic waves [106, 107], light scattering [108], and shock waves [109, 110]. The application of the theory to heat conduction was considered as well [111]. See Part V for details.

1.9.2 Singular Limit from Polyatomic Gas to Monatomic Gas

Let us consider the limiting process from polyatomic gas to monatomic gas when we let D approach 3 (i.e., $D \to 3$) from above, where D is assumed to be a continuous variable. The limit is singular in the sense that the system for rarefied polyatomic gases with 14 independent variables seems to converge to the system with only 13 independent variables for rarefied monatomic gases. What is going on with the remaining one equation?

It is possible to prove that the 14 equations and the solutions of the system (1.33) converge to the 13 equations and solutions of monatomic gas (1.28) with $\Pi = 0$, respectively, provided that the data are compatible with a monatomic gas, i.e., $\Pi(\mathbf{x}, 0) = 0$ [112]. Details will be given in Sect. 6.5.

1.9.3 MEP Closure and the Molecular Approach for the 14-Moment Theory

Concerning the kinetic counterpart, a crucial step towards the development of the theory of rarefied polyatomic gases was made by Borgnakke and Larsen [113]. The distribution function is assumed to depend on an additional continuous variable representing the energy of the internal modes of a molecule in order to take into account the exchange of energy (other than translational one) in binary collisions. This model was initially used for Monte Carlo simulations of polyatomic gases, and later it was applied to the derivation of the generalized Boltzmann equation by Bourgat et al. [114].

As a consequence of the introduction of one additional parameter I, the velocity distribution function $f(t, \mathbf{x}, \boldsymbol{c}, I)$ is defined on the extended domain $[0, \infty) \times \mathbb{R}^3 \times \mathbb{R}^3 \times [0, \infty)$. Its rate of change is determined by the Boltzmann equation which has the same form as the one of monatomic gases (1.20) but the collision integral $Q(f)$ takes into account the influence of the internal degrees of freedom through the collisional cross section.

Pavić, Ruggeri, and Simić proved [115] that, by means of the maximum entropy principle (MEP), the kinetic model for rarefied polytropic polyatomic gases presented in [113] and [114] yields appropriate macroscopic balance laws. This is a natural generalization of the classical procedure of MEP from monatomic gases to polyatomic gases. They considered the case of 14 moments, and showed the complete agreement with the binary hierarchy (1.31). The moments are defined by

$$\begin{pmatrix} F \\ F_{i_1} \\ F_{i_1 i_2} \end{pmatrix} = \int_{\mathbb{R}^3} \int_0^{\infty} m \begin{pmatrix} 1 \\ c_{i_1} \\ c_{i_1} c_{i_2} \end{pmatrix} f(t, \mathbf{x}, \mathbf{c}, I)\, \varphi(I)\, dI\, d\mathbf{c},$$
$$\begin{pmatrix} G_{pp} \\ G_{ppk_1} \end{pmatrix} = \int_{\mathbb{R}^3} \int_0^{\infty} m \begin{pmatrix} c^2 + 2\frac{I}{m} \\ \left(c^2 + 2\frac{I}{m}\right) c_{k_1} \end{pmatrix} f(t, \mathbf{x}, \mathbf{c}, I)\, \varphi(I)\, dI\, d\mathbf{c}. \tag{1.34}$$

The weighting function $\varphi(I)$ is determined in such a way that it recovers the caloric equation of state in equilibrium for polyatomic gases.

Later, using the MEP, Ruggeri studied the closure in the most complicated case of non-polytropic gas, in which the internal energy is a nonlinear function of the temperature [116]. Therefore, also for rarefied polyatomic gases, the closure procedures of RET and MEP give the same result for any non-polytropic gas. And, in the polytropic case, also the Grad procedure is equivalent to the above two!

Remark 1.3 For a model of polyatomic gas, Gamba and Pavić-Čolić recently established existence and uniqueness theory in the space homogeneous setting for the full nonlinear case under suitable conditions [117]. Global existence and asymptotic behavior of classical solutions in the ellipsoidal BGK model for polyatomic molecules with the initial data given sufficiently close to a global Maxwellian were studied by Yun [118].

1.9.4 Molecular ET of Polyatomic Gases

For polyatomic gases, we can also carry out the closure of the system with generic number of moments. In this case, we use the same idea introduced in [115]. We consider a distribution function depending on the additional parameter I that takes into account the internal degrees of freedom of a molecule, and adopt the generalized hierarchy structure (1.34).

We can define a binary hierarchy of the moments in the following way [115, 119]: We consider now the same binary hierarchy of 14 moments but for a generic number of moments truncated, for the F-series, at the index of truncation N and, for the G-series, at the index M:

$$\partial_t F + \partial_i F_i = 0,$$
$$\partial_t F_{k_1} + \partial_i F_{ik_1} = 0,$$
$$\partial_t F_{k_1 k_2} + \partial_i F_{ik_1 k_2} = P_{k_1 k_2}, \qquad \partial_t G_{kk} + \partial_i G_{ikk} = 0,$$

$$
\begin{array}{ll}
\vdots & \partial_t G_{kkj_1} + \partial_i G_{kkij_1} = Q_{kkj_1}, \\
\vdots & \vdots \\
\partial_t F_{k_1k_2\dots k_N} + \partial_i F_{ik_1k_2\dots k_N} = P_{k_1k_2\dots k_N}, & \vdots \\
 & \partial_t G_{kkj_1j_2\dots j_M} + \partial_i G_{kkij_1j_2\dots j_M} = Q_{kkj_1j_2\dots j_M}.
\end{array} \tag{1.35}
$$

The truncation order N of the F-hierarchy and the order M of the G-hierarchy are a priori independent of each other. It is worth noting that the first and the second equations of the F-hierarchy represent the conservation laws of mass and momentum, respectively, while the first equation of the G-hierarchy represents the conservation law of energy, and that, in each of the two hierarchies, the flux in one equation appears as the density in the following equation—a feature in common with the single hierarchy of monatomic gases.

The Euler 5-moment system is a particular case of (1.35) with $N = 1, M = 0$. And the 14-moment system (1.31) is another particular case of (1.35) with $N = 2, M = 1$.

It was proved that the truncation indexes N and M of the two hierarchies are, in reality, not independent of each other because of physical reasons: (1) Galilean invariance of field equations and (2) the fact that the characteristic velocities depend on the degrees of freedom of a molecule. We arrived at the conclusion that the relation $M = N - 1$ should be satisfied [119].

The closure of the system was achieved by means of the maximum entropy principle, and it was also proved that, in the present case, this closure is equivalent to the closure by the entropy principle with concave entropy density. In this way, the system becomes symmetric when it is written in terms of the *main field* components, definition of which will be shown in Chap. 2.

The characteristic velocities in an equilibrium state are analyzed. These velocities play an important role in the following cases: the propagation of acceleration waves [120, 121], the determination of the phase velocity of linear waves in the high-frequency limit [122, 123], and the *subshock* formation [124].

With regard to this, it will be discussed how the characteristic velocities of the system depend on the internal degrees of freedom and on the order of the truncation of hierarchies. In particular, the two limit cases of monatomic gases and of a gas with infinite internal degrees of freedom will be investigated.

Finally, using the convexity arguments and the subcharacteristic conditions for the principal subsystems, the lower-bound estimate for the maximum characteristic velocity was obtained as follows:

$$
\frac{\lambda^{E,\max}_{(N)}}{c_0} \geqslant \sqrt{\frac{6}{5}\left(N - \frac{1}{2}\right)}, \qquad \left(c_0 = \sqrt{\frac{5}{3}\frac{k_B}{m}T}\right).
$$

This is independent of the degrees of freedom of a molecule. Noteworthy point is that this estimate is exactly the same as that for monatomic gases established by Boillat and Ruggeri [90]. This was recently used by Slemrod in his analysis of the hydrodynamic limit of the Boltzmann equation and Hilbert's sixth problem [10]. Therefore, also for polyatomic gases, the maximum characteristic velocity tends to be unbounded when the order of the hierarchies tends to infinity!

As in the case of 14 fields [112], it is possible to prove [125] that, in the limit $D \to 3$, the solutions of the system for rarefied polyatomic gases, which is composed of $\frac{1}{6}(N+1)(N+2)(2N+3)$ equations, converge to those of the system of rarefied monatomic gases, which is composed of $\frac{1}{6}(N+1)(N^2+8N+6)$ equations. This subject will be the contents of Chap. 10.

1.9.5 Six-Field Theory and the Meixner Theory of a Relaxation Process

The 14-field RET theory (ET_{14}) gives us a complete phenomenological model but its differential system is rather complex. For this reason we construct a simplified RET theory with 6 fields. This simplified theory preserves the main physical properties of the more complex 14-field theory, when the bulk viscosity plays more important role than the shear viscosity and the heat conductivity. This situation is observed in many polyatomic gases at some temperature ranges [126]. This model is particularly interesting because, as seen later, it is also valid in a situation far from equilibrium.

In the 14-field theory, there exist three relaxation times τ_σ, τ_Π, and τ_q that characterize the relaxation of the shear stress, the dynamic pressure, and the heat flux, respectively. The relaxation times depend on the mass density and the temperature, and their magnitudes are usually comparable with each other. As mentioned above, the shear and bulk viscosities and the heat conductivity in the Navier–Stokes Fourier theory can be expressed in terms of these relaxation times.

It was revealed recently, however, by studying the dispersion relation of linear harmonic waves [106] and the shock wave structure [109, 127] that, in an appropriate temperature range of some polyatomic gases such as hydrogen gas or carbon dioxide gas, the relaxation time τ_Π is several orders larger than the other two relaxation times τ_σ and τ_q. In such a situation, the dynamic pressure relaxes very slowly compared with the relaxation of the shear stress and the heat flux. And the effect of the shear stress and the heat flux on the relaxation process is negligibly small.

In order to focus our attention on such slow relaxation phenomena, a simplified version of the ET_{14} theory, that is, a RET theory with 6 independent fields (ET_6) of the mass density, the velocity, the temperature, and the dynamic pressure was proposed [128, 132]. It was shown that the well-known Meixner theory of a relaxation process [129, 130] can be understood by the ET_6 theory.

Due to its simplicity, the ET_6 theory with a nonlinear constitutive equation was also developed [131–133]. It is noteworthy that the molecular ET_6 theory is consistent with a theory in the different approach where a polyatomic gas with discrete internal energy is treated as a sort of mixture of monatomic gases [134].

Details of the ET_6 theory will be presented in Chaps. 12 and 13.

1.9.6 ET_7 and ET_{15} of Polyatomic Gases with Two Molecular Relaxation Processes

The RET theory of polyatomic gases with the binary hierarchy has the limitation of its applicability, although the theory has been successfully utilized to analyze various nonequilibrium phenomena (see Part V). In fact, we have many experimental data showing that the relaxation times of the rotational mode and of the vibrational mode of a molecule are quite different to each other. In such a case, more than one molecular relaxation processes should be taken into account to make the RET theory more precise.

In order to describe the relaxation processes of rotational and vibrational modes separately, we decompose the energy of internal modes I as the sum of the energy of rotational mode I^R and the energy of vibrational mode I^V:

$$I = I^R + I^V.$$

Generalizing the Borgnakke-Larsen idea [113], we assume the same form of the Boltzmann equation (1.20) with a velocity distribution function that depends on these additional parameters, i.e., $f \equiv f\left(t, \mathbf{x}, \mathbf{c}, I^R, I^V\right)$. And we also take into account the effect of the parameters I^R and I^V on the collision term $Q(f)$. In this way we have a triple hierarchy. In the papers [135] and [136] we presented more refined versions, that is, ET_7 and ET_{15} respectively. The details of these new RET will be explained in Chaps. 8 and 14.

1.9.7 Applications of the RET Theories of Polyatomic Gases

Applications of the RET theories of a polyatomic gas to specific nonequilibrium phenomena have been made. In this section we review briefly some of them. Their detailed exposition will be made in Part V.

1.9.7.1 Dispersion Relation for Sound in Rarefied Diatomic Gases

The dispersion relation for sound in rarefied diatomic gases; hydrogen, deuterium, and hydrogen deuteride gases based on the ET_{14} theory was studied in detail [106]. The dispersion relation was compared with those obtained by experiments and by the Navier–Stokes and Fourier (NSF) theory. As is expected, the applicable frequency-range of the RET theory was shown to be much wider than that of the NSF theory. The values of the bulk viscosity and the relaxation times involved in nonequilibrium processes were evaluated. It was found that the relaxation time related to the dynamic pressure has a possibility to become much larger than the other relaxation times related to the shear stress and the heat flux. The isotope effects on sound propagation were also clarified [106, 107]. The analysis was made in the temperature range where the internal modes in a molecule play an important role.

Using the ET_{15} theory, we revisit the dispersion relation of a gas in a wider temperature range where molecular rotational and vibrational modes play significant roles individually. The cross effect among the molecular relaxations, shear viscosity, and heat conduction on the dispersion and absorption of sound wave was studied. The ET_7 theory was also utilized for the study in a low-frequency region because the difference between the results derived from the ET_{15} and ET_7 theories is useful for understanding the molecular relaxation phenomena.

1.9.7.2 Shock Wave Structure in a Rarefied Polyatomic Gas

The shock wave structure in a rarefied polyatomic gas is, under some conditions, quite different from the shock wave structure in a rarefied monatomic gas due to the presence of the microscopic internal modes in a polyatomic molecule such as the rotational and vibrational modes [137, 138]. For examples: (1) The shock wave thickness in a rarefied monatomic gas is of the order of the mean free path. On the other hand, owing to the slow relaxation process involving the internal modes, the thickness of a shock wave in a rarefied polyatomic gas is several orders larger than the mean free path; (2) As the Mach number increases from unity, the profile of the shock wave structure in a rarefied polyatomic gas changes from the nearly symmetric profile (Type A) to the asymmetric profile (Type B), and then changes further to the profile composed of thin and thick layers (Type C) [139–144]. Schematic profiles of the mass density are shown in Fig. 1.2. Such change of the shock wave profile with the Mach number cannot be observed in a monatomic gas.

In order to explain the shock wave structure in a rarefied polyatomic gas, there have been two well-known approaches. One was proposed by Bethe and Teller [145] and the other is proposed by Gilbarg and Paolucci [146]. Although the Bethe-Teller theory can describe qualitatively the shock wave structure of Type C, its theoretical basis is not clear enough. Indeed, from a quantitative viewpoint, there exist experimental data that show the deviation from the exponential decay of the mass density in the thick layer Ψ predicted by the Bethe-Teller theory. The Gilbarg-

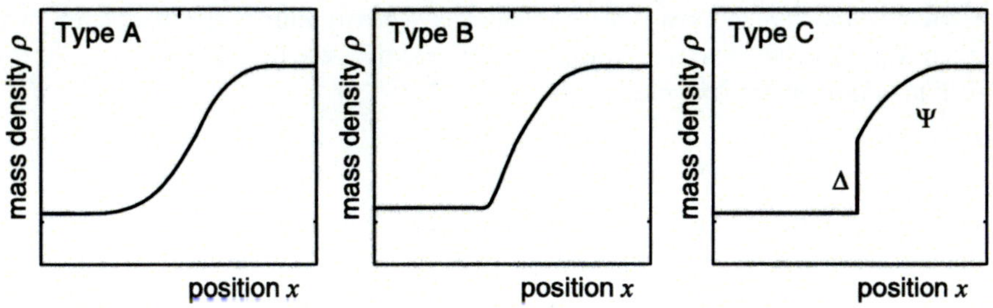

Fig. 1.2 Schematic representation of three types of shock wave structure in a rarefied polyatomic gas, where ρ and x are the mass density and the position, respectively. As the Mach number increases from unity, the profile of the shock wave structure changes from Type A to Type B, and then to Type C that consists of the thin layer Δ and the thick layer Ψ

Paolucci theory, on the other hand, cannot explain asymmetric shock wave structure (Type B) nor thin layer (Type C).

Recently it was shown that the ET_{14} theory can describe the shock wave structure of all Types A to C in a rarefied polyatomic gas [109, 110] in a consistent way. In other words the ET_{14} theory has overcome the difficulties encountered in the previous two approaches. This new approach indicates clearly the usefulness of the RET theory for the analysis of shock wave phenomena. These basic studies will be useful for various practical applications, for example, re-entry of a satellite into the atmosphere of a planet.

The shock wave structure was also analyzed by both linear and non-linear ET_6 theories. In this analysis, the thin layer in Type C (see Fig. 1.2) with finite thickness described by the ET_{14} theory is replaced by a discontinuous jump, which we call subshock. And, by using the nonlinear ET_6 theory, the temperature overshoot at a subshock in terms of Meixner's temperature was studied.

The successful results obtained from the RET theories have also attracted the researchers working in the field of the kinetic theory. When the bulk viscosity (or the relaxation time for the dynamic pressure) has a large value, the modified version of the ellipsoidal statistical (ES) model for polyatomic gas was shown to be able to explain the above-mentioned features 1) and 2). This model supports the predictions by the RET theories quantitatively [147–149].

1.9.7.3 Some Other Applications

Other applications of the RET theory of rarefied polyatomic gases explained in Part V are as follows:

- *Steady flow*, in particular, in a nozzle, analyzed by ET_7 and ET_6.
- *Acceleration waves*, analyzed by ET_6
- *Light scattering*, analyzed by ET_{14}.

- *Heat conduction* in a gas at rest confined between two infinite parallel plates, two coaxial cylinders, and two concentric spheres, analyzed by ET_{14}.
- *Fluctuating hydrodynamics*, analyzed by ET_{14}.

1.10 RET of Dense Polyatomic Gas

Compared with RET of rarefied gases, *RET of dense gases* is not well developed at present. Therefore one of the big challenges in the study of RET is to construct a RET theory of dense gases that is valid in a wider region in the state space. In this section, we make a preliminary discussion on a RET theory of dense polyatomic gases.

From a microscopic point of view, the Hamiltonian of a rarefied polyatomic gas is given in the form:

$$\mathscr{H} = \mathscr{H}^K + \mathscr{H}^I,$$

where $\mathscr{H}^K$ is the *kinetic energy* of molecular translational motion and $\mathscr{H}^I$ is the *energy of the internal motion* of molecules such as molecular rotation and vibration. Instead the Hamiltonian of a dense polyatomic gas is given in the form:

$$\mathscr{H} = \mathscr{H}^{K+U} + \mathscr{H}^I,$$

where $\mathscr{H}^{K+U}$ is composed of the kinetic energy of molecular translational motion and the *potential energy* among molecules, and $\mathscr{H}^I$ is the energy of the internal motion of molecules.

By considering the parallelism between rarefied gas and dense gas as shown above, we proposed the duality principle [150]: *The differential system of a moderately dense gas can be obtained from the system of a rarefied gas by the following substitution:*

$$\left(p^K, \varepsilon^K, s^K\right) \quad \rightarrow \quad \left(p^{K+U}, \varepsilon^{K+U}, s^{K+U}\right). \tag{1.36}$$

Here p^K, ε^K and s^K are, respectively, the equilibrium pressure p, equilibrium specific energy density of the translational mode K, and equilibrium specific entropy of K for a rarefied gas. The quantities p^{K+U}, ε^{K+U}, and s^{K+U} have similar physical meanings for a dense gas.

In this way, from the knowledge of ET_6 of rarefied polyatomic gases, we can construct a new theory of dense polyatomic gases (ET_6^D) in which the internal motion is treated as a unit [150]. Instead, in the paper [151] studying a theory of dense polyatomic gases (ET_7^D), rotation and vibration modes of a molecule are treated individually:

$$\mathscr{H}^{\mathscr{I}} = \mathscr{H}^R + \mathscr{H}^V,$$

where $\mathscr{H}^R$ and $\mathscr{H}^V$ are the energies of molecular rotation and vibration, respectively. Here we can see another parallelism between the pair $\{ET_6, ET_7\}$ for rarefied polyatomic gases and the pair $\{ET_6^D, ET_7^D\}$ for dense polyatomic gases. Details of these theories ET_6^D and ET_7^D will be explained in Chaps. 24 and 25, respectively.

1.11 Relativistic RET of Rarefied Polyatomic Gas

In relativistic fluid dynamics, it is mandatory to adopt hyperbolic systems so as to satisfy the condition that perturbations propagate with finite velocities less than the light velocity. For this reason, Müller [49] and Israel [50] gave the first tentative of a causal relativistic phenomenological theory in contrast with the classical parabolic system due to Eckart [152]. A successive modern approach being compatible, at mesoscopic scale, with the kinetic theory was done by Liu, Müller, and Ruggeri (LMR) [25, 55]. A recent theory on this subject was given by Pennisi and Ruggeri [153] with the aim to include polyatomic gases.

The field equations of the model proposed in [153] are given by

$$\partial_\alpha V^\alpha = 0, \qquad \partial_\alpha T^{\alpha\beta} = 0, \qquad \partial_\alpha A^{\alpha\langle\beta\gamma\rangle} = I^{\langle\beta\gamma\rangle}, \tag{1.37}$$

where V^α and $T^{\alpha\beta}$ are, respectively, the particle flux vector and the energy-momentum tensor. The first two equations represent as usual the conservation laws of particle number and energy-momentum. The last equation represents the extended balance law with fluxes $A^{\alpha\langle\beta\gamma\rangle}$ and productions $I^{\langle\beta\gamma\rangle}$.

The motivation of the system (1.37) comes from the generalized Boltzmann-Chernikov equation

$$p^\alpha \partial_\alpha f = Q, \tag{1.38}$$

where the distribution function f depends not only on the space-time coordinates x^α and the four-momentum of a molecule p^β, but also on a quantity I which takes into account the internal energy of a molecule. According with [153] we have the following generalized moments as in the classical case:

$$\begin{aligned}
V^\alpha &= mc \int_{\mathbb{R}^3} \int_0^{+\infty} f p^\alpha \varphi(I)\, d\, I\, d\mathbf{P}, \\
T^{\alpha\beta} &= \frac{1}{mc} \int_{\mathbb{R}^3} \int_0^{+\infty} f \left(mc^2 + I \right) p^\alpha p^\beta\, \varphi(I)\, d\, I\, d\mathbf{P}, \\
A^{\alpha\beta\gamma} &= \frac{1}{m^2 c} \int_{\mathbb{R}^3} \int_0^{+\infty} f \left(mc^2 + 2I \right) p^\alpha p^\beta p^\gamma \varphi(I)\, d\, I\, d\mathbf{P}, \\
I^{\beta\gamma} &= \frac{1}{m^2 c} \int_{\mathbb{R}^3} \int_0^{+\infty} Q \left(m\, c^2 + 2I \right) p^\beta p^\gamma\, \varphi(I)\, d\, I\, d\mathbf{P},
\end{aligned}$$

where $\varphi(I)$ is the state density of the internal mode and

$$d\mathbf{P} = \frac{dp^1\, dp^2\, dp^3}{p^0}.$$

By using the *maximum entropy principle* [56], the hierarchy of moments was closed and the following expression for the triple tensor was obtained [153]:

$$\begin{aligned} A^{\alpha\beta\gamma} &= A_1^0 U^\alpha U^\beta U^\gamma + 3A_{11}^0 h^{(\alpha\beta} U^{\gamma)} \\ &- \frac{3}{c^2}\frac{N_1^\pi}{D_1^\pi} \Pi U^\alpha U^\beta U^\gamma - 3\frac{N_{11}^\pi}{D_1^\pi} \pi h^{(\alpha\beta} U^{\gamma)} + \frac{3}{c^2}\frac{N_3}{D_3} q^{(\alpha} U^\beta U^{\gamma)} \\ &+ \frac{3}{5}\frac{N_{31}}{D_3} h^{(\alpha\beta} q^{\gamma)} + 3C_5 t^{(\langle\alpha\beta\rangle} U^{\gamma)}, \end{aligned}$$

where U^α is the four-velocity, $h^{\alpha\beta}$ is the projector tensor, $t^{\langle\alpha\beta\rangle}$ is the viscous deviatoric stress, Π is the dynamic pressure, and q^α is the heat flux four-vector with the orthogonality constraints:

$$t^{\langle\alpha\beta\rangle} U_\beta = 0, \qquad q^\alpha U_\alpha = 0. \tag{1.39}$$

The scalar coefficients A_1^0, A_{11}^0, D_1^π, N_1^π, N_{11}^π, N_3, D_3, N_{31}, C_5 were introduced in [153] and are functions of

$$\gamma = \frac{m\,c^2}{k_B\, T}, \tag{1.40}$$

(m and c are, respectively, the rest mass of a particle and the light velocity). The present model contains the monatomic LMR theory as a limiting case [154]. More details are given in Chap. 26.

1.12 Nonequilibrium Temperature

If our study is restricted within nonequilibrium thermodynamics under the local equilibrium assumption, there exists no conceptual difficulty in the temperature in nonequilibrium. Therefore, TIP does no suffer from such a difficulty.

However, if we go beyond the local equilibrium assumption to study highly nonequilibrium phenomena, we encounter an extremely difficult problem. The establishment of a suitable definition of the nonequilibrium temperature in such a situation has always been a big challenge. In RET (and also the kinetic theory) of monatomic gases, the so-called kinetic temperature, which is defined through the thermal average of the kinetic energy of a molecule, has usually been adopted as a nonequilibrium temperature.[2] While, in the papers [155, 156], for example, the so-called thermodynamic temperature was also introduced in the framework of RET. The thermodynamic temperature is defined by the zeroth law of thermodynamics, that is, the continuity conditions of the heat flux and of the entropy flux, especially, at the boundary of a system [5].

For a survey about the nonequilibrium temperature in more general context, see the review paper [157].

In Chap. 15, we will study the temperature and also the chemical potential in polyatomic gases in nonequilibrium. In addition to the kinetic temperature, we introduce another well-defined nonequilibrium temperature and chemical potential on the basis of the generalized Gibbs relation in RET where the *main field* plays an essential role. And subsequently these quantities are examined explicitly in the RET theories with 6 and 14 fields. In the ET_6 theory, in particular, it will be shown in Chap. 12 that their definitions correspond exactly to the definitions of the temperature and chemical potential in the Meixner theory of relaxation processes [129, 130]. An example in the case of shock wave will also be shown.

1.13 Mixture of Gases with Multi-Temperature

In Part VIII, we will study some models of a mixture of compressible fluids. In particular, we will discuss the most general model of a mixture in which each component has its own temperature (multi-temperature, or MT).

We will firstly compare the solutions of this model with those with a unique common temperature (single temperature, or ST) [158]. In the case of Eulerian fluids, it will be shown that the corresponding ST differential system is a principal subsystem of the MT system [158]. Global behavior of smooth solutions for large time for both systems will also be discussed by applying the Shizuta–Kawashima K-condition (see Sect. 2.6.2).

Secondly, we introduce the concept of the average temperature of a mixture based on the consideration that the internal energy of a mixture with multi-temperature is the same as that of a single-temperature mixture [159, 160]. As a consequence, it is shown that the entropy of a mixture reaches a local maximum in equilibrium. Through the procedure of the Maxwellian iteration, new constitutive equations for nonequilibrium temperatures of the components are obtained in a classical limit, together with Fick's law for the diffusion flux.

In order to justify the Maxwellian iteration, we will present, for dissipative fluids, a possible approach to a classical theory of a mixture with multi-temperature. We will prove that the differences of the temperatures between the components imply

[2] In computer simulations by the molecular-dynamics method, the kinetic temperature has been exclusively adopted as the temperature in nonequilibrium.

the existence of the dynamic pressure even if the fluids have zero bulk viscosity [159].

In the paper [161], Ha and Ruggeri observed an analogy between the model of a mixture of fluids with multi-temperature and the famous Cucker-Smale (CS) model for collective behavior of many-body systems, e.g., aggregation of bacteria, flocking of birds, swarming of fish, herding of sheep [162]. This was the strategy to construct a thermo-mechanical CS (TCS) model in Chap. 30. Recently in [163], the authors constructed a model of a relativistic mixture of gases with multi-temperature with explicit production terms, and then using the technique of the main field theory [164] and the principal subsystem [165] they derived a "mechanical" relativistic model that can be regarded as a relativistic counterpart of the CS model. Moreover, they showed that the relativistic CS model reduces to the CS model in the classical limit. This part will be discussed in Chap. 32.

In the previous models, the components of a mixture are Eulerian gases. In Chap. 31, we present a mixture of dissipative fluids in which we take into account only the dynamical pressure of each component and we neglect shear viscosity and heat conductivity. We present also a general mathematical framework defining the system of mixture type in which the main part of the operator has the same form as the one for a single component and the interaction between components is only due to production terms.

1.14 Qualitative Analysis

In RET, the differential system is closed by the universal principles: the objectivity principle, the entropy principle, and the principle of causality and stability. This procedure permits an intimate connection between RET and the mathematical theory of hyperbolic systems with convex extension. Then there exists a privileged field (*main field*) such that the differential system becomes a symmetric hyperbolic system with the well posedness of the local (in time) Cauchy problem [164, 166, 167].

Moreover the requirement that the balance laws are invariant with respect to the Galilean transformation permits to fix, in a unique way, the velocity dependence in the field equations. The entropy principle becomes a constraint only for the constitutive equations [167].

If the system is nonlinear and hyperbolic, global smooth solutions can exist due to the interrelationship between the first five conservation laws and the remaining dissipative ones. In fact, for generic hyperbolic systems of balance laws, endowed with a convex entropy law and dissipation, the Shizuta–Kawashima condition (K-condition) [168] becomes a sufficient condition for the existence of global smooth solutions, provided that the initial data are sufficiently smooth [169–172]. Lou and Ruggeri [173] observed that there exists a weaker K-condition that is a necessary (but unfortunately not sufficient) condition for the global existence of smooth solutions. It was proved that the assumptions of the previous theorems are fulfilled in both classical [174] and relativistic [175, 176] monatomic RET, and also in the case of mixture of gases with multi-temperature [158]. The same property exists also in the case of polyatomic rarefied gas. Details will be explained in Chap. 20 for polyatomic gas with 6 fields and in Chap. 28 for the model of mixture with multi-temperature. In Chap. 33, we will discuss the parabolic limit of the RET theories via the Maxwellian iteration.

1.15 About this Book

The present book is composed of nine Parts I-IX with an additional chapter: "Introduction and Overview" at the beginning.

In Part I, we present the mathematical structure of RET and its intimate connection with hyperbolic nonlinear systems of first order. In particular, on the basis of the entropy principle and the convexity of entropy, we can transform the original system into a system in a symmetric form in terms of the main field. The main field is crucial also for clarifying the nesting structure of RET. Moreover symmetric hyperbolic systems have good properties concerning the well-posedness of the Cauchy problem. We discuss the theorems that guarantee the existence of global smooth solutions for all time and, as is expected from a physical point of view, the asymptotic tendency to an equilibrium state. The Galilean invariance dictates the explicit dependence of the field variables on the velocity, while the entropy principle gives a selection rule for admissible constitutive equations. One chapter is devoted to explain the general properties of wave propagation phenomena like linear wave, acceleration wave, shock wave, and shock wave structure. These are important to understand RET deeply.

Part II is a brief summary of the main results of RET of rarefied monatomic gases in both classical and relativistic frameworks, details of which are explained in the book of Müller and Ruggeri [25]. However, the method for deriving the closed system of field equations adopted here is new and more systematic than the previous one. This Part is necessary for understanding the new progress that will be explained in the subsequent Parts.

From Part III, we start to explain the newly obtained results after the book of Müller and Ruggeri [25]. In Part III, firstly, we study the RET theory of rarefied polyatomic gases with 14 independent fields. The finding that the system of field equations should have the structure of binary hierarchy is the main breakthrough in the new progress. We also study the molecular ET with the maximum entropy principle. This is compared with the above phenomenological approach, and the consistency between two approaches is shown. Secondly, we study the mathematical structure of the RET theory with binary hierarchy of arbitrary number of independent densities. Nesting theory of many moments is studied. Thirdly, we study RET of rarefied polyatomic gases with molecular relaxation processes, where molecular rotational and vibrational modes are treated individually. Lastly, the monatomic-gas limit of RET of polyatomic gases is studied.

Part IV is devoted to the RET theories of rarefied polyatomic gases with six and seven independent fields that are applicable to the phenomena far from equilibrium. The correspondence relation between the RET theory and the Meixner theory of relaxation processes is established. The concept of the nonequilibrium temperature is revisited.

Part V deals with some applications of the RET theories. Linear harmonic wave and shock wave are studied. Comparison of the theoretical predictions with experimental data shows the remarkable superiority of the RET theory to the classical Navier–Stokes and Fourier theory. Some other typical applications are also briefly discussed.

In Part VI, the RET theories of dense polyatomic gases are presented. Firstly, we study the RET theory with 6 fields, where we neglect shear viscosity and heat conductivity and we treat the internal (rotational and vibrational) motion of a molecule as a unit. We postulate a principle of duality between rarefied gas and dense gas. Then we also construct the RET theory with 7 fields where molecular rotational and vibrational modes are treated individually.

In Part VII, the relativistic RET theory of a rarefied polyatomic gas with 14 fields is presented. Then relativistic RET of a polyatomic gas with many moments associated with the relativistic Boltzmann-Chernikov kinetic equation is discussed.

In Part VIII, multi-temperature mixture of Euler fluids is studied. Then we study the shock structure in a mixture. We study, in particular, profile of shock structure, subshock regions, and the temperature overshoot in a shock wave. We discuss also the Cucker-Smale model from the viewpoint of the mixture theory. Two advanced topics such as dissipative mixture of polyatomic gases and relativistic mixture are also explained.

In Part IX, the parabolic limit of the hyperbolic RET systems obtained via the Maxwellian iteration is analyzed. In particular, we study the question whether the entropy principle is preserved in this limit, and we discuss the range of validity of the so-called regularized system.

Part I
Mathematical Structure and Waves

Chapter 2
Mathematical Structure

Abstract In this chapter, we make a survey on the mathematical structure of the system of rational extended thermodynamics, which is strictly related to the mathematical problems of hyperbolic systems in balance form with a convex entropy density. We summarize the main results: The proof of the existence of the *main field* in terms of which a system becomes symmetric, and several properties derived from the qualitative analysis concerning symmetric hyperbolic systems. In particular, the Cauchy problem is well-posed locally in time, and if the so-called *K-condition* is satisfied, there exist global smooth solutions provided that the initial data are sufficiently small. Moreover the main field permits us to identify natural subsystems and in this way we have a structure of nesting theories. The main property of these subsystems is that the characteristic velocities satisfy the so-called *subcharacteristic* conditions that imply, in particular, that the maximum characteristic velocity does not decrease when the number of equations increases. Another beautiful general property is the compatibility of the balance laws with the Galilean invariance that dictates the precise dependence of the field equations on the velocity.

2.1 System of Balance Laws

Rational extended thermodynamics (RET) [25] is based on the assumptions that the density $\mathbf{F}^0$, the flux $\mathbf{F}^i$, and the production $\mathbf{f}$ in the balance-law system (1.2) depend locally on the field variable $\mathbf{u}$, and that the quasi-linear dissipative system of balance laws:

$$\frac{\partial \mathbf{F}^0(\mathbf{u})}{\partial t} + \frac{\partial \mathbf{F}^i(\mathbf{u})}{\partial x^i} = \mathbf{f}(\mathbf{u}) \tag{2.1}$$

is hyperbolic with respect to the t-direction. The variable $\mathbf{u} \equiv \mathbf{u}(\mathbf{x}, t)$ is the unknown column field-vector with N components to be determined in a problem under consideration.

T. Ruggeri, M. Sugiyama, *Classical and Relativistic Rational Extended Thermodynamics of Gases*, https://doi.org/10.1007/978-3-030-59144-1_2

It is convenient to rewrite the system (2.1). The most compact form of the hyperbolic system in the space-time is to use the relativistic notation: $x^0 = t$; $\partial_\alpha = \partial/\partial x^\alpha \quad (\alpha = 0, 1, 2, 3)$. Then we have

$$\partial_\alpha \mathbf{F}^\alpha(\mathbf{u}) = \mathbf{f}(\mathbf{u}). \tag{2.2}$$

2.1.1 Hyperbolicity in the t-Direction

Let us consider the quasi-linear first order system of PDE's:

$$\mathbf{A}^\alpha(\mathbf{u})\partial_\alpha \mathbf{u} = \mathbf{f}(\mathbf{u}). \tag{2.3}$$

Note that the system (2.2) is a particular case of (2.3) where $\mathbf{A}^\alpha$ is given by

$$\mathbf{A}^\alpha = \frac{\partial \mathbf{F}^\alpha}{\partial \mathbf{u}}.$$

Definition 2.1 The system (2.3) is called *hyperbolic in the t-direction*, if it has the following two properties:

- $\det \mathbf{A}^0 \neq 0$,
- for all unit vectors $\mathbf{n} \equiv (n_i)$, the eigenvalue problem:

$$\left(\mathbf{A}^i n_i - \lambda \mathbf{A}^0\right) \mathbf{d} = 0 \tag{2.4}$$

  admits only real eigenvalues λ and a set of linearly independent right eigenvectors $\mathbf{d}$.

The λ's are called *characteristic velocities* and the polynomial

$$\det \left(\mathbf{A}^i n_i - \lambda \mathbf{A}^0\right) = 0 \tag{2.5}$$

is called *characteristic polynomial.* If all λ are distinct the system is called *strictly hyperbolic*. The left eigenvectors $\mathbf{l}$ are defined by

$$\mathbf{l}\left(\mathbf{A}^i n_i - \lambda \mathbf{A}^0\right) = 0,$$

and may be chosen in such a way that

$$\mathbf{l}^{(I)} \cdot \mathbf{d}^{(J)} = \delta^{IJ}, \qquad \text{for all} \qquad I, J = 1, 2, \ldots N.$$

2.1.2 Symmetric Hyperbolic System

Definition 2.2 The system (2.3) is called *symmetric hyperbolic* (in the t-direction), or briefly *symmetric*—by the definition of Friedrichs—, if

- the matrices $\mathbf{A}^\alpha$ are symmetric;
- the matrix $\mathbf{A}^0$ is positive definite.

By linear algebra every symmetric system is hyperbolic, but the reverse statement is not true.

Symmetric systems play an important role in RET because the RET theory uses hyperbolic equations and also because the entropy principle, together with the convexity of entropy density, ensures that the equations form a *symmetric* hyperbolic system.

2.1.3 Covariant Definition of Hyperbolicity

The previous definition can be given in a covariant formalism that is necessary in the relativistic case (see e.g. [164]). Let V^4 be a four-dimensional C^∞ manifold and $\mathbf{x}$ be a point of V^4 with x^α being local coordinates of $\mathbf{x}$. The manifold is supposed to be endowed with a pseudo-Riemannian metric. In the local coordinates, $g_{\alpha\beta}$ represents the components of the metric tensor with signature $(+, -, -, -)$. On V^4 we consider the quasi-linear first order partial differential system (2.3) for the unknown N-vector $\mathbf{u}(x^\alpha) \in \mathbb{R}^N$. The components of $\mathbf{u}$ are the blocks of contravariant tensors, and now ∂_α indicates the covariant derivative.

Definition 2.3 The system (2.3) is said to be hyperbolic if a timelike covector $\xi_\alpha (\xi_\alpha \xi^\alpha > 0)$ exists such that the following two statements hold:

$$\det(\mathbf{A}^\alpha \xi_\alpha) \neq 0, \tag{2.6}$$

and, for any covector ζ_α of spacelike $(\zeta_\alpha \zeta^\alpha < 0)$, the following eigenvalue problem

$$\mathbf{A}^\alpha \left(\zeta_\alpha - \lambda \xi_\alpha\right) \mathbf{d} = 0 \tag{2.7}$$

has only real eigenvalues λ and N linearly independent eigenvectors $\mathbf{d}$. The covectors $\zeta_\alpha - \lambda \xi_\alpha$ are called *characteristic*, while ξ_α fulfilling (2.6) and (2.7) are called *subcharacteristic*.

Definition 2.4 A system (2.3) is called symmetric if all $\mathbf{A}^\alpha$ are symmetric matrices and a timelike covector ξ_α exists such that $\mathbf{A}^\alpha \xi_\alpha$ is positive definite for any $\mathbf{u} \in \mathscr{D}$, where $\mathscr{D}$ is a convex open subset of $\mathbb{R}^N$.

Remark 2.1 Without loss of generality, it is possible to choose the timelike covector ξ_α and the spacelike covector ζ_α such that $\xi^\alpha \zeta_\alpha = 0$ and $\xi_\alpha \xi^\alpha = 1$, $\zeta_\alpha \zeta^\alpha = -1$.

In a particular local frame in which $\xi_\alpha \equiv (1, 0, 0, 0)$ and $\zeta_\alpha \equiv (0, n_1, n_2, n_3)$, the Definitions 2.3 and 2.4 coincide with the classical Definitions 2.1 and 2.2, respectively.

2.2 Axioms of Rational Extended Thermodynamics

The main axioms of RET are summarized as follows:

- The balance laws (2.2) must satisfy the relativity principle, i.e., the balance laws are invariant under Galilean transformation (or Lorentz transformation in a relativistic case) and the proper constitutive equations are invariant under any change of observer.
- The entropy principle requires that constitutive equations must be selected so that all thermodynamic processes are compatible with the second law of thermodynamics, where a thermodynamic process is defined as a solution of the total system (balance laws plus constitutive equations). More precisely, the principle consists in the following axioms:

  (i) *There exists an additive and objective scalar, which we call entropy*;
  (ii) *The entropy density and the flux of the entropy are constitutive functions to be determined*;
  (iii) *The entropy production is non-negative for all thermodynamic processes.*

  The axiom (i) says that a physical system possesses an additive quantity called entropy, like mass and energy. As for any additive quantity, there exists a balance law of entropy:

  $$\partial_\alpha h^\alpha(\mathbf{u}) = \Sigma(\mathbf{u}), \tag{2.8}$$

  where h^0, h^i, and Σ are, respectively, the entropy density, the entropy flux, and the entropy production. The entropy production depends also on $\mathbf{u}$. The second axiom (ii) says that both h^0 and h^i are constitutive quantities to be determined as functions of the field variables. Finally, the third axiom (iii) requires that the entropy production is non-negative for all thermodynamic processes:

  $$\Sigma \geqslant 0. \tag{2.9}$$

  We have seen, in Sect. 1.6, the genesis of this entropy principle. In the context of hyperbolic systems of conservation laws, this formulation is due to Friedrichs and Lax [177].
- The causality and the thermodynamic stability are required, i.e., the entropy density must be a concave function of the density fields. When $\mathbf{u} \equiv \mathbf{F}^0$, we

have the condition:

$$\frac{\partial^2 h^0}{\partial \mathbf{u}\,\partial \mathbf{u}} \quad : \text{negative definite,}$$

i.e.,

$$\delta^2 h^0 = \delta \mathbf{u} \cdot \frac{\partial^2 h^0}{\partial \mathbf{u}\,\partial \mathbf{u}} \delta \mathbf{u} < 0 \quad \forall\ \delta \mathbf{u} \neq 0. \tag{2.10}$$

In the mathematical community, the sign of the entropy density h^0 is usually opposite to the one adopted in physics. Therefore following the tradition we still use the word, convexity, instead of the more appropriate word, concavity.

2.3 Entropy Principle and Symmetric System

2.3.1 General Discussions

First we prove the following important theorem by Ruggeri and Strumia [164]:[1]

Theorem 2.1 (Main Field and Symmetric Form) *The compatibility between the system of balance laws (2.2) and the supplementary balance law (2.8) with h^0 being a convex function of $\mathbf{u} \equiv \mathbf{F}^0$ implies the existence of the* main field $\mathbf{u}'$ *that satisfies*

$$dh^\alpha = \mathbf{u}' \cdot d\mathbf{F}^\alpha, \qquad \Sigma = \mathbf{u}' \cdot \mathbf{f} \geqslant 0. \tag{2.11}$$

If we choose the components of $\mathbf{u}'$ as field variables, the original system (2.2) can be rewritten in a symmetric form with Hessian matrices:

$$\partial_\alpha \left(\frac{\partial h'^\alpha}{\partial \mathbf{u}'} \right) = \mathbf{f} \quad \iff \quad \frac{\partial^2 h'^\alpha}{\partial \mathbf{u}' \partial \mathbf{u}'} \partial_\alpha \mathbf{u}' = \mathbf{f}, \tag{2.12}$$

where h'^α is the four-potential defined by

$$h'^\alpha = \mathbf{u}' \cdot \mathbf{F}^\alpha - h^\alpha. \tag{2.13}$$

Proof Let us firstly notice the following two points: the first is that (2.2) and (2.8) are quasi-linear equations (linear in the derivative), and the second is that the system composed of (2.2) and (2.8) is overdetermined, i.e., $N + 1$ equations for N unknowns. Therefore the compatibility between (2.2) and (2.8) implies that the

[1] The proof in [164] was done in a covariant formalism. Instead, here, we use a classical formalism for simplicity.

supplementary law (2.8) must be a linear combination of the equations of the system (2.2). In other words, there exists a vector $\mathbf{u}' \equiv \mathbf{u}'(\mathbf{u})$ such that

$$\partial_\alpha h^\alpha - \Sigma \equiv \mathbf{u}' \cdot \left(\partial_\alpha \mathbf{F}^\alpha - \mathbf{f}\right). \tag{2.14}$$

As a consequence of this identity, we have (2.11). Differentiating (2.13), taking into account $(2.11)_1$ and choosing $\mathbf{u}'$ as a field, we obtain the following relation:

$$\mathbf{F}^\alpha = \frac{\partial h'^\alpha}{\partial \mathbf{u}'}. \tag{2.15}$$

Inserting (2.15) into (2.2), we obtain (2.12). The change of variables from the density field $\mathbf{u} = \mathbf{F}^0$ to the main field $\mathbf{u}'$ is ensured because of the convexity condition of the entropy density. In fact, when $\alpha = 0$, it is easy to see from $(2.11)_1$ and (2.15) that $\mathbf{u}$ and $\mathbf{u}'$ are dual fields in the Legendre transform (2.13):

$$\mathbf{u} = \frac{\partial h'^0}{\partial \mathbf{u}'}, \qquad \mathbf{u}' = \frac{\partial h^0}{\partial \mathbf{u}} \tag{2.16}$$

and the convexity condition (2.10) implies the convexity of h'^0 with respect to the main field $\mathbf{u}'$:

$$\delta^2 h'^0 = \delta\mathbf{u}' \cdot \frac{\partial^2 h'^0}{\partial \mathbf{u}' \partial \mathbf{u}'} \delta\mathbf{u}' < 0 \qquad \forall\ \delta\mathbf{u}' \neq 0. \tag{2.17}$$

Therefore the system (2.12) is symmetric, and the proof is completed.[2]

We can verify that the inverse is also true. Every system of the form (2.12) satisfies automatically the entropy principle (2.8) and (2.9) with h^α being given by

$$h^\alpha = \mathbf{u}' \cdot \left(\frac{\partial h'^\alpha}{\partial \mathbf{u}'}\right) - h'^\alpha, \tag{2.18}$$

provided that the production term $\mathbf{f}$, which is the function of $\mathbf{u}'$, satisfies the inequality $(2.11)_2$.

Because $\mathbf{u}'$ is the only privileged field with respect to which the system assumes the symmetric Hessian form (2.12), Ruggeri and Strumia [164] proposed to call it *main field*. If we introduce the inequality

$$Q = \delta\mathbf{u} \cdot \delta\mathbf{u}' < 0, \tag{2.19}$$

[2] We recall that, due to the different sign of the entropy density in Definition 2.2, the word "positive definite" is changed into "negative definite".

then it is easy to verify that the inequality (2.19) coincides with the convexity condition of h^0 (2.10) and the convexity condition of h'^0 (2.17). In fact from (2.16) we have

$$Q = \delta \mathbf{u} \cdot \delta \mathbf{u}' = \delta \mathbf{u} \cdot \delta \left(\frac{\partial h^0}{\partial \mathbf{u}} \right) = \delta^2 h^0 < 0 \tag{2.20}$$

or

$$Q = \delta \mathbf{u} \cdot \delta \mathbf{u}' = \delta \left(\frac{\partial h'^0}{\partial \mathbf{u}'} \right) \cdot \delta \mathbf{u}' = \delta^2 h'^0 < 0.$$

2.3.2 Direct and Inverse Methods for Exploiting the Entropy Principle

This theorem provides two complementary methods for exploiting the entropy principle:

- **Direct method**: We use the physical field $\mathbf{u} \equiv \mathbf{F}^0$. From $(2.11)_1$, we find the integrability conditions. We deduce the main field $\mathbf{u}'$, the entropy density h^0, the entropy flux h^i, and the compatible fluxes $\mathbf{F}^i$. Then we obtain the entropy production Σ from $(2.11)_2$ and the four-potential h'^α from (2.13).
- **Inverse method** [32]: We use the main field $\mathbf{u}'$. We write down the four-potential h'^α as the function of the main field. Then, without addressing any integrability conditions, we have h^α and $\mathbf{F}^\alpha$ from (2.18) and (2.15), respectively. Although this method is very powerful, there exists one disadvantage: Since the closure is made in term of the main field, it is necessary for us to have the physical variables $\mathbf{u} \equiv \mathbf{u}(\mathbf{u}')$ by carrying out a nonlinear inversion of (2.15) with $\alpha = 0$.

2.3.3 Convexity and Symmetrization in Covariant Formalism

Theorem 2.1 is valid for both classical and relativistic frameworks. Nevertheless, in a covariant formalism being typical of relativity, there is a subtle point concerning the convexity of entropy and the symmetric systems, which was realized by Ruggeri [178]. In fact, the symmetry condition requires, according to the Definition 2.4, that $\mathbf{A}^\alpha \xi_\alpha$ must be negative definite for a given timelike covector ξ_α. In the present case, from $(2.12)_2$, we have

$$\mathbf{A}^\alpha \xi_\alpha = \frac{\partial^2 h'^\alpha}{\partial \mathbf{u}' \partial \mathbf{u}'} \xi_\alpha \qquad \text{negative definite.} \tag{2.21}$$

By taking into account (2.15), condition (2.21) is equivalent to the inequality:

$$Q = \delta \mathbf{u}' \cdot \delta \mathbf{F}^{\alpha}\, \xi_{\alpha} < 0. \tag{2.22}$$

In the covariant formalism, the quantities $h^0, \mathbf{u} = \mathbf{F}^0$, and h'^0 are, respectively, expressed as

$$h = h^{\alpha}\xi_{\alpha}, \quad \mathbf{u} = \mathbf{F}^{\alpha}\xi_{\alpha}, \quad h' = h'^{\alpha}\xi_{\alpha}.$$

If the congruence ξ_{α} is constant and we locally choose $\xi_{\alpha} \equiv (1, 0, 0, 0)$, all relations above become the classical ones. In particular, (2.21) is equivalent to (2.17), Q in (2.22) coincides with (2.19), and the symmetry condition (2.22) is equivalent to the convexity condition $\delta^2 h < 0$ (see (2.20)).

Instead, a problem arises when the time congruence ξ_{α} depends on some field variables. Indeed, in relativistic fluids it is common to choose

$$\xi_{\alpha} = U_{\alpha}, \tag{2.23}$$

where U^{α} is the four-velocity. For non-constant congruence, it was proved in [178] that

$$Q = \delta^2 h + h'^{\alpha} \delta^2 \xi_{\alpha}. \tag{2.24}$$

Therefore the symmetry condition (2.22) does not coincide with the convexity $\delta^2 h < 0$ of entropy $h = h^{\alpha}\xi_{\alpha}$. But in the same paper it was proved that there exists a privileged congruence $\bar{\xi}_{\alpha}$ such that the second term in (2.24) is not positive and therefore the convexity condition $\delta^2 h < 0$ implies the symmetry condition $Q < 0$. It is remarkable that the privileged congruence is collinear with the four-potential h'^{α}, i.e.,

$$\bar{\xi}_{\alpha} = -\frac{c\, h'^{\alpha}}{\sqrt{h'^{\beta} h'_{\beta}}}, \tag{2.25}$$

where c is the light speed. It is also remarkable that, for Euler relativistic fluid (see (5.9)) and for non-degenerate relativistic RET of monatomic gases with 14 moments, the privileged congruence (2.25) coincides exactly with (2.23) for the usual common congruence.[3]

[3] We recall that, in the paper [178], the sign of the quantities h^{α} and h'^{α} is opposite to the one adopted here.

2.3.4 Historical Remarks

The history concerning the symmetrization seems to be not so clear, and it sometimes does stir up debate. We here want to list some remarks on it to our knowledge:

- In 1961, Godunov wrote a short paper "An interesting class of quasi-linear systems" [179] in which he was able to prove, in the case of Euler fluid, that if we choose $\left(\left(g + \frac{v^2}{2}\right)/T, -\mathbf{v}/T, 1/T\right)$ as variables where g is the chemical potential (see (2.28)), the original system becomes symmetric. Moreover he proved that all systems that come from a variational principle can be put in a symmetric form.
- In 1971, Friedrichs and Lax proved [177] that all systems that are compatible with the entropy principle are symmetrizable. This means that, for an original system that is not symmetric, a new one after a pre-multiplication of a matrix $\mathbf{H}(\mathbf{u})$ becomes symmetric. As a consequence the symmetric system is different from the previous system. Therefore the disadvantage of this method is that weak solutions of the original system are not weak solutions of the new one. It seems that they ignored completely Godunov's work!
- In 1974, Boillat [166] introduced the field $\mathbf{u}' = \partial h^0/\partial \mathbf{F}^0$, and he realized that the original system can be put in a symmetric form. Therefore, to our knowledge, he is the first person who symmetrized the original hyperbolic system that is compatible with the entropy principle. At that time, he did not know Godunov's paper, but after he discovered the paper of Godunov, he has called this system *Godunov system* (and Godunov has always quoted Boillat). The reader can find more details in the Lecture Notes in Mathematics n.1640 (1986) of a Cime Course by Boillat, Dafermos, Lax, and Liu. In the part written by Boillat [180], there are several physical examples of symmetrization by using this technique: non-linear elasticity, Born-Infeld non-linear electro-dynamics, magneto-fluid dynamics, etc. in which the main field was used to symmetrize these systems. See also the recent paper [181].
- In 1981, Ruggeri and Strumia [164] were interested in extending this technique to a relativistic case by using a covariant formulation. But they realized that it is impossible to define $\mathbf{u}'$ in the same manner as before because h^0 is the temporal part of the four-vector h^α and therefore is not a scalar invariant. Also $\mathbf{F}^0$ is not a vector. Therefore they introduced $\mathbf{u}'$ as a multiplier such that if we multiply the balance laws by $\mathbf{u}'$ we obtain identically the supplementary entropy law (see (2.14)). This has a benefit that all components of $\mathbf{u}'$ are tensors and, in the classical case, reduce to the Boillat field. They realized the importance of this change of variables and for this reason they proposed for the first time the name: *main field*. The technique of Lagrange multipliers to explore the entropy principle is similar to the one used by Ruggeri and Strumia, and it was given first by I-Shi Liu [182]. However, it should be noticed that the Lagrange multipliers that define the main field are obtained only when the system is written in a balance

form. While, if we use a differential system in a different form (for example, the system expressed by using the material derivative), the Lagrange multipliers are not independent variables (see for more details Sect. 2.7).

- In 1982, Boillat extended the symmetrization also to the case with constraints [183]. This problem was considered also by Dafermos [184].
- In 1983, in the first tentative to construct RET, Ruggeri realized that it is possible to carry out the symmetrization also for parabolic systems, and he wrote down for the first time the expression of the main field for Navier–Stokes Fourier fluids [53].
- In 1989, in the paper [167], Ruggeri proved that symmetrization is compatible with the Galilean invariance.
- In a relativistic case, there is a subtle point because the entropy depends on the choice of the temporal congruence. In 1990, a discussion about the convexity and the symmetrization, and a choice of temporal congruence were the subjects of the paper [178]. See also Sect. 2.3.3.
- In some recent papers, Gouin and coworkers payed the attention to the symmetrization of some hyperbolic-parabolic systems [185–187].

2.3.5 Example of Symmetric form of Euler Fluids

As an example of the previous general results, let us consider the simple case of Euler fluid. We have

$$\mathbf{F}^0 = \begin{pmatrix} \rho \\ \rho v_k \\ \frac{1}{2}\rho v^2 + \rho\varepsilon \end{pmatrix}, \quad \mathbf{F}^i = \begin{pmatrix} \rho v_i \\ \rho v_i v_k + p\delta_{ik} \\ (\frac{1}{2}\rho v^2 + \rho\varepsilon + p)v_i \end{pmatrix}, \quad \mathbf{f} = \begin{pmatrix} 0 \\ 0_k \\ 0 \end{pmatrix},$$

$$h^0 = \rho s, \qquad h^i = \rho s v_i, \qquad \Sigma = 0. \tag{2.26}$$

Let $(\lambda, \lambda_i, \zeta)$ be the components of $\mathbf{u}'$, then, from $(2.11)_1$, we have

$$d(\rho s) = \lambda d\rho + \lambda_i d(\rho v_i) + \zeta d\left(\rho \frac{v^2}{2} + \rho\varepsilon\right).$$

As ds is given by the Gibbs equation (1.8), we have

$$\lambda = \frac{1}{T}\left(-g + \frac{v^2}{2}\right), \qquad \lambda_i = -\frac{v_i}{T}, \qquad \zeta = \frac{1}{T}, \tag{2.27}$$

where

$$g = \varepsilon + \frac{p}{\rho} - Ts \tag{2.28}$$

is the chemical potential. It is easy to verify that (2.11)$_1$ with $\alpha = i = 1, 2, 3$ are automatically satisfied. Therefore the main field coincides with the one derived by Godunov [179]:

$$\mathbf{u}' = \frac{1}{T}\left(-g + \frac{v^2}{2}, -v_i, 1\right). \tag{2.29}$$

From (2.13) we have also the potentials:

$$h'^0 = -\frac{p}{T}, \quad h'^i = -v_i \frac{p}{T}. \tag{2.30}$$

The convexity condition is always satisfied under the usual thermodynamic stability condition, that is, the positivity of the specific heat and the compressibility:

$$\left(\frac{\partial \varepsilon}{\partial T}\right)_\rho > 0, \qquad \left(\frac{\partial p}{\partial \rho}\right)_T > 0. \tag{2.31}$$

2.4 Principal Subsystems

Importance of the main field was again recognized when Boillat and Ruggeri discovered that this field permits us to define *nesting theories* through the concept of principal subsystems [165].

Let us split the main field $\mathbf{u}' \in \mathbb{R}^N$ into two parts $\mathbf{u}' \equiv (\mathbf{v}', \mathbf{w}')$ where $\mathbf{v}' \in \mathbb{R}^M$ and $\mathbf{w}' \in \mathbb{R}^{N-M}$ $(0 < M < N)$, then the system (2.12) with $\mathbf{f} \equiv (\mathbf{r}, \mathbf{g})$ reads:

$$\partial_\alpha \left(\frac{\partial h'^\alpha(\mathbf{v}', \mathbf{w}')}{\partial \mathbf{v}'}\right) = \mathbf{r}(\mathbf{v}', \mathbf{w}'), \tag{2.32}$$

$$\partial_\alpha \left(\frac{\partial h'^\alpha(\mathbf{v}', \mathbf{w}')}{\partial \mathbf{w}'}\right) = \mathbf{g}(\mathbf{v}', \mathbf{w}'). \tag{2.33}$$

Definition 2.5 (Principal Subsystem) Given some assigned constant value $\mathbf{w}'_*$ of $\mathbf{w}'$, we call the system:[4]

$$\partial_\alpha \left(\frac{\partial h'^\alpha(\mathbf{v}', \mathbf{w}'_*)}{\partial \mathbf{v}'}\right) = \mathbf{r}(\mathbf{v}', \mathbf{w}'_*)$$

[4]The definition and the properties remain valid for prescribed values of $\mathbf{w}'_*$ that depend on x^α in an arbitrary manner. In this case the principal subsystem is not autonomous [165].

principal subsystem of (2.12). In other words, a principal subsystem (there are $2^N - 2$ such subsystems) coincides with the first block of the system putting $\mathbf{w}' = \mathbf{w}'_*$.

The principal subsystems have two important properties: they admit also a convex subentropy law, and the spectrum of the characteristic velocities is contained in the spectrum of the full system (subcharacteristic conditions). In fact it is possible to prove the following theorems [165]:

Theorem 2.2 *Solutions of a principal subsystem satisfy also a supplementary law (subentropy law):*

$$\partial_\alpha \overline{h}^\alpha = \overline{\Sigma}, \tag{2.34}$$

where the entropy four-vector $\overline{h}^\alpha(\mathbf{v}', \mathbf{w}'_*)$ and the entropy production $\overline{\Sigma}$ are related to the restrictions of the entropy four-vector $h^\alpha(\mathbf{v}', \mathbf{w}'_*)$ and of the entropy production $\Sigma(\mathbf{v}', \mathbf{w}'_*)$ of the full system:

$$\overline{h}^\alpha(\mathbf{v}', \mathbf{w}'_*) = h^\alpha(\mathbf{v}', \mathbf{w}'_*) - \mathbf{w}'_* \cdot \left(\frac{\partial h'^\alpha}{\partial \mathbf{w}'}\right)_{\mathbf{w}' \equiv \mathbf{w}'_*}, \tag{2.35}$$

$$\overline{\Sigma} = \Sigma(\mathbf{v}', \mathbf{w}'_*) - \mathbf{w}'_* \cdot \mathbf{g}(\mathbf{v}', \mathbf{w}'_*). \tag{2.36}$$

The subentropy is convex and therefore every principal subsystem are also symmetric hyperbolic.

Let $\lambda^{(k)}(\mathbf{v}', \mathbf{w}', \mathbf{n})$ and $\overline{\lambda}^{(\overline{k})}(\mathbf{v}', \mathbf{w}'_*, \mathbf{n})$ be the characteristic velocities of the total system and of the subsystem, respectively, where $\mathbf{n}$ is the unit normal to the wave front. In general, solutions of the subsystem are not particular solutions of the system (for $\mathbf{w}' = \mathbf{w}'_*$) and the spectrum of the $\overline{\lambda}$'s is not part of the spectrum of the λ's. However, we have the following theorem when we define

$$\lambda_{\max} = \max_{k=1,2,\ldots,N} \lambda^{(k)}, \qquad \overline{\lambda}_{\max} = \max_{\overline{k}=1,2,\ldots,M} \overline{\lambda}^{(\overline{k})}$$

and also define in a similar way for the minima:

Theorem 2.3 (Subcharacteristic Conditions) *Under the assumption that h^0 is a convex function, the following subcharacteristic conditions hold for every principal subsystem:*

$$\lambda_{\max}(\mathbf{v}', \mathbf{w}'_*, \mathbf{n}) \geqslant \overline{\lambda}_{\max}(\mathbf{v}', \mathbf{w}'_*, \mathbf{n}); \qquad \lambda_{\min}(\mathbf{v}', \mathbf{w}'_*, \mathbf{n}) \leqslant \overline{\lambda}_{\min}(\mathbf{v}', \mathbf{w}'_*, \mathbf{n}), \tag{2.37}$$

$\forall\, \mathbf{v}' \in \mathbb{R}^M$ *and* $\forall\, \mathbf{n} \in \mathbb{R}^3 : \|\, \mathbf{n} \,\| = 1.$

The proofs of the above two theorems are given in [165].

2.4.1 Example of Euler Principal Subsystem

Usually the principal subsystems are obtained by letting some relaxation times go to zero (see e.g. Sect. 4.1.2). But the concept is more general. In fact, for Euler system:

$$
\begin{aligned}
&\frac{\partial \rho}{\partial t}+\frac{\partial}{\partial x_k}\left(\rho v_k\right)=0, \\
&\frac{\partial \rho v_i}{\partial t}+\frac{\partial}{\partial x_k}\left(\rho v_i v_k+p \delta_{ik}\right)=0, \\
&\frac{\partial}{\partial t}\left(\rho v^2+2 \rho \varepsilon\right)+\frac{\partial}{\partial x_k}\left\{\rho v^2 v_k+2(\rho \varepsilon+p) v_k\right\}=0,
\end{aligned}
\tag{2.38}
$$

where no relaxation time exists, we can still consider ET$_4$ principal subsystem requiring to freeze the last component of the main field (2.27):

$$\zeta = \text{const.}, \qquad \rightarrow \qquad T = T^* = \text{const.},$$

and eliminating the last equation in (2.38). We obtain the isothermal Euler system:

$$
\begin{aligned}
&\frac{\partial}{\partial t}\rho+\frac{\partial}{\partial x_i}\left(\rho v_i\right)=0, \\
&\frac{\partial}{\partial t}\left(\rho v_j\right)+\frac{\partial}{\partial x_i}\left(\rho v_i v_j+p^* \delta_{ij}\right)=0,
\end{aligned}
\tag{2.39}
$$

where

$$p^* \equiv p(\rho, T^*).$$

From (2.36), we have

$$
\begin{aligned}
&\bar{h}^0=\rho s(\rho, T^*)-\frac{1}{T^*}\left(\rho \varepsilon\left(\rho, T^*\right)+\frac{1}{2} \rho v^2\right), \\
&\bar{h}^i=\rho s(\rho, T^*) v^i-\frac{1}{T^*}\left(\rho \varepsilon\left(\rho, T^*\right)+\frac{1}{2} \rho v^2+p\left(\rho, T^*\right)\right) v^i, \\
&\Sigma=0.
\end{aligned}
\tag{2.40}
$$

The Gibbs relation (1.8) evaluated at $T = T^*$ is given as

$$T^* ds\left(\rho, T^*\right) = d\varepsilon\left(\rho, T^*\right) - \frac{p\left(\rho, T^*\right)}{\rho^2} d\rho.$$

By integration, we have

$$s\left(\rho, T^*\right) = \frac{1}{T^*}\left(\varepsilon\left(\rho, T^*\right) - \mathscr{E}\left(\rho, T^*\right)\right), \quad \mathscr{E}\left(\rho, T^*\right) = \int \frac{p\left(\rho, T^*\right)}{\rho^2} d\rho. \tag{2.41}$$

Substituting (2.41) into (2.40), we can see that the supplementary equation (2.34) of the isothermal Euler system (2.39) is nothing but the mechanical energy balance law:

$$\frac{\partial}{\partial t}\left(\rho\,\mathscr{E}\left(\rho, T^*\right) + \frac{1}{2}\rho v^2\right) + \frac{\partial}{\partial x_i}\left[\left(\rho\mathscr{E}\left(\rho, T^*\right) + \frac{1}{2}\rho v^2 + p\left(\rho, T^*\right)\right) v^i\right] = 0.$$

Therefore the supplementary law becomes the energy equation in the isothermal case! In the case of non-isothermal Euler system, the entropy principle is required, when a weak solution evolves, in such a way that the entropy grows. Instead, as is natural, in the mechanical isothermal case, we observe a decay of energy in a weak solution, and, in particular, it is true across a shock.

This example clearly illustrates the power of the concept of principal subsystem. We can identify a proper approximated system even though the system does not have relaxation times.

Concerning the subcharacteristic condition, it is automatically satisfied according with Theorem 2.3. In fact, in the case of Euler system, we have the following inequality:

$$\lambda_{\max} = \sqrt{\left(\frac{\partial p}{\partial \rho}\right)_s} > \bar{\lambda}_{\max} = \sqrt{\left(\frac{\partial p}{\partial \rho}\right)_T}.$$

2.5 Conservation and Balance Laws, and Equilibrium Subsystem

A particular case of (2.32) and (2.33) is the case where the first M equations are conservation laws, i.e., $\mathbf{r} \equiv 0$. This is the case of all RET theories. Then the block of conservation laws is expressed as

$$\partial_\alpha \mathbf{V}^\alpha = 0 \quad \Longleftrightarrow \quad \partial_\alpha\left(\frac{\partial h'^\alpha(\mathbf{v}', \mathbf{w}')}{\partial \mathbf{v}'}\right) = 0, \tag{2.42}$$

and a block of balance laws as

$$\partial_\alpha \mathbf{W}^\alpha = \mathbf{g} \quad \Longleftrightarrow \quad \partial_\alpha \left(\frac{\partial h'^\alpha(\mathbf{v}', \mathbf{w}')}{\partial \mathbf{w}'} \right) = \mathbf{g}(\mathbf{v}', \mathbf{w}'). \tag{2.43}$$

In this case, it is possible to define, as in usual thermodynamics, the equilibrium state:

Definition 2.6 (Equilibrium State) An equilibrium state is a state for which the entropy production $\Sigma|_E$ vanishes and hence attains its minimum value. The suffix E indicates that a quantity is evaluated at an equilibrium state.

It is possible to prove the following theorem [124, 165]:

Theorem 2.4 (Equilibrium Manifold) *In an equilibrium state, under the assumption of dissipative productions i.e., the assumption that*

$$\mathbf{D} = \frac{1}{2} \left\{ \frac{\partial \mathbf{g}}{\partial \mathbf{w}'} + \left(\frac{\partial \mathbf{g}}{\partial \mathbf{w}'} \right)^T \right\} \Bigg|_E \quad \textit{is negative definite}, \tag{2.44}$$

the production vanishes and the main field components vanish except for the first M components. Thus

$$\mathbf{g}|_E = 0, \qquad \mathbf{w}'|_E = 0. \tag{2.45}$$

Therefore, in an equilibrium state, all the components of the main field corresponding to the Lagrange multipliers of the balance laws (2.43) vanish, and only the Lagrange multipliers corresponding to the conservation laws (2.42) survive. This confirms again the importance of the main field!

We have another important characteristic property of the equilibrium state [172, 188]:

Theorem 2.5 (Maximum of the Entropy) *In equilibrium, the entropy density h^0 is maximal, i.e.,*

$$h^0 < h^0|_E \quad \forall \mathbf{u}' \neq \mathbf{u}'|_E, \quad \textit{where} \quad h^0|_E = h^0\left(\mathbf{v}'|_E, 0\right),$$

and $\mathbf{v}'|_E$ denotes the restriction in equilibrium of the components of the main field corresponding to the Lagrange multipliers of conservation laws.

Therefore we have found, at this general level, also the well-known property of the entropy in thermodynamics, i.e., the property of maximal entropy in equilibrium.

If we limit our study within one-dimensional space, the system (2.32) and (2.33) assumes the form:

$$\begin{cases} \mathbf{V}_t + \left(k'_{\mathbf{v}'}\right)_x = 0, \\ \mathbf{W}_t + \left(k'_{\mathbf{w}'}\right)_x = -\mathbf{G}\left(\mathbf{v}', \mathbf{w}'\right)\mathbf{w}' \end{cases} \tag{2.46}$$

with $\mathbf{V} = h'_{\mathbf{v}'}$, $\mathbf{W} = h'_{\mathbf{w}'}$ and $h' = h'^0$, $k' = h'^1$. The matrix $\mathbf{G}$ is a positive definite $(N - M) \times (N - M)$ matrix.

2.6 Qualitative Analysis

In this section, we discuss the importance of the entropy principle for the qualitative analysis of the Cauchy problem.

2.6.1 Competition Between Hyperbolicity and Dissipation

In the general theory of hyperbolic conservation laws and hyperbolic-parabolic conservation laws, the existence of a strictly convex entropy function is a basic condition for the well-posedness. In fact, if the fluxes $\mathbf{F}^i$ and the production $\mathbf{f}$ are smooth enough in a suitable convex open set $\in \mathbb{R}^N$, it is well known that system (2.2) has a unique local (in time) smooth solution for smooth initial data [177, 189, 190].

However, in a general case, even for arbitrarily small and smooth initial data, there is no global continuation for these smooth solutions, which may develop singularities, shocks, or blowup, in a finite time, see for instance [191, 192].

On the other hand, in many physical examples, thanks to the interplay between the source term and the hyperbolicity, there exist global smooth solutions for a suitable set of initial data. This is the case, for example, of the isentropic Euler system with damping [193, 194]. Roughly speaking, for such a system, the relaxation term induces a dissipative effect. This effect then competes with the hyperbolicity. If the dissipation is sufficiently strong so as to dominate the hyperbolicity, the system is *dissipative*, and we expect that the classical solution exists for all time and converges to a constant state. While, if the dissipation and the hyperbolicity are equally important, we expect that only part of the perturbation diffuses. In the latter case the system is called *of composite type* by Zeng [195].

2.6.1.1 A Simple Example: Burgers' Equation

The simplest example of the problem is presented by Burgers' equation:

$$u_t + uu_x = 0, \qquad u(x,0) = u_0(x), \tag{2.47}$$

which, by using the method of characteristics, can be rewritten as

$$\begin{cases} \dfrac{du}{dt} = 0, & u(0) = u_0(x_0), \\ \dfrac{dx}{dt} = u, & x(0) = x_0, \end{cases}$$

and therefore admits the general solution:

$$\begin{cases} u(x,t) = u_0(x_0), \\ x = x_0 + u_0(x_0)t. \end{cases}$$

Here u is a function of (x, t) through the parameter x_0.

The invertibility for x_0 as a function of (x, t) is lost, for each value of x_0, at the time

$$t_c(x_0) = -\frac{1}{u_0'(x_0)}.$$

The critical time is defined by the smallest positive value of $t_c(x_0)$:

$$t_{cr} = \inf_{x_0} \{t_c(x_0) > 0\}.$$

If, instead of (2.47), we take into account a dissipative production term like

$$u_t + uu_x = -\nu u, \qquad u(x,0) = u_0(x)$$

with ν = constant > 0, then we have

$$\begin{cases} \dfrac{du}{dt} = -\nu u, & u(0) = u_0(x_0), \\ \dfrac{dx}{dt} = u, & x(0) = x_0, \end{cases}$$

which admits the solution:

$$\begin{cases} u(x,t) = u_0(x_0)e^{-\nu t}, \\ x = x_0 + \dfrac{u_0(x_0)}{\nu}\left(1 - e^{-\nu t}\right), \end{cases}$$

and we have

$$t_c(x_0) = -\frac{1}{\nu}\log\left(1+\frac{\nu}{u_0'(x_0)}\right), \qquad t_{cr} = \inf_{x_0}\{t_c(x_0) > 0\}.$$

Therefore, if

$$\nu > \nu^*$$

with $\nu^* = \max_{x_0}|u_0'(x_0)|$, the classical solution exists any time: *the dissipation wins over the hyperbolicity*. While, if

$$\nu \leqslant \nu*,$$

the hyperbolicity dominates the dissipation and, in general, we cannot expect global existence of a smooth solution. See Fig. 2.1 for these cases.

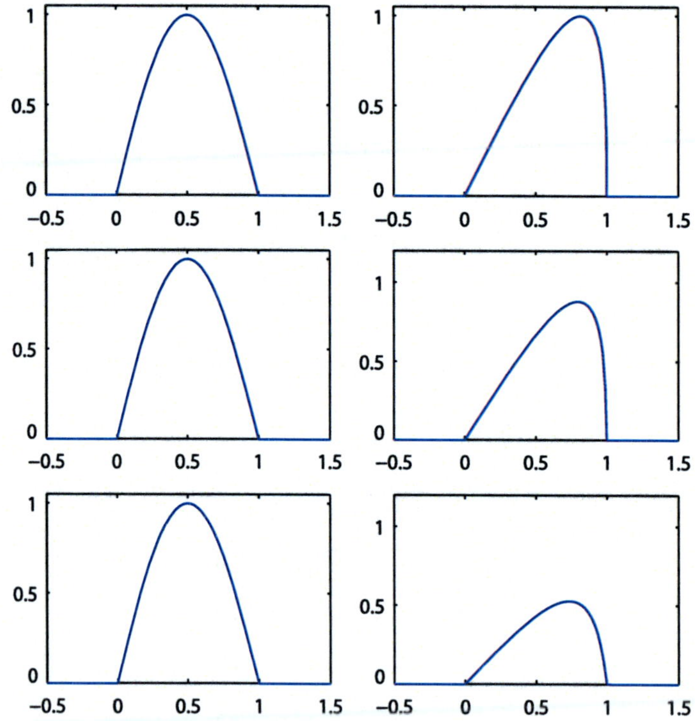

Fig. 2.1 Burgers' equation: Evolution of the profile from the initial time (left) to the critical time (right) for $\nu = 0$, $0 < \nu \leqslant \nu^*$, and for $\nu > \nu^*$, respectively, from top to bottom

2.6.2 Shizuta–Kawashima K-Condition

In general, there are several ways to identify whether a hyperbolic system with relaxation is of dissipative type or of composite type. One way, which is completely parallel to the case of the hyperbolic-parabolic system, was discussed first by Shizuta and Kawashima [168]. It is known in the literature as *K-condition* or *genuine coupling*.

Definition 2.7 (K-Condition) A system (2.3) satisfies the K-condition if, in the equilibrium manifold, any right characteristic eigenvectors $\mathbf{d}$ of (2.4) are not in the null space of $\nabla \mathbf{f}$, where $\nabla \equiv \partial/\partial \mathbf{u}$:

$$\left(\nabla \mathbf{f}\, \mathbf{d}^I\right)\Big|_E \neq 0 \quad \forall \mathbf{d}^I, \quad I = 1, 2, \dots N. \tag{2.48}$$

2.6.3 Global Existence and Stability of Constant State

For dissipative one-dimensional systems (2.46) satisfying the K-condition, it is possible to prove the following global existence theorem by Hanouzet and Natalini [169]:

Theorem 2.6 (Global Existence) *Assume that the system (2.46) is strictly dissipative: (2.44), and that the K-condition is satisfied. Then there exists $\delta > 0$, such that, if $\left\|\mathbf{u}'(x, 0)\right\|_2 \leqslant \delta$, there is a unique global smooth solution, which verifies*

$$\mathbf{u}' \in C^0([0, \infty); H^2(R) \cap C^1([0, \infty); H^1(R)).$$

This global existence theorem was generalized to a higher-dimensional case by Yong [170] and successively by Bianchini, Hanouzet, and Natalini [171].

Moreover Ruggeri and Serre [172] proved that the constant equilibrium state is stable:

Theorem 2.7 (Stability of an Equilibrium State) *Under natural hypotheses of strongly convex entropy, strict dissipativeness, genuine coupling, and "zero mass" initial for the perturbation of the equilibrium variables, the constant solution stabilizes in such a way that*

$$\|\mathbf{u}\,(t)\|_2 = O\left(t^{-1/2}\right).$$

In [169], the authors reported several examples of dissipative systems satisfying the K-condition: the p-system with damping, the Suliciu model for the isothermal viscoelasticity, the Kerr-Debye model in nonlinear electromagnetism, and the Jin-Xin relaxation model. Dafermos showed the existence and long time behavior of spatially periodic BV solutions [196].

2.6.3.1 An Example: Global Existence Without the K-Condition

Zeng [195] considered a toy model of vibrational nonequilibrium gas in Lagrangian variables. She proved that, if the system is of composite type, the global existence holds. Therefore the K-condition is only a sufficient condition for the global existence of smooth solutions.

An intriguing open problem is to make clear the following questions: Does a weaker K-condition that is necessary to ensure global solutions exist? If such a condition exists, does it have a physical meaning that gives a possible new principle of RET in addition to the convexity condition of entropy?

Lou and Ruggeri [173] observed that there indeed exists a weaker K-condition that is a necessary (but unfortunately not sufficient) condition for the global existence of smooth solutions. Instead of the condition that the right eigenvectors are not in the null space of $\nabla \mathbf{f}$, they posed this condition only on the right eigenvectors corresponding to genuine nonlinear eigenvalues. It was proved that the assumptions of the previous theorems are fulfilled in both classical [174] and relativistic [175, 176] RET theories of monatomic gases, and also in the theory of mixtures of gases with multi-temperature [158].

2.7 Galilean Invariance

The Galilean invariance of a general system of balance laws was studied by Ruggeri [167]. It was proved that the Galilean invariance imposes the specific velocity dependence of the density, flux, and production.

More precisely, if we split the field $\mathbf{u}$ into two fields $(\mathbf{v}, \mathbf{w})$, i.e., $\mathbf{u} \equiv (\mathbf{v}, \mathbf{w})$ where $\mathbf{v} \in \mathbb{R}^3$ is the velocity and $\mathbf{w} \in \mathbb{R}^{N-3}$ is the other objective field, we have the following theorem:

Theorem 2.8 (Galilean Invariance) *The system of balance laws (1.2) is invariant under the Galilean transformation if there exists $N \times N$ matrix $\mathbf{X}(\mathbf{v})$ such that:*

$$\begin{cases} \mathbf{F}^0\,(\mathbf{v}, \mathbf{w}) \;= \mathbf{X}(\mathbf{v})\, \hat{\mathbf{F}}^0(\mathbf{w}), \\ \boldsymbol{\Phi}^i\,(\mathbf{v}, \mathbf{w}) = \mathbf{X}(\mathbf{v})\, \hat{\boldsymbol{\Phi}}^i(\mathbf{w}), \\ \mathbf{f}\,(\mathbf{v}, \mathbf{w}) \quad = \mathbf{X}(\mathbf{v})\, \hat{\mathbf{f}}(\mathbf{w}), \end{cases} \tag{2.49}$$

where

$$\boldsymbol{\Phi}^i = \mathbf{F}^i - \mathbf{F}^0 v^i, \qquad \hat{\boldsymbol{\Phi}}^i = \hat{\mathbf{F}}^i. \tag{2.50}$$

and $\mathbf{X}(\mathbf{v})$ *is an exponential matrix:*

$$\mathbf{X}(\mathbf{v}) = e^{\mathbf{A}^r v_r} = \mathbf{I} + \mathbf{A}^r v_r + \frac{1}{2}\mathbf{A}^r\mathbf{A}^s v_r v_s + \dots \tag{2.51}$$

with $\mathbf{A}^r$ *being three* $(N \times N)$ *constant matrices such that*

$$\mathbf{A}^r\mathbf{A}^s = \mathbf{A}^s\mathbf{A}^r, \quad \forall\, r, s = 1, 2, 3.$$

The hat on a quantity indicates, here and hereafter, the corresponding quantity evaluated at zero velocity (intrinsic quantities):

$$\hat{\mathbf{F}}^0(\mathbf{w}) = \mathbf{F}^0\,(0, \mathbf{w})\,, \quad \hat{\boldsymbol{\Phi}}^i\,(\mathbf{w}) = \boldsymbol{\Phi}^i(0, \mathbf{w}), \quad \hat{\mathbf{f}}\,(\mathbf{w}) = \mathbf{f}(0, \mathbf{w}).$$

From (2.51) we have:

$$\frac{\partial \mathbf{X}(\mathbf{v})}{\partial v_s} = \mathbf{X}\mathbf{A}^s = \mathbf{A}^s\mathbf{X} \qquad \rightarrow \qquad \mathbf{A}^s = \left.\frac{\partial \mathbf{X}}{\partial v_s}\right|_{\mathbf{v}=0}. \tag{2.52}$$

2.7.1 Remark and Example of Galilean Invariance for Balance Laws of a Fluid

It is important to observe that Theorem 2.8 is valid for any systems of balance laws irrespective of constitutive equations. As an example, let us consider the system of a fluid (1.3), which can be rewritten in the form (1.2) with (1.4). In this case we have

$$\mathbf{F}^0 = \begin{pmatrix} \rho \\ \rho v_k \\ \frac{1}{2}\rho v^2 + \rho\varepsilon \end{pmatrix}, \quad \boldsymbol{\Phi}^i = \begin{pmatrix} 0 \\ -t_{ik} \\ -t_{ik}v_k + q_i \end{pmatrix}, \quad \mathbf{f} = \begin{pmatrix} 0 \\ \rho b_k \\ \rho b_j v_j + r \end{pmatrix},$$

$$\hat{\mathbf{F}}^0 = \begin{pmatrix} \rho \\ 0_j \\ \rho\varepsilon \end{pmatrix}, \quad \hat{\boldsymbol{\Phi}}^i = \begin{pmatrix} 0 \\ -t_{ij} \\ q_i \end{pmatrix}, \quad \hat{\mathbf{f}} = \begin{pmatrix} 0 \\ \rho b_j \\ r \end{pmatrix}.$$

The matrices $\mathbf{A}^r$ and $\mathbf{X}(\mathbf{v})$ in the present case read

$$\mathbf{A}^r = \begin{bmatrix} 0 & 0 & 0 \\ \delta_k^r & 0 & 0 \\ 0 & \delta_j^r & 0 \end{bmatrix} \quad \text{and} \quad \mathbf{X}(\mathbf{v}) = \begin{bmatrix} 1 & 0 & 0 \\ v_k & \delta_{kj} & 0 \\ \frac{1}{2}v^2 & v_j & 0 \end{bmatrix}. \tag{2.53}$$

We can appreciate, even if this simple example, the validity of (2.49) for any constitutive equations.

2.7.2 Compatibility Between Entropy Principle and Galilean Invariance for the System with Local Constitutive Equations

Now we return to the system with local constitutive equations. We discuss the following question: Because the dependence of field variables on the velocity is prescribed by the Galilean invariance (2.49), the natural question is whether the relations (2.49) are compatible with the constraints (2.15) due to the entropy principle.

We observe first that, for the supplementary entropy law, we have

$$h^0 = h = \hat{h}, \qquad h^i = \hat{h} v_i + \varphi_i, \qquad \varphi_i = \hat{h}^i. \tag{2.54}$$

In [167], it was proved that the compatibility exists and that, as we expect, the entropy principle becomes a constraint only for the objective quantities $\mathbf{w}$. In fact, it was proved that

$$\mathbf{u}' = \hat{\mathbf{u}}'\mathbf{X}^{-1}(\mathbf{v}) = \hat{\mathbf{u}}'\mathbf{X}\,(-\mathbf{v})\,. \tag{2.55}$$

Then from (2.13) and (2.54) we have:

$$\begin{aligned} &h'^0 = h' = \hat{h}', \qquad h'^i = \hat{h}' v_i + \hat{h}'^i, \\ &\text{where} \qquad \hat{h}'^\alpha = \hat{\mathbf{u}}' \cdot \hat{\mathbf{F}}^\alpha - \hat{h}^\alpha. \end{aligned} \tag{2.56}$$

By taking into account of (2.49), (2.50), (2.51), and (2.55), $(2.11)_1$ becomes

$$\hat{\mathbf{u}}' \cdot d\hat{\mathbf{F}}^0 = d\hat{h}, \qquad \hat{\mathbf{u}}' \cdot d\hat{\mathbf{F}}^i = d\hat{h}^i, \tag{2.57}$$

together with the constraints:

$$\hat{\mathbf{u}}' \cdot \mathbf{A}^r \hat{\mathbf{F}}^0 = 0, \qquad \hat{\mathbf{u}}' \cdot \mathbf{A}^r \hat{\mathbf{F}}^i = -\hat{h}'^0 \delta^{ir}. \tag{2.58}$$

While, from (2.13), we obtain

$$\begin{gathered} \hat{\mathbf{F}}^0 \cdot d\hat{\mathbf{u}}' = d\hat{h}'^0, \qquad \hat{\mathbf{F}}^i \cdot d\hat{\mathbf{u}}' = d\hat{h}'^i, \\ h'^0 = \hat{h}'^0, \quad h'^i = \hat{h}'^0 v^i + \hat{h}'^i. \end{gathered}$$

The presence of the three constraints $(2.58)_1$ is not surprising because, while $\mathbf{u}'$ is an $\mathbb{R}^N$ field, $\hat{\mathbf{u}}'$ are not independent variables. There are only $N - 3$ independent components of $\hat{\mathbf{u}}'$, which, together with the 3 components of the velocity, forms a field. Therefore the entropy principle is full compatible with the Galilean invariance and it becomes constraints for the proper constitutive functions (see (2.57)).

The convexity condition (2.19), in particular, implies (see for details [167]):

$$\hat{Q} = \delta\hat{\mathbf{u}}' \cdot \delta\hat{\mathbf{F}} < 0. \tag{2.59}$$

An alternative procedure to satisfy the Galilean invariance and the entropy principle avoiding the constraints (2.58) was made in [197].

2.7.3 Field Equations in Terms of Intrinsic Quantities

If we introduce the material derivative (1.7) and take into account (2.49) and (2.52), we can prove [167] that, within classical solutions, every solution of (2.2) is also the solution of

$$\frac{d\hat{\mathbf{F}}^0}{dt} + \hat{\mathbf{F}}^0 \frac{\partial v_i}{\partial x_i} + \frac{\partial \hat{\mathbf{F}}^i}{\partial x_i} + \mathbf{A}^r \left\{ \hat{\mathbf{F}}^0 \frac{d v_r}{dt} + \hat{\mathbf{F}}^i \frac{\partial v_r}{\partial x_i} \right\} = \hat{\mathbf{f}}.$$

These equations are much more useful for practical purposes. However, for the study of weak solutions, in particular, shock waves, we need the system of field equations in balance form.

2.7.4 Diagonal Structure in RET

In the case of moment theory, the density $\mathbf{F}^0$, the (non-convective or intrinsic) flux $\boldsymbol{\Phi}^i$, and the production $\mathbf{f}$ are expressed as

$$\mathbf{F}^0 \equiv \begin{pmatrix} F \\ F_{k_1} \\ F_{k_1 k_2} \\ \vdots \\ F_{k_1 k_2 \dots k_N} \end{pmatrix}, \quad \boldsymbol{\Phi}^i \equiv \begin{pmatrix} \Phi_i \\ \Phi_{ik_1} \\ \Phi_{ik_1 k_2} \\ \vdots \\ \Phi_{ik_1 k_2 \dots k_N} \end{pmatrix}, \quad \mathbf{f} \equiv \begin{pmatrix} f \\ f_{k_1} \\ f_{k_1 k_2} \\ \vdots \\ f_{k_1 k_2 \dots k_N} \end{pmatrix}, \tag{2.60}$$

where $F_{k_1 \dots k_j}$, $\Phi_{ik_1 \dots k_j}$, and $f_{k_1 \dots k_j}$ are symmetric tensors. Thus the system of balance equations (2.2) has a natural order with increasing tensorial rank. Each block of tensorial equations of rank j governs the evolution of a new quantity $F_{k_1 k_2 \dots k_j}$.

The field equations have the form:

$$\frac{\partial F_{i_1 \dots i_l}}{\partial t} + \frac{\partial (F_{i_1 \dots i_l} v_i + \Phi_{i i_1 \dots i_l})}{\partial x^i} = f_{i_1 \dots i_l} \quad (l = 0, 1, 2, \dots, N).$$

We assume that all subsystems that result from (2.60) by ignoring the tensor equation of rank N, or (N and $N-1$), or (N, $N-1$ and $N-2$), etc. have the Galilean invariance. Thus the matrix $\mathbf{X}(\mathbf{v})$ must be a sub-triangular block matrix. It is possible to prove (see [25, 167]) that the matrix $\mathbf{X}(\mathbf{v})$ is a polynomial matrix in $\mathbf{v}$ of order N whose diagonal elements are blocks of Kronecker delta. It reads

$$\mathbf{X}(\mathbf{v}) \equiv \begin{bmatrix} 1 & & & & & \\ v_{k_1} & \delta^{h_1}_{k_1} & & & & \\ v_{k_1} v_{k_2} & 2\delta^{h_1}_{(k_1} v_{k_2)} & \cdots & & & \\ v_{k_1} v_{k_2} v_{k_3} & 3\delta^{h_1}_{(k_1} v_{k_2} v_{k_3)} & \cdots & \delta^{h_1}_{k_1}\delta^{h_2}_{k_2}\delta^{h_3}_{k_3} & & \\ \vdots & \vdots & \vdots & \vdots & & \\ v_{k_1} v_{k_2} \cdots v_{k_n} & \binom{n}{1}\delta^{h_1}_{(k_1} v_{k_2} \cdots v_{k_n)} & \cdots & \vdots & & \\ \vdots & \vdots & \vdots & \vdots & & \vdots \quad \vdots \\ v_{k_1} v_{k_2} \cdots v_{k_N} & \binom{N}{1}\delta^{h_1}_{(k_1} v_{k_2} \cdots v_{k_N)} & \cdots & \binom{N}{3}\delta^{h_1}_{(k_1}\delta^{h_2}_{k_2}\delta^{h_3}_{k_3} v_{k_4} \cdots v_{k_N)} & \cdots & \delta^{h_1}_{k_1}\delta^{h_2}_{k_2} \cdots \delta^{h_N}_{k_N} \end{bmatrix}, \tag{2.61}$$

while the matrices A^r are nil-potent matrices: $\mathbf{A}^{k_1}\mathbf{A}^{k_2}\ldots\mathbf{A}^{k_{N+1}} = 0$ for all $k_1, k_2, \ldots, k_{N+1}$ over $1, 2, 3$, and have the following expression:

$$\mathbf{A}^r = \begin{bmatrix} 0 & 0 & 0 & & 0 \\ \delta^r_{k_1} & 0 & 0 & & 0 \\ 0 & 2\delta^{(rh_1)}_{(k_1 k_2)} & 0 \cdots & & 0 \\ \vdots & \vdots & \vdots & & \vdots \\ 0 & 0 & 0 \cdots & N\delta^{(rh_1 h_2 \ldots h_{N-1})}_{(k_1 k_2 \ldots k_N)} & 0 \end{bmatrix}.$$

Thus we are able to decompose tensors of arbitrary rank into the velocity-dependent part and intrinsic part. In particular, the decomposition of an arbitrary tensorial density is given by

$$F_{i_1 \ldots i_l} = \hat{F}_{i_1 \ldots i_l} + \binom{l}{1} \hat{F}_{(i_1 \ldots i_{l-1}} v_{i_l)} + \binom{l}{2} \hat{F}_{(i_1 \ldots i_{l-2}} v_{i_{l-1}} v_{i_l)} + \ldots$$
$$+ \binom{l}{l-1} \hat{F}_{(i_1} v_{i_2 \ldots} v_{i_l)} + \hat{F} v_{i_1 \ldots} v_{i_l}.$$

The same decompositions can be made for $\boldsymbol{\Phi}^i$ and $\mathbf{f}$.

2.7.4.1 Example of Euler Fluid

As an example, we consider the Euler fluid with no external force and zero heat supply. The Galilean invariance was proved in Sect. 2.7.1 for general fluids. In the present case, $t_{ik} = -p\delta_{ik}$, $q_i = 0$, $\hat{h}^0 = \rho s$, $\hat{h}^i = 0$, and $\hat{\Sigma} = 0$. The main field in the present case is given by (2.29) and therefore the main field evaluated for zero velocity is:

$$\hat{\mathbf{u}}' \equiv \frac{1}{T}\left(-g, 0_j, 1\right).$$

From (2.26) and (2.53), it is easy to check the validity of the relation (2.55) in the present case. And, from (2.30), the relation $(2.57)_1$ reduces to the Gibbs equation (1.8), while $(2.57)_2$ and (2.58) are identically satisfied.

Chapter 3
Waves in Hyperbolic Systems

Abstract Wave propagation phenomena give us an important mean to check the validation of a nonequilibrium thermodynamics theory. In this chapter, we present a short review on the modern theory of wave propagation for hyperbolic systems. Firstly, we present the theory of linear waves emphasizing the role of the dispersion relation. The high frequency limit in the dispersion relation is also studied. Secondly, nonlinear acceleration waves are discussed together with the transport equation and the critical time. Thirdly, we present the main results concerning shock waves as a particular class of weak solutions and the admissibility criterion to select physical shocks (Lax condition, entropy growth condition, and Liu condition). Fourth, we discuss traveling waves, in particular, shock waves with structure. The subshock formation is particularly interesting. The Riemann problem and the problem of the large-time asymptotic behavior are also discussed. Lastly, we present toy models to show explicitly some interesting features obtained here.

3.1 Linear Wave

A typical method to test a theory of nonequilibrium thermodynamics is to study plane harmonic waves and to compare the theoretical prediction of the dispersion relation with the experimental data. Let us consider here some fundamental properties of the solutions for a general linear hyperbolic system. We therefore limit our analysis within the one-space-dimensional problem.

3.1.1 Plane Harmonic Wave and the Dispersion Relation

Let us consider the quasi-linear system (2.3). Because of the hyperbolicity, we may write it in the normal form:

$$\frac{\partial \mathbf{u}}{\partial t} + \mathbf{A}(\mathbf{u})\frac{\partial \mathbf{u}}{\partial x} = \mathbf{f}(\mathbf{u}),$$

T. Ruggeri, M. Sugiyama, *Classical and Relativistic Rational Extended Thermodynamics of Gases*, https://doi.org/10.1007/978-3-030-59144-1_3

where we have taken into account the condition that all quantities depend on only the position x in addition to the time t. We linearize this equation by setting

$$\mathbf{u} = \tilde{\mathbf{u}} + \bar{\mathbf{u}},$$

where $\bar{\mathbf{u}}$ is a small perturbation of an equilibrium state $\tilde{\mathbf{u}}$ for which $\mathbf{f}(\tilde{\mathbf{u}}) = 0$. Thus we obtain the linearized equation with respect to $\bar{\mathbf{u}}$:

$$\frac{\partial \bar{\mathbf{u}}}{\partial t} + \tilde{\mathbf{A}}\frac{\partial \bar{\mathbf{u}}}{\partial x} = \tilde{\mathbf{B}}\bar{\mathbf{u}}, \qquad \text{where} \qquad \tilde{\mathbf{A}} = \mathbf{A}(\tilde{\mathbf{u}}), \quad \tilde{\mathbf{B}} = (\nabla \mathbf{f})_{\tilde{\mathbf{u}}}, \quad \nabla \equiv \frac{\partial}{\partial \mathbf{u}}. \tag{3.1}$$

We look for the solution of the form:

$$\bar{\mathbf{u}} = \mathbf{w}e^{i(\omega t - kx)}, \tag{3.2}$$

which represents a plane harmonic wave with real (angular) frequency ω, complex wave number $k = k_r + ik_i$, and complex amplitude $\mathbf{w}$ traveling in the x-direction.

Substitution of (3.2) into (3.1) provides a homogeneous algebraic linear system of the form:

$$\left(\mathbf{I} - z\tilde{\mathbf{A}} + \frac{i}{\omega}\tilde{\mathbf{B}}\right)\mathbf{w} = 0,$$

where $\mathbf{I}$ is the unit matrix, and z stands for k/ω.

For non-trivial solutions, the *dispersion relation* (from now we omit the tilde)

$$\det\left(\mathbf{I} - z\mathbf{A} + \frac{i}{\omega}\mathbf{B}\right) = 0 \tag{3.3}$$

must be satisfied. The dispersion relation permits the calculation of the phase velocity v_{ph} and of the attenuation factor α in terms of the frequency ω:

$$v_{ph} = \frac{1}{\mathscr{R}(z)} = \frac{\omega}{k_r}, \qquad \alpha = -\omega\mathscr{I}(z) = -k_i. \tag{3.4}$$

For linear stability, $\alpha(\omega)$ must be positive (negative) for waves traveling to the right (left). When the perturbation $\bar{\mathbf{u}}$ starts to evolve from $x = 0$ at time $t = 0$, we have the linear stability condition: $\alpha(\omega)x > 0$. Since the path of the perturbation is given by $x = \lambda t$ (where λ denotes an eigenvalue of the matrix $\mathbf{A}$), we may write the condition of linear stability as

$$\alpha(\omega)\lambda > 0. \tag{3.5}$$

3.1.2 High Frequency Limit

Rigorous expressions of v_{ph} and α in the limit of high frequency (i.e., $\omega \to \infty$) were given by Muracchini, Ruggeri, and Seccia in the case of simple characteristic eigenvalue λ [122], and by Banach, Larecki, and Ruggeri in the case of multiple eigenvalue [123]. In the derivation process, these authors considered the formal power-series expansions of z and $\mathbf{w}$ with respect to $1/\omega$, that is,

$$z = \sum_{\alpha \geqslant 0} \frac{z_\alpha}{\omega^\alpha}, \qquad \mathbf{w} = \sum_{\beta \geqslant 0} \frac{\mathbf{w}_\beta}{\omega^\beta}, \tag{3.6}$$

where z_α and $\mathbf{w}_\beta$ are expansion coefficients. Insertion of (3.6) into (3.3) provides a recurrence formula.

In the case that the eigenvalue λ has multiplicity 1, it was proved that [122]

$$\lim_{\omega \to \infty} v_{ph}(\omega) = \lambda \qquad \text{and} \qquad \lim_{\omega \to \infty} \alpha(\omega)\lambda = -\mathbf{l}\,\mathbf{B}\,\mathbf{d}. \tag{3.7}$$

The first equation says that the phase velocity coincides with the characteristic velocity in the limit of high frequency. The second result in (3.7) furnishes the condition of linear stability because, from (3.5), we have the condition:

$$\mathbf{l}\,\mathbf{B}\,\mathbf{d} < 0. \tag{3.8}$$

Later we shall see that the condition (3.8) guarantees the nonlinear stability as well.

In the case that the eigenvalue λ has multiplicity $m > 1$, the condition $(3.7)_1$ remains the same, while $(3.7)_2$ becomes [123]:

$$\lim_{\omega \to \infty} \alpha_K(\omega)\lambda = -\mu_K, \qquad K = 1, 2, \ldots, m$$

where μ_K's are m eigenvalues of the matrix:

$$b_{IJ} = \mathbf{l}_I\,\mathbf{B}\,\mathbf{d}_J, \qquad I, J = 1, 2, \ldots, m \tag{3.9}$$

with $\mathbf{l}_I$ and $\mathbf{d}_J$ being, respectively, m linearly independent left and right eigenvectors of the matrix $\mathbf{A}$ corresponding to the eigenvalue λ. In the present case, the stability condition (3.8) is substituted by the condition that the μ_K's are real and negative for all $K(= 1, 2, \ldots, m)$.

3.2 Acceleration Wave

For a generic quasi-linear hyperbolic system, it is possible to consider a particular class of solutions that characterizes the so-called *weak discontinuity waves* or, in

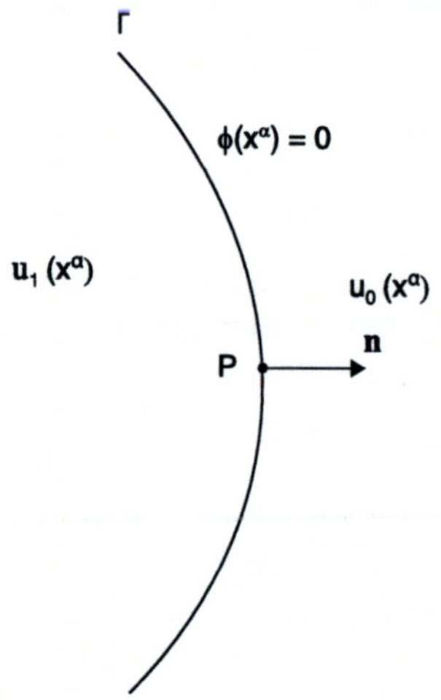

Fig. 3.1 Front of an acceleration wave. $\mathbf{u}_0$ and $\mathbf{u}_1$ denote, respectively, the unperturbed and perturbed state

the language of continuum mechanics, *acceleration waves*. Let us study a moving surface (wave front) Γ prescribed by the Cartesian equation $\phi(x,t) = 0$ that separates the space into two subspaces (see Fig. 3.1). Ahead of the wave front we have a known unperturbed field $\mathbf{u}_0(x,t)$, and behind the wave front we have an unknown perturbed field $\mathbf{u}(x,t)$. Both the fields $\mathbf{u}_0$ and $\mathbf{u}$ are supposed to be regular solutions of (2.1) and to be continuous across the surface Γ, but to be discontinuous in the normal derivative, i.e.,

$$[\![\mathbf{u}]\!] = 0, \quad \left[\!\!\left[\frac{\partial \mathbf{u}}{\partial \phi}\right]\!\!\right] = \boldsymbol{\mathcal{A}} \neq 0, \tag{3.10}$$

where the square brackets indicate the jump at the wave front:[1]

$$[\![\cdot]\!] = (\cdot)_{\phi=0^-} - (\cdot)_{\phi=0^+}.$$

We have the following well-known results [120, 121, 198]:

1. The normal speed $U = -\dfrac{\phi_t}{|\nabla\phi|}$ is equal to a characteristic speed evaluated at $\mathbf{u}_0$:

$$U = \lambda(\mathbf{u}_0).$$

[1] For simplicity, we use the symbols g and g_0 for the values of a generic quantity g evaluated at Γ with the condition that $\phi \to 0^-$ and $\phi \to 0^+$, respectively.

2. The jump vector $\boldsymbol{\mathcal{A}}$ is proportional to the right eigenvector $\mathbf{d}$ of the eigenvalue λ evaluated at $\mathbf{u}_0$:

$$\boldsymbol{\mathcal{A}} = \mathscr{A}\,\mathbf{d}(\mathbf{u}_0). \tag{3.11}$$

3. The *amplitude* $\mathscr{A}$ satisfies the *Bernoulli equation* along the characteristic line:

$$\frac{d\mathscr{A}}{dt} + a(t)\mathscr{A}^2 + b(t)\mathscr{A} = 0, \tag{3.12}$$

where d/dt indicates the time derivative along the bicharacteristics lines, and $a(t)$ and $b(t)$ are known functions of the time through $\mathbf{u}_0$.

For an example in the case of one space-dimension, we have [121]:

$$\begin{aligned}
&\frac{d}{dt} = \partial_t + \lambda_0 \partial_x, \quad \frac{dx}{dt} = \lambda_0 \ (characteristic), \quad \lambda_0 = \lambda(\mathbf{u}_0),\\
&a(t) = \phi_x (\nabla\lambda \cdot \mathbf{d})_0, \qquad \nabla = \frac{\partial}{\partial \mathbf{u}},\\
&b(t) = \left\{ d_j \left(\frac{\partial l_i}{\partial u_j} - \frac{\partial l_j}{\partial u_i} \right) \frac{du_i}{dt} + (\mathbf{l} \cdot \mathbf{u}_x)(\nabla\lambda \cdot \mathbf{d}) - \frac{\partial(\mathbf{l} \cdot \mathbf{f})}{\partial u_j} d_j \right\}_0,\\
&\frac{d\phi_x}{dt} + (\nabla\lambda \cdot \mathbf{u}_x)_0\, \phi_x = 0, \qquad \phi_x(0) = 1.
\end{aligned}$$

The solution of (3.12) is expressed as

$$\mathscr{A}(t) = \frac{\mathscr{A}(0) exp\left(-\int_0^t b(\xi) d\xi\right)}{1 + \mathscr{A}(0) \int_0^t a(\zeta) exp\left(-\int_0^\zeta b(\xi) d\xi\right) d\zeta}. \tag{3.13}$$

In order to make an analysis of the evolution of amplitude, we recall that, in the theory of hyperbolic systems, a wave associated to a characteristic velocity λ is called:

- *genuinely nonlinear*, if

$$\nabla\lambda \cdot \mathbf{d} \neq 0 \quad \forall \mathbf{u}, \tag{3.14}$$

- *linearly degenerate* (or *exceptional*), if

$$\nabla\lambda \cdot \mathbf{d} \equiv 0 \quad \forall \mathbf{u}, \tag{3.15}$$

- *locally linearly degenerate* (or *locally exceptional*), if

$$\nabla\lambda \cdot \mathbf{d} = 0 \ \text{ for some } \mathbf{u}. \tag{3.16}$$

If a wave is genuinely nonlinear, there exists, in general, a critical time t_{cr} such that the denominator of (3.13) tends to zero and the discontinuity becomes unbounded. This instant usually corresponds to the emergence of a *strong discontinuity*, i.e., a shock wave, and the field itself presents a discontinuity across the wave front. If a wave satisfies (3.14), the coefficient $a(t) \neq 0$ and, without any loss of generality, it can always be chosen to be positive by an appropriate choice of the right eigenvector.

The qualitative analysis of the Bernoulli equation (3.12) was made by Ruggeri [121]. In particular, the stability of the zero solution of (3.12) (λ-*stability*) was proved under the conditions:

$$\int_0^\infty a(\xi) e^{-\int_0^\xi b(\zeta) d\zeta} d\xi = K < \infty, \tag{3.17}$$

$$\exists \text{ a constant } m \text{ such that } \int_0^t b(\xi) d\xi > m, \ \ \forall\ t > 0. \tag{3.18}$$

In fact, if (3.17) and (3.18) are fulfilled and if

$$|\mathscr{A}(0)| < \mathscr{A}_{cr}, \qquad \mathscr{A}_{cr} = 1/K, \tag{3.19}$$

the solution $\mathscr{A}(t)$ exists for all time and remains to be bounded. Moreover, if

$$\int_0^\infty b(\xi) d\xi = +\infty, \tag{3.20}$$

then $\lim_{t\to\infty} |\mathscr{A}(t)| = 0$ and the zero solution is asymptotically stable.

If (3.19) is not satisfied and

$$\mathscr{A}(0) < 0, \qquad |\mathscr{A}(0)| > \mathscr{A}_{cr}, \tag{3.21}$$

then, from (3.13), there exists a positive critical time t_{cr} given by

$$1 + \mathscr{A}(0) \int_0^{t_{cr}} a(\zeta) exp\left(-\int_0^\zeta b(\xi) d\xi\right) d\zeta = 0.$$

When the unperturbed state is an equilibrium constant state: $\mathbf{u}_0 = \mathbf{u}_E =$ constant, we have

$$\mathbf{f}(\mathbf{u}_E) = 0,$$

$$\phi_x = 1, \quad a = (\nabla\lambda \cdot \mathbf{d})_0 = \text{const.}, \quad b = -(\mathbf{l}\, \nabla \mathbf{f}\, \mathbf{d})_0 = \text{const.}, \tag{3.22}$$

and (3.13) becomes

$$\mathscr{A}(t) = \frac{\mathscr{A}(0)\, e^{-bt}}{1 - \mathscr{A}(0)\, \frac{a}{b}\left(e^{-bt} - 1\right)}. \tag{3.23}$$

In this case, $\mathscr{A}_{cr} = b/a$, and the conditions (3.17) and (3.20) reduce to

$$\mathbf{l}_0\, \mathbf{B}\, \mathbf{d}_0 < 0 \quad \text{with} \quad \mathbf{B} = (\nabla \mathbf{f})_0\,. \tag{3.24}$$

This is a necessary and sufficient condition for the zero solution of the Bernoulli equation to be asymptotically stable. From the definition of the K-condition (2.48), we see that, for genuinely nonlinear waves, the K-condition must be satisfied in order to satisfy (3.24) [173]. Moreover we see that the linear and nonlinear λ-stability conditions, namely (3.8) and (3.24), coincide.

In the present case, if the initial amplitude satisfies the inequalities (3.21), then we have a critical time:

$$t_{cr} = -\frac{1}{b} \log\left(1 + \frac{b}{a\mathscr{A}(0)}\right). \tag{3.25}$$

We notice that if the system is not dissipative and $b = 0$ the amplitude (3.13) becomes

$$\mathscr{A}(t) = \frac{\mathscr{A}(0)}{1 + \mathscr{A}(0)\, a\, t},$$

and if $\mathscr{A}(0) < 0$ the critical time always exists and is given by

$$t_{cr} = -\frac{1}{a\mathscr{A}(0)}.$$

Instead, if $a = (\nabla\lambda \cdot \mathbf{d})_E \equiv 0$ (linearly degenerate wave), the Bernoulli equation becomes linear:

$$\frac{d\mathscr{A}}{dt} + b\,\mathscr{A} = 0, \tag{3.26}$$

and it does not have the critical time. Furthermore, if $b = 0$, we have $\mathscr{A}(t) = \mathscr{A}(0)$. Hence the K-condition is not necessary in this case. Therefore, we have the weaker Lou-Ruggeri K-condition:

Definition 3.1 (Weak K-Condition) A system (2.3) satisfies the weak K-condition if, in the equilibrium manifold, the right characteristic eigenvectors $\mathbf{d}$ corresponding to the genuinely nonlinear eigenvalues are not in the null space of $\nabla \mathbf{f}$.

This condition together with the dissipation condition ($b > 0$ if $a \neq 0$) is a necessary and sufficient condition such that the discontinuity wave solution exists for all time for a small initial perturbation.

Finally we observe that if we introduce the operator $\delta = [\![\partial/\partial\phi]\!]$, the characteristic velocities λ and the right eigenvectors (2.4) can be obtained by the system (2.3) with the operators chain rule:

$$\frac{\partial}{\partial t} \to -\lambda\delta, \qquad \frac{\partial}{\partial x_i} \to n_i\delta, \quad \mathbf{f} \to 0, \tag{3.27}$$

for which $\delta\mathbf{u} \propto \mathbf{d}$. And the weak K-condition reads

$$\delta\mathbf{f}|_E = (\nabla\mathbf{f} \cdot \delta\mathbf{u})\,|_E \propto (\nabla\mathbf{f} \cdot \mathbf{d})\,|_E \neq 0 \tag{3.28}$$

for all genuinely nonlinear waves. That is, the weak K-condition requires that, if the production vanishes in equilibrium: $\mathbf{f}|_E = 0$, all the genuinely nonlinear discontinuity waves transport the disturbance of the normal derivative of the production: $\delta\mathbf{f}|_E \neq 0$.

Remark 3.1 This weak K-condition is, in general, only necessary for global existence of smooth solutions. In fact, it is satisfied for completely linearly degenerate systems (all waves are exceptional) or for semi-linear systems. But we know that, in general, smooth solutions of these cases cannot exist for all time. Therefore we need to add more conditions to the weak K-condition in order to ensure the global existence. This problem is, however, still open!

Nevertheless, it is important to have the necessary condition in order to select physically admissible productions and also to know the possibility of the global existence of the solution when the K-condition is violated. For example, for a mixture of Euler fluids with single temperature, it was proved that the weak K-condition is violated, while for a mixture with multi-temperature the condition is satisfied [158]. This implies an important fact that the model with multi-temperature is physically more realistic than the model with single temperature!

Remark 3.2 If λ has multiplicity $m > 1$ and the system (2.3) is a system of balance laws, then, according to the theorem by Boillat [199], the corresponding wave is exceptional, i.e., it satisfies (3.15) for all right eigenvectors $\mathbf{d}_I$ $(I = 1, 2, \ldots, m)$. In the present case, the jump vector (3.11) and the transport law for the amplitude (3.26) become, respectively,

$$\mathscr{A} = \sum_{I=1}^{m} \mathscr{A}_I \mathbf{d}_I$$

and

$$\frac{d\mathscr{A}_I}{dt} - \sum_{J=1}^{m} b_{IJ}\,\mathscr{A}_J = 0, \qquad I = 1, 2, \cdots, m$$

where the matrix b_{IJ} is the same as in the case of linear wave (3.9). Therefore, also in this case, conditions of linear and nonlinear stability coincide with each other, i.e., μ_K, eigenvalues of b_{IJ}, must be negative.

Remark 3.3 There exists a huge literature concerning physical applications of acceleration waves. Interested readers can find some references in the papers quoted in this chapter. Moreover, other interesting examples can be seen in [180, 198] and in the book of Sharma [200].

3.3 Shock Wave

Shock wave phenomena are fascinating and important from both mathematical and physical points of view. The phenomena have been studied intensively for long years [201]. And we have also many scientific and practical applications of shock waves in various fields, some of which are exemplified as follows:

- Aerodynamics: The knowledge of shock waves is essential for designing optimal geometry of an object moving faster than the sound velocity, for example, supersonic aircrafts, spacecrafts that reenter into the atmosphere of a planet [201].
- Material science: Shock waves are used to induce phase transitions or the chemical reactions in a material. An interesting example is the transformation of graphite to diamond induced by a shock wave [202].
- Astrophysics: Supernova explosion is a typical example involving shock waves. The analysis of supernova remnants gives us important information about its mechanism [203].
- Medical science: The kidney stone in a human body can be fragmented by a weak shock wave generated in the Extra-corporeal Shock-Wave Lithotripsy (ESWL) [204].

A recent survey on shock waves can be found in [205].

3.3.1 Rankine-Hugoniot Relations

If the field $\mathbf{u}$ itself experiences a jump across the wave front instead of the jump of its first derivative (see Fig. 3.1), we say that it is a *shock wave*. Shock waves are possible only for systems of balance laws, and they belong to a particular class of *weak solutions*. In fact, it is well known that a shock wave solution is a weak solution of (2.1) if and only if it satisfies the so-called *Rankine-Hugoniot relations* (RH relations, in short) or *Rankine-Hugoniot conditions* (RH conditions) across the shock front:

$$-s\left[\!\left[\mathbf{F}^0\right]\!\right]+\left[\!\left[\mathbf{F}^i\right]\!\right]n_i=0, \tag{3.29}$$

where $[\![g(\mathbf{u})]\!] = g(\mathbf{u}_1) - g(\mathbf{u}_0)$ for a generic function g and $(\mathbf{u}_1, \mathbf{u}_0)$ are, respectively, the values of the rear (perturbed) side and the front (unpertubed) side of the wave surface. And s stands for the normal speed of a shock wave with the unit normal $\mathbf{n} \equiv (n_i)$.

The set of all perturbed states $\mathbf{u}_1$ satisfying (3.29) for a given unperturbed state $\mathbf{u}_0$ is called *Hugoniot locus* for the point $\mathbf{u}_0$ and is denoted as $\mathscr{H}(\mathbf{u}_0)$.

If the unperturbed field $\mathbf{u}_0$ is known and we consider plane shocks with $\mathbf{n} =$ constant, the RH relations furnish a system of N equations for the $N+1$ unknowns $\mathbf{u}_1$ and s. Thus *any one* among the $(N+1)$-tuple $(\mathbf{u}_1, s)$ may be chosen as the *shock parameter* i.e., the quantity that characterizes the *strength* of a shock. Therefore, as the shock parameter, we may choose the speed s, or any component of $\mathbf{u}_1$, or a combination of these, which will be generically denoted by ξ hereafter.

If we introduce the mapping:

$$\boldsymbol{\Psi}_s(\mathbf{u}) = -s\mathbf{F}^o(\mathbf{u}) + \mathbf{F}^i(\mathbf{u})n_i \,,$$

the RH relations can be written as $\boldsymbol{\Psi}_s(\mathbf{u}_1) = \boldsymbol{\Psi}_s(\mathbf{u}_0)$. Therefore the mapping $\boldsymbol{\Psi}_s(\mathbf{u})$ must be locally non-invertible in the neighborhood of $\mathbf{u}_0$:

$$\det\left(\frac{\partial \Psi_s}{\partial \mathbf{u}}\right)\bigg|_{\mathbf{u}=\mathbf{u}_0} = 0.$$

On the other hand, the Jacobian of the mapping:

$$\det\left(\frac{\partial \Psi_s}{\partial \mathbf{u}}\right)\bigg|_{\mathbf{u}=\mathbf{u}_0} = \det\left(\mathbf{A}^i n_i - s\mathbf{A}^0\right)\bigg|_{\mathbf{u}=\mathbf{u}_0}$$

becomes singular for $s = \lambda(\mathbf{u}_0)$ as seen from (2.5). Therefore the local non-invertibility occurs when s is equal to the characteristic speed.

In a schematic way, we may represent the values $\mathbf{u}_1$ for a given value $\mathbf{u}_0$ as the function of s. See Fig. 3.2, which indicates that *shock solutions may be seen as bifurcating curves from the trivial solution* $\mathbf{u}_1 = \mathbf{u}_0$ of (3.29) at the points where s is equal to the one of the unperturbed characteristic speeds λ_0.

3.3.2 Admissibility of Shock Wave

According to the theory of hyperbolic systems, not every solution of the RH relations corresponds to a physically meaningful shock wave. Thus, we need a criterion to select the perturbed states $\mathbf{u}_1$ that, together with $\mathbf{u}_0$, form *admissible* shocks. Since admissible shocks propagate with no change in shape when they evolve from the initial data of a Riemann problem, these solutions are sometimes called *stable shocks*.

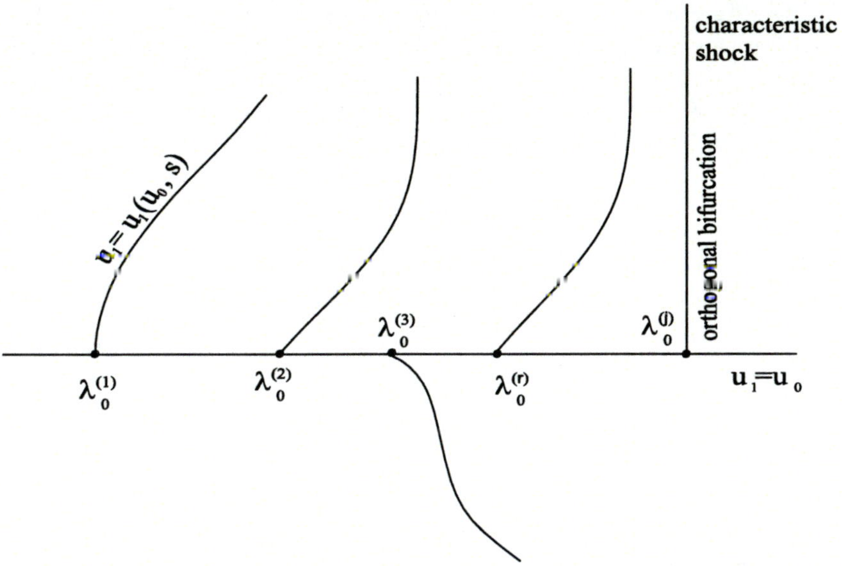

Fig. 3.2 Shocks as bifurcating branches of the trivial solution $\mathbf{u}_1 = \mathbf{u}_0$ with s being the shock parameter

The issue of shock admissibility, when genuinely nonlinear and linearly degenerate waves are involved, has been largely and deeply investigated in the past decades. For example, the Euler system of an ideal gas features only waves belonging to these two types. On the contrary, the Euler system of a van der Waals fluid features linearly degenerate and locally linearly degenerate waves. A comprehensive analysis of the shock admissibility in this kind of fluids was the subject of the paper [206].

3.3.2.1 Admissibility Conditions: Lax, Entropy Growth, and Liu

The selection rule being useful to study the admissibility of shock waves depends on the type of the nonlinear waves involved. Thus, it is necessary to discuss separately the cases of genuinely nonlinear, linearly degenerate, and locally linearly degenerate waves.

- When we deal with genuinely nonlinear waves, the selection rule is given by the *Lax condition* [207], according to which a shock wave is admissible if the shock velocity satisfies the inequality:

$$\lambda_0 < s < \lambda_1,$$

where $\lambda_0 \equiv \lambda(\mathbf{u}_0)$ and $\lambda_1 \equiv \lambda(\mathbf{u}_1)$ are the unperturbed and perturbed characteristic velocities, respectively. Such a shock wave is called *k–shock* (being

λ the kth eigenvalue of the system). The Lax condition turns out to be equivalent (at least for *weak shock waves*) to the condition of entropy growth across the shock:

$$\eta = s[\![h]\!] - [\![h_n]\!] \geqslant 0 \quad \rightleftarrows \quad \eta = s\left(h\left(\mathbf{u}_1\right) - h\left(\mathbf{u}_0\right)\right) - \left(h_n(\mathbf{u}_1) - h_n(\mathbf{u}_0)\right) \geqslant 0, \tag{3.30}$$

where $h = h^0$ and $h_n = h^i n_i$.

- When we deal with linearly degenerate waves, admissible *k-shocks* are called *characteristic shocks* and they propagate with the speed $s = \lambda_0 = \lambda_1$. In this case, there is no entropy growth across the shock, i.e., $\eta = 0$. The admissibility of a characteristic shock may thus be studied by means of the so-called *generalized Lax condition*:

$$\lambda_0 \leqslant s \leqslant \lambda_1,$$

  A characteristic shock depends on as many parameters as the multiplicity of the eigenvalue λ [199]; the system of equations of an Euler fluid in three space-dimensions, for example, has an eigenvalue of multiplicity three. And the shock wave associated to this eigenvalue is thus a characteristic shock depending on three parameters.
- When the system features locally linearly degenerate waves, the selection rule is given by the *Liu condition* [208, 209], which states that a shock wave is admissible if

$$s \geqslant s_*, \ \forall s_* \in \{s_* : s_*\left(\mathbf{u}_* - \mathbf{u}_0\right) = \mathbf{F}\left(\mathbf{u}_*\right) - \mathbf{F}\left(\mathbf{u}_0\right), \ \mathbf{u}_* \in \mathscr{H}\left(\mathbf{u}_0\right) \text{ between } \mathbf{u}_0 \text{ and } \mathbf{u}_1\}.$$

  This means that a shock is admissible if its speed, s, is not smaller than the speed of any other shock with the same unperturbed state $\mathbf{u}_0$ and with perturbed state $\mathbf{u}_*$ lying on the Hugoniot locus for $\mathbf{u}_0$ between $\mathbf{u}_0$ and $\mathbf{u}_1$ (see Fig. 3.3).

The entropy growth in this case is not sufficient for the admissibility, and we need a superposition principle (see [210]).

Concerning shock waves, we remark shock-induced phase transitions in real gases. For details, see the references [206, 211] for a van der Waals gas, and [212–216] for a hard-sphere gas.

Moreover there exists also a general theory concerning the interaction between shocks and discontinuity waves given in the work of Boillat and Ruggeri [217]. See also Ruggeri [218], and Pandey and Sharma [219] for its applications to fluids, and Mentrelli et al. [220] for a complete analysis with numerical simulations in the case of van der Waals fluid.

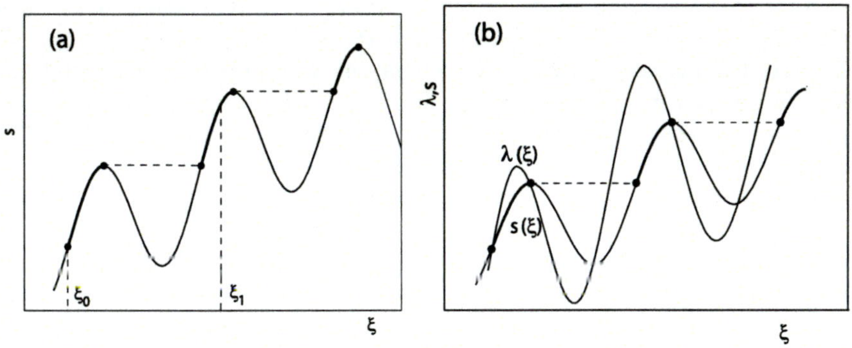

Fig. 3.3 Shock speed s as a function of the shock parameter ξ and the range of the admissible shocks (bold curve). λ is the characteristic speed

3.4 Shock Structure

A shock wave is, in reality, not a discontinuous surface. It has a continuous structure with sharp transition from an unperturbed state to a perturbed state. Typically these states are two different equilibrium states. As the shock profile depends on the shock parameter, say, the unperturbed Mach number M_0, the thickness of a shock changes with this parameter. The experiments of Alsemeyer [221] in monatomic gases show that the thickness decreases until $M_0 \sim 3.1$ and then increases with the Mach number. Several authors tried to explain this behavior of the shock thickness [146, 222–225]. The satisfactory result obtained by molecular dynamics simulation was first presented by Bird [226] in 1970. But until now no complete satisfactory phenomenological approach exists.

New interest on this subject was aroused by Ruggeri in 1993 [227]. The general mathematical structure of RET [56] was recognized, and the problem of the shock wave structure was analyzed in the context of hyperbolic systems of balance laws. With this analysis, as mentioned above, Ruggeri noticed that a singularity appears each time when the shock speed equals a characteristic eigenvalue. As a consequence, the singularity, which occurred at $M_0 = 1.65$ in the 13-moment theory, moves closer and closer to $M_0 = 1$ as the number of moments increases. Therefore the subshock seems to emerge sooner after $M_0 = 1$, and no smooth shock structure can be expected or, at best, it exists only very close to $M_0 = 1$. This interpretation is, however, not correct. In fact Weiss [228], through making a numerical analysis, showed that at least up to 35 moments all but one of these singularities are regular ones. The singular one corresponds to the highest characteristic speed evaluated in an equilibrium state in front of the shock, $\lambda_0^{(\max)}$. Thus he calculated smooth shock structures up to $s = \lambda_0^{(\max)}$. Beyond this value of s, a subshock appears. This numerical evidence of the subshock formation when $s > \lambda_0^{(\max)}$ was confirmed by a theorem due to Boillat and Ruggeri [124]. Therefore, as the maximum characteristic velocity increases with the number of moments [90, 165], according to the Weiss conjecture, the quantitative features of the shock structure improve with more and more moments. In conclusion, RET with many moments is the natural theory that is able to explain the shock wave structure. For a comprehensive survey on this topic, see, for example, the book of RET by Müller and Ruggeri [56]. In this book the chapter on shock wave structure in RET was written by W. Weiss.

As was explained in Chap. 1, the shock wave structure in polyatomic gases are quite different from that in monatomic gases. This is one of the main topics to be explained in the present book.

In this section, we summarize the mathematical background knowledge on shock structure solutions for a generic hyperbolic dissipative system of balance laws, to which the models of RET belong.

3.4.1 Shock Wave Structure and Subshock Formation

The shock wave structure is a regular solution depending on one variable φ:

$$\mathbf{u} \equiv \mathbf{u}(\varphi), \qquad \varphi = x^i n_i - st, \qquad s = \text{const.}, \quad \mathbf{n} \equiv (n_i) = \text{const.}, \tag{3.31}$$

such that

$$\lim_{\varphi \to \pm\infty} \mathbf{u}(\varphi) = \begin{cases} \mathbf{u}_0 \\ \mathbf{u}_1 \end{cases}; \qquad \lim_{\varphi \to \pm\infty} \frac{d\mathbf{u}}{d\varphi} = 0, \tag{3.32}$$

i.e., a plane wave solution connecting the two constant states.[2]

Substituting (3.31) into (2.1), we obtain the ordinary differential system:

$$\frac{d}{d\varphi}\left\{-s\mathbf{F}^0(\mathbf{u}) + \mathbf{F}_n(\mathbf{u})\right\} = \mathbf{f}(\mathbf{u}) \tag{3.33}$$

with the boundary conditions (3.32), where $\mathbf{F}_n = \mathbf{F}^i n_i$. This system is equivalent to the following ODE system:

$$\left(-s\mathbf{A}^0 + \mathbf{A}_n\right)\frac{d\mathbf{u}}{d\varphi} = \mathbf{f}(\mathbf{u}), \tag{3.34}$$

($\mathbf{A}_n = \mathbf{A}^i n_i$). As was noticed by Ruggeri [227], when s approaches a characteristic eigenvalue λ (see (2.4)), the solution may have a breakdown.

[2]As the balance laws satisfy the Galilean invariance, it is possible to adopt the frame moving with the shock velocity. For such an observer the wave appears stationary: $\varphi = x$.

In physics, especially in RET, M field equations are conservation laws and $N-M$ field equations are balance laws, i.e., these have the structure given in (2.42) and (2.43). We rewrite these by separating space and time as follows:

$$\begin{cases} \dfrac{\partial \mathbf{V}(\mathbf{u})}{\partial t} + \dfrac{\partial \mathbf{P}^i(\mathbf{u})}{\partial x^i} = 0, \\ \\ \dfrac{\partial \mathbf{W}(\mathbf{u})}{\partial t} + \dfrac{\partial \mathbf{R}^i(\mathbf{u})}{\partial x^i} = \mathbf{g}(\mathbf{u}), \end{cases} \tag{3.35}$$

where $\mathbf{V},\ \mathbf{P}^i \in \mathbb{R}^M$, while $\mathbf{W},\ \mathbf{R}^i$, and $\mathbf{g}$ are vectors of $\mathbb{R}^{N-M}$:

$$\mathbf{F}^0 \equiv \begin{bmatrix} \mathbf{V} \\ \mathbf{W} \end{bmatrix}, \qquad \mathbf{F}^i \equiv \begin{bmatrix} \mathbf{P}^i \\ \mathbf{R}^i \end{bmatrix}, \qquad \mathbf{f} \equiv \begin{bmatrix} 0 \\ \mathbf{g} \end{bmatrix}. \tag{3.36}$$

In what follows, we choose, as field variables, the components of the main field, i.e., $\mathbf{u} \equiv \mathbf{u}'$. Let $\mathbf{u}$ be expressed by a pair: $\mathbf{u} \equiv (\mathbf{v}, \mathbf{w})$ with $\mathbf{v} \in \mathbb{R}^M$ and $\mathbf{w} \in \mathbb{R}^{N-M}$.

As seen in (2.45), $\mathbf{w} = 0$ in equilibrium. Then, from $(3.35)_1$ with $\mathbf{w} \equiv 0$, we obtain the associated system:

$$\frac{\partial \mathbf{V}(\mathbf{v}, 0)}{\partial t} + \frac{\partial \mathbf{P}^i(\mathbf{v}, 0)}{\partial x^i} = 0. \tag{3.37}$$

According to the discussion made in Sect. 2.4 [165], (3.37) is a *principal equilibrium subsystem* of (3.35). Then (3.37) satisfies the entropy principle together with the following *subcharacteristic conditions*:

$$\max_{k=1,2,\dots,N} \lambda^{(k)}(\mathbf{v}, 0) \geqslant \max_{J=1,2,\dots,M} \mu^{(J)}(\mathbf{v}), \tag{3.38}$$

and similar inequalities for the minimum. The λ's are the characteristic speeds of the total system (2.1) while the μ's are the ones of the equilibrium subsystem (3.37). In RET, the latter system corresponds to the Euler fluid system.

Going back to (3.33), we rewrite it as follows:

$$\begin{cases} \dfrac{d}{d\varphi}\left\{-s\mathbf{V}(\mathbf{v}, \mathbf{w}) + \mathbf{P}_n(\mathbf{v}, \mathbf{w})\right\} = 0, \\ \\ -s\dfrac{d}{d\varphi}\mathbf{W}(\mathbf{v}, \mathbf{w}) + \dfrac{d}{d\varphi}\mathbf{R}_n(\mathbf{v}, \mathbf{w}) = \mathbf{g}(\mathbf{v}, \mathbf{w}). \end{cases} \tag{3.39}$$

From $(3.39)_2$ evaluated at $\varphi \to \pm\infty$ with the condition $(3.32)_2$, we obtain

$$\mathbf{g}(\mathbf{v}_0, \mathbf{w}_0) = \mathbf{g}(\mathbf{v}_1, \mathbf{w}_1) = 0,$$

which implies that the solutions at infinity on both sides are equilibrium solutions (see [124, 165]):

$$\mathbf{w}_0 = \mathbf{w}_1 = 0.$$

By $(3.39)_1$, we have

$$-s\mathbf{V}(\mathbf{v}, \mathbf{w}) + \mathbf{P}_n(\mathbf{v}, \mathbf{w}) = \mathbf{c} = \text{const.}$$

That is, the quantity on the left-hand side is conserved along the process. In particular, for $\varphi \to \pm\infty$ (see (2.45)), we have

$$\mathbf{c} = -s\mathbf{V}(\mathbf{v}_0, 0) + \mathbf{P}_n(\mathbf{v}_0, 0) = -s\mathbf{V}(\mathbf{v}_1, 0) + \mathbf{P}_n(\mathbf{v}_1, 0). \tag{3.40}$$

Equation (3.40) represents the Rankine-Hugoniot conditions for shocks of the *equilibrium subsystem* (3.37), and therefore

$$\mathbf{v}_1 \equiv \mathbf{v}_1(\mathbf{v}_0, s). \tag{3.41}$$

The problem now is to find a C^1 solution of (3.33) connecting the two equilibrium states $\mathbf{u}_0 \equiv (\mathbf{v}_0, 0)$ and $\mathbf{u}_1 \equiv (\mathbf{v}_1(\mathbf{v}_0, s), 0)$ for the shock velocity s with the prescribed values of $\mathbf{v}_0$ and $\mathbf{n}$.

3.4.2 Non-existence of Smooth Shock When $s > \lambda^{\max}(\mathbf{u}_0)$

We now prove the following theorem due to Boillat and Ruggeri [124]:

Theorem 3.1 (Subshock Formation) *We consider a system of N balance laws (2.2), $M(<N)$ equations of which are conservation laws, as shown in (3.35). We assume that the system satisfies the entropy principle with convex entropy, (2.8) and (2.9). Then, if the shock velocity s is greater than the maximum characteristic velocity evaluated in the equilibrium state in front of the shock, C^1 shock-structure cannot exist. That is, smooth solutions may exist only if $s \leqslant \lambda^{\max}(\mathbf{u}_0)$.*

Proof Since any smooth solution of (2.2) is also a solution of (2.8), we must have the condition for shock structure solutions:

$$\frac{d}{d\varphi}\{-s\,h(\mathbf{u}) + h_n(\mathbf{u})\} = \Sigma(\mathbf{u}) \geqslant 0. \tag{3.42}$$

This implies that the bracketed part is an increasing function of φ. Therefore we have

$$-s\,h(\mathbf{u}_1) + h_n(\mathbf{u}_1) \leqslant -s\,h(\mathbf{u}) + h_n(\mathbf{u}) \leqslant -s\,h(\mathbf{u}_0) + h_n(\mathbf{u}_0).$$

In particular, the entropy grows across the shock (3.30). The idea of the proof is to verify that the entropy inequality (3.42) is violated when $s > \lambda^{\max}(\mathbf{u}_0)$.

We remark also that, at least up to moderately strong shock waves, the entropy growth condition (3.30) is equivalent to the *Lax condition* for admissibility of shock waves: For a fixed eigenvalue $\mu(\mathbf{v}_0) \in \{\mu^{(J)}(\mathbf{v}_0),\ J = 1, 2, \ldots, M\}$ of the equilibrium subsystem (3.37), a shock satisfies

$$\mu(\mathbf{v}_0) < s < \mu(\mathbf{v}_1); \qquad \lim_{s \to \mu(\mathbf{v}_0)} \mathbf{v}_1(\mathbf{v}_0, s) = \mathbf{v}_0. \tag{3.43}$$

Therefore we study shock waves satisfying the Lax condition (3.43), where we increase s starting from the trivial shock with $s = \mu(\mathbf{v}_0)$.

First we observe that the production term $\mathbf{f}(\mathbf{u})$ is always orthogonal to the main field in any constant equilibrium state $\mathbf{u}_0$:

$$\mathbf{u}_0 \cdot \mathbf{f}(\mathbf{u}) \equiv 0, \qquad \forall \mathbf{u}.$$

In fact, the first M components of $\mathbf{f}(\mathbf{u})$ are null as shown in $(3.36)_3$ and an equilibrium state $\mathbf{u}_0$ corresponds to the pair $(\mathbf{v}_0, 0)$ (see $(2.45)_2$). The residual inequality $(2.11)_3$ can be written as

$$\Sigma = \mathbf{u} \cdot \mathbf{f}(\mathbf{u}) = (\mathbf{u} - \mathbf{u}_0) \cdot \mathbf{f}(\mathbf{u}) \geqslant 0. \tag{3.44}$$

Multiplying (3.33) by $\mathbf{u} - \mathbf{u}_0$, and taking into account (2.11) and (3.44), we have

$$\frac{d}{d\varphi} \{-sh(\mathbf{u}) + h_n(\mathbf{u}) - \mathbf{u}_0 \{-s\mathbf{F}(\mathbf{u}) + \mathbf{F}_n(\mathbf{u})\}\} = (\mathbf{u} - \mathbf{u}_0) \cdot \mathbf{f} \geqslant 0. \tag{3.45}$$

Therefore the outermost-bracketed part on the left-hand side is a increasing function of φ. Then we obtain the inequality:

$$\begin{aligned} & - s\,(h(\mathbf{u}) - h(\mathbf{u}_0)) + h_n(\mathbf{u}) - h_n(\mathbf{u}_0) \\ & \quad - \mathbf{u}_0 \{-s \{\mathbf{F}(\mathbf{u}) - \mathbf{F}(\mathbf{u}_0)\} + \mathbf{F}_n(\mathbf{u}) - \mathbf{F}_n(\mathbf{u}_0)\} \leqslant 0. \end{aligned} \tag{3.46}$$

From (2.13) and (2.15), we have

$$h = \mathbf{u} \cdot \frac{\partial h'}{\partial \mathbf{u}} - h', \quad h_n = \mathbf{u} \cdot \frac{\partial h'_n}{\partial \mathbf{u}} - h'_n, \tag{3.47}$$

where $h' = h'^0,\ h'_n = h'^i n_i$. Inserting (3.47) into (3.46), we have

$$s\left(h'(\mathbf{u}) - h'(\mathbf{u}_0)\right) - h'_n(\mathbf{u}) + h'_n(\mathbf{u}_0) + (\mathbf{u} - \mathbf{u}_0) \cdot \left(-s\frac{\partial h'(\mathbf{u})}{\partial \mathbf{u}} + \frac{\partial h'_n(\mathbf{u})}{\partial \mathbf{u}}\right) \geqslant 0.$$

We can rewrite this as follows:

$$(\mathbf{u}-\mathbf{u}_0)^T \cdot \left(\frac{\partial^2 h'_n(\mathbf{u})}{\partial \mathbf{u}\partial \mathbf{u}} - s\frac{\partial^2 h'(\mathbf{u})}{\partial \mathbf{u}\partial \mathbf{u}}\right)_{\mathbf{u}^*} (\mathbf{u}-\mathbf{u}_0) \leqslant 0$$

or

$$(\mathbf{u}-\mathbf{u}_0)^T \cdot \left(\mathbf{A}_n(\mathbf{u}^*) - s\mathbf{H}(\mathbf{u}^*)\right) (\mathbf{u}-\mathbf{u}_0) \geqslant 0, \tag{3.48}$$

where

$$\mathbf{u}^* = \mathbf{u}_0 + \tau(\mathbf{u}-\mathbf{u}_0); \;\; \tau \in [0,1], \quad \mathbf{A}_n = -\frac{\partial^2 h'_n}{\partial \mathbf{u}\partial \mathbf{u}}, \quad \mathbf{H} = -\frac{\partial^2 h'}{\partial \mathbf{u}\partial \mathbf{u}}.$$

For $\mathbf{u}$ near $\mathbf{u}_0$:

$$\mathbf{u} = \mathbf{u}_0 + \varepsilon \mathbf{q}$$

with ε being a small parameter, (3.48) becomes

$$\mathbf{q}^T \cdot (\mathbf{A}_n(\mathbf{u}_0) - s\mathbf{H}(\mathbf{u}_0))\,\mathbf{q} + \mathcal{O}(\varepsilon) \geqslant 0. \tag{3.49}$$

It is obviously impossible to have any $s > \lambda_0^{\max}$ (where $\lambda_0^{\max} = \lambda^{\max}(\mathbf{u}_0)$) that satisfies (3.49), because, in this case, the quadratic form in (3.49) becomes negative definite due to the fact that two matrices are symmetric and $\mathbf{H}$ is positive definite. Therefore smooth solutions may exist only for $s \leqslant \lambda_0^{\max}$. The proof is now completed.

Remark 3.4 It is worth noting that the impossibility to have smooth shocks with velocity greater than $\lambda_0^{\max}$ is a strict consequence of all the three aspects in the entropy principle: (1) the existence of a supplementary balance law (2.8) that implies (3.45), (2) the convexity of entropy density (convexity of h) that permits to put the system in a symmetric form with $\mathbf{H}$ being positive definite, and (3) the residual inequality that determines the sign of the left-hand side of (3.49).

Remark 3.5 As seen just above, the upper bound for the velocity of a continuous shock is determined only by the maximum characteristic velocity in the equilibrium state *in front of* the shock $\lambda_0^{\max}$. Why does the maximum characteristic velocity in the equilibrium state *behind* the shock $\lambda_1^{\max}$ have no role? The answer is simple: it is just impossible for s to meet $\lambda_1^{\max}$ because of the stability reason (entropy condition). In fact, let us firstly remind that shock waves, which have the parameter s starting from $s = \mu_0^{\max}$, satisfy the Lax condition $(3.43)_1$. Then we obtain the following inequality from the subcharacteristic conditions (3.38):

$$\mu_0^{\max} \leqslant s \leqslant \mu_1^{\max} \leqslant \lambda_1^{\max}.$$

We note that $\mu_1^{max} \equiv \mu_1^{max}(\mathbf{u}_1) = \mu_1^{max}(\mathbf{v}_1, 0)$ and $\lambda_1^{max} \equiv \lambda^{max}(\mathbf{u}_1) = \lambda^{max}(\mathbf{v}_1, 0)$ are the functions of s because the perturbed field $\mathbf{v}_1$ is a function of $\mathbf{v}_0$ and s that is prescribed by the Rankine-Hugoniot equations of the equilibrium subsystem (3.41).

Remark 3.6 After the shock-speed exceeds the maximum characteristic eigenvalue in the unperturbed state, a *subshock* arises necessarily.

Remark 3.7 Due to the isotropy of the space, if there exists a wave with velocity λ_0, there exists also a wave with the velocity $-\lambda_0$, i.e., we have two waves propagating in opposite directions to each other. In this case, replacing $\mathbf{n}$ by $-\mathbf{n}$, we can easily verify that smooth shocks cannot propagate with the velocity less than λ_0^{min}.

Remark 3.8 In the above, we have always considered only shocks that are not characteristic ones. In the exceptional case of a characteristic shock, which propagates with a characteristic velocity, it was proved that shock structure solutions never exist [229].

Remark 3.9 An open problem is to see whether there exists a physical case for which a subshock arises at the eigenvalues less than the maximum one. Until now, in all RET theories for any number of moments, a shock is continuous until it reaches the maximum eigenvalue. If this is a general property, it is very favorable because the increase of the number of moments implies the increase of the maximum characteristic velocity and also implies the increase of the critical shock velocity. We will discuss this question in Sect. 3.6. On the other hand, in the case of mixture, we will prove that this is possible (see Chap. 29).

Remark 3.10 Concerning the existence of shock structures, we mention, among others, the classical paper by Gilbarg [230] in the context of parabolic continuum theory of fluids, and the paper by Yong and Zumbrun [231] for hyperbolic systems with relaxation. Interesting analysis of stationary points in the context of stability and bifurcation was made by Simić [232, 233].

3.5 Riemann Problem

Georg Bernhard Riemann raised the famous question about the time-evolution of a gas under the initial condition that the gas is divided into two regions by a thin diaphragm. Each region is filled with the same gas, but with different values of thermodynamic quantities such as the pressure, the density, and the temperature. This problem is of fundamental importance and has many applications. An experimental apparatus called shock tube has been used in a wide variety of aerodynamic or ballistic topics like supersonic aircraft flight, gun performance, asteroid impacts, shuttle atmospheric entry, etc.

Then the Riemann problem is as follows: What happens when the diaphragm is put away? In literature, by extension of this problem, the Riemann problem deals with every solution of a system of conservation laws in one-space dimension along the x axis when the initial data composed of two different constant states $(\mathbf{u}_1, \mathbf{u}_0)$ are connected with a jump at $x = 0$.

This problem, at least in one-space dimension, has been completely solved for systems of conservation laws (see e.g. [192, 207, 234–237]). It was shown that the solution of the Riemann problem for hyperbolic systems of conservation laws is a combination of rarefaction waves, contact waves, and shock waves. A huge literature of the Riemann problem exists. In particular, many numerical results have been obtained by using the Riemann solver (see e.g., [237]). Concerning the numerical approach for hyperbolic systems see also the recent paper of Dumbser, Boscheri, Semplice, and Russo [238], and concerning approximate Riemann solvers see the contribution of Klingenberg [239].

3.5.1 Riemann Problem with Structure

Liu [240, 241] noticed that Riemann initial data can be regarded as a rough approximation of continuous initial data containing steep variations. Therefore this problem takes into account the presence of shock thickness, oscillations, noises, and continuous (although very steep) changes, since initial data coincide with the Riemann problem only for large $|x|$ (that is to say, they are constant equilibrium states for large $|x|$). This problem is called *Riemann problem with structure* (see Fig. 3.4). In the case of conservation laws of hyperbolic type, Liu proved, roughly speaking, that the solution of this problem converges, for large time, to the solution of the corresponding Riemann problem.

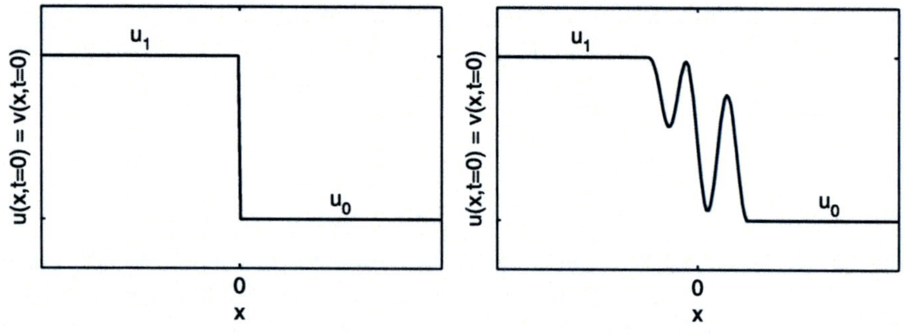

Fig. 3.4 Riemann problem (left) and the Riemann problem with structure (right)

3.6 Toy Models: 2 x 2 Hyperbolic System of Balance Laws

For understanding explicitly the previous questions concerning the subshocks and the Riemann problem for a system of balance laws, we consider a toy model, that is, the following 2×2 dissipative hyperbolic system of balance laws proposed by Mentrelli and Ruggeri [242]. This system is equipped with all the characteristics of RET systems: a block of conservation laws and a block of balance laws, entropy law, convexity of entropy, and K-condition:

$$\begin{aligned} &\frac{\partial u}{\partial t}+\frac{\partial}{\partial x}\left(\frac{\partial K}{\partial u}\right)=-\frac{1}{\tau}(u-v), \\ &\frac{\partial v}{\partial t}+\frac{\partial}{\partial x}\left(\frac{\partial K}{\partial v}\right)=-\frac{1}{\tau}(v-u), \end{aligned} \tag{3.50}$$

or, alternatively,

$$\begin{aligned} &\frac{\partial}{\partial t}(u+v)+\frac{\partial}{\partial x}\left(\frac{\partial K}{\partial u}+\frac{\partial K}{\partial v}\right)=0, \\ &\frac{\partial u}{\partial t}+\frac{\partial}{\partial x}\left(\frac{\partial K}{\partial u}\right)=-\frac{1}{\tau}(u-v) \end{aligned} \tag{3.51}$$

for the unknown field $\mathbf{U}=(u, v)^T$, which is a function of space x and time t. Here $K \equiv K(u, v)$ is an arbitrary smooth function of the variables u and v, and $\tau > 0$ stands for a constant relaxation time. The equilibrium state is achieved when $u = v$.

The system (3.50) (or, (3.51)) satisfies all the requirement of RET. In fact, a solution of the balance equations (3.50) (or, (3.51)) satisfies the following entropy inequality:[3]

$$\frac{\partial h}{\partial t}+\frac{\partial h^1}{\partial x}=\Sigma \geqslant 0,$$

where h, h^1, and Σ are, respectively, the entropy density, the entropy flux, and the entropy production density given by

$$h=-\frac{1}{2}\left(u^2+v^2\right), \quad h^1=-u\frac{\partial K}{\partial u}-v\frac{\partial K}{\partial v}+K, \quad \Sigma=\frac{1}{\tau}(u-v)^2. \tag{3.52}$$

Moreover, the convexity of the entropy density h with respect to the field $(u, v)^T$ is automatically satisfied (see $(3.52)_1$). By the formal substitution (3.27), we obtain a linear system of two equations where λ represents the characteristic velocity and

[3] We adopt different definition of the sign of the entropy from the one adopted in [242].

$(\delta u, \delta v)^T$ is proportional to the characteristic eigenvector of the system associated with λ:

$$\left(-\lambda + \frac{\partial^2 K}{\partial u^2}\right)\delta u + \frac{\partial^2 K}{\partial u \partial v}\delta v = 0,$$
$$\frac{\partial^2 K}{\partial u \partial v}\delta u + \left(-\lambda + \frac{\partial^2 K}{\partial u^2}\right)\delta v = 0. \tag{3.53}$$

Therefore the characteristic velocities $\lambda^{(1)}$ and $\lambda^{(2)}$ are obtained as the solutions of the characteristic polynomial $P(\lambda) = 0$, where

$$P(\lambda) = \lambda^2 - \left\{\frac{\partial^2 K}{\partial u^2} + \frac{\partial^2 K}{\partial v^2}\right\}\lambda + \frac{\partial^2 K}{\partial u^2}\frac{\partial^2 K}{\partial v^2} - \left(\frac{\partial^2 K}{\partial u \partial v}\right)^2.$$

In particular, the equilibrium characteristic velocities $\lambda_E^{(1)}$ and $\lambda_E^{(2)}$ are the roots of $P_E(\lambda_E) = 0$, where

$$P_E(\lambda_E) = \lambda_E^2 - \left\{\frac{\partial^2 K}{\partial u^2} + \frac{\partial^2 K}{\partial v^2}\right\}\bigg|_E \lambda_E + \left\{\frac{\partial^2 K}{\partial u^2}\frac{\partial^2 K}{\partial v^2} - \left(\frac{\partial^2 K}{\partial u \partial v}\right)^2\right\}\bigg|_E. \tag{3.54}$$

Here the quantities with subscript E represent the quantities evaluated in the equilibrium state in which $v = u$.

In the present case, the equilibrium subsystem associated with the system (3.51) is obtained by putting $v = u$ into the Eq. $(3.51)_1$:

$$\frac{\partial u}{\partial t} + \frac{1}{2}\frac{\partial}{\partial x}\left(\frac{d\bar{K}}{du}\right) = 0, \tag{3.55}$$

where $\bar{K} = \bar{K}(u)$ is defined by $\bar{K}(u) = K(u, u)$. The characteristic velocity μ of the equilibrium subsystem (3.55) is given by

$$\mu = \frac{1}{2}\frac{d^2\bar{K}}{du^2}.$$

Taking into account the following identities:

$$\frac{d\bar{K}}{du} = \left(\frac{\partial K}{\partial u} + \frac{\partial K}{\partial v}\right)\bigg|_E,$$
$$\frac{d^2\bar{K}}{du^2} = \left(\frac{\partial^2 K}{\partial u^2} + 2\frac{\partial^2 K}{\partial u \partial v} + \frac{\partial^2 K}{\partial v^2}\right)\bigg|_E,$$

we have

$$P_E(\mu) = -\frac{1}{4}\left\{\left(\frac{\partial^2 K}{\partial u^2} - \frac{\partial^2 K}{\partial v^2}\right)\Big|_E\right\}^2 \leqslant 0. \tag{3.56}$$

Therefore, we have the subcharacteristic condition

$$\lambda_E^{(1)} \leqslant \mu \leqslant \lambda_E^{(2)}, \tag{3.57}$$

which is automatically satisfied according with Theorem 2.3. In the present case, the K-condition (2.48) requires

$$\delta\mathbf{f}|_E \neq 0 \quad \Longleftrightarrow \quad (\delta u - \delta v)|_E \neq 0. \tag{3.58}$$

We need to consider two possible cases separately:

- If

$$\frac{\partial^2 K}{\partial u \partial v}\Big|_E = 0,$$

  from (3.54) and (3.53), we have

$$\lambda_E^{(1)} = \frac{\partial^2 K}{\partial u^2}\Big|_E, \quad (\delta u)|_E = 1, \; (\delta v)|_E = 0,$$
$$\lambda_E^{(2)} = \frac{\partial^2 K}{\partial v^2}\Big|_E, \quad (\delta u)|_E = 0, \; (\delta v)|_E = 1,$$

  and (3.58) is automatically satisfied for both eigenvectors.
- If

$$\frac{\partial^2 K}{\partial u \partial v}\Big|_E \neq 0,$$

  from (3.53), we have

$$(\delta u)|_E = -\frac{\partial^2 K}{\partial u \partial v}\Big|_E, \qquad (\delta v)|_E = -\lambda_E + \frac{\partial^2 K}{\partial u^2}\Big|_E,$$

  and therefore the K-condition (3.58) is satisfied when

$$\lambda_E \neq \omega \quad \text{with} \quad \omega = \frac{\partial^2 K}{\partial u^2}\Big|_E + \frac{\partial^2 K}{\partial u \partial v}\Big|_E. \tag{3.59}$$

From (3.54), we have

$$P_E(\omega) = \left.\frac{\partial^2 K}{\partial u \partial v}\right|_E \left.\left(\frac{\partial^2 K}{\partial u^2} - \frac{\partial^2 K}{\partial v^2}\right)\right|_E .$$

As the K-condition (3.59) implies $P_E(\omega) \neq 0$, we have

$$\left.\left(\frac{\partial^2 K}{\partial u^2} - \frac{\partial^2 K}{\partial v^2}\right)\right|_E \neq 0. \tag{3.60}$$

We notice that, if (3.60) holds, the equilibrium characteristic velocities for the full system have different values from those for the equilibrium subsystem and the inequalities in (3.56) and (3.57) become strict.

According with Theorem 2.6, for any smooth function $K(u, v)$ such that

$$\left.\frac{\partial^2 K}{\partial u \partial v}\right|_{u=v} = 0, \quad \text{or} \quad \left.\left(\frac{\partial^2 K}{\partial u^2} - \frac{\partial^2 K}{\partial v^2}\right)\right|_{u=v} \neq 0, \tag{3.61}$$

and for sufficiently small initial data, the system (3.50) has global smooth solutions for all time.

3.6.1 Subshock Formation with $s < \lambda^{\max}(\mathbf{u}_0)$ and Multiple Subshock

The RET theory of monatomic gases predicts the subshock formation only when the shock velocity is greater than the maximum characteristic velocity in the unperturbed state. Based on this fact, it was conjectured that the subshock arises only when $s > \lambda^{\max}(\mathbf{u}_0)$. Furthermore, also in the case of polyatomic gas, ET_{14} predicts the subshock formation only when $s > \lambda^{\max}(\mathbf{u}_0)$. This fact reinforces the conjecture.

In contrast to these results, the subshock formation with $s < \lambda^{\max}(\mathbf{u}_0)$ and the multiple subshock formation were numerically observed in a binary mixture of the Eulerian gases [243] (see Chap. 29) and the Grad system of a mixture of monatomic gases, [244, 245]. However, the system of binary mixture has special properties such that (1) the form of balance equation for each component is the same as the one of a single fluid, and (2) the coupling effect between the components comes only through the production terms.

To understand this intriguing problem more deeply, let us consider a particular case of the 2×2 system (3.50) by choosing $K = u^4/4 + v^6/6$ [246]:

$$\frac{\partial u}{\partial t} + \frac{\partial}{\partial x}\left(\frac{u^3}{3}\right) = -\frac{u - v}{\tau},$$

$$\frac{\partial v}{\partial t} + \frac{\partial}{\partial x}\left(\frac{v^5}{5}\right) = -\frac{v - u}{\tau}.$$

The above system can be rewritten as a combination of a conservation law and a balance law:

$$\frac{\partial}{\partial t}(u + v) + \frac{\partial}{\partial x}\left(\frac{u^3}{3} + \frac{v^5}{5}\right) = 0,$$

$$\frac{\partial u}{\partial t} + \frac{\partial}{\partial x}\left(\frac{u^3}{3}\right) = -\frac{u - v}{\tau}.$$

Figure 3.5 shows typical shock structures predicted by this model. As the velocity of shock wave increases, the shock structure changes from continuous one to the structure with a subshock ($s < \lambda^{\max}(\mathbf{u}_0)$), and furthermore changes to the structure with multiple subshock ($s > \lambda^{\max}(\mathbf{u}_0)$). This is a clear counter example of the conjecture mentioned above. It is noticeable that, as will be seen in Sect. 3.6.2, this simple toy model satisfies all important requirements of the system of RET, namely, the entropy principle, convexity of the entropy, dissipative character satisfying

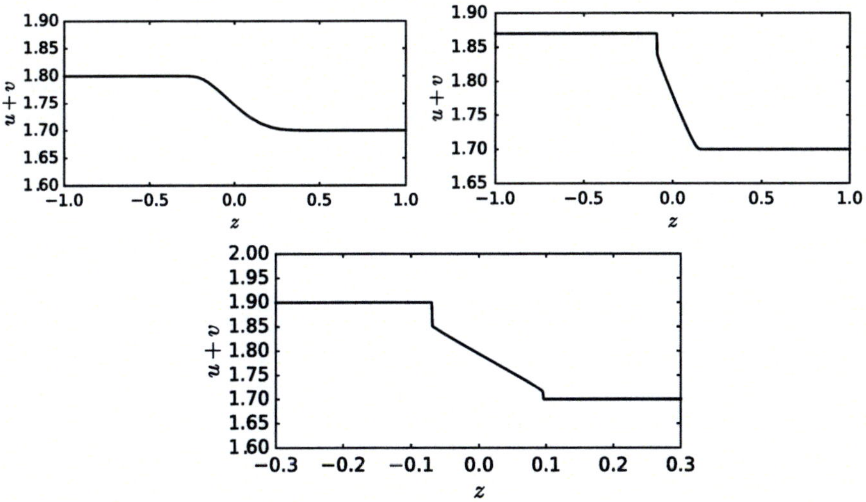

Fig. 3.5 Continuous shock structure with $s = 0.677$ (top–left), subshock formation with $s = 0.717$ ($< \lambda^{\max}(\mathbf{u}_0)$) (top–right), and multiple subshock with $s = 0.735$ ($> \lambda^{\max}(\mathbf{u}_0)$) (bottom) [246]. In this case, $\lambda^{\max}(\mathbf{u}_0) = 0.723$. $z \equiv x - st$

Shizuta-Kawashima condition [168]. Therefore there is still an open and important problem: What property is hidden in RET, according to which a subshock arises only when the shock velocity is greater than the maximum characteristic velocity?

3.6.2 Conjecture Concerning Large-Time Asymptotic Behavior of Shock Structure for System of Balance Laws

For balance laws, few mathematical theories of Riemann problem exist because a system of balance laws cannot admit rarefaction wave solutions depending on x/t. Recently, however, Ruggeri and coworkers [242, 247, 248]—following Liu [249]—proposed a conjecture about the large-time behaviors of Riemann problem and Riemann problem with structure for a system of balance laws. According to this conjecture, solutions of both Riemann problems with and without structure converge, for large time, to a solution that represents the combination of shock structures (with and without subshocks) and rarefactions of the equilibrium subsystem.

More precisely, let us consider Riemann data (or Riemann data with structure) for the system of balance laws (3.35) between two equilibrium constant states $(\mathbf{u}_0, \mathbf{u}_1)$ with $\mathbf{g}(\mathbf{u}_1) = \mathbf{g}(\mathbf{u}_0) = 0$. First of all we note that, in the case of the corresponding Riemann problem for the equilibrium subsystem (3.37), we have a combination of rarefactions $\mathscr{R}$ and shocks $\mathscr{S}$ plus contact discontinuities, and the constant states, according with the general theory. Now the conjecture is as follows:

Conjecture 3.1 A solution of the Riemann problem for the full system converges, after large time elapsed, to a solution composed of the same rarefactions $\mathscr{R}$ of the equilibrium subsystem and a shock-structure of the full system $\mathscr{S}_{struc}$ (including subshock) corresponding to $\mathscr{S}$.

Remark 3.11 In particular, if the Riemann initial data correspond to a particular shock family $\mathscr{S}$, then, for large time, a solution of the Riemann problem of the full system converges to a corresponding shock structure solution. This means that the numerical study of the shock structure, instead of using a complex mathematical solver of ODE, can be used as a Riemann solvers [237] if we wait enough time after the initial time. This strategy seems to be useful in RET as will be discussed in the following chapters.

The conjecture was tested numerically for the Grad 13-moment system and a system for mixture of fluids [247, 248]. In the followings, we test the conjecture through studying a simple 2×2 dissipative model considered by Mentrelli and Ruggeri [242]. We can calculate analytically the shock structures and the rarefactions of its equilibrium subsystem.

Among many possible choices for the function K in (3.50) satisfying K-condition (3.61), we discuss in detail a toy model with K being a function of u and v given in [242]

$$K(u, v) = uv^2 \qquad \text{with} \quad u > 0 \text{ and } v > 0.$$

In this particular case, two equations in the system (3.50) become

$$\begin{aligned}\frac{\partial u}{\partial t}+\frac{\partial}{\partial x}\left(v^2\right)&=\frac{1}{\tau}\left(v-u\right),\\ \frac{\partial v}{\partial t}+\frac{\partial}{\partial x}\left(2uv\right)&=\frac{1}{\tau}\left(u-v\right).\end{aligned} \tag{3.62}$$

Or equivalently,

$$\begin{aligned}\frac{\partial}{\partial t}(u+v)+\frac{\partial}{\partial x}\left(v^2+2uv\right)&=0,\\ \frac{\partial u}{\partial t}+\frac{\partial}{\partial x}\left(v^2\right)&=\frac{1}{\tau}\left(v-u\right).\end{aligned} \tag{3.63}$$

System (3.62) (or (3.63)) has a supplementary law:

$$\frac{\partial}{\partial t}\left(\frac{1}{2}\left(u^2+v^2\right)\right)+\frac{\partial}{\partial x}\left(2uv^2\right)=-\frac{1}{\tau}\left(u-v\right)^2,$$

and an equilibrium subsystem:

$$\frac{\partial u}{\partial t}+3u\frac{\partial u}{\partial x}=0. \tag{3.64}$$

The characteristic eigenvalues of (3.62) (or (3.63)) are given by

$$\begin{aligned}&\lambda^{(1)}=u-\sqrt{u^2+4v^2}, &&\lambda^{(2)}=u+\sqrt{u^2+4v^2},\\ &\lambda^{(1)}\Big|_E=u\left(1-\sqrt{5}\right), &&\lambda^{(2)}\Big|_E=u\left(1+\sqrt{5}\right),\end{aligned} \tag{3.65}$$

which are genuinely nonlinear for any $u>0$ and $v>0$. While the eigenvalue of the equilibrium subsystem (3.64) is given by

$$\mu=3u.$$

It is easy to see that the subcharacteristic condition (3.57) is satisfied.

3.6.2.1 Exact Solutions

In this section, we present some exact solutions of the system (3.62). We also show exact solutions of the associate equilibrium subsystem (3.64) that, following the conjecture, are the asymptotic limit of the Riemann and of the Riemann problem with structure.

If the two constant equilibrium states characterizing the boundary data are $\mathbf{u}_1 \equiv (u_1, v_1 = u_1)^T$ and $\mathbf{u}_0 \equiv (u_0, v_0 = u_0)^T$, three different cases are possible depending on the value of the ratio:

$$\gamma = \frac{u_1}{u_0}.$$

In the case (a) where

$$1 < \gamma \leqslant \gamma_*, \qquad \gamma_* = \frac{2\sqrt{5}-1}{3}, \tag{3.66}$$

we have a continuous shock structure solution of the system (3.62). The case (b), where

$$\gamma > \gamma_*,$$

corresponds to a shock structure solution of the system (3.62) with a subshock in the structure. Finally the case (c), where

$$\gamma < 1,$$

is characterized by a rarefaction observed in the equilibrium subsystem (3.64).

In the following subsections, we deduce exact solutions of the three cases (a)–(c), and then, in Sect. 3.6.2.5, we validate numerically that solutions for both Riemann problems without and with structure converge, for large time, to one of the cases (a)–(c) depending on the value of γ.

3.6.2.2 Case (a): Continuous Shock Structure Solution

The system (3.62) (or (3.63)) admits a shock structure solution, i.e., a solution of the type $\mathbf{u} = \mathbf{u}(\varphi)$ where $\varphi = x - st$ ($s =$ constant) under the boundary conditions (3.32). In the present case $\mathbf{u}_0 \equiv (u_0, u_0)^T$ and $\mathbf{u}_1 \equiv (u_1, u_1)^T$. From $(3.63)_1$, we have

$$\begin{aligned} -s\,(u+v) + v^2 + 2uv &= C = \text{const.} \\ &= -2su_0 + 3u_0^2 \\ &= -2su_1 + 3u_1^2. \end{aligned} \tag{3.67}$$

And there remains one equation $(3.63)_2$:

$$-s\frac{du}{d\varphi} + 2v\frac{dv}{d\varphi} = \frac{1}{\tau}\,(v-u)\,. \tag{3.68}$$

The last equality in (3.67) corresponds to the Rankine–Hugoniot equation of the equilibrium subsystem (3.64), and it gives us the velocity s of the shock structure

$$s = \frac{3}{2}\left(u_0 + u_1\right), \tag{3.69}$$

as well as the value of the constant C:

$$C = -3u_0u_1.$$

From (3.67), the curve $\mathscr{C}$ of shock structure in the configuration space (u, v) is prescribed by

$$\mathscr{C}: \qquad u = \frac{6u_0u_1 - 3\left(u_0 + u_1\right)v + 2v^2}{3\left(u_0 + u_1\right) - 4v}. \tag{3.70}$$

While, from (3.68), taking into account also (3.69) and (3.70), we deduce the differential equation for $v\left(\varphi\right)$:

$$\left(\frac{-9\left(u_0 + u_1\right)\left(3u_0^2 - 2u_0u_1 + 3u_1^2\right) + 72\left(u_0 + u_1\right)v^2 - 64v^3}{12\left(u_0 - v\right)\left(u_1 - v\right)\left(3\left(u_0 + u_1\right) - 4v\right)}\right)\frac{dv}{d\varphi} = \frac{1}{\tau}. \tag{3.71}$$

It is now necessary to recall that, according to Theorem 3.1, a smooth shock structure solution may exist only for shock velocity s being less than or equal to the maximum characteristic speed evaluated in the equilibrium unperturbed state $\mathbf{u}_0$:

$$s \leqslant \lambda^{max}\left(\mathbf{u}_0\right). \tag{3.72}$$

By taking into account (3.65), the condition (3.72) turns into

$$\frac{3}{2}\left(u_0 + u_1\right) \leqslant u_0\left(1 + \sqrt{5}\right),$$

or, equivalently,

$$u_1 \leqslant \left(\frac{2\sqrt{5} - 1}{3}\right)u_0. \tag{3.73}$$

The Lax condition for the equilibrium subsystem $\mu(u_1) > \mu(u_0)$ implies

$$u_1 > u_0.$$

To sum up, the existence of a C^1 shock structure solution for the system (3.62) is possible only when the following inequalities between the constant states u_0 and u_1

hold:

$$u_0 < u_1 \leqslant \left(\frac{2\sqrt{5}-1}{3}\right) u_0,$$

which is equivalent to (3.66).

In what follows, it is convenient to introduce the new variables:

$$\hat{s} = \frac{s}{u_0}, \qquad U = \frac{u}{u_0}, \qquad V = \frac{v}{u_0}, \qquad z = \frac{\varphi}{\tau u_0}.$$

The curve $\mathscr{C}$ in the plane (U, V) is expressed as

$$\mathscr{C}: \qquad U = \frac{6\gamma - 3(1+\gamma)V + 2V^2}{3(1+\gamma) - 4V}, \tag{3.74}$$

while (3.71) becomes

$$F(V, \gamma)\frac{dV}{dz} = 1 \tag{3.75}$$

with

$$F(V, \gamma) = \frac{64V^3 - 9(1+\gamma)\left(8V^2 + 2\gamma - 3\left(1+\gamma^2\right)\right)}{12(V-1)(4V - 3(1+\gamma))(V-\gamma)}.$$

Equation (3.75) admits an implicit solution that can be written in the form:

$$-\frac{3(1+\gamma)}{2}\log|3(1+\gamma) + 4V| + \frac{19\gamma^2 - 6\gamma - 9}{12(\gamma-1)}\log|\gamma - V| + \frac{9\gamma^2 + 6\gamma - 19}{12(\gamma-1)}\log|V-1| + \frac{4}{3}V = z + C_1. \tag{3.76}$$

The presence of the constant C_1 is due to the fact that the shock structure solution has the translational symmetry.

In conclusion, if γ satisfies the inequalities (3.66), there exists a continuous shock structure solution given by (3.74) and (3.76).

3.6.2.3 Case (b): Shock Structure Solution with Subshock

When $\gamma > \gamma_*$, the condition (3.73) is not satisfied and a C^1 shock structure solution cannot exist. In this case a subshock appears, but an analytical representation of the solution may still be given.

Let (u_*, v_*) be the breaking point, where the shock appears; the Rankine–Hugoniot relations associated to the system (3.63) must be satisfied between this point and the unperturbed equilibrium state $(u_0, u_0)^T$. In the dimensionless variables we have

$$\begin{aligned} & -\hat{s}\,(U_* + V_* - 2) + \left(V_*^2 + 2U_*V_* - 3\right) = 0, \\ & -\hat{s}\,(U_* - 1) + \left(V_*^2 - 1\right) = 0. \end{aligned} \tag{3.77}$$

From $(3.77)_1$, we obtain that (U_*, V_*) is a point of the curve $\mathscr{C}$ given by (3.74). Eliminating U_* from (3.77), we obtain

$$2V_*^2 + 2V_* + \hat{s}(2 - \hat{s}) = 0.$$

Then we obtain

$$V_* = \frac{1}{2}\left(-1 + \sqrt{1 + 2\,(\hat{s} - 2)\,\hat{s}}\right).$$

To sum up, the coordinates of the breaking point are given by

$$\begin{aligned} U_* &= \frac{9\gamma^2 + 18\gamma + 5 - 2\sqrt{2\,(9\gamma^2 + 6\gamma - 1)}}{12\,(1 + \gamma)}, \\ V_* &= \frac{-2 + \sqrt{2\,(9\gamma^2 + 6\gamma - 1)}}{4}. \end{aligned}$$

Therefore, if $\gamma > \gamma_*$, there is a shock structure with subshock that is made up by a continuous solution connecting the states (u_1, u_1) and (u_*, v_*), and a shock connecting (u_*, v_*) and (u_0, u_0). The whole profile moves as a traveling wave with velocity s given by (3.69).

The analytical representation of the solution connecting the states (u_1, u_1) and (u_*, v_*) with $\gamma \neq 3$ is given, in the dimensionless variables, by (3.74) and (3.76). In the case $\gamma = 3$, Eqs. (3.74) and (3.76) should be replaced by

$$\begin{aligned} & 2\gamma \log(\gamma - U) = z + C_2, \qquad U \in [U_* = \tfrac{7}{3}, \gamma = 3[\\ & V = \gamma. \end{aligned}$$

It should be noticed that, for $\gamma_* < \gamma < 3$, the function V connecting the states u_1 and v_* is monotonically decreasing, while for $\gamma > 3$ the function becomes monotonically increasing and it presents a *overshoot* in the region close to the breaking point. When $\gamma = 3$, the behavior of V is simply given by the constant states $V = \gamma$.

3.6.2.4 Case (c): Rarefaction Solution of the Equilibrium Subsystem

The equilibrium subsystem (3.64), when Lax condition is not satisfied (when $\gamma < 1$ in the present case), admits a rarefaction solution:

$$u(x,t) = v(x,t) = \begin{cases} u_1 & \frac{x}{t} \leqslant 3u_1, \\ \frac{x}{3t} & 3u_1 < \frac{x}{t} < 3u_0, \\ u_0 & \frac{x}{t} \geqslant 3u_0. \end{cases} \tag{3.78}$$

3.6.2.5 Results of the Numerical Simulations

In order to test the Conjecture 3.1, solutions of the system (3.62) with $\tau = 10^{-2}$ have been numerically calculated under different Cauchy initial data of both Riemann problems with and without structure. In the first case, the initial data are given by

$$u(x,0) = v(x,0) = \begin{cases} u_1 & x < 0, \\ u_0 & x > 0, \end{cases}$$

while in the second case the discontinuity in the initial profile is replaced by a smooth function connecting the two constant states. Both types of initial data are represented in Fig. 3.4. Simulations have been carried out for different values of u_1 and u_0, which give rise to different values of γ, so as to investigate the behavior of the system in the three cases (a)–(c).

3.6.2.6 Case (a): Continuous Shock Structure Solutions

The results of the numerical simulation of the Riemann problem with initial data

$$u_1 = 1.1, \qquad u_0 = 1$$

are presented in Fig. 3.6 and in Fig. 3.7. In this case the condition (3.66) is satisfied and we expect the system to show asymptotically the shock structure solution discussed in Sect. 3.6.2.2. The behavior of the numerical solution is shown in the phase plane (Fig. 3.6) for different times t, along with the curve representing the shock structure solution analytically calculated by means of (3.70) and (3.76). It comes out that the curve representing the solution in the phase plane approaches, as time increases, the one representing the continuous shock structure.

The behavior of the solution in terms of U and V as functions of z is shown in Fig. 3.7, for the same times t as those in Fig. 3.6.

In this case, in order to appreciate the convergence of the numerical solution to the traveling wave, the numerical results have been plotted together with one

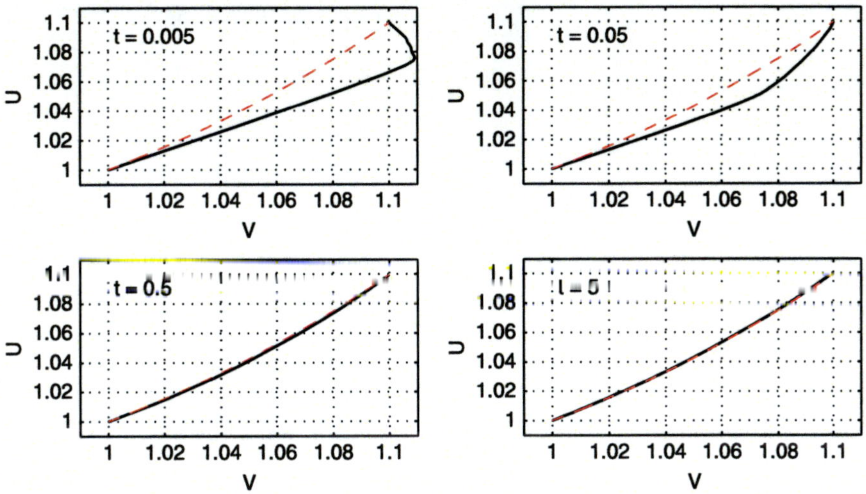

Fig. 3.6 Representation in the phase plane $U - V$ of the numerical solution of the Riemann problem with $u_1 = 1.1$ and $u_0 = 1$ (solid line) and of the analytical continuous shock structure solution (dashed line) at different times t

particular traveling wave belonging to the family of continuous shock structure. The shift constant C_1 in (3.76), which characterizes the traveling waves plotted in Fig. 3.7, is determined by minimizing the norm of the difference between the two curves when the time is large enough.

It is easy to check that the behavior of both U and V seems to asymptotically get closer and closer to the expected results. It is interesting to note that the continuous shock structure is obtained asymptotically as the result of a shock wave moving forward and a perturbed "rarefaction" wave moving backward.

The solution of the Riemann problem with structure, for the same values of u_1 and u_0, is shown in Fig. 3.8 and in Fig. 3.9. From both the phase plane representation of the solution and the behavior of U and V as functions of z, we see that the solution of the system asymptotically approaches the continuous shock structure solution. Even in this case, in Fig. 3.9, the numerical solution of the system has been plotted together with the particular traveling wave chosen from the family of shock structure solutions that represents the expected asymptotic behavior of the solution.

We can also observe that, with these particular initial Cauchy data, both Figs. 3.8 and 3.9 suggest that in the case of Riemann problem with structure the convergence to the continuous shock structure solution is *slower* than that achieved in the Riemann problem, i.e., longer time is required before the solution becomes similar to the traveling wave.

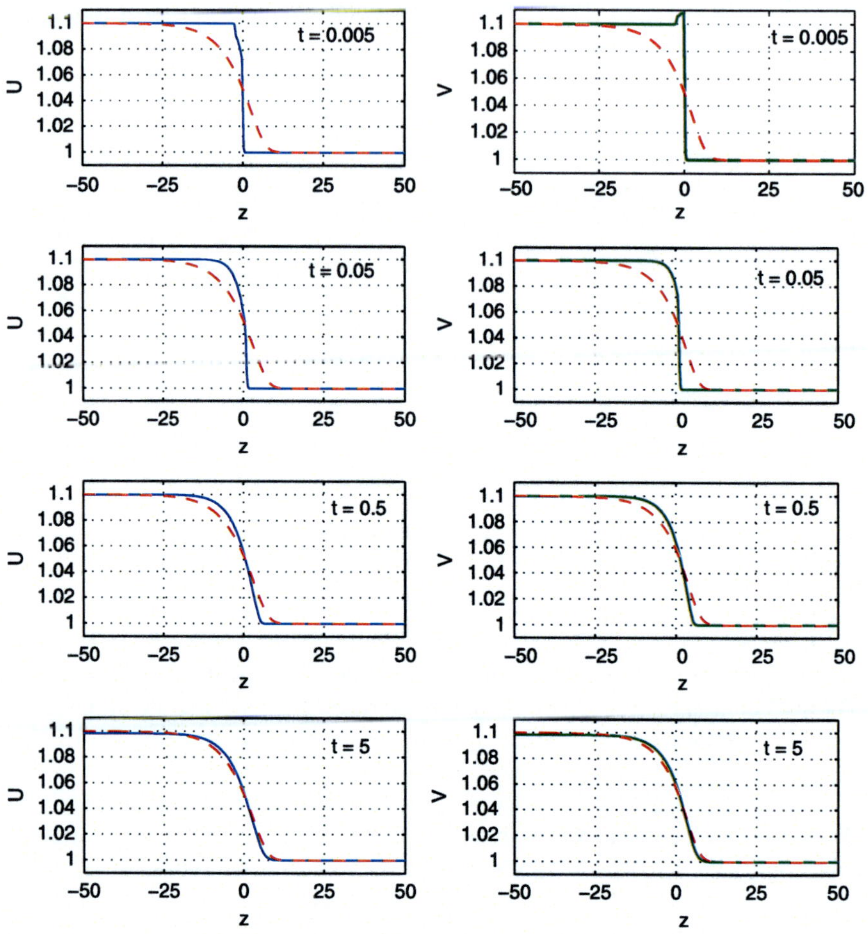

Fig. 3.7 Numerical solution of the Riemann problem with $u_1 = 1.1$ and $u_0 = 1$ (solid line) and analytical continuous shock structure solution (dashed line) at different times t. U and V are represented as functions of $z = x - st$

3.6.2.7 Case (b): Shock Structure Solutions with Subshock

The behaviors of the solutions for Riemann problem and Riemann problem with structure under the condition:

$$u_1 = 2, \qquad u_0 = 1$$

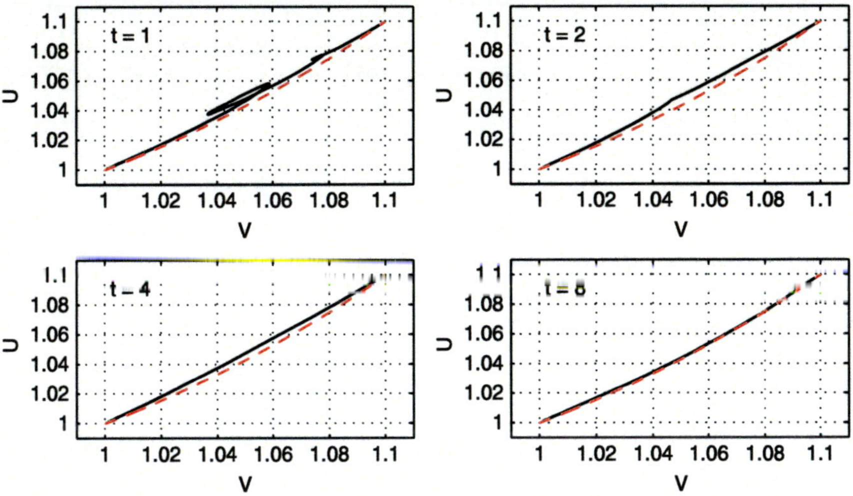

Fig. 3.8 Representation in the phase plane $U - V$ of the numerical solution of the Riemann problem with structure with $u_1 = 1.1$ and $u_0 = 1$ (solid line) and of the analytical continuous shock structure solution (dashed line) at different times t

are shown in Fig. 3.10 and in Fig. 3.11, respectively. In this case, the value of γ satisfies the condition

$$\gamma > \frac{2\sqrt{5} - 1}{3}.$$

According to the conjecture, the solution of the system is expected to converge to a shock structure solution with a subshock, as explained in Sect. 3.6.2.3.

In this case, in contrast to case (a), the wave traveling forward has a velocity that tends to a value larger than the velocity of the shock structure. Therefore a continuous solution may not exist. As expected, a sharp discontinuity is met in the numerical results, and the position of the discontinuity point fit very well with the expectation, as shown in Fig. 3.10.

The numerical solution is plotted together with the discontinuous shock structure evaluated as described in Sect. 3.6.2.3. In this case, the choice of the particular traveling wave has been done in such a way that the subshock is located at the same spatial coordinate.

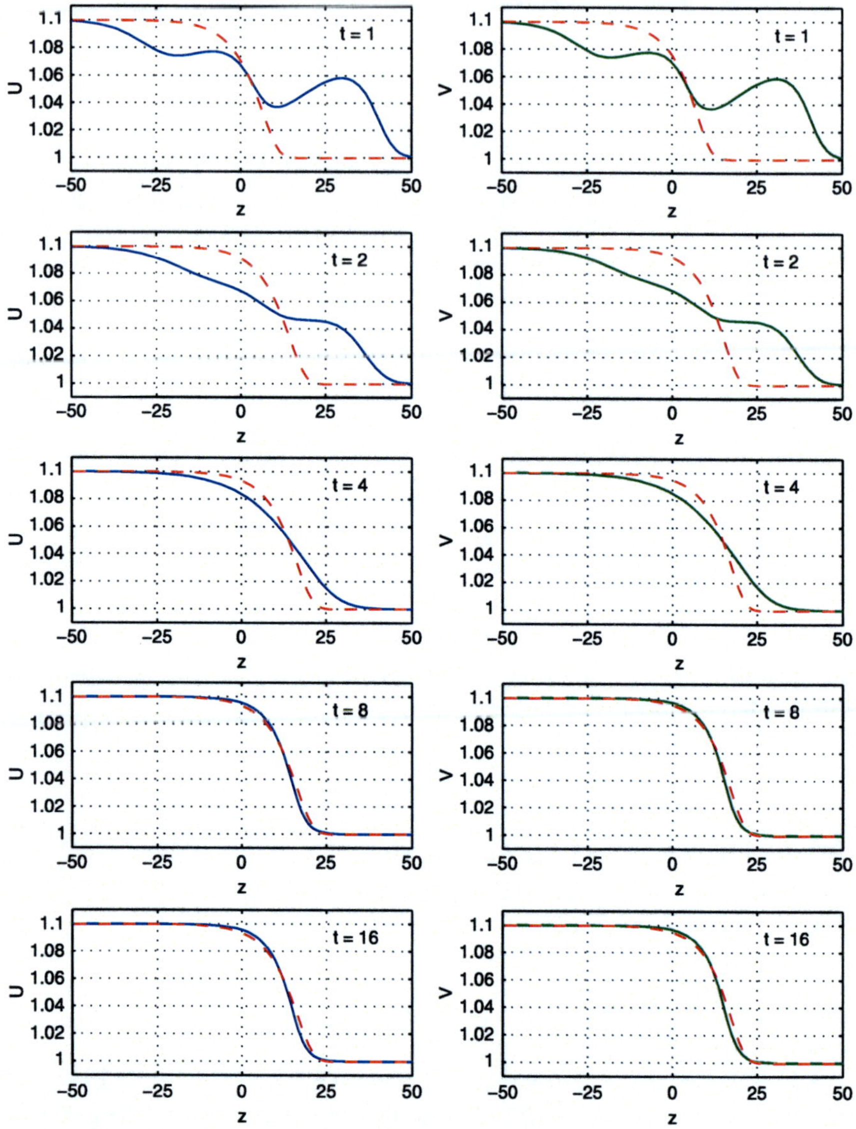

Fig. 3.9 Numerical solution of the Riemann problem with structure with $u_1 = 1.1$ and $u_0 = 1$ (solid line) and analytical continuous shock structure solution (dashed line) at different times t. U and V are represented as functions of $z = x - st$

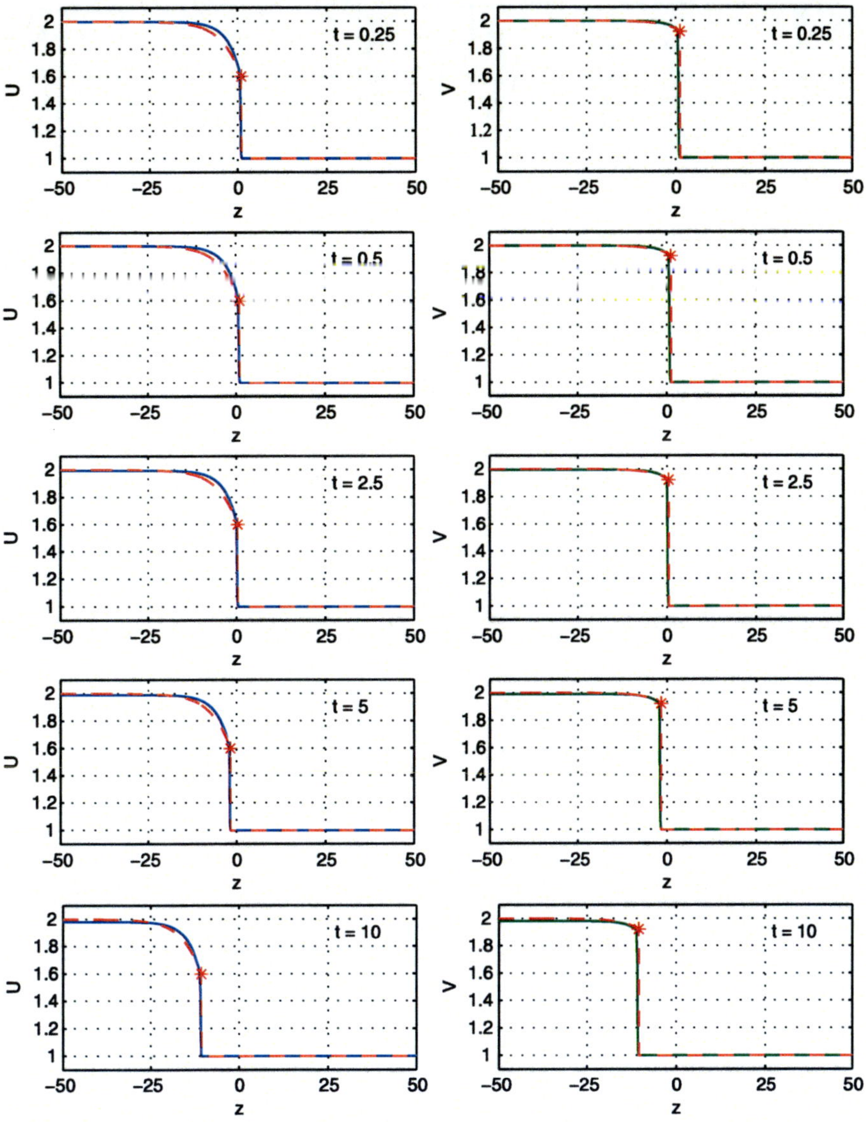

Fig. 3.10 Numerical solution of the Riemann problem with $u_1 = 2$ and $u_0 = 1$ (solid line) and analytical shock structure solution with subshock (dashed line) at different times t. U and V are represented as functions of $z = x - st$. The symbol "*" indicates the point (U_*, V_*) where the continuity of the analytical shock structure solution is lost

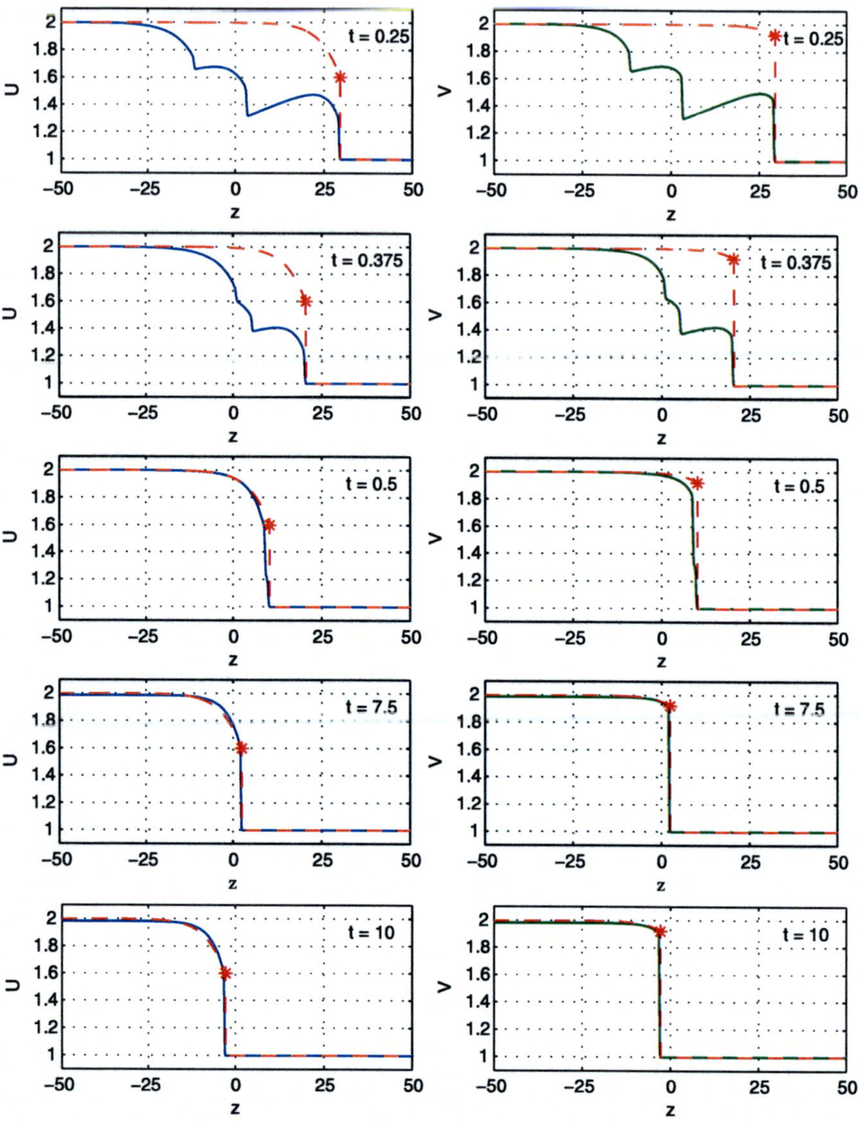

Fig. 3.11 Numerical solution of the Riemann problem with structure with $u_1 = 2$ and $u_0 = 1$ (solid line) and analytical shock structure solution with subshock (dashed line) at different times t. U and V are represented as functions of $z = x - st$. The symbol "*" indicates the point (U_*, V_*) where the continuity of the analytical shock structure solution is lost

3.6.2.8 Case (c): Equilibrium Rarefaction Solutions

Finally, in the case $\gamma < 1$ with $u_1 = 1$ and $u_0 = 2$, we can see from Figs. 3.12 and 3.13 that solutions of both Riemann problem and Riemann problem with structure converge soon to the equilibrium rarefaction wave given by (3.78).

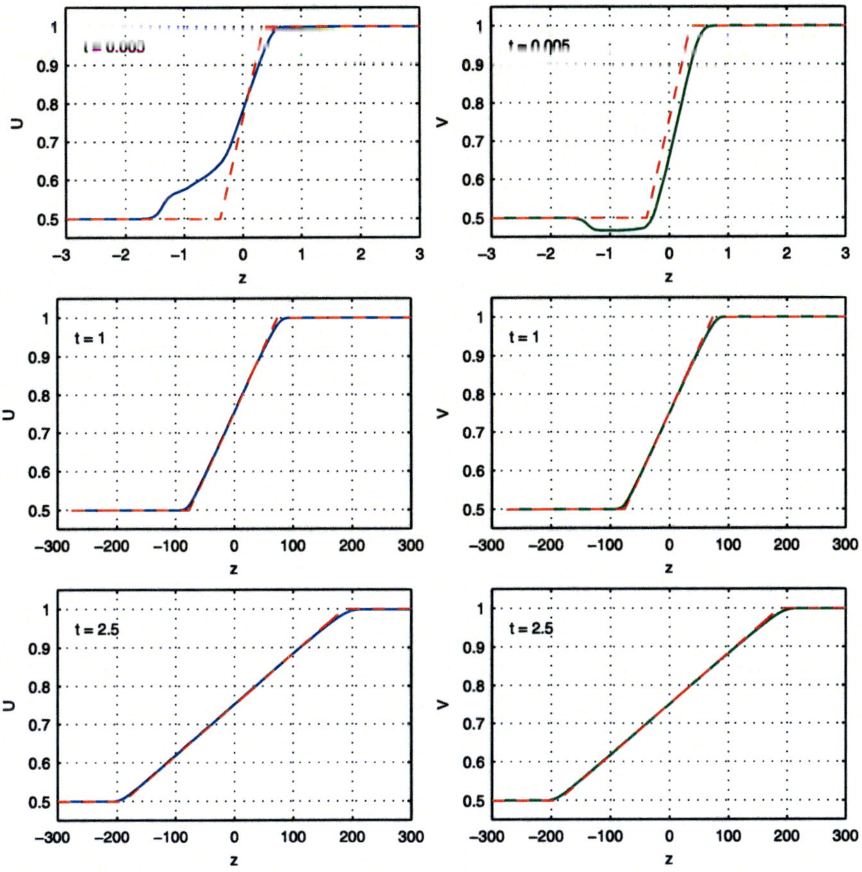

Fig. 3.12 Numerical solution of the Riemann problem with $u_1 = 1$ and $u_0 = 2$ (solid line) and rarefaction wave of the equilibrium subsystem (dashed line) at different times t. U and V are represented as functions of $z = x - st$

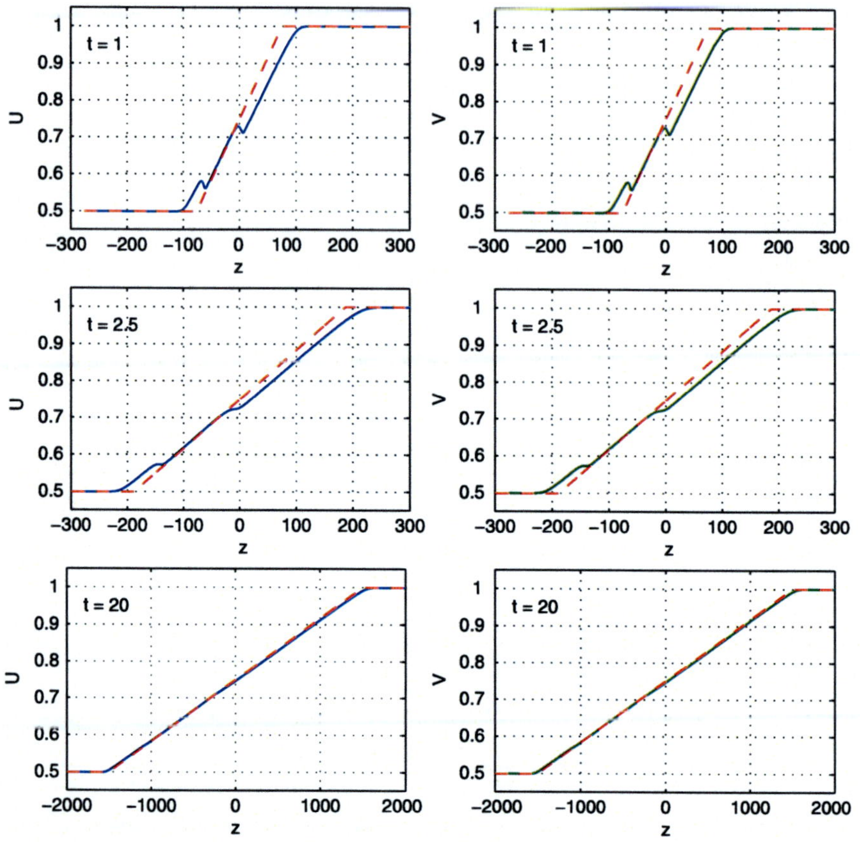

Fig. 3.13 Numerical solution of the Riemann problem with structure with $u_1 = 1$ and $u_0 = 2$ (solid line) and rarefaction wave of the equilibrium subsystem (dashed line) at different times t. U and V are represented as functions of $z = x - st$

Part II
Classical and Relativistic Rational Extended Thermodynamics of Rarefied Monatomic Gas

Chapter 4
RET of Rarefied Monatomic Gas: Non-relativistic Theory

Abstract We make a survey about RET of rarefied monatomic gases. In addition to some results that have been already given in the Müller-Ruggeri book (Müller and Ruggeri: Rational Extended Thermodynamics, 2nd edn. Springer, New York (1998)), many others obtained recently are presented.

We start from the phenomenological RET theory with 13 fields. We prove that the closure of RET coincides with both the closure proposed by Grad using the kinetic-theoretical arguments and the closure by the MEP procedure. The closure here is newly adopted by restyling the closure used in the Liu-Müller paper (Liu and Müller (Arch. Rat. Mech. Anal. 83:285, 1983)).

The RET theory with m moments obtained by the MEP closure is also presented together with the nesting theory that emerges from the concept of principal subsystem. We present the ET_m^α theory, i.e., m-moment theory where the closure by using the terms up to order α with respect to the nonequilibrium variables is adopted. The domain of hyperbolicity is studied. We discuss, in particular, the results due to Brini and Ruggeri (Continuum Mech. Thermodyn. 32:23, 2020) concerning the extension of the hyperbolicity domain when we move our viewpoint from ET_{13}^1 to ET_{13}^2.

A problematic point concerning a bounded domain in RET is also discussed. A simple example in heat conduction is explained to show explicitly that the prediction of the RET theory is appreciably different from the counterpart of the Navier-Stokes and Fourier theory.

A lower bound for the maximum characteristic velocity is obtained as a function of the truncation tensor index N. The velocity increases as the number of moments grows, and it becomes unbounded when $N \to \infty$.

The chapter contains also a brief comparison of the RET predictions with experimental data concerning sound wave and light scattering.

T. Ruggeri, M. Sugiyama, *Classical and Relativistic Rational Extended Thermodynamics of Gases*, https://doi.org/10.1007/978-3-030-59144-1_4

4.1 RET of Rarefied Monatomic Gas with 13 Fields

We saw, in Sect. 1.8.1, that RET of rarefied monatomic gases is intimately related to the moment theory (1.22) associated with the Boltzmann equation. As emphasized there, the 13-field RET (ET_{13}) theory is a purely phenomenological theory [54] in which the system of field equations in balance type (1.27) is adopted and the closure of the system with the use of the universal principles of physics is accomplished. We saw also, in Sect. 1.8.2, the problem of closure of the 13-moment theory.

We have three closure methods by using the universal principles of macroscopic RET [54], the perturbative method of Grad [84] at the kinetic level, and the maximum entropy principle (MEP) [88]. The vital point is that three different and apparently uncorrelated closure methods give the same system (1.28).

First we explain the macroscopic closure by restyling the corresponding part in the paper of Liu and Müller [54].

4.1.1 *Macroscopic Closure of Monatomic* ET_{13}

We adopt the system of balance equations as follows [54]:

$$\begin{aligned}
&\frac{\partial F}{\partial t} + \frac{\partial F_k}{\partial x_k} = 0, \\
&\frac{\partial F_i}{\partial t} + \frac{\partial F_{ik}}{\partial x_k} = 0, \\
&\frac{\partial F_{ij}}{\partial t} + \frac{\partial F_{ijk}}{\partial x_k} = P_{\langle ij \rangle}, \\
&\frac{\partial F_{lli}}{\partial t} + \frac{\partial F_{llik}}{\partial x_k} = P_{lli},
\end{aligned} \tag{4.1}$$

where F is the mass density, F_i is the momentum density, F_{ij} is the momentum flux, and F_{lli} is the energy flux. And F_{ijk} and F_{llik} are the fluxes of F_{ij} and F_{lli}, respectively, and $P_{\langle ij \rangle}$ and P_{lli} are the productions with respect to F_{ij} and F_{lli}, respectively. All tensors are symmetric. The balance equations of $(4.1)_{1,2}$ and the trace part of $(4.1)_3$ represent the conservation laws of mass, momentum, and energy, respectively.

4.1.1.1 Thermal and Caloric Equations of State

The thermal and caloric equations of state for a rarefied monatomic gas are, respectively, expressed as

$$p = \frac{k_B}{m}\rho T, \qquad \varepsilon = \frac{3}{2}\frac{k_B}{m}T. \tag{4.2}$$

In this case the specific heat at constant volume is constant:

$$c_v = \frac{3}{2}\frac{k_B}{m}.$$

4.1.1.2 Closure Quantities

Choosing the field **u** composed of the densities:

$$\mathbf{u} \equiv (F,\ F_i,\ F_{ij},\ F_{lli})^T,$$

we assume that the other fluxes and productions $\psi \equiv \{F_{ijk},\ F_{llik},\ P_{\langle ij\rangle},\ P_{lli}\}$ depend on the field **u** at that point and time, i.e., local and instantaneous:

$$\psi \equiv \psi(\mathbf{u}). \tag{4.3}$$

The entropy density h^0 and the entropy flux h^k (2.54) are also assumed to be in the form (4.3).

4.1.1.3 Exploitation of the Galilean Invariance

The matrices $\mathbf{X}$ and $\mathbf{A}^r$ of the Galilean invariance (see (2.49) and (2.51)) are given by

$$\mathbf{X} = \begin{pmatrix} 1 & 0 & 0 & 0 \\ v_i & \delta_i^{h_1} & 0 & 0 \\ v_i v_j & 2\delta_{(i}^{h_1} v_{j)} & \delta_i^{h_1}\delta_j^{h_2} & 0 \\ v^2 v_i & 3v_{(l}v_l\delta_{i)}^{h_1} & 3\delta_{(ih_1}v_{h_2)} & \delta_i^{h_1} \end{pmatrix} \tag{4.4}$$

and

$$\mathbf{A}^r = \begin{pmatrix} 0 & 0 & 0 & 0 \\ \delta_i^r & 0 & 0 & 0 \\ 0 & 2\delta_{(i}^{h_1}\delta_{j)}^r & 0 & 0 \\ 0 & 0 & 3\delta_{(ih_1}\delta_{h_2)r} & 0 \end{pmatrix} \qquad (r = 1, 2, 3).$$

Therefore, according with Theorem 2.8, the velocity dependence of field variables is completely prescribed by the Galilean invariance as follows (see (2.49)):

$$\mathbf{F}^0 = \begin{pmatrix} F \\ F_i \\ F_{ij} \\ F_{lli} \end{pmatrix} = \begin{pmatrix} \rho \\ \rho v_i \\ \rho v_i v_j + \hat{F}_{ij} \\ \rho v^2 v_i + 3\hat{F}_{(il}v_{l)} + \hat{F}_{lli} \end{pmatrix},$$

$$\boldsymbol{\Phi}^k = \begin{pmatrix} \Phi_k \\ \Phi_{ik} \\ \Phi_{ijk} \\ \Phi_{llik} \end{pmatrix} = \begin{pmatrix} 0 \\ \hat{F}_{ik} \\ 2\hat{F}_{k(i}v_{j)} + \hat{F}_{ijk} \\ 3\hat{F}_{k(i}v_l v_{l)} + 3\hat{F}_{k(il}v_{l)} + \hat{F}_{llik} \end{pmatrix}, \tag{4.5}$$

$$\mathbf{f} = \begin{pmatrix} 0 \\ 0 \\ P_{\langle ij\rangle} \\ P_{lli} \end{pmatrix} = \begin{pmatrix} 0 \\ 0 \\ \hat{P}_{\langle ij\rangle} \\ 2\hat{P}_{\langle il\rangle}v_l + \hat{P}_{lli} \end{pmatrix},$$

where we recall $\hat{\boldsymbol{\Phi}}^k = \hat{\mathbf{F}}^k$ (see (2.50)).

Since the balance equations of F, F_i, and F_{ll} represent, respectively, the conservation laws of mass, momentum, and energy, the intrinsic quantities $\hat{F}_{ij}$, $\hat{F}_{ll}$, and $\hat{F}_{lli}$ have the following conventional meanings:

stress tensor: $$t_{ij} = -\hat{F}_{ij} \ (= -p\delta_{ij} + \sigma_{\langle ij\rangle}), \tag{4.6}$$

specific internal energy: $$\varepsilon = \frac{1}{2\rho}\hat{F}_{ll}, \tag{4.7}$$

heat flux: $$q_i = \frac{1}{2}\hat{F}_{lli}. \tag{4.8}$$

From (4.6) and (4.7), we have the relation:

$$p = \frac{2}{3}\rho\varepsilon,$$

which is consistent with the equations of state (4.2).

Then the **constitutive equations of ET$_{13}$** are expressed by the Galilean objective variables in the form (see the closure equations (4.3)):

$$\hat{\psi} = \hat{\psi}(\rho, \varepsilon, \sigma_{\langle ij\rangle}, q_i).$$

Since the entropy density $h(= h^0)$ and the intrinsic entropy flux $\varphi_k(= \hat{h}^k)$ are independent of the velocity, these are also expressed as follows:

$$h = h(\rho,\ \varepsilon,\ \sigma_{\langle ij\rangle},\ q_i),$$
$$\varphi_k = \varphi_k(\rho,\ \varepsilon,\ \sigma_{\langle ij\rangle},\ q_i).$$

4.1.1.4 Exploitation of the Entropy Principle

The entropy principle requires the following relations (see (2.11)):

$$\begin{aligned} dh &= \lambda dF + \lambda_i dF_i + \lambda_{\langle ij\rangle} dF_{ij} + \mu dF_{ll} + \mu_i dF_{lli}, \\ dh_k &= \lambda dF_k + \lambda_i dF_{ik} + \lambda_{\langle ij\rangle} dF_{ijk} + \mu dF_{llk} + \mu_i dF_{llik}, \\ \Sigma &= \lambda_{\langle ij\rangle} P_{\langle ij\rangle} + \mu_i P_{lli}, \end{aligned} \tag{4.9}$$

where

$$\mathbf{u}' \equiv \left(\lambda,\ \lambda_i,\ \lambda_{\langle ij\rangle},\ \mu,\ \mu_i\right) \tag{4.10}$$

is the main field composed of Lagrange multipliers. Therefore, taking (2.57) into account, we have

$$\begin{aligned} dh &= \hat{\lambda} d\rho + \hat{\lambda}_{\langle ij\rangle} d\hat{F}_{ij} + \hat{\mu} d\hat{F}_{ll} + \hat{\mu}_i d\hat{F}_{lli}, \\ d\varphi_k &= \hat{\lambda}_i d\hat{F}_{ik} + \hat{\lambda}_{\langle ij\rangle} d\hat{F}_{ijk} + \hat{\mu} d\hat{F}_{llk} + \hat{\mu}_i d\hat{F}_{llik}, \\ \Sigma &= \hat{\lambda}_{\langle ij\rangle} \hat{P}_{\langle ij\rangle} + \hat{\mu}_i \hat{P}_{lli}. \end{aligned} \tag{4.11}$$

The dependence of the Lagrange multipliers on the velocity is dictated by the relation (2.55). In the present case, taking (4.4) into account, we have the following relations:

$$\begin{aligned} \lambda &= \hat{\lambda} - \hat{\lambda}_i v_i + \hat{\lambda}_{\langle ij\rangle} v_i v_j + \hat{\mu} v^2 - \hat{\mu}_i v^2 v_i, \\ \lambda_i &= \hat{\lambda}_i - 2\hat{\lambda}_{\langle il\rangle} v_l - 2\hat{\mu} v_i + 2\hat{\mu}_l v_i v_l + \hat{\mu}_i v^2, \\ \lambda_{\langle ij\rangle} &= \hat{\lambda}_{\langle ij\rangle} - \left(\hat{\mu}_i v_j + \hat{\mu}_j v_i - \frac{2}{3}\hat{\mu}_k v_k \delta_{ij}\right), \\ \mu &= \hat{\mu} - \hat{\mu}_l v_l, \\ \mu_i &= \hat{\mu}_i. \end{aligned} \tag{4.12}$$

The constraints (2.58) are expressed as

$$\begin{gathered}\hat{\lambda}_i = -\frac{1}{\rho}\left(2\hat{F}_{il}\hat{\mu}_l + \hat{F}_{ll}\hat{\mu}_i\right),\\ \left(\hat{\lambda}\rho + \hat{\lambda}_{\langle ij\rangle}\hat{F}_{ij} + \hat{\mu}\hat{F}_{ll} + \hat{\mu}_k\hat{F}_{llk} - h\right)\delta_{ir} + 2\hat{\lambda}_{\langle rl\rangle}\hat{F}_{il} + 2\hat{\mu}\hat{F}_{ir} + 2\hat{\mu}_l\hat{F}_{lri} + \hat{\mu}_r\hat{F}_{lli} = 0.\end{gathered} \tag{4.13}$$

(i) Equilibrium State: From Definition 2.6 of an equilibrium state, which says that the entropy production Σ becomes minimum and vanishes in equilibrium, we notice that the productions $\hat{P}_{\langle ij\rangle}$ and $\hat{P}_{lli}$ must vanish together with all Lagrange multipliers corresponding to the balance laws (see (2.45)):

$$\hat{\lambda}^E_{\langle ij\rangle} = 0, \qquad \hat{\mu}^E_i = 0, \tag{4.14}$$

where superscript E indicates that quantities are evaluated in an equilibrium state. Since the Lagrange multipliers $\hat{\lambda}_{\langle ij\rangle}$ and $\hat{\mu}_i$ are non-zero only in nonequilibrium, these are called nonequilibrium variables. Taking the representation theorem for isotropic vectors and tensors into account, we can show the following fact: the above requirement for the intrinsic productions and the intrinsic nonequilibrium Lagrange multipliers in equilibrium implies that q_i and $\sigma_{\langle ij\rangle}$ must also vanish in equilibrium.

(ii) Entropy: Let us rewrite the entropy density in the form:

$$h = \rho s + \rho k, \tag{4.15}$$

where $s(\rho, \varepsilon)$ is the equilibrium specific entropy density and $k(\rho, \varepsilon, \sigma_{\langle ij\rangle}, q_i)$ denotes the nonequilibrium part of the entropy density. Therefore, from the discussion (i) above, we have the condition:

$$k(\rho, \varepsilon, 0, 0) = 0.$$

The equilibrium entropy density s obeys the Gibbs equation (1.8). From $(4.11)_1$, we obtain

$$\begin{aligned}\hat{\lambda} &= -\frac{g}{T} + k + \rho k_\rho - \varepsilon k_\varepsilon,\\ \hat{\lambda}_{\langle ij\rangle} &= -\rho k_{\sigma_{\langle ij\rangle}},\\ \hat{\mu} &= \frac{1}{2T} + \frac{1}{2}k_\varepsilon,\\ \hat{\mu}_i &= \frac{\rho}{2}k_{q_i},\end{aligned} \tag{4.16}$$

where $g = g(\rho, \varepsilon) = \varepsilon + p/\rho - Ts$ is the chemical potential. A subscript in the right-hand side of (4.16) stands for a partial differentiation with respect to the quantity, for example, $k_\rho \equiv \partial k/\partial \rho$. In an equilibrium state, we have the condition (4.14) and the following relations:

$$\hat{\lambda}^E = -\frac{g}{T}, \qquad \hat{\lambda}_j^E = 0, \qquad \hat{\mu}^E = \frac{1}{2T}, \tag{4.17}$$

Remark 4.1 We have introduced the quantity T in (4.16) through the Gibbs equation (1.8). And, from (4.17), we notice that this quantity T is just the absolute temperature when the system is in equilibrium. Therefore we use the same symbol T in both nonequilibrium and equilibrium cases, and call it (local equilibrium) temperature. For general discussions on nonequilibrium temperature and chemical potential, see Chap. 15.

(iii) Constitutive Equations Near Equilibrium: As usual in RET, we study processes not far from equilibrium, which, however, may be out of local equilibrium. Then we adopt the constitutive equations that are linear with respect to the nonequilibrium variables $\{\sigma_{\langle ij\rangle}, q_i\}$. Then, we have k and φ_k in the following forms:

$$k = k_1\sigma_{\langle ij\rangle}\sigma_{\langle ij\rangle} + k_2 q_i q_i + O(3), \tag{4.18}$$

$$\varphi_k = \beta_1 q_k + \beta_2 \sigma_{\langle ki\rangle} q_i + O(3), \tag{4.19}$$

where coefficients k_1, k_2 and β_1, β_2 are functions of ρ and ε.

From the symmetry of F_{ijk} with respect to the suffixes i, j, k, we also represent fluxes as follows:

$$\hat{F}_{ijk} = \frac{2}{5}(q_i\delta_{jk} + q_k\delta_{ij} + q_j\delta_{ki}) + O(2), \qquad (\hat{F}_{lli} = 2q_i)$$

$$\hat{F}_{llij} = g_1\delta_{ij} - g_2\sigma_{\langle ij\rangle} + O(2),$$

where g_1 and g_2 are functions of ρ and ε.

From (4.18), we obtain

$$dk = \sigma_{\langle ij\rangle}\sigma_{\langle ij\rangle}dk_1 + q_i q_i dk_2 + 2k_1\sigma_{\langle ij\rangle}d\sigma_{\langle ij\rangle} + 2k_2 q_i dq_i. \tag{4.20}$$

Therefore, substituting (4.20) into (4.16), we obtain the Lagrange multipliers as follows:

$$\hat{\lambda} = -\frac{g}{T} + (k_1 + \rho k_{1\rho} - \varepsilon k_{1\varepsilon})\sigma_{\langle ij\rangle}\sigma_{\langle ij\rangle} + (k_2 + \rho k_{2\rho} - \varepsilon k_{2\varepsilon})q_i q_i,$$

$$\hat{\lambda}_{\langle ij\rangle} = -2\rho k_1\sigma_{\langle ij\rangle},$$

$$\hat{\mu} = \frac{1}{2T} + \frac{1}{2}k_{1\varepsilon}\sigma_{\langle ij\rangle}\sigma_{\langle ij\rangle} + \frac{1}{2}k_{2\varepsilon}q_i q_i,$$
$$\hat{\mu}_i = \rho k_2 q_i, \tag{4.21}$$

where, from (4.13), $\hat{\lambda}_i$ is related to $\hat{\mu}_i$.

Let us evaluate the coefficients k_1 and k_2 by using the constraint $(4.13)_2$ and the expressions of the Lagrange multipliers (4.21). We decompose the constraint $(4.13)_2$ into the trace part and traceless part as follows:

$$\begin{aligned} h &= \rho\hat{\lambda} - \frac{5}{3}\hat{\lambda}_{\langle ij\rangle}\sigma_{\langle ij\rangle} + 5p\hat{\mu} + 4q_i\hat{\mu}_i, \\ &\hat{\lambda}_{\langle l\langle r\rangle}\hat{F}_{i\rangle l} + 2\hat{\mu}_{\langle r}q_{i\rangle} = 0. \end{aligned} \tag{4.22}$$

From (4.22), we obtain

$$k_1 = -\frac{1}{4\rho pT}, \qquad k_2 \propto \frac{1}{\rho^2\varepsilon^3}. \tag{4.23}$$

Next we analyze the entropy flux. From (4.19), we have

$$d\varphi_k = q_k d\beta_1 + \sigma_{\langle ki\rangle}q_i d\beta_2 + \beta_2 q_i d\sigma_{\langle ki\rangle} + (\beta_1\delta_{ki} + \beta_2\sigma_{\langle ki\rangle})dq_i.$$

On the other hand, $d\varphi_k$ is expressed as $(4.11)_2$ with $(4.13)_1$. By comparing these expressions with each other, we obtain the following relations:

$$\beta_1 = 2\mu = \frac{1}{T}, \qquad \beta_2 = -\frac{8}{5}\rho k_1 = \frac{2}{5pT},$$

and

$$g_{1\rho} = \frac{5p}{\rho}p_\rho, \quad \rho k_2\left(g_{1\varepsilon} - \frac{5p}{\rho}p_\varepsilon\right) = \beta_{1\varepsilon}, \quad \rho k_2\left(g_2 - \frac{5p}{\rho}\right) = -\beta_2. \tag{4.24}$$

We now analyze the relations (4.24). From the integrability condition of g_1, we obtain the proportionality coefficient in (4.23). Then we can easily obtain the expressions of k_2, g_1, and g_2 as follows:

$$k_2 = -\frac{1}{5p^2T}, \qquad g_1 = \frac{10}{3}p\varepsilon, \qquad g_2 = \frac{14}{3}\varepsilon.$$

Here we have used the condition that the integration constant in g_1 vanishes because g_1 must become null when $\rho \to 0$.

4.1.1.5 Linear Constitutive Equations

Using the expressions of the expansion coefficients obtained above, we have the linear constitutive equations:

$$\begin{aligned}\hat{F}_{ijk} &= \frac{2}{5}(q_i\delta_{jk} + q_k\delta_{ij} + q_j\delta_{ki}) + O(2), \qquad (\hat{F}_{lli} = 2q_i) \\ \hat{F}_{llij} &= \frac{10}{3}p\varepsilon\delta_{ij} - \frac{14}{3}\varepsilon\sigma_{\langle ij\rangle} + O(2).\end{aligned} \tag{4.25}$$

4.1.1.6 Productions

The productions are also expanded with respect to the nonequilibrium variables $\{\sigma_{\langle ij\rangle}, q_i\}$ around an equilibrium state. In the linear approximation, we have

$$\hat{P}_{\langle ij\rangle} = \frac{1}{\tau_\sigma}\sigma_{\langle ij\rangle}, \qquad \hat{P}_{lli} = -\frac{2}{\tau_q}q_i, \tag{4.26}$$

where τ_σ and τ_q are the expansion coefficients, meaning of which will be understood in Sect. 4.1.1.10.

4.1.1.7 Main Field

From (4.21), the intrinsic Lagrange multipliers in the linear approximation are obtained as follows:

$$\begin{aligned}&\hat{\lambda} = -\frac{g}{T}, \quad \hat{\lambda}_i = \frac{1}{pT}q_i, \quad \hat{\lambda}_{\langle ij\rangle} = \frac{1}{2pT}\sigma_{\langle ij\rangle}, \\ &\hat{\mu} = \frac{1}{2T}, \quad \hat{\mu}_i = -\frac{\rho}{5p^2T}q_i.\end{aligned} \tag{4.27}$$

Taking into account (4.12), we can evaluate the main field components (4.10):

$$\begin{aligned}\lambda &= -\frac{1}{T}\left(g - \frac{v^2}{2} + \frac{1}{2p}\sigma_{\langle ij\rangle}v_i v_j - \frac{\rho}{5p^2}q_i v_i v^2\right), \\ \lambda_i &= -\frac{1}{T}\left\{v_i - \frac{1}{p}\sigma_{\langle ij\rangle}v_j + \frac{\rho}{5p^2}\left(v^2 q_i + 2q_j v_j v_i\right) - \frac{q_i}{p}\right\}, \\ \mu &= \frac{1}{2T}\left(1 - \frac{2\rho}{3p^2}q_k v_k\right),\end{aligned} \tag{4.28}$$

$$\lambda_{\langle ij\rangle} = \frac{1}{T}\left\{\frac{1}{2p}\sigma_{\langle ij\rangle} + \frac{\rho}{5p^2}\left(v_i\, q_j + v_j\, q_i - \frac{2}{3}v_k\, q_k\, \delta_{ij}\right)\right\},$$
$$\mu_i = -\frac{\rho}{5Tp^2}q_i,$$

in terms of which the differential system becomes symmetric hyperbolic.

4.1.1.8 Entropy Density, Entropy Flux, and Entropy Production

The entropy density (4.15) and entropy flux (4.19) are expressed as

$$h = \rho s - \frac{1}{4pT}\sigma_{\langle ij\rangle}\sigma_{\langle ij\rangle} - \frac{\rho}{5p^2T}q_i q_i + O(3),$$
$$\varphi_k = \frac{1}{T}q_k + \frac{2}{5pT}q_i\sigma_{\langle ik\rangle} + O(3).$$

From (4.9), (4.27), and (4.26), the entropy production is given by

$$\Sigma = \frac{1}{2pT\tau_\sigma}\sigma_{\langle ij\rangle}\sigma_{\langle ij\rangle} + \frac{2\rho}{5p^2T\tau_q}q_i q_i \geqq 0.$$

Then we have two conditions:

$$\tau_\sigma > 0, \qquad \tau_q > 0.$$

4.1.1.9 Convexity of the Entropy Density

The convexity condition (2.59) evaluated in an equilibrium state becomes

$$\hat{Q}_E = \hat{Q}_{5E} - \frac{1}{4pT}\delta\sigma_{\langle ij\rangle}\delta\sigma_{\langle ij\rangle} - \frac{\rho}{5p^2T}\delta q_i \delta q_i < 0,$$

where $\hat{Q}_{5E}$ is the corresponding quantity for Euler fluids.

From the condition $\hat{Q}_{5E} < 0$, we obtain the usual thermodynamic inequalities, that is, the positivity of the heat capacity and the compressibility (see (2.31)). Therefore, the convexity condition is satisfied in the neighborhood of equilibrium state. The potential $h' = h'^0$ is expressed as

$$\begin{aligned} h' &= \rho\hat{\lambda} + 2\rho\varepsilon\hat{\mu} - \sigma_{\langle ij\rangle}\hat{\lambda}_{\langle ij\rangle} + 2q_i\hat{\mu}_i - h \\ &= -\frac{p}{T}\left(1 + \frac{\sigma_{\langle ij\rangle}\sigma_{\langle ij\rangle}}{4p^2} + \frac{\rho q_i q_i}{5p^3}\right). \end{aligned} \tag{4.29}$$

This is also a convex function with respect to the main field because of the convexity of the entropy density. While from (2.56) we have

$$h'^i = -\frac{p}{T}\left(1 + \frac{\sigma_{\langle ij\rangle}\sigma_{\langle ij\rangle}}{4p^2} + \frac{\rho q_i q_i}{5p^3}\right) v_i + \frac{q_j \sigma_{\langle ij\rangle}}{4pT}. \tag{4.30}$$

The potentials h' (4.29) and h'^i (4.30) as functions of the main field (4.28) symmetrize the system of ET_{13}, and, in equilibrium, they reduce to the ones of Euler fluids (2.30).

4.1.1.10 Closed System of Field Equations

We can close the system of field equations by substituting the constitutive equations (4.25) and (4.26) into (4.1) with (4.5) and (4.6)–(4.8). Then the closed system with the independent fields $\{\rho, v_i, T, \sigma_{\langle ik\rangle}, q_i\}$ is given by (1.28), where p and ε are the functions of ρ and T given by the thermal and caloric equations of state (4.2).

By using the material derivative, the system (1.28) can be rewritten as

$$\begin{aligned}
&\dot{\rho} + \rho\frac{\partial v_k}{\partial x_k} = 0,\\
&\rho\dot{v}_i + \frac{\partial(p\delta_{ij} - \sigma_{\langle ij\rangle})}{\partial x_j} = 0,\\
&\rho\frac{3}{2}\frac{k_B}{m}\dot{T} + p\frac{\partial v_k}{\partial x_k} - \frac{\partial v_i}{\partial x_k}\sigma_{\langle ik\rangle} + \frac{\partial q_k}{\partial x_k} = 0,\\
&\dot{\sigma}_{\langle ij\rangle} - 2p\frac{\partial v_{\langle i}}{\partial x_{j\rangle}} + \sigma_{\langle ij\rangle}\frac{\partial v_k}{\partial x_k} + 2\frac{\partial v_{\langle i}}{\partial x_k}\sigma_{\langle j\rangle k\rangle} - \frac{4}{5}\frac{\partial q_{\langle i}}{\partial x_{j\rangle}} = -\frac{1}{\tau_\sigma}\sigma_{\langle ij\rangle},\\
&\dot{q}_i + \frac{7}{5}q_i\frac{\partial v_k}{\partial x_k} + \frac{2}{5}q_k\frac{\partial v_k}{\partial x_i} + \frac{7}{5}q_k\frac{\partial v_i}{\partial x_k} + \frac{1}{2}\frac{k_B}{m}\left(5p\delta_{ki} - 7\sigma_{\langle ki\rangle}\right)\frac{\partial T}{\partial x_k}\\
&\quad + \frac{\sigma_{\langle ki\rangle}}{\rho}\frac{\partial p}{\partial x_k} - \frac{1}{\rho}\left(p\delta_{ki} + \sigma_{\langle ki\rangle}\right)\frac{\partial \sigma_{\langle kl\rangle}}{\partial x_l} = -\frac{1}{\tau_q}q_i.
\end{aligned} \tag{4.31}$$

We now understand that τ_σ and τ_q can be regarded as the relaxation times of the shear stress and the heat flux, respectively.

4.1.1.11 Relationship Between RET Theory and Navier-Stokes Fourier Theory

We carry out the Maxwellian iteration (for more details see Chap. 33, Sects. 33.2.1 and 33.3) for the system (4.31): The first iterates $\sigma^{(1)}_{\langle ij\rangle}$ and $q_i^{(1)}$ are obtained by the substitution of the 0th iterates $\sigma^{(0)}_{\langle ij\rangle} = 0$ and $q^{(0)}_i = 0$ into the left-hand side of $(4.31)_{4,5}$. Then we obtain

$$\sigma^{(1)}_{\langle ij\rangle} = 2p\tau_\sigma \frac{\partial v_{\langle i}}{\partial x_{j\rangle}}, \qquad q^{(1)}_i = -\frac{5}{2}\frac{k_B}{m} p\tau_q \frac{\partial T}{\partial x_i}.$$

On the other hand, we have the laws of Navier-Stokes and Fourier expressed by (1.11). Their comparison reveals that

$$\mu = p\tau_\sigma, \qquad \kappa = \frac{5}{2}\frac{k_B}{m} p\tau_q.$$

We can therefore estimate the values of the relaxation times τ_σ and τ_q from the experimental data on the phenomenological coefficients μ and κ.

The second iterates are obtained by substituting the first iterates into the left-hand side of $(4.31)_{4,5}$, and higher iterates are obtained in a similar way.

In conclusion, the system can be certainly closed by the universal principles, provided that we know the thermal and caloric equations of state and the viscosity and heat conductivity coefficients. The consistency of the present theory to the theory derived from the kinetic theory will be shown in Sect. 4.4.1.

4.1.2 *Principal Subsystems of* ET_{13}

The maximum characteristic velocity in equilibrium of the system (1.28) is $\lambda_{\max} \simeq 1.65\, c_0$, where c_0 is the equilibrium sound velocity for a monatomic gas:

$$c_0 = \sqrt{\frac{5}{3}\frac{k_B}{m} T}.$$

Let us consider possible principal subsystems of ET_{13} following the general theory presented in Sect. 2.4.

4.1.2.1 ET_{10} Principal Subsystem

From (4.28), the 10-field system is a principal subsystem of the 13-field (ET_{13}) system when

$$\mu_i = 0 \quad \to \quad q_i = 0.$$

Neglecting the last block of the corresponding equation of (1.28) and inserting $q_i = 0$ in the remaining equations, we have

$$
\begin{aligned}
&\frac{\partial \rho}{\partial t} + \frac{\partial}{\partial x_k}(\rho v_k) = 0, \\
&\frac{\partial \rho v_i}{\partial t} + \frac{\partial}{\partial x_k}\left(\rho v_i v_k + p\delta_{ik} - \sigma_{\langle ik\rangle}\right) = 0, \\
&\frac{\partial}{\partial t}(\rho v^2 + 2\rho\varepsilon) + \frac{\partial}{\partial x_k}\left\{\rho v^2 v_k + 2(\rho\varepsilon + p)v_k - 2\sigma_{\langle kl\rangle} v_l\right\} = 0, \qquad (4.32)\\
&\frac{\partial}{\partial t}\left\{\rho v_i v_j + p\delta_{ij} - \sigma_{\langle ij\rangle}\right\} + \\
&\quad + \frac{\partial}{\partial x_k}\left\{\rho v_i v_j v_k + p(v_i\delta_{jk} + v_j\delta_{ki} + v_k\delta_{ij}) - \sigma_{\langle ij\rangle} v_k - \sigma_{\langle jk\rangle} v_i - \sigma_{\langle ki\rangle} v_j\right\} = \frac{\sigma_{\langle ij\rangle}}{\tau_\sigma}.
\end{aligned}
$$

In this case, in equilibrium, $\lambda_{\max} \simeq 1.34\, c_0$. This principal subsystem can be obtained formally from the general system (1.28) by taking the limit for the relaxation time: $\tau_q \to 0$.

4.1.2.2 Euler ET$_5$ Principal Subsystem

The Euler system is the 5-field principal subsystem of the 13- and 10-field systems when

$$
\mu_i = 0, \quad \lambda_{\langle ij\rangle} = 0 \;\rightarrow\; q_i = 0, \quad \sigma_{\langle ij\rangle} = 0.
$$

Neglecting the last block of the corresponding equation of (4.32), and inserting $\sigma_{\langle ij\rangle} = 0$ in the remaining equations, we have

$$
\begin{aligned}
&\frac{\partial \rho}{\partial t} + \frac{\partial}{\partial x_k}(\rho v_k) = 0, \\
&\frac{\partial \rho v_i}{\partial t} + \frac{\partial}{\partial x_k}(\rho v_i v_k + p\delta_{ik}) = 0, \qquad (4.33)\\
&\frac{\partial}{\partial t}(\rho v^2 + 2\rho\varepsilon) + \frac{\partial}{\partial x_k}\left\{\rho v^2 v_k + 2(\rho\varepsilon + p)v_k\right\} = 0.
\end{aligned}
$$

The maximum velocity in equilibrium is now $\lambda_{\max} = c_0$. This principal subsystem can be obtained formally from the general system (1.28) by taking the limits for the relaxation times: both τ_q and τ_σ approach zero.

4.1.2.3 ET$_4$ Principal Subsystem

Up to here we have seen that the principal subsystems are obtained by letting some relaxation times go to zero. Now, with Euler system (4.33) where no relaxation time exists, we can still consider ET$_4$ principal subsystem requiring

$$\mu_i = 0, \quad \lambda_{\langle ij\rangle} = 0, \quad \mu = \text{const.} \quad \rightarrow \quad q_i = 0, \quad \sigma_{\langle ij\rangle} = 0, \quad T = T^* = \text{const.},$$

and eliminating the last equation in (4.33). The field equations are given by the isothermal Euler system:

$$\begin{aligned} &\frac{\partial}{\partial t}\rho + \frac{\partial}{\partial x_i}(\rho v_i) = 0, \\ &\frac{\partial}{\partial t}\left(\rho v_j\right) + \frac{\partial}{\partial x_i}\left(\rho v_i v_j + p^* \delta_{ij}\right) = 0, \end{aligned} \tag{4.34}$$

where

$$p^* \equiv p(\rho, T^*).$$

In the present case, in equilibrium, $\lambda_{\max} \simeq 0.7746\, c_0$. This case was presented in Sect. 2.4.1.

4.1.2.4 ET$_1$ Principal Subsystem

Finally, the transport equation:

$$\frac{\partial}{\partial t}\rho + \frac{\partial}{\partial x_i}\left(\rho v_i^*\right) = 0$$

is the 1-field principal subsystem of all the previous ones:

$$\begin{aligned} &\mu_i = 0, \ \lambda_{\langle ij\rangle} = 0, \ \mu = \text{const.}, \ \lambda_j = \text{const.} \ \rightarrow \\ &q_i = 0, \ \sigma_{\langle ij\rangle} = 0, \ T = T^* = \text{const.}, \ v_i = v_i^* = \text{const.} \end{aligned}$$

We have, in equilibrium, $\lambda_{\max} = 0$.

According with the general result, the maximum characteristic velocity increases with the number of fields [90].

4.2 Molecular ET

For rarefied gases, we have discovered that, in highly nonequilibrium phenomena such as sound waves with high frequencies, light scattering with large scattering angle, shock waves with large Mach number, predictions of the 13-moment RET theory are superior to those of the Navier-Stokes and Fourier theory. However, we also notice that the 13-moment theory is still not quite satisfactory when we compare its predictions with experimental data. For such phenomena, we need a theory with more moments. But it is too difficult to proceed within a purely macroscopic theory like the 13-field RET theory. Therefore let us recall that the fields F's can be regarded as the moments of a distribution function f.

In order to explain this approach by using the moments F's, we first rewrite the hierarchy of balance laws in more compact notation:

$$\partial_t F_A + \partial_i F_{iA} = P_A, \qquad 0 \leqslant A \leqslant N \tag{4.35}$$

with

$$F_A = \int_{\mathbb{R}^3} m f c_A d\mathbf{c}, \quad F_{iA} = \int_{\mathbb{R}^3} m f c_i\, c_A d\mathbf{c}, \tag{4.36}$$

$$P_A = \int_{\mathbb{R}^3} Q c_A d\mathbf{c}, \tag{4.37}$$

where

$$c_A = \begin{cases} 1 & \text{for } A = 0, \\ c_{i_1} c_{i_2} \cdots c_{i_A} & \text{for } 1 \leqslant A \leqslant N, \end{cases}$$

and

$$F_A = \begin{cases} F & \text{for } A = 0, \\ F_{i_1 i_2 \cdots i_A} & \text{for } 1 \leqslant A \leqslant N, \end{cases} \qquad F_{iA} = \begin{cases} F_i & \text{for } A = 0, \\ F_{i i_1 i_2 \cdots i_A} & \text{for } 1 \leqslant A \leqslant N. \end{cases}$$

The indexes i and $i_1 \leqslant i_2 \leqslant \cdots \leqslant i_A$ are defined over 1, 2, 3. In the followings, similar notations will be adopted.

Now we adopt the following definition:

Definition 4.1 **($N^{(i,j,k,\dots)}$-System)** A system of moment equations (4.35) truncated at the tensorial index N is called (N)-system. The system (4.35) is called $N^{(1)}$-system if, for the last balance equation, we consider only the trace with respect to the two indexes, $F_{k_1 k_2 \dots k_{N-2} ll}$, instead of the full N-order tensor $F_{k_1 k_2 \dots k_{N-2} k_{N-1} k_N}$. If, instead of a pair of 2 indexes, we have the contraction with respect to 2 pairs of 2 indexes, we write $N^{(2)}$. If we have the contraction with respect to i pairs of 2 indexes for the last tensorial balance equation, j pairs of 2 indexes for the second to last balance equation, k pairs of 2 indexes for the third to last balance equation, and so on, the system is called $N^{(i,j,k,\cdots)}$-system.

According with this definition the Euler fluid (2.26) is $2^{(1)}$-system and the Grad system (1.27) is $3^{(1)}$-system.

Since all tensors are symmetric, the number n of moments, for (N)-system, is given by

$$n_N = \frac{1}{6}(N+1)(N+2)(N+3), \tag{4.38}$$

and, for $N^{(1)}$-system

$$n_{N^{(1)}} = \frac{1}{6}N(N^2+6N-1). \tag{4.39}$$

Other examples to clarify the definition are given in this table:

(N)	Densities	$n_{(N)}$
$(2^{(1)})$	F, F_i, F_{ll}	5
$(3^{(1)})$	F, F_i, F_{ij}, F_{lli}	13
$(4^{(2,1)})$	$F, F_i, F_{ij}, F_{lli}, F_{llmm}$	14
$(6^{(3,2,1)})$	$F, F_i, F_{ij}, F_{ijk}, F_{llij}, F_{llmmi}, F_{llmmnn}$	30

We will see in Sect. 5.6 that the optimal choice of moments obtained as the limit of a relativistic theory corresponds to a special case of this kind of moments.

Sometimes it is more convenient to separate the tensor $F_{i_1 i_2 \cdots i_n}$ into the deviatoric part and the trace part with respect to a pair or pairs of indexes. Let us introduce the symmetric traceless tensor $F_{\langle i_1 i_2 \cdots i_n \rangle}$ as follows (see for example [76]):

$$F_{\langle i_1 i_2 \cdots i_n \rangle} = F_{i_1 i_2 \cdots i_n} + \sum_{k=1}^{[\frac{n}{2}]} a_{nk} \delta_{(i_1 i_2} \cdots \delta_{i_{2k-1} i_{2k}} F_{i_{2k+1} \cdots i_n) j_1 \cdots j_k j_1 \cdots j_k}$$

with

$$a_{nk} = (-1)^k \frac{n!(2n-2k-1)!!}{(n-2k)!(2n-1)!!(2k)!!},$$

where the following notations are adopted:

$$\left[\frac{n}{2}\right] = \begin{cases} \frac{n}{2} & \text{if } n \text{ is even.} \\ \frac{n-1}{2} & \text{if } n \text{ is odd.} \end{cases}, \qquad n!! = \prod_{j=0}^{\left[\frac{n-1}{2}\right]} (n-2j).$$

Then the system (4.35) in the case of $N^{(1)}$-system can be written as

$$\begin{aligned} &\partial_t F_{\langle A\rangle} + \partial_i F_{i\langle A\rangle} = P_{\langle A\rangle}, \\ &\quad (0 \leqslant A \leqslant N-1) \qquad\qquad \partial_t F_{llA'} + \partial_i F_{lliA'} = P_{llA'}, \\ &\qquad\qquad\qquad\qquad\qquad\qquad\quad (0 \leqslant A' \leqslant N-2) \end{aligned} \tag{4.40}$$

where

$$F_{\langle A\rangle} = \begin{cases} F & \text{for } A = 0 \\ F_{i_1} & \text{for } A = 1 \\ F_{\langle i_1 \cdots i_A\rangle} & \text{for } 2 \leqslant A \leqslant N-1 \end{cases}, \qquad F_{i\langle A\rangle} = \begin{cases} F_i & \text{for } A = 0 \\ F_{ii_1} & \text{for } A = 1 \\ F_{i\,\langle i_1 \cdots i_A\rangle} & \text{for } 2 \leqslant A \leqslant N-1, \end{cases}$$

$$P_{\langle A\rangle} = \begin{cases} 0 & \text{for } A = 0 \\ 0 & \text{for } A = 1 \\ P_{\langle i_1 \cdots i_A\rangle} & \text{for } 2 \leqslant A \leqslant N-1, \end{cases} \tag{4.41}$$

and

$$F_{llA'} = \begin{cases} F_{ll} & \text{for } A' = 0 \\ F_{lli_1\cdots i_{A'}} & \text{for } 1 \leqslant A' \leqslant N-2, \end{cases} \qquad F_{lliA'} = \begin{cases} F_{lli} & \text{for } A' = 0 \\ F_{llii_1\cdots i_{A'}} & \text{for } 1 \leqslant A' \leqslant N-2, \end{cases}$$

$$P_{llA'} = \begin{cases} 0 & \text{for } A' = 0 \\ P_{i_1\cdots i_{A'}} & \text{for } 1 \leqslant A' \leqslant N-2. \end{cases} \tag{4.42}$$

It is simple to calculate the number of the components of $F_{\langle A\rangle}$ with $0 \leqslant A \leqslant N-1$:

$$n_{\langle A\rangle} = N^2, \tag{4.43}$$

while the number of the components of $F_{llA'}$ with $0 \leqslant A' \leqslant N-2$ is given by

$$n_{llA'} = \frac{1}{6}N(N^2-1). \tag{4.44}$$

As an example, we take the Grad system (ET_{13}) (see (1.27)). In this case $N = 3$, and

$$F_{\langle A\rangle} \equiv \begin{pmatrix} F \\ F_{i_1} \\ F_{\langle i_1 i_2\rangle} \end{pmatrix}, \qquad F_{i\langle A\rangle} \equiv \begin{pmatrix} F_i \\ F_{ii_1} \\ F_{i\langle i_1 i_2\rangle} \end{pmatrix}, \qquad P_{\langle A\rangle} \equiv \begin{pmatrix} 0 \\ 0 \\ P_{\langle i_1 i_2\rangle}, \end{pmatrix},$$

$$F_{llA'} \equiv \begin{pmatrix} F_{ll} \\ F_{i_1 ll} \end{pmatrix}, \qquad F_{illA'} \equiv \begin{pmatrix} F_{ill} \\ F_{ii_1 ll} \end{pmatrix}, \qquad P_{llA'} \equiv \begin{pmatrix} 0 \\ P_{i_1 ll} \end{pmatrix}.$$

According with (4.43) and (4.44), we have $n_{\langle A\rangle} = 9$ and $n_{llA'} = 4$.

4.2.1 Closure via the Entropy Principle

We require the compatibility of the truncated system (4.35) with the entropy law, i.e., all solutions of (4.35) must satisfy also the supplementary entropy balance law (2.8) where h^0, h^i, and Σ are functionals of f (see (1.23)) through the moments (4.36) with $A = 0, \ldots, N$. This is a strong restriction on the distribution function f. Now the problem to be solved is as follows: Determine the distribution function f_N under the condition that any classical solution of (4.35) with (4.36) and (4.37) is also the solution of (2.8).

For a generic entropy functional, which is valid not only for classical gases but also for degenerate gases such as Bose and Fermi gases, the following theorem was proved by Boillat and Ruggeri [90]:

Theorem 4.1 *Necessary and sufficient condition such that the truncated system of moments (4.35) satisfies the entropy principle (2.8) is that the truncated distribution function f_N depends on $(\mathbf{x}, t, \mathbf{c})$ only through a single variable:*

$$f_N \equiv f_N(\chi_N),$$

where

$$\chi_N = \sum_{A=0}^{N} u'_A(\mathbf{x}, t) c_A$$

is a polynomial in $\mathbf{c}$ with the coefficients u'_A:

$$u'_A = \begin{cases} u' & \text{for } A = 0, \\ u'_{i_1 i_2 \cdots i_A} & \text{for } 1 \leqslant A \leqslant N. \end{cases}$$

The entropy density, flux, and production have the following expressions:

$$h^0 = m \int_{\mathbb{R}^3} \left(\chi_N \Omega'(\chi_N) - \Omega(\chi_N)\right) d\mathbf{c}, \tag{4.45}$$

$$h^i = m \int_{\mathbb{R}^3} c^i \left(\chi_N \Omega'(\chi_N) - \Omega(\chi_N)\right) d\mathbf{c}, \qquad \Sigma = m \int_{\mathbb{R}^3} Q\, \chi_N \, d\mathbf{c},$$

where the partition function $\Omega(\chi_N)$ satisfies the relation:

$$\Omega' = \frac{d\Omega}{d\chi_N} = f_N(\chi_N).$$

The system (4.35) becomes symmetric in the form (2.12) with the main field u'_A and the potentials:

$$h'^0 = m\int_{\mathbb{R}^3} \Omega(\chi_N)\,d\mathbf{c}, \quad h'^i = m\int_{\mathbb{R}^3} \Omega(\chi_N)c^i\,d\mathbf{c}, \tag{4.46}$$

provided $\Omega''(\chi_N) < 0$.

The proof can be made by following Theorem 2.1. In fact (2.15) becomes in this case

$$F_A = \frac{\partial h'^0}{\partial u'_A}, \qquad F_{iA} = \frac{\partial h'^i}{\partial u'_A}.$$

Then

$$dh'^0 = \sum_{A=0}^{N} F_A du'_A = m\int_{\mathbb{R}^3} f_N \sum_{A=0}^{N} c_A du'_A d\mathbf{c} = m\int_{\mathbb{R}^3} f_N d\chi_N d\mathbf{c} = d\int_{\mathbb{R}^3} m\Omega(\chi_N)\,d\mathbf{c}$$

and $(4.46)_1$ holds. From (2.13), we have (4.45). Analogous considerations are also valid for h'^i and h^i.

Therefore the original system of the moments is closed. It is symmetric hyperbolic in terms of the main-field components (2.12):

$$J_{AB}\partial_t u'_B + J_{iAB}\partial_i u'_B = P_A(u'_C), \qquad A = 0, \dots, N \tag{4.47}$$

where

$$J_{AB}(u'_C) = \frac{\partial^2 h'^0}{\partial u'_A \partial u'_B} = \int_{\mathbb{R}^3} m\Omega''(\chi_N) c_A c_B d\mathbf{c},$$

$$J_{iAB}(u'_C) = \frac{\partial^2 h'_i}{\partial u'_A \partial u'_B} = \int_{\mathbb{R}^3} m\Omega''(\chi_N) c_i\, c_A c_B\, d\mathbf{c}.$$

Here and hereafter, the summation symbol with respect to A and/or B is omitted for simplicity. Indeed, provided that $\Omega'' < 0$ holds, the matrix J_{AB} is negative definite since

$$J_{AB} X_A X_B = \int_{\mathbb{R}^3} m\Omega''\,(c_A X_A)^2\,d\mathbf{c} < 0 \quad \forall X_A \neq 0.$$

If we require that h^0 is the usual entropy density for non-degenerate gases, that is,

$$h^0 = -k_B \int_{\mathbb{R}^3} f_N \ln f_N \, d\mathbf{c}, \tag{4.48}$$

we obtain, from (4.45),

$$\left(\chi_N + \frac{k_B}{m} \ln \Omega'\right) \Omega' - \Omega = 0.$$

And, by differentiation, we obtain

$$f_N(\chi_N) = e^{-1-m\chi_N/k_B}. \tag{4.49}$$

In the present case we have

$$J_{AB}\left(u'_C\right) = -\frac{m^2}{k_B} \int_{\mathbb{R}^3} f_N c_A c_B \, d\mathbf{c}, \qquad J_{iAB}\left(u'_C\right) = -\frac{m^2}{k_B} \int_{\mathbb{R}^3} f_N c_i c_A c_B \, d\mathbf{c}.$$

In an equilibrium state, (4.49) is reduced to the well-known Maxwellian distribution function.

We note that f_N is not a solution of the Boltzmann equation. But we may have the conjecture (open problem) that, for $N \to \infty$, f_N tends to a solution of the Boltzmann equation.

4.2.2 *Closure via the Maximum Entropy Principle*

Instead of adopting the entropy principle, there is an alternative in the RET theory of moments for the determination of the phase density f_N. This is the method of maximization of the entropy under some constraints.

We have discussed the MEP in Sect. 1.8.5. In this subsection, we summarize the results in the case of monatomic gas (see [90] for more details). Let us treat firstly a general case where the entropy h^0 is a generic functional of f:

$$h^0 = \int_{\mathbb{R}^3} \psi(f) d\mathbf{c}\,. \tag{4.50}$$

We ask for the phase density f_N that provides the maximum of h^0 under the constraints of fixed values F_A of the moments:

$$F_A = \int_{\mathbb{R}^3} m f c_A d\mathbf{c}.$$

With the Lagrange multipliers λ_A, we introduce the functional $\mathscr{L}$:

$$\mathscr{L} = \int_{\mathbb{R}^3} \psi(f) d\mathbf{c} + \lambda_A \left(F_A - m \int c_A f d\mathbf{c} \right).$$

This can be substituted by the equivalent one:

$$\mathscr{L} = \int_{\mathbb{R}^3} (\psi(f) - m\lambda_A c_A f)\, d\mathbf{c}. \tag{4.51}$$

And we impose the extremum condition:

$$\delta\mathscr{L} = \int_{\mathbb{R}^3} \left(\frac{d\psi}{df} - m\lambda_A c_A \right) \delta f d\mathbf{c} = 0.$$

Thus we have

$$\frac{d\psi}{df} = m\lambda_A c_A$$

as a necessary condition for an extremum. Hence it follows that f is a function of

$$\chi = \lambda_A c_A, \tag{4.52}$$

and that $\psi(f)$ has the form:

$$\psi(f) = m \left(\chi f - \int f d\chi \right). \tag{4.53}$$

Insertion of (4.53) into (4.50) gives exactly the same result as that from the entropy principle (4.45). Thus we conclude that *the maximization of the entropy leads to the same result as that from the entropy principle in molecular ET of moments* [90]. In particular, the Lagrange multipliers λ_A are identical to the main field components u'_A.

Remark 4.2 If, instead of the system (4.35), we consider the equivalent system (4.40), we have, instead of χ given by (4.52),

$$\chi = \Lambda_{\langle A \rangle}\, c_{\langle A \rangle} + \Lambda_{llA'} c^2 c_{A'},$$

where $\Lambda_{\langle A \rangle}$, $(A = 0, \cdots N-1)$ and $\Lambda_{llA'}$, $(A' = 0, \cdots N-2)$ denote the Lagrange multipliers corresponding to the densities $F_{\langle A \rangle}$ and $F_{llA'}$. The main field components $u'_A = \lambda_A$ $(A = 0, \cdots N)$ for the system with full moments such as (4.35) are also expressed by the linear combinations of the main field components $\Lambda_{\langle A \rangle}$ and $\Lambda_{llA'}$.

Remark 4.3 Therefore we have no RET theory other than nesting theories depending on the truncation indexes. We can pass from a general system to simpler systems through the concept of principal subsystem. This illustrates the power and the beauty of this definition (see Sect. 2.4). From a physical point of view, a principal subsystem in the nesting theory is obtained by neglecting some relaxation times. In the limit where N tends to infinity (assuming that this limit exists and f_N converges to f), we may say roughly that all nesting theories of RET are principal subsystems of the Boltzmann equation!

4.3 Maximum Characteristic Velocity

Characteristic velocities λ of the symmetric hyperbolic system (4.47) with (4.49) in the propagation direction with unit vector $\mathbf{n} \equiv (n^i)$ are eigenvalues of

$$K_{AB} = J_{iAB}n^i - \lambda J_{AB} = -\frac{m^2}{k_B}\int_{\mathbb{R}^3} f(\chi)(\mathbf{c}\cdot\mathbf{n} - \lambda)c_A c_B\, d\mathbf{c}.$$

In particular, the wave speeds for disturbances propagating in an equilibrium state are eigenvalues of

$$\int_{\mathbb{R}^3} f^{(M)}(\mathbf{c}\cdot\mathbf{n} - \lambda)c_A c_B\, d\mathbf{c}, \tag{4.54}$$

where $f^{(M)}$ is the Maxwellian distribution function. As the integrals in (4.54) are known, it is easy to evaluate the maximum eigenvalues for increasing N.

Numerical analysis was made by Weiss [251] who obtained increasing value of the maximum characteristic velocity for increasing number of moments n that depends on the truncation index N through (4.38). For instance, for $n = 20$, $\lambda_{\max} = 1.8c_0$ and for $n = 15{,}180$, $\lambda_{\max} = 9.36c_0$, where $\lambda_{\max}$ is the maximum characteristic velocity in equilibrium with c_0 being the sound wave velocity. An interesting problem is to answer the following question: what is the limit of $\lambda_{\max}$ as $N \to \infty$?

4.3.1 Lower Bound Estimate and Characteristic Velocities for Large Number of Moments

In the previous examples, we have seen the validity of the subcharacteristic conditions. Now we are able to prove the behavior of $\lambda_{\max}$ when $N \to \infty$. The $(k+1)(k+2)/2$ components of the main field of order k:

$$u'_{i_1 i_2 \dots i_k}, \qquad i_1 \leqslant i_2 \leqslant \cdots \leqslant i_k$$

can be mapped in the corresponding variables:

$$u'_{pqr}, \qquad p+q+r=k,$$

where p, q, r are, respectively, the numbers of indexes over 1, 2, 3. With this notation, χ is expressed as

$$\chi = \sum_{p,q,r} u'_{pqr} c_1^p c_2^q c_3^r, \qquad 0 \leqslant p+q+r \leqslant N.$$

We can prove the following theorem by Boillat and Ruggeri [90]:

Theorem 4.2 *For any N we have the lower bound condition:*

$$\frac{\lambda_{\max}}{c_0} \geqslant \sqrt{\frac{6}{5}\left(N - \frac{1}{2}\right)}, \tag{4.55}$$

where c_0 is the sound velocity. Therefore, $\lambda_{\max}$ becomes unbounded when $N \to \infty$.

Sketch of the Proof By the use of the variable u'_{pqr}, the components of the matrix (4.54) are given by

$$\int_{\mathbb{R}^3} f^{(M)}(c_i n^i - \lambda) c_1^{p+s} c_2^{q+t} c_3^{r+u} d\mathbf{c}.$$

The matrix is negative semi-definite, if λ is the largest eigenvalue $\lambda_{\max}$. Since the elements a_{ij} of a semi-definite matrix satisfy the inequality:

$$a_{ii} a_{jj} \geqslant a_{ij}^2, \tag{4.56}$$

we have

$$\int_{\mathbb{R}^3} f^{(M)}(c_i n^i - \lambda_{\max}) c_1^{2p} c_2^{2q} c_3^{2r} d\mathbf{c} \int_{\mathbb{R}^3} f^{(M)}(c_i n^i - \lambda_{\max}) c_1^{2s} c_2^{2t} c_3^{2u} d\mathbf{c}$$
$$\geqslant \left(\int_{\mathbb{R}^3} f^{(M)}(c_i n^i - \lambda_{\max}) c_1^{p+s} c_2^{q+t} c_3^{r+u} d\mathbf{c} \right)^2. \tag{4.57}$$

In this case, (4.57) is reduced to

$$\lambda_{\max}^2 \int_{\mathbb{R}^3} f^{(M)} c_1^{2p} c_2^{2q} c_3^{2r} d\mathbf{c} \int_{\mathbb{R}^3} f^{(M)} c_1^{2s} c_2^{2t} c_3^{2u} d\mathbf{c}$$
$$\geqslant \left(\int_{\mathbb{R}^3} f^{(M)}(c_i n^i - \lambda_{\max}) c_1^{p+s} c_2^{q+t} c_3^{r+u} d\mathbf{c} \right)^2.$$

With the choice: $p = N, s = N-1, q = r = t = u = 0, \mathbf{n} \equiv (1,0,0)$, this inequality becomes

$$\lambda_{\max}^2 \geqslant \frac{\int_0^\infty f^{(M)} c_1^{2N} dc_1}{\int_0^\infty f^{(M)} c_1^{2(N-1)} dc_1} = 2\frac{k_B}{m}T\frac{\Gamma(N+1/2)}{\Gamma(N-1/2)} = \frac{6}{5}c_0^2\left(N-\frac{1}{2}\right).$$

and the proof is completed.

Therefore, we have

$$\lim_{N\to\infty} \lambda_{\max} = \infty.$$

In Fig. 4.1, the numerical values of $\lambda_{\max}/c_0$ given by Weiss [251] are compared with the lower bound in the right-hand side of (4.55). This is a surprising result because the original motivation of RET was to repair the paradox of infinite velocity of the Navier-Stokes and Fourier classical approach. Therefore, for any finite N, we have symmetric hyperbolic systems with finite characteristic velocities. But when we take infinite moments we have a parabolic behavior.

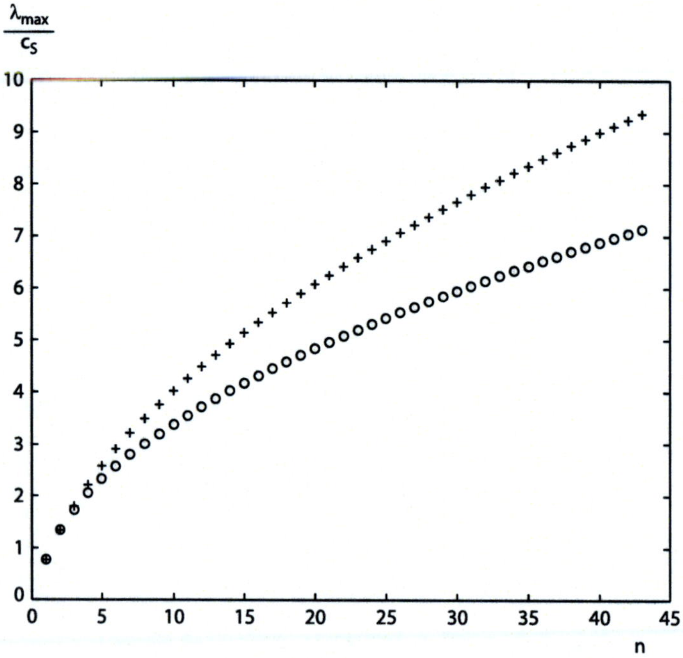

Fig. 4.1 The behavior of the maximum characteristic velocity (cross) versus the truncation number N and the lower bound estimate (4.55) (circle)

Instead, in the relativistic context, it was proved that the limit of the maximum characteristic velocity for $N \to \infty$ is the light velocity [252–254] (See Sect. 5.5.2).

4.4 Convergence Problem, Junk Observation, and RET Near an Equilibrium State of Order α: ET^{α}_{M}

All results explained above are valid also for a case far from equilibrium provided that the integrals in (4.36) and (4.37) are convergent. The problem of the convergence of the moments is one of the main questions in a far-from-equilibrium case. In particular, as we will see, the index of truncation N must be even. This implies, in particular, that a theory with 13 moments is not allowed when we study phenomena far from equilibrium!

Moreover, if the conjecture—the distribution function f_N, when $N \to \infty$, tends to the distribution function f that satisfies the Boltzmann equation—is true, we need another convergence requirement.

These problems were studied by Boillat and Ruggeri in [90]. As before, χ_N is expressed as

$$\chi_N = \sum_{p,q,r} u'_{pqr} c_1^p c_2^q c_3^r, \quad 0 \leqslant p+q+r \leqslant N.$$

Since

$$\left| \sum_{p,q,r} u'_{pqr} c_1^p c_2^q c_3^r \right| \leqslant a_N c^N$$

with

$$a_N = \max_{|\mathbf{t}|=1} \nu_N(\mathbf{t}), \qquad \nu_N(\mathbf{t}) = \sum_{p,q,r} u'_{pqr} t_1^p t_2^q t_3^r,$$

the series is absolutely convergent, when $p+q+r = N \to \infty$, for any $\mathbf{c}$ provided that

$$u'_{pqr} \to 0, \qquad \frac{a_{N+1}}{a_N} \to 0.$$

Hence the components of the main field become smaller and smaller when N increases. This justifies the truncation of the system. On the other hand, when N is finite, the integrals of moments must also be convergent. When c is large, $\chi_N \simeq |c|^N \nu_N$. Therefore it is easy to see, by using the spherical coordinates, that the integrals of moments converge provided that $\nu_N(\mathbf{t}) < 0$ for any unit vector $\mathbf{t}$. But, since $\nu_N(-\mathbf{t}) = (-1)^N \nu_N(\mathbf{t})$, we can conclude that N must be even and $\max_{|\mathbf{t}|=1} \nu_N(\mathbf{t}) < 0$.

The question about the convergence is not the only one in the full nonlinear case. Indeed, Junk and co-workers have shown [95, 96] that even the complete RET theory, which is obtained through the maximum entropy principle (MEP) without any approximation, raises a problem: there exist physically admissible states that correspond to singularities. A possible way to bypass numerically the Junk problem was given in an interesting paper by McDonald and Torrilhon [255].

When we study nonequilibrium phenomena of ideal gas in the neighborhood of a local equilibrium state, we do not encounter the above-mentioned problems. In fact, taking into account the fact that all the equilibrium components of the main field vanish except for those corresponding to the first 5 moments (see Theorem 2.4) and that the first 5 are given in (2.29), then the f_N in equilibrium coincides with the Maxwellian $f^{(M)}$. Therefore, near equilibrium, the distribution function, which is obtained as a solution of the variational problem (4.49), is formally expanded around an equilibrium Maxwellian distribution $f^{(M)}$:

$$f_N = e^{-1-\frac{m}{k_B}u'_A c_A} = e^{-1-\frac{m}{k_B}\hat{u}'_A C_A} \approx f^{(M)}\left(1 - \frac{m}{k_B}\tilde{u}'_A C_A\right), \tag{4.58}$$

where

$$\tilde{u}'_A = \left(\hat{u}'_A - \hat{u}'^E_A\right)$$

and $\hat{u}'^E_A$ are the main-field components evaluated at a local equilibrium state with zero velocity. Plugging (4.58) into $(4.36)_1$, we obtain a linear algebraic system that permits to evaluate the nonequilibrium main field $\tilde{u}'_A$ in terms of the intrinsic densities $\hat{F}_A$:

$$J^{\mathcal{M}}_{AB}\,\tilde{u}'_B = \left(\hat{F}_A - \hat{F}^E_A\right), \tag{4.59}$$

where $\hat{F}^E_A$ denotes the moment $\hat{F}_A$ evaluated at the equilibrium state with zero velocity and

$$J^{\mathcal{M}}_{AB} = -\frac{m^2}{k_B}\int_{\mathbb{R}^3} f^{(M)}\, C_A C_B\, d\mathbf{C}. \tag{4.60}$$

By inserting $\tilde{u}'_A$ just obtained above into $(4.36)_2$ and (4.37), the explicit dependences of the truncated fluxes and source terms on the densities are elucidated. In this way, the (N)–system becomes a closed system for the densities F_A.

The generalization of this procedure to higher order expansion of the distribution function was made in [256]. For nonequilibrium processes close to an equilibrium state, the distribution function f_N can be approximated by $f_N^{(\alpha)}$ through a Taylor-expansion up to order $\alpha > 0$ ($\alpha \in \mathbb{N}$):

$$f_N^{(\alpha)} = f^{(M)} \left\{ 1 - \frac{m}{k_B} \left(\tilde{u}'_{A_1} c_{A_1} + \frac{1}{2} \tilde{u}'_{A_1} \tilde{u}'_{A_2} c_{A_1} c_{A_2} + \cdots + \right.\right.$$
$$\left.\left. + \frac{1}{\alpha!} \tilde{u}'_{A_1} \tilde{u}'_{A_2} \ldots \tilde{u}'_{A_\alpha} c_{A_1} c_{A_2} \ldots c_{A_\alpha} \right) \right\}. \tag{4.61}$$

The components $F_{ik_1k_2k_3\ldots k_N}$ that are not in the list of the densities are also approximated as

$$F_{ik_1k_2k_3\ldots k_N}^{(\alpha)} = m \int_{\mathbb{R}^3} f_N^{(\alpha)}(\mathbf{x}, t, \mathbf{c}) c_i c_{k_1} c_{k_2} c_{k_3} \ldots c_{k_N} d\mathbf{c}.$$

This is a RET theory with M moments of order α (ET_M^α). In particular, ET_{13}^1 corresponds to the usual Grad system [84].

It must be emphasized that, when we use an approximated distribution function, we must be careful about its validity range. In other words, we must check the convexity condition of the entropy and the hyperbolicity condition. Usually these conditions hold only in a neighborhood of an equilibrium state.

An alternative interesting method of MEP with a similar spirit of the Taylor expansion (4.61) was given by Abdelmalik and van Brummelen [257] based on the observation that the exponential in the distribution function (4.49) can be rewritten as $\exp(\bullet) = \lim_{n\to\infty}(1 + \bullet/n)$.

Remark 4.4 Although the closure of the phenomenological theory gives the same results as those obtained through the principle of maximum entropy, there is a profound conceptual difference. In fact, in the first case, the fields are not moments of a distribution function and therefore they have no inequalities to satisfy. On the other hand, in the case of MEP, the fields are moments of a positive distribution function and therefore some conditions of *realizability* that must be considered. There is the question of physical realizability, which asks which moment states can be reached by an arbitrary positive distribution function. Second, one must consider which moment states can be reached by distribution functions of the assumed form used to close the system. This problematic question is the so-called Hamburger moment problem [258]. On this delicate problem, there exists a huge literature, but concerning the MEP we refer, for example, to the papers [255, 259, 260]. Another big difference between the two approaches is as follows: In macroscopic theory, no condition of integrability is required and any number of moments is, in principle, admissible. Instead, as we have seen before, in the kinetic approach of MEP, we encounter the problem of integrability, and, as a necessary condition, the truncation index of the moment N must be even.

4.4.1 Example: Molecular Closure for ET_{13}

To understand the previous procedure in a more specific way, we study the molecular ET closure for ET_{13}. We consider non-degenerate and non-polytropic gas in which $\varepsilon = \varepsilon(T)$ is not necessarily a linear function of T.

We now want to use the MEP in the case of the hierarchy with 13 fields (2.49). For the entropy defined by (4.48), the MEP poses the following variational problem: Determine the distribution function $f(t, \mathbf{x}, \mathbf{c})$ such that $h = h^0 \to$ max under the constraints of prescribed 13 moments. The solution of the problem is given by the following theorem:

Theorem 4.3 *The distribution function, which maximizes the entropy (4.48) under the constraints of prescribed* 13 *moments and under the assumption that processes are not so far from equilibrium, is the Grad distribution function* (1.30).

In this case, the functional for the variational problem with the constraints, (4.51), is given by

$$\mathscr{L} = \int_{\mathbb{R}^3} \left\{ -k_B f \log f - m \left[\lambda + \lambda_i c_i + \lambda_{\langle ij \rangle} c_i c_j + (\mu + \mu_i c_i) c^2 \right] f \right\} d\mathbf{c}. \tag{4.62}$$

By the Galilean invariance, the functional is the same as in the case with zero hydrodynamic velocity ($\mathbf{v} = 0$). Therefore, we have

$$\mathscr{L} = \int_{\mathbb{R}^3} \left\{ -k_B f \log f - m \left[\hat{\lambda} + \hat{\lambda}_i C_i + \hat{\lambda}_{\langle ij \rangle} C_i C_j + (\hat{\mu} + \hat{\mu}_i C_i) C^2 \right] f \right\} d\mathbf{C}. \tag{4.63}$$

Comparison between (4.62) and (4.63) gives the velocity dependence of the Lagrange multipliers:

$$\begin{aligned}
\lambda &= \hat{\lambda} - \hat{\lambda}_i v_i + \hat{\lambda}_{\langle ij \rangle} v_i v_j + \hat{\mu} v^2 - \hat{\mu}_i v^2 v_i, \\
\lambda_i &= \hat{\lambda}_i - 2\hat{\lambda}_{\langle il \rangle} v_l - 2\hat{\mu} v_i + \hat{\mu}_i v^2 + 2\hat{\mu}_l v_i v_l, \\
\lambda_{\langle ij \rangle} &= \hat{\lambda}_{\langle ij \rangle} - \left(\hat{\mu}_i v_j + \hat{\mu}_j v_i - \frac{2}{3} \hat{\mu}_k v_k \delta_{ij} \right), \\
\mu &= \hat{\mu} - \hat{\mu}_l v_l, \\
\mu_i &= \hat{\mu}_i.
\end{aligned}$$

These relations are perfectly in agreement with the general Galilean invariance theorem (2.55). In the present case, $\mathbf{X}$ is given by (4.4).

From (4.63), the solution of the Euler-Lagrange equation $\delta\mathscr{L}/\delta f = 0$ is given by

$$f = \exp\left(-1 - \frac{m}{k_B}\chi\right)$$

with

$$\chi = \hat{\lambda} + \hat{\lambda}_i C_i + \hat{\lambda}_{\langle ij\rangle} C_i C_j + (\hat{\mu} + \hat{\mu}_i C_i)C^2. \tag{4.64}$$

As in the general case, there is the problem of the convergence of moments also in ET_{13}. We, therefore, try to obtain the approximate solution in the form of expansion around the local equilibrium:

$$f = f^{(M)}\left\{1 - \frac{m}{k_B}\left(\tilde{\lambda} + \tilde{\lambda}_i C_i + \tilde{\lambda}_{\langle ij\rangle} C_i C_j + (\tilde{\mu} + \tilde{\mu}_i C_i)C^2\right)\right\}, \tag{4.65}$$

where $f^{(M)}$ is the equilibrium Maxwellian distribution (1.29) and

$$\begin{aligned} &\tilde{\lambda} = \hat{\lambda} - \hat{\lambda}^E, \quad \tilde{\lambda}_i = \hat{\lambda}_i - \hat{\lambda}_i^E, \quad \tilde{\lambda}_{\langle ij\rangle} = \hat{\lambda}_{\langle ij\rangle} - \hat{\lambda}_{\langle ij\rangle}^E, \\ &\tilde{\mu} = \hat{\mu} - \hat{\mu}^E, \quad \tilde{\mu}_i = \hat{\mu}_i - \hat{\mu}_i^E. \end{aligned} \tag{4.66}$$

Here quantities with the superscript E are the Lagrange multipliers at a local equilibrium state. $\hat{\lambda}^E$, $\hat{\lambda}_i^E$, and $\hat{\mu}^E$ coincide with the main field components of the Euler fluid (2.29) at zero velocity, while the remainders are null according to (2.45), i.e.,

$$\hat{\lambda}^E = -\frac{g}{T}, \quad \hat{\lambda}_i^E = 0, \quad \hat{\mu}^E = \frac{1}{2T}, \quad \hat{\lambda}_{\langle ij\rangle}^E = 0, \quad \hat{\mu}_i^E = 0.$$

Inserting (4.65) into (4.59) and taking into account (4.64), we obtain

$$\begin{aligned} &\tilde{\lambda}\hat{F}^E + \tilde{\mu}\,\hat{F}_{ll}^E = 0, \\ &\tilde{\lambda}_k\hat{F}_{ik}^E + \tilde{\mu}_k\,\hat{F}_{llik}^E = 0, \\ &\tilde{\lambda}\hat{F}_{ll}^E + \tilde{\mu}\,\hat{F}_{llkk}^E = 0, \\ &\tilde{\lambda}_{\langle rs\rangle}\hat{F}_{ijrs}^E = \frac{k_B}{m}\sigma_{\langle ij\rangle}, \\ &\tilde{\lambda}_k\hat{F}_{llkj}^E + \tilde{\mu}_k\,\hat{F}_{llsskj}^E = -2\frac{k_B}{m}q_j. \end{aligned} \tag{4.67}$$

By using the Maxwellian distribution (1.29), expressions of the moments in equilibrium are easily obtained as follows:

$$\begin{aligned}&\hat{F}^E=\rho,\quad \hat{F}^E_{ij}=p\delta_{ij},\quad \hat{F}^E_{ijrs}=\frac{p^2}{\rho}\left(\delta_{ij}\delta_{rs}+\delta_{ir}\delta_{js}+\delta_{is}\delta_{jr}\right),\\&\hat{F}^E_{llijrs}=7\frac{p^3}{\rho^2}\left(\delta_{ij}\delta_{rs}+\delta_{ir}\delta_{js}+\delta_{is}\delta_{jr}\right).\end{aligned}\tag{4.68}$$

Therefore, from (4.67) with (4.68), we have

$$\tilde{\lambda}=0,\quad \tilde{\lambda}_i=\frac{1}{pT}q_i,\quad \tilde{\mu}=0,\quad \tilde{\lambda}_{\langle ij\rangle}=\frac{1}{2pT}\sigma_{\langle ij\rangle},\quad \tilde{\mu}_i=-\frac{\rho}{5p^2T}q_i.\tag{4.69}$$

Inserting (4.69) into (4.65), we obtain the nonequilibrium Grad distribution function (1.30), and the theorem is proved. This theorem is due to the pioneering study of Dreyer [88]. The expressions (4.69) are in perfect agreement with those of the phenomenological theory (4.27).

Once we have the distribution function, we can calculate the moments that are not in the list of densities, and the system is closed. In the present case, using the Grad distribution function (1.30), we have

$$\hat{F}_{ijk}=\frac{2}{5}\left(q_j\delta_{ik}+q_i\delta_{jk}+q_k\delta_{ij}\right),\quad \hat{F}_{llik}=\frac{p}{\rho}\left(5p\delta_{ik}-7\sigma_{\langle ik\rangle}\right).$$

Then, we derive explicit expressions of F_{ijk} and F_{llk} from Theorem 2.8 and Grad distribution function:

$$\begin{aligned}F_{ijk}&=\int_{\mathbb{R}^3} mfc_ic_jc_k d\mathbf{c}\\&=\rho v_iv_jv_k+p(v_i\delta_{jk}+v_j\delta_{ki}+v_k\delta_{ij})-\sigma_{\langle ij\rangle}v_k-\sigma_{\langle jk\rangle}v_i-\sigma_{\langle ki\rangle}v_j\\&\quad+\frac{2}{5}(q_i\delta_{jk}+q_j\delta_{ki}+q_k\delta_{ij}),\\F_{llij}&=\int_{\mathbb{R}^3} mfc_ic_jc^2 d\mathbf{c}\\&=\rho v^2v_iv_j+p(v^2\delta_{ij}+7v_iv_j)-\sigma_{\langle ij\rangle}v^2-2\sigma_{\langle li\rangle}v_lv_j-2\sigma_{\langle lj\rangle}v_lv_i\\&\quad+\frac{4}{5}q_lv_l\delta_{ij}+\frac{14}{5}q_iv_j+\frac{14}{5}q_jv_i+5\frac{p^2}{\rho}\delta_{ij}-7\frac{p}{\rho}\sigma_{\langle ij\rangle}.\end{aligned}$$

These relations are exactly the same as those obtained by the method of macroscopic closure (see (4.25)).

4.5 Hyperbolicity Region

The analysis on the determination of the hyperbolicity region was made firstly by Müller and Ruggeri [56] in the case of the 13-moment theory. Brini [261] carried on the study of a one-dimensional system with linear expansion and also extended it to the corresponding one-dimensional 14-moment theory of monatomic gases. Ruggeri and Trovato [262] studied the 13-moment model for degenerate monatomic gases with both linear and quadratic expansions.

The characteristic polynomial associated to the system (4.35) has the form.

$$\mathscr{P}(\lambda) = \sum_{k=0}^{n} a_k \lambda^k, \quad \text{with } a_k = a_k(\mathbf{u}). \tag{4.70}$$

The hyperbolicity region is a set of points $\mathbf{u}$ in which all roots of the equation $\mathscr{P}(\lambda) = 0$ are real and the right eigenvectors form a basis.

For many models, however, it is difficult to make an analytical study of the roots because of the high degree of the characteristic polynomial. We need to employ a different approach to overcome the difficulty.

Here, in order to identify the boundaries of the hyperbolicity region, we adopt the following procedure, which can be seen as a generalization of the procedure proposed in [56]: The boundary is a set of values of $\mathbf{u} = \mathbf{u}_B$ at which at least two real eigenvalues coincide with each other. In fact, by continuity, the transition from real roots to complex conjugate roots takes place at such a point. Therefore we require that, when $\mathbf{u} = \mathbf{u}_B$, the characteristic polynomial with $n \geqslant 2$ can be factorized as

$$\mathscr{P}(\lambda) = (\lambda - \lambda_*)^2 \sum_{j=0}^{n-2} b_j \lambda^j, \tag{4.71}$$

where λ_* denotes the double root and b_j are coefficients determined from (4.70). The requirement that there exists a double root implies two constraints for the coefficients a_k in terms of the parameter λ_* (see (4.71)):

$$\begin{aligned} \mathscr{P}(\lambda_*) = 0 \quad &\Leftrightarrow \quad \sum_{k=0}^{n} a_k\, \lambda_*^k = 0, \\ \frac{d\mathscr{P}}{d\lambda}(\lambda_*) = 0 \quad &\Leftrightarrow \quad \sum_{k=0}^{n} k\, a_k\, \lambda_*^{k-1} = 0. \end{aligned} \tag{4.72}$$

It is possible to verify that the coefficients b_j $(0 \leqslant j \leqslant n-2)$ can be expressed in terms of λ_* and a_k $(0 \leqslant k \leqslant n)$:

$$b_j = \sum_{l=j+2}^{n} (l-j-1)\lambda_*^{l-j-2} a_l, \qquad 0 \leqslant j \leqslant n-2.$$

Because the coefficients a_k are functions of the field $\mathbf{u}$, the conditions (4.72) constitute the parametric equations for the boundaries of the hyperbolicity region with the parameter λ_*.

4.5.1 Hyperbolicity Region of ET$_{13}$ for the First Order Approximation

In this subsection, we will focus on the model that describes one-dimensional phenomena depending only on one spatial variable $x = x_1$. The system (2.1) or equivalently (2.3) can be rewritten as

$$\partial_t \mathbf{F}^0 + \partial_x \mathbf{F}^1 = \mathbf{f} \qquad \Longrightarrow \qquad \mathbf{A}(\mathbf{u})\partial_t \mathbf{u} + \mathbf{B}(\mathbf{u})\partial_1 \mathbf{u} = \mathbf{f}, \tag{4.73}$$

where $\mathbf{A}(\mathbf{u}) = \partial_{\mathbf{u}} \mathbf{F}^0$ and $\mathbf{B}(\mathbf{u}) = \partial_{\mathbf{u}} \mathbf{F}^1$ are square matrices. The hyperbolicity of the system (4.73) is guaranteed if $\mathbf{A}$ is a non-singular matrix and the generalized eigenvalue problem $(\mathbf{B} - \lambda \mathbf{A})\,\mathbf{r} = 0$ presents all real characteristic velocities λ and a complete set of eigenvectors.

Now we consider the system of ET$_{13}$ given by (1.28) under the assumption that both the velocity and the heat flux are parallel to the x-direction, i.e., $\mathbf{v} = (v_1, 0, 0)$ and $\mathbf{q} = (q_1, 0, 0)$. We assume also that $\sigma_{12} = \sigma_{13} = \sigma_{23} = 0$ and $\sigma_{22} = \sigma_{33} = -\sigma_{11}/2$, i.e., the viscous tensor is traceless. Thus, we deal with only five field variables $\mathbf{u} = (\rho, v_1, T, \sigma_{11}, q_1)$.

The structure of the hyperbolicity region for a monatomic gas described by the one-dimensional ET$^1_{13}$ (i.e., Grad system) was studied in the literature [25, 56]. In what follows, we reconstruct the model and summarize the results as a benchmark for the next step of the study.

By using the explicit expressions of matrices $\mathbf{A}$ and $\mathbf{B}$, we notice that $\mathbf{A}$ is non-singular: $\det(\mathbf{A}) = -6k_B\rho^2/m < 0$. For convenience, we use dimensionless quantities referring to the velocity $c_s = \sqrt{k_B T/m}$.[1] In particular, we introduce the following dimensionless quantities:

$$\tilde{\sigma}_{ij} = \frac{\sigma_{ij}}{\rho c_s^2}, \qquad \tilde{q}_i = \frac{q_i}{\rho c_s^3},$$

[1]We remark that the dimensionless quantities introduced here are different from those in [25, 56]. In the previous papers $c = \sqrt{(5/3)}\, c_s$ was used instead of c_s.

and

$$\tilde{\lambda} = \frac{\lambda - v_1}{c_s}.$$

The characteristic polynomial, which is firstly obtained in [56], is given by

$$\mathscr{P}^{(1)}(\tilde{\lambda}) = 6\frac{k_B}{m}\sqrt{\frac{k_B}{m}T}p^2\tilde{\lambda}\left[a_4^{(1)}\tilde{\lambda}^4 + a_2^{(1)}\tilde{\lambda}^2 + a_1^{(1)}\tilde{\lambda} + a_0^{(1)}\right]$$

with

$$\begin{aligned} &a_4^{(1)} = 1, \quad a_2^{(1)} = -\tfrac{2}{15}(39 - 31\tilde{\sigma}_{11}), \\ &a_1^{(1)} = -\tfrac{96}{25}\tilde{q}_1, \quad a_0^{(1)} = \tfrac{3}{5}(5 - 10\tilde{\sigma}_{11} + 7\tilde{\sigma}_{11}^2). \end{aligned} \tag{4.74}$$

The superscript (1) indicates that we are now studying the first-order expansion of the theory. At equilibrium, the values of $\tilde{\lambda}$ are all real and distinct [25, 56]:

$$\tilde{\lambda}_1\Big|_E = 0, \qquad \tilde{\lambda}_{2,3}\Big|_E = \pm\sqrt{\frac{13 - \sqrt{94}}{5}}, \qquad \tilde{\lambda}_{4,5}\Big|_E = \pm\sqrt{\frac{13 + \sqrt{94}}{5}}.$$

This fact guarantees the existence of a basis of eigenvectors and the hyperbolicity property at the equilibrium state. By continuity arguments, such a property has to be maintained in a neighborhood of the equilibrium state.

The analytic equations for the boundaries of the hyperbolicity region can be deduced [25, 56] through the procedure described above for $n = 4$. We obtain, from (4.74) and (4.72), the parametric equations for the hyperbolic region:

$$\begin{cases} 9\lambda_*^4 + \frac{2}{5}(-39 + 31\tilde{\sigma}_{11})\lambda_*^2 - \frac{9}{5}(5 - 10\tilde{\sigma}_{11} + 7\tilde{\sigma}_{11}^2) = 0, \\ 3\lambda_*^3 + \frac{1}{5}(-39 + 31\tilde{\sigma}_{11})\lambda_* - \frac{72\tilde{q}_1}{25} = 0. \end{cases}$$

More explicitly, we can rewrite these equations as follows:

$$\tilde{q}_1 = \pm\frac{\sqrt{5}}{648}\sqrt{\gamma_1 + \sqrt{\gamma_2}}\,(-2\gamma_1 + \sqrt{\gamma_2}),$$

where

$$\gamma_1 = 39 - 31\tilde{\sigma}_{11}, \quad \gamma_2 = 3546 + 4\tilde{\sigma}_{11}(-1617 + 949\tilde{\sigma}_{11}).$$

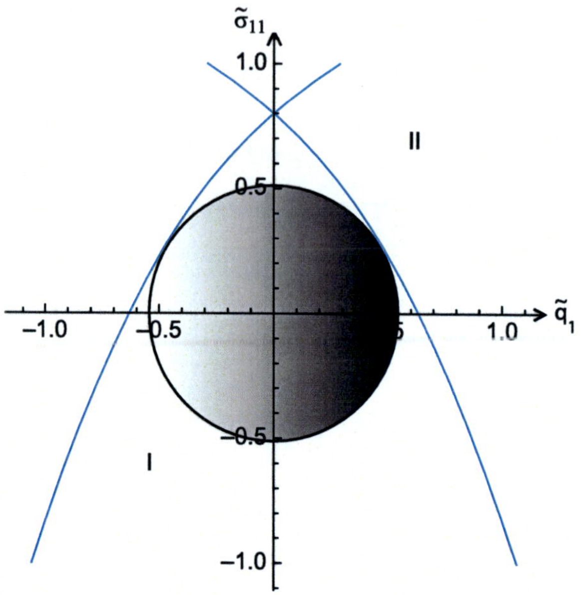

Fig. 4.2 Boundaries of the hyperbolicity region for the one-dimensional ET^1_{13} in the $(\tilde{q}_1, \tilde{\sigma}_{11})$ space. The hyperbolicity region is denoted by 'I', while the zones that do not satisfy the hyperbolicity property are labeled with 'II'. The grey area represents the circle with the maximum radius, centered in the equilibrium point and inscribed within the hyperbolicity region

The hyperbolicity region is the domain enclosed in the two curves for $\tilde{\sigma}_{11} \leqslant 3(89 - \sqrt{7545})/8$ as shown in Fig. 4.2. This result is exactly the same as the one in [25] but with different dimensionless variables.

4.5.2 *Hyperbolicity Region of ET_{13} for the Second Order Approximation*

Concerning the hyperbolicity region, an interesting problem was raised by Cai, Fan, and Li [263]. They showed that a thermodynamic equilibrium state is just on the boundary of the hyperbolicity region if the heat flux and velocity are no longer parallel to a fixed direction. And they concluded that an equilibrium state in such a situation is unstable since any small perturbation can bring an equilibrium state out of the hyperbolicity region. The same authors also suggested a different tool in an ingenious mathematical way to overcome the problem [263, 264]. They proposed an ad hoc modified system, called a new globally hyperbolic system, to 'let' the equilibrium point be inside the hyperbolicity region.

However, their modified system cannot be cast into a system in balance form because the only system of balance equations that is compatible with the entropy principle in the first order approximation is the Grad system. Therefore there is no place for the entropy principle in the modified system. This means that the modified system cannot be used for studying the phenomena where weak solutions play roles such as shock wave phenomena. And a continuum theory without the entropy principle has no sound physical basis.

The hyperbolicity region was revisited by Brini and Ruggeri [250] by constructing a RET model with quadratic expansion with respect to the nonequilibrium variables. They again used the technique explained in Sect. 4.4. We will see that, in the one-dimensional variables, the radius of the hyperbolicity increases when we go on to the quadratic expansion. Moreover we can avoid the above-mentioned problem about the instability of an equilibrium state in the case of three-dimensional variables, where we deal with a local-hyperbolic model in balance form that is compatible with the entropy principle.

After some cumbersome calculations, the system of ET_{13}^2 was determined [250]:

$$
\begin{aligned}
&\partial_t \rho + \partial_k(\rho v_k) = 0,\\
&\partial_t(\rho v_i) + \partial_k(\rho v_i v_k + p\delta_{ik} - \sigma_{\langle ik\rangle}) = 0,\\
&\partial_t(\rho v_i v_j + p\delta_{ij} - \sigma_{\langle ij\rangle}) + \partial_k \Big\{\rho v_i v_j v_k + p(v_i\delta_{jk} + v_j\delta_{ik} + v_k\delta_{ij}) - \sigma_{\langle ij\rangle}v_k -\\
&- \sigma_{\langle jk\rangle}v_i - \sigma_{\langle ik\rangle}v_j + \frac{2}{5}(q_i\delta_{jk} + q_j\delta_{ik} + q_k\delta_{ij}) + \underline{\Big[\frac{8}{25p}(\sigma_{\langle il\rangle}q_l\delta_{jk} +}\\
&\underline{+\sigma_{\langle jl\rangle}q_l\delta_{ik} + \sigma_{\langle kl\rangle}q_l\delta_{ij}) - \frac{4}{5p}(q_i\sigma_{\langle jk\rangle} + q_j\sigma_{\langle ik\rangle} + q_k\sigma_{\langle ij\rangle})\Big]}\Big\} = P_{\langle ij\rangle},\\
&\partial_t(\rho v^2 v_i + 2(\rho\varepsilon + p)v_i - 2\sigma_{\langle il\rangle}v_l + 2q_i) + \partial_k\Big\{\rho v^2 v_i v_k + 2\rho\varepsilon v_i v_k +\\
&+ p(v^2\delta_{ik} + 4v_i v_k) - \sigma_{\langle ik\rangle}v^2 - 2\sigma_{\langle il\rangle}v_l v_k - 2\sigma_{\langle kl\rangle}v_i v_l +\\
&+\frac{4}{5}q_l v_l\delta_{ik} + \frac{14}{5}q_i v_k + \frac{14}{5}q_k v_i + 5\frac{p^2}{\rho}\delta_{ik} - 7\frac{p}{\rho}\sigma_{\langle ik\rangle} + \underline{\Big[\frac{2}{\rho}\sigma_{\langle il\rangle}\sigma_{\langle kl\rangle} +}\\
&\underline{+\frac{1}{25p}(112 q_i q_k + 36 q^2\delta_{ik} + 16(\sigma_{\langle il\rangle}q_l v_k + \sigma_{\langle kl\rangle}q_l v_i + \sigma_{\langle lm\rangle}q_l v_m\delta_{ik}) -}\\
&\underline{-40(\sigma_{\langle il\rangle}q_k v_l + \sigma_{\langle kl\rangle}q_i v_l + \sigma_{\langle ik\rangle}q_l v_l))\Big]}\Big\} = P_{ill},
\end{aligned}
\tag{4.75}
$$

where the underlined terms are the quadratic ones. The conservation law of energy is given by the trace of Eq. (4.75)$_3$.

The characteristic polynomial associated with the system (4.75) is given by

$$\mathscr{P}^{(2)}(\tilde{\lambda}) = 6\frac{k_B}{m}\sqrt{\frac{k_B}{m}T}p^2\Big[a_5^{(2)}\tilde{\lambda}^5 + a_4^{(2)}\tilde{\lambda}^4 + +a_3^{(2)}\tilde{\lambda}^3 + a_2^{(2)}\tilde{\lambda}^2 + a_1^{(2)}\tilde{\lambda} + a_0^{(2)}\Big], \tag{4.76}$$

where the superscript (2) indicates the approximation with second-order expansion, and

$$\begin{aligned}
a_5^{(2)} &= 1, \qquad a_4^{(2)} = -\frac{184}{25}\tilde{q}_1,\\
a_3^{(2)} &= -\frac{1}{1875}(150(\tilde{\sigma}_{11}-1)(18\tilde{\sigma}_{11}-65) - 19{,}684\tilde{q}_1^2),\\
a_2^{(2)} &= -\frac{12}{625}\tilde{q}_1(148\tilde{q}_1^2 + 8\tilde{\sigma}_{11}^2 + 945\tilde{\sigma}_{11} - 975),\\
a_1^{(2)} &= \frac{1}{1875}(5625(-1+\tilde{\sigma}_{11})^2 - 36\tilde{q}_1^2(76\tilde{\sigma}_{11}^2 - 619\tilde{\sigma}_{11} + 580)),\\
a_0^{(2)} &= \frac{12}{15}\tilde{q}_1(\tilde{\sigma}_{11}-1)^2(2\tilde{\sigma}_{11}-5).
\end{aligned} \tag{4.77}$$

Obviously, at an equilibrium state, the roots of $\mathscr{P}^{(2)} = 0$ coincide with those in the first-order approximation, and the same considerations about hyperbolicity remain valid. While, out of equilibrium, we deal with the fifth-degree polynomial (4.76), which cannot be factorized analytically. However, using (4.72) with the coefficients (4.77), one can determine the parametric equations of the boundaries. In Fig. 4.3, such boundaries are plotted in the $(\tilde{q}_1, \tilde{\sigma}_{11})$-plane. The hyperbolicity region is composed of different zones with the symbol 'I'. We stress that the results presented in Fig. 4.3 are compatible with those obtained in a different approach [262], if one takes the non-degeneracy limit of a gas.

The hyperbolicity regions corresponding to ET_{13}^1 and ET_{13}^2 present completely different structures from each other. In order to compare them quantitatively, we determine the radius of the maximum circle centered in an equilibrium state and inscribed in the region. In Fig. 4.2 and in Fig. 4.4, we see the maximum circles for ET_{13}^1 and ET_{13}^2. The region corresponding to the quadratic expansion turns out to be larger in the neighborhood of the equilibrium state. In fact, for the Grad system the maximum radius[2] is $\tilde{r}^{(1)} \simeq 0.5432$, while for the second-order expansion the radius is $\tilde{r}^{(2)} \simeq 0.8220$.

[2]The values of the dimensionless radius $\tilde{r}^{(1)}$ calculated in [25, 56] is different from the present one since, as mentioned above, different dimensionless variables are used.

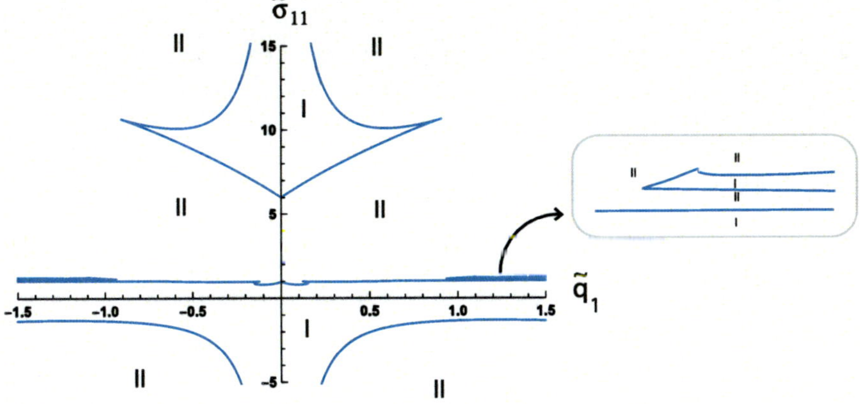

Fig. 4.3 Boundaries of the hyperbolicity region in the $(\tilde{q}_1, \tilde{\sigma}_{11})$-plane for the ET_{13}^2 model when one-dimensional variables are taken into account. The hyperbolicity region is denoted by 'I', while the zones that do not satisfy the hyperbolicity property are labeled with 'II'. A zoom is also presented to show the details of the structure of the hyperbolicity region

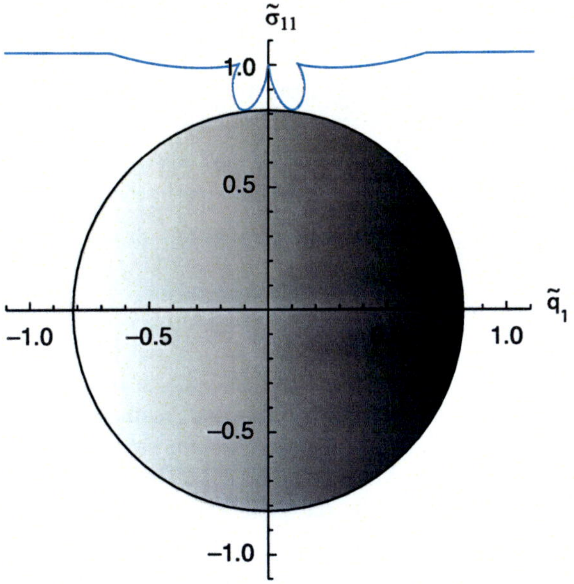

Fig. 4.4 Circle with the maximum radius centered in an equilibrium state and inscribed in the hyperbolicity region for ET_{13}^2 in the $(\tilde{q}_1, \tilde{\sigma}_{11})$-plane

Lastly let us take into account all the 13 components: $\mathbf{u} = (\rho, v_j, T, \sigma_{jk}, q_j)$ for $j, k = 1, 2, 3$ ($\sigma_{ll} = 0$), but we still assume that the variables depend on t and only one space variable x. The corresponding system of equations is in the form of (4.73) with 13 equations. In this case, the problem in ET^1_{13} presented by Cai, Fan, and Li [263] does not occur in ET^2_{13}. In fact, Brini and Ruggeri [250] proved that, in the quadratic approximation, there exists always a neighborhood of an equilibrium point within the hyperbolicity region. The instability problem has disappeared!

4.6 Bounded Domain: Heat Conduction and Problem of Boundary Data

Up to here, we have considered phenomena evolving in an unbounded domain. We have seen that RET is successful because (1) it satisfies explicitly the universal principles of physics, (2) it has a desired mathematical structure of symmetric hyperbolic system, (3) it is perfectly consistent with the kinetic theory, and (4) the results derived from it are in good agreement with experimental data. The analysis of phenomena in a bounded domain is, however, not quite satisfactory when the number of moments is more than 13 as we will see below.

In this subsection, we summarize some problems encountered in the study of nonequilibrium phenomena in a bounded domain. We start with the heat-conduction problem by using the 13-moment RET theory. We will see that the results in non-planar geometry are very different from the results predicted by the classical Navier-Stokes and Fourier (NSF) theory.

4.6.1 Heat Conduction Analyzed by the 13-Moment RET Theory

Müller and Ruggeri [265] studied one-dimensional heat conduction in a gas at rest in planar, cylindrical, and spherical geometries by using the 13-moment RET theory. It turns out that, in the radially symmetric cases, the stress tensor does not reduce to a scalar pressure and that the heat flux depends on the normal components of the deviatoric stress tensor. As a result, the singularities of temperature on the axis of the cylinder and at the center of the sphere—which are characteristic for the Navier-Stokes and Fourier solution—disappear. In this subsection, we explain this result.

For the present argument the details of molecular interaction are irrelevant. Therefore we choose the simplest model and consider the system of equations based on the BGK model with the relations $\tau_\sigma = \tau_q = \tau$. Furthermore, we employ the covariant derivatives by using the Christoffel symbols. Then we obtain a more specific, but more complex version of the field equations (1.28), which, in a stationary case, becomes [265]:

$$
\begin{aligned}
&g^{ik}\frac{\partial p}{\partial x^k}-\frac{\partial \sigma^{\langle ik\rangle}}{\partial x^k}-\Gamma^i_{kl}\,\sigma^{\langle kl\rangle}-\Gamma^k_{kl}\,\sigma^{\langle il\rangle}=0,\\
&\frac{\partial q^k}{\partial x^k}+\Gamma^k_{kl}\,q^l=0,\\
&\frac{2}{5}\left(g^{ik}\frac{\partial q^j}{\partial x^k}+g^{jk}\frac{\partial q^i}{\partial x^k}+g^{ik}\Gamma^j_{kl}\,q^l+g^{jk}\Gamma^i_{kl}\,q^l\right)=\frac{1}{\tau}\sigma^{\langle ij\rangle}, \qquad (4.78)\\
&5pg^{ik}\frac{k_B}{m}\frac{\partial T}{\partial x^k}-7\sigma^{\langle ik\rangle}\frac{k_B}{m}\frac{\partial T}{\partial x^k}-2\frac{k_B}{m}T\left(\frac{\partial \sigma^{\langle ik\rangle}}{\partial x^k}+\Gamma^i_{kl}\;\sigma^{\langle kl\rangle}+\Gamma^k_{kl}\sigma^{\langle il\rangle}\right)=-\frac{2}{\tau}q^i,
\end{aligned}
$$

where g_{ik} is the metric tensor and Γ^i_{jk} is the Christoffel symbol appropriate to the coordinates x^k.

For the planar case with rectangular Cartesian coordinates, g^{ik} is the Kronecker tensor and all Christoffel symbols vanish. In the cylindrical case with coordinates $(x^1, x^2, x^3) = (r, \vartheta, z)$, we have

$$
g^{ik}=\begin{pmatrix}1 & & 0\\ & \frac{1}{r^2} & \\ 0 & & 1\end{pmatrix} \text{ and } \Gamma^1_{22}=-r,\;\; \Gamma^2_{21}=\Gamma^2_{12}=\frac{1}{r},\;\; \Gamma^m_{kn}=0 \text{ else.}
$$

And, in the spherical case with $(x^1, x^2, x^3) = (r, \vartheta, \varphi)$, we have

$$
g^{ik}=\begin{pmatrix}1 & & 0\\ & \frac{1}{r^2} & \\ 0 & & \frac{1}{r^2\sin^2\vartheta}\end{pmatrix} \text{ and } \Gamma^3_{31}=\Gamma^3_{13}=\Gamma^2_{21}=\Gamma^2_{12}=\tfrac{1}{r},\, \Gamma^1_{22}=-r,
$$

$$
\Gamma^1_{33}=-r\sin^2\vartheta,\;\; \Gamma^3_{32}=\Gamma^3_{23}=ctg\;\vartheta,
$$

$$
\Gamma^2_{33}=-\sin\vartheta\cos\vartheta,
$$

$$
\Gamma^m_{kn}=0 \quad \text{else.}
$$

Note that, in all the three cases, the coordinate lines are orthogonal so that the metric tensors are diagonal.

4.6.2 Comparison with the Navier-Stokes and Fourier Theory

We investigate solutions under the conditions that (1) all fields depend only on x^1 and (2) q^1 is the only non-vanishing component of the heat flux. In such a case, the

Eqs. (4.78) readily imply that shear stress vanishes and the deviatoric normal stress is related to q^1 by

$$\sigma^{\langle 11\rangle} = -\frac{4}{5}\Gamma^k_{k1}\tau q^1,\ \sigma^{\langle 22\rangle} = \frac{4}{5}g^{22}\Gamma^2_{21}\tau q^1,\ \sigma^{\langle 33\rangle} = \frac{4}{5}g^{33}\Gamma^3_{31}\tau q^1. \tag{4.79}$$

We also obtain

$$\frac{dq^1}{dx^1} = -\Gamma^k_{k1}q^1.$$

We conclude from (4.78) that $p = const.$, and the 1-component provides a relation between the heat flux and the temperature gradient, namely

$$q^1 = -\kappa\left(1 - \frac{7}{5}\frac{\sigma^{\langle 11\rangle}}{p}\right)\frac{dT}{dx^1}, \tag{4.80}$$

where κ is the heat conductivity related to the relaxation time τ by

$$\kappa = \frac{5}{2}\frac{k_B}{m}\tau p.$$

From (4.79), we understand that, in spherical coordinates, the physical components of the shear stress are not zero. This result is particular to RET, because with the Navier-Stokes constitutive equations we have $\sigma^{\langle ij\rangle} = 0$ in any geometry. Thus we conclude, from (4.79), that in the planar case where $\sigma^{\langle 11\rangle}$ vanishes, Fourier's law with a constant heat conductivity is recovered because the heat flux is proportional to the gradient of the temperature. In the cylindrical and the spherical cases, Fourier's law is not valid because of the second term on the right hand side of (4.79).

We now solve the system of equations and obtain the fields $\sigma^{\langle 11\rangle}$, q^1, and T. The only geometric quantity left in that system is Γ^k_{k1} which reads

$$\Gamma^k_{k1} = \frac{j}{x^1} \quad \text{with } j = \begin{bmatrix} 0 & & \text{planar} \\ 1 & \text{for the} & \text{cylindrical case.} \\ 2 & & \text{spherical} \end{bmatrix}$$

The general integral is easily obtained. In the planar case we have the simple solution

$$q^1 = c_1, \quad \sigma^{\langle 11\rangle} = 0, \quad T = c_2 - \frac{c_1}{\kappa}x^1.$$

This represents a linear temperature profile, which is also predicted by the Fourier theory (c_1,c_2 are integration constants). For $j = 1, 2$ we have (we write now

$r,\ q^r,\ \sigma^{\langle rr\rangle}$ instead of $x^1,\ q^1,\ \sigma^{\langle 11\rangle}$)

$$q^r = \frac{c_1}{r^j} \qquad \text{and} \qquad \sigma^{\langle rr\rangle} = -\frac{8j}{25}\frac{m}{k_B}\frac{\kappa}{p}\frac{c_1}{r^{j+1}}.$$

And (4.80) becomes

$$\frac{dT}{dr} = -\frac{c_1}{\kappa}\frac{r}{\frac{56}{125}j\frac{m}{k_B}\frac{\kappa}{p^2}c_1 + r^{j+1}}. \tag{4.81}$$

The first term in the denominator is absent in the Navier-Stokes and Fourier theory, because $\sigma^{\langle rr\rangle} = 0$ in that theory.

Thus we can see a first difference between RET and the Fourier theory: The derivative of the temperature tends to zero for $r \to 0$, while in the Fourier case it diverges. The general solution of (4.81) reads:

In the cylindrical case, i.e., for $j = 1$,

$T = c_2 - \frac{c_1}{2\kappa}\log(b + r^2)$.

In the spherical case, i.e., for $j = 2$,

$$T = c_2 + \frac{c_1}{3\, 2^{\frac{1}{3}}\, b^{\frac{1}{3}}\kappa}\left\{\sqrt{3}\arctan\left(\frac{b^{\frac{1}{3}} - 2^{\frac{2}{3}}r}{\sqrt{3}\, b^{\frac{1}{3}}}\right) + \right.$$
$$\left. + \log\left(2\, b^{\frac{1}{3}} + 2^{\frac{2}{3}}r\right) - \frac{1}{2}\log\left(2\, b^{\frac{2}{3}} - 2^{\frac{2}{3}}\, b^{\frac{1}{3}}r + 2^{\frac{1}{3}}r^2\right)\right\}.$$

While, according to the Fourier law, we have

$$T = c_2 - \frac{c_1}{\kappa}\log r \quad \text{for } j = 1 \quad \text{and} \quad T = c_2 + \frac{c_1}{\kappa}\frac{1}{r} \quad \text{for } j = 2.$$

For abbreviation we have set $b = \frac{56}{125}\frac{m}{k_B}\frac{\kappa}{p^2}c_1$.

Figure 4.5 illustrates the difference between these solutions in the spherical case for c_1 arbitrarily assigned and c_2 such that the temperature vanishes for large values of r. We see that the temperature is finite in RET, while it diverges in the Fourier theory. We also observe that the solutions coincide with each other when the gradient is small, while they differ significantly where the curves become steeper. As a matter of course, RET becomes relevant when gradients are large.

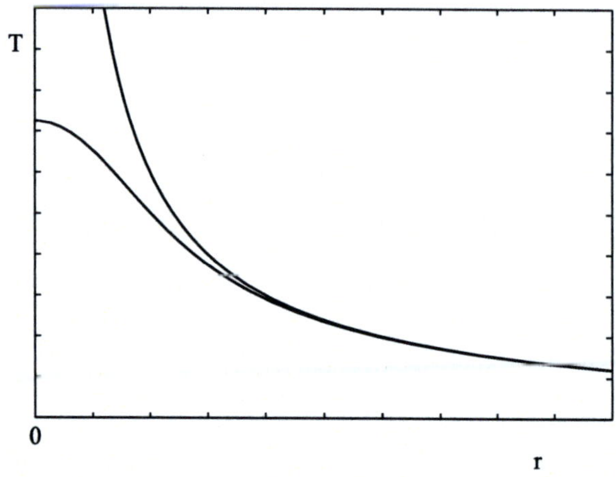

Fig. 4.5 The behavior of the general integral of the temperature in the spherical symmetry. The divergent curve represents the solution of the Fourier theory while the bounded one represents the 13-moment solution

4.6.3 Solution of a Boundary Value Problem

In the planar case, the result from the Fourier theory is identical to the one from RET. Indeed, from (4.11)$_3$, we conclude that the temperature is linear in x^1. The radially symmetric cases, in particular the cylindrically symmetric one, are more interesting.

We consider a gas between two co-axial cylinder or between two concentric spheres with inner radius r_i and outer radius r_e. We heat the gas at the inner radius with a prescribed heat flux q and the temperature at the outer radius is kept to a value T_e. Thus we solve the boundary value problem:

$$q^r(r_i) = q \quad \text{and} \quad T(r_e) = T_e.$$

In the cylindrical case we obtain the solution:

$$q^r = \frac{q\,r_i}{r}, \quad \sigma^{\langle rr\rangle} = \frac{8}{25}\frac{m}{k_B}\frac{\kappa}{p}\frac{q\,r_i}{r^2}, \quad T = T_e - \frac{q\,r_i}{2\kappa}\log\left(\frac{\frac{56}{125}\frac{m}{k_B}\frac{\kappa}{p^2}q r_i + r^2}{\frac{56}{125}\frac{m}{k_B}\frac{\kappa}{p^2}q r_i + r_e^2}\right). \tag{4.82}$$

Figure 4.6 shows this solution for T in the case of argon with the relative atomic mass $m = 40$ and for the data:

$$r_i = 10^{-3}[\text{m}],\ r_e = 10^{-2}[\text{m}],\ p = 10^2[\text{N/m}^2],\ \tau = 10^{-5}[\text{s}],\ q = 10^4[\text{W/m}^2],\ T_e = 300[\text{K}]. \tag{4.83}$$

The curve marked by continuous line in Fig. 4.6 represents the solution of RET and the one with dashed line is the solution of the Fourier theory. At the inner cylinder, the values of the two theories differ by 7.5K.

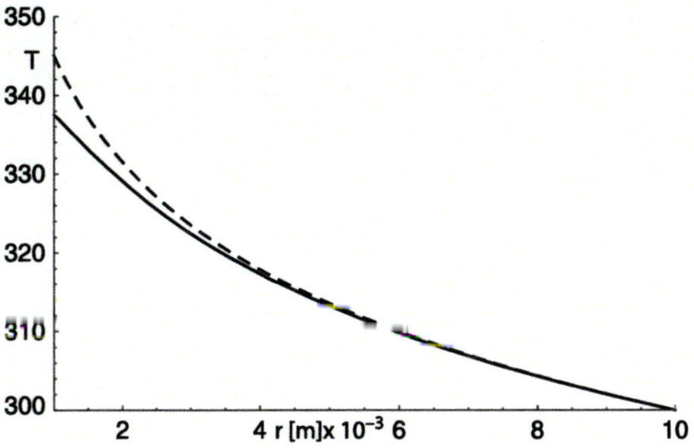

Fig. 4.6 The behavior of the temperatures in the cylindrical case with the boundary data (4.83). Dashed line: Fourier theory. Continuous line: 13-moment RET theory

In the spherical case we obtain with $b = \frac{56}{125}\frac{m}{k_B}\frac{\kappa}{p^2} q r_i^2$:

$$
\begin{aligned}
q^r &= \frac{q\, r_i^2}{r^2}, \quad \sigma^{\langle rr\rangle} = \frac{16}{25}\frac{m}{k_B}\frac{\kappa}{p}\frac{q\, r_i^2}{r^3}, \\
T &= T_e + \frac{c_1}{3\, 2^{\frac{1}{3}}\, b^{\frac{1}{3}}\kappa}\left\{\sqrt{3}\left[\arctan\left(\frac{b^{\frac{1}{3}} - 2^{\frac{2}{3}} r}{\sqrt{3}\, b^{\frac{1}{3}}}\right) - \arctan\left(\frac{b^{\frac{1}{3}} - 2^{\frac{2}{3}} r_e}{\sqrt{3}\, b^{\frac{1}{3}}}\right)\right]\right. \\
&\quad \left. + \log\left(\frac{2\, b^{\frac{1}{3}} + 2^{\frac{2}{3}} r}{2\, b^{\frac{1}{3}} + 2^{\frac{2}{3}} r_e}\right) - \frac{1}{2}\log\left(\frac{2\, b^{\frac{2}{3}} - 2^{\frac{2}{3}}\, b^{\frac{1}{3}} r + 2^{\frac{1}{3}} r^2}{2\, b^{\frac{2}{3}} - 2^{\frac{2}{3}}\, b^{\frac{1}{3}} r_e + 2^{\frac{1}{3}} r_e^2}\right)\right\}.
\end{aligned}
\tag{4.84}
$$

Both solutions (4.82) and (4.84) hold in the interval $r_i \leqslant r \leqslant r_e$.

The case of co-axial cylinders should be easy to set up experimentally. Indeed, the inner cylinder could be realized by a wire and the heating may be effected by letting an electric current run through that wire. On the other hand, the corresponding case of two concentric spheres may be quite difficult to realize experimentally.

4.6.4 Difficulty in the RET Theory in a Bounded Domain When the Number of Fields Is More than 13

In the above, we have seen that the 13-moment RET theory predicts the new and interesting results that show an appreciable difference from the results predicted by the Navier-Stokes and Fourier theory. We understand that, when there exists a steep gradient (and/or a rapid change), the RET theory is superior to the conventional classical TIP theory. Therefore it was natural that several authors tried to understand more deeply nonequilibrium phenomena in a bounded domain by using the RET theory with more than 13 moments. However, in such studies, the authors encountered a conceptually difficult problem. This is the problem of boundary conditions.

All quantities in the 13-moment theory have concrete physical meanings and are observable. However, the quantities expressed by higher moments have not definite physical meanings and are, in general, impossible to be measured in experiments. Therefore we cannot pose the values of these quantities at the boundary of a domain, that is, we cannot pose the boundary conditions for the RET theory with more than 13 moments appropriately.

We may have a question: Is it really necessary to know all the boundary values? Indeed let us consider, for example, heat conduction in a gas filled in between two parallel plates. In this case, we need not know all boundary values except for, say, the temperatures at the two plates in order to study the phenomena experimentally. This consideration seems to lead us to the following conjecture: if the temperatures are fixed at the both sides, the other quantities adjust by themselves and have appropriate boundary values that are consistent with the experimental data. Such quantities are sometimes called uncontrollable quantities.

If this conjecture is true, we further want to know the mechanism of the self-adjustment. Several studies have been done. Struchtrup and Weiss [266] proposed the minimax principle for the entropy production, according to which the boundary values of the uncontrollable quantities are fixed. Barbera, Müller, Reitebuch, and Zhao also studied this problem [267]. Expecting that the uncontrollable quantities fluctuate around the most probable values, they assume that the boundary values of the quantities are given by these most probable values.

Another strategy was proposed by Ruggeri and Lou [268]. Their approach is purely phenomenological. Let us consider, as an example, the heat conduction in a mixture of gases. In order to obtain the temperature profile, they impose not only the boundary conditions but also some conditions inside a gas that should be given by experiments. The latter conditions are, for example, the temperatures at several points inside a gas (see Sect. 28.6.2).

Moreover, as pointed out by Brini and Ruggeri [256], there is another subtle problem. They proved that, if nonequilibrium variables have small values of the same order, then some derivatives of these variables (critical derivatives) are not necessarily of the same order. As a consequence, the solutions violate the entropy principle and the system becomes inconsistent with the near-equilibrium approximation adopted.

Kinetic-theoretical study by using the concept of the accommodation factor may be helpful to solve this problem (see, for example, references [76, 269–272]). The accommodation factor is defined as the ratio between the effects of specular reflection and thermalized (diffuse) reflection of an incident molecule at the boundary wall. The introduction of this factor implies that the interaction between a gas and a boundary wall should be properly taken into consideration in order to fix the boundary conditions.

In conclusion, except for the 13-moment RET theory, the problem of the boundary conditions in RET in general still remains as a big issue to be solved even in the case of monatomic gas.

4.7 Comparison with Experimental Data and with Solutions of the Boltzmann Equation

The RET theory is successful when its results are compared with experimental data and also with the solutions of the Boltzmann equation.

4.7.1 Sound Waves, Light Scattering, and Shock Waves: Comparison with Experimental Data

Concerning the experimental data, the RET results are very good for sound waves with high frequencies and light scattering. In Fig. 4.7 taken from the book [25], we can see that the dynamic structure factor $S(x, y)$ obtained by the RET theory fits very well the experimental data on light scattering (represented in the figure by dots) when the number of moments N is sufficiently large. For the definition of $S(x, y)$ and for the details on the comparison with experiments, see Ref. [25]. See also Chap. 21.

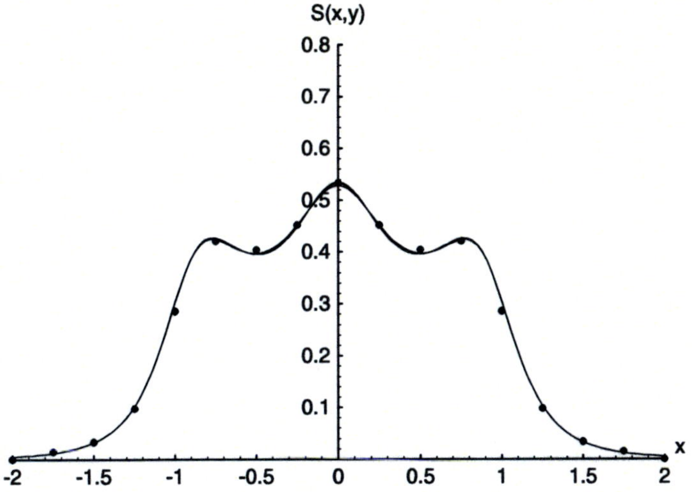

Fig. 4.7 Dynamic structure factor: We see an excellent agreement between the prediction of RET and the experimental data

Concerning shock structure a chapter written by Weiss is contained in the book [25]. Here the situation is more complex for several reasons. First, according with Boillat-Ruggeri Theorem 3.1, there is always a subshock when the shock speed exceeds the non-perturbed equilibrium maximum characteristic velocity. These appear as artificial discontinuities, at least for a monatomic gas, since no evidence of such discontinuities is seen in experimental results. Second, it is very hard, even numerically, to obtain the shock structure in the theory with many moments. In principle, we expect better results when we increase the number of moments because, according with Theorem 4.2, the maximum characteristic velocity increases with the number of moments and therefore the subshock arises at the bigger critical Mach number, which grows with the number of moments.

4.7.2 *Shock Waves: Comparison Between RET and Kinetic Theory*

An interesting contribution was recently given by Mentrelli and Ruggeri [273] adopting the second order closure explained in Sect. 4.5.2. They proved that the strength of the subshock that appears in the shock structure profile for large enough Mach number is remarkably reduced compared to what is found in the case of the first order closure (see Fig. 4.8), and the overall profile of shock structure solution is in better agreement with the results obtained with the kinetic theoretical approach. In order to correctly interpret Fig. 4.8, we recall that the critical Mach number for ET_{13} is $M_0^{cr} \simeq 1.65$, and only above this value a jump in the value of the density ahead and before the newly formed shock appears.

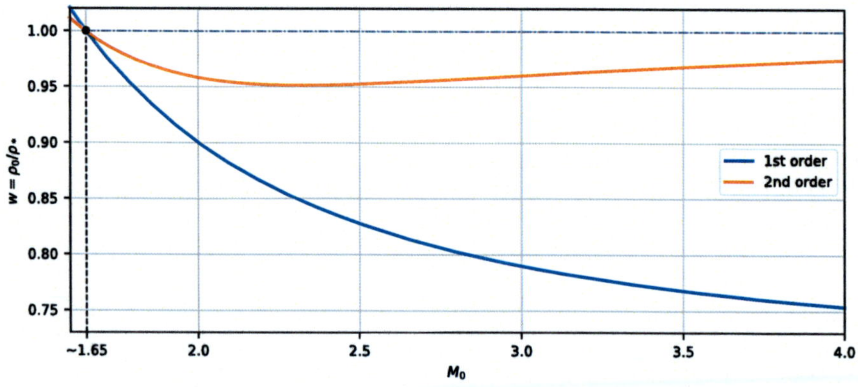

Fig. 4.8 Compression factor w, i.e. ratio of the mass density ahead and behind the fast subshock front ($w = \rho_0/\rho_*$), predicted for the 13-moment system for a monatomic gas with first and second order maximum entropy principle closure (respectively, blue curve and orange curve), as a function of the unperturbed Mach number M_0

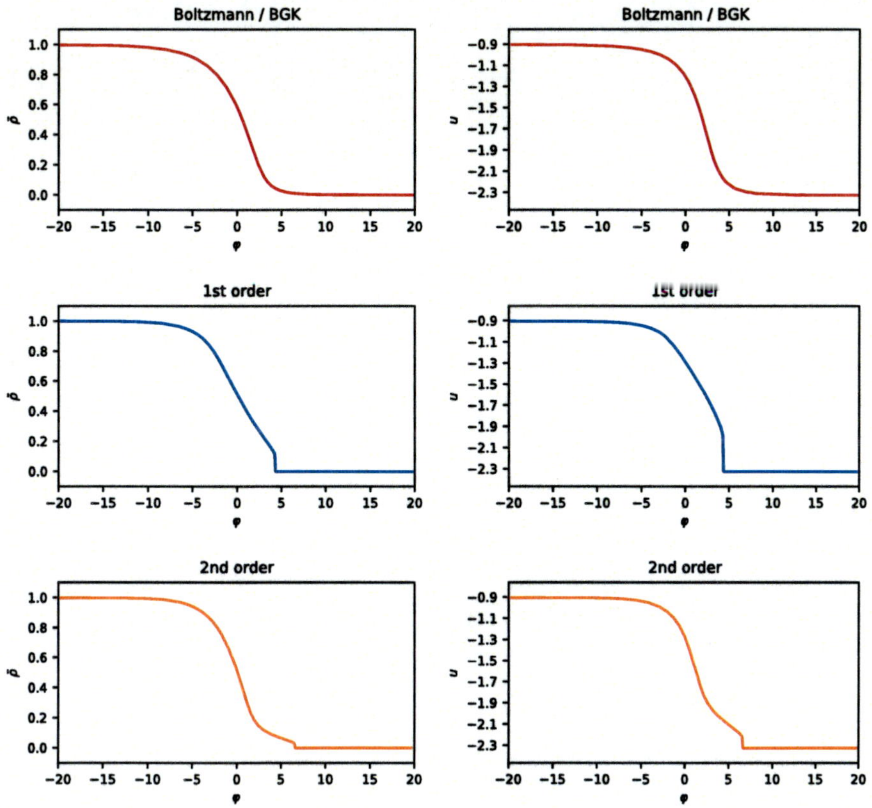

Fig. 4.9 Rescaled mass density profile $\bar{\rho} = (\rho - \rho_0) / (\rho_1 - \rho_0)$ (on the left) and relative velocity profile $u = v - s$ (on the right) of the shock structure solution obtained for $M_0 = 2.3$ in a monatomic gas with the Boltzmann/BGK model (top row), and with the 13 moment system with the first order MEP closure (middle row) and second order MEP closure (bottom row)

In Figs. 4.9 and 4.10, we can see the benefit of the second order closure in the case of two different values of the Mach number ($M_0 = 2.3$ and $M_0 = 4$, respectively).

In the analysis of these results, it is important to consider the region of hyperbolicity of the process. For the first order closure, it is known since the early '90s [56, 274] that for a Mach number greater than $M_0 \simeq 2.7$, the perturbed non-equilibrium state $\mathbf{u}_*$ associated to a subshock is not in the hyperbolicity region (see Fig. 4.11) and it is necessary to keep in mind that the numerical results obtained for larger Mach numbers might be compromised by a partial lack of hyperbolicity along the process, being the model not valid outside the hyperbolicity region.

More interestingly, a close analysis of the results obtained adopting a second order closure reveals that the process never leaves the hyperbolicity region, at least for Mach number up to 4 (and beyond). In Fig. 4.12 it is shown that all the characteristic velocities in the nonequilibrium perturbed state $\mathbf{u}_*$ behind the shock are always real in this range, and the peculiar topology of the hyperbolicity region

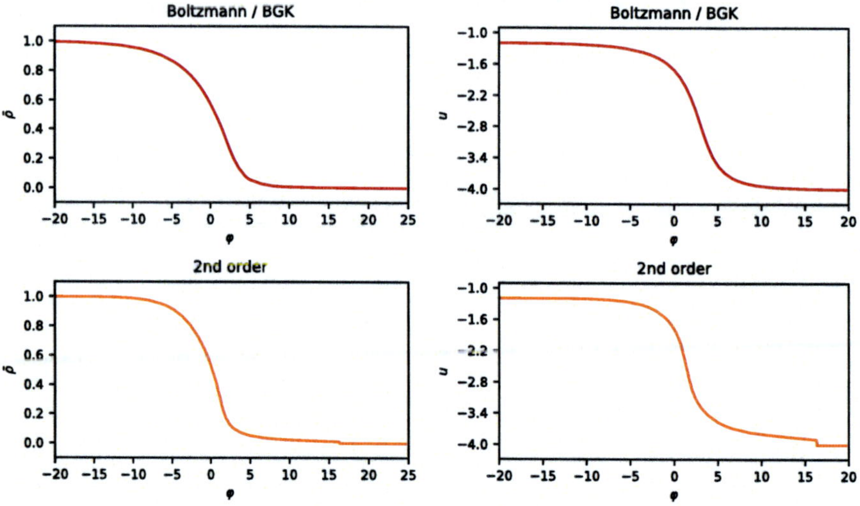

Fig. 4.10 Rescaled mass density profile $\bar{\rho} = (\rho - \rho_0)/(\rho_1 - \rho_0)$ (on the left) and relative velocity profile $u = v - s$ (on the right) of the shock structure solution obtained for $M_0 = 4.0$ in a monatomic gas with the Boltzmann/BGK model (top row), and with the 13 moment system with the first order MEP closure (middle row) and second order MEP closure (bottom row)

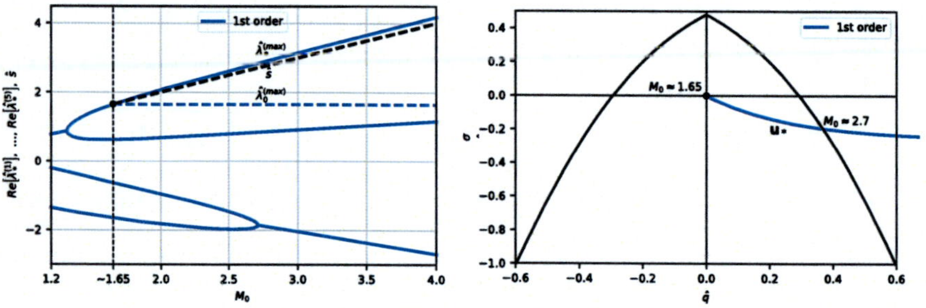

Fig. 4.11 (left) Real part of the dimensionless characteristic velocities $\hat{\lambda}_*^{(k)}$ $(k = 1, \ldots 5)$ evaluated in the state $\mathbf{u}_*$ behind the shock, predicted by the 13-Moment system for a monatomic gas in the one-dimensional case with first order MEP closures; (right) Graphical representation of the hyperbolicity region and of the perturbed state $\mathbf{u}_*$ as the Mach number varies. When $M_0 \simeq 2.7$ the hyperbolicity is locally lost [56, 274]

discussed in Fig. 4.3 is such that the process along the shock structure happens to be always inside the hyperbolicity region. In Fig. 4.12b the process along the shock structure is depicted together with the hyperbolicity region illustrated in Fig. 4.3. Therefore, in the results shown in Fig. 4.10, only the results obtained by means of the second order closure are reported, along with the results obtained by a kinetic theory approach, since the results obtained with a first order closure are not, strictly speaking, meaningful for such a large Mach number.

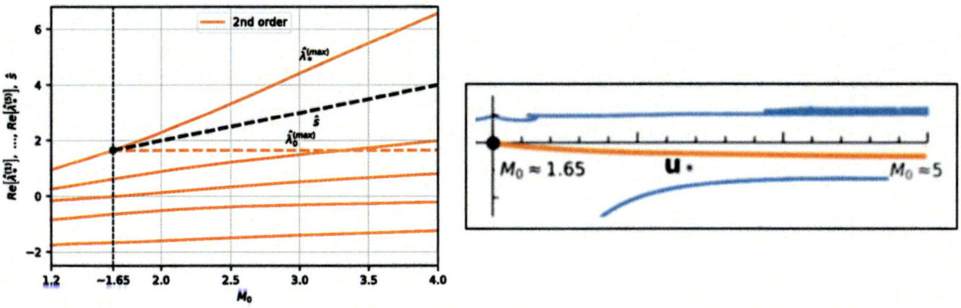

Fig. 4.12 (left) Real part of the dimensionless characteristic velocities $\hat{\lambda}_*^{(k)}$ ($k = 1, \ldots 5$) evaluated in the state $\mathbf{u}_*$ behind the shock, predicted by the 13-Moment system for a monatomic gas in the one-dimensional case with second order MEP closures; (right) Graphical representation of the hyperbolicity region (see Fig. 4.3) and of the perturbed state $\mathbf{u}_*$ as the Mach number varies. It was numerically shown that even for *large* Mach number ($M_0 = 5$ and well beyond) the state $\mathbf{u}_*$ never leaves the hyperbolicity region calculated in [256]

For more details concerning RET of monatomic gases, readers can consult the book [25] and the survey papers [275] and [276].

Finally we want to recall that many applications of RET not only to rarefied gases but also to other materials such as semiconductor (for example, fluid-dynamic models of semiconductor [92, 277–279]) have been made.

Chapter 5
Relativistic RET of Rarefied Monatomic Gas

Abstract In this chapter, firstly, the relativistic Euler model of a gas is presented with the proof that the system is symmetric hyperbolic for general constitutive equations. In the case of rarefied monatomic gas, by using the relativistic Boltzmann-Chernikov equation, appropriate constitutive equations and, in particular, the Synge energy are discussed. The two limits in classical and ultra-relativistic cases are also studied.

Then, the modern approach to a relativistic gas with dissipation, that is, RET of a relativistic gas given by Liu, Müller, and Ruggeri (LMR) is summarized. The LMR theory is an improvement to the previous casual theories of Müller and Israel.

The RET with many moments, say N moments, is also studied together with the evaluation of the characteristic velocities for increasing number N. In this framework, the maximum characteristic velocity is bounded for any number N. It converges to the light velocity from below when $N \to \infty$.

In the last section, the classical limit is studied. Then it is proved that, for a fixed N, there exists a unique choice of the moments in classical case. This is probably an answer to the longstanding problem about the optimal choice of moments in the classical case.

5.1 Introduction

In Chap. 4, the non-relativistic (classical) RET theory of rarefied monatomic gases is explained. However, in the case when γ given by (1.40) is very small, a non-relativistic RET theory becomes unsatisfactory to describe a fluid. Then we need a relativistic RET theory. The smallness of γ means that (1) a fluid is very hot so that the thermal kinetic energy of a particle surpasses its rest energy to a large extent or (2) the particle mass is extremely small. Therefore a suitable relativistic thermodynamic theory of fluids is of considerable important in several areas such as astrophysics, nuclear physics. In particular, since the state of matter in the emerging universe is believed to be characterized by an extremely high temperature, relativistic thermodynamics is required for describing dynamics of the universe once

T. Ruggeri, M. Sugiyama, *Classical and Relativistic Rational Extended Thermodynamics of Gases*, https://doi.org/10.1007/978-3-030-59144-1_5

the recombination has occurred and the system has been enough cooled down. In such a situation, we need not worry about its detailed features at a microscopic scale. See, for example, [5, 277, 280, 281].

5.2 Relativistic Euler Fluid

The relativistic Euler fluid is the simplest model, where all dissipative effects are ignored. As in the classical counterpart, the balance equations come from the conservation of particle number and the conservation of energy-momentum:

$$\partial_\alpha V^\alpha = 0, \qquad \partial_\alpha T^{\alpha\beta} = 0, \tag{5.1}$$

where x^α ($\alpha = 0, 1, 2, 3$) are the space-time coordinates and $\partial_\alpha = \partial/\partial x^\alpha$. The particle flux vector V^α and the energy-momentum tensor $T^{\alpha\beta}$ are expressed as

$$V^\alpha = nmU^\alpha, \qquad T^{\alpha\beta} = ph^{\alpha\beta} + \frac{e}{c^2}\,U^\alpha U^\beta, \tag{5.2}$$

where n is the particle number density ($\rho = nm$ is the mass density), U^α is the four-velocity ($U^\alpha U_\alpha = c^2$), p is the pressure, e is the energy density, and $h^{\alpha\beta}$ is the projector:

$$h^{\alpha\beta} = -g^{\alpha\beta} + \frac{1}{c^2} U^\alpha U^\beta$$

with $g^{\alpha\beta} = diag(1\,,\,-1\,,\,-1\,,\,-1)$ being the metric tensor. In this case, the energy e and the four-vector ρU^α constitute 5 independent variables. To close the system we need, as in the classical case, one constitutive equation, for example,

$$p \equiv p(\rho, e). \tag{5.3}$$

5.2.1 *Symmetrization of the Relativistic Euler System*

Following the symmetrization technique described in Theorem 2.1, Ruggeri and Strumia [164] proved for the first time that the hyperbolic system (5.1) can be rewritten in a symmetric form by changing the independent fields into the main field:

$$\mathbf{u}' \equiv \frac{1}{T}\left(-g_r, U_\beta\right)^T. \tag{5.4}$$

The components of the main field are the Lagrange multipliers in such a way that, if we multiply equation (5.1)$_1$ by $-g_r/T$ and (5.1)$_2$ by U_β/T and add these up, we obtain the supplementary balance law of entropy:

$$\partial_\alpha h^\alpha = 0, \quad h^\alpha = \rho S U^\alpha, \tag{5.5}$$

where S is the entropy density. The quantity g_r in (5.4) is given by

$$g_r = \frac{e+p}{\rho} - TS. \tag{5.6}$$

This is the chemical potential except for c^2. In fact, taking into account that the energy density e is composed of the internal energy density, $\rho\varepsilon$, and the rest-energy density:

$$e = \rho(\varepsilon + c^2), \tag{5.7}$$

we have

$$g_r = g + c^2, \qquad g = \varepsilon + \frac{p}{\rho} - TS, \tag{5.8}$$

where g is the usual chemical potential. The compatibility with the entropy principle requires only the condition: the Gibbs equation (1.8), which remains unchanged in the relativistic framework, is valid.

In the present case, the potential four-vector (2.13) that symmetrizes the system (2.12) is given by

$$h'^\alpha = -\frac{p}{T}U^\alpha. \tag{5.9}$$

By choosing (e, S) as independent variables and then by considering $p \equiv p(e, S)$ instead of (5.3), the convexity of entropy density was proved in [164]. It requires:

$$p_e = \left.\frac{\partial p}{\partial e}\right|_S < 1, \qquad c_p = \frac{k_B}{m} + c_v > 0, \tag{5.10}$$

where $c_v = d\varepsilon/dT$ is the specific heat at constant volume. The first condition means that the maximum characteristic velocity in the rest frame is less than the light velocity c. In fact, it is well known that the maximum characteristic velocity in the rest frame $\hat{\lambda}$ is given by

$$\hat{\lambda} = \sqrt{p_e}\, c.$$

And the second condition in (5.10) requires that the specific heat at constant pressure c_p is positive. This implies that $p_e > 0$, and, as a consequence, the system is hyperbolic. Therefore, the convexity condition is more restrictive than the mere condition of hyperbolicity. The two conditions in (5.10) are equivalent to the condition of hyperbolicity and the requirement that the signal propagates with velocity less than c:

$$0 < p_e < 1. \tag{5.11}$$

To sum up, the Euler system of a relativistic gas is symmetric hyperbolic provided that the inequalities (5.11) are satisfied.

5.3 Space-Time Decomposition

We rewrite the system of balance laws (5.1) in the form of space-time decomposition as follows:

$$\partial_\alpha V^\alpha = 0 \qquad \leftrightarrows \qquad \frac{1}{c}\partial_t V^0 + \partial_i V^i = 0, \tag{5.12}$$

$$\partial_\alpha T^{\alpha\beta} = 0 \qquad \leftrightarrows \qquad \begin{cases} \frac{1}{c}\partial_t T^{0j} + \partial_i T^{ij} = 0, \\ \frac{1}{c}\partial_t \left(cT^{00} - c^2 V^0\right) + c\,\partial_i \left(T^{0i} - cV^i\right) = 0. \end{cases} \tag{5.13}$$

Taking into account the following relations with τ being the proper time:

$$U^\alpha = \frac{dx^\alpha}{d\tau} = \frac{dx^\alpha}{dt}\frac{dt}{d\tau} \equiv (\Gamma c, \Gamma v^i),$$

$$\text{where} \quad \Gamma = \frac{1}{\sqrt{1 - \frac{v^2}{c^2}}} \quad \text{is the Lorentz factor,}$$

and also taking (5.2) and (5.7) into account, we rewrite (5.12) and (5.13) as follows:

$$\begin{aligned}
&\partial_t(\rho\Gamma) + \partial_i(\rho\Gamma v^i) = 0,\\
&\partial_t \left(\rho\Gamma^2 v^j + (p + \rho\varepsilon)\frac{\Gamma^2}{c^2}v^j\right) + \partial_i \left(p\delta^{ij} + \Gamma^2\rho v^i v^j + (p + \rho\varepsilon)\frac{\Gamma^2 v^i v^j}{c^2}\right) = 0,\\
&\partial_t \left(p(\Gamma^2 - 1) + \rho\varepsilon\Gamma^2 + \rho c^2\Gamma(\Gamma - 1)\right) + \partial_i \left\{\left((p + \rho\varepsilon)\Gamma^2 + \rho c^2\Gamma(\Gamma - 1)\right)v^i\right\} = 0.
\end{aligned} \tag{5.14}$$

Equations (5.14) are the relativistic counterparts of the conservation laws of mass, momentum, and energy in the classical theory. In fact, with the use of the following expansion forms:

$$\Gamma = 1 + \frac{v^2}{2c^2} + \frac{3}{8}\frac{v^4}{c^4} + \dots,$$
$$\Gamma^2 = 1 + \frac{v^2}{c^2} + \frac{v^4}{c^4} + \dots,$$
$$c^2\Gamma(\Gamma - 1) = \frac{v^2}{2} + \frac{5}{8}\frac{v^4}{c^2} + \dots,$$

we see that the limits of (5.14) for $c \to \infty$ are just the usual conservation laws of mass, momentum, and energy for classical Euler fluids (4.33).

Note that the main field of the system (5.14) is slightly different from that of (5.4). This is because the energy equation $(5.14)_3$ is not

$$\partial_\alpha T^{\alpha 0} = 0, \tag{5.15}$$

but (see (5.13))

$$\partial_\alpha \left(cT^{\alpha 0} - c^2 V^\alpha \right) = 0. \tag{5.16}$$

Equation (5.15) diverges when $c \to \infty$ due to the contribution ρc^2 to the energy in the rest frame: $e = \rho c^2 + \rho\varepsilon$. While (5.16) converges, in the classical limit, to the energy conservation law $(4.33)_3$. Let us introduce the main field (ξ, λ_i, ζ) associated with the system (5.14) such that the linear combination of the equations in (5.14) gives the entropy law (5.5). Comparing two expressions for dh^0 in terms of the two different main fields:

$$dh^0 = \xi dV^0 + \lambda_i dT^{0i} + \zeta d\left(T^{00} - cV^0\right) = -\frac{g_r}{T} dV^0 + \frac{u_\beta}{T} dT^{0\beta},$$

we have the explicit expression of the main field that symmetrizes the system (5.14):

$$\xi = -\frac{1}{T}\left(g - c^2(\Gamma - 1)\right), \quad \lambda_i = -\Gamma\frac{v^i}{T}, \quad \zeta = \frac{\Gamma}{T}. \tag{5.17}$$

In the classical limit when $c \to \infty$, the main field (5.17) converges to the Godunov main field (2.29).

5.3.1 Kinetic Relativistic Theory and Synge Energy

Equations (5.1) (or equivalently (5.14)) are valid for any kind of relativistic Euler fluid. We now focus our attention on a rarefied monatomic gas with specific constitutive equations. And we can obtain the equations of state by using the kinetic theory. The distribution function $f(x^\alpha, p^\alpha)$ $(\alpha = 0, 1, 2, 3)$ satisfies the Boltzmann-Chernikov equation [280, 282, 283]:

$$p^\alpha \partial_\alpha f = Q, \tag{5.18}$$

where p^α is the four-momentum of a molecule. We have $p_\alpha p^\alpha = (p^0)^2 - \mathbf{p}^2 = m^2c^2$ with $\mathbf{p}^2 = (p^1)^2 + (p^2)^2 + (p^3)^2$. The quantity Q expresses the effect due to the collision between molecules.

We can easily derive the Euler system (5.1) from (5.18) using the first five moments defined by

$$V^\alpha = mc \int_{\mathbb{R}^3} f p^\alpha \, d\mathbf{P}\,, \quad T^{\alpha\beta} = c \int_{\mathbb{R}^3} f p^\alpha p^\beta \, d\mathbf{P}, \tag{5.19}$$

where

$$d\mathbf{P} = \frac{dp^1\, dp^2\, dp^3}{p^0}.$$

In equilibrium, the distribution function reduces to the Jüttner distribution function:

$$f_J = \frac{n\gamma}{K_2(\gamma)} \frac{1}{4\pi m^3 c^3} e^{-\frac{\gamma}{mc^2} U_\beta p^\beta}, \tag{5.20}$$

where $K_n(\gamma)$ is the modified Bessel functions:

$$K_n(\gamma) = \int_0^{+\infty} \cosh ns \quad e^{-\gamma \cosh s}\, d\,s.$$

Inserting (5.20) into (5.19) and taking (5.2) into account, we obtain

$$\begin{aligned} p &= \frac{mnc^2}{\gamma} = \frac{k_B}{m} \rho T\,, \\ e &= \frac{nmc^2}{K_2(\gamma)} \left[K_3(\gamma) - \frac{1}{\gamma} K_2(\gamma) \right]. \end{aligned} \tag{5.21}$$

The energy expressed by (5.21) is called Synge energy [277, 283]. In the classical limit ($\gamma \to \infty$), by taking into account the expansion forms of the Bessel functions:

$$\begin{aligned} K_3(\gamma) &= \sqrt{\frac{\pi}{2}} \gamma^{-1/2} e^{-\gamma} \left[1 + \frac{35}{8} \frac{1}{\gamma} + o\left(\frac{1}{\gamma^2}\right) \right], \\ K_2(\gamma) &= \sqrt{\frac{\pi}{2}} \gamma^{-1/2} e^{-\gamma} \left[1 + \frac{15}{8} \frac{1}{\gamma} + o\left(\frac{1}{\gamma^2}\right) \right], \end{aligned} \tag{5.22}$$

the Synge energy converges to

$$\frac{e}{n} = mc^2 + m\varepsilon, \quad \text{with} \quad \varepsilon = \frac{3}{2}\frac{k_B}{m}T.$$

These expressions show that both classical and relativistic kinetic theories are valid only for rarefied monatomic gases.

Many papers are dedicated to the relativistic Euler system. We quote the book by Rezzolla and Zanotti [281] for numerical problems and other references.

Nonlinear waves and the Riemann problem for the system (5.14) with Synge energy was the subject in a paper by Ruggeri et al. [284]. The classical and ultra-relativistic limits of this Riemann problem were studied in [285]. Interesting properties of symmetry for the system were obtained in [286, 287]. In the ultra-relativistic case, there is an interesting paper by Freistühler [288] regarding the symmetric form. And, for the Riemann problem, we quote the classical paper by Smoller and Temple [289].

5.3.2 Principal Subsystem of Relativistic Euler Fluid

We can see that the system (5.14) is composed of two blocks of equations: the first block consists of Eqs. $(5.14)_{1,2}$ and the second block is the energy equation $(5.14)_3$. Now we are ready to discuss the principal subsystem (see Sect. 2.4) consisting of the first four equations in (5.14), i.e., the first block.

According to the theory of principal subsystem, we need to freeze the last component of the main field (5.17), that is, we require that it is constant:

$$\frac{\Gamma}{T} = \frac{1}{T^*} = \text{const.}$$

Unlike the classical case, this requirement does not imply that the temperature is homogeneously constant because the temperature now depends on the velocity:

$$T = T^*\Gamma = \frac{T^*}{\sqrt{1 - \frac{v^2}{c^2}}}. \tag{5.23}$$

This Eq. (5.23) looks like the transformation law of the temperature, which is firstly proposed by Ott [290] in 1963 and is revisited by Arzelés [291]. Here T^* has the meaning of the temperature in the rest frame. It is also well known that there is another famous formula proposed by Einstein and Planck in 1907, in which the square root is a multiplying factor. Numerous textbooks on the relativity theory refer the latter formula. There has been a longstanding debate concerning the legitimacy of the two formulas. However, some authors have yet another opinion that both formulas are wrong because T must be a scalar, and therefore it is an invariant being independent of the velocity. In particular, supporting this idea, Müller talks about "*Ott-Plank imbroglio*" in his book on a history of thermodynamics [4]. For a recent survey on this topic, see for example [292].

If we accept the idea that the temperature is a scalar being independent of the velocity, we conclude that we cannot have the "relativistic isothermal" principal subsystem. This conclusion probably reflects the fact that, in a relativistic case, the energy (temperature) and momentum are integrated into a single physical quantity, i.e., the energy-momentum tensor. Therefore, it is, in general, intrinsically impossible to deal with the energy and the momentum separately.

On the other hand, we have already appreciated the concept of the principal subsystem in RET. We have the nesting theories, each of which provides us a RET theory with different order of approximation. Therefore, it must be meaningful to study the principal subsystem in a relativistic context with the hope that we shed new light on the formula (5.23).

Substituting (5.23) into the first four equations of (5.14), we obtain the following principal subsystem with the field variables ρ and $\mathbf{v}$:

$$\begin{aligned} &\partial_t(\rho\Gamma) + \partial_i(\rho\Gamma v^i) = 0, \\ &\partial_t\left(\rho\Gamma^2 v^j\left(1+\frac{\bar{p}+\rho\bar{\varepsilon}}{\rho c^2}\right)\right) + \partial_i\left(\bar{p}\delta^{ij} + \rho\Gamma^2 v^i v^j\left(1+\frac{\bar{p}+\rho\bar{\varepsilon}}{\rho c^2}\right)\right) = 0, \\ &\bar{p} \equiv p(\rho, T^*\Gamma), \quad \bar{\varepsilon} = \varepsilon(\rho, T^*\Gamma). \end{aligned} \tag{5.24}$$

In this framework also pressure and internal energies are not scalars but depending of the velocity according the laws given by $(5.24)_3$.

This principal subsystem was firstly presented by Boillat [180] and was revisited in [163] to construct a relativistic Cucker Smale model as a particular case of mixture of gases. The system (5.24) can be viewed as the *relativistic mechanical Euler system*. And, as is expected, when $c \to \infty$ (non-relativistic limit), the system (5.24) converges to the isothermal principal subsystem (4.34). Obviously further studies on this subject are required.

5.4 Relativistic Dissipative Gas with 14 Fields

A pioneer of relativistic thermodynamics is Carl Eckart [152] who as early as 1940 established thermodynamics of irreversible processes (TIP) as seen in Sect. 1.3. Eckart's theory provides a relativistic counterpart of the Navier-Stokes and Fourier system. It gives the expression of the deviatoric stress and a generalization of Fourier's law of heat conduction. The latter permits a heat flux to be generated by an acceleration. That is, a temperature gradient can be equilibrated by a gravitational field. However, Eckart's theory has a serious draw-back: it leads to parabolic system of equations for the temperature and velocity. Therefore it predicts infinite pulse speeds in contradiction to the relativity principle.

As is well known, in the pioneering works of Müller [49] and Israel [50], the first tentative of a causal relativistic phenomenological theory was made. They proposed a system of equations of hyperbolic type such that wave speeds are finite in accordance with the relativity principle. This approach is based substantially on the idea of the modified Gibbs relation in a nonequilibrium state. This method has been fundamental for long time because of its simplicity. But a more refined analyses reveal that there remains arbitrariness in the theory. The assumptions adopted do not seem to be completely justified from a "rational" point of view. Moreover the equations so obtained suffer from a mathematical and physical problems. For details, see [53–55].

Because of these reasons, Liu et al. [55] (see also [25]) explored a new theory that starts with a few natural assumptions and uses only some universal principles. The theory obtained was motivated by the kinetic theory and is based on the following general assumptions:

- The objective of RET of relativistic fluids is the determination of the 14 fields:

$$V^{\alpha}(x^{\mu}) \quad - \quad \text{particle flux vector,}$$
$$T^{\alpha\beta}(x^{\mu}) \quad - \quad \text{energy momentum tensor.}$$

- For the determination of the 14 variables, we need the field equations, i.e., the conservation laws of particle number and energy-momentum (5.1), and the extended balance law of fluxes,

$$\partial_{\alpha} V^{\alpha} = 0, \qquad \partial_{\alpha} T^{\alpha\beta} = 0, \qquad \partial_{\alpha} A^{\alpha\beta\gamma} = I^{\beta\gamma}. \tag{5.25}$$

  It is assumed that $T^{\alpha\beta}$, $A^{\alpha\beta\gamma}$, and $I^{\beta\gamma}$ are completely symmetric tensors and moreover that

$$I^{\alpha}{}_{\alpha} = 0, \qquad A^{\alpha\beta}{}_{\beta} = c^2 V^{\alpha}. \tag{5.26}$$

- The constitutive equations are assumed to be of local type, i.e.,

$$A^{\alpha\beta\gamma} \equiv A^{\alpha\beta\gamma}(V^{\mu}, T^{\mu\nu}), \qquad I^{\alpha\beta} \equiv I^{\alpha\beta}(V^{\mu}, T^{\mu\nu}).$$

- The closure of RET is achieved by means of the universal principles of physics: *objectivity principle*, *entropy principle*, and *principle of causality and stability*.

The balance laws (5.25) and the trace conditions (5.26) (that guarantee that the number of the independent field equations is 14) are suggested by the kinetic theory. In fact, we have (5.19) and

$$A^{\alpha\beta\gamma} = \frac{c}{m} \int_{\mathbb{R}^3} f p^{\alpha} p^{\beta} p^{\gamma} \, d\mathbf{P}, \qquad I^{\beta\gamma} = \frac{c}{m} \int_{\mathbb{R}^3} Q \, p^{\beta} p^{\gamma} \, d\mathbf{P}.$$

By using the familiar nonequilibrium variables, we can rewrite the quantities V^α and $T^{\alpha\beta}$ as follows:

$$V^\alpha = nmU^\alpha,$$
$$T^{\alpha\beta} = t^{\langle\alpha\beta\rangle} + (p+\Pi)h^{\alpha\beta} + \frac{1}{c^2}(U^\alpha q^\beta + U^\beta q^\alpha) + \frac{e}{c^2}U^\alpha U^\beta,$$

where the deviatoric stress tensor $t^{\langle\alpha\beta\rangle}$ and the heat flux four-vectors q^α satisfy the orthogonality conditions (1.39).

The closure obtained by Liu et al. [55] gives the following expression for the triple tensor:

$$\begin{aligned}A^{\alpha\beta\gamma} = &(C_1 + C_2\Pi)U^\alpha U^\beta U^\gamma + \frac{c^2}{6}(nm - C_1 - C_2\Pi)\left(g^{\alpha\beta}U^\gamma + g^{\beta\gamma}U^\alpha + g^{\alpha\gamma}U^\beta\right)\\ &+ C_3\left(g^{\alpha\beta}q^\gamma + g^{\beta\gamma}q^\alpha + g^{\alpha\gamma}q^\beta\right) - \frac{6}{c^2}C_3\left(U^\alpha U^\beta q^\gamma + U^\gamma U^\beta q^\alpha + U^\alpha U^\gamma q^\beta\right)\\ &+ C_4\left(t^{\langle\alpha\beta\rangle}U^\gamma + t^{\langle\beta\gamma\rangle}U^\gamma + t^{\langle\alpha\gamma\rangle}U^\beta\right).\end{aligned}$$

The main field calculated in [55] is given by

$$u' = -\frac{g_r}{T} - \tau_0\Pi,$$
$$u'_\alpha = \frac{U_\alpha}{T} - \tau_1\Pi U_\alpha - \tau_2 q_\alpha,$$
$$u'_{\alpha\beta} = -\tau_3 t_{\langle\alpha\beta\rangle} - \tau_4\Pi h_{\alpha\beta} - \frac{1}{c^2}\tau_5(U_\alpha q_\beta + U_\beta q_\alpha) - \frac{3}{c^2}\tau_6\Pi U_\alpha u_\beta,$$

where we omit the explicit expressions of the coefficients $\tau_0, \ldots, \tau_6$ for simplicity. For more details, see [55] or [25].

If we impose the condition:

$$u'_{\alpha\beta} = 0,$$

we have

$$t_{\langle\alpha\beta\rangle} = 0, \qquad q_\alpha = 0, \qquad \Pi = 0.$$

This corresponds to the equilibrium principal subsystem, i.e., the Euler relativistic fluid system (5.1) and (5.2). The convexity of entropy is guaranteed at least in the limit of a neighborhood of equilibrium and the system is symmetric hyperbolic. In the papers [175, 176], it was proved that the system satisfies also the K-condition, and therefore global existence for smooth solutions is guaranteed provided that the initial data are sufficiently small.

Remark on the Einstein Equation From the viewpoint of RET, we can see the Einstein equation from a different viewpoint. In fact, in the classical approach, the number of balance laws is usually five. Only one component (for example, the internal energy) in the energy-momentum tensor is considered as a field variable, and the remaining ones are prescribed by the constitutive equations. While, in RET, it is assumed that all the components of the energy-momentum tensor are field variables. Therefore, in RET, the Einstein equation (as usual $R_{\mu\nu}$, R, and G are, respectively, the Ricci tensor, the scalar curvature, and Newton's gravitational constant):

$$R_{\mu\nu} - \frac{1}{2} g_{\mu\nu} R = -\frac{8\pi G}{c^4} T_{\mu\nu}$$

is regarded as a *universal* equation irrespective of the constitution of materials [293]. Concerning the qualitative analysis of Einstein equation, interested reader can refer to the monumental book by Y. Choquet Bruhat [294].

5.5 Relativistic Theory with Many Moments

The relativistic system of moment equations with the truncation index N is expressed as

$$\partial_\alpha A^{\alpha\alpha_1\cdots\alpha_n} = I^{\alpha_1\cdots\alpha_n} \quad \text{with} \quad n = 0, \cdots, N \tag{5.27}$$

with

$$A^{\alpha\alpha_1\cdots\alpha_n} = \frac{c}{m^{n-1}} \int_{\mathbb{R}^3} f\, p^\alpha p^{\alpha_1} \cdots p^{\alpha_n}\, d\mathbf{P}, \quad I^{\alpha_1\cdots\alpha_n} = \frac{c}{m^{n-1}} \int_{\mathbb{R}^3} Q\, p^{\alpha_1} \cdots p^{\alpha_n}\, d\mathbf{P}. \tag{5.28}$$

When $N = 1$ we have the relativistic Euler system (5.1), and when $N = 2$ we have the Liu-Müller-Ruggeri system (5.25). As in the classical case, we may use a multi-index, and rewrite (5.27) and (5.28) as follows:

$$\partial_\alpha A^{\alpha B} = I^B \quad \text{with} \quad B = 0, \cdots, N \tag{5.29}$$

with

$$A^{\alpha B} = \begin{cases} A^\alpha & \text{for } B = 0 \\ A^{\alpha\alpha_1\cdots\alpha_B} & \text{for } B \geqslant 1 \end{cases}, \quad I^B = \begin{cases} 0 & \text{for } B = 0, 1 \\ I^{\alpha_1\cdots\alpha_B} & \text{for } B \geqslant 2 \end{cases}.$$

We introduce a quantity:

$$h^\alpha = -k_B \int_{\mathbb{R}^3} p^\alpha \left\{ (s^2 - 1 + \ln f) f + s(1 - s f) \ln(1 - s f) \right\} d\mathbf{P},$$

where $s = 0, -1, 1$ correspond, respectively, to the non-degenerate gas, the Fermi gas, and the Bose gas. It is well known that we can obtain, from (5.18), the supplementary inequality:

$$\partial_\alpha h^\alpha = \Sigma \geqslant 0. \tag{5.30}$$

This expression suggests the balance of entropy if we identify h^α and Σ with the four-entropy vector and the entropy production, respectively. Then we can say that (5.30) expresses the H-theorem in the relativistic framework.

5.5.1 Closed System of Moment Equations

The closure of the system by using the entropy principle or the maximum entropy principle [252, 253] gives us the distribution function in the following form:

$$f_N = \frac{1}{e^{-\chi/k_B} + s},$$

where

$$\chi = \sum_{B=0}^{N} u'_B p^B = u'_B p^B \tag{5.31}$$

with

$$p^B = \begin{cases} 1 & \text{for } B = 0 \\ p^{\alpha_1} p^{\alpha_2} \ldots p^{\alpha_B} & \text{for } B \geqslant 1 \end{cases}, \quad u'_B = \begin{cases} u' & \text{for } B = 0 \\ u'_{\alpha_1} u'_{\alpha_2} \cdots u'_{\alpha_B} & \text{for } B \geqslant 1 \end{cases}.$$

The closed system in terms of the main field components u'_B is given by

$$H^{\alpha AB}(u'_C) \partial_\alpha u'_B = P_A(u'_C), \quad A = 0, 1 \ldots N \tag{5.32}$$

with the symmetric matrices:

$$H^{\alpha AB} = \int_{\mathbb{R}^3} \frac{\exp(-\chi/k_B)}{k_B \left(1 + s \exp(-\chi/k_B)\right)^2} p^\alpha p^A p^B d\mathbf{P}.$$

We note that $H^{\alpha AB}\zeta_\alpha$ is negative definite for any timelike vector ($\zeta_\alpha\zeta^\alpha > 0$). This implies that the system (5.32) is symmetric hyperbolic (in the sense of Friedrichs) and the Cauchy problem is well posed (local in time).

5.5.2 *Wave Propagation in an Equilibrium State and the Maximum Characteristic Velocity*

The wave surface $\phi(x^\alpha) = 0$ is a solution of the characteristic equation:

$$\det(H^{\alpha AB}\partial_\alpha\phi) = 0.$$

As a consequence, the four gradient $\partial_\alpha\phi$ cannot be timelike. Therefore the velocities of waves cannot exceed the velocity of light, i.e., $\lambda_{\max} \leqslant c$. When the number of equations increases, as was already shown, the maximum wave velocity cannot decrease. Now a question is: Does this velocity tend to c when N tends to infinity? The present subsection is devoted to the resolution of this question.

A thermodynamic equilibrium state is defined as the state for which the productions vanish and the entropy production Σ reaches its minimum value, i.e., zero. According with the general result, all the main-field variables except for the first five variables are zero. Then we have

$$u' = -\frac{g_r}{T}, \quad u'_\alpha = \frac{U_\alpha}{T}.$$

Therefore, for any truncation index N, the quantity χ, which is given by (5.31), reduces to $\chi/k_B = (-g_r + U_\alpha p^\alpha)/(k_B T)$ in an equilibrium state. And, in a case at rest where $U^i = 0,\ U^0 = c,\ p^0 = \sqrt{m^2c^2 + \mathbf{p}^2}$, we have

$$\frac{\chi}{k_B} = -a + \gamma\sqrt{1 + \frac{\mathbf{p}^2}{m^2c^2}},$$

where $a = g_r/k_B T$ and $\gamma = mc^2/k_B T$. In this case, the distribution function reduces to the Jüttner equilibrium distribution (5.20). We recall that a can assume all real values for a Fermi gas, while $a + \gamma > 0$ for a Bose gas.

The wave velocity λ in the normal direction to the wave front is given by an eigenvalue of

$$\det(H^{iAB}n_i - \lambda H^{0AB}) = 0, \tag{5.33}$$

where $\mathbf{n}$ is the unit normal vector to the wave front. The matrix in (5.33) is negative semidefinite for the maximum eigenvalue. Then, if we take $\mathbf{n} \equiv (1, 0, 0)$, the components of the matrix:

$$H^{1AB} - \lambda_{\max} H^{0AB} = \int_{\mathbb{R}^3} \frac{df}{d\chi} p^A p^B (p^1 - \lambda_{\max} p^0) d\mathbf{P}$$

satisfy the inequality in the form $a_{ii}a_{jj} - a_{ij}^2 \geqslant 0$ (see (4.56)). Therefore, choosing $p^A = (p^1)^N$, $p^B = (p^1)^{N-1}$, we have

$$\frac{\lambda_{\max}^2}{c^2} \int_{\mathbb{R}^3} \frac{df}{d\chi} \left(p^1\right)^{2N} d\mathbf{p} \int_{\mathbb{R}^3} \frac{df}{d\chi} \left(p^1\right)^{2(N-1)} d\mathbf{p} \geqslant \left(\int_{\mathbb{R}^3} \frac{df}{d\chi} \left(p^1\right)^{2N} \frac{d\mathbf{p}}{p^0} \right)^2,$$

since the integrals of odd functions vanish ($d\mathbf{p} = dp^1 dp^2 dp^3$).

By introducing the spherical coordinates, the above inequality, after some straightforward calculations, yields

$$\frac{\lambda_{\max}^2}{c^2} \geqslant \frac{2N-1}{2N+1} \frac{J_{N+1}^2}{I_N I_{N+1}},$$

where

$$I_n = \int_0^\infty \phi(r) r^{2n} dr, \quad J_n = \int_0^\infty \frac{\phi(r) r^{2n}}{\sqrt{1+r^2}} dr,$$

$$\phi(r) = \frac{e^{\psi(r)}}{(1 + s e^{\psi(r)})^2},$$

$$\psi(r) = \frac{\chi_e}{k_B} = -a - \gamma\sqrt{1+r^2}, \quad s = 0, \pm 1.$$

Therefore we conclude that: *For any type of gas including the degenerate gases of fermions and bosons, the largest wave velocity has the lower and upper bounds.*

In particular, the previous integrals in the case of non-degenerate gas can be expressed in terms of the Bessel function of second kind. Then we have

$$\frac{\lambda_{\max}^2}{c^2} \geqslant \frac{(2N-1)}{\gamma} \frac{K_{N+1}(\gamma)}{K_{N+2}(\gamma)}. \tag{5.34}$$

Using the recurrence relation for Bessel functions $K_{N+2} - K_N = 2(N+1)K_{N+1}/\gamma$, we obtain that, for $N \to \infty$, the limit value of $\lambda_{\max}$ in (5.34) is the light velocity c, since it has already been proved that it cannot be larger than c. Thus, *when the number of moments tends to infinity, the maximum velocity in equilibrium tends to the light velocity.*

This result can be proved also for degenerate gases because our proof is completely independent of the collision term Q of (5.18).

In the *ultrarelativistic case* with small γ, taking into account the properties of the Bessel functions for $\gamma \to 0$, we obtain the simple inequality:

$$\frac{\lambda_{\max}^2}{c^2} \geqslant \frac{(2N-1)}{2(N+1)}.$$

5.6 Classical Limit of Relativistic Moments and Optimal Choice of Moments

In the previous sections, we used generically the notation N as the truncation index. In this section, since we consider the classic limit of relativistic moments, we use $\bar{N}$ as the truncation index for classical moments and N as the one for relativistic moments.

In Sect. 4.2, we have studied RET with many moments in the classical framework. The Grad system with 13 independent moments $(F, F_i, F_{ij}, F_{lli})$ is a special case of (4.35) with $\bar{N} = 3$. We notice that, instead of the full triple tensor F_{ijk}, the trace of two indexes F_{llk} is adopted, and also notice that all the 13 moments have concrete physical meanings. As a next generalization of the Grad system, Kremer [295] studied a theory with $\bar{N} = 4$, that is, RET with 14 moments: $(F, F_i, F_{ij}, F_{lli}, F_{llkk})$. However, instead of going to $\bar{N} = 4$, we may study alternative RET with 20 moments by choosing all index-free tensors until $\bar{N} = 3$: $(F, F_i, F_{ij}, F_{ijk})$. Therefore, the generalization of the Grad system is obviously not unique.

One may naturally have the following question: Is there a guiding principle to choose the optimal truncated system of moments? In this section, we discuss an interesting viewpoint to answer this question.

Of course, there are some general constraints in choosing the moments: (1) The system of balance laws must be Galilean invariant. This implies several restrictions on the system. In particular, we cannot neglect some blocks of equations with respect to the tensorial order. (2) In the nonlinear closure, there is the requirement of the integrability of moments. As a necessary condition, $\bar{N}$ must be even (see Sect. 4.4). (3) The nesting theory requires that, for a given system of moments, any subsystems of moments must be principal subsystems. This implies, in particular, that the maximum characteristic velocity increases with the increase of the number of moments (see Sect. 4.3).

Pennisi and Ruggeri [296] found an intriguing viewpoint to the above-mentioned question. They proved that the classical limit of the relativistic moments may give a natural hierarchy of the classic moments. This hierarchy satisfies automatically all the constraints mentioned above. Furthermore it fixes the moments for a given truncation index N in a unique way. This assertion is valid for both cases of monatomic gas and polyatomic gas. For the case of monatomic gas, we have the following theorem [296]:

Theorem 5.1 *For a prescribed truncation index N, any integers $0 \leqslant s \leqslant N$ and multi-index $0 \leqslant A \leqslant N - s$, the relativistic moment system for a monatomic gas (5.27) (or equivalently (5.29)) converges, when $c \to \infty$, to the classical moments in the form:*

$$\partial_t F_A + \partial_i F_{iA} = P_A, \quad \textit{if} \quad s = 0, \quad 0 \leqslant A \leqslant N, \quad \textit{and}$$

$$\partial_t F_{j_1 j_1 \ldots j_s j_s i_1 i_2 \ldots i_{N-s}} + \partial_i F_{i j_1 j_1 \ldots j_s j_s i_1 i_2 \ldots i_{N-s}} = P_{j_1 j_1 \ldots j_s j_s i_1 i_2 \ldots i_{N-s}}, \quad \textit{with} \quad 1 \leqslant s \leqslant N.$$

where the moments F are given by (4.36) and the productions P given by (4.37). In particular, for $s = 0$ we have the F moments with all free indexes until index of truncation N and for $1 \leqslant s \leqslant N$ there is a single block of F moments with increasing number of pairs of contracted indexes.

Some interesting examples are given as follows:

- $N = 1$: Relativistic hierarchy of moment equations (Relativistic Euler system):

$$\partial_\alpha A^\alpha = 0, \quad \partial_\alpha A^{\alpha\alpha_1} = 0$$

converges to the classical hierarchy of moment equations:

$$(s = 0): \quad \partial_t F + \partial_i F_i = 0, \qquad \partial_t F_i + \partial_i F_{ij} = 0,$$
$$(s = 1): \quad \partial_t F_{j_1 j_1} + \partial_i F_{i j_1 j_1} = 0.$$

This is the classical Euler system ET$_5$.

- $N = 2$: Relativistic hierarchy of moment equations (Liu, Müller, and Ruggeri-model):

$$\partial_\alpha A^\alpha = 0, \quad \partial_\alpha A^{\alpha\alpha_1} = 0, \quad \partial_\alpha A^{\alpha\alpha_1\alpha_2} = I^{\alpha_1\alpha_2}$$

converges to the classical hierarchy of moment equations:

$$(s = 0): \quad \partial_t F + \partial_i F_i = 0, \qquad \partial_t F_{i_1} + \partial_i F_{i i_1} = 0, \qquad \partial_t F_{i_1 i_2} + \partial_i F_{i i_1 i_2} = P_{\langle i_1 i_2 \rangle},$$
$$(s = 1): \quad \partial_t F_{j_1 j_1 i_1} + \partial_i F_{i j_1 j_1 i_1} = P_{j_1 j_1 i_1},$$
$$(s = 2): \quad \partial_t F_{j_1 j_1 j_2 j_2} + \partial_i F_{i j_1 j_1 j_2 j_2} = P_{j_1 j_1 j_2 j_2}.$$

In this case, we have the Kremer theory ET$_{14}$ of monatomic gases [295]. Dreyer and Weiss [297] firstly proved this limit.

- $N = 3$: Relativistic hierarchy of moment equations:

$$\partial_\alpha A^\alpha = 0, \quad \partial_\alpha A^{\alpha\alpha_1} = 0, \quad \partial_\alpha A^{\alpha\alpha_1\alpha_2} = I^{\alpha_1\alpha_2}, \quad \partial_\alpha A^{\alpha\alpha_1\alpha_2\alpha_3} = I^{\alpha_1\alpha_2\alpha_3}$$

converges to the classical hierarchy of moment equations:

$$\begin{aligned}
(s=0): \quad & \partial_t F + \partial_i F_i = 0, \qquad \partial_t F_{i_1} + \partial_i F_{ii_1} = 0, \\
& \partial_t F_{i_1 i_2} + \partial_i F_{ii_1 i_2} = P_{\langle i_1 i_2\rangle}, \quad \partial_t F_{i_1 i_2 i_3} + \partial_i F_{ii_1 i_2 i_3} = P_{i_1 i_2 i_3} \\
(s=1): \quad & \partial_t F_{j_1 j_1 i_1 i_2} + \partial_i F_{ij_1 j_1 i_1 i_2} = P_{j_1 j_1 i_1 i_2}, \\
(s=2): \quad & \partial_t F_{j_1 j_1 j_2 j_2 i_1} + \partial_i F_{ij_1 j_1 j_2 j_2 i_1} = P_{j_1 j_1 j_2 j_2 i_1}, \\
(s=3): \quad & \partial_t F_{j_1 j_1 j_2 j_2 j_3 j_3} + \partial_i F_{ij_1 j_1 j_2 j_2 j_3 j_3} = P_{j_1 j_1 j_2 j_2 j_3 j_3}.
\end{aligned}$$

This is the ET_{30} theory of monatomic gases.

From the theorem and also from the previous examples, it is easy to see that the hierarchies of the classical moments thus constructed satisfy all the requirements:

1. The system is Galilean invariant.
2. The truncation index $\bar{N} = 2N$ obtained when $s = N$ is even. Therefore all classical moments obtained by the MEP must be integrable.
3. The classical system corresponding to the relativistic system with index N is a principal subsystem of the classical system corresponding to the relativistic system with index $N+1$.

For given N, the previous system is, according with the Definition 4.1, a $2N^{(N,N-1,\cdots,1)}$-system, and the total number of the components of the classical moments, $\mathcal{N}$, is determined by the following relation:

$$\mathcal{N} = \frac{1}{6}(N+1)(N+2)(2N+3). \tag{5.35}$$

Therefore the number $\mathcal{N}$ is not anymore arbitrary, but is prescribed by the formula (5.35). Some typical numbers are shown below:

$$\begin{pmatrix} N \\ \mathcal{N} \end{pmatrix} = \begin{pmatrix} 1 & 2 & 3 & 4 & 5 & 6 & 7 & 8 & 9 & 10 \\ 5 & 14 & 30 & 55 & 91 & 140 & 204 & 285 & 385 & 506 \end{pmatrix}.$$

Then the best choice of RET theories in monatomic gas is given by ET_5, ET_{14}, ET_{30}, ET_{55}, ET_{91}, etc. According with the Definition 4.1 the best choice of systems is as follows:

$$2^{(1)},\ 4^{(2,1)},\ 6^{(3,2,1)},\ 8^{(4,3,2,1)},\ 10^{(5,4,3,2,1)},\ 12^{(6,5,4,3,2,1)},\ 14^{(7,6,5,4,3,2,1)}$$

$$16^{(8,7,6,5,4,3,2,1)},\ 18^{(9,8,7,6,5,4,3,2,1)},\ 20^{(10,9,8,7,6,5,4,3,2,1)},\ \ldots$$

Part III
Rational Extended Thermodynamics of Rarefied Polyatomic Gas

Chapter 6
Macroscopic Theory of Rarefied Polyatomic Gas with 14 Fields

Check for updates

Abstract The objective of the present chapter is to explain the phenomenological RET theory (ET_{14}) of rarefied polyatomic gases with 14 independent fields, that is, mass density, velocity, temperature, shear stress, dynamic pressure, and heat flux. A gas is assumed to be *non-polytropic*, that is, the internal energy density of a gas has the nonlinear dependence on the temperature. The system of field equations has a binary hierarchy structure. We show that, by exploiting the universal principles explained in Sect. 2.2, the constitutive equations can be determined completely by the caloric and thermal equations of state as in the ET_{13} theory of rarefied monatomic gases. We obtain the closed system of field equations and the main field explicitly. The relationship between the ET_{14} theory and the Navier-Stokes and Fourier theory is discussed by using the Maxwellian iteration method. For completeness, as a special case, ET_{14} of rarefied polyatomic gases with *polytropic* caloric equation of state is also presented.

6.1 Previous Tentatives

As seen in the previous chapters, the Navier-Stokes and Fourier theory emerges from RET as a limiting case by carrying out the Maxwellian iteration [25, 104]. In this respect, the Navier-Stokes and Fourier theory can be seen as an approximation of RET where the relaxation times of dissipative fluxes (viscous stress and heat flux) are very small. We call this Navier-Stokes Fourier limit.

On the other hand, within its validity range, the classical Navier-Stokes and Fourier theory is applicable to any fluids that are not necessarily limited to rarefied gases nor to monatomic gases. Therefore, after the successful establishment of RET for rarefied monatomic gases, there appeared many studies of RET for rarefied polyatomic gases [298–300] and also for dense gases [100–102, 301–303]. In such gases in nonequilibrium, the dynamic pressure Π does not vanish identically and, in general, no simple relationship between the pressure p and the specific internal energy ε exists.

T. Ruggeri, M. Sugiyama, *Classical and Relativistic Rational Extended Thermodynamics of Gases*, https://doi.org/10.1007/978-3-030-59144-1_6

Previous authors tried to establish RET of dense gases by postulating single hierarchy structure similar to (1.27), but with 14 densities by introducing a new *fourth-rank tensorial density* such as F_{kkll} [102, 301, 303]. However, in their theories, the other important feature of the system of field equations—a flux in an equation becomes a density in the next equation—was abandoned. Because of this generality, constitutive equations could not be fully determined from the knowledge of the equilibrium properties of gases. There remain many phenomenological constants in the constitutive equations that are impossible to be evaluated experimentally or theoretically. Moreover, when the Navier-Stokes Fourier limit is taken, the postulation of the fourth-rank tensorial density seems to be not well justified because such a density does not have any straightforward counterpart in the Navier-Stokes and Fourier theory.

In this chapter, we explain the phenomenological RET theory (ET_{14}) of *rarefied polyatomic gases* with 14 independent fields, that is, mass density, velocity, temperature, shear stress, dynamic pressure, and heat flux. We will see that the ET_{14} theory overcomes the previous difficulties mentioned above.

The RET theory of dense gases will be studied in Chaps. 24 and 25.

6.2 Binary Hierarchy in RET of Rarefied Polyatomic Gas: Heuristic Viewpoint

It is interesting to reconsider the structure of the classical Navier-Stokes and Fourier system where, in addition to the usual conservation laws of mass, momentum, and energy (1.3), we have the constitutive equations (1.11). We observe that the constitutive equations can be rewritten in the following form [53]:

$$\begin{aligned}
&\frac{\partial}{\partial x_k}\left(v_i\delta_{jk} + v_j\delta_{ik} - \frac{2}{3}v_k\delta_{ij}\right) = \frac{\sigma_{\langle ij\rangle}}{\mu},\\
&\frac{\partial v_k}{\partial x_k} = -\frac{\Pi}{\nu},\\
&\frac{\partial T}{\partial x_k} = -\frac{q_k}{\kappa}.
\end{aligned} \tag{6.1}$$

The system composed of Eqs. (1.3) and (6.1) can be seen as a system of 14 equations for the 14 unknown fields: ρ, v_i, ε, $\sigma_{\langle ij\rangle}$, Π, and q_i. Its mathematical structure is in the form of balance type, but, in (6.1), we have no term with time derivative. Therefore the system is not hyperbolic but parabolic.

Taking seriously this viewpoint, we adopt the mathematical structure of balance laws in the following type [103]:

$$
\begin{aligned}
&\frac{\partial F}{\partial t}+\frac{\partial F_k}{\partial x_k}=0, \\
&\frac{\partial F_i}{\partial t}+\frac{\partial F_{ik}}{\partial x_k}=0, \\
&\frac{\partial F_{ij}}{\partial t}+\frac{\partial F_{ijk}}{\partial x_k}=P_{ij}, \qquad && \frac{\partial G_{ll}}{\partial t}+\frac{\partial G_{llk}}{\partial x_k}=0, \\
& && \frac{\partial G_{lli}}{\partial t}+\frac{\partial G_{llik}}{\partial x_k}=Q_{lli},
\end{aligned}
\tag{6.2}
$$

where F is the mass density, F_i is the momentum density, G_{ll} is the energy density, F_{ik} is the momentum flux, and G_{llk} is the energy flux. And F_{ijk} and G_{llik} are the fluxes of F_{ij} and G_{lli}, respectively, and P_{ij} and Q_{lli} are the productions with respect to F_{ij} and G_{lli}, respectively. In order to justify this structure, we admit that Eqs. (1.3) correspond to $(6.2)_{1,2,4}$ with the condition that F_{ll} is different from G_{ll} because, except for rarefied monatomic gases, no simple relation exists between the pressure and the internal energy. Equation $(6.2)_3$ can be split into the deviatoric and trace parts that have the mathematical structure of $(6.1)_{1,2}$ when the terms with time derivatives are neglected. While Eq. $(6.2)_5$ in a steady case has the mathematical structure of the type of Fourier law $(6.1)_3$.

To sum up, as a new RET theory, we adopt 14 independent fields:

mass density:	F,
momentum density:	F_i,
energy density:	G_{ii},
momentum flux:	F_{ij},
energy flux:	G_{lli}.

And we adopt the system (6.2) that is composed of two parallel hierarchical series: The one is the series starting from the mass and momentum balance equations (F-series) and the other is from the energy balance equation (G-series). In each series, the flux in one equation becomes the density in the next equation. This binary hierarchy will be justified from the kinetic-theoretical considerations in Chap. 7.

The results in this Chapter was firstly obtained in the paper by Arima et al. [103]. However, the analytical procedure here is different from the previous method. Furthermore, the considerations on dense gases in the paper [103] will not be explained in this book because, as shown in Part VI, Chaps. 24 and 25, we have now more satisfactory RET theories of dense gases.

6.3 RET 14-Field Theory of Rarefied Polyatomic Gas with Non-polytropic Equation of State

In this section, we present the phenomenological RET 14-field theory (ET_{14}) of rarefied polyatomic gases with non-polytropic caloric equation of state by imposing the universal physical principles explained in Sect. 2.2.

6.3.1 *Non-polytropic Gas*

For rarefied non-polytropic gases in equilibrium, the internal energy is a non-linear function of the temperature. The functional form $\varepsilon_E(T)$ is given usually by experimental data or by a statistical-mechanical calculation. For example, since $c_v(T) = d\varepsilon_E(T)/dT$ can be measured by experiments as a function of the temperature T, we can obtain ε by using the inverse relation:

$$\varepsilon = \varepsilon_E(T) = \frac{k_B}{m}\int_{T_0}^{T} \hat{c}_v(x)\,\mathrm{d}x, \tag{6.3}$$

where $\hat{c}_v$ is the dimensionless specific heat at constant volume:

$$\hat{c}_v = \frac{c_v}{k_B/m}, \tag{6.4}$$

and T_0 is an inessential reference temperature.

6.3.2 *Local Dependence of Unknown Quantities*

We proceed as in Sect. 4.1. In this case, the field $\mathbf{u}$ is given by

$$\mathbf{u} \equiv (F,\ F_i,\ F_{ij},\ G_{ll},\ G_{lli})^T.$$

The unknown fluxes and productions: $\psi \equiv \{F_{ijk},\ G_{llik},\ P_{ij},\ Q_{lli}\}$, together with the entropy density, entropy flux, and entropy production, depend locally on the field $\mathbf{u}$.

6.3.3 Exploitation of the Galilean Invariance

The matrices $\mathbf{X}$ and $\mathbf{A}^r$ of the Galilean invariance (see (2.49) and (2.51)) are given by

$$\mathbf{X} = \begin{pmatrix} 1 & 0 & 0 & 0 & 0 \\ v_i & \delta_i^{h_1} & 0 & 0 & 0 \\ v_i v_j & 2\delta_{(i}^{h_1} v_{j)} & \delta_i^{h_1}\delta_j^{h_2} & 0 & 0 \\ v^2 & 2v_i & 0 & 1 & 0 \\ v^2 v_i & 3v_{(l} v_l \delta_{i)}^{h_1} & 2\delta_i^{(h_1} v_{h_2)} & v_i & \delta_i^{h_1} \end{pmatrix} \tag{6.5}$$

and

$$\mathbf{A}^r = \begin{pmatrix} 0 & 0 & 0 & 0 & 0 \\ \delta_i^r & 0 & 0 & 0 & 0 \\ 0 & 2\delta_{(i}^{h_1}\delta_{j)}^r & 0 & 0 & 0 \\ 0 & 2\delta_{h_1}^r & 0 & 0 & 0 \\ 0 & 0 & 2\delta_i^{(h_1}\delta_r^{h_2)} & \delta_i^r & 0 \end{pmatrix}, \qquad (r = 1, 2, 3).$$

Therefore, the velocity dependence of field variables is completely prescribed as follows:

$$\mathbf{F}^0 = \begin{pmatrix} F \\ F_i \\ F_{ij} \\ G_{ll} \\ G_{lli} \end{pmatrix} = \begin{pmatrix} \rho \\ \rho v_i \\ \rho v_i v_j + \hat{F}_{ij} \\ \rho v^2 + \hat{G}_{ll} \\ \rho v^2 v_i + 2\hat{F}_{li} v_l + \hat{G}_{ll} v_i + \hat{G}_{lli} \end{pmatrix},$$

$$\boldsymbol{\Phi}^k = \begin{pmatrix} \Phi_k \\ \Phi_{ik} \\ \Phi_{ijk} \\ \Psi_k \\ \Psi_{ik} \end{pmatrix} = \begin{pmatrix} 0 \\ \hat{F}_{ik} \\ 2\hat{F}_{k(i} v_{j)} + \hat{F}_{ijk} \\ 2\hat{F}_{kl} v_l + \hat{G}_{llk} \\ 3\hat{F}_{k(i} v_l v_{l)} + 2\hat{F}_{ikl} v_l + \hat{G}_{llk} v_i + \hat{G}_{llik} \end{pmatrix}, \tag{6.6}$$

$$\mathbf{f} = \begin{pmatrix} 0 \\ 0 \\ P_{ij} \\ 0 \\ Q_{lli} \end{pmatrix} = \begin{pmatrix} 0 \\ 0 \\ \hat{P}_{ij} \\ 0 \\ 2\hat{P}_{li} v_l + \hat{Q}_{lli} \end{pmatrix}.$$

Since the balance equations of F, F_i, and G_{ll} represent the conservation laws of mass, momentum, and energy, the intrinsic quantities $\hat{F}_{ij}$, $\hat{G}_{ll}$, and $\hat{G}_{lli}$ have the following conventional meanings:

$$\begin{aligned}
&\text{stress tensor:} \quad t_{ij} = -\hat{F}_{ij} = -\mathscr{P}\,\delta_{ij} + \sigma_{\langle ij\rangle} = -\,(p+\Pi)\,\delta_{ij} + \sigma_{\langle ij\rangle},\\
&\text{specific internal energy:} \quad \varepsilon = \frac{1}{2\rho}\hat{G}_{ll},\\
&\text{heat flux:} \quad q_i = \frac{1}{2}\hat{G}_{lli},
\end{aligned} \tag{6.7}$$

where $\mathscr{P}$ and p are, respectively, the nonequilibrium and equilibrium pressures, and the so-called dynamic pressure Π is introduced by the difference between them:

$$\Pi = \mathscr{P} - p.$$

The equilibrium pressure for a rarefied gas is given by the thermal equation of state:

$$p = \frac{k_B}{m}\rho T. \tag{6.8}$$

Note that, in the case of monatomic gas, $\Pi = 0$, that is, $\mathscr{P} = p$ and the viscous stress $\sigma_{\langle ij\rangle}$ is symmetric and traceless. Therefore $\hat{P}_{ij}$ is symmetric and $\hat{F}_{ijk}$ is symmetric a priori only with respect to the first two indexes. Then the constitutive equations of ET$_{14}$ are expressed by the Galilean objective variables in the form :

$$\begin{aligned}
&\hat{\psi} \equiv \hat{\psi}(\rho,\,\varepsilon,\,\Pi,\,\sigma_{\langle ij\rangle},\,q_i),\\
&h \equiv h(\rho,\,\varepsilon,\,\Pi,\,\sigma_{\langle ij\rangle},\,q_i), \qquad \varphi_k \equiv \varphi_k(\rho,\,\varepsilon,\,\Pi,\,\sigma_{\langle ij\rangle},\,q_i).
\end{aligned}$$

6.3.4 Exploitation of the Entropy Principle

The entropy principle requires the following relations (see (2.11)):

$$\begin{aligned}
&dh = \lambda dF + \lambda_i dF_i + \lambda_{ij} dF_{ij} + \mu dG_{ll} + \mu_i dG_{lli},\\
&dh_k = \lambda dF_k + \lambda_i dF_{ik} + \lambda_{ij} dF_{ijk} + \mu dG_{llk} + \mu_i dG_{llik},\\
&\Sigma = \lambda_{ij} P_{ij} + \mu_i Q_{lli},
\end{aligned} \tag{6.9}$$

where $\mathbf{u}' \equiv (\lambda, \lambda_i, \lambda_{ij}, \mu, \mu_i)$ is the main field with the components of the Lagrange multipliers. Therefore, taking (2.57) into account, we have

$$
\begin{aligned}
dh &= \hat{\lambda} d\rho + \hat{\lambda}_{ij} d\hat{F}_{ij} + \hat{\mu} d\hat{G}_{ll} + \hat{\mu}_i d\hat{G}_{lli},\\
d\varphi_k &= \hat{\lambda}_i d\hat{F}_{ik} + \hat{\lambda}_{il} d\hat{F}_{ilk} + \hat{\mu} d\hat{G}_{llk} + \hat{\mu}_i d\hat{G}_{llik},\\
\Sigma &= \hat{\lambda}_{ij}\hat{P}_{ij} + \hat{\mu}_i \hat{Q}_{lli}.
\end{aligned}
\tag{6.10}
$$

The dependence of the Lagrange multipliers on the velocity is dictated by the relation (2.55). In the present case, from (6.5), we have the following relations:

$$
\begin{aligned}
\lambda &= \hat{\lambda} - \hat{\lambda}_i v_i + \hat{\lambda}_{ij} v_i v_j + \hat{\mu} v^2 - \hat{\mu}_i v^2 v_i,\\
\lambda_i &= \hat{\lambda}_i - 2\hat{\lambda}_{il} v_l - 2\hat{\mu} v_i + \hat{\mu}_i v^2 + 2\hat{\mu}_l v_i v_l,\\
\lambda_{ij} &= \hat{\lambda}_{ij} - 2\hat{\mu}_{(i} v_{j)},\\
\mu &= \hat{\mu} - \hat{\mu}_l v_l,\\
\mu_i &= \hat{\mu}_i.
\end{aligned}
\tag{6.11}
$$

The constraints (2.58), which come from the fact that the entropy density and the intrinsic entropy flux are independent of the velocity, are now expressed as

$$
\begin{aligned}
&\hat{\lambda}_i = -\frac{1}{\rho}\left(\hat{G}_{ll}\hat{\mu}_i + 2\hat{F}_{il}\hat{\mu}_l\right),\\
&\left(\hat{\lambda}\rho + \hat{\lambda}_{kj}\hat{F}_{kj} + \hat{\mu}\hat{G}_{ll} + \hat{\mu}_k \hat{G}_{llk} - h\right)\delta_{ir} + 2\hat{\lambda}_{rl}\hat{F}_{il} + 2\hat{\mu}\hat{F}_{ir} + 2\hat{\mu}_l \hat{F}_{lri} + \hat{\mu}_r \hat{G}_{lli} = 0.
\end{aligned}
\tag{6.12}
$$

6.3.4.1 Equilibrium State

According to the general Theorem 2.4, the productions $\hat{P}_{ii}$, $\hat{P}_{\langle ij\rangle}$, and $\hat{Q}_{lli}$ must vanish together with all Lagrange multipliers corresponding to the balance laws (see (2.45)):

$$
\hat{\lambda}^E_{ll} = 0, \qquad \hat{\lambda}^E_{\langle ij\rangle} = 0, \qquad \hat{\mu}^E_i = 0,
$$

where superscript E indicates that quantities are evaluated in an equilibrium state. Since the Lagrange multipliers $\hat{\lambda}_{ll}$, $\hat{\lambda}_{\langle ij\rangle}$, and $\hat{\mu}_i$ are non-zero only in nonequilibrium, these are called nonequilibrium variables. Taking the representation theorem for isotropic vectors and tensors into account, we can show the following fact: the above requirement for the intrinsic productions and the intrinsic nonequilibrium Lagrange multipliers in equilibrium implies that q_i, $\sigma_{\langle ij\rangle}$, and Π must also vanish in equilibrium.

6.3.4.2 Intrinsic Lagrange Multipliers

Let us rewrite the entropy density in the form (4.15) with the condition

$$k(\rho, \varepsilon, 0, 0, 0) = 0. \tag{6.13}$$

From $(6.10)_1$ and the Gibbs equation (1.8), we obtain

$$\begin{aligned}
\hat{\lambda} &= -\frac{g}{T} + k + \rho k_\rho - \varepsilon k_\varepsilon - (\rho p_\rho - \varepsilon p_\varepsilon) k_\Pi, \\
\hat{\mu} &= \frac{1}{2T} + \frac{1}{2} k_\varepsilon - \frac{1}{2} p_\varepsilon k_\Pi, \\
\hat{\lambda}_{ll} &= \rho k_\Pi, \\
\hat{\lambda}_{\langle ij \rangle} &= -\rho k_{\sigma_{\langle ij \rangle}}, \\
\hat{\mu}_i &= \frac{\rho}{2} k_{q_i}.
\end{aligned} \tag{6.14}$$

A subscript in the right-hand side of (6.14) stands for a partial differentiation with respect to the quantity, for example, $k_\Pi \equiv \partial k / \partial \Pi$. In an equilibrium state, we have the condition (6.13) and the following relations:

$$\hat{\lambda}^E = -\frac{g}{T}, \qquad \hat{\mu}^E = \frac{1}{2T}. \tag{6.15}$$

Remark 6.1 In a similar way in the case of rarefied monatomic gas, we have introduced the quantity T in (6.14) through the Gibbs equation (1.8). And, from (6.15), we notice that this quantity T is just the absolute temperature when the system is in equilibrium. Therefore we use the same symbol T in both nonequilibrium and equilibrium cases, and call it (local equilibrium) temperature. In the next Chap. 7, we will study the temperature T from a different point of view with the help of the kinetic theory. Similar argument is valid also for nonequilibrium pressure and nonequilibrium chemical potential, see Chap. 15.

6.3.4.3 Constitutive Equations Near Equilibrium

As usual in RET, we study processes not far from equilibrium, which, however, may be out of local equilibrium. Then we adopt the constitutive equations that are linear with respect to the nonequilibrium variables $\{\Pi, \sigma_{\langle ij \rangle}, q_i\}$. As we will see in the analysis below, in order to fix all the expansion coefficients, we need to represent the potentials h'^α and therefore the entropy and the entropy flux (see (2.13)) in an expansion form with respect to the nonequilibrium fields up to third order. Then, we have k and φ_k in the following forms:

$$
\begin{aligned}
k =& k_1 \Pi^2 + k_2 \sigma_{\langle ij\rangle}\sigma_{\langle ij\rangle} + k_3 q_i q_i + k_4 \Pi^3 + k_5 \Pi \sigma_{\langle ij\rangle}\sigma_{\langle ij\rangle} + k_6 \sigma_{\langle ij\rangle}\sigma_{\langle ni\rangle}\sigma_{\langle jn\rangle} \\
&+ k_7 \Pi q_i q_i + k_8 \sigma_{\langle ij\rangle} q_i q_j + O(4), \qquad (6.16)\\
\varphi_k =& (\beta_1 + \beta_2 \Pi) q_k + \beta_3 \sigma_{\langle ki\rangle} q_i + O(3), \qquad (6.17)
\end{aligned}
$$

where coefficients $k_1, \cdots, k_8$ and $\beta_1, \beta_2, \beta_3$ are functions of ρ and ε. Note that, in (6.16), there is no linear term with respect to Π. This is justified by the convexity condition as seen in Sect. 6.3.9.

We also represent fluxes as follows:

$$
\begin{aligned}
\hat{F}_{ijk} &= f_1 q_k \delta_{ij} + f_2 q_{\langle i}\delta_{j\rangle k} + O(2),\\
\hat{G}_{llij} &= (g_1 + g_2 \Pi)\delta_{ij} - g_3 \sigma_{\langle ij\rangle} + O(2),
\end{aligned}
\qquad (6.18)
$$

where f_1, f_2, g_1, g_2, and g_3 are functions of ρ and ε.

From (6.16), we obtain

$$
\begin{aligned}
dk =& \Pi^2 dk_1 + \sigma_{\langle ij\rangle}\sigma_{\langle ij\rangle} dk_2 + q_i q_i dk_3 + \Pi^3 dk_4 \\
&+ \Pi \sigma_{\langle ij\rangle}\sigma_{\langle ij\rangle} dk_5 + \sigma_{\langle ij\rangle}\sigma_{\langle in\rangle}\sigma_{\langle nj\rangle} dk_6 + \Pi q_i q_i dk_7 + q_i q_j \sigma_{\langle ij\rangle} dk_8 \\
&+ \left(2k_1 \Pi + 3k_4 \Pi^2 + k_5 \sigma_{\langle ij\rangle}\sigma_{\langle ij\rangle} + k_7 q_i q_i\right) d\Pi \\
&+ \left(2k_2 \sigma_{\langle ij\rangle} + 2k_5 \Pi \sigma_{\langle ij\rangle} + 3k_6 \sigma_{\langle n\langle i\rangle}\sigma_{\langle j\rangle n\rangle} + k_8 q_{\langle i} q_{j\rangle}\right) d\sigma_{\langle ij\rangle} \\
&+ \left(2k_3 q_i + 2k_7 \Pi q_i + 2k_8 \sigma_{\langle ij\rangle} q_j\right) dq_i . \qquad (6.19)
\end{aligned}
$$

Therefore, substituting (6.19) into (6.14), we obtain the Lagrange multipliers as follows:

$$
\begin{aligned}
\hat{\lambda} =& -\frac{g}{T} - 2k_1 \left(\rho p_\rho - \varepsilon p_\varepsilon\right) \Pi + \left\{k_1 + \rho k_{1\rho} - \varepsilon k_{1\varepsilon} - 3k_4 \left(\rho p_\rho - \varepsilon p_\varepsilon\right)\right\} \Pi^2 \\
&+ \left\{k_2 + \rho k_{2\rho} - \varepsilon k_{2\varepsilon} - k_5 \left(\rho p_\rho - \varepsilon p_\varepsilon\right)\right\} \sigma_{\langle ij\rangle}\sigma_{\langle ij\rangle} \\
&+ \left\{k_3 + \rho k_{3\rho} - \varepsilon k_{3\varepsilon} - k_7 \left(\rho p_\rho - \varepsilon p_\varepsilon\right)\right\} q_i q_i + O(3),\\
\hat{\mu} =& \frac{1}{2T} - p_\varepsilon k_1 \Pi + \frac{1}{2}\left(k_{1\varepsilon} - 3p_\varepsilon k_4\right) \Pi^2 + \frac{1}{2}\left(k_{2\varepsilon} - p_\varepsilon k_5\right) \sigma_{\langle ij\rangle}\sigma_{\langle ij\rangle} \\
&+ \frac{1}{2}\left(k_{3\varepsilon} - p_\varepsilon k_7\right) q_i q_i + O(3),\\
\hat{\lambda}_{ll} =& 2\rho k_1 \Pi + 3\rho k_4 \Pi^2 + \rho k_5 \sigma_{\langle ij\rangle}\sigma_{\langle ij\rangle} + \rho k_7 q_i q_i + O(3),\\
\hat{\lambda}_{\langle ij\rangle} =& -2\rho k_2 \sigma_{\langle ij\rangle} - 2\rho k_5 \Pi \sigma_{\langle ij\rangle} - 3\rho k_6 \sigma_{\langle n\langle i\rangle}\sigma_{\langle j\rangle n\rangle} - \rho k_8 q_{\langle i} q_{j\rangle} + O(3),\\
\hat{\mu}_i =& \rho k_3 q_i + \rho k_7 \Pi q_i + \rho k_8 \sigma_{\langle ij\rangle} q_j + O(3).
\end{aligned}
\qquad (6.20)
$$

Let us study the coefficients $k_1, \cdots, k_8$ by using the constraint (6.12)$_2$ and the expressions of the Lagrange multipliers (6.20). We decompose the constraint (6.12)$_2$ into the trace part, traceless part, and antisymmetric part as follows:

$$\begin{aligned}
&h = \rho\hat{\lambda} + \frac{5}{3}(p+\Pi)\hat{\lambda}_{ll} - \frac{5}{3}\hat{\lambda}_{\langle ij\rangle}\sigma_{\langle ij\rangle} + 2(\rho\varepsilon + p + \Pi)\hat{\mu} + \frac{8}{3}q_i\hat{\mu}_i + \frac{2}{3}\hat{\mu}_l\hat{F}_{lpp},\\
&2\hat{\lambda}_{l\langle r}\hat{F}_{i\rangle l} - 2\hat{\mu}\sigma_{\langle ir\rangle} + 2\hat{\mu}_l\hat{F}_{l\langle ri\rangle} + 2\hat{\mu}_{\langle r}q_{i\rangle} = 0,\\
&2\hat{\lambda}_{l[r}\hat{F}_{i]l} + 2\hat{\mu}_{[r}q_{i]} + 2\hat{\mu}_l\hat{F}_{l[ri]} = 0,
\end{aligned} \tag{6.21}$$

where the brackets [ ] stand for the antisymmetry with respect to the suffixes inside it. From (6.21)$_{1,2}$, we obtain

$$\begin{aligned}
k_1 &= -\frac{1}{2\rho T\Gamma},\\
k_2 &= -\frac{1}{4\rho Tp},\\
k_4 &= -\frac{1}{3\Gamma}\left\{\left(\frac{10}{3} - 2\frac{p_\varepsilon}{\rho}\right)k_1 + \rho k_{1\rho} + \frac{p}{\rho}k_{1\varepsilon}\right\},\\
k_5 &= -\frac{1}{\Gamma}\left\{\frac{10}{3}k_2 + \rho k_{2\rho} + \frac{p}{\rho}k_{2\varepsilon}\right\}\\
&= -\frac{1}{p}\left\{\left(\frac{1}{3} - \frac{p_\varepsilon}{2\rho}\right)k_1 + k_2\right\},\\
k_6 &= -\frac{1}{6\rho Tp^2},\\
k_7 &= -\frac{1}{\Gamma}\left\{\frac{2}{3}(4+3f_1)k_3 + \rho k_{3\rho} + \frac{p}{\rho}k_{3\varepsilon}\right\},\\
k_8 &= \frac{1}{p}(1+f_2)k_3,
\end{aligned} \tag{6.22}$$

where Γ is a function of ρ and ε defined by

$$\Gamma = \frac{5}{3}p - \rho p_\rho - \frac{p}{\rho}p_\varepsilon = \frac{5}{3}p - \rho\left(\frac{\partial p}{\partial \rho}\right)_s.$$

We notice that the coefficients k_1, k_2, k_4, k_5, k_6 are expressed explicitly as functions of ρ and ε since we know the functional forms of the equilibrium thermodynamic quantities, while, for k_7 and k_8, we need the expressions of k_3, f_1, and f_2. In the below we will obtain also their expressions.

From the antisymmetric part of the constraint (6.21)$_3$, we obtain

$$\hat{F}_{lri} = \hat{F}_{lir}.$$

Since $\hat{F}_{lri}$ is symmetric with respect to l and r, we understand that $\hat{F}_{lri}$ is symmetric with respect to all indexes. Therefore instead of the expression (6.18)$_1$, we can adopt

$$\hat{F}_{ijk} = \hat{F}_{\langle ijk\rangle} + \frac{3}{5}\hat{F}_{ll(i}\delta_{jk)}$$

with

$$\hat{F}_{\langle ijk\rangle} = O(2), \quad \text{and} \quad \hat{F}_{lli} = 3f_1 q_i + O(2).$$

This concludes, from (6.18)$_1$, that

$$f_1 = \frac{5}{6}f_2. \tag{6.23}$$

Next we analyze the entropy flux. From (6.17), we have

$$\begin{aligned} d\varphi_k =& q_k d\beta_1 + \Pi q_k d\beta_2 + \sigma_{\langle ki\rangle} q_i d\beta_3 + \beta_2 q_k d\Pi + \beta_3 q_i d\sigma_{\langle ki\rangle} \\ &+ \left\{(\beta_1 + \beta_2\Pi)\delta_{ki} + \beta_3\sigma_{\langle ki\rangle}\right\} dq_i. \end{aligned}$$

On the other hand, $d\varphi_k$ is expressed as (6.10)$_2$ with (6.12)$_1$. By comparing these expressions with each other, we obtain the following relations:

$$\begin{aligned}
\beta_{1_\rho} &= \rho\left\{g_{1_\rho} - 2\left(\varepsilon + \frac{p}{\rho}\right)p_\rho\right\}k_3, \\
\beta_{2_\rho} &= \rho\left\{2k_1 f_{1_\rho} + k_3\left(g_{2_\rho} - 2\frac{p_\rho}{\rho}\right) + k_7\left(g_{1_\rho} - 2\left(\varepsilon + \frac{p}{\rho}\right)p_\rho\right)\right\}, \\
\beta_{3_\rho} &= -\rho\left\{2k_2 f_{2_\rho} + k_3\left(g_{3_\rho} - 2\frac{p_\rho}{\rho}\right) - k_8\left(g_{1_\rho} - 2\left(\varepsilon + \frac{p}{\rho}\right)p_\rho\right)\right\}, \\
\beta_{1_\varepsilon} &= \rho\left\{g_{1_\varepsilon} - 2\left(\varepsilon + \frac{p}{\rho}\right)p_\varepsilon\right\}k_3, \\
\beta_{2_\varepsilon} &= \rho\left\{2k_1 f_{1_\varepsilon} + k_3\left(g_{2_\varepsilon} - 2\frac{p_\varepsilon}{\rho}\right) + k_7\left(g_{1_\varepsilon} - 2\left(\varepsilon + \frac{p}{\rho}\right)p_\varepsilon\right)\right\}, \\
\beta_{3_\varepsilon} &= -\rho\left\{2k_2 f_{2_\varepsilon} + k_3\left(g_{3_\varepsilon} - 2\frac{p_\varepsilon}{\rho}\right) - k_8\left(g_{1_\varepsilon} - 2\left(\varepsilon + \frac{p}{\rho}\right)p_\varepsilon\right)\right\}, \\
\beta_1 &= \frac{1}{T},
\end{aligned}$$

$$\beta_2 = \rho k_3 \left\{ g_2 - 2\left(\varepsilon + \frac{p}{\rho}\right)\right\},$$
$$\beta_3 = -\rho k_3 \left\{ g_3 - 2\left(\varepsilon + \frac{p}{\rho}\right)\right\}, \tag{6.24}$$

and

$$f_1 = \frac{\beta_2}{2\rho k_1} + \frac{p_\varepsilon}{\rho},$$
$$f_2 = -\frac{\beta_3}{2\rho k_2}.$$

From (6.24)$_{1,2,\cdots,7}$ and (6.22), we obtain the relations:

$$\begin{aligned}
dg_1 &= -\frac{1}{\rho k_3 T^2} dT + 2\left(\varepsilon + \frac{p}{\rho}\right) dp,\\
d(T\Gamma g_2) &= \Gamma\left\{2\left(\varepsilon + \frac{p}{\rho}\right) + \frac{k_7}{T\rho k_3^2}\right\} dT + 2\left(\varepsilon + \frac{p}{\rho}\right) T d\Gamma + \frac{2T\Gamma}{\rho} dp + \frac{1}{\rho k_3} d\left(\frac{p_\varepsilon}{\rho}\right),\\
d(pg_3) &= 2\left(\varepsilon + 2\frac{p}{\rho}\right) dp + \frac{1}{T}\left\{pg_3 - \frac{1}{T\rho k_3} - 2p\left(\varepsilon + \frac{p}{\rho}\right)\right\} dT.
\end{aligned} \tag{6.25}$$

Finally, using the integrability condition for g_1 and the relation:

$$g_2 = \frac{1}{\Gamma}\left\{\frac{5}{3} p g_3 + \frac{p_\varepsilon}{\rho^2 T k_3} + 2\left(\varepsilon + \frac{p}{\rho}\right)\left(\Gamma - \frac{5}{3} p\right)\right\},$$

which comes from (6.23), we can prove that (6.25)$_2$ and (6.25)$_3$ are equivalent.

In conclusion, if we know the explicit expressions of g_1 and g_3 as functions of ρ and ε, we can also determine explicitly the expressions of all the other coefficients. For determining g_1 and g_3, it is more convenient to change independent variables from $\{\rho, \varepsilon\}$ to $\{\rho, T\}$.

6.3.4.4 Determination of the Expansion Coefficients in Terms of ρ and T

In order to use the equations of states (6.3) and (6.8) more explicitly, we adopt $\{\rho, T\}$ as independent variables. Since ε depends only on T, we have the following relations for a generic quantity f:

$$f_\rho \equiv \left(\frac{\partial f}{\partial \rho}\right)_\varepsilon = \left(\frac{\partial f}{\partial \rho}\right)_T, \qquad f_\varepsilon \equiv \left(\frac{\partial f}{\partial \varepsilon}\right)_\rho = \frac{1}{c_v}\left(\frac{\partial f}{\partial T}\right)_\rho. \tag{6.26}$$

For example, we have the expression of the quantity Γ:

$$\Gamma = \frac{2\hat{c}_v - 3}{3\hat{c}_v} p.$$

Using (6.26), we can easily rewrite all the previous results in terms of ρ and T. For simplicity, we only rewrite the following relations, which come from $(6.25)_{1,3}$:

$$\begin{aligned} \left(\frac{\partial g_1}{\partial \rho}\right)_T &= 2\left(\varepsilon + \frac{p}{\rho}\right)\left(\frac{\partial p}{\partial \rho}\right)_T, \\ \left(\frac{\partial\, p g_3}{\partial \rho}\right)_T &= 2\left(\varepsilon + 2\frac{p}{\rho}\right)\left(\frac{\partial p}{\partial \rho}\right)_T, \end{aligned} \tag{6.27}$$

and other relations are not explicitly rewritten here.

We now determine the explicit expression of g_1 by integrating $(6.27)_1$ with respect to ρ. The arbitrary function of T arose in the integration vanishes because g_1 must become null at any value of T when $\rho \to 0$. Therefore we obtain the relation:

$$g_1 = 2\left(\varepsilon + \frac{k_B}{m}T\right)p.$$

In a similar way, we have

$$g_3 = 2\left(\varepsilon + 2\frac{k_B}{m}T\right).$$

Then we have the expressions of all the other coefficients. In particular, we have

$$f_2 = \frac{2}{1+\hat{c}_v}, \quad g_2 = g_3, \quad k_3 = -\frac{1}{2p^2T(1+\hat{c}_v)}.$$

6.3.5 Linear Constitutive Equations

Using the expressions of the expansion coefficients obtained above, we have the linear constitutive equations:

$$\begin{aligned} \hat{F}_{ijk} &= \frac{1}{\hat{c}_V + 1}\left(q_i\delta_{jk} + q_j\delta_{ki} + q_k\delta_{ij}\right), \\ \hat{G}_{llij} &= 2\left(\varepsilon + \frac{k_B}{m}T\right)p\delta_{ij} + 2\left(\varepsilon + 2\frac{k_B}{m}T\right)\left(\Pi\delta_{ij} - \sigma_{\langle ij\rangle}\right). \end{aligned} \tag{6.28}$$

6.3.6 Productions

The productions are also expanded with respect to the nonequilibrium variables $\{\Pi,\ \sigma_{\langle ij\rangle},\ q_i\}$ around an equilibrium state. In the linear approximation, we have

$$\hat{P}_{ll} = -\frac{3}{\tau_\Pi}\Pi, \quad \hat{P}_{\langle ij\rangle} = \frac{1}{\tau_\sigma}\sigma_{\langle ij\rangle}, \quad \hat{Q}_{lli} = -\frac{2}{\tau_q}q_i, \tag{6.29}$$

where τ_Π, τ_σ, and τ_q are relaxation times as will be shown in Sect. 6.3.10.

6.3.7 Main Field

From (6.20), the intrinsic Lagrange multipliers in the linear approximation are obtained as follows:

$$\begin{aligned}
\hat{\lambda} &= -\frac{g}{T} - \frac{3(\varepsilon - \hat{c}_v \frac{k_B}{m}T)}{pT(2\hat{c}_v - 3)}\Pi, & \hat{\lambda}_i &= \frac{\rho(\varepsilon + \frac{k_B}{m}T)}{p^2T(\hat{c}_v + 1)}q_i, \\
\hat{\lambda}_{\langle ij\rangle} &= \frac{1}{2pT}\sigma_{\langle ij\rangle}, & \hat{\lambda}_{ll} &= -\frac{3\hat{c}_v}{pT(2\hat{c}_v - 3)}\Pi, \\
\hat{\mu} &= \frac{1}{2T} + \frac{3}{2pT(2\hat{c}_v - 3)}\Pi, & \hat{\mu}_i &= -\frac{\rho}{2p^2T(\hat{c}_v + 1)}q_i.
\end{aligned} \tag{6.30}$$

From (6.11), we can evaluate the main field components, by using which the differential system becomes symmetric hyperbolic.

6.3.8 Entropy Density, Entropy Flux, and Entropy Production

From (6.16) and (6.17) with the coefficients determined above, the entropy density and entropy flux are expressed as

$$\begin{aligned}
h &= \rho s - \frac{3\hat{c}_v}{2(2\hat{c}_v - 3)pT}\Pi^2 - \frac{1}{4pT}\sigma_{\langle ij\rangle}\sigma_{\langle ij\rangle} - \frac{\rho}{2p^2T\left(1 + \hat{c}_v\right)}q_iq_i + O(3), \\
\varphi_k &= \frac{1}{T}q_k - \frac{1}{pT\left(1 + \hat{c}_v\right)}\Pi q_k + \frac{1}{pT\left(1 + \hat{c}_v\right)}q_i\sigma_{\langle ik\rangle} + O(3).
\end{aligned}$$

From (6.9), (6.30), and (6.29), the entropy production is given by

$$\Sigma = \frac{3\hat{c}_v}{pT(2\hat{c}_v - 3)\tau_\Pi}\Pi^2 + \frac{1}{2pT\tau_\sigma}\sigma_{\langle ij\rangle}\sigma_{\langle ij\rangle} + \frac{\rho}{p^2T(1+\hat{c}_v)\tau_q}q_iq_i \geqslant 0.$$

Then we have three conditions:

$$(2\hat{c}_v - 3)\tau_\Pi > 0, \quad \tau_\sigma > 0, \quad (1+\hat{c}_v)\tau_q > 0. \tag{6.31}$$

6.3.9 Convexity of the Entropy Density

The convexity condition (2.59) evaluated in an equilibrium state becomes

$$\hat{Q}_E = \hat{Q}_{5E} - \frac{3\hat{c}_v}{2(2\hat{c}_v - 3)pT}(\delta\Pi)^2 - \frac{1}{4pT}\delta\sigma_{\langle ij\rangle}\delta\sigma_{\langle ij\rangle} - \frac{\rho}{2p^2T(1+\hat{c}_v)}\delta q_i\delta q_i < 0,$$

where $\hat{Q}_{5E}$ is the corresponding quantity for Euler fluids.

From the condition $\hat{Q}_{5E} < 0$, we obtain the usual thermodynamic inequalities, that is, the positivity of the heat capacity and of the compressibility (see (2.31)). Therefore, the convexity condition is satisfied when

$$2\hat{c}_v - 3 > 0. \tag{6.32}$$

This condition is always satisfied for a polyatomic gas since $\hat{c}_v > 3/2$. (As will be discussed in Sect. 6.5, we should be careful in the case of monatomic gas with $\hat{c}_v = 3/2$.) From (6.31) and (6.32), we obtain the following inequalities:

$$\tau_\Pi > 0, \quad \tau_\sigma > 0, \quad \tau_q > 0.$$

The potentials h'^α (2.56) in the present case become:

$$h' = -\frac{p}{T} - \frac{3\hat{c}_v\Pi^2}{2(2\hat{c}_v - 3)pT} - \frac{\sigma_{\langle ij\rangle}\sigma_{\langle ij\rangle}}{4pT} - \frac{\rho q_iq_i}{2(\hat{c}_v + 1)p^2T},$$

$$h'^i = h'v_i + \frac{1}{(2\hat{c}_v + 1)pT}\left(\frac{\hat{c}_v + 3}{2\hat{c}_v - 3}\Pi q_i + q_j\sigma_{\langle ij\rangle}\right).$$

Since the convexity of the entropy density is proved, the potential h' is also a convex function with respect to the main field components.

6.3.10 Closed System of Field Equations

We can close the system of field equations by substituting the constitutive equations (6.28) and (6.29) into (6.2) with (6.6) and (6.7). Then the closed system with the independent fields $\{\rho, v_i, T, \Pi, \sigma_{\langle ik\rangle}, q_i\}$ is given by

$$
\begin{aligned}
&\frac{\partial \rho}{\partial t}+\frac{\partial}{\partial x_k}(\rho v_k)=0,\\
&\frac{\partial \rho v_i}{\partial t}+\frac{\partial}{\partial x_k}\left\{\rho v_i v_k+(p+\Pi)\delta_{ik}-\sigma_{\langle ik\rangle}\right\}=0,\\
&\frac{\partial}{\partial t}(\rho v^2+2\rho\varepsilon)+\frac{\partial}{\partial x_k}\left\{\rho v^2 v_k+2(\rho\varepsilon+p+\Pi)v_k-2\sigma_{\langle kl\rangle}v_l+2q_k\right\}=0,\\
&\frac{\partial}{\partial t}\left\{\rho v_i v_j+(p+\Pi)\delta_{ij}-\sigma_{\langle ij\rangle}\right\}+\\
&\quad+\frac{\partial}{\partial x_k}\left\{\rho v_i v_j v_k+(p+\Pi)(v_i\delta_{jk}+v_j\delta_{ki}+v_k\delta_{ij})-\sigma_{\langle ij\rangle}v_k-\sigma_{\langle jk\rangle}v_i-\sigma_{\langle ki\rangle}v_j\right.\\
&\qquad\left.+\frac{1}{1+\hat{c}_v}(q_i\delta_{jk}+q_j\delta_{ki}+q_k\delta_{ij})\right\}=-\frac{\Pi\delta_{ij}}{\tau_\Pi}+\frac{\sigma_{\langle ij\rangle}}{\tau_\sigma},\qquad(6.33)\\
&\frac{\partial}{\partial t}\left\{\rho v^2 v_i+2(\rho\varepsilon+p+\Pi)v_i-2\sigma_{\langle li\rangle}v_l+2q_i\right\}+\\
&\quad+\frac{\partial}{\partial x_k}\left\{\rho v^2 v_i v_k+2\rho\varepsilon v_i v_k+(p+\Pi)(v^2\delta_{ik}+4v_i v_k)\right.\\
&\qquad-\sigma_{\langle ik\rangle}v^2-2\sigma_{\langle li\rangle}v_l v_k-2\sigma_{\langle lk\rangle}v_l v_i+\frac{2(2+\hat{c}_v)}{1+\hat{c}_v}(q_i v_k+q_k v_i)+\frac{2}{1+\hat{c}_v}q_l v_l\delta_{ik}\\
&\qquad\left.+2p\left(\varepsilon+\frac{k_B}{m}T\right)\delta_{ik}+2\left(\varepsilon+2\frac{k_B}{m}T\right)(\Pi\delta_{ik}-\sigma_{\langle ik\rangle})\right\}\\
&\quad=-2\frac{q_i}{\tau_q}-2\left(\frac{\Pi\delta_{ij}}{\tau_\Pi}-\frac{\sigma_{\langle ij\rangle}}{\tau_\sigma}\right)v_j.
\end{aligned}
$$

By using the material derivative, the closed system above can be rewritten as

$$
\begin{aligned}
&\dot{\rho}+\rho\frac{\partial v_k}{\partial x_k}=0,\\
&\rho\dot{v}_i+\frac{\partial\left\{(p+\Pi)\delta_{ij}-\sigma_{\langle ij\rangle}\right\}}{\partial x_j}=0,\\
&\rho\frac{k_B}{m}\hat{c}_v\dot{T}+(p+\Pi)\frac{\partial v_k}{\partial x_k}-\frac{\partial v_i}{\partial x_k}\sigma_{\langle ik\rangle}+\frac{\partial q_k}{\partial x_k}=0,\\
&\dot{\Pi}+\left(\frac{2\hat{c}_v-3}{3\hat{c}_v}p+\frac{5\hat{c}_v-3}{3\hat{c}_v}\Pi\right)\frac{\partial v_k}{\partial x_k}-\frac{2\hat{c}_v-3}{3\hat{c}_v}\frac{\partial v_{\langle i}}{\partial x_{k\rangle}}\sigma_{\langle ik\rangle}
\end{aligned}
$$

$$-\frac{5}{3}\frac{1}{\left(1+\hat{c}_v\right)^2}\frac{d\hat{c}_v}{dT}q_k\frac{\partial T}{\partial x_k}+\frac{2\hat{c}_v-3}{3\hat{c}_v(1+\hat{c}_v)}\frac{\partial q_k}{\partial x_k}=-\frac{1}{\tau_\Pi}\Pi,$$

$$\dot{\sigma}_{\langle ij\rangle}-2(p+\Pi)\frac{\partial v_{\langle i}}{\partial x_{j\rangle}}+\sigma_{\langle ij\rangle}\frac{\partial v_k}{\partial x_k}+2\frac{\partial v_{\langle i}}{\partial x_k}\sigma_{\langle j\rangle k\rangle} \tag{6.34}$$

$$+\frac{2}{\left(1+\hat{c}_v\right)^2}\frac{d\hat{c}_v}{dT}\frac{\partial T}{\partial x_k}q_{\langle i}\delta_{j\rangle k}-\frac{2}{1+\hat{c}_v}\frac{\partial q_{\langle i}}{\partial x_{j\rangle}}=-\frac{1}{\tau_\sigma}\sigma_{\langle ij\rangle},$$

$$\dot{q}_i+\frac{2+\hat{c}_v}{1+\hat{c}_v}q_i\frac{\partial v_k}{\partial x_k}+\frac{1}{1+\hat{c}_v}q_k\frac{\partial v_k}{\partial x_i}+\frac{2+\hat{c}_v}{1+\hat{c}_v}q_k\frac{\partial v_i}{\partial x_k}$$

$$+\frac{k_B}{m}\left\{\left(1+\hat{c}_v\right)p\delta_{ki}+\left(2+\hat{c}_v\right)\left(\Pi\delta_{ki}-\sigma_{\langle ki\rangle}\right)\right\}\frac{\partial T}{\partial x_k}$$

$$-\frac{1}{\rho}(\Pi\delta_{ki}-\sigma_{\langle ki\rangle})\frac{\partial p}{\partial x_k}+\frac{1}{\rho}\left\{(p-\Pi)\delta_{ki}+\sigma_{\langle ki\rangle}\right\}\frac{\partial}{\partial x_l}(\Pi\delta_{kl}-\sigma_{\langle kl\rangle})=-\frac{1}{\tau_q}q_i.$$

We now understand that τ_Π, τ_σ, and τ_q can be regarded as the relaxation times of the dynamic pressure, the shear stress, and the heat flux, respectively.

6.3.11 *Relationship Between RET Theory and Navier-Stokes and Fourier Theory*

We carry out the Maxwellian iteration (for more detail see Chap. 33, Sects. 33.2.1 and 33.3) for the system (6.34): The first iterates $\Pi^{(1)}$, $\sigma^{(1)}_{\langle ij\rangle}$, and $q^{(1)}_i$ are obtained by the substitution of the 0th iterates $\Pi^{(0)}=0$, $\sigma^{(0)}_{\langle ij\rangle}=0$, and $q^{(0)}_i=0$ into the left-hand side of $(6.34)_{4,5,6}$. Then we obtain

$$\Pi^{(1)}=-\frac{2\hat{c}_v-3}{3\hat{c}_v}p\tau_\Pi\frac{\partial v_k}{\partial x_k},\quad \sigma^{(1)}_{\langle ij\rangle}=2p\tau_\sigma\frac{\partial v_{\langle i}}{\partial x_{j\rangle}},\quad q^{(1)}_i=-\frac{k_B}{m}(1+\hat{c}_v)p\tau_q\frac{\partial T}{\partial x_i}. \tag{6.35}$$

On the other hand, we have the laws of Navier-Stokes and Fourier expressed by (1.11). Their comparison reveals that

$$\nu=\frac{2\hat{c}_v-3}{3\hat{c}_v}p\tau_\Pi,\qquad \mu=p\tau_\sigma,\qquad \kappa=\frac{k_B}{m}(1+\hat{c}_v)p\tau_q. \tag{6.36}$$

We can therefore estimate the values of the relaxation times τ_Π, τ_σ, and τ_q from the experimental data on the phenomenological coefficients ν, μ, and κ.

In conclusion, the system can be certainly closed by the universal principles, provided that we know the thermal and caloric equations of state, and the viscosity and heat conductivity coefficients. The consistency of the present theory to the theory derived from the molecular ET will be shown in Chap. 7.

6.4 ET$_{14}$ Theory of Polytropic Gas

In this section, we show the closed system of the ET$_{14}$ theory in the case of polytropic gas. We will show also its hyperbolicity region.

6.4.1 *Closed System of the ET$_{14}$ Theory*

For a rarefied polytropic gas, the thermal and caloric equations of state are given in (1.32). Therefore the dimensionless specific heat is expressed as

$$\hat{c}_v = \frac{D}{2}.$$

The system (6.33) in this case is already given in (1.33). The convexity condition (6.32) is given by

$$D > 3. \tag{6.37}$$

Using the material derivative, we have the system with independent fields $\{\rho, v_i, T, \Pi, \sigma_{\langle ik\rangle}, q_i\}$:

$$\dot{\rho} + \rho \frac{\partial v_k}{\partial x_k} = 0,$$

$$\rho \dot{v}_i + \frac{\partial \left\{(p+\Pi)\delta_{ij} - \sigma_{\langle ij\rangle}\right\}}{\partial x_j} = 0,$$

$$\rho \frac{D}{2}\frac{k_B}{m}\dot{T} + (p+\Pi)\frac{\partial v_k}{\partial x_k} - \frac{\partial v_i}{\partial x_k}\sigma_{\langle ik\rangle} + \frac{\partial q_k}{\partial x_k} = 0,$$

$$\dot{\Pi} + \left(\frac{2(D-3)}{3D}p + \frac{5D-6}{3D}\Pi\right)\frac{\partial v_k}{\partial x_k} - \frac{2(D-3)}{3D}\frac{\partial v_{\langle i}}{\partial x_{k\rangle}}\sigma_{\langle ik\rangle} + \frac{4(D-3)}{3D(D+2)}\frac{\partial q_k}{\partial x_k} = -\frac{1}{\tau_\Pi}\Pi,$$

$$\dot{\sigma}_{\langle ij\rangle} - 2p\frac{\partial v_{\langle i}}{\partial x_{j\rangle}} + \sigma_{\langle ij\rangle}\frac{\partial v_k}{\partial x_k} - 2\Pi\frac{\partial v_{\langle i}}{\partial x_{j\rangle}} + 2\frac{\partial v_{\langle i}}{\partial x_k}\sigma_{\langle j\rangle k} - \frac{4}{D+2}\frac{\partial q_{\langle i}}{\partial x_{j\rangle}} = -\frac{1}{\tau_\sigma}\sigma_{\langle ij\rangle},$$

$$\dot{q}_i + \frac{D+4}{D+2}q_i\frac{\partial v_k}{\partial x_k} + \frac{2}{D+2}q_k\frac{\partial v_k}{\partial x_i} + \frac{D+4}{D+2}q_k\frac{\partial v_i}{\partial x_k}$$

$$
-\frac{k_B}{m}T\frac{\partial p}{\partial x_i}+\frac{1}{2}\frac{k_B}{m}\left((D+2)\,p\delta_{ki}+(D+4)\left(\Pi\delta_{ki}-\sigma_{\langle ki\rangle}\right)\right)\frac{\partial T}{\partial x_k}
$$

$$
+\frac{1}{\rho}\left((p-\Pi)\delta_{ki}+\sigma_{\langle ki\rangle}\right)\frac{\partial}{\partial x_l}\left((p+\Pi)\delta_{kl}-\sigma_{\langle kl\rangle}\right)=-\frac{1}{\tau_q}q_i, \tag{6.38}
$$

Then relaxation times are related to the phenomenological coefficients:

$$
\mu=p\tau_\sigma, \qquad \nu=\frac{2(D-3)}{3D}p\tau_\Pi, \qquad \kappa=\frac{D+2}{2}\frac{p^2}{\rho T}\tau_q. \tag{6.39}
$$

In this case the entropy density and the entropy flux are expressed as

$$
h=h^E-\frac{3D}{4(D-3)pT}\Pi^2-\frac{1}{4pT}\sigma_{\langle ij\rangle}\sigma_{\langle ij\rangle}-\frac{\rho}{(D+2)p^2T}q_iq_i+O(3),
$$

$$
\varphi_k=\frac{1}{T}q_k-\frac{2}{(D+2)pT}\Pi q_k+\frac{2}{(D+2)pT}q_i\sigma_{\langle ik\rangle}+O(3).
$$

The convexity condition is always satisfied for a polyatomic gas since $D>3$. The intrinsic Lagrange multipliers (6.30) are given by

$$
\begin{aligned}
\hat{\lambda}&=-\frac{g}{T}, & \hat{\lambda}_i&=\frac{1}{pT}q_i,\\
\hat{\lambda}_{\langle ij\rangle}&=\frac{1}{2pT}\sigma_{\langle ij\rangle}, & \hat{\lambda}_{ll}&=-\frac{3D}{2pT(D-3)}\Pi,\\
\hat{\mu}&=\frac{1}{2T}+\frac{3}{2pT(D-3)}\Pi, & \hat{\mu}_i&=-\frac{\rho}{p^2T(D+2)}q_i.
\end{aligned}
$$

A remarkable point is that $\hat{\lambda}$ is independent of the dynamic pressure and coincides with the equilibrium value.

We will see in Chap. 7 that the system (6.38) is completely equivalent to the one obtained via MEP. Moreover, for diatomic gases with $D=5$, field equations (6.38) coincide with those derived by Mallinger using a Grad procedure [304] except for the expressions of the relaxation times.

6.4.2 *Hyperbolicity Region in the Case of Polyatomic Gas*

In a similar way as in the case of monatomic gas explained in Sect. 4.5, it is possible to construct the hyperbolicity region associated with the system (6.34). In the present case, the nonequilibrium characteristic polynomial is evaluated as [305]:

$$
\mathscr{P}(\tilde{\lambda})=-6D\frac{k_B}{m}\sqrt{\frac{k_B}{m}T}\,p^2\tilde{\lambda}^2\left(a_4\tilde{\lambda}^4+a_3\tilde{\lambda}^3+a_2\tilde{\lambda}^2+a_1\tilde{\lambda}+a_0\right) \tag{6.40}
$$

with (λ denotes now the characteristic velocity)

$$\tilde{\lambda} = \frac{\lambda - v_1}{c_s}, \quad \sigma_{11} = \sigma_{\langle 11 \rangle} - \Pi, \quad \tilde{\sigma}_{11} = \frac{\sigma_{11}}{\rho c_s^2}, \quad \tilde{q}_1 = \frac{q_1}{\rho c_s^3}, \quad c_s = \sqrt{\frac{k_B}{m} T}$$

and

$$a_4 = 1, \quad a_3 = 0, \quad a_2 = \frac{2((2D^2 + 4D + 3)\sigma_{11} - (2D^2 + 7D))}{D(D+2)},$$
$$a_1 = -\frac{12(D+5)\hat{q}_1}{(D+2)^2}, \quad a_0 = \frac{3((D+4)\sigma_{11}^2 - (2D+4)\sigma_{11} + D + 2)}{D+2}.$$

At equilibrium, the roots of the characteristic polynomial are given by

$$\tilde{\lambda}^E_{1,2} = 0, \qquad \tilde{\lambda}^E_{3,4} = \pm\sqrt{\frac{2D + 7 - \sqrt{37 + D(16 + D)}}{D+2}},$$
$$\tilde{\lambda}^E_{5,6} = \pm\sqrt{\frac{2D + 7 + \sqrt{37 + D(16 + D)}}{D+2}}.$$

Since all the characteristic velocities are real and it is possible to verify that there exists a basis of eigenvectors, the hyperbolicity requirement is satisfied at equilibrium. By continuity, it is possible to show that the hyperbolicity property is valid at least in a neighborhood of the equilibrium state.

Due to the decomposition of the characteristic polynomial (6.40), an explicit expression of the boundary equation is obtained through (4.72):

$$\tilde{q}_1 = \pm \frac{(D+2)^2 (2a_2 + \sqrt{a_2^2 + 12a_0})\sqrt{-a_2 + \sqrt{a_2^2 + 12a_0}}}{18\sqrt{6}(D+5)} \quad \forall\ \tilde{\sigma}_{11} \le s_{11}, \tag{6.41}$$

where the maximum value of the 11-component of the dimensionless viscous tensor reads[1]

$$s_{11} = \tilde{\sigma}_{11}|_{\max} = \frac{b_1 - \sqrt{6 b_2}}{(D+1)^2 (D-4)^2}, \tag{6.42}$$

where

$$b_1 = D^4 + 8D^3 + 17D^2 + 28D, \quad b_2 = D^7 + 26D^6 + 85D^5 + 50D^4 - 32D^3 + 32D^2.$$

[1]The ostensible singularity for $D = 4$ in (6.42) can be easily overcome. In fact, it holds that $\lim_{D \to 4} s_{11} = 13/16$.

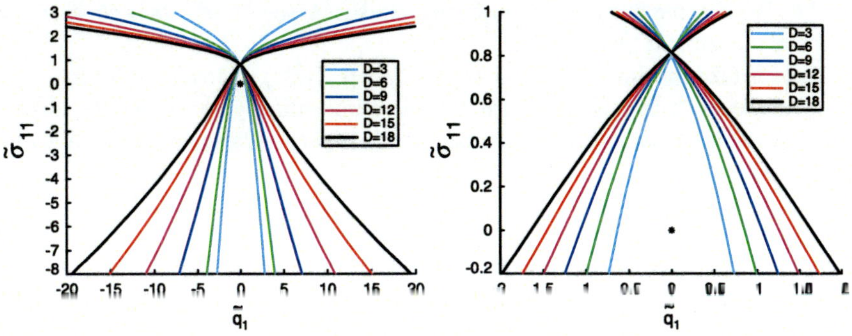

Fig. 6.1 Hyperbolicity region in the case of one-dimensional field variables, for different values of the number of degrees of freedom, D. On the right, a zoom of the figure on the left in the neighborhood of the equilibrium point (denoted by a star)

The hyperbolicity region, bounded by two curves described in (6.41), is described in Fig. 6.1 for different values of D, the star indicates the local equilibrium point.

From Fig. 6.1, we can see clearly that the domain of hyperbolicity and the radius around the equilibrium point increase with D and that the domain is minimal in the case of monatomic gas.

In [305], the hyperbolicity region for the second-order theory was also evaluated.

6.5 Singular Limit from Polyatomic Gas to Monatomic Gas

In this section, we show that a rarefied monatomic gas, where there exists no dynamic pressure, can be identified as a singular limit of a rarefied polyatomic gas. We confine our discussion within the singular limit from ET_{14} of a rarefied polyatomic gas to ET_{13} of a rarefied monatomic gas [112]. The gas is assumed to be polytropic.

Let us discuss the limiting process in the system (6.38) from polyatomic to monatomic rarefied gases when we let D approach 3 from above, where D is assumed to be a continuous variable. The limit is singular in the sense that the system for a rarefied polyatomic gas with 14 independent fields needs to converge to the system with only 13 independent fields for a rarefied monatomic gas.

The singularity can be seen also by the inequalities required for the symmetric hyperbolicity in equilibrium (6.37). The condition is obviously satisfied only for polyatomic gases with $D > 3$, and the case of monatomic gases with $D = 3$ is not admissible. Therefore only the limit of D toward 3 from above is meaningful.

In the present case, the relaxation times τ_σ, τ_Π, and τ_q are, respectively, related to the shear viscosity μ, the bulk viscosity ν, and the heat conductivity κ as seen in

(6.39). We observe that the bulk viscosity vanishes when $D \to 3$ as is consistent in rarefied monatomic gases.

Let us take the limit $D \to 3$ of the system (1.33), that is, the limit from a polyatomic gas to a monatomic gas. Then we immediately notice that the limit of the system exists, but it still has 14 equations. However, we also notice the following three points (I)–(III):

(I) The limit of the equation for Π, $(1.33)_5$, is given by

$$\dot{\Pi} = -\left(\frac{1}{\tau_\Pi} + \frac{\partial v_k}{\partial x_k}\right)\Pi, \quad \longleftrightarrow \quad \left(\frac{\Pi}{\rho}\right)^{\bullet} = -\frac{1}{\tau_\Pi}\frac{\Pi}{\rho}. \tag{6.43}$$

This is the first-order quasi-linear partial differential equation with respect to the dynamic pressure Π.

As the limit case is the case of monatomic gas, the initial condition for (6.43) must be compatible with monatomic gases. We therefore should impose the following initial condition:

$$\Pi(\mathbf{x}, 0) = 0. \tag{6.44}$$

Then, by assuming the uniqueness of the solution, the only possible solution of (6.43) under the initial condition (6.44) is given by

$$\Pi(\mathbf{x}, t) = 0 \qquad \forall t > 0. \tag{6.45}$$

Therefore the dynamic pressure in a monatomic gas vanishes identically for any time once we impose the initial condition (6.44).

(II) If we insert the solution (6.45) into the remaining equations in (6.38) with $D \to 3$, we confirm that the resulting equations are the same as the ones of ET_{13} for rarefied monatomic gases. This means that the limiting system with 14 equations is essentially equivalent to that of ET_{13} of monatomic gases.

(III) The violation of the symmetric hyperbolicity condition ($D > 3$) disappears because this inequality comes out from the non-vanishing dynamic pressure. Therefore the condition for the symmetric hyperbolicity is the same as the one in the ET_{13} theory, which is always satisfied near equilibrium.

In the reference [112], two illustrative numerical results in the process of the singular limit, that is, the linear waves and the shock waves are shown in order to grasp the asymptotic behavior of the physical quantities, in particular, of the dynamic pressure.

To sum up, we may conclude that the ET_{14} theory is applicable also to rarefied monatomic gases if we impose the initial or boundary condition of zero dynamic pressure.

Chapter 7
Molecular ET of Rarefied Polyatomic Gas with 14 Fields

Abstract In this chapter, we prove, in the case of rarefied polyatomic gas with non-polytropic caloric equation of state, that the maximum entropy principle (MEP) gives the same closed system as that obtained in the phenomenological RET theory with 14 fields discussed in Chap. 6. The key idea for the study of polyatomic gases with MEP is to adopt a generalized distribution function. This is the function not only of the usual variables, i.e., time, position, and velocity, but also of an extra variable that connects with the internal degrees of freedom of a constituent molecule. We can obtain the same binary hierarchy introduced in Chap. 6 in a natural way: the one is the momentum-type, F-hierarchy, and the other is the energy-type, G-hierarchy. The extra variable plays a role in the G-hierarchy. The coincidence of the systems between the phenomenological RET theory and the molecular ET theory in the case of more fields will be proved in Chap. 9.

7.1 Kinetic Theory of Rarefied Polyatomic Gas

As seen in the previous chapters, one of the most important results in the study of rarefied monatomic gases is that the closure of the system obtained by the macroscopic principles of RET, i.e., phenomenological RET closure is exactly the same as both the closures of MEP and of the Grad procedure. Therefore, a question naturally arises: Is this remarkable result still valid in the case of polyatomic gas? To answer this question is the main purpose of this chapter. We will see below that the answer is affirmative.

The kinetic theory of rarefied polyatomic gases was developed by Borgnakke and Larsen [113], and successively its mathematical aspect, in particular, was reconsidered by Bourgat et al. [114]. Their idea is to adopt the generalized distribution function f that depends not only on $(t, \mathbf{x}, \mathbf{c})$ defined in $([0, \infty) \times \mathbb{R}^3 \times \mathbb{R}^3)$ but also on an additional continuous variable I defined in $[0, \infty)$. The quantity I represents the energy of the internal mode of a constituent molecule.

The time-evolution of the distribution function is determined by the Boltzmann equation (1.38) as in the case of monatomic gas. However, the collision term

T. Ruggeri, M. Sugiyama, *Classical and Relativistic Rational Extended Thermodynamics of Gases*, https://doi.org/10.1007/978-3-030-59144-1_7

$Q(f)$ takes into account the influence of the molecular internal degrees of freedom through the collision cross-section.

7.2 Equilibrium Distribution Function of a Rarefied Polyatomic Gas and Caloric Equation of State

It is well known that, in the case of monatomic gas, the Maxwellian distribution function $f^{(M)}$ given in (1.29) plays an important role, and that it can be derived under the condition that it maximizes the entropy. Therefore, in the case of polyatomic gas, we also need to know firstly the equilibrium distribution function. For this aim, Pavić et al. [115] determined the equilibrium distribution function for a polyatomic gas $f^{(E)}$ considering the first five equations in the binary hierarchy (6.2):

$$\begin{aligned} &\frac{\partial F}{\partial t} + \frac{\partial F_k}{\partial x_k} = 0, \\ &\frac{\partial F_i}{\partial t} + \frac{\partial F_{ik}}{\partial x_k} = 0, \\ &\frac{\partial G_{ll}}{\partial t} + \frac{\partial G_{llk}}{\partial x_k} = 0. \end{aligned} \tag{7.1}$$

These equations express the conservation laws of mass, momentum, and energy. Therefore all production terms vanish identically.

The system (7.1) comes from the existence of the collision invariants:

$$m\left(1,\ c_i,\ c^2 + 2\frac{I}{m}\right)^T.$$

In fact, the densities (F, F_i, G_{ll}) corresponding to the invariants in a generic state described by the distribution function $f(t, \mathbf{x}, \mathbf{c}, I)$ are given by

$$\begin{pmatrix} F \\ F_i \\ G_{ll} \end{pmatrix} = \begin{pmatrix} \rho \\ \rho v_i \\ \rho v^2 + 2\rho\varepsilon \end{pmatrix} = \int_{\mathbb{R}^3}\int_0^\infty m \begin{pmatrix} 1 \\ c_i \\ c^2 + \frac{2}{m}I \end{pmatrix} f(t, \mathbf{x}, \mathbf{c}, I)\varphi(I)\, dI\, d\mathbf{c}, \tag{7.2}$$

where the non-negative measure (state density) $\varphi(I)$ is introduced so as to recover the classical caloric equation of state for polyatomic gases in equilibrium.

Whereas, the fluxes F_{ik} and G_{llk} that a priori are not known are given by

$$F_{ik} = \int_{\mathbb{R}^3}\int_0^\infty m c_i c_k f\, \varphi(I)\, dI\, d\mathbf{c} \tag{7.3}$$

and

$$G_{llk} = \int_{\mathbb{R}^3} \int_0^\infty m(c^2 + 2I/m) c_k f \, \varphi(I) \, dI \, d\mathbf{c}. \tag{7.4}$$

Introducing the peculiar velocity $\mathbf{C}$, we have the following relations from (7.2):

$$\begin{pmatrix} \rho \\ 0_i \\ 2\rho\varepsilon \end{pmatrix} = \int_{\mathbb{R}^3} \int_0^\infty m \begin{pmatrix} 1 \\ C_i \\ C^2 + \frac{2}{m} I \end{pmatrix} f(t, \mathbf{x}, \mathbf{C}, I) \varphi(I) \, dI \, d\mathbf{C}. \tag{7.5}$$

Note that the internal energy density $\rho\varepsilon$ can be divided into the energy density due to the translational motion $\rho\varepsilon^K$ and the energy density related to the internal degrees of freedom $\rho\varepsilon^I$:

$$\begin{aligned} \rho\varepsilon &= \rho\varepsilon^K + \rho\varepsilon^I; \\ \rho\varepsilon^K &= \int_{\mathbb{R}^3} \int_0^\infty \frac{1}{2} m C^2 f(t, \mathbf{x}, \mathbf{C}, I) \varphi(I) \, dI \, d\mathbf{C}, \\ \rho\varepsilon^I &= \int_{\mathbb{R}^3} \int_0^\infty I f(t, \mathbf{x}, \mathbf{C}, I) \varphi(I) \, dI \, d\mathbf{C}. \end{aligned} \tag{7.6}$$

For later use, we define here also the entropy density h in a generic state with f as follows:

$$h = -k_B \int_{\mathbb{R}^3} \int_0^\infty f \log f \varphi(I) \, dI \, d\mathbf{c}. \tag{7.7}$$

7.2.1 *Equilibrium Distribution Function*

The equilibrium distribution function $f^{(E)}(t, \mathbf{x}, \mathbf{C}, I)$ is determined in such a way that it is the function with the maximum value of the entropy h under the constraints (7.2), or equivalently, due to the Galilean invariance, under the constraints (7.5). Then we have the following theorem [115]:

Theorem 7.1 *The equilibrium distribution function that maximizes the entropy (7.7) under the constraints (7.5) has the form:*

$$f^{(E)} = \frac{\rho}{m\,A(T)} \left(\frac{m}{2\pi k_B T} \right)^{3/2} \exp\left\{ -\frac{1}{k_B T} \left(\frac{1}{2} m C^2 + I \right) \right\} = f^{(M)} f^{(I)}, \tag{7.8}$$

where $f^{(M)}$ denotes the Maxwellian distribution (1.29), $f^{(I)}$ is the distribution function for the internal mode I:

$$f^{(I)} = \frac{1}{A(T)} \exp\left(-\frac{I}{k_B T}\right),$$

and $A(T)$ is its normalization factor:

$$A(T) = \int_0^\infty \exp\left(-\frac{z}{k_B T}\right) \varphi(z)\, dz \tag{7.9}$$

in such a way that

$$\int_0^\infty f^{(I)} \varphi(I)\, dI = 1. \tag{7.10}$$

Proof We use the method of Lagrange multipliers. Denoting the multipliers $(\lambda, \lambda_i, \mu)$, we define the functional of the distribution function f:

$$\begin{aligned}
\mathscr{L} = & -\int_{\mathbb{R}^3}\int_0^\infty k_B f \log f \varphi(I)\, dI\, d\mathbf{c} + \lambda\left(\rho - \int_{\mathbb{R}^3}\int_0^\infty m f \varphi(I)\, dI\, d\mathbf{c}\right) \\
& + \lambda_i \left(\rho v_i - \int_{\mathbb{R}^3}\int_0^\infty m f c_i \varphi(I)\, dI\, d\mathbf{c}\right) \\
& + \mu\left(\rho v^2 + 2\rho\varepsilon - \int_{\mathbb{R}^3}\int_0^\infty m\left(c^2 + 2\frac{I}{m}\right) f \varphi(I)\, dI d\mathbf{c}\right).
\end{aligned}$$

Now we look for the condition that this functional $\mathscr{L}$ becomes maximum with respect to f. However, as the macroscopic quantities are fixed, the above functional can be substituted by the following one:

$$\mathscr{L} = \int_{\mathbb{R}^3}\int_0^\infty \left\{-k_B f \log f - m\left[\lambda + \lambda_i c_i + \mu\left(c^2 + 2\frac{I}{m}\right)\right] f\right\} \varphi(I)\, dI\, d\mathbf{c}. \tag{7.11}$$

Furthermore, since $\mathscr{L}$ is a scalar, it must retain the value in the case of zero hydrodynamic velocity $\mathbf{v} = \mathbf{0}$ due to the Galilean invariance. Therefore, it can be rewritten as follows:

$$\mathscr{L} = \int_{\mathbb{R}^3}\int_0^\infty \left\{-k_B f \log f - m\left[\hat{\lambda} + \hat{\lambda}_i C_i + \hat{\mu}\left(C^2 + 2\frac{I}{m}\right)\right] f\right\} \varphi(I)\, dI\, d\mathbf{C} \tag{7.12}$$

with the intrinsic Lagrange multipliers $\left(\hat{\lambda}, \hat{\lambda}_i, \hat{\mu}\right)$. Comparison between (7.11) and (7.12) yields the relations (2.55) (with matrix $\mathbf{X}$ given by (2.53)):

$$\lambda = \hat{\lambda} - \hat{\lambda}_i v_i + \hat{\mu} v^2, \quad \lambda_i = \hat{\lambda}_i - 2\hat{\mu} v_i, \quad \mu = \hat{\mu}. \tag{7.13}$$

These relations dictate the velocity dependence of the Lagrange multipliers, which is consistent with the general results of the Galilean invariance. (See Sect. 2.7.)

The Euler-Lagrange equation $\delta\mathscr{L}/\delta f = 0$ leads to the following form of the equilibrium distribution function:

$$f^{(E)} = \exp\left\{-1 - \frac{m}{k_B}\left[\hat{\lambda} + \hat{\lambda}_i C_i + \hat{\mu}\left(C^2 + 2\frac{I}{m}\right)\right]\right\}.$$

Plugging this into the constraints (7.5) and noting that $\varepsilon_E^K(T) = (3k_B)/(2m)T$ where $\varepsilon_E^K(T)$ is the specific energy density ε^K in an equilibrium state with temperature T, one determines the intrinsic Lagrange multipliers in terms of the hydrodynamic variables:

$$\exp\left(-1 - \frac{m}{k_B}\hat{\lambda}\right) = \frac{\rho}{m\,A(T)}\left(\frac{m}{2\pi k_B T}\right)^{3/2}, \quad \hat{\lambda}_i = 0 \quad \hat{\mu} = \frac{1}{2T}, \tag{7.14}$$

with $A(T)$ being defined by (7.9). The proof is now completed.

The distribution function (7.8) is the generalization of the Maxwellian distribution function in the case of polyatomic gas. It was obtained also in [114] with different arguments.

Remark 7.1 $A(T)$ can be regarded as the partition function for the molecular internal mode in statistical mechanics.

7.2.2 *Internal Energy Density in Equilibrium and the Measure $\varphi(I)$ of Internal Mode*

For ideal non-polytropic gases in equilibrium, the internal energy is a non-linear function of the temperature. Its functional form $\varepsilon_E(T)$ is given usually by experimental data or by a statistical-mechanical calculation. For example, since the specific heat at constant volume c_v can be measured by experiments as a function of the temperature T, we can obtain ε_E by using the relation (6.3).

From (7.6), inserting the equilibrium distribution (7.8) and taking into account (7.9), we obtain the energy due to the internal mode in equilibrium:

$$\varepsilon_E^I(T) = \frac{1}{m}\int_0^\infty I f^{(I)} \varphi(I)\, dI = \frac{k_B}{m} T^2 \frac{d \log A(T)}{dT}. \tag{7.15}$$

Therefore, if we know the caloric equation of state (6.3), we know the expression of $\varepsilon_E^I(T) = \varepsilon(T) - \varepsilon_E^K(T)$, and then, from (7.15), we can obtain $A(T)$:

$$A(T) = A_0 \exp\left(\frac{m}{k_B}\int_{T_0}^{T}\frac{\varepsilon_E^I(x)}{x^2}dx\right), \tag{7.16}$$

where A_0 is an inessential constant. As was observed in [306] and in [134], the function A is, according to (7.9), the Laplace transform of φ:

$$A(s) = \mathscr{L}\left[\varphi(I)\right](s) = \int_0^\infty e^{-sI}\varphi(I)dI, \qquad s = \frac{1}{k_B T}.$$

Then we can obtain the measure φ as the inverse Laplace transform of A:

$$\varphi(I) = \mathscr{L}^{-1}\left[A(s)\right].$$

Differentiating (7.15) with respect to T, we have

$$\frac{1}{m^2}\int_0^\infty I^2 f^{(I)}\varphi(I)\,dI = \frac{p^2}{\rho^2}\hat{c}_V^I + \varepsilon_E^{I2}, \tag{7.17}$$

where

$$c_V^I = \frac{d\varepsilon_E^I(T)}{dT}, \qquad \hat{c}_V^I = \frac{m}{k_B}c_V^I.$$

7.2.2.1 Polytropic Gas

In the polytropic case, c_v is constant and the specific internal energy is linear with respect to the temperature:

$$\varepsilon_E(T) = \frac{D}{2}\frac{k_B}{m}T, \tag{7.18}$$

where D is the degrees of freedom of a molecule. This is a particular case of the non-polytropic case, and we obtain

$$\varepsilon_E^I = \frac{D-3}{2}\frac{k_B}{m}T, \qquad \hat{c}_V^I = \frac{D-3}{2}, \tag{7.19}$$

$$\varphi(I) = I^\alpha, \quad \text{with} \quad \alpha = \frac{D-5}{2} > -1$$

and

$$A(T) = (k_B T)^{1+\alpha} \Gamma(1+\alpha),$$

where Γ is the gamma function.

7.3 Euler System of a Rarefied Polyatomic Gas

We have the following theorem:

Theorem 7.2 *The system of a rarefied polyatomic gas (7.1) closed by using the equilibrium distribution function (7.8) is the system of an Euler gas with non-polytropic caloric equation of state.*

Proof Inserting the equilibrium distribution function (7.8) into the expressions of the fluxes (7.3) and (7.4), we have, as closure:

$$F_{ik} = \int_{\mathbb{R}^3} \int_0^\infty m c_i c_k f^{(E)} \varphi(I) \, dI \, d\mathbf{c} = \rho v_i v_k + p\delta_{ik},$$

where p is the equilibrium pressure:

$$p = \frac{1}{3} m \int_{\mathbb{R}^3} \int_0^\infty C^2 f^{(E)} \varphi(I) \, dI \, d\mathbf{C} = \frac{1}{3} m \int_{\mathbb{R}^3} \int_0^\infty C^2 f^{(M)} f^{(I)} \varphi(I) \, dI \, d\mathbf{C}$$
$$= \frac{1}{3} m \int_{\mathbb{R}^3} C^2 f^{(M)} \, d\mathbf{C}.$$

Therefore we notice the relation:

$$p = \frac{2}{3} \rho \varepsilon_E^K.$$

And

$$G_{llk} = \int_{\mathbb{R}^3} \int_0^\infty m(c^2 + 2I/m) c_k f^{(E)} \varphi(I) \, dI \, d\mathbf{c} = \left(\rho v^2 + 2\rho\varepsilon + 2p\right) v_k,$$

where ε is related to the temperature T by the relation: $\varepsilon = \varepsilon_E(T)$. These expressions are exactly the same as (2.26).

Note that, using (7.13) and (7.14), we can have the explicit form of the Lagrange multipliers (i.e., the main field) that symmetrize the system:

$$\lambda = \frac{1}{T}\left(-g + \frac{1}{2} v^2\right), \quad \lambda_i = -\frac{v_i}{T}, \quad \mu = \frac{1}{2T}.$$

Naturally, these coincide exactly with the Godunov variables (2.29).

7.4 Molecular ET 14-Field Theory of Rarefied Polyatomic Gas

7.4.1 *System of Field Equations and Nonequilibrium Distribution Function*

Let us now study the 14-moment theory of a rarefied polyatomic gas with non-polytropic equation of state. We adopt the same binary hierarchy proposed in the macroscopic approach (6.2):

$$
\begin{aligned}
&\partial_t F + \partial_k F_k = 0, \\
&\partial_t F_i + \partial_k F_{ik} = 0, \\
&\partial_t F_{\langle ij\rangle} + \partial_k F_{\langle ij\rangle k} = P_{\langle ij\rangle}, \\
&\partial_t F_{ll} + \partial_k F_{llk} = P_{ll}, \qquad \partial_t G_{ll} + \partial_k G_{llk} = 0, \\
&\qquad\qquad\qquad\qquad\qquad\quad\; \partial_t G_{lli} + \partial_k G_{llik} = Q_{lli}.
\end{aligned}
\tag{7.20}
$$

In the present case, the densities, fluxes, and productions are given by the moments of the generalized distribution function $f(t, \mathbf{x}, \mathbf{c}, I)$ as follows:

$$
\begin{aligned}
\begin{pmatrix} F \\ F_{i_1} \\ F_{i_1 i_2} \\ F_{i_1 i_2 i_3} \end{pmatrix} &= \int_{\mathbb{R}^3} \int_0^\infty m \begin{pmatrix} 1 \\ c_{i_1} \\ c_{i_1} c_{i_2} \\ c_{i_1} c_{i_2} c_{i_3} \end{pmatrix} f(t, \mathbf{x}, \mathbf{c}, I)\, \varphi(I)\, dI\, d\mathbf{c}, \\
P_{ij} &= \int_{\mathbb{R}^3} \int_0^\infty m c_i c_j\, Q(f)\, \varphi(I)\, dI\, d\mathbf{c}, \\
\begin{pmatrix} G_{ll} \\ G_{llk_1} \\ G_{llk_1 k_2} \end{pmatrix} &= \int_{\mathbb{R}^3} \int_0^\infty m \begin{pmatrix} c^2 + 2\frac{I}{m} \\ \left(c^2 + 2\frac{I}{m}\right) c_{k_1} \\ \left(c^2 + 2\frac{I}{m}\right) c_{k_1} c_{k_2} \end{pmatrix} f(t, \mathbf{x}, \mathbf{c}, I)\, \varphi(I)\, dI\, d\mathbf{c}, \\
Q_{lli} &= \int_{\mathbb{R}^3} \int_0^\infty m \left(c^2 + 2\frac{I}{m}\right) c_i\, Q(f)\, \varphi(I)\, dI\, d\mathbf{c}.
\end{aligned}
\tag{7.21}
$$

The densities of the system (7.20) are related to the conventional quantities as summarized below:

$$
\text{mass density}: \qquad \rho \equiv \int_{\mathbb{R}^3} \int_0^\infty m f\, \varphi(I)\, dI d\mathbf{c},
$$

$$
\text{velocity}: \qquad v_i \equiv \frac{1}{\rho} \int_{\mathbb{R}^3} \int_0^\infty m c_i f\, \varphi(I)\, dI d\mathbf{c},
$$

specific translational energy density : $\varepsilon^K \equiv \frac{1}{2\rho}\int_{\mathbb{R}^3}\int_0^\infty mC^2 f\,\varphi(I)\; dI d\mathbf{C},$

specific internal energy density : $\varepsilon^I \equiv \frac{1}{\rho}\int_{\mathbb{R}^3}\int_0^\infty If\,\varphi(I)\; dI d\mathbf{C},$

nonequilibrium pressure : $\mathscr{P} = \frac{2}{3}\rho\varepsilon^K = \frac{m}{3}\int_{\mathbb{R}^3}\int_0^\infty C^2 f\,\varphi(I)\; dI d\mathbf{C},$

dynamic pressure : $\Pi = \mathscr{P} - p = \frac{m}{3}\int_{\mathbb{R}^3}\int_0^\infty C^2(f - f^{(E)})\,\varphi(I)\; dI d\mathbf{C},$

shear stress : $\sigma_{\langle ij\rangle} \equiv -\int_{\mathbb{R}^3}\int_0^\infty mC_{\langle i}C_{j\rangle} f\,\varphi(I)\; dI d\mathbf{C},$

heat flux : $q_i \equiv \frac{1}{2}\int_{\mathbb{R}^3}\int_0^\infty (mC^2 + 2I)C_i f\,\varphi(I)\; dI d\mathbf{C}.$

And the trace part of the momentum flux F_{ll} is related to the *pressure* p and the *dynamic pressure* Π in continuum mechanics as follows:

$$F_{ll} = 3\mathscr{P} + \rho v^2 = 3(p + \Pi) + \rho v^2.$$

In particular, for a gas in equilibrium at temperature T, using the equilibrium distribution function $f^{(E)}$, we have

$$\begin{aligned} \varepsilon_E^K &= \frac{m}{2\rho}\int_{\mathbb{R}^3}\int_0^\infty C^2 f^{(E)}\varphi(I)\, dI\, d\mathbf{C} = \frac{3p}{2\rho} = \frac{3}{2}\frac{k_B}{m}T, \\ \varepsilon_E^I &= \frac{1}{\rho}\int_{\mathbb{R}^3}\int_0^\infty I f^{(E)}\varphi(I)\, dI\, d\mathbf{C} = \int_0^\infty I f^{(I)}\varphi(I)\, dI. \end{aligned} \tag{7.22}$$

Remark 7.2 Because of the constraints in the derivation of $f^{(E)}$ in the last section, we have, in particular, the relation:

$$\varepsilon = \varepsilon^K + \varepsilon^I = \varepsilon_E^K + \varepsilon_E^I, \tag{7.23}$$

where ε^K and ε^I are defined in (7.6), while ε_E^K and ε_E^I are defined in (7.22) with the temperature T. However, it is important to keep in mind that the nonequilibrium pressure $\mathscr{P}$ is not equal to the equilibrium pressure p at temperature T due to the non-zero dynamic pressure Π, where $\mathscr{P}$ is defined just above and p is given in $(7.22)_1$. Similarly, in contrast to the relation (7.23), ε^K and ε^I in nonequilibrium are individually not equal to $\varepsilon_E^K(T)$ and $\varepsilon_E^I(T)$, respectively.

With the entropy defined by (7.7), the MEP poses the following variational problem: Determine the distribution function $f(t, \mathbf{x}, \mathbf{c}, I)$ such that h becomes maximum under the constraints of the prescribed moments $(F, F_i, F_{ij}, G_{ll}, G_{lli})$. The solution of the problem is given by the following theorem [116]:

Theorem 7.3 *The distribution function obtained by the MEP has the following form, if processes are not so far from equilibrium:*

$$
\begin{aligned}
f =f^{(E)} \Bigg\{ 1 + \frac{3}{2p}\left(\frac{\varepsilon_E^I}{Tc_V^I} - 1\right)\Pi + \frac{\rho}{p^2}\Bigg[& -\frac{2\rho\varepsilon_E^I + 5p}{p(5+2\hat{c}_V^I)} q_i C_i \\
& + \left(-\sigma_{\langle ij\rangle} + \left(1 + \frac{3}{2\hat{c}_V^I}\right)\Pi\delta_{ij}\right) C_i C_j - \frac{3}{4\hat{c}_V^I}\Pi\left(C^2 + 2\frac{I}{m}\right) \\
& + \frac{\rho}{p(5+2\hat{c}_V^I)}\left(C^2 + 2\frac{I}{m}\right) q_i C_i \Bigg]\Bigg\},
\end{aligned}
\tag{7.24}
$$

where $f^{(E)}$ is the equilibrium distribution function (7.8).

Proof We introduce the Lagrange multipliers $(\lambda, \lambda_i, \lambda_{\langle ij\rangle}, \nu, \mu, \mu_{ill})$ associated with the moments $(F, F_i, F_{\langle ij\rangle}, F_{ll}, G_{ll}, G_{ill})$, respectively. The functional for the constrained variational problem (4.51) in this case is given by

$$
\begin{aligned}
\mathscr{L} = \int_{\mathbb{R}^3}\int_0^\infty \Bigg\{ -k_B f \log f - m \Bigg[& \lambda + \lambda_i c_i + \lambda_{\langle ij\rangle} c_i c_j + \nu c^2 \\
& + (\mu + \mu_i c_i)\left(c^2 + 2\frac{I}{m}\right)\Bigg] f \Bigg\} \varphi(I)\, dI\, d\mathbf{c}.
\end{aligned}
\tag{7.25}
$$

Due to the Galilean invariance, the functional can be replaced by the one in the case of zero hydrodynamic velocity ($\mathbf{v} = \mathbf{0}$). Therefore, we have

$$
\begin{aligned}
\mathscr{L} = \int_{\mathbb{R}^3}\int_0^\infty \Bigg\{ -k_B f \log f - m \Bigg[& \hat{\lambda} + \hat{\lambda}_i C_i + \hat{\lambda}_{\langle ij\rangle} C_i C_j + \hat{\nu} C^2 \\
& + (\hat{\mu} + \hat{\mu}_i C_i)\left(C^2 + 2\frac{I}{m}\right)\Bigg] f \Bigg\} \varphi(I)\, dI\, d\mathbf{C}
\end{aligned}
\tag{7.26}
$$

with the intrinsic Lagrange multipliers $(\hat{\lambda}, \hat{\lambda}_i, \hat{\lambda}_{\langle ij\rangle}, \hat{\nu}, \hat{\mu}, \hat{\mu}_{ill})$. The comparison between (7.25) and (7.26) gives the velocity dependence of the Lagrange multipliers, which is in agreement with the general result given in Sect. 4.2. In particular, this coincides exactly with the velocity dependence for the main field in the macroscopic theory (6.11).

The solution of the Euler-Lagrange equation $\delta\mathscr{L}/\delta f = 0$ is given by

$$f = \exp\left(-1 - \frac{m}{k_B}\chi\right)$$

with

$$\chi = \hat{\lambda} + \hat{\lambda}_i C_i + \hat{\lambda}_{\langle ij\rangle} C_i C_j + \hat{\nu}\, C^2 + (\hat{\mu} + \hat{\mu}_i C_i)\left(C^2 + 2\frac{I}{m}\right).$$

As in a monatomic gas, there is the problem of the convergence of moments. Therefore, we study the solution in the form of expansion around the local equilibrium:

$$f = f^{(E)}\left\{1 - \frac{m}{k_B}\left(\tilde{\lambda} + \tilde{\lambda}_i C_i + \tilde{\lambda}_{\langle ij\rangle} C_i C_j + \tilde{\nu}\, C^2 + (\tilde{\mu} + \tilde{\mu}_i C_i)\left(C^2 + 2\frac{I}{m}\right)\right)\right\}, \tag{7.27}$$

where the quantities with a tilde are defined in (4.66). Inserting (7.27) into (7.21) evaluated at zero velocity, we obtain the algebraic system (see the general approach to the algebraic system (4.59)):

$$\begin{aligned}
&\tilde{\lambda}\hat{F}^E + \tilde{\nu}\,\hat{F}^E_{ll} + \tilde{\mu}\,\hat{G}^E_{ll} = 0,\\
&\tilde{\lambda}_k \hat{F}^E_{ik} + \tilde{\mu}_k\,\hat{G}^E_{llik} = 0,\\
&\tilde{\lambda}\hat{G}^E_{ll} + \tilde{\nu}\,\hat{G}^E_{llkk} + \tilde{\mu}\,\hat{H}^E_{llkk} = 0,\\
&\tilde{\lambda}_{\langle rs\rangle}\hat{F}^E_{ijrs} = \frac{k_B}{m}\sigma_{\langle ij\rangle},\\
&\tilde{\lambda}\hat{F}^E_{ll} + \tilde{\nu}\,\hat{F}^E_{llkk} + \tilde{\mu}\,\hat{G}^E_{llkk} = -3\frac{k_B}{m}\Pi,\\
&\tilde{\lambda}_k\hat{G}^E_{llkj} + \tilde{\mu}_k\,\hat{H}^E_{llsskj} = -2\frac{k_B}{m}q_j,
\end{aligned} \tag{7.28}$$

where $\hat{F}^E_{k_1k_2\ldots k_n}$ and $\hat{G}^E_{llk_1k_2\ldots k_n}$ are the $\hat{F}$ and $\hat{G}$ moments in an equilibrium state, i.e.,

$$\begin{aligned}
\hat{F}^E_{k_1k_2\ldots k_n} &= m\int_{\mathbb{R}^3}\int_0^\infty f^{(M)} f^{(I)} C_{k_1} C_{k_2}\ldots C_{k_n}\varphi(I)\, dI\, d\mathbf{C},\\
\hat{G}^E_{llk_1k_2\ldots k_r} &= m\int_{\mathbb{R}^3}\int_0^\infty f^{(M)} f^{(I)}\left(C^2 + 2\frac{I}{m}\right) C_{k_1} C_{k_2}\ldots C_{k_r}\varphi(I)\, dI\, d\mathbf{C},
\end{aligned} \tag{7.29}$$

and

$$\hat{H}^E_{llssk_1k_2\ldots k_s} = m \int_{\mathbb{R}^3} \int_0^\infty f^{(M)} f^{(I)} \left(C^2 + 2\frac{I}{m} \right)^2 C_{k_1} C_{k_2} \ldots C_{k_s} \varphi(I)\, dI\, d\mathbf{C}. \tag{7.30}$$

From (7.29) and (7.30), taking into account the relations (7.10), (7.15), and (7.17), it is easy to verify that the moments of $\hat{F}$, $\hat{G}$, and $\hat{H}$ in an equilibrium state—$\hat{F}^E$, $\hat{G}^E$, and $\hat{H}^E$—can be expressed in terms of the moments of $\hat{F}$ calculated by using the Maxwellian—$\hat{F}^M$—, i.e., the moments of $\hat{F}$ for an equilibrium monatomic gas:

$$\begin{aligned}
\hat{F}^E_{k_1k_2\ldots k_n} &= \hat{F}^M_{k_1k_2\ldots k_n}, \\
\hat{G}^E_{llk_1k_2\ldots k_r} &= \hat{F}^M_{llk_1k_2\ldots k_r} + 2\varepsilon^I_E \hat{F}^M_{k_1k_2\ldots k_r}, \\
\hat{H}^E_{lljjk_1k_2\ldots k_s} &= \hat{F}^M_{lljjk_1k_2\ldots k_s} + 4\varepsilon^I_E \hat{F}^M_{jjk_1k_2\ldots k_s} + 4\left(\frac{p^2}{\rho^2}\hat{c}^I_V + \varepsilon^{I2}_E \right) \hat{F}^M_{k_1k_2\ldots k_s}.
\end{aligned} \tag{7.31}$$

Moments ($\hat{F}^M$, $\hat{F}^M_{ij}$, $\hat{F}^M_{ijrs}$, $\hat{F}^M_{llijrs}$) are the same as those of monatomic gas (4.68), and therefore, from (7.28) with (7.31), we have the following relations:

$$\begin{aligned}
&\tilde{\lambda} = \frac{3}{2}\frac{k_B}{m}\left(1 - \frac{\varepsilon^I_E}{Tc^I_V}\right)\frac{\Pi}{p}, \quad \tilde{\nu} = -\frac{1}{2T}\left(1 + \frac{3}{2\hat{c}^I_V}\right)\frac{\Pi}{p}, \quad \tilde{\mu} = \frac{3}{4T\hat{c}^I_V}\frac{\Pi}{p}, \\
&\tilde{\lambda}_i = \frac{2\rho\varepsilon^I_E + 5p}{p^2T\left(2\hat{c}^I_V + 5\right)} q_i, \quad \tilde{\mu}_i = -\frac{\rho}{p^2T\left(2\hat{c}^I_V + 5\right)} q_i \\
&\tilde{\lambda}_{\langle ij \rangle} = \frac{1}{2pT}\sigma_{\langle ij \rangle}.
\end{aligned} \tag{7.32}$$

Inserting (7.32) into (7.27), we obtain the nonequilibrium distribution function (7.24). The proof of the theorem is now completed.

7.4.2 Non-convective Fluxes

Now, as we have the the nonequilibrium distribution function (7.24), we can evaluate the fluxes and the productions that are not in the list of density variables, and then we can close the system. The non-convective fluxes $\hat{F}_{ijk}$ and $\hat{G}_{llij}$ are defined by

$$\hat{F}_{ijk} = m \int_{\mathbb{R}^3} \int_0^\infty C_i C_j C_k f(t, \mathbf{x}, \mathbf{C}, I)\, \varphi(I)\, dI\, d\mathbf{C}$$

and

$$\hat{G}_{llij} = m \int_{\mathbb{R}^3} \int_0^\infty \left(C^2 + 2\frac{I}{m} \right) C_i C_j f(t, \mathbf{x}, \mathbf{C}, I) \varphi(I) \, dI \, d\mathbf{C}.$$

These are expressed, with the use of the distribution function (7.24) and by taking (7.31) into account, as follows:

$$\begin{aligned} \hat{F}_{ijk} &= \frac{2}{2\hat{c}_V^I + 5} \left(q_i \delta_{jk} + q_j \delta_{ki} + q_k \delta_{ij} \right), \\ \hat{G}_{llij} &= p \left(5\frac{p}{\rho} + 2\varepsilon_E^I \right) \delta_{ij} + \left(7\frac{p}{\rho} + 2\varepsilon_E^I \right) \left(\Pi \delta_{ij} - \sigma_{\langle ij \rangle} \right). \end{aligned} \tag{7.33}$$

The expressions (7.33) coincide with those of the macroscopic closure (6.28).

In this way, we can see that the closed system studied here is completely equivalent to the one obtained by the macroscopic approach as far as the principal part of the operator (left-hand side of the system) is concerned.

7.4.3 Polytropic Gas

In the case of polytropic gas (7.19), the nonequilibrium distribution function (7.24) is expressed as

$$\begin{aligned} f = & f^{(E)} \left\{ 1 - \frac{\rho}{p^2} \left[q_i C_i + \left(-\sigma_{\langle ij \rangle} + \frac{D}{D-3} \Pi \delta_{ij} \right) C_i C_j \right. \right. \\ & \left. \left. - \frac{3}{2(D-3)} \Pi \left(C^2 + 2\frac{I}{m} \right) + \frac{\rho}{p(D+2)} \left(C^2 + 2\frac{I}{m} \right) q_i C_i \right] \right\}. \end{aligned}$$

This result coincides with the one obtained by Pavić et al. [115].

7.4.4 Nonequilibrium Quantities

Using the nonequilibrium distribution function f given in (7.24), we can evaluate the nonequilibrium pressure $\mathscr{P}$ and the nonequilibrium energy densities ε^K and ε^I. These are expressed as follows:

$$\mathscr{P} = p + \Pi, \quad \rho\varepsilon^K = \rho\varepsilon_E^K + \frac{3}{2}\Pi, \quad \rho\varepsilon^I = \rho\varepsilon_E^I - \frac{3}{2}\Pi. \tag{7.34}$$

The relations above are quite impressive because, in contrast to the case of monatomic gas, the nonequilibrium pressure $\mathscr{P}$ and the specific energy densities ε^K, ε^I for a polyatomic gas are different from the corresponding equilibrium ones due to the presence of the dynamic pressure Π. Moreover the dynamic pressure can be interpreted in three different ways: (1) difference between nonequilibrium pressure and equilibrium one, (2) difference between $\rho\varepsilon^K$ and $\rho\varepsilon_E^K$ except for the factor 2/3, and (3) difference between $\rho\varepsilon^I$ and $\rho\varepsilon_E^I$ except for the factor $-2/3$. But, as pointed out in Remark 7.2 above, the sum $\varepsilon^K + \varepsilon^I$, i.e., total energy ε is equal to the sum $\varepsilon_E^K + \varepsilon_E^I$, i.e., equilibrium total energy ε_E!

It is, therefore, natural to introduce two nonequilibrium temperatures θ^K and θ^I in an indirect way:

$$\varepsilon^K = \varepsilon_E^K(\theta^K), \qquad \varepsilon^I = \varepsilon_E^I(\theta^I). \tag{7.35}$$

In other words, the nonequilibrium temperatures are the ones at which the corresponding equilibrium functions give the values in nonequilibrium. In particular, the temperature θ^K is the temperature for the translational mode, i.e., so-called kinetic temperature.

Comparing (7.35) with (7.34), we have

$$\mathscr{P} = p(\rho, \theta^K) = \frac{2}{3}\rho\varepsilon_E^K(\theta^K) = \Pi + p(\rho, T), \tag{7.36}$$

i.e.,

$$\theta^K = T\left(1 + \frac{\Pi}{p(\rho, T)}\right),$$

and

$$\varepsilon^I = \varepsilon_E^I(\theta^I) = \varepsilon(T) - \varepsilon_E^K(\theta^K). \tag{7.37}$$

with the identity:

$$\varepsilon_E^I(\theta^I) - \varepsilon_E^I(T) = \varepsilon_E^K(T) - \varepsilon_E^K(\theta^K).$$

Therefore, as independent variables, we can use, instead of the pair (T, Π), the pair $(\varepsilon^K, \varepsilon^I)$ or (θ^K, θ^I). There are one-to-one correspondences among these pairs through Eqs. (7.36) and (7.37) thanks to the monotonicity of the energy densities.

7.5 Generalized BGK Model

The MEP approach has an advantage over the phenomenological one. At least in principle, we can derive the explicit expressions of the production terms, while in the phenomenological approach we know only the sign of the production terms appearing in the second member (right-hand side) of (7.20). In fact we have the expressions $(7.21)_{2,4}$ for the productions P_{ij} and Q_{lli}. The main problem is, however, that, in order to have explicit expressions for the productions, we need a model for the collision term $Q(f)$ in the Boltzmann equation. It is usually very difficult to obtain it in the case of polyatomic gas. Concerning the collision term, Struchtrup [307], and Rahimi and Struchtrup [308] proposed a variant of the BGK-model [309] to take into account a relaxation of the energy of the internal mode. In a recent paper [310], an alternative BGK expression is studied with the aim to use Chapman-Enskog procedure for a high temperature rarefied gas.

In this section, we introduce, as in [136], a simple collision term with two relaxation times.

7.5.1 Two Relaxation Times

In polyatomic gases, we may introduce two characteristic times corresponding to two relaxation processes (see also [311–314]):

1. Relaxation time τ_K: This characterizes the relaxation process within the translational mode (mode K) of molecules. The process shows the tendency to approach an equilibrium state of the mode K with the distribution function $f_K^{(E)}$ having the temperature θ^K, explicit expression of which is shown below. However, the internal mode I remains, in general, in nonequilibrium. This process exists also in monatomic gases.
2. Relaxation time τ of the second stage: After the relaxation process of the translational mode K, two modes, K and I, eventually approach a local equilibrium state characterized by $f^{(E)}$ with a common temperature T. Naturally we have assumed the condition: $\tau > \tau_K$.

7.5.2 Generalized BGK Collision Term

The generalized BGK collision term is proposed as follows:

$$Q(f) = -\frac{1}{\tau_K}(f - f_K^{(E)}) - \frac{1}{\tau}(f - f^{(E)}), \tag{7.38}$$

where the distribution functions $f_K^{(E)}$ is given by

$$f_K^{(E)} = \frac{\rho^I(I)}{m}\left(\frac{m}{2\pi k_B \theta^K}\right)^{3/2} \exp\left(-\frac{mC^2}{2k_B\theta^K}\right) \tag{7.39}$$

with $\rho^I(I)$ being defined as

$$\rho^I(I) = \int_{\mathbb{R}^3} mf d\mathbf{c}.$$

This is the equilibrium function with respect to the K-mode with the temperature θ^K and with the "frozen" energy I. In other words, $f_K^{(E)}$ is a Maxwellian with the mass density $\rho^I(I)$ and temperature θ^K. Therefore $f_K^{(E)}$ given in (7.39) is obtained by maximizing the entropy with the fixed energy I:

$$h^I(I) = -k_B \int_{\mathbb{R}^3} f \log f \, d\mathbf{c}$$

under the constraints:

$$\begin{pmatrix} \rho^I(I) \\ \rho^I(I)v_i \\ 2\rho^I(I)\varepsilon_E^K(\theta^K) \end{pmatrix} = \int_{\mathbb{R}^3} \begin{pmatrix} m \\ mc_i \\ mC^2 \end{pmatrix} f d\mathbf{c}.$$

7.5.3 H-theorem

From the definition of the distribution functions $f_K^{(E)}$ and $f^{(E)}$, it is easy to verify the following relations:

$$\int_{\mathbb{R}^3} \int_0^\infty (f - f_K^{(E)}) \log f_K^{(E)} \varphi(I) \, dI \, d\mathbf{C} = 0,$$

$$\int_{\mathbb{R}^3} \int_0^\infty (f - f^{(E)}) \log f^{(E)} \varphi(I) \, dI \, d\mathbf{C} = 0.$$

Then it is easy to show that the entropy production

$$\Sigma = -k_B \int_{\mathbb{R}^3} \int_0^\infty Q(f) \log f \, \varphi(I) \, dI d\mathbf{c}$$

is positive:

$$\Sigma = k_B \int_{\mathbb{R}^3} \int_0^\infty \left\{ \frac{f - f_K^{(E)}}{\tau^K} \log \frac{f}{f_K^{(E)}} + \frac{f - f^{(E)}}{\tau} \log \frac{f}{f^{(E)}} \right\} \varphi(I) \, dI \, d\mathbf{C} \geqslant 0,$$

and the H-theorem holds.

7.6 Production Terms in the Generalized BGK Model

Adopting the generalized BGK model (7.38) and taking into account (7.36), we obtain the explicit expressions of the production terms $\hat{P}_{ll}$, $\hat{P}_{\langle ij\rangle}$, and $\hat{Q}_{lli}$:

$$\hat{P}_{ll} = -\frac{3}{\tau}\Pi, \quad \hat{P}_{\langle ij\rangle} = \left(\frac{1}{\tau_K} + \frac{1}{\tau}\right)\sigma_{\langle ij\rangle}, \quad \hat{Q}_{lli} = -2\left(\frac{1}{\tau_K} + \frac{1}{\tau}\right)q_i. \tag{7.40}$$

It should be noted that, as usual in the BGK model, the Prandtl number predicted by the present model is not satisfactory. One possible way to avoid this difficulty is that we use the procedure adopted in the phenomenological RET theory of rarefied polyatomic gases. That is, instead of $(7.40)_{2,3}$, we adopt the following production terms

$$\hat{P}_{\langle ij\rangle} = \frac{1}{\tau_\sigma}\sigma_{\langle ij\rangle}, \qquad \hat{Q}_{lli} = -\frac{2}{\tau_q}q_i,$$

where τ_σ and τ_q (and also $\tau = \tau_\Pi$) are functions of ρ and T, which are estimated by using the experimental data on bulk viscosity ν, shear viscosity μ, and heat conductivity κ. It is possible to verify that these production terms satisfy the entropy principle.

7.7 Closed System of Field Equations: ET$_{14}$

Taking into account that $c_i = C_i + v_i$, from (7.21), we have the following relations that are in perfect agreement with the general theorem of Galilean invariance given in Sect. 2.7:

$$\begin{aligned}
&F = \hat{F}, \quad F_i = \hat{F}v_i, \quad F_{ij} = \hat{F}_{ij} + \hat{F}v_iv_j, \\
&F_{ijk} = \hat{F}_{ijk} + \hat{F}_{ij}v_k + \hat{F}_{ik}v_j + \hat{F}_{kj}v_i + \hat{F}v_iv_jv_k, \\
&G_{ll} = \hat{G}_{ll} + \hat{F}v^2, \quad G_{lli} = \hat{G}_{lli} + 2\hat{F}_{ij}v_j + \hat{G}_{ll}v_i + \hat{F}v^2v_i, \\
&G_{llik} = \hat{G}_{llik} + 2\hat{F}_{iks}v_s + \hat{F}_{ik}v^2 + \hat{G}_{lli}v_k + \hat{G}_{llk}v_i + 2\hat{F}_{is}v_kv_s + 2\hat{F}_{ks}v_iv_s + \hat{G}_{ll}v_iv_k + \hat{F}v^2v_iv_k.
\end{aligned} \tag{7.41}$$

Inserting (7.41) into (7.20), we obtain the closed system of field equations for non-polytropic gases in terms of the 14 variables $(\rho, v_i, T, \Pi, \sigma_{\langle ij\rangle}, q_i)$. This system is coincident with the one obtained in the macroscopic theory (6.33). In this way we have now proved the perfect matching between the phenomenological ET$_{14}$ and the molecular ET$_{14}$ of rarefied polyatomic gases with non-polytropic caloric equation of state.

In the particular case of polytropic gas, we have the specific internal energy given by (7.18), and $\hat{c}_v = D/2$. The results above reduce to the ones obtained in [115]. The closed system of field equations is already given explicitly in (1.33).

Remark 7.3 Application of MEP is not exhausted in the moment closure problem of hierarchies of macroscopic equations as we have seen before but was used for the estimation of transport coefficients in macroscopic models. In particular, in [315], it was applied by Brull and Schneider in the context of the ellipsoidal statistical model for polyatomic gases and exploited in the analysis of transport coefficients in the BGK approximation.

Chapter 8
Relaxation Processes of Molecular Rotation and Vibration: ET_{15}

Abstract In the previous Chaps. 6 and 7, we have studied RET of rarefied polyatomic gases under the assumption that molecular internal modes can be treated as a whole in terms of the single variable I. However, by dividing the internal modes into the molecular rotational and vibrational modes, we can obtain a more refined version of RET of rarefied polyatomic gases with a wider applicability range. To study this subject is the main purpose of the present chapter. In this case, we need to introduce (at least) two variables instead of I, and as a result we have a triple (or multiple) hierarchy of the moment system. As the simplest model, we derive explicitly the closed system of field equations with 15 variables (ET_{15}) by using the maximum entropy principle (MEP). Three different types of the production terms, which are suggested by a generalized BGK-type collision term in the Boltzmann equation, are adopted. The Navier-Stokes and Fourier (NSF) theory is again derived from the ET_{15} theory as a limiting case of small relaxation times via the Maxwellian iteration. The relaxation times introduced in ET_{15} are related to the shear and bulk viscosities and heat conductivity.

8.1 Introduction

The RET theory of rarefied polyatomic gases with the binary hierarchy explained in Chaps. 6 and 7 has the limitation of its applicability, although the theory has been successfully utilized to analyze various nonequilibrium phenomena as will be seen in Part V. In fact, we have many experimental data showing that the relaxation times of the rotational mode and of the vibrational mode are quite different to each other. In such a case, more than one molecular relaxation processes should be taken into account to make the RET theory more precise.

The aim of the present chapter is to establish such a RET theory with much wider applicability range for rarefied polyatomic gases, that is, the ET_{15} theory with 15 independent fields: mass density, velocity, translational energy density, rotational energy density, vibrational energy density, shear stress, and heat flux [136].

T. Ruggeri, M. Sugiyama, *Classical and Relativistic Rational Extended Thermodynamics of Gases*, https://doi.org/10.1007/978-3-030-59144-1_8

Firstly, we explain molecular ET of rarefied polyatomic gases with two internal relaxation processes by using two parameters expressing the rotational and vibrational energies of a molecule. The equilibrium distribution function, the expressions of thermal and caloric equations of state, and the entropy density in equilibrium are also shown. Secondly, we define three kinds of moments, and we derive a triple hierarchy of moment equations from the Boltzmann equation. Then we study the truncated system of balance equations, which is closed by using the nonequilibrium distribution function derived from MEP. Thirdly, we introduce a simple collision term with three relaxation times, which is a generalization of the BGK model. Lastly, the relationship between the ET$_{15}$ theory and the Navier-Stokes and Fourier theory is discussed.

8.2 Distribution Function with Molecular Rotational and Vibrational Energies

In order to describe the relaxation processes of molecular rotational and vibrational modes separately, we decompose the energy of internal modes I into the energy of rotational mode I^R and the energy of vibrational mode I^V:

$$I = I^R + I^V. \tag{8.1}$$

Generalizing the Borgnakke-Larsen idea explained in Chap. 7, we assume the same form of the Boltzmann equation (1.20) but with a velocity distribution function depending on these additional parameters, i.e., $f \equiv f\left(t, \mathbf{x}, \mathbf{c}, I^R, I^V\right)$. We also take into account the effect of the parameters I^R and I^V on the collision term $Q(f)$.

Remark 8.1 Since a state near the dissociation temperature, where the molecular vibration is highly anharmonic, is out of the scope of the present study, the relation (8.1) can be safely assumed. In a harmonic approximation of the molecular vibration, we may further divide I^V into the energies of several harmonic modes. However, because we focus, in the present study, on the contribution from the rotational or vibrational mode, each of which is treated as a whole, we do not enter into such details although the generalization in this direction is in principle straightforward.

8.2.1 Equilibrium Distribution Function

We derive the equilibrium distribution function $f^{(E)}$ by means of MEP. We remark that the collision invariants of the present model are m, mc_i, and $mc^2 + 2I^R + 2I^V$. These quantities correspond to the hydrodynamics variables, i.e., the mass density

$F(=\rho)$, the momentum density $F_i(=\rho v_i)$, and twice the energy density $G_{ll}(= 2\rho\varepsilon + \rho v^2)$ through the following relations:

$$
\begin{aligned}
F &= \int_{\mathbb{R}^3}\int_0^\infty\int_0^\infty mf\,\varphi\left(I^R\right)\psi\left(I^V\right)dI^R dI^V d\mathbf{c},\\
F_i &= \int_{\mathbb{R}^3}\int_0^\infty\int_0^\infty mc_i f\,\varphi\left(I^R\right)\psi\left(I^V\right)dI^R dI^V d\mathbf{c},\\
G_{ll} &= \int_{\mathbb{R}^3}\int_0^\infty\int_0^\infty \left(mc^2+2I^R+2I^V\right) f\,\varphi\left(I^R\right)\psi\left(I^V\right)dI^R dI^V d\mathbf{c}.
\end{aligned}
\tag{8.2}
$$

Here $\varphi\left(I^R\right)$ and $\psi\left(I^V\right)$ are the state densities corresponding to I^R and I^V. And it is easy to see from (8.2)$_3$, that the specific internal energy ε is composed of three parts, that is, the kinetic part ε^K and the parts of rotational mode ε^R and of vibrational mode ε^V, i.e.,

$$\varepsilon = \varepsilon^K + \varepsilon^R + \varepsilon^V.$$

Expressions of ε^K, ε^R, and ε^V will be given explicitly in (8.16), (8.17), and (8.18), respectively.

The entropy density h, the entropy flux h_i, and the entropy production Σ are defined by

$$
\begin{aligned}
h &= -k_B\int_{\mathbb{R}^3}\int_0^\infty\int_0^\infty f\log f\,\varphi\left(I^R\right)\psi\left(I^V\right)dI^R dI^V d\mathbf{c},\\
h_i &= -k_B\int_{\mathbb{R}^3}\int_0^\infty\int_0^\infty c_i f\log f\,\varphi\left(I^R\right)\psi\left(I^V\right)dI^R dI^V d\mathbf{c},\\
\Sigma &= -k_B\int_{\mathbb{R}^3}\int_0^\infty\int_0^\infty Q(f)\log f\,\varphi\left(I^R\right)\psi\left(I^V\right)dI^R dI^V d\mathbf{c}.
\end{aligned}
\tag{8.3}
$$

The H-theorem requires that the collision term Q must satisfy the condition: $\Sigma \geqslant 0$.

Theorem 8.1 *The equilibrium distribution function $f^{(E)}$, which maximizes the entropy density* (8.3)$_1$ *under the constraints* (8.2), *is given by*

$$f^{(E)} = \frac{\rho}{m\,A^R(T)\,A^V(T)}\left(\frac{m}{2\pi k_B T}\right)^{3/2}\exp\left\{-\frac{1}{k_B T}\left(\frac{1}{2}mC^2 + I^R + I^V\right)\right\}, \tag{8.4}$$

where $A^R(T)$ and $A^V(T)$ are normalization factors:

$$A^R(T) = \int_0^\infty \varphi\left(I^R\right) e^{-\frac{I^R}{k_B T}} dI^R, \qquad A^V(T) = \int_0^\infty \psi\left(I^V\right) e^{-\frac{I^V}{k_B T}} dI^V. \tag{8.5}$$

The proof is omitted here for simplicity because it is quite similar to the one shown in Chap. 7.

The equilibrium distribution function can be expressed by the product of the equilibrium distribution functions of the three modes:

$$f^{(E)} = f^{(M)} f^{(R)} f^{(V)}, \tag{8.6}$$

where

$$\begin{aligned} f^{(M)} &= \frac{\rho}{m}\left(\frac{m}{2\pi k_B T}\right)^{3/2} \exp\left(-\frac{mC^2}{2k_B T}\right), \\ f^{(R)} &= \frac{1}{A^R(T)} \exp\left(-\frac{I^R}{k_B T}\right), \\ f^{(V)} &= \frac{1}{A^V(T)} \exp\left(-\frac{I^V}{k_B T}\right). \end{aligned} \tag{8.7}$$

8.2.2 Thermal and Caloric Equations of State

By using the equilibrium distribution function $f^{(E)}$, the caloric equation of state is expressed by $\varepsilon_E(T)$:

$$\varepsilon_E(T) = \varepsilon_E^K(T) + \varepsilon_E^R(T) + \varepsilon_E^V(T), \tag{8.8}$$

and, proceeding in a similar way as shown in Chap. 7, we have

$$\begin{aligned} \varepsilon_E^K(T) &\equiv \frac{3}{2}\frac{k_B}{m}T, \\ \varepsilon_E^R(T) &\equiv \frac{k_B}{m}T^2 \frac{d \log A^R(T)}{dT}, \\ \varepsilon_E^V(T) &\equiv \frac{k_B}{m}T^2 \frac{d \log A^V(T)}{dT}. \end{aligned} \tag{8.9}$$

Therefore if we know the normalization factors $A^R(T)$ and $A^V(T)$, which can be regarded as the partition functions for the molecular rotational and vibrational modes in statistical mechanics, we can derive the equilibrium energies of rotational and vibrational modes from (8.9). Vice versa if we know, at the macroscopic phenomenological level, the constitutive equations $\varepsilon_E^R(T)$ and $\varepsilon_E^V(T)$, we can obtain by integration of $(8.9)_{2,3}$:

$$A^R(T) = A_0^R \exp\left(\frac{m}{k_B}\int_{T_0}^{T}\frac{\varepsilon_E^R(x)}{x^2}dx\right),$$

$$A^V(T) = A_0^V \exp\left(\frac{m}{k_B}\int_{T_0}^{T}\frac{\varepsilon_E^V(x)}{x^2}dx\right),$$

where A_0^R, A_0^V, and T_0 are inessential constants. The functions A^R and A^V are, according to (8.5), the Laplace transforms of φ and ψ, respectively:

$$A^R(T) = \mathscr{L}\left[\varphi(I^R)\right](s), \quad A^V(T) = \mathscr{L}\left[\psi\left(I^V\right)\right](s), \quad s = \frac{1}{k_B T},$$

and then we obtain the state functions $\varphi\left(I^R\right)$ and $\psi\left(I^V\right)$ as the inverse Laplace transforms of $A^R(T)$ and $A^V(T)$, respectively:

$$\varphi(I^R) = \mathscr{L}^{-1}\left[A^R(T)\right](I^R),\ \psi\left(I^V\right) = \mathscr{L}^{-1}\left[A^V(T)\right](I^V),\ T = \frac{1}{k_B s}.$$

We also notice the relation:

$$p(\rho, T) = \frac{2}{3}\rho\varepsilon_E^K(T), \tag{8.10}$$

and the specific entropy density $s = h_E/\rho$ in equilibrium is given by

$$s = s_E^K(\rho, T) + s_E^R(T) + s_E^V(T),$$

where

$$\begin{aligned} s_E^K(\rho, T) &\equiv -\frac{k_B}{\rho}\int_{\mathbb{R}^3}\int_0^\infty\int_0^\infty f^{(E)}\log f^{(M)}\varphi\left(I^R\right)\psi\left(I^V\right)dI^R dI^V d\mathbf{c} \\ &= \frac{k_B}{m}\log\left(\frac{T^{3/2}}{\rho}\right) + \frac{\varepsilon_E^K(T)}{T} - \frac{k_B}{m}\log\left[\frac{1}{m}\left(\frac{m}{2\pi k_B}\right)^{3/2}\right], \\ s_E^R(T) &\equiv -\frac{k_B}{\rho}\int_{\mathbb{R}^3}\int_0^\infty\int_0^\infty f^{(E)}\log f^{(R)}\varphi\left(I^R\right)\psi\left(I^V\right)dI^R dI^V d\mathbf{c} \\ &= \frac{k_B}{m}\log A^R(T) + \frac{\varepsilon_E^R(T)}{T}, \end{aligned}$$

$$
\begin{aligned}
s_E^V(T) &\equiv -\frac{k_B}{\rho}\int_{\mathbb{R}^3}\int_0^\infty\int_0^\infty f^{(E)}\log f^{(V)}\varphi\left(I^R\right)\psi\left(I^V\right)dI^R dI^V d\mathbf{c} \\
&= \frac{k_B}{m}\log A^V(T) + \frac{\varepsilon_E^V(T)}{T}.
\end{aligned} \tag{8.11}
$$

The Gibbs relations of the three modes are given by

$$
\begin{aligned}
&T\mathrm{d}s_E^K(\rho, T) = \mathrm{d}\varepsilon_E^K(T) - \frac{p(\rho, T)}{\rho^2}\mathrm{d}\rho, \\
&T\mathrm{d}s_E^R(T) = \mathrm{d}\varepsilon_E^R(T), \qquad T\mathrm{d}s_E^V(T) = \mathrm{d}\varepsilon_E^V(T).
\end{aligned} \tag{8.12}
$$

Proceeding as was done for (7.15) and (7.17), we have the following relations, which will be useful later:

$$
\begin{aligned}
&\int_0^\infty f^{(R)}\varphi(I^R)\,dI^R = 1, \qquad \frac{1}{m}\int_0^\infty I^R f^{(R)}\varphi(I^R)\,dI^R = \varepsilon_E^R(T), \\
&\frac{1}{m^2}\int_0^\infty (I^R)^2 f^{(R)}\varphi(I^R)\,dI^R = \frac{k_B}{m}T^2 c_v^R + (\varepsilon_E^R)^2 = \frac{p^2}{\rho^2}\hat{c}_v^R + (\varepsilon_E^R)^2, \\
&\int_0^\infty f^{(V)}\psi\left(I^V\right)dI^V = 1, \qquad \frac{1}{m}\int_0^\infty I^V f^{(V)}\psi\left(I^V\right)dI^V = \varepsilon_E^V(T), \\
&\frac{1}{m^2}\int_0^\infty (I^V)^2 f^{(V)}\psi\left(I^V\right)dI^V = \frac{k_B}{m}T^2 c_v^V + (\varepsilon_E^V)^2 = \frac{p^2}{\rho^2}\hat{c}_v^V + (\varepsilon_E^V)^2,
\end{aligned}
$$

where we have introduced two specific heats at constant volume: $c_v^R = d\varepsilon_E^R/dT = (k_B/m)\hat{c}_v^R$ and $c_v^V = d\varepsilon_E^V/dT = (k_B/m)\hat{c}_v^V$.

8.3 Nonequilibrium Theory with Two Molecular Relaxation Processes: ET$_{15}$

Since we are now considering the rotational and vibrational modes separately, we assume 15 balance laws as follows:

$$
\begin{aligned}
&\frac{\partial F}{\partial t} + \frac{\partial F_i}{\partial x_i} = 0, \\
&\frac{\partial F_j}{\partial t} + \frac{\partial F_{ij}}{\partial x_i} = 0, \\
&\frac{\partial F_{ij}}{\partial t} + \frac{\partial F_{ijk}}{\partial x_k} = P_{ij}^K, \qquad \frac{\partial H_{ll}^R}{\partial t} + \frac{\partial H_{lli}^R}{\partial x_i} = P_{ll}^R, \qquad \frac{\partial H_{ll}^V}{\partial t} + \frac{\partial H_{lli}^V}{\partial x_i} = P_{ll}^V, \\
&\qquad\qquad\qquad\qquad \frac{\partial G_{lli}}{\partial t} + \frac{\partial G_{llik}}{\partial x_k} = Q_{lli},
\end{aligned} \tag{8.13}
$$

where F, F_i are defined in (8.2), and the other moments are defined as

$$\begin{aligned}
F_{ij} &= \int_{\mathbb{R}^3}\int_0^\infty\int_0^\infty m\, c_i c_j f \varphi\left(I^R\right)\psi\left(I^V\right)\, dI^R dI^V d\mathbf{c},\\
H_{ll}^R &= \int_{\mathbb{R}^3}\int_0^\infty\int_0^\infty 2I^R f \varphi\left(I^R\right)\psi\left(I^V\right)\, dI^R dI^V d\mathbf{c},\\
H_{ll}^V &= \int_{\mathbb{R}^3}\int_0^\infty\int_0^\infty 2I^V f \varphi\left(I^R\right)\psi\left(I^V\right)\, dI^R dI^V d\mathbf{c},\\
G_{lli} &= \int_{\mathbb{R}^3}\int_0^\infty\int_0^\infty m\, c_i \left(c^2 + \frac{2}{m}I^R + \frac{2}{m}I^V\right) f \varphi\left(I^R\right)\psi\left(I^V\right)\, dI^R dI^V d\mathbf{c}.
\end{aligned} \tag{8.14}$$

The fluxes are defined by

$$\begin{aligned}
F_{ijk} &= \int_{\mathbb{R}^3}\int_0^\infty\int_0^\infty m\, c_i c_j c_k f \varphi\left(I^R\right)\psi\left(I^V\right)\, dI^R dI^V d\mathbf{c},\\
H_{lli}^R &= \int_{\mathbb{R}^3}\int_0^\infty\int_0^\infty 2I^R c_i f \varphi\left(I^R\right)\psi\left(I^V\right)\, dI^R dI^V d\mathbf{c},\\
H_{lli}^V &= \int_{\mathbb{R}^3}\int_0^\infty\int_0^\infty 2I^V c_i f \varphi\left(I^R\right)\psi\left(I^V\right)\, dI^R dI^V d\mathbf{c},\\
G_{llik} &= \int_{\mathbb{R}^3}\int_0^\infty\int_0^\infty m\, c_i c_k \left(c^2 + \frac{2}{m}I^R + \frac{2}{m}I^V\right) f \varphi\left(I^R\right)\psi\left(I^V\right)\, dI^R dI^V d\mathbf{c},
\end{aligned}$$

and the productions by

$$\begin{aligned}
P_{ij}^K &= \int_{\mathbb{R}^3}\int_0^\infty\int_0^\infty m\, c_i c_j Q(f) \varphi\left(I^R\right)\psi\left(I^V\right)\, dI^R dI^V d\mathbf{c},\\
P_{ll}^R &= \int_{\mathbb{R}^3}\int_0^\infty\int_0^\infty 2I^R Q(f) \varphi\left(I^R\right)\psi\left(I^V\right)\, dI^R dI^V d\mathbf{c},\\
P_{ll}^V &= \int_{\mathbb{R}^3}\int_0^\infty\int_0^\infty 2I^V Q(f) \varphi\left(I^R\right)\psi\left(I^V\right)\, dI^R dI^V d\mathbf{c},\\
Q_{lli} &= \int_{\mathbb{R}^3}\int_0^\infty\int_0^\infty m\, c_i \left(c^2 + \frac{2}{m}I^R + \frac{2}{m}I^V\right) Q(f) \varphi\left(I^R\right)\psi\left(I^V\right)\, dI^R dI^V d\mathbf{c}.
\end{aligned} \tag{8.15}$$

Since the sum of the balance equations for F_{ll}, H_{ll}^R, H_{ll}^V expresses the conservation law of energy, we have

$$P_{ll}^K + P_{ll}^R + P_{ll}^V = 0.$$

It should be noted also that the balance equation of the energy flux G_{lli} is obtained by summing the balance equations of F_{lli}, H^R_{lli}, and H^V_{lli} that appear in (8.13), then we have

$$G_{llik} = F_{llik} + H^R_{llik} + H^V_{llik}, \qquad Q_{lli} = P^K_{lli} + P^R_{lli} + P^V_{lli}.$$

It is worth emphasizing again that the ET_{14} theory adopts $F, F_i, F_{ij}, G_{ll}(= F_{ll} + H^R_{ll} + H^V_{ll})$ and G_{lli} as independent fields and the internal modes of a molecule are treated as a unit. On the other hand, ET_{15} describes the rotational and vibrational modes individually.

8.3.1 Galilean Invariance and Intrinsic Variable

The densities of the system (8.13) are related to the following conventional field variables:

mass density:

$$\rho = \int_{\mathbb{R}^3}\int_0^\infty\int_0^\infty mf\varphi\left(I^R\right)\psi\left(I^V\right)dI^R dI^V d\mathbf{c},$$

velocity:

$$v_i = \frac{1}{\rho}\int_{\mathbb{R}^3}\int_0^\infty\int_0^\infty mc_i f\varphi\left(I^R\right)\psi\left(I^V\right)dI^R dI^V d\mathbf{c},$$

specific translational energy density:

$$\varepsilon^K = \frac{1}{2\rho}\int_{\mathbb{R}^3}\int_0^\infty\int_0^\infty mC^2 f\varphi\left(I^R\right)\psi\left(I^V\right)dI^R dI^V d\mathbf{C}, \tag{8.16}$$

specific rotational energy density:

$$\varepsilon^R = \frac{1}{\rho}\int_{\mathbb{R}^3}\int_0^\infty\int_0^\infty I^R f\varphi\left(I^R\right)\psi\left(I^V\right)dI^R dI^V d\mathbf{C}, \tag{8.17}$$

specific vibrational energy density:

$$\varepsilon^V = \frac{1}{\rho}\int_{\mathbb{R}^3}\int_0^\infty\int_0^\infty I^V f\varphi\left(I^R\right)\psi\left(I^V\right)dI^R dI^V d\mathbf{C}, \tag{8.18}$$

shear stress:

$$\sigma_{\langle ij\rangle} = -\int_{\mathbb{R}^3}\int_0^\infty\int_0^\infty mC_{\langle i}C_{j\rangle} f\varphi\left(I^R\right)\psi\left(I^V\right)dI^R dI^V d\mathbf{C},$$

nonequilibrium pressure:

$$\mathscr{P} = \frac{2}{3}\rho\varepsilon^K = \frac{1}{3}\int_{\mathbb{R}^3}\int_0^\infty\int_0^\infty mC^2 f \varphi\left(I^R\right)\psi\left(I^V\right) dI^R dI^V d\mathbf{C}, \tag{8.19}$$

heat flux:

$$q_i = \frac{1}{2}\int_{\mathbb{R}^3}\int_0^\infty\int_0^\infty (mC^2 + 2I^R + 2I^V)C_i f \varphi\left(I^R\right)\psi\left(I^V\right) dI^R dI^V d\mathbf{C}.$$

We have seen in Chap. 7 that there is a one-to-one correspondence between (T, Π) and the nonequilibrium energies $(\varepsilon^K, \varepsilon^I)$, or between (T, Π) and the nonequilibrium temperatures (θ^K, θ^I), where the dynamic pressure Π is given by $\mathscr{P} - p$. While, in the present case, we have three new nonequilibrium variables θ^K, θ^R, and θ^V associated with the nonequilibrium specific energies ε^K, ε^R, and ε^V in (8.16), (8.17), and (8.18) through following relations:

$$\varepsilon^K = \varepsilon^K_E(\theta^K), \qquad \varepsilon^R = \varepsilon^R_E(\theta^R), \qquad \varepsilon^V = \varepsilon^V_E(\theta^V).$$

Because of the monotonicity of $\varepsilon_E^{K,R,V}$, these are one-to-one relationships among the new variables $\theta^{K,R,V}$ and the specific energies $\varepsilon^{K,R,V}$. From (8.6), the temperature T is determined in an implicit way by the relation:

$$\varepsilon_E(T) = \varepsilon^K + \varepsilon^R + \varepsilon^V \quad \text{in terms of nonequilibrium energies,} \tag{8.20}$$

or by the relation:

$$\varepsilon_E(T) = \varepsilon^K_E(\theta^K) + \varepsilon^R_E(\theta^R) + \varepsilon^V_E(\theta^V) \quad \text{in terms of nonequilibrium temperatures.} \tag{8.21}$$

Another possible choice of nonequilibrium variables is to adopt the following energy differences between nonequilibrium and local equilibrium energies:

$$\begin{aligned}
\Delta^K &= \varepsilon^K_E(\theta^K) - \varepsilon^K_E(T),\\
\Delta^R &= \varepsilon^R_E(\theta^R) - \varepsilon^R_E(T),\\
\Delta^V &= \varepsilon^V_E(\theta^V) - \varepsilon^V_E(T).
\end{aligned}$$

These energy differences indicate the energy exchange among the three molecular modes, i.e., translational, rotational, and vibrational modes. But we need to observe that only 2 of these are independent. In fact we have

$$\Delta^K + \Delta^R + \Delta^V = 0. \tag{8.22}$$

From (8.19), we notice that, also in this case, the nonequilibrium pressure is different from the equilibrium one. In fact, we have the following relation:

$$\mathscr{P} = p + \frac{2}{3}\rho\Delta^K.$$

On the other hand, the difference between $\mathscr{P}$ and p is nothing but the dynamical pressure Π. Therefore we have the following relations:

$$\Pi = \mathscr{P} - p = \frac{2}{3}\rho\Delta^K = \frac{2}{3}\rho\left(\varepsilon^K - \varepsilon_E^K\right) = \frac{2}{3}\rho\left(\varepsilon_E^K(\theta^K) - \varepsilon_E^K(T)\right). \quad (8.23)$$

To sum up, in addition to the 12 variables $(\rho, v_i, \sigma_{\langle ij\rangle}, q_i)$, we have several possible choices for the remaining 3 variables, among which there are one-to-one relationships. Three typical choices are as follows:

- $(\varepsilon^K, \varepsilon^R, \varepsilon^V)$, in this case the local equilibrium temperature T is obtained by (8.20);
- $(\theta^K, \theta^R, \theta^V)$, in this case the local equilibrium temperature T is obtained by (8.21);
- T and two of $(\Delta^K, \Delta^R, \Delta^V)$.

Remark 8.2 In the general framework of RET, nonequilibrium temperatures are defined in terms of the main field. For details, see Chap. 15. It was proved in [136] that the variables $(\theta^K, \theta^R, \theta^V)$ defined above can be regarded as the nonequilibrium temperatures at least not so far from local equilibrium.

Remark 8.3 An advantage of using the variables $(\theta^K, \theta^R, \theta^V)$ may be understood well when we want to study the following practical problem: We firstly prepare a gas in equilibrium with the temperature T_0 in a region Ω. Then we excite only the rotational mode (or vibrational mode) of a gas in a small region $\Omega_0 (\subset \Omega)$ from the temperature T_0 to the temperature θ_0^R $(> T_0)$ (or θ_0^V $(> T_0)$) instantaneously at the initial time. And we analyze, by using the ET$_{15}$ theory, the time-evolution of relaxations among the modes from such a nonequilibrium initial state. A numerical simulation of this problem is being prepared and its results will be shown elsewhere.

Remark 8.4 There is a possibility that rotational and vibrational nonequilibrium temperatures defined above may be measured directly, for example, by using infrared spectroscopy.

Since the intrinsic variables are the moments in terms of the peculiar velocity C_i instead of c_i, we have

$$\begin{aligned}
&F = \rho, \quad F_i = \rho v_i, \quad F_{ll} = 2\rho\varepsilon^K + \rho v^2, \quad F_{\langle ij\rangle} = -\sigma_{\langle ij\rangle} + \rho v_{\langle i}v_{j\rangle},\\
&H_{ll}^R = 2\rho\varepsilon^R, \quad H_{ll}^V = 2\rho\varepsilon^V,\\
&G_{lli} = 2q_i + \{2(\rho\varepsilon + \mathscr{P})\}v_i - 2\sigma_{\langle li\rangle}v_l + \rho v^2 v_i.
\end{aligned}$$

Similarly, the velocity dependence of the fluxes and productions is obtained as follows:

$$
\begin{aligned}
&F_{ijk} = \hat{F}_{ijk} + 3\hat{F}_{(ij}v_{k)} + \rho v_i v_j v_k, \\
&H^R_{lli} = \hat{H}^R_{ll} v_i + \hat{H}^R_{lli}, \qquad H^V_{lli} = \hat{H}^V_{ll} v_i + \hat{H}^V_{lli}, \\
&G_{llik} = \hat{G}_{llik} + 2\hat{G}_{ll(i}v_{k)} + 2v_l\hat{F}_{lik} + 6v_{(i}v_l\hat{F}_{lk)} + v_i v_k\left(\hat{H}^R_{ll} + \hat{H}^V_{ll}\right) + \rho v^2 v_i v_k, \\
&P^K_{ij} = \hat{P}^K_{ij}, \qquad P^R_{ll} = \hat{P}^R_{ll}, \qquad P^V_{ll} = \hat{P}^V_{ll}, \\
&Q_{lli} = 2v_l\hat{P}^K_{il} + \hat{Q}_{lli},
\end{aligned}
$$

where a hat on a quantity indicates the intrinsic part of the quantity.

8.3.2 *Nonequilibrium Distribution Function Derived from MEP*

To close the system (8.13), we need the nonequilibrium distribution function f, which is derived from the MEP under the constraints (8.2) and (8.14). We have the following theorem:

Theorem 8.2 *The nonequilibrium distribution function f in ET$_{15}$ obtained by using the MEP is in the following form, if a process is not so far from equilibrium:*

$$
\begin{aligned}
f =& f^{(E)}\left\{1 - \frac{\rho^2}{p^2}\left[\frac{\varepsilon^K_E}{\hat{c}^K_v}\Delta^K + \frac{\varepsilon^R_E}{\hat{c}^R_v}\Delta^R + \frac{\varepsilon^V_E}{\hat{c}^V_v}\Delta^V - \frac{2\varepsilon^I_E + 5p/\rho}{p(5+2\hat{c}^I_V)}q_i C_i - \frac{1}{\rho}\sigma_{\langle ij\rangle}C_iC_j\right.\right. \\
&\left.\left. - \frac{\Delta^K}{2\hat{c}^K_v}C^2 - \frac{I^R}{m}\frac{\Delta^R}{\hat{c}^R_v} - \frac{I^V}{m}\frac{\Delta^V}{\hat{c}^V_v} + \frac{q_iC_i}{p(5+2\hat{c}^I_V)}\left(C^2 + 2\frac{I^R + I^V}{m}\right)\right]\right\},
\end{aligned}
\tag{8.24}
$$

where $f^{(E)}$ is the (local) equilibrium distribution function (8.4) *with* (8.5)*, $\varepsilon^I_E = \varepsilon^R_E + \varepsilon^V_E$, $\hat{c}^I_v = \hat{c}^R_v + \hat{c}^V_v$, and Δ's obey the constraint* (8.22).

Proof The most suitable distribution function f of the truncated system (8.13) is obtained by maximizing the functional defined by

$$
\begin{aligned}
\mathscr{L}(f) =& -k_B \int_{\mathbb{R}^3}\int_0^\infty\int_0^\infty f\log f\,\varphi\left(I^R\right)\psi\left(I^V\right)dI^R dI^V d\mathbf{c} \\
&+ \lambda\left(F - \int_{\mathbb{R}^3}\int_0^\infty\int_0^\infty mf\varphi\left(I^R\right)\psi\left(I^V\right)dI^R dI^V d\mathbf{c}\right)
\end{aligned}
$$

$$+\lambda_i\left(F_i-\int_{\mathbb{R}^3}\int_0^\infty\int_0^\infty mc_i f\varphi\left(I^R\right)\psi\left(I^V\right)dI^R dI^V d\mathbf{c}\right)$$
$$+\lambda_{ij}\left(F_{ij}-\int_{\mathbb{R}^3}\int_0^\infty\int_0^\infty mc_ic_j f\varphi\left(I^R\right)\psi\left(I^V\right)dI^R dI^V d\mathbf{c}\right)$$
$$+\mu^R\left(H^R_{ll}-\int_{\mathbb{R}^3}\int_0^\infty\int_0^\infty 2I^R f\varphi\left(I^R\right)\psi\left(I^V\right)dI^R dI^V d\mathbf{c}\right)$$
$$+\mu^V\left(H^V_{ll}-\int_{\mathbb{R}^3}\int_0^\infty\int_0^\infty 2I^V f\varphi\left(I^R\right)\psi\left(I^V\right)dI^R dI^V d\mathbf{c}\right)$$
$$+\mu_i\left(G_{lli}-\int_{\mathbb{R}^3}\int_0^\infty\int_0^\infty m\,c_i\left(c^2+\frac{2}{m}I^R+\frac{2}{m}I^V\right)f\varphi\left(I^R\right)\psi\left(I^V\right)dI^R dI^V d\mathbf{c}\right),$$

where λ, λ_i, λ_{ij}, μ^R, μ^V, and μ_i are the Lagrange multipliers. Their velocity dependences are determined as follows:

$$\begin{aligned}
\lambda &= \hat{\lambda} - v_i\hat{\lambda}_i + v_iv_j\hat{\lambda}_{ij} - v^2 v_i\hat{\mu}_i,\\
\lambda_i &= \hat{\lambda}_i - 2v_j\hat{\lambda}_{ji} + 2v_iv_j\hat{\mu}_j + v^2\hat{\mu}_i,\\
\lambda_{ij} &= \hat{\lambda}_{ij} - v_l\hat{\mu}_l\delta_{ij} - v_i\hat{\mu}_j - v_j\hat{\mu}_i,\\
\mu^R &= \hat{\mu}^R - v_i\hat{\mu}_i,\\
\mu^V &= \hat{\mu}^V - v_i\hat{\mu}_i,\\
\mu_i &= \hat{\mu}_i.
\end{aligned}$$

The distribution function f, which satisfies $\delta\mathscr{L}/\delta f = 0$, is given by

$$f = \exp\left(-1-\frac{m}{k_B}\chi\right),$$
$$\chi = \hat{\lambda} + C_i\hat{\lambda}_i + C_iC_j\hat{\lambda}_{ij} + \frac{2I^R}{m}\hat{\mu}^R + \frac{2I^V}{m}\hat{\mu}^V + \left(C^2+\frac{2I^R}{m}+\frac{2I^V}{m}\right)C_i\hat{\mu}_i. \tag{8.25}$$

Taking into account that, in equilibrium, f_{15} coincides with the equilibrium distribution function (8.6), we can easily see that the equilibrium components of the Lagrange multipliers are given by

$$\lambda_E = -\frac{g}{T}+\frac{v^2}{2T},\quad \lambda_{i_E} = -\frac{v_i}{T},\quad \frac{\lambda_{ll_E}}{3} = \mu^R_E = \mu^V_E = \frac{1}{2T},\quad \lambda_{\langle ij\rangle_E} = 0.\quad \mu_{i_E} = 0,$$

Because of the reason explained in Chap. 7, we focus on the processes near equilibrium. Then we expand (8.25) around an equilibrium state in the following form:

$$
\begin{aligned}
f &= f^{(E)}\left(1-\frac{m}{k_B}\tilde{\chi}\right),\\
\tilde{\chi} &= \tilde{\lambda}+C_i\tilde{\lambda}_i+C_iC_j\tilde{\lambda}_{ij}+\frac{2I^R}{m}\tilde{\mu}^R+\frac{2I^V}{m}\tilde{\mu}^V+\left(C^2+\frac{2I^R}{m}+\frac{2I^V}{m}\right)C_i\tilde{\mu}_i,
\end{aligned}
\tag{8.26}
$$

where a tilde on a quantity indicates its nonequilibrium part.

Inserting (8.26) into (4.59), and proceeding as in Chap. 7, we obtain the intrinsic nonequilibrium Lagrange multipliers as functions of the field variables:

$$
\begin{aligned}
\tilde{\lambda} &= \frac{\varepsilon_E^K(T)}{T^2c_v^K(T)}\Delta^K+\frac{\varepsilon_E^R(T)}{T^2c_v^R(T)}\Delta^R+\frac{\varepsilon_E^V(T)}{T^2c_v^V(T)}\Delta^V,\\
\tilde{\lambda}_i &= \frac{p+\rho\varepsilon(T)}{p^2T\left(1+\hat{c}_v(T)\right)}q_i,\\
\tilde{\nu} &= \frac{\tilde{\lambda}_{ll}}{3}=-\frac{\Delta^K}{2T^2c_v^K(T)},\\
\tilde{\mu}^R &= -\frac{\Delta^R}{2T^2c_v^R(T)},\quad \tilde{\mu}^V=-\frac{\Delta^V}{2T^2c_v^V(T)},\\
\tilde{\lambda}_{\langle ij\rangle} &= \frac{\sigma_{\langle ij\rangle}}{2pT},\quad \tilde{\mu}_i=-\frac{\rho q_i}{2p^2T\left(1+\hat{c}_v(T)\right)},
\end{aligned}
\tag{8.27}
$$

where $c_v^{K,R,V}(T) = \mathrm{d}\varepsilon_E^{K,R,V}(T)/\mathrm{d}T = (k_B/m)\hat{c}_v^{K,R,V}(T)$ are the specific heats at constant volume of the three modes. Finally, by inserting (8.27) into (8.26), the theorem is proved.

8.3.3 *Closure of the System*

Using the distribution function (8.24), we obtain the constitutive equations for the fluxes up to the first order with respect to the nonequilibrium variables as follows:

$$
\begin{aligned}
\hat{F}_{ijk} &= \int_{\mathbb{R}^3}\int_0^\infty\int_0^\infty mC_iC_jC_kf\,\varphi\left(I^R\right)\psi\left(I^V\right)dI^RdI^Vd\mathbf{C}\\
&= \frac{1}{1+\hat{c}_v}(q_k\delta_{ij}+q_j\delta_{ik}+q_i\delta_{jk}),
\end{aligned}
$$

$$
\begin{aligned}
\hat{H}^R_{llk} &= \int_{\mathbb{R}^3}\int_0^\infty\int_0^\infty 2I^R C_k f\,\varphi\left(I^R\right)\psi\left(I^V\right)\,dI^R dI^V d\mathbf{C} \\
&= \frac{2\,\hat{c}^R_v}{1+\hat{c}_v}q_k, \\
\hat{H}^V_{llk} &= \int_{\mathbb{R}^3}\int_0^\infty\int_0^\infty 2I^V C_k f\,\varphi\left(I^R\right)\psi\left(I^V\right)\,dI^R dI^V d\mathbf{C} \\
&= \frac{2\,\hat{c}^V_v}{1+\hat{c}_v}q_k, \\
\hat{G}_{llik} &= \int_{\mathbb{R}^3}\int_0^\infty\int_0^\infty (mC^2+2I^R+2I^V)C_iC_k f\,\varphi\left(I^R\right)\psi\left(I^V\right)\,dI^R dI^V d\mathbf{C} \\
&= -2\left\{\frac{p^2}{\rho}-\frac{2}{3}\rho\left(\varepsilon+2\frac{p}{\rho}\right)\left(\varepsilon^K_E+\Delta^K\right)\right\}\delta_{ik}-2\left(\varepsilon+2\frac{p}{\rho}\right)\sigma_{\langle ik\rangle}.
\end{aligned}
$$

With these constitutive equations, we obtain the closed system of field equations for the 15 independent fields, ρ, v_i, T, $\sigma_{\langle ij\rangle}$, q_i and two of the fields Δ^K, Δ^R, Δ^V:

$$
\begin{aligned}
&\frac{\partial\rho}{\partial t}+\frac{\partial}{\partial x_i}(\rho v_i)=0, \\
&\frac{\partial\rho v_j}{\partial t}+\frac{\partial}{\partial x_i}\left\{(p+\Pi)\delta_{ij}-\sigma_{\langle ij\rangle}+\rho v_i v_j\right\}=0, \\
&\frac{\partial}{\partial t}\left\{2\rho\left(\varepsilon^K_E(T)+\Delta^K\right)+\rho v^2\right\}+\frac{\partial}{\partial x_i}\left\{\frac{5}{1+\hat{c}_v}q_i+5(p+\Pi)v_i\right. \\
&\qquad\qquad\left.-2\sigma_{\langle li\rangle}v_l+\rho v^2 v_i\right\}=\hat{P}^K_{ll}, \\
&\frac{\partial}{\partial t}\left\{2\rho\left(\varepsilon^R_E(T)+\Delta^R\right)\right\}+\frac{\partial}{\partial x_i}\left\{2\rho\left(\varepsilon^R_E(T)+\Delta^R\right)v_i+\frac{2\hat{c}^R_v}{1+\hat{c}_v}q_i\right\}=\hat{P}^R_{ll}, \\
&\frac{\partial}{\partial t}\left\{2\rho\left(\varepsilon^V_E(T)+\Delta^V\right)\right\}+\frac{\partial}{\partial x_i}\left\{2\rho\left(\varepsilon^V_E(T)+\Delta^V\right)v_i+\frac{2\hat{c}^V_v}{1+\hat{c}_v}q_i\right\}=\hat{P}^V_{ll}, \\
&\frac{\partial}{\partial t}\left\{-\sigma_{\langle ij\rangle}+\rho v_{\langle i}v_{j\rangle}\right\} \\
&+\frac{\partial}{\partial x_k}\left\{\frac{2}{1+\hat{c}_v}q_{\langle i}\delta_{j\rangle k}+2(p+\Pi)v_{\langle i}\delta_{j\rangle k}-\sigma_{\langle ij\rangle}v_k-2\sigma_{\langle k\langle i\rangle}v_{j\rangle}+\rho v_{\langle i}v_{j\rangle}v_k\right\}=\hat{P}^K_{\langle ij\rangle}, \\
&\frac{\partial}{\partial t}\left\{2q_i+2\left(\rho\varepsilon+p+\Pi\right)v_i-2\sigma_{\langle li\rangle}v_l+\rho v^2 v_i\right\}+ \\
&\quad+\frac{\partial}{\partial x_k}\left\{2\left[p\left(\varepsilon+\frac{p}{\rho}\right)+\Pi\left(\varepsilon+2\frac{p}{\rho}\right)\right]\delta_{ik}-2\left(\varepsilon+2\frac{p}{\rho}\right)\sigma_{\langle ik\rangle}\right.
\end{aligned}
$$

$$+\frac{2}{1+\hat{c}_v}q_l v_l \delta_{ik} + 2\left(1+\frac{1}{1+\hat{c}_v}\right)(q_i v_k + q_k v_i) + (p+\Pi)v^2\delta_{ik}$$
$$+2(\rho\varepsilon + 2p + 2\Pi)v_i v_k - v^2\sigma_{\langle ik\rangle} - 2v_l v_i \sigma_{\langle lk\rangle} - 2v_l v_k \sigma_{\langle il\rangle}$$
$$\left. + \rho v^2 v_i v_k \right\} = 2v_l \hat{P}^K_{il} + \hat{Q}_{lli} \tag{8.28}$$

with Π being given by (8.23) and Δ's being constrained by (8.22).

8.4 Entropy Density, Flux, and Production

The nonequilibrium part of the entropy density $h - \rho s$ and the intrinsic entropy flux (non-convective entropy flux) $\varphi_i (= h_i - h v_i)$ up to the second order with respect to the dissipative fluxes for the truncated system (8.13) are obtained, from (8.3) with the distribution function (8.26), as follows:

$$\begin{aligned} h - \rho s =& -\frac{\rho}{2T^2}\frac{(\Delta^K)^2}{c_v^K} - \frac{\rho}{2T^2}\frac{(\Delta^R)^2}{c_v^R} - \frac{\rho}{2T^2}\frac{(\Delta^V)^2}{c_v^V} \\ & - \frac{\sigma_{\langle ik\rangle}\sigma_{\langle ik\rangle}}{4pT} - \frac{\rho q_i q_i}{2p^2 T(1+\hat{c}_v)}, \\ \varphi_i =& \frac{q_i}{T} - \frac{1}{pT(1+\hat{c}_v)}\left(\frac{2\rho}{3}\Delta^K q_i - q_j \sigma_{\langle ij\rangle}\right). \end{aligned}$$

And the entropy production is given by

$$\Sigma = \hat{P}^K_{ij}\tilde{\lambda}_{ij} + \hat{P}^R_{ll}\tilde{\mu}^R + \hat{P}^V_{ll}\tilde{\mu}^V + \hat{Q}_{lli}\tilde{\mu}_i. \tag{8.29}$$

8.5 Generalized BGK Model

Concerning the collision term, as explained in Chap. 7, Struchtrup [307] and Rahimi and Struchtrup [308] proposed a variant of the BGK model [309] to take into account a relaxation of the energy of the internal mode. In this section we introduce a simple collision term with three relaxation times in order to describe a more refined model in which rotational and vibrational modes are treated individually.

8.5.1 Three Relaxation Times

In polyatomic gases, we may introduce three characteristic times corresponding to three relaxation processes caused by the molecular collision (see also [311–314]):

(i) Relaxation time τ_K: This characterizes the relaxation process within the translational mode (mode K) of molecules. The process shows the tendency to approach an equilibrium state of the mode K with the distribution function $f_K^{(E)}$ having the temperature θ^K, explicit expression of which is shown below. However, the rotational and vibrational modes are, in general, in nonequilibrium. This process is observable also in monatomic gases.

(ii) Relaxation time $\tau_{\mathfrak{bc}}$: There are energy exchanges among the three modes: mode K, rotational mode (mode R), and vibrational mode (mode V). The relaxation process occurs in such a way that two of the three modes, say $(\mathfrak{bc})$ (= (KR), (KV), (RV)) approach, after the relaxation time $\tau_{\mathfrak{bc}}$, an equilibrium state characterized by the distribution function $f_{\mathfrak{bc}}^{(E)}$ with a common temperature $\theta^{\mathfrak{bc}}$, explicit expression of which is shown below. Because of the lack of experimental data, we have no reliable magnitude-relationship between τ_K and $\tau_{\mathfrak{bc}}$. However it seems natural to adopt the relation: $O(\tau_{\mathfrak{bc}}) \gtrsim O(\tau_K)$, which we assume hereafter. In Table 8.1, possible three cases are summarized depending on the choice of $\mathfrak{b}$ and $\mathfrak{c}$.

(iii) Relaxation time τ of the last stage: After the relaxation process between $\mathfrak{b}$ and $\mathfrak{c}$, all modes, K, R, and V, eventually approach a local equilibrium state characterized by $f^{(E)}$ with a common temperature T among K, R, and V-modes, which is given by (8.4). Naturally we have the relation: $\tau > \tau_{\mathfrak{bc}}$.

Diagrams of the possible relaxation processes are shown in Fig. 8.1.

8.5.2 Generalized BGK Collision Term

The generalized BGK collision term for $(\mathfrak{bc})$-process ($(\mathfrak{bc})$ = (KR), (KV), (RV)) is proposed as follows:

$$Q^{\mathfrak{bc}}(f) = -\frac{1}{\tau_K}(f - f_K^{(E)}) - \frac{1}{\tau_{\mathfrak{bc}}}(f - f_{\mathfrak{bc}}^{(E)}) - \frac{1}{\tau}(f - f^{(E)}), \tag{8.30}$$

Table 8.1 Three possible relaxation processes in the second stage (ii)

$(\mathfrak{bc})$-Process	$(\mathfrak{a}, \mathfrak{b}, \mathfrak{c})$	Relaxation time	Collision term
(KR)-process	(V, K, R)	τ_{KR}	$Q^{KR}(f)$
(KV)-process	(R, K, V)	τ_{KV}	$Q^{KV}(f)$
(RV)-process	(K, R, V)	τ_{RV}	$Q^{RV}(f)$

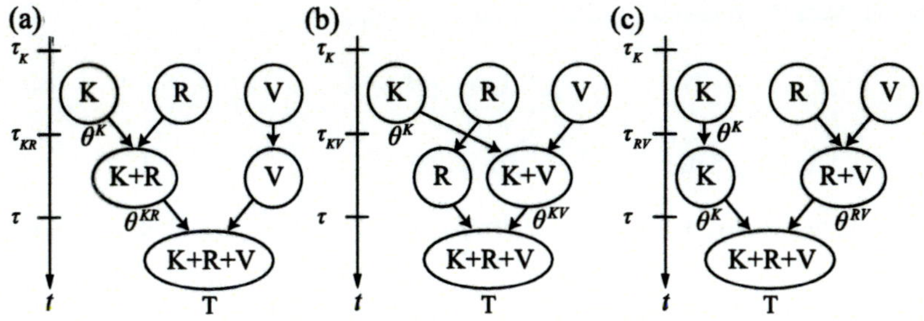

Fig. 8.1 Diagram of the three possible relaxation processes of the translational mode (K), rotational mode (R), and vibrational mode (V) for (**a**) (KR)-process, (**b**) (KV)-process, and (**c**) (RV)-process. The symbols θ^K, $\theta^{\mathfrak{bc}}$ (($\mathfrak{bc}$) = (KR), (KV), (RV)) are partial equilibrium temperatures and T is the local equilibrium temperature. A mode without attaching a symbol of the temperature is not necessarily in partial equilibrium

where the distribution functions $f_K^{(E)}$ and $f_{\mathfrak{bc}}^{(E)}$ are given as follows:

Distribution Function $f_K^{(E)}$ This is given by

$$f_K^{(E)} = \frac{\rho^{RV}(I^R, I^V)}{m}\left(\frac{m}{2\pi k_B \theta^K}\right)^{3/2} \exp\left(-\frac{mC^2}{2k_B\theta^K}\right), \tag{8.31}$$

where

$$\rho^{RV}(I^R, I^V) = \int_{\mathbb{R}^3} mf d\mathbf{c}. \tag{8.32}$$

This is the equilibrium function with respect to the K-mode with the temperature θ^K and with the "frozen" energies, I^R and I^V. In other words, $f_K^{(E)}$ is a Maxwellian with the mass density $\rho^{RV}(I^R, I^V)$ and temperature θ^K. Therefore $f_K^{(E)}$ given in (8.31) is obtained by maximizing the entropy with the frozen energies, which is not the true entropy (8.3):

$$h^{RV}(I^R, I^V) = -k_B \int_{\mathbb{R}^3} f \log f \, d\mathbf{c}$$

under the constraints:

$$\begin{pmatrix} \rho^{RV}(I^R, I^V) \\ \rho^{RV}(I^R, I^V)v_i \\ 2\rho^{RV}(I^R, I^V)\varepsilon_E^K(\theta^K) \end{pmatrix} = \int_{\mathbb{R}^3} \begin{pmatrix} m \\ mc_i \\ mC^2 \end{pmatrix} f d\mathbf{c}.$$

Then we have the relation (8.32), and the relation:

$$\varepsilon^K = \varepsilon_E^K(\theta^K),$$

from which we can determine the temperature θ^K.

Distribution Function $f_{KR}^{(E)}$ Let us study the process in which the K and R-modes reach their common equilibrium with the temperature θ^{KR} and the vibrational energy I^V can be considered as frozen. In this case, we have the distribution function:

$$f_{KR}^{(E)} = \frac{\rho^V(I^V)}{mA^R(\theta^{KR})}\left(\frac{m}{2\pi k_B\theta^{KR}}\right)^{3/2}\exp\left\{-\frac{1}{k_B\theta^{KR}}\left(\frac{mC^2}{2}+I^R\right)\right\}, \tag{8.33}$$

where

$$\rho^V(I^V) = m\int_{\mathbb{R}^3}\int_0^\infty f\varphi\left(I^R\right)dI^R d\mathbf{c}. \tag{8.34}$$

Since the equilibrium state is described by the mass density $\rho^V(I^V)$ with the frozen vibrational energy I^V and the internal energy $\varepsilon_E^{K+R}(\theta^{KR}) \equiv \varepsilon_E^K(\theta^{KR}) + \varepsilon_E^R(\theta^{KR})$, we obtain $f_{KR}^{(E)}$ by using the MEP for the entropy:

$$h^V(I^V) = -k_B\int_{\mathbb{R}^3}\int_0^\infty f\log f\,\varphi\left(I^R\right)dI^R\,d\mathbf{c}$$

under the constraints:

$$\begin{pmatrix}\rho^V(I^V)\\ \rho^V(I^V)v_i\\ 2\rho^V(I^V)\varepsilon_E^{K+R}(\theta^{KR})\end{pmatrix} = \int_{\mathbb{R}^3}\int_0^\infty\begin{pmatrix}m\\ mc_i\\ mC^2+2I^R\end{pmatrix}f\varphi\left(I^R\right)dI^R\,d\mathbf{c}.$$

Therefore we have the relation (8.34), and the relation:

$$\varepsilon^{K+R} \equiv \varepsilon^K + \varepsilon^R = \varepsilon_E^K(\theta^{KR}) + \varepsilon_E^R(\theta^{KR}) \equiv \varepsilon_E^{K+R}(\theta^{KR}), \tag{8.35}$$

from which we can determine the temperature θ^{KR}.

Distribution Function $f_{KV}^{(E)}$ In a similar way, we have

$$f_{KV}^{(E)} = \frac{\rho^R(I^R)}{mA^V(\theta^{KV})}\left(\frac{m}{2\pi k_B\theta^{KV}}\right)^{3/2}\exp\left\{-\frac{1}{k_B\theta^{KV}}\left(\frac{mC^2}{2}+I^V\right)\right\}, \tag{8.36}$$

where

$$\rho^R(I^R) = \int_{\mathbb{R}^3}\int_0^\infty mf\psi\left(I^V\right)dI^V d\mathbf{c}.$$

And we have the relation, from which we can determine the temperature θ^{KV}:

$$\varepsilon^{K+V} \equiv \varepsilon^K + \varepsilon^V = \varepsilon_E^K(\theta^{KV}) + \varepsilon_E^V(\theta^{KV}) \equiv \varepsilon_E^{K+V}(\theta^{KV}). \tag{8.37}$$

Distribution Function $f_{RV}^{(E)}$ In this case the K-mode is in equilibrium with the temperature θ^K, and R and V-modes are also in equilibrium but with the different temperature θ^{RV}. Then we have the expression similar to (8.4):

$$f_{RV}^{(E)} = \frac{\rho}{mA^R(\theta^{RV})A^V(\theta^{RV})}\left(\frac{m}{2\pi k_B\theta^K}\right)^{3/2}\exp\left(-\frac{mC^2}{2k_B\theta^K} - \frac{I^R + I^V}{k_B\theta^{RV}}\right), \tag{8.38}$$

where the temperature θ^{RV} is determined by the relation:

$$\varepsilon^{R+V} \equiv \varepsilon^R + \varepsilon^V = \varepsilon_E^R(\theta^{RV}) + \varepsilon_E^V(\theta^{RV}) \equiv \varepsilon_E^{R+V}(\theta^{RV}). \tag{8.39}$$

Distribution Function $f^{(E)}$ This is the local equilibrium distribution function given by (8.4), in which the temperature T is given by the condition:

$$\varepsilon = \varepsilon_E(T). \tag{8.40}$$

8.5.3 H-theorem

From the definition of the distribution functions $f_K^{(E)}$, $f_{KR}^{(E)}$, $f_{KV}^{(E)}$, $f_{RV}^{(E)}$, and $f^{(E)}$, it is easy to verify the following relations ((𝔟𝔠) = (KR), (KV), (RV)):

$$\int_{\mathbb{R}^3}\int_0^\infty\int_0^\infty (f - f_K^{(E)})\log f_K^{(E)}\varphi\left(I^R\right)\psi\left(I^V\right)dI^R dI^V d\mathbf{c} = 0,$$

$$\int_{\mathbb{R}^3}\int_0^\infty\int_0^\infty (f - f_{\mathfrak{bc}}^{(E)})\log f_{\mathfrak{bc}}^{(E)}\varphi\left(I^R\right)\psi\left(I^V\right)dI^R dI^V d\mathbf{c} = 0,$$

$$\int_{\mathbb{R}^3}\int_0^\infty\int_0^\infty (f - f^{(E)})\log f^{(E)}\varphi\left(I^R\right)\psi\left(I^V\right)dI^R dI^V d\mathbf{c} = 0.$$

Then the entropy production (8.3)$_3$ can easily be shown to be positive:

$$\Sigma = k_B \int_{\mathbb{R}^3}\int_0^\infty\int_0^\infty \left\{ \frac{f - f_K^{(E)}}{\tau^K} \log\frac{f}{f_K^{(E)}} + \frac{f - f_{\mathfrak{bc}}^{(E)}}{\tau_{\mathfrak{bc}}} \log\frac{f}{f_{\mathfrak{bc}}^{(E)}} \right.$$

$$\left. + \frac{f - f^{(E)}}{\tau} \log\frac{f}{f^{(E)}} \right\} \varphi\left(I^R\right) \psi\left(I^V\right) dI^R dI^V d\mathbf{c} \geqslant 0,$$

and the H-theorem holds.

8.6 Production Terms in the Generalized BGK Model

We derive the expressions of the production terms by adopting the generalized BGK model explained in Sect. 8.5.2.

Firstly we notice that, in the case of ET$_{15}$, the mass densities of equilibrium states of the stages (i) and (ii) explained in Sect. 8.5.2, i.e., ρ^{RV}, ρ^V, and ρ^R, which appear, respectively, in (8.31), (8.33), and (8.36), are given as follows:

$$\rho^{RV}(I^R, I^V) = \rho f^{(R)} f^{(V)} \left\{ 1 + \frac{\Delta^R}{\frac{k_B}{m} T^2 c_v^R} \left(\frac{I^R}{m} - \varepsilon^R \right) + \frac{\Delta^V}{\frac{k_B}{m} T^2 c_v^V} \left(\frac{I^V}{m} - \varepsilon^V \right) \right\},$$

$$\rho^V(I^V) = \rho f^{(V)} \left\{ 1 + \frac{\Delta^V}{\frac{k_B}{m} T^2 c_v^V} \left(\frac{I^V}{m} - \varepsilon^V \right) \right\},$$

$$\rho^R(I^R) = \rho f^{(R)} \left\{ 1 + \frac{\Delta^R}{\frac{k_B}{m} T^2 c_v^R} \left(\frac{I^R}{m} - \varepsilon^R \right) \right\},$$

where $f^{(R)}$ and $f^{(V)}$ are defined in (8.7).

Inserting (8.26) with (8.27) into the corresponding components of (8.15), the production terms for ($\mathfrak{bc}$)-process (($\mathfrak{a}$, $\mathfrak{b}$, $\mathfrak{c}$) = (V,K,R), (R,K,V), (K,R,V)) are expressed as follows:

$$\begin{aligned}
\hat{P}_{ll}^{\mathfrak{a}} &= -\frac{2\rho}{\tau} \left(\varepsilon_E^{\mathfrak{a}}(\theta^{\mathfrak{a}}) - \varepsilon_E^{\mathfrak{a}}(T) \right), \\
\hat{P}_{ll}^{\mathfrak{b}} &= -\frac{2\rho}{\tau_{\mathfrak{bc}}} \left(\varepsilon_E^{\mathfrak{b}}(\theta^{\mathfrak{b}}) - \varepsilon_E^{\mathfrak{b}}(\theta^{\mathfrak{bc}}) \right) - \frac{2\rho}{\tau} \left(\varepsilon_E^{\mathfrak{b}}(\theta^{\mathfrak{b}}) - \varepsilon_E^{\mathfrak{b}}(T) \right), \\
\hat{P}_{ll}^{\mathfrak{c}} &= -\frac{2\rho}{\tau_{\mathfrak{bc}}} \left(\varepsilon_E^{\mathfrak{c}}(\theta^{\mathfrak{c}}) - \varepsilon_E^{\mathfrak{c}}(\theta^{\mathfrak{bc}}) \right) - \frac{2\rho}{\tau} \left(\varepsilon_E^{\mathfrak{c}}(\theta^{\mathfrak{c}}) - \varepsilon_E^{\mathfrak{c}}(T) \right), \\
\hat{P}_{\langle ij \rangle}^K &= \left(\frac{1}{\tau_K} + \frac{1}{\tau_{\mathfrak{bc}}} + \frac{1}{\tau} \right) \sigma_{\langle ij \rangle}, \\
\hat{Q}_{lli} &= -2 \left(\frac{1}{\tau_K} + \frac{1}{\tau_{\mathfrak{bc}}} + \frac{1}{\tau} \right) q_i.
\end{aligned} \tag{8.41}$$

For the expression (8.41), it may be useful to introduce the following quantities, i.e., the energy exchanges among the three modes:

$$\delta \equiv \varepsilon_E^{\mathfrak{b}}(\theta^{\mathfrak{b}}) - \varepsilon_E^{\mathfrak{b}}(\theta^{\mathfrak{bc}}) = -\varepsilon_E^{\mathfrak{c}}(\theta^{\mathfrak{c}}) + \varepsilon_E^{\mathfrak{c}}(\theta^{\mathfrak{bc}}),$$
$$\Delta \equiv \Delta^{\mathfrak{a}} = \varepsilon_E^{\mathfrak{a}}(\theta^{\mathfrak{a}}) - \varepsilon_E^{\mathfrak{a}}(T) = -\varepsilon_E^{\mathfrak{b}+\mathfrak{c}}(\theta^{\mathfrak{bc}}) + \varepsilon_E^{\mathfrak{b}+\mathfrak{c}}(T).$$

Expanding the energy exchanges with respect to the nonequilibrium temperatures around a temperature T up to the first order, we obtain

$$\delta = c_v^{\mathfrak{b}}(\theta^{\mathfrak{b}} - \theta^{\mathfrak{bc}}) = -c_v^{\mathfrak{c}}(\theta^{\mathfrak{c}} - \theta^{\mathfrak{bc}}), \qquad \Delta = c_v^{\mathfrak{a}}(\theta^{\mathfrak{a}} - T) = -c_v^{\mathfrak{b}+\mathfrak{c}}(\theta^{\mathfrak{bc}} - T).$$

Here and hereafter, for simplicity, we use the notation $c_v^{\mathfrak{a}}$ instead of $c_v^{\mathfrak{a}}(T)$ and so on. Inversely, the nonequilibrium temperatures are expressed as follows:

$$\theta^{\mathfrak{a}} - T = \frac{\Delta}{c_v^{\mathfrak{a}}}, \qquad \theta^{\mathfrak{bc}} - T = -\frac{\Delta}{c_v^{\mathfrak{b}+\mathfrak{c}}},$$
$$\theta^{\mathfrak{b}} - T = \frac{\delta}{c_v^{\mathfrak{b}}} - \frac{\Delta}{c_v^{\mathfrak{b}+\mathfrak{c}}}, \qquad \theta^{\mathfrak{c}} - T = -\frac{\delta}{c_v^{\mathfrak{c}}} - \frac{\Delta}{c_v^{\mathfrak{b}+\mathfrak{c}}}.$$

For the $(\mathfrak{bc})$-process, δ and Δ are expressed in terms of $\{\Delta^R, \Delta^V\}$ as summarized in Table 8.2.

The production terms are now expressed as

$$P_{ll}^{\mathfrak{a}} = -2\rho\frac{\Delta}{\tau}, \qquad P_{ll}^{\mathfrak{b}} = -2\rho\frac{\delta}{\tau_\delta} + 2\rho\frac{c_v^{\mathfrak{b}}}{c_v^{\mathfrak{b}+\mathfrak{c}}}\frac{\Delta}{\tau}, \qquad P_{ll}^{\mathfrak{c}} = 2\rho\frac{\delta}{\tau_\delta} + 2\rho\frac{c_v^{\mathfrak{c}}}{c_v^{\mathfrak{b}+\mathfrak{c}}}\frac{\Delta}{\tau},$$

where τ_δ is defined by

$$\frac{1}{\tau_\delta} \equiv \frac{1}{\tau_{\mathfrak{bc}}} + \frac{1}{\tau}.$$

Table 8.2 Relation between $\{\delta, \Delta\}$ and $\{\Delta^R, \Delta^V\}$

$(\mathfrak{bc})$	δ	Δ
(KR)	$-\frac{c_v^R}{c_v^{K+R}}\Delta^V - \Delta^R$	Δ^V
(KV)	$-\frac{c_v^V}{c_v^{K+V}}\Delta^R - \Delta^V$	Δ^R
(RV)	$\frac{c_v^V\Delta^R - c_v^R\Delta^V}{c_v^{R+V}}$	$-\Delta^R - \Delta^V$

It is possible to verify that the production terms give a non-negative entropy production (8.29) ($\Sigma \geqslant 0$). In fact, we have

$$\Sigma = \frac{\rho}{\tau T^2} \frac{c_v}{c_v^{\mathfrak{a}} c_v^{\mathfrak{b}+\mathfrak{c}}} \Delta^2 + \frac{\rho}{\tau_\delta T^2} \frac{c_v^{\mathfrak{b}+\mathfrak{c}}}{c_v^{\mathfrak{b}} c_v^{\mathfrak{c}}} \delta^2 + \frac{1}{2pT}\left(\frac{1}{\tau_K} + \frac{1}{\tau_\delta}\right)\left(\sigma_{\langle ij\rangle}\sigma_{\langle ij\rangle} + \frac{2}{T\left(\frac{k_B}{m} + c_v\right)} q_i q_i\right) \geqslant 0,$$

since the relaxation times and specific heats are positive.

8.7 Closed System of Field Equations: ET$_{15}$

The balance equations (8.28) are rewritten in terms of the independent fields $\{\rho, v_i, T, \delta, \Delta, \sigma_{\langle ij\rangle}, q_i\}$ with material derivative as follows:

$$\dot{\rho} + \rho \frac{\partial v_k}{\partial x_k} = 0,$$

$$\rho \dot{v}_i + \frac{\partial \left\{(p+\Pi)\delta_{ij} - \sigma_{\langle ij\rangle}\right\}}{\partial x_j} = 0,$$

$$\rho c_v \dot{T} + (p+\Pi)\frac{\partial v_k}{\partial x_k} - \frac{\partial v_i}{\partial x_k}\sigma_{\langle ik\rangle} + \frac{\partial q_k}{\partial x_k} = 0,$$

$$\dot{\delta} + \frac{1}{\rho}\left\{A_1 + \frac{1}{c_v}\frac{\mathrm{d}}{\mathrm{d}T}\left(\frac{c_v^{\mathfrak{b}}}{c_v^{\mathfrak{b}+\mathfrak{c}}}\right)\Delta\right\}\left\{(p+\Pi)\delta_{kl} - \sigma_{\langle kl\rangle}\right\}\frac{\partial v_l}{\partial x_k} + \frac{1}{\rho}\left\{\frac{\frac{k_B}{m}A_1}{\frac{k_B}{m} + c_v} + \frac{1}{c_v}\frac{\mathrm{d}}{\mathrm{d}T}\left(\frac{c_v^{\mathfrak{b}}}{c_v^{\mathfrak{b}+\mathfrak{c}}}\right)\Delta\right\}\frac{\partial q_k}{\partial x_k}$$
$$+ \frac{1}{\rho}\left\{\frac{c_v^{\mathfrak{b}}}{c_v^{\mathfrak{b}+\mathfrak{c}}}\frac{\mathrm{d}}{\mathrm{d}T}\left(\frac{\frac{k_B}{m}\frac{A_2}{c_v} + c_v^{\mathfrak{a}}}{\frac{k_B}{m} + c_v}\right) + \frac{\mathrm{d}}{\mathrm{d}T}\left(\frac{\frac{k_B}{m}\left(1 - \frac{A_2}{c_v}\right) + c_v^{\mathfrak{b}}}{\frac{k_B}{m} + c_v}\right)\right\} q_k \frac{\partial T}{\partial x_k} = -\frac{\delta}{\tau_\delta},$$

$$\dot{\Delta} + \frac{1}{\rho}\frac{A_2 - c_v^{\mathfrak{a}}}{c_v}\left\{(p+\Pi)\delta_{kl} - \sigma_{\langle kl\rangle}\right\}\frac{\partial v_l}{\partial x_k} + \frac{1}{\rho}\frac{\mathrm{d}}{\mathrm{d}T}\left(\frac{\frac{k_B}{m}\frac{A_2}{c_v} + c_v^{\mathfrak{a}}}{\frac{k_B}{m} + c_v}\right) q_k \frac{\partial T}{\partial x_k}$$
$$+ \frac{1}{\rho}\frac{A_2 - c_v^{\mathfrak{a}}}{(\frac{k_B}{m} + c_v)c_v}\frac{k_B}{m}\frac{\partial q_k}{\partial x_k} = -\frac{\Delta}{\tau},$$

Table 8.3 Explicit expression of A_1, A_2 and Π

(bc)	A_1	A_2	Π
(KR) or (KV)	$\frac{c_v^c}{c_v^{b+c}}$	0	$\frac{2}{3}\rho\left(\delta - \frac{c_v^b}{c_v^{b+c}}\Delta\right)$
(RV)	0	c_v	$\frac{2}{3}\rho\Delta$

$$
\begin{aligned}
&\dot{\sigma}_{\langle ij\rangle} - 2(p+\Pi)\frac{\partial v_{\langle i}}{\partial x_{j\rangle}} + \sigma_{\langle ij\rangle}\frac{\partial v_k}{\partial x_k} + 2\frac{\partial v_{\langle i}}{\partial x_k}\sigma_{\langle j\rangle k\rangle} + \frac{2\frac{k_B}{m}}{\left(\frac{k_B}{m}+c_v\right)^2}\frac{dc_v}{dT}\frac{\partial T}{\partial x_k}q_{\langle i}\delta_{j\rangle k} \\
&\quad - \frac{2\frac{k_B}{m}}{\frac{k_B}{m}+c_v}\frac{\partial q_{\langle i}}{\partial x_{j\rangle}} = -\frac{1}{\tau_\sigma}\sigma_{\langle ij\rangle}, \\
&\dot{q}_i + \frac{2\frac{k_B}{m}+c_v}{\frac{k_B}{m}+c_v}q_i\frac{\partial v_k}{\partial x_k} + \frac{\frac{k_B}{m}}{\frac{k_B}{m}+c_v}q_k\frac{\partial v_k}{\partial x_i} + \frac{2\frac{k_B}{m}+c_v}{\frac{k_B}{m}+c_v}q_k\frac{\partial v_i}{\partial x_k} \\
&\quad + \left\{\left(\frac{k_B}{m}+c_v\right)p\delta_{ki} + \left(2\frac{k_B}{m}+c_v\right)(\Pi\delta_{ki} - \sigma_{\langle ki\rangle})\right\}\frac{\partial T}{\partial x_k} \\
&\quad - \frac{p}{\rho}\frac{\partial p}{\partial x_i} + \frac{1}{\rho}\left\{(p-\Pi)\delta_{ki} + \sigma_{\langle ki\rangle}\right\}\frac{\partial}{\partial x_l}\left\{(p+\Pi)\delta_{kl} - \sigma_{\langle kl\rangle}\right\} = -\frac{1}{\tau_q}q_i,
\end{aligned}
\tag{8.42}
$$

where A_1, A_2 and Π are given in Table 8.3.

8.8 Maxwellian Iteration and Phenomenological Coefficients

When we carry out the Maxwellian iteration (for more detail see Chap. 33, Sects. 33.2.1 and 33.3) on (8.42) and retain the first order terms with respect to the relaxation times τ, τ_σ, and τ_q, we obtain the system of the Navier-Stokes and Fourier theory with

$$
\begin{aligned}
\delta &= -\tau_\delta \frac{p}{\rho}A_1\frac{\partial v_i}{\partial x_i}, & \Delta &= -\tau\frac{p}{\rho}\frac{A_2 - c_v^a}{c_v}\frac{\partial v_i}{\partial x_i}, \\
\sigma_{\langle ij\rangle} &= 2p\tau_\sigma\frac{\partial v_{\langle i}}{\partial x_{j\rangle}}, & q_i &= -\left(\frac{k_B}{m}+c_v\right)p\tau_q\frac{\partial T}{\partial x_i},
\end{aligned}
$$

where, from (8.41),

$$
\frac{1}{\tau_\sigma} = \frac{1}{\tau_q} = \frac{1}{\tau_K} + \frac{1}{\tau_{bc}} + \frac{1}{\tau}.
$$

If $\tau_K \ll \tau_{\mathfrak{bc}}$, we have the relation: $\tau_\sigma = \tau_q \sim \tau_K$. In particular, for ($\mathfrak{bc}$)-process (($\mathfrak{bc}$) =(KR) or (KV)), we have

$$\Pi = \Pi^{\mathfrak{bc}} + \Pi^{\mathfrak{a}} \qquad \text{with}$$

$$\Pi^{\mathfrak{bc}} = p(\rho, \theta^K) - p(\rho, \theta^{\mathfrak{bc}}) = -\tau_\delta p \frac{k_B}{m} \frac{c_v^{\mathfrak{c}}}{c_v^{\mathfrak{b}} c_v^{\mathfrak{b}+\mathfrak{c}}} \frac{\partial v_i}{\partial x_i}$$

$$\Pi^{\mathfrak{a}} = p(\rho, \theta^{\mathfrak{bc}}) - p(\rho, T) = -\tau p \frac{k_B}{m} \frac{c_v^{\mathfrak{a}}}{c_v^{\mathfrak{b}+\mathfrak{c}} c_v} \frac{\partial v_i}{\partial x_i},$$

and for (RV)-process, we have

$$\Pi = -\tau p \frac{k_B}{m} \frac{c_v^{R+V}}{c_v^K c_v} \frac{\partial v_i}{\partial x_i}.$$

Recalling the definition (1.11) of the bulk viscosity ν, shear viscosity μ, and heat conductivity κ, we obtain

$$\begin{aligned}
\nu &= \begin{cases} \frac{k_B}{m} \frac{c_v^{\mathfrak{c}}}{c_v^{\mathfrak{b}} c_v^{\mathfrak{b}+\mathfrak{c}}} p\tau_\delta + \frac{k_B}{m} \frac{c_v^{\mathfrak{a}}}{c_v^{\mathfrak{b}+\mathfrak{c}} c_v} p\tau & \text{for } (\mathfrak{bc})\text{-process } ((\mathfrak{bc}) = (KR) \text{ or } (KV)) \\ \frac{k_B}{m} \frac{c_v^{R+V}}{c_v^K c_v} p\tau & \text{for (RV)-process} \end{cases}, \\
\mu &= p\tau_\sigma, \\
\kappa &= \left(\frac{k_B}{m} + c_v\right) p\tau_q.
\end{aligned} \tag{8.43}$$

It should be noted that, as usual in the BGK model, the Prandtl number predicted by the present generalized BGK model is not satisfactory. One possible way to avoid this difficulty is that we use the procedure adopted in the phenomenological RET theory of rarefied polyatomic gases. That is, the relaxation times τ, τ_δ, τ_σ, and τ_q are regarded as functions of ρ and T that are estimated by using the experimental data on ν, μ, and κ. In this respect, for (RV)-process, the relations (8.43) are the same as those derived from ET$_{14}$ (see (6.36)). On the other hand, for ($\mathfrak{bc}$)-process (($\mathfrak{bc}$) =(KR) or (KV)), only the relation for ν is different from the one derived from ET$_{14}$.

Chapter 9
Nesting Theory of Many Moments and Maximum Entropy Principle

Abstract We consider molecular extended thermodynamics (molecular ET) of rarefied polyatomic gases with the system composed of two hierarchies of balance equations for the moments of a distribution function. The internal degrees of freedom of a molecule are properly taken into account in the distribution function. By the reasoning of physical relevance, the truncation orders of the two hierarchies are shown to be dependent on each other. And the two closure procedures based on the maximum entropy principle and on the entropy principle are also proved to be equivalent to each other.

Characteristic velocities of a hyperbolic system of balance equations for a polyatomic gas are compared to those obtained for a monatomic gas. The lower bound estimate for the maximum equilibrium characteristic velocity established for a monatomic gas is proved to be valid also for a rarefied polyatomic gas, that is, the estimate is independent of the degrees of freedom of a molecule. As a consequence, also for polyatomic gases, when the number of moments increases the maximum characteristic velocity becomes unbounded.

9.1 Introduction

We discussed, in Sect. 1.8.5, the role of the maximum entropy principle (MEP) and, in Sect. 4.2, we made a survey of molecular ET. We proved the equivalence of the closures via MEP and via the entropy principle in the case of monatomic gas. In Chap. 7, we described the closure via MEP of the 14-field RET theory of rarefied polyatomic gases. We proved that the binary hierarchy of moment equations obtained by the distribution function is consistent with the binary hierarchy presented in Chap. 6. In particular, it was shown that the *momentum-like* hierarchy (F-hierarchy) is related to the usual moments of the distribution function, and the *energy-like* hierarchy (G-hierarchy) is related to the moments of an additional continuous variable representing the internal energy of a molecule.

The purpose of the present chapter is to establish the RET theory of rarefied polyatomic gases with any number of moments. These results are based on the papers of Arima et al. [119, 316].

T. Ruggeri, M. Sugiyama, *Classical and Relativistic Rational Extended Thermodynamics of Gases*, https://doi.org/10.1007/978-3-030-59144-1_9

In particular, by physical arguments (namely, Galilean invariance and the requirement that the characteristic velocities depend on the degrees of freedom of a molecule), the relation between the truncation orders of the *momentum-like* and *energy-like* hierarchies is investigated. The conclusion is that once the truncation order of one hierarchy is chosen, the truncation order of the other one is automatically prescribed.

The closure of the system is achieved by means of MEP, and it is proved that, also in the present general case, the MEP approach is equivalent to the approach with the requirement that the truncated system satisfies the entropy principle with convex entropy density. In this way, it is shown that the system is symmetric in terms of the *main field* components.

The characteristic velocities in an equilibrium state are analyzed. These velocities play an important role in processes such as propagation of an acceleration wave [120, 121] (see Sect. 3.2), determination of the phase velocity of linear wave in the high-frequency limit [122, 123] (see Sect. 3.1), and *subshock* formation [124] (see Theorem 3.1). With regard to this, the following problem will be discussed: how do the characteristic velocities of the system depend on the internal degrees of freedom of a molecule and on the truncation orders of hierarchies? In particular, the two limit cases, that is, the monatomic-gas limit and the limit of a gas with infinite internal degrees of freedom are investigated.

Finally, using the convexity arguments and the subcharacteristic conditions for principal subsystems (see Theorem 2.3), we prove that the lower bound estimate for the maximum characteristic velocity established for monatomic gases by Boillat and Ruggeri [90] (see (4.55)) still holds also for polyatomic gases, that is, the estimate is independent of the degrees of freedom of a molecule. Therefore, also for polyatomic gases, the maximum characteristic velocity tends to be unbounded when the orders of the hierarchies tend to infinity!

9.2 MEP Closure for Rarefied Polyatomic Gas with Many Moments

The discussion is focused on gases characterized by the thermal and caloric equations of state (1.32), that is, rarefied polytropic and polyatomic gases.

We consider now the same binary hierarchy of the 14-moment theory but for a generic number of moments truncated for the F-hierarchy at the order of truncation N and for the G-hierarchy at the order M:

$$
\begin{aligned}
&\partial_t F + \partial_i F_i = 0, \\
&\partial_t F_{k_1} + \partial_i F_{ik_1} = 0, \\
&\partial_t F_{k_1k_2} + \partial_i F_{ik_1k_2} = P_{k_1k_2}, \qquad \partial_t G_{ll} + \partial_i G_{lli} = 0, \\
&\vdots \qquad\qquad\qquad\qquad\qquad \partial_t G_{llj_1} + \partial_i G_{llij_1} = Q_{llj_1}, \\
&\vdots \qquad\qquad\qquad\qquad\qquad \vdots \\
&\partial_t F_{k_1k_2\ldots k_N} + \partial_i F_{ik_1k_2\ldots k_N} = P_{k_1k_2\ldots k_N} \quad \vdots \\
&\qquad\qquad\qquad\qquad \partial_t G_{llj_1j_2\ldots j_M} + \partial_i G_{llij_1j_2\ldots j_M} = Q_{llj_1j_2\ldots j_M}.
\end{aligned}
\tag{9.1}
$$

Definition 9.1 ((N, M)-System) The above system can be rewritten in a simple form by using the multi-index notations defined below:

$$
\begin{aligned}
&\partial_t F_A + \partial_i F_{iA} = P_A, \\
&(0 \leqslant A \leqslant N) \qquad\qquad \partial_t G_{llA'} + \partial_i G_{lliA'} = Q_{llA'}, \\
&\qquad\qquad\qquad\qquad\qquad (0 \leqslant A' \leqslant M)
\end{aligned}
\tag{9.2}
$$

which we call (N, M)-system.

The moments of the F-hierarchy are the usual ones:

$$
\begin{aligned}
F_A &= \int_{\mathbb{R}^3}\int_0^\infty m f c_A I^\alpha \, dI d\mathbf{c}, \qquad F_{iA} = \int_{\mathbb{R}^3}\int_0^\infty m f c_i c_A I^\alpha \, dI d\mathbf{c}, \\
P_A &= \int_{\mathbb{R}^3}\int_0^\infty m Q(f) c_A I^\alpha \, dI d\mathbf{c},
\end{aligned}
\tag{9.3}
$$

while the moments of the G-hierarchy are expressed with the additional variable I as

$$
\begin{aligned}
G_{llA'} &= \int_{\mathbb{R}^3}\int_0^\infty m f \left(c^2 + \frac{2I}{m}\right) c_{A'} I^\alpha \, dI d\mathbf{c}, \\
G_{lliA'} &= \int_{\mathbb{R}^3}\int_0^\infty m f \left(c^2 + \frac{2I}{m}\right) c_i c_{A'} I^\alpha \, dI d\mathbf{c}, \\
Q_{llA'} &= \int_{\mathbb{R}^3}\int_0^\infty m Q(f) \left(c^2 + \frac{2I}{m}\right) c_{A'} I^\alpha \, dI d\mathbf{c}.
\end{aligned}
\tag{9.4}
$$

The following multi-index notations are introduced for the sake of compactness:

$$F_A = \begin{cases} F & \text{for } A = 0 \\ F_{i_1 \cdots i_A} & \text{for } 1 \leqslant A \leqslant N \end{cases}, \qquad F_{iA} = \begin{cases} F_i & \text{for } A = 0 \\ F_{i\, i_1 \cdots i_A} & \text{for } 1 \leqslant A \leqslant N \end{cases},$$

$$P_A = \begin{cases} 0 & \text{for } A = 0, 1 \\ P_{i_1 \cdots i_A} & \text{for } 2 \leqslant A \leqslant N, \end{cases}$$

$$G_{llA'} = \begin{cases} G_{ll} & \text{for } A' = 0 \\ G_{lli_1 \cdots i_{A'}} & \text{for } 1 \leqslant A' \leqslant M \end{cases}, \qquad G_{lliA'} = \begin{cases} G_{lli} & \text{for } A' = 0 \\ G_{llii_1 \cdots i_{A'}} & \text{for } 1 \leqslant A' \leqslant M \end{cases},$$

$$Q_{llA'} = \begin{cases} 0 & \text{for } A' = 0 \\ Q_{lli_1 \cdots i_{A'}} & \text{for } 1 \leqslant A' \leqslant M \end{cases},$$

and

$$c_A = \begin{cases} 1 & \text{for } A = 0 \\ c_{i_1} \cdots c_{i_A} & \text{for } 1 \leqslant A \leqslant N \end{cases},$$

where the indexes i and $i_1 \leqslant i_2 \leqslant \cdots \leqslant i_A$ assume the values $1, 2, 3$. The truncation index N of the F-hierarchy and the index M of the G-hierarchy are a priori independent of each other. It is worth noting that the first and second equations of the F-hierarchy represent the conservation laws of mass and momentum, respectively ($P \equiv 0, P_i \equiv 0$), while the first equation of the G-hierarchy represents the conservation law of energy ($Q_{ll} \equiv 0$), and that, in each of the two hierarchies, the flux in one equation appears as the density in the following equation—a feature in common with the single hierarchy of monatomic gases.

The Euler 5-moment system (4.33) is a particular case of (9.2) with $N = 1$, $M = 0$, and the 14-moment system (6.2) is another particular case of (9.2) with $N = 2$, $M = 1$.

Instead of the (N, M)-system that includes all tensors of F_A and $G_{llA'}$, it is possible to construct other systems. Then similar to the Definition 4.1, we have the following:

Definition 9.2 A system (9.1) is called $(N^{(1)}, M)$-system if, for the last balance equation in the F-hierarchy, we consider only the trace with respect to the two indexes, $F_{k_1 k_2 \ldots k_{N-2} ll}$, instead of the full N-order tensor $F_{k_1 k_2 \ldots k_{N-2} k_{N-1} k_N}$. Similar definition is valid for the G-hierarchy and we have $(N, M^{(1)})$-system. If, instead of 2 indexes, we have the contraction with respect to 2 pairs of 2 indexes, we write: $(N^{(2)}, M)$ etc.

For example, in the classical kinetic approach for rarefied polyatomic gases, the 17-moment theory is proposed [317, 318]. This theory corresponds to the RET theory with the densities; $F, F_i, F_{ij}, G_{ll}, F_{lli}, G_{lli}$. In this case, in the last tensor of F-hierarchy, only the trace is adopted as an independent field, and therefore all components of the G-hierarchy have the corresponding components in the F-hierarchy. Since this 17-moment system adopts the trace with respect to the indexes i_{N-1} and i_N of the N-order tensorial equation with $N = 3$, the 17-moment system is denoted as $(3^{(1)}, 1)$-system. Similarly, the monatomic 13-moment system is denoted as $(3^{(1)})$-system.

Remark 9.1 Definition 9.2 does not cover all possible cases of physically plausible systems. For example, the system that adopts the traceless part of the highest order tensor of F- and/or G-hierarchies is also possible. Another possibility is that the system adopts the trace part not only of the highest-order tensor of F- and/or G-hierarchies but also of the second-highest-order tensor. However, in the present study, we skip such systems and pay attention solely to (N, M)-system with or without trace of two indexes.

From (4.38) and (4.39), the number of moments for (N, M)-system is given by

$$n_{(N,M)} = \frac{1}{6}(N+1)(N+2)(N+3) + \frac{1}{6}(M+1)(M+2)(M+3), \tag{9.5}$$

and, for $(N^{(1)}, M)$-system,

$$n_{(N^{(1)},M)} = \frac{1}{6}N(N^2+6N-1) + \frac{1}{6}(M+1)(M+2)(M+3). \tag{9.6}$$

In Chap. 27, we will see the optimal choice of moments for polyatomic gases with the use of classical limit of the relativistic theory.

9.2.1 Galilean Invariance

An important issue to be addressed concerns the relation between the orders N and M of the two hierarchies (9.2). In the treatment outlined above, the orders N and M have been considered to be independent of each other. However, from the physical point of view, one should clarify whether some restrictions on the truncation procedure of the two hierarchies exist or not.

In the spirit of RET, the application of the universal principles—in the present case, the Galilean invariance—suggests that in order to have a physically acceptable model, the orders of truncation N and M cannot be chosen independently. Indeed, the following theorem gives a first restriction on the relation between the two orders N and M:

Theorem 9.1 *In order for the* (N, M)*-system* (9.2) *to be Galilean invariant, it must be* $M \leqslant N - 1$.

Proof Recalling the definition of the peculiar velocity $C_i = c_i - v_i$, and defining the velocity-independent *internal* moments (so-called Galilean tensors) $\hat{F}_A$ and $\hat{G}_{llA'}$ as follows:

$$\hat{F}_A = \int_{\mathbb{R}^3} \int_0^\infty mf\, C_A I^\alpha\, dI d\mathbf{C},$$
$$\hat{G}_{llA'} = \int_{\mathbb{R}^3} \int_0^\infty mf \left(C^2 + \frac{2I}{m}\right) C_{A'} I^\alpha\, dI d\mathbf{C},$$

we can write

$$\begin{aligned} F_A &= \sum_{k=0}^{A} \binom{A}{k} \hat{F}_{(i_1 \cdots i_k} v_{i_{k+1}} \cdots v_{i_A)}, \\ &= X_{AB} \hat{F}_B, \end{aligned} \tag{9.7}$$

where X_{AB} is a component of the matrix (2.61), and

$$\begin{aligned} G_{llA'} &= \sum_{k=0}^{A'} \binom{A'}{k} \\ &\quad \times \left(\hat{G}_{ll(i_1 \cdots i_k} v_{i_{k+1}} \cdots v_{i_{A'})} + 2 v_l \hat{F}_{l(i_1 \cdots i_k} v_{i_{k+1}} \cdots v_{i_{A'})} + v^2 \hat{F}_{(i_1 \cdots i_k} v_{i_{k+1}} \cdots v_{i_{A'})} \right) \\ &= X_{A'B'} \left(\hat{G}_{llB'} + 2 v_l \hat{F}_{lB'} + v^2 \hat{F}_{B'} \right). \end{aligned} \tag{9.8}$$

It is noticeable that $G_{llA'}$ depends not only on the Galilean tensors of the G-hierarchy, but also on those of the F-hierarchy. Similarly, the fluxes F_{iA}, $G_{lliA'}$ and the productions P_A, $Q_{llA'}$ are expressed, respectively, by these internal quantities $\hat{F}_{iA}$, $\hat{G}_{lliA'}$ and $\hat{P}_A$, $\hat{Q}_{llA'}$, which are evaluated by imposing $v_{i_k} = 0$ for all k.

The structure of the terms (9.7) and (9.8) guarantees that the results proved in [167] can be extended to the present case and, when $M > N - 1$, unknown moments of the F-hierarchy appear in a higher order moment of the G-hierarchy. For example, in the case with $M = N$, $G_{lli_1 \cdots i_N}$, which is the highest-order moment of the G-hierarchy, includes $\hat{F}_{li_1 \cdots i_N}$, and the corresponding moment $F_{li_1 \cdots i_N}$ is not included in the F-hierarchy. Therefore we conclude that the system cannot be Galilean invariant whenever $M > N - 1$. The 14 moment theory, namely (2, 1)-system, and the Euler $(1, 0)$-system are of course Galilean invariant.

It should be noted that the requirement of the Galilean invariance for (N, M)-system; $M \leqslant N - 1$ is satisfied also for $(N^{(1)}, M)$, $(N, M^{(1)})$, and $(N^{(1)}, M^{(1)})$-systems except for $(N^{(1)}, N - 1)$-system.

9.2.2 *Closure of* (N, M)*-system via the Maximum Entropy Principle*

To close the (N, M)-system (9.2) with $M \leqslant N-1$, the maximum entropy principle is applied: The distribution function $f_{(N,M)}$ is the one that maximizes the entropy h^0 defined by

$$h^0 = -k_B \int_{\mathbb{R}^3} \int_0^{\infty} f \log f \, I^{\alpha} \, dI d\mathbf{c}. \tag{9.9}$$

under the constraints that the moments F_A and $G_{llA'}$ are given by $(9.3)_1$ and $(9.4)_1$.

The variational problem, from which the distribution function $f_{(N,M)}$ is obtained, is connected to the functional:

$$\begin{aligned} &\mathcal{L}_{(N,M)}(f) \\ &= -k_B \int_{\mathbb{R}^3} \int_0^{\infty} f \log f \, I^{\alpha} \, dI d\mathbf{c} + \sum_{A=0}^{N} \lambda_A \left(F_A - \int_{\mathbb{R}^3} \int_0^{\infty} m f c_A I^{\alpha} \, dI d\mathbf{c} \right) \\ &\quad + \sum_{A'=0}^{M} \mu_{A'} \left(G_{llA'} - \int_{\mathbb{R}^3} \int_0^{\infty} m f \left(c^2 + \frac{2I}{m} \right) c_{A'} I^{\alpha} dI d\mathbf{c} \right), \end{aligned}$$

where λ_A and $\mu_{A'}$ are the Lagrange multipliers. Then the distribution function is obtained as follows [119, 316]:

Theorem 9.2 *The distribution function* $f_{(N,M)}$ *that maximizes the functional* $\mathcal{L}_{(N,M)}$ *is (for simplicity, we omit from now on the symbol of summation in* A *and in* A'*):*

$$f_{(N,M)} = \exp\left(-1 - \frac{m}{k_B} \chi_{(N,M)}\right), \qquad \chi_{(N,M)} = \lambda_A c_A + \left(c^2 + \frac{2I}{m}\right) \mu_{A'} c_{A'}. \tag{9.10}$$

By inserting (9.10) into $(9.3)_1$ and $(9.4)_1$, the Lagrange multipliers λ_A and $\mu_{A'}$ are evaluated in terms of the densities F_A and $G_{llA'}$. Finally, plugging (9.10) into the last flux and production terms, the system can be closed. We will give more details in the following sections.

9.2.3 *Closure of* (N, M)*-system via the Entropy Principle*

As discussed above, an alternative approach to achieve the closure of the system makes use of the entropy principle. In this case, it is required that all the solutions of (9.2) satisfy the entropy principle (2.8). The condition (2.14) can now be written as

$$\partial_t h^0 + \partial_i h^i - u'_A \left(\partial_t F_A + \partial_i F_{iA} - P_A\right) - v'_{A'} \left(\partial_t G_{llA'} + \partial_i G_{lliA'} - Q_{llA'}\right)$$
$$= \Sigma \geqslant 0,$$

where u'_A and $v'_{A'}$ are the main field components. Treating h^0, h^i, and Σ as constitutive functions of F_A and $G_{llA'}$, we obtain (see (2.11))

$$dh^0 = u'_A dF_A + v'_{A'} dG_{llA'}, \quad dh_i = u'_A dF_{iA} + v'_{A'} dG_{lliA'},$$
$$\Sigma = u'_A P_A + v'_{A'} Q_{llA'} \geqslant 0.$$

Recalling (9.3) and (9.4), we can write

$$dh^0 = \int_{\mathbb{R}^3}\int_0^\infty m\chi_{(N,M)} I^\alpha \, df \, dI d\mathbf{c}, \quad dh^i = \int_{\mathbb{R}^3}\int_0^\infty mc_i\, \chi_{(N,M)} I^\alpha \, df \, dI d\mathbf{c},$$
$$\Sigma = \int_{\mathbb{R}^3}\int_0^\infty m\chi_{(N,M)}\, Q(f)\, I^\alpha \, dI d\mathbf{c}, \quad \chi_{(N,M)} = u'_A c_A + \left(c^2 + \frac{2I}{m}\right) v'_{A'} c_{A'}. \tag{9.11}$$

On the other hand, recalling that the entropy density h^0 is expressed as (9.9) by using the distribution function, and similarly the entropy flux and production can be expressed as

$$h^i = -k_B \int_{\mathbb{R}^3}\int_0^\infty c_i f \log f \, I^\alpha \, dI d\mathbf{c}, \qquad \Sigma = -k_B \int_{\mathbb{R}^3}\int_0^\infty Q(f) \log f \, I^\alpha \, dI d\mathbf{c},$$

we can write the quantities dh^0 and dh^i as

$$\begin{aligned} dh^0 &= -k_B \int_{\mathbb{R}^3}\int_0^\infty (\log f + 1)\, I^\alpha \, df \, dI d\mathbf{c}, \\ dh_i &= -k_B \int_{\mathbb{R}^3}\int_0^\infty c_i(\log f + 1)\, I^\alpha \, df \, dI d\mathbf{c}. \end{aligned} \tag{9.12}$$

By comparing (9.11) to (9.12), it is easy to have

Theorem 9.3 *The distribution function $f_{(N,M)}$ that satisfies the entropy principle is given by*

$$f_{(N,M)} = \exp\left(-1 - \frac{m}{k_B}\chi_{(N,M)}\right). \tag{9.13}$$

As a consequence of Theorems 9.2 and 9.3, it is concluded that:

Statement 9.1 *The MEP and the entropy principle are equivalent with respect to the closure also for rarefied polyatomic gases. In addition, the Lagrange multipliers of MEP coincide with the main field of the entropy principle:* $u'_A \equiv \lambda_A$ *and* $v'_{A'} \equiv \mu_{A'}$.

For the case of a generic entropy functional $h^0 = h^0(f)$ valid for any gas including degenerate gas, the equivalence of the entropy principle and MEP was proved in [316] in a similar way in the case of monatomic gas (see Sect. 4.2).

9.2.4 Closure and Symmetric Hyperbolic Form

As discussed in Sect. 4.2 in the case of rarefied monatomic gas, the system (9.2) can be written in a symmetric hyperbolic form in terms of the main field $\left(u'_A, v'_{A'}\right)$. In fact, we have the following theorem:

Theorem 9.4 *The system* (9.2) *may be written as follows:*

$$\begin{pmatrix} J^0_{AB} & J^1_{AB'} \\ J^1_{A'B} & J^2_{A'B'} \end{pmatrix} \partial_t \begin{pmatrix} u'_B \\ v'_{B'} \end{pmatrix} + \begin{pmatrix} J^0_{iAB} & J^1_{iAB'} \\ J^1_{iA'B} & J^2_{iA'B'} \end{pmatrix} \partial_i \begin{pmatrix} u'_B \\ v'_{B'} \end{pmatrix} = \begin{pmatrix} P_A \\ Q_{llA'} \end{pmatrix}. \tag{9.14}$$

The coefficient matrix of time-derivatives of the fields is negative-definite, and all other matrices are symmetric. Therefore the system is symmetric hyperbolic.

Proof From (9.13) dropping the subscript (N, M) for simplicity, we have

$$\partial_t f = -\frac{m}{k_B} f \, \partial_t \chi = -\frac{m}{k_B} f \left(c_A \partial_t u'_A + \left(c^2 + \frac{2I}{m} \right) c_{A'} \partial_t v'_{A'} \right),$$

and then

$$\partial_t F_A = J^0_{AB} \partial_t u'_B + J^1_{AB'} \partial_t v'_{B'}, \qquad \partial_t G_{llA'} = J^1_{BA'} \partial_t u'_B + J^2_{A'B'} \partial_t v'_{B'}, \tag{9.15}$$

$$\partial_i F_{iA} = J^0_{iAB} \partial_i u'_B + J^1_{iAB'} \partial_i v'_{B'}, \quad \partial_i G_{lliA'} = J^1_{iBA'} \partial_i u'_B + J^2_{iA'B'} \partial_i v'_{B'},$$

where

$$
\begin{aligned}
J^0_{AB} &= -\frac{m^2}{k_B}\int_{\mathbb{R}^3}\int_0^\infty f c_A c_B I^\alpha \, dI d\mathbf{c}, \\
J^0_{iAB} &= -\frac{m^2}{k_B}\int_{\mathbb{R}^3}\int_0^\infty f c_i c_A c_B I^\alpha \, dI d\mathbf{c}, \\
J^1_{AB'} &= -\frac{m^2}{k_B}\int_{\mathbb{R}^3}\int_0^\infty f c_A c_{B'} \left(c^2 + \frac{2I}{m}\right) I^\alpha \, dI d\mathbf{c}, \\
J^1_{iAB'} &= -\frac{m^2}{k_B}\int_{\mathbb{R}^3}\int_0^\infty f c_i c_A c_{B'} \left(c^2 + \frac{2I}{m}\right) I^\alpha \, dI d\mathbf{c}, \\
J^2_{A'B'} &= -\frac{m^2}{k_B}\int_{\mathbb{R}^3}\int_0^\infty f c_{A'} c_{B'} \left(c^2 + \frac{2I}{m}\right)^2 I^\alpha \, dI d\mathbf{c}, \\
J^2_{iA'B'} &= -\frac{m^2}{k_B}\int_{\mathbb{R}^3}\int_0^\infty f c_i c_{A'} c_{B'} \left(c^2 + \frac{2I}{m}\right)^2 I^\alpha \, dI d\mathbf{c}.
\end{aligned}
\tag{9.16}
$$

Inserting (9.15) into (9.2), we obtain (9.14), and this completes the proof.

9.3 Closure in the Neighborhood of a Local Equilibrium State and Principal Subsystems

As seen in the monatomic gas case (Sect. 4.4), we have also in this case the problem of the convergence of the moment integrals. Therefore the distribution function (9.10) obtained as a solution of the variational problem is expanded in the neighborhood of a local equilibrium state:

$$
\begin{aligned}
& f \approx f^{(E)}\left[1 - \frac{m}{k_B}\left(\tilde{u}'_A c_A + \left(c^2 + \frac{2I}{m}\right)\tilde{v}'_{A'} c_{A'}\right)\right], \\
& \tilde{u}'_A = u'_A - u'^E_A, \ \tilde{v}'_{A'} = v'_{A'} - v'^E_{A'},
\end{aligned}
\tag{9.17}
$$

where u'^E_A and $v'^E_{A'}$ are the main field components evaluated in the local equilibrium state. The equilibrium distribution function $f^{(E)}$, obtained by Pavić et al. [115] (see also [319]), is given by (7.8).

All the Lagrange multipliers (main field) in equilibrium vanish except those corresponding to the hydrodynamic variables ρ, v_i, ε. By inserting (9.17) into $(9.3)_1$ and $(9.4)_1$, a linear algebraic system that permits to evaluate the main field $(u'_A, v'_{A'})$ in terms of the densities F_A and $G_{llA'}$ is obtained:

$$
\begin{pmatrix} J^{0|E}_{AB} & J^{1|E}_{AB'} \\ J^{1|E}_{A'B} & J^{2|E}_{A'B'} \end{pmatrix}
\begin{pmatrix} \tilde{u}'_B \\ \tilde{v}'_{B'} \end{pmatrix}
= \begin{pmatrix} F_A - F^E_A \\ G_{llA'} - G^E_{llA'} \end{pmatrix},
\tag{9.18}
$$

where the superscript "E" denotes the quantities evaluated in an equilibrium state by using the local equilibrium distribution function $f^{(E)}$ given by (7.8). Plugging (7.8) into (9.16), we obtain the following useful relations that express the quantities $J^{0|E}_{AB}$, $J^{1|E}_{AB'}$, and $J^{2|E}_{A'B'}$ by the corresponding quantities $J^{\mathscr{M}}_{AB}$ of the monatomic gas case defined in (4.60) and the parameter $\alpha = \frac{D-5}{2}$ (D: degrees of freedom of a molecule):

$$\begin{aligned} &J^{0|E}_{AB} = J^{\mathscr{M}}_{AB}, \qquad J^{1|E}_{AB'} = J^{\mathscr{M}}_{iiAB'} + 2c_s^2(1+\alpha)J^{\mathscr{M}}_{AB'}, \\ &J^{2|E}_{A'B'} = J^{\mathscr{M}}_{iijjA'B'} + 4c_s^2(1+\alpha)\left(J^{\mathscr{M}}_{iiA'B'} + c_s^2(2+\alpha)J^{\mathscr{M}}_{A'B'}\right), \end{aligned} \tag{9.19}$$

where

$$c_s = \sqrt{\frac{k_B}{m}T}.$$

Once $\left(\tilde{u}'_B, \tilde{v}'_{B'}\right)$ is calculated as a solution of (9.18) in terms of the densities F_A and $G_{llA'}$, the main field $\left(u'_A, v'_{A'}\right)$ is obtained from $(9.17)_2$ by considering the fact that, from the general Theorem 2.4, all the components of main field in equilibrium vanish except for the first 5 components corresponding to the Lagrange multipliers of the conservation laws of mass, momentum, and energy (u', u'_i, v'_{ll}). Then inserting the solution of (9.18) $\left(\tilde{u}'_B, \tilde{v}'_{B'}\right)$ into $(9.17)_1$, we obtain all the fluxes $(9.3)_{2,3}$ and the productions $(9.4)_{2,3}$ in terms of the densities. Thus the closure is completed.

9.3.1 14-Moment System and Its Principal Subsystems

In the case of the 14-moment system ($N = 2$, $M = 1$), the above results coincide with the ones described in Chap. 7 and the closed system in physical variables is given in (1.33).

According to the theory of principal subsystems by Boillat and Ruggeri [165] presented in Sect. 2.4, the following systems can be obtained as *principal subsystems* of the 14-moment system:

The 11-Moment System ($N = 2$, $M = 0$) Setting $q_i = 0$ and neglecting the equation of q_j in (1.33), the following principal subsystem is obtained

$$\begin{aligned}
&\partial_t \rho + \partial_i (\rho v_i) = 0,\\
&\partial_t \left\{\rho v_j\right\} + \partial_i \left\{\rho v_i v_j + (p+\Pi)\delta_{ij} - \sigma_{\langle ij\rangle}\right\} = 0,\\
&\partial_t \left\{\rho v^2 + 3(p+\Pi)\right\} + \partial_i \left\{\rho v^2 v_i + 5v_i(p+\Pi) - 2v_l\sigma_{\langle li\rangle}\right\} = -\frac{3\Pi}{\tau_\Pi},\\
&\partial_t \left\{\rho v_{\langle j}v_{k\rangle} - \sigma_{\langle jk\rangle}\right\} + \partial_i \left\{\rho v_{\langle j}v_{k\rangle}v_i + 2v_{\langle j}\delta_{k\rangle i}(p+\Pi) - 2v_{\langle j}\sigma_{k\rangle i} - v_i\sigma_{\langle jk\rangle}\right\} = \frac{\sigma_{\langle jk\rangle}}{\tau_\sigma},\\
&\partial_t \left\{\rho v^2 + 2\rho\varepsilon\right\} + \partial_i \left\{(\rho v^2 + 2\rho\varepsilon + 2p + 2\Pi)v_i - 2\sigma_{\langle li\rangle}v_l\right\} = 0,
\end{aligned} \tag{9.20}$$

This is the system in which F, F_i, F_{ij} and G_{ll} are retained as field variables.

The 6-Moment System ($N = 2^{(1)}$, $M = 0$) Setting $\sigma_{\langle ij\rangle} = 0$ and neglecting the equation of $\sigma_{\langle ij\rangle}$ in (9.20), the following principal subsystem is obtained

$$\begin{aligned}
&\partial_t \rho + \partial_i (\rho v_i) = 0,\\
&\partial_t \left\{\rho v_j\right\} + \partial_i \left\{\rho v_i v_j + (p+\Pi)\delta_{ij}\right\} = 0,\\
&\partial_t \left\{\rho v^2 + 3(p+\Pi)\right\} + \partial_i \left\{\rho v^2 v_i + 5v_i(p+\Pi)\right\} = -\frac{3\Pi}{\tau_\Pi},\\
&\partial_t \left\{\rho v^2 + 2\rho\varepsilon\right\} + \partial_i \left\{(\rho v^2 + 2\rho\varepsilon + 2p + 2\Pi)v_i\right\} = 0,
\end{aligned}$$

This is the system in which F, F_i, F_{ll} and G_{ll} appear as field variables.

The 5-Moment System ($N = 1$, $M = 0$) The 5-moment system (well-known as the Euler system), which retains F, F_i, and G_{ll} as field variables, is obtained as a principal subsystem of the 14-moment system by setting $\Pi = 0$ in the 6-moment principal subsystem. The system reads as follows:

$$\begin{aligned}
&\partial_t \rho + \partial_i (\rho v_i) = 0,\\
&\partial_t \left(\rho v_j\right) + \partial_i \left(\rho v_i v_j + p\delta_{ij}\right) = 0,\\
&\partial_t \left(\rho v^2 + 2\rho\varepsilon\right) + \partial_i \left((\rho v^2 + 2\rho\varepsilon + 2p)v_i\right) = 0.
\end{aligned}$$

9.3.2 Closure for Higher-Order Systems

As examples of the higher order system than 14-moment system, the closed system with 17, 18, and 30 moments are shown. For simplicity, the one-dimensional variables are displayed with the following notation:[1]

$$\begin{aligned} F_{p,q} &= \int_{\mathbb{R}^3}\int_0^\infty mf(c_1)^p(c^2)^q I^\alpha \, dI d\mathbf{c}, \\ G_{p',q'} &= \int_{\mathbb{R}^3}\int_0^\infty mf\left(c^2+\frac{2I}{m}\right)(c_1)^{p'}(c^2)^{q'} I^\alpha \, dI d\mathbf{c}, \end{aligned} \tag{9.21}$$

where the indexes p, q, p', and q' are the non-negative integers satisfying:

$$0 \leqslant p+2q \leqslant N, \qquad 0 \leqslant p'+2q' \leqslant M \qquad (M \leqslant N-1).$$

In these systems, the first 14 moments ($F_{0,0}, F_{1,0}, F_{2,0}, F_{0,1}, G_{0,0}, G_{1,0}$) are common, and these moments have the following form:

$$\begin{aligned} &F_{0,0} = \rho, \quad F_{1,0} = \rho v, \quad F_{2,0} = p + \Pi - \sigma + \rho v^2, \quad F_{0,1} = 3(p+\Pi) + \rho v^2, \\ &G_{0,0} = 2\rho\varepsilon + \rho v^2, \quad G_{1,0} = 2q + v((D+2)p + 2\Pi - 2\sigma) + \rho v^3. \end{aligned}$$

In the following, we show the explicit form of the remaining densities and fluxes. The nonequilibrium parts of the Galilean tensors of higher order are denoted as $\tilde{F}_{p,q} = \hat{F}_{p,q} - \hat{F}_{p,q}|_E$ and $\tilde{G}_{p',q'} = \hat{G}_{p',q'} - \hat{G}_{p',q'}|_E$.

9.3.2.1 17-Moment System ($N = 3^{(1)}, M = 1$)

The independent variables are $F_{0,0}, F_{1,0}, F_{2,0}, F_{0,1}, F_{1,1}, G_{0,0}, G_{1,0}$ and the constitutive functions are $F_{3,0}, F_{2,1}, G_{2,0}$. The densities and fluxes are obtained as follows:

$$\begin{aligned} F_{3,0} &= \frac{3}{5}\tilde{F}_{1,1} + 3v(p+\Pi-\sigma) + \rho v^3, \\ F_{1,1} &= \tilde{F}_{1,1} + v(5p + 5\Pi - 2\sigma) + \rho v^3, \\ F_{2,1} &= \frac{p}{\rho}(5p + 10\Pi - 7\sigma) + \frac{16}{5} v\tilde{F}_{1,1} + v^2(8p + 8\Pi - 5\sigma) + \rho v^4, \end{aligned}$$

[1] Upon inspection it can be seen that in the one-dimensional case, for any N and M, $\{F_{i_1 i_2 \ldots i_A}, 0 \leqslant A \leqslant N\}$ is mapped into $\{F_{p,q}, 0 \leqslant p+2q \leqslant N\}$, and $\{G_{ll i_1 i_2 \ldots i_{A'}}, 0 \leqslant A' \leqslant M\}$ is mapped into $\{G_{p',q'}, 0 \leqslant p'+2q' \leqslant M\}$.

$$G_{2,0} = \frac{p}{\rho}\{(D+2)p + (D+4)(\Pi - \sigma)\} + v\left(\frac{6}{5}\tilde{F}_{1,1} + 4q\right)$$
$$+ v^2\{(D+5)p + 5\Pi - 5\sigma\} + \rho v^4.$$

9.3.2.2 18-Moment System ($N = 3^{(1)}$, $M = 2^{(1)}$)

The independent variables are $F_{0,0}$, $F_{1,0}$, $F_{2,0}$, $F_{0,1}$, $F_{1,1}$, $G_{0,0}$, $G_{1,0}$, $G_{0,1}$ and the constitutive functions are $F_{3,0}$, $F_{2,1}$, $G_{2,0}$, $G_{1,1}$. The densities and fluxes are obtained as follows:

$$F_{3,0} = \frac{3}{5}\tilde{F}_{1,1} + 3v(p + \Pi - \sigma) + \rho v^3,$$

$$F_{1,1} = \tilde{F}_{1,1} + v(5p + 5\Pi - 2\sigma) + \rho v^3,$$

$$F_{2,1} = \frac{10\tilde{G}_{0,1} + 3\frac{p}{\rho}\{5(D+7)p + 30\Pi - 7(D+7)\sigma\}}{3(D+7)}$$
$$+ \frac{16}{5}v\tilde{F}_{1,1} + v^2(8p + 8\Pi - 5\sigma) + \rho v^4,$$

$$G_{2,0} = \frac{1}{3}\tilde{G}_{0,1} + v\left(\frac{6}{5}\tilde{F}_{1,1} + 4q\right)$$
$$+ \frac{p}{\rho}\{(D+2)p - (D+4)\sigma\} + v^2\{(D+5)p + 5\Pi - 5\sigma\} + \rho v^4,$$

$$G_{0,1} = \tilde{G}_{0,1} + 3(D+2)\frac{p^2}{\rho} + 2v(\tilde{F}_{1,1} + 2q) + v^2\{(D+7)p + 7\Pi - 4\sigma\} + \rho v^4,$$

$$G_{1,1} = \frac{p}{\rho}\left\{(D+6)\tilde{F}_{1,1} + 10q\right\}$$
$$+ v\frac{5(D+11)\tilde{G}_{0,1} + 3\frac{p}{\rho}\{5(D+7)(D+4)p + 60\Pi - 2(D+7)(D+11)\sigma\}}{3(D+7)}$$
$$+ v^2\left(\frac{27}{5}\tilde{F}_{1,1} + 6q\right) + v^3\{(D+11)p + 11\Pi - 8\sigma\} + \rho v^5.$$

The 18-field system was studied from a phenomenological point of view [320].

9.3.2.3 30-Moment System ($N = 3$, $M = 2$)

The independent variables are $F_{0,0}$, $F_{1,0}$, $F_{2,0}$, $F_{0,1}$, $F_{3,0}$, $F_{1,1}$, $G_{0,0}$, $G_{1,0}$, $G_{2,0}$, $G_{0,1}$ and the constitutive functions are $F_{4,0}$, $F_{2,1}$, $G_{3,0}$, $G_{1,1}$. The densities and fluxes are obtained as follows:

$$F_{3,0} = \tilde{F}_{3,0} + 3v(p + \Pi - \sigma) + \rho v^3,$$

$$F_{1,1} = \tilde{F}_{1,1} + v(5p + 5\Pi - 2\sigma) + \rho v^3,$$

$$
\begin{aligned}
F_{4,0} &= \frac{6\left\{(D+7)\tilde{G}_{2,0} - \tilde{G}_{0,1}\right\} + 3\frac{p}{\rho}(D+4)\{(D+7)p + 6\Pi\}}{(D+4)(D+7)} \\
&\quad + 4v\tilde{F}_{3,0} + 6v^2(p + \Pi - \sigma) + \rho v^4, \\
F_{2,1} &= \frac{(D-3)\tilde{G}_{0,1} + 7(D+7)\tilde{G}_{2,0} + 5\frac{p}{\rho}(D+4)\{(D+7)p + 6\Pi\}}{(D+4)(D+7)} \\
&\quad + 2v(\tilde{F}_{1,1} + \tilde{F}_{3,0}) + v^2(8p + 8\Pi - 5\sigma) + \rho v^4, \\
G_{2,0} &= \tilde{G}_{2,0} + (D+2)\frac{p^2}{\rho} + 2v(\tilde{F}_{3,0} + 2q) + v^2\{(D+5)p + 5\Pi - 5\sigma\} + \rho v^4, \\
G_{0,1} &= \tilde{G}_{0,1} + 3(D+2)\frac{p^2}{\rho} + 2v(\tilde{F}_{1,1} + 2q) + v^2\{(D+7)p + 7\Pi - 4\sigma\} + \rho v^4, \\
G_{3,0} &= \frac{p}{\rho}\left\{(D+6)\tilde{F}_{3,0} + 6q\right\} \\
&\quad + 3v\frac{(D+7)(D+8)\tilde{G}_{2,0} - 4\tilde{G}_{0,1} + \frac{p}{\rho}(D+4)\{(D+4)(D+7)p + 12\Pi\}}{(D+4)(D+7)} \\
&\quad + v^2(7\tilde{F}_{3,0} + 6q) + v^3((D+9)p + 9\Pi - 9\sigma) + \rho v^5, \\
G_{1,1} &= \frac{p}{\rho}\left\{(D+6)\tilde{F}_{1,1} + 10q\right\} \\
&\quad + v\frac{(D+11)\left\{(D+2)\tilde{G}_{0,1} + 2(D+7)\tilde{G}_{2,0}\right\} + 5\frac{p}{\rho}(D+4)\{(D+4)(D+7)p + 12\Pi\}}{(D+4)(D+7)} \\
&\quad + v^2(3\tilde{F}_{1,1} + 4\tilde{F}_{3,0} + 6q) + v^3((D+11)p + 11\Pi - 8\sigma) + \rho v^5.
\end{aligned}
$$

Remark 9.2 Also in the case of polyatomic gas, we have the nesting theories of RET that can be identified at any level of approximation via the procedure of principal subsystem. Similar comment to the one in Remark 9.3 of Chap. 4 for monatomic gas is valid.

9.4 Characteristic Velocities of (N, M)-System

The set of the characteristic velocities $\lambda_{(N,M)}$ of the system (9.14) in the propagation direction with a unit vector $\mathbf{n} \equiv (n^i)$ is the set of the roots of the characteristic polynomial $T_{(N,M)}$:

$$
T_{(N,M)} = \det\left[\begin{pmatrix} J^0_{iAB} & J^1_{iAB'} \\ J^1_{iA'B} & J^2_{iA'B'} \end{pmatrix} n^i - \lambda_{(N,M)} \begin{pmatrix} J^0_{AB} & J^1_{AB'} \\ J^1_{A'B} & J^2_{A'B'} \end{pmatrix}\right] = 0.
$$

In particular, the wave velocities for disturbances propagating in an equilibrium state are the solutions of the characteristic polynomial $T^E_{(N,M)}$:

$$T^E_{(N,M)} = \det\left[\begin{pmatrix} J^{0|E}_{iAB} & J^{1|E}_{iAB'} \\ J^{1|E}_{iA'B} & J^{2|E}_{iA'B'} \end{pmatrix} n^i - \lambda^E_{(N,M)} \begin{pmatrix} J^{0|E}_{AB} & J^{1|E}_{AB'} \\ J^{1|E}_{A'B} & J^{2|E}_{A'B'} \end{pmatrix}\right] = 0.$$

Hereafter, the analysis is restricted to the one-dimensional case for the sake of simplicity. Under this assumption, in the same manner as (9.21), the relations (9.19) can be rewritten as

$$\begin{aligned} J^{0|E}_{p+r,q+s} &= -\frac{m^2}{k_B}\int_{\mathbb{R}^3}\int_0^\infty f^E c_1^{p+r}(c^2)^{q+s} I^\alpha \, dI d\mathbf{c} \\ &= J^{\mathcal{M}}_{p+r,q+s}, \end{aligned} \tag{9.22}$$

$$\begin{aligned} J^{1|E}_{p'+r',q'+s'} &= -\frac{m^2}{k_B}\int_{\mathbb{R}^3}\int_0^\infty f^E c_1^{p'+q'}(c^2)^{r'+s'}\left(c^2+\frac{2I}{m}\right) I^\alpha \, dI d\mathbf{c} \\ &= J^{\mathcal{M}}_{p'+r',q'+s'+1} + 2c_s^2(1+\alpha) J^{\mathcal{M}}_{p'+r',q'+s'}, \end{aligned} \tag{9.23}$$

$$\begin{aligned} J^{2|E}_{p'+r',q'+s'} &= -\frac{m^2}{k_B}\int_{\mathbb{R}^3}\int_0^\infty f^E c_1^{p'+r'}(c^2)^{q'+s'}\left(c^2+\frac{2I}{m}\right)^2 I^\alpha \, dI d\mathbf{c} \\ &= J^{\mathcal{M}}_{p'+r',q'+s'+2} + 4c_s^2(1+\alpha)\left(J^{\mathcal{M}}_{p'+r',q'+s'+1} + c_s^2(2+\alpha) J^{\mathcal{M}}_{p'+r',q'+s'}\right), \end{aligned} \tag{9.24}$$

where

$$0 \leqslant r+2s \leqslant N, \qquad 0 \leqslant r'+2s' \leqslant M \qquad (M \leqslant N-1)$$

and

$$\begin{aligned} J^{\mathcal{M}}_{p,q} &= -\frac{m^2}{k_B}\int_{\mathbb{R}^3} f^{\mathcal{M}} c_1^p (c^2)^q \, d\mathbf{c} \\ &= -\frac{m}{k_B}\rho\left(\frac{k_B T}{m}\right)^{\frac{p+2q}{2}} \frac{2^{\frac{p}{2}+q}}{p+1}\Gamma\left(\frac{p+3}{2}+q\right)\frac{1+(-1)^p}{\sqrt{\pi}}. \end{aligned}$$

Introducing the notation:

$$\tilde{J}^{a|E}_{p,q} = -\lambda J^{a|E}_{p,q} + J^{a|E}_{p+1,q} \qquad (a = 0, 1, 2),$$

we rewrite the characteristic polynomial $T^E_{(N,M)}$ in one-dimensional case as

$$T^E_{(N,M)} = \det \begin{pmatrix} \tilde{J}^{0|E}_{p+r,q+s} & \tilde{J}^{1|E}_{p'+r,q'+s} \\ \tilde{J}^{1|E}_{p+r',q+s'} & \tilde{J}^{2|E}_{p'+r',q'+s'} \end{pmatrix} =$$
$$= \det \begin{pmatrix} \tilde{J}^{\mathcal{M}}_{p+r,q+s} & \tilde{J}^{\mathcal{M}}_{p'+r,q'+s+1} \\ \tilde{J}^{\mathcal{M}}_{p+r',q+s'+1} & \tilde{J}^{\mathcal{M}}_{p'+r',q'+s'+2} + 4c_s^4(1+\alpha)\tilde{J}^{\mathcal{M}}_{p'+r',q'+s'} \end{pmatrix} = 0, \tag{9.25}$$

where we make use of the properties of the determinant and of the relations (9.22)–(9.24).

9.4.1 *Characteristic Velocities of the 14-, 11-, 6-, and 5-Moment Systems*

For simplicity, the set $\hat{\lambda}^E_{(N,M)}$ of the dimensionless characteristic velocities is introduced, starting from the set $\lambda^E_{(N,M)}$, as follows:

$$\hat{\lambda}^E_{(N,M)} = \left\{ \frac{\lambda}{c_0}, \ \forall \lambda \in \lambda^E_{(N,M)} \right\},$$

where

$$c_0 = \sqrt{\frac{5}{3} \frac{k_B}{m} T_0}$$

is the sound velocity of a monatomic gas at the temperature T_0.

The explicit form of the dimensionless characteristic velocities in the case of the 14-, 11-, 6-, and 5-moment systems is derived.

The 14-Moment System ($N = 2$, $M = 1$) The characteristic polynomial (9.25) is

$$T^E_{(2,1)} = \det \mathcal{J}^E = 0, \qquad \mathcal{J}^E = \begin{pmatrix} \tilde{J}^{0|E}_{0,0} & \tilde{J}^{0|E}_{1,0} & \tilde{J}^{0|E}_{0,1} & \tilde{J}^{0|E}_{2,0} & \tilde{J}^{1|E}_{0,0} & \tilde{J}^{1|E}_{0,1} \\ \tilde{J}^{0|E}_{1,0} & \tilde{J}^{0|E}_{2,0} & \tilde{J}^{0|E}_{1,1} & \tilde{J}^{0|E}_{3,0} & \tilde{J}^{1|E}_{1,0} & \tilde{J}^{1|E}_{1,1} \\ \tilde{J}^{0|E}_{0,1} & \tilde{J}^{0|E}_{1,1} & \tilde{J}^{0|E}_{0,2} & \tilde{J}^{0|E}_{2,1} & \tilde{J}^{1|E}_{0,1} & \tilde{J}^{1|E}_{0,2} \\ \tilde{J}^{0|E}_{2,0} & \tilde{J}^{0|E}_{3,0} & \tilde{J}^{0|E}_{2,1} & \tilde{J}^{0|E}_{4,0} & \tilde{J}^{1|E}_{2,0} & \tilde{J}^{1|E}_{2,1} \\ \tilde{J}^{1|E}_{0,0} & \tilde{J}^{1|E}_{1,0} & \tilde{J}^{1|E}_{0,1} & \tilde{J}^{1|E}_{2,0} & \tilde{J}^{2|E}_{0,0} & \tilde{J}^{2|E}_{0,1} \\ \tilde{J}^{1|E}_{0,1} & \tilde{J}^{1|E}_{1,1} & \tilde{J}^{1|E}_{0,2} & \tilde{J}^{1|E}_{2,1} & \tilde{J}^{2|E}_{0,1} & \tilde{J}^{2|E}_{0,2} \end{pmatrix},$$

and the equilibrium characteristic velocities in one-dimensional case are calculated as follows:

$$\hat{\lambda}^E_{(2,1)} = \left\{ 0 \text{ (multiplicity 2)}, \ \pm\sqrt{\frac{3}{5}}\sqrt{\frac{2D+7 \pm \sqrt{D^2+16D+37}}{D+2}} \right\}. \tag{9.26}$$

The 11-Moment System ($N = 2$, $M = 0$) The characteristic polynomial of the 11-moment system is obtained as the determinant of the matrix $\mathscr{J}^E$ after removing the last row and the last column. The characteristic velocities are given by

$$\hat{\lambda}^E_{(2,0)} = \left\{ 0 \text{ (multiplicity 3)}, \ \pm\sqrt{\frac{9}{5}} \right\}. \tag{9.27}$$

The 6-Moment System ($N = 2^{(1)}$, $M = 0$) In this case, the characteristic polynomial is obtained as the determinant of the matrix $\mathscr{J}^E$ after removing the third and last rows and columns. In this case the characteristic velocities are given by

$$\hat{\lambda}^E_{(2^{(1)},0)} = \{0 \text{ (multiplicity 2)}, \ \pm 1\}. \tag{9.28}$$

The 5-Moment System ($N = 1$, $M = 0$) The characteristic polynomial of the 5-moment system (Euler system) is the determinant of the matrix $\mathscr{J}^E$ after removing the last three rows and columns. The characteristic velocities are given by

$$\hat{\lambda}^E_{(1,0)} = \left\{ 0, \ \pm\sqrt{\frac{3}{5}}\sqrt{\frac{D+2}{D}} \right\}. \tag{9.29}$$

9.4.2 Systems with D-independent Characteristic Velocities

By analyzing the features of the matrix $\mathscr{J}^E$ by means of (9.19), it seems that, as expected, the equilibrium characteristic velocities $\lambda^E_{(N,M)}$ depend on the parameter α, and thus on the degrees of freedom of a molecule D. Nonetheless, as seen, for example, in (9.27) for the case of 11-moment system and in (9.28) for the case of 6-moment system, this is not always the case. Upon inspection of the features of the matrix $\mathscr{J}^E$, the following theorem holds:

Theorem 9.5 *If $M < N - 1$, the equilibrium characteristic velocities of (N, M)-system, $\lambda^E_{(N,M)}$, are independent of the degrees of freedom D of a molecule. More specifically, the equilibrium characteristic velocities coincide with those of (N)-system, $\lambda^E_{(N)}$,* and *those of (M)-system, $\lambda^E_{(M)}$, of monatomic gases:*

$$\lambda^E_{(N,M)} = \lambda^E_{(N)} \cup \lambda^E_{(M)}.$$

In particular, the maximum equilibrium characteristic velocity of (N, M)*-system,* $\lambda^{E,\max}_{(N,M)}$*, is independent of the order* M*, and coincides with the one of* (N)*-systems for monatomic gases, i.e.,*

$$\lambda^{E,\max}_{(N,M)} = \lambda^{E,\max}_{(N)}.$$

Proof In the characteristic polynomial $T^E_{(N,M)}$ described as (9.25), all components of $\tilde{J}^{\mathcal{M}}_{p+r',q+s'+1}$ are included in $\tilde{J}^{\mathcal{M}}_{p+r,q+s}$ since $M < N - 1$, therefore these are subtracted from the determinant. Similarly, all components of $\tilde{J}^{\mathcal{M}}_{p'+r',q'+s'+2}$ are subtracted by $\tilde{J}^{\mathcal{M}}_{p'+r,q'+s+1}$. Then, $T^E_{(N,M)}$ may be manipulated as follows:

$$\begin{aligned} T^E_{(N,M)} &= \det\begin{pmatrix} \tilde{J}^{\mathcal{M}}_{p+r,q+s} & 0_{p'+r,q'+s+1} \\ 0_{p+r',q+s'+1} & 4c_s^4(1+\alpha)\tilde{J}^{\mathcal{M}}_{p'+r',q'+s'} \end{pmatrix} \\ &= \left(4c_s^4(1+\alpha)\right)^d \det\left(\tilde{J}^{\mathcal{M}}_{p+r,q+s}\right) \det\left(\tilde{J}^{\mathcal{M}}_{p'+r',q'+s'}\right), \end{aligned} \tag{9.30}$$

with

$$d = \frac{M+1+\mathrm{mod}(M+1,2)}{2} = \begin{cases} \frac{M+1}{2} & \text{if } M \text{ is odd,} \\ \frac{M+2}{2} & \text{if } M \text{ is even,} \end{cases}$$

where 0 with indexes denotes the zero block matrix. From (9.30), it becomes evident that all the characteristic velocities that are obtained as the roots of the polynomial $T^E_{(N,M)}$ are independent of the degrees of freedom D of a molecule, and these are identified as the characteristic velocities of (N)-system and of (M)-system of rarefied monatomic gases (Fig. 9.1).

Finally, the claim that the maximum characteristic velocity coincides with those of (N)-system for the monatomic gas is a consequence of the subcharacteristic conditions (2.37) (see Sect. 2.4).

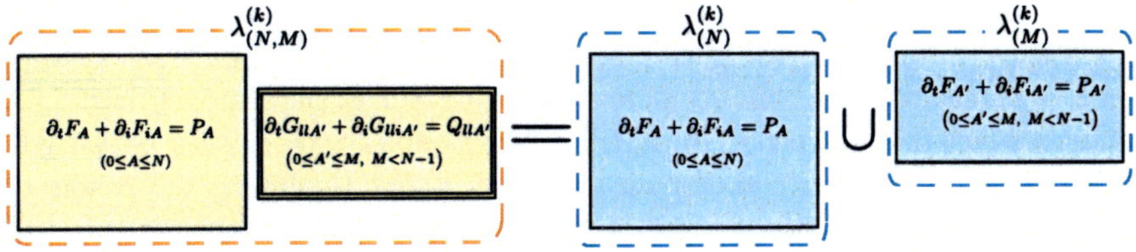

Fig. 9.1 Graphical representation of the collapse of the characteristic velocities derived from (N, M)-system (on the left) into those derived from (N)-system and (M)-system (on the right) when $M < N - 1$

It is worth pointing out that Theorem 9.5 is true for $(N^{(1)}, M)$-system with $M < N-1$, since the proof comes from the corresponding relationship between $F_{llA'}$ and $G_{llA'}$ $(0 \leqslant A' \leqslant M)$.

As already noticed, the equilibrium characteristic velocities of the 11-moment and of the 6-moment systems do not depend on D. In these cases it is easily seen that the set of the equilibrium characteristic velocities are obtainable according to Theorem 9.5 as follows:

The (2, 0)*-system (11-Moment System)* In this case the equilibrium characteristic velocities of the (2)-system and of the (0)-system of monatomic gases are:

$$\hat{\lambda}^E_{(2)} = \left\{0 \text{ (multiplicity 2)}, \ \pm\sqrt{\frac{9}{5}}\right\}, \qquad \hat{\lambda}^E_{(0)} = \{0\},$$

which, compared to (9.27), show that $\lambda^E_{(2,0)} = \lambda^E_{(2)} \cup \lambda^E_{(0)}$ and $\lambda^{E,\max}_{(2,0)} = \lambda^{E,\max}_{(2)}$.

The $(2^{(1)}, 0)$*-system (6-Moment System)* Also in this case it is easy to see that the equilibrium characteristic velocities (9.28) are those of the $(2^{(1)})$-system *and* those of the (0)-system of monatomic gases, i.e., $\lambda^E_{(2^{(1)},0)} = \lambda^E_{(2^{(1)})} \cup \lambda^E_{(0)}$ with

$$\hat{\lambda}^E_{(2^{(1)})} = \{0, \ \pm 1\}, \qquad \hat{\lambda}^E_{(0)} = \{0\},$$

and $\lambda^{E,\max}_{(2^{(1)},0)} = \lambda^{E,\max}_{(2^{(1)})}$.

Remark 9.3 Theorem 9.5 is satisfied also for the systems: $(N, M^{(1)})$, $(N^{(1)}, M)$, and $(N^{(1)}, M^{(1)})$ because its proof is true when each G-moment has the correspondence to F-moments. From the requirement of the Galilean invariance, the possible $(N^{(1)}, M)$-system has $M < N-1$, therefore $(N^{(1)}, M)$-system is always independent of D.

Remark 9.4 When $M < N-1$ and the trace part of the N-tensorial equation in the F-hierarchy, i.e., $F_{lli_1\cdots i_{N-2}}$ is absent, the characteristic velocities can depend on D. The 10-moment system in which F, F_i, $F_{\langle ij\rangle}$ and G_{ll} are retained as field variables is an example.

In the following, according with the physical case of 5 moments (Euler system) and 14 moments (hyperbolic counterpart of Navier-Stokes and Fourier system), the case in which the trace part of the N-tensorial equation always exists is considered. Therefore the above-remarked case is excluded, and combining the results of Theorems 9.1 and 9.5, it is possible to draw the following conclusion:

Theorem 9.6 *(N, M)-system satisfies the relevant features of Galilean invariance and has equilibrium characteristic velocities depending on the degrees of freedom D if and only if*

$$M = N - 1.$$

Table 9.1 Densities, number of moments, and the dependence of $\hat{\lambda}^E$ on D for some (N, M)-systems

(N,M)	Densities	Number of moments	Dependence of $\hat{\lambda}^E$ on D
$(1, 0)$	(F, F_i, G_{ll})	5	Yes
$(2^{(1)}, 0)$	(F, F_i, F_{ll}, G_{ll})	6	No
$(2, 0)$	(F, F_i, F_{ij}, G_{ll})	11	No
$(2,1)$	$(F, F_i, F_{ij}, G_{ll}, G_{lli})$	14	Yes
$(3^{(1)},1)$	$(F, F_i, F_{ij}, F_{lli}, G_{ll}, G_{lli})$	17	No
$(3^{(1)},2^{(1)})$	$(F, F_i, F_{ij}, F_{lli}, G_{ll}, G_{lli}, G_{lljj})$	18	Yes
$(3,2^{(1)})$	$(F, F_i, F_{ij}, F_{ijk}, G_{ll}, G_{lli}, G_{lljj})$	25	Yes
$(3,2)$	$(F, F_i, F_{ij}, F_{ijk}, G_{ll}, G_{lli}, G_{llij})$	30	Yes

This conclusion is also true for $(N, (N-1)^{(1)})$ and $(N^{(1)}, (N-1)^{(1)})$-systems, while $(N^{(1)}, (N-1))$-system is excluded.

From (9.5) and (9.6), the number of moments for $(N, N-1)$-system, is given by

$$n_{(N,N-1)} = \frac{1}{6}(N+1)(N+2)(2N+3).$$

For $(N, (N-1)^{(1)})$-system,

$$n_{(N,(N-1)^{(1)})} = \frac{1}{6}(2N^3 + 9N^2 + N + 12),$$

and, for $(N^{(1)}, (N-1)^{(1)})$-system,

$$n_{(N^{(1)},(N-1)^{(1)})} = 1 + \frac{1}{6}N(N-1)(11+2N).$$

As examples, we present several systems with the list of the densities, the number of moments, and the dependence of $\hat{\lambda}^E$ on D in Table 9.1.

9.5 Characteristic Velocities of $(N, N-1)$-System and the Analysis of the Cases: $D \to 3$ and $D \to \infty$

In this section, an analysis of the equilibrium characteristic velocities of the physically relevant system, i.e., $(N, N-1)$-system, is presented. In particular, the equilibrium maximum characteristic velocity is discussed.

The maximum equilibrium characteristic velocity of $(N, N-1)$-system of rarefied polyatomic gases is limited by that of monatomic gases, as shown below:

Theorem 9.7 *For any truncation order N, the maximum equilibrium characteristic velocity of $(N, N-1)$-system of a rarefied polyatomic gas, $\lambda^{\max}_{(N,N-1)}$, is bounded by the maximum equilibrium characteristic velocities of $(N+1)^{(1)}$-system and of (N)-system of a rarefied monatomic gas as follows:*

$$\lambda^{\max}_{(N)} \leqslant \lambda^{\max}_{(N,N-1)} \leqslant \lambda^{\max}_{(N+1)^{(1)}}.$$

In order to prove this theorem, it suffices to notice that both the upper and lower bounds are obtained from Theorem 9.5 and the subcharacteristic conditions:

$$\lambda^{\max}_{(N+1)^{(1)}} = \lambda^{\max}_{((N+1)^{(1)},M)} = \lambda^{\max}_{((N+1)^{(1)},N-1)} \geqslant \lambda^{\max}_{(N,N-1)}, \qquad \forall M < N$$

$$\lambda^{\max}_{(N,N-1)} \geqslant \lambda^{\max}_{(N,M)} = \lambda^{\max}_{(N)}, \qquad \forall M < N-1.$$

Hereafter, the influences of the degrees of freedom D of a molecule on the characteristic velocities are studied.

9.5.1 Characteristic Velocities in the Limit Case: $D \to 3$

It was proved [112] that, in the limit case $D \to 3$, the solutions of the 14-moment system of a polyatomic gas converge to those of the 13-moment system of a monatomic gas (see Sect. 6.5).

A similar proof can be done for a system with generic number of moments. In [112], it was proved that $(N, N-1)$-system for a rarefied polyatomic gas (which is composed of $n_{(N,N-1)} = \frac{1}{6}(N+1)(N+2)(2N+3)$ equations) converges to the system of order $(N+1)^{(1)}$ for a rarefied monatomic gas (which is composed of $n_{(N+1)^{(1)}} = \frac{1}{6}(N+1)(N^2+8N+6)$ equations). Details of this limit will be considered in Sect. 10.3 of the next Chapter. Therefore, in this limit, the binary-hierarchy can be regarded as the single-hierarchy of rarefied monatomic gases. A graphical representation of the collapse of the $(N, N-1)$-system into the $(N+1)^{(1)}$-system is presented in Fig. 9.2.

As a consequence of the convergence of the solutions of polyatomic gases towards those of monatomic gases, the characteristic velocities follow the same rule. It should be noted that when the characteristic polynomial (9.25) is calculated in this limit, we may obtain the characteristic velocities that are composed of not only those of $(N+1)^{(1)}$-system but also those of $(N-2)$-system:

$$\lim_{D\to 3} \lambda^{E}_{(N,N-1)} = \lambda^{E}_{(N+1)^{(1)}} \cup \lambda^{E}_{(N-2)}. \tag{9.31}$$

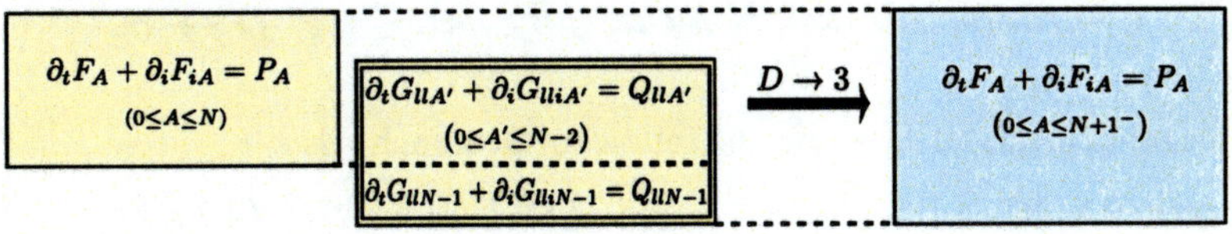

Fig. 9.2 Graphical representation of the collapse of $(N, N-1)$-system (on the left) into $(N+1)^{(1)}$-system (on the right) when $D \to 3$. $(N+1)^{(1)}$-system is (N)-system augmented by the trace part of order N, i.e., augmented by the balance laws for the moments $F_{lli_1 \cdots i_{N-1}}$

However, the characteristic velocities of $(N-2)$-system are related to the balance equations which vanish in this limit.

Moreover (see Theorem 9.5), it is noticeable that, in the limit case $D \to 3$, the maximum characteristic velocity of $(N, N-1)$-system coincides with the one of $(N+1)^{(1)}$-system for rarefied monatomic gases, which in turn coincides with that of $((N+1)^{(1)}, M)$-system for any $M < N$, i.e.,

$$\lim_{D \to 3} \lambda^{\max}_{(N,N-1)} = \lambda^{\max}_{(N+1)^{(1)}} = \lambda^{\max}_{((N+1)^{(1)},M)}, \qquad \forall M < N. \tag{9.32}$$

It is easily proved that for the Euler system (9.29) and the 14-moment system (9.26), the relations (9.31) and (9.32) are confirmed as follows:

Euler System For this system, there is no I_A. Therefore, in the limit case $D \to 3$, it is enough to consider the Euler system for a monatomic gas, i.e., $(2^{(1)})$-system. This corresponds to the fact that the Euler system can be applicable for any fluids. The characteristic velocities are given by

$$\lim_{D \to 3} \hat{\lambda}^E_{(1,0)} = \hat{\lambda}^E_{(2^{(1)})} = \{0, \ \pm 1\}.$$

The 14-Moment System For this system, in the limit case $D \to 3$, the dynamic pressure $\Pi \propto \lim_{D \to 3}(F_{ll} - G_{ll}) = I_0$ vanishes, then the solutions of the 14-moment system coincide with those of the 13-moment rarefied monatomic gas system, i.e., $3^{(1)}$-system. The characteristic velocities, converging to those of the 13-moment system of rarefied monatomic gases, are the following:

$$\lim_{D \to 3} \hat{\lambda}^E_{(2,1)} = \hat{\lambda}^E_{3^{(1)}} \cup \hat{\lambda}^E_{(0)} = \left\{ 0 \, (\text{multiplicity} \, 2), \ \pm \sqrt{\frac{3}{5}} \sqrt{\frac{13 \pm \sqrt{94}}{5}} \right\},$$

where

$$\hat{\lambda}^E_{3^{(1)}} = \left\{ 0, \ \pm \sqrt{\frac{3}{5}} \sqrt{\frac{13 \pm \sqrt{94}}{5}} \right\}$$

are the characteristic velocities of the 13-moment system of a monatomic gas.

9.5.2 Characteristic Velocities in the Limit Case: $D \to \infty$

On the other hand, for $D \to \infty$, the following theorem holds:

Theorem 9.8 *When $D \to \infty$, $n_{(N,N-1)} = \frac{1}{6}(N+1)(N+2)(2N+3)$ characteristic velocities of $(N, N-1)$-system coincide with $n_{(N)} = \frac{1}{6}(N+1)(N+2)(N+3)$ characteristic velocities of (N)-system and $n_{(N-1)} = \frac{1}{6}N(N+1)(N+2)$ characteristic velocities of $(N-1)$-system.*

$$\lim_{D\to\infty} \lambda^E_{(N,N-1)} = \lambda^E_{(N)} \cup \lambda^E_{(N-1)}.$$

In particular, the maximum characteristic velocity of $(N, N-1)$-system coincides with the one of (N)-system for monatomic gases that is in turn coincident with the one of (N, M)-system for any $M < N-1$, i.e.,

$$\lim_{D\to\infty} \lambda^{\max}_{(N,N-1)} = \lambda^{\max}_{(N)} = \lambda^{\max}_{(N,M)}, \qquad \forall M < N-1. \tag{9.33}$$

Proof The characteristic polynomial $T^E_{(N,N-1)}$, as seen from (9.25), can be written in the form:

$$T^E_{(N,N-1)} = \left(4c_s^4(1+\alpha)\right)^d \det\begin{pmatrix} \tilde{J}^{\mathcal{M}}_{p+r,q+s} & \tilde{J}^{\mathcal{M}}_{p'+r,q'+s+1} \\ \frac{\tilde{J}^{\mathcal{M}}_{p+r',q+s'+1}}{4c_s^4(1+\alpha)} & \frac{\tilde{J}^{\mathcal{M}}_{p'+r',q'+s'+2}}{4c_s^4(1+\alpha)} + \tilde{J}^{\mathcal{M}}_{p'+r',q'+s'} \end{pmatrix} = 0,$$

therefore

$$\begin{aligned}\lim_{\alpha\to\infty} \frac{T^E_{(N,N-1)}}{\left(4c_s^4(1+\alpha)\right)^d} &= \det\begin{pmatrix} \tilde{J}^{\mathcal{M}}_{p+r,q+s} & \tilde{J}^{\mathcal{M}}_{p'+r,q'+s+1} \\ 0_{p+r',q+s'} & \tilde{J}^{\mathcal{M}}_{p'+r',q'+s'} \end{pmatrix} \\ &= \det\left(\tilde{J}^{\mathcal{M}}_{p+r,q+s}\right)\det\left(\tilde{J}^{\mathcal{M}}_{p'+r',q'+s'}\right) = 0.\end{aligned} \tag{9.34}$$

From (9.34), it is shown that the characteristic velocities of $(N, N-1)$-system are those of (N)-system and of $(N-1)$-system of monatomic gases. Since $(N-1)$-system is a principal subsystem of (N)-system, the subcharacteristic conditions guarantee that the maximum characteristic velocity of $(N, N-1)$-system coincides with the maximum characteristic velocity of (N)-system. Moreover, from the Theorem 9.5, the maximum characteristic velocity coincides with that of (N, M)-system for any $M < N-1$. This means that, in the limit $D \to \infty$, the maximum characteristic velocity is determined only by the order of truncation of the F-hierarchy.

The collapse of the characteristic velocities of $(N, N-1)$-system into those derived from (N)-system and $(N-1)$-system as $D \to \infty$ is graphically shown in Fig. 9.3.

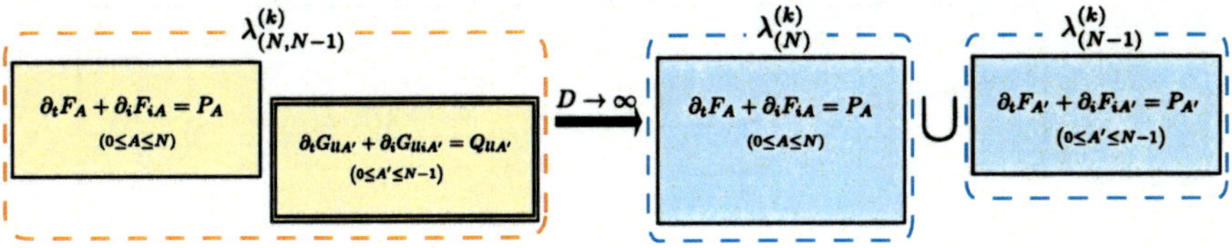

Fig. 9.3 Graphical representation of the collapse of the characteristic velocities derived from $(N, N-1)$-system (on the left) into those derived from (N)-system and $(N-1)$-system (on the right) when $D \to \infty$

For the Euler system (9.29) and the 14-moment system (9.26), in agreement with Theorem 9.8, the following results are obtained:

The 14-Moment System The characteristic velocities of the 14-moment system for rarefied polyatomic gases (9.26) converge to those of the (2)-system for monatomic gases (i.e., the 10-moment system retaining F, F_i, and F_{ij} as independent fields) and those of the (1)-system (i.e., the 4-moment system retaining F and F_i as independent fields):

$$\lim_{D\to\infty} \lambda^E_{(2,1)} = \hat{\lambda}^E_{(2)} \cup \hat{\lambda}^E_{(1)} = \left\{0\,(\text{multiplicity}\,2),\ \pm\sqrt{\frac{9}{5}}\right\} \cup \left\{\pm\sqrt{\frac{3}{5}}\right\} = \left\{0\,(\text{multiplicity}\,2),\ \pm\sqrt{\frac{3}{5}},\ \pm\sqrt{\frac{9}{5}}\right\}.$$

Euler System The characteristic velocities of the Euler system for rarefied polyatomic gases (9.29) converge to those of the (1)-system (i.e., the 4-moment system retaining F and F_i as independent fields) and that of the (0)-system (i.e., the system retaining only F as independent field):

$$\lim_{D\to\infty} \lambda^E_{(1,0)} = \hat{\lambda}^E_{(1)} \cup \hat{\lambda}^E_{(0)} = \left\{\pm\sqrt{\frac{3}{5}}\right\} \cup \{0\} = \left\{0,\ \pm\sqrt{\frac{3}{5}}\right\}.$$

9.5.3 The Case: $3 < D < \infty$

In the case: $3 < D < \infty$, as a corollary of Theorem 9.7, considering (9.32) and (9.33), we obtain the following result:

Theorem 9.9 *For any truncation order N, the maximum equilibrium characteristic velocity of (N, M)-system of polyatomic gases is bounded as follows:*

$$\lim_{D\to\infty} \lambda^{\max}_{(N,N-1)} \leqslant \lambda^{\max}_{(N,N-1)} \leqslant \lim_{D\to 3} \lambda^{\max}_{(N,N-1)}.$$

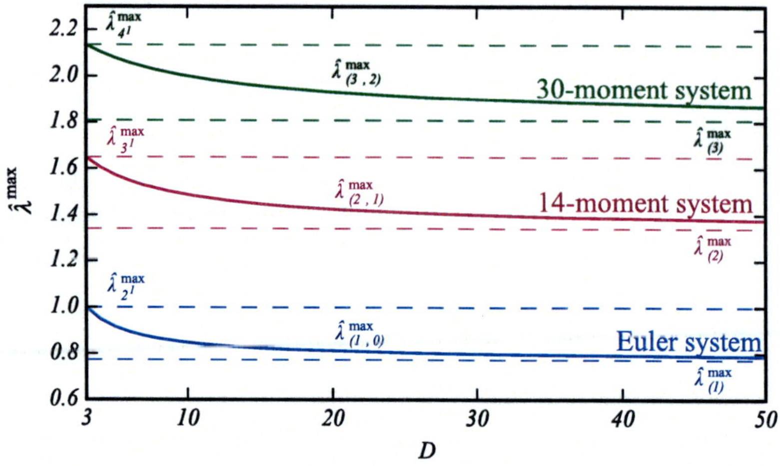

Fig. 9.4 Nondimensional maximum characteristic velocity (at equilibrium) of $(N, N-1)$-system, $\hat{\lambda}^{\max}_{(N,N-1)} = \lambda^{\max}_{(N,N-1)}/c_0$, for three different values of N ($N = 1$: Euler system; $N = 2$: 14-moment system; $N = 3$: 30-moment system) as a function of the degrees of freedom D of a molecule. The limit values of the nondimensional characteristic velocities for $D \to 3$ and $D \to \infty$ (respectively, $\hat{\lambda}^{\max}_{(N+1)^{(1)}} = \lambda^{\max}_{(N+1)^{(1)}}/c_0$ and $\hat{\lambda}^{\max}_{(N)} = \lambda^{\max}_{(N)}/c_0$) are indicated with dashed lines (c_0 being the sound velocity in a monatomic gas)

In Fig. 9.4, dependence of $\lambda^{\max}_{(N,N-1)}$ on D (in the one-dimensional case) for the $(1, 0)$-system (5-moment Euler system), for the $(2, 1)$-system (14-moment system), and for the $(3, 2)$-system (30-moment system) is plotted.

9.6 Dependence of the Maximum Characteristic Velocity on the Truncation Order N

In this section, dependence of the maximum equilibrium characteristic velocity of $(N, N-1)$-system for rarefied polyatomic gases, $\lambda^{\max}_{(N,N-1)}$, on the order of truncation N is analyzed. From Theorem 9.5, the following result can be obtained:

Theorem 9.10 *The maximum equilibrium characteristic velocity of* $(N, N-1)$-*system has the same lower bound as that for the maximum characteristic velocity of* (N)-*system of monatomic gases, i.e.,*

$$\hat{\lambda}^{\max}_{(N,N-1)} \geqslant \sqrt{\frac{6}{5}\left(N - \frac{1}{2}\right)}. \tag{9.35}$$

In particular,

$$\lim_{N\to\infty} \lambda^{\max}_{(N,N-1)} = \infty.$$

Dependence of the maximum equilibrium characteristic velocity on the order N of $(N, N-1)$-system has been numerically calculated for various values of the degrees of freedom D of a molecule and the results are shown in Table 9.2.

In Fig. 9.5, the dependence is shown for two representative values: $D = 10$ and $D = 50$, together with the values obtained for the limit cases $D \to 3$ and $D \to \infty$, and the lower bound given by (9.35).

Remark 9.5 In a special case, the systems characterized by $M < N - 1$ play an interesting role. $(N^{(1)}, N-2)$-system has a special feature that each moment of G-hierarchy has one-to-one correspondence to the trace part of the F-hierarchy. Then, as the total energy can be decomposed into the translational and internal energy, each moment of $G_{llA'}$ can be decomposed into two moments; one relating to the translational mode, i.e., $F_{llA'}$, and the other relating to the internal mode, i.e., $G_{llA'} - F_{llA'}$. From this feature, although equilibrium characteristic velocities are independent of D, $(N^{(1)}, N-2)$-system is physically and mathematically interesting. One of the examples is the RET theory with 6 fields, F, F_i, F_{ll}, G_{ll}, $((2^{(1)}, 0)$-system) [107, 127, 128], which is the simplest theory next to the Euler system. The RET theory with 17 fields $((3^{(1)}, 1)$-system) that adopts $F, F_i, F_{ij}, F_{lli}, G_{ll}, G_{lli}$ as independent fields is another example. This theory is in full agreement with the kinetic theory of polyatomic gases [317, 318] that adopts 17 moments where the internal mode of the energy and the internal heat flux are introduced. Moreover, recently, Rahimi and Struchtrup [308] deduced Grad's 36-moment system based on the kinetic model with the two collision terms of BGK-type, and adopted the method of regularization. The RET theory of 6 moments will be the object in Chaps. 12 and 13.

Remark 9.6 Recently Pennisi and Ruggeri [296] discovered that, taking into account the fact that a RET theory of polyatomic gases can be obtained as a limiting case of a relativistic theory, the proper choice of the moments in the classical theory is completely determined by the index of truncation in the relativistic moments. In particular, according to Theorem 27.1, the index N in the classical case is always equal to $M - 1$, and the optimal RET theories are ET_5, ET_{15}, ET_{35}, ET_{70}, etc. See Chap. 27 for details.

Table 9.2 Nondimensional maximum characteristic velocity (at equilibrium) of $(N, N-1)$-system, $\hat{\lambda}^{\max}_{(N,N-1)} = \lambda^{\max}_{(N,N-1)}/c_0$, for different values of the degrees of freedom D of a molecule and for different values of the truncation order N. The lower bound of the maximum characteristic velocity is given in the last column)

N	Number of moments	$D = 3.000001$	$D = 5$	$D = 7$	$D = 9$	$D = 10^9$	$\sqrt{\frac{6}{5}\left(N-\frac{1}{2}\right)}$
1	5	1.00000000	0.91651514	0.87831007	0.85634884	0.77459667	0.77459667
2	14	1.65028578	1.57434001	1.52907918	1.49884345	1.34164079	1.34164079
3	30	2.13542008	2.07965378	2.04104554	2.01252658	1.80822948	1.73205081
4	55	2.53553413	2.49590924	2.46579529	2.44196686	2.21299946	2.04939015
5	91	2.88470819	2.85640788	2.83350669	2.81447218	2.57495874	2.32379001
6	140	3.19976009	3.17923587	3.16186390	3.14688733	2.90507811	2.56904652
7	204	3.49004595	3.47488600	3.46161633	3.44985116	3.21035245	2.79284801
8	285	3.76126186	3.74984774	3.73959608	3.73030519	3.49555791	3.00000000
9	385	4.01712781	4.00836824	4.00034138	3.99293937	3.76412372	3.19374388
10	506	4.26021705	4.25336891	4.24699414	4.24103382	4.01860847	3.37638860
11	650	4.49239843	4.48694972	4.48181455	4.47696002	4.26098014	3.54964787
12	819	4.71508666	4.71068005	4.70648638	4.70248678	4.49279023	3.71483512
13	1015	4.92939054	4.92577317	4.92230397	4.91897186	4.71528716	3.87298335
14	1240	5.13620410	5.13319435	5.13029015	5.12748479	4.92949284	4.02492236
15	1496	5.33626485	5.33373027	5.33127253	5.32888746	5.13625617	4.17133072
16	1785	5.53019228	5.52803475	5.52593429	5.52388824	5.33629130	4.31277173
17	2109	5.71851433	5.71666004	5.71484891	5.71307921	5.53020569	4.44971909
18	2470	5.90168619	5.90007879	5.89850458	5.89696241	5.71852112	4.58257569
19	2870	6.08010411	6.07869999	6.07732178	6.07596866	5.90168962	4.71168760
20	3311	6.25411586	6.25288081	6.25166626	6.25047163	6.08010585	4.83735465

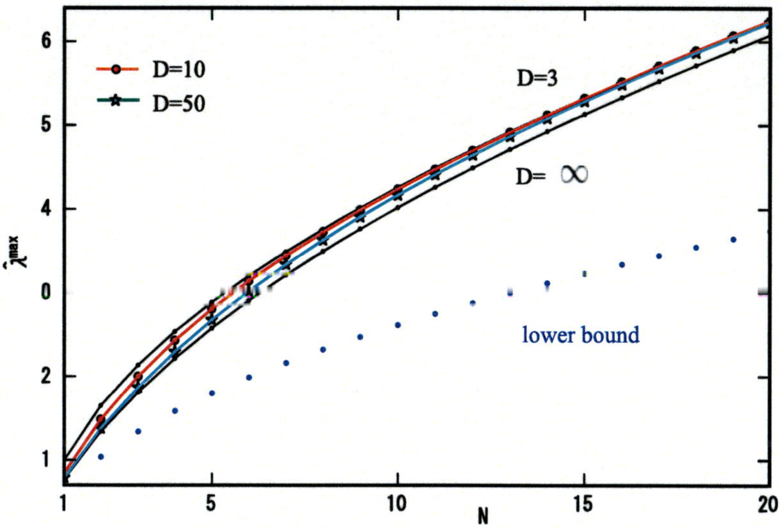

Fig. 9.5 Nondimensional maximum characteristic velocity (at equilibrium) of $(N, N-1)$-system, $\hat{\lambda}^{max}_{(N,N-1)} = \lambda^{max}_{(N,N-1)}/c_0$, for two different values of the degrees of freedom D of a molecule ($D = 10$: line with circle-shaped markers; $D = 50$: line with star-shaped markers) as a function of the order N. The limit values of the nondimensional characteristic velocities for $D \to 3$ and $D \to \infty$ (respectively, $\hat{\lambda}^{max}_{(N+1)^{(1)}} = \lambda^{max}_{(N+1)^{(1)}}/c_0$ and $\hat{\lambda}^{max}_{(N,M)} = \hat{\lambda}^{max}_{(N)} = \lambda^{max}_{(N)}/c_0$ for $M < N - 1$) are indicated with the thin upper and lower lines. The lower bound of the maximum characteristic velocity, given by (9.35), is represented by the dots (c_0 being the sound velocity in a monatomic gas)

Chapter 10
Monatomic-Gas Limit in Molecular ET of Polyatomic Gas

Abstract The difference in the theoretical structure between monatomic gas and polyatomic gas in highly nonequilibrium states is discussed from the viewpoint of molecular extended thermodynamics (molecular ET) of rarefied gases, which is free from the local equilibrium assumption. The molecular ET theories of these two types of gas are based on the moment balance equations with different hierarchy structures due to whether the internal degrees of freedom of a molecule are incorporated in their distribution functions or not. In particular, the number of balance equations in the molecular ET theory of polyatomic gases is greater than the number in the corresponding theory of monatomic gases. In this chapter we prove that the solutions for polyatomic gases converge, in the limit where the degrees of freedom of a molecule D tend to 3, to the ones for monatomic gases provided that we impose appropriate initial conditions compatible with monatomic gases. Thus a molecular ET theory of rarefied monatomic gases can be identified as a singular limit of the corresponding molecular ET theory of rarefied polyatomic gases. As illustrative examples, the asymptotic behaviors when $D \to 3$ in the dispersion relation of ultrasonic sound and in the shock wave structure are shown.

10.1 Introduction

Now we have grasped an important point that, from the viewpoint of RET, the theoretical difference between polyatomic gas and monatomic gas resides in the difference in the hierarchy structure of balance equations in addition to the caloric equation of state. Therefore, one of the essential problems is to clarify the relationship between the molecular extended thermodynamics (molecular ET) theories of monatomic and polyatomic gases when D tends to 3, where D is assumed to be a continuous variable.

It was proved in [112] and discussed in Sect. 6.5 that, in the limit $D \to 3$, the system of 14-field equations for polyatomic gases has the same solutions as the system of 13-field equations for monatomic gases with null dynamic pressure Π in the following sense: The limit should be regarded as a singular limit with an appropriate initial condition compatible with a property of monatomic gases, i.e., in

T. Ruggeri, M. Sugiyama, *Classical and Relativistic Rational Extended Thermodynamics of Gases*, https://doi.org/10.1007/978-3-030-59144-1_10

this case, $\Pi = 0$ at an initial time. In this sense, we can say that the binary hierarchy of polyatomic gases collapses into the single hierarchy of monatomic gases in the limit $D \to 3$.

What happens if the molecular ET theory has a number of independent fields larger than 14? In this case, there appear many new nonequilibrium variables and equations which characterize polyatomic gases in addition to the dynamic pressure. The aim of this chapter is to obtain the relationship between the molecular ET theories of monatomic and polyatomic gases when D tends to 3 in such general cases. In particular, we will study the singular limit of the RET theory with many independent fields, where such nonequilibrium variables vanish. We will prove that the solutions of the system of polyatomic gases coincide with those of monatomic gases in this singular limit for appropriate initial data. In this way, we will have a unified and general molecular ET theory that is valid for both monatomic and polyatomic gases. In this chapter we summarize the results obtained in the paper [125].

10.2 Characteristic Variables of Polyatomic Gas

We try to single out the variables that characterize polyatomic gases. In the RET theory with 14 fields, this problem was studied by recalling that a polyatomic gas is characterized by the following relation:

$$G_{ll} \neq F_{ll} \qquad \text{if} \qquad D > 3,$$

and by the existence of the dynamic pressure, which is expressed by

$$\Pi = \frac{1}{3}\left(F_{ll} - G_{ll}\right) - \frac{1}{3}\left(F_{ll}^{E} - G_{ll}^{E}\right), \tag{10.1}$$

where a quantity with the superscript E is the quantity evaluated at an equilibrium state. If we split the F's hierarchy into the deviatoric part $F_{\langle A\rangle}$ and the trace part F_{llA} as for a monatomic gas in (4.41) and (4.42), we pay attention to the difference between F- and G-series as the *characteristic nonequilibrium variables typical of polyatomic gases*:

$$\Pi_{A''} = \left(F_{llA''} - G_{llA''}\right) - \left(F_{llA''}^{E} - G_{llA''}^{E}\right), \tag{10.2}$$

where $0 \leqslant A'' \leqslant N-2$ and we have put:

$$F_{llA''} = \int_{\mathbb{R}^3}\int_0^\infty mfc^2 c_{A''} I^\alpha \, dI d\mathbf{c}, \quad G_{llA''} = \int_{\mathbb{R}^3}\int_0^\infty mf\left(c^2 + 2\frac{I}{m}\right) c_{A''} I^\alpha \, dI d\mathbf{c}. \tag{10.3}$$

The dynamic pressure Π is the first component of $\Pi_{A''}$ except for a factor $1/3$. Therefore the multi-index tensors $\Pi_{A''}$ ($0 \leqslant A'' \leqslant N-2$) play the role like the dynamic pressure when the moments are more than 14, and we will prove in the following that $\Pi_{A''}$ vanish in the limit of monatomic gas provided that these are zero at the initial time. For this reason we call $\Pi_{A''}$ *the dynamical multi-index pressure tensors.*

Let us introduce variables $I_{A''}$ defined by

$$I_{A''} = F_{llA''} - G_{llA''} = -2 \int_{\mathbb{R}^3} \int_0^\infty f c_{A''} I^{\alpha+1} \, dI d\mathbf{c}, \tag{10.4}$$

and let $I_{iA''}$ and $R_{A''}$ be given by

$$I_{iA''} = F_{lliA''} - G_{lliA''} = -2 \int_{\mathbb{R}^3} \int_0^\infty f c_i c_{A''} I^{\alpha+1} \, dI d\mathbf{c},$$

$$R_{A''} = P_{llA''} - Q_{llA''} = -2 \int_{\mathbb{R}^3} \int_0^\infty Q(f) c_{A''} I^{\alpha+1} \, dI d\mathbf{c}.$$

Then the dynamical multi-index pressure tensors $\Pi_{A''}$ are the nonequilibrium part of $I_{A''}$. If we adopt $I_{A''}$, instead of $F_{llA''}$, as independent variables for the closed $(N, N-1)$-system, the system (9.2) can be rewritten in an equivalent way as follows:

$$\begin{gathered} \partial_t F_{\langle A \rangle} + \partial_i F_{i\langle A \rangle} = P_{\langle A \rangle}, \\ (0 \leqslant A \leqslant N) \qquad \partial_t I_{A''} + \partial_i I_{iA''} = R_{A''}, \quad \partial_t G_{llA'} + \partial_i G_{lliA'} = Q_{llA'}, \\ \left(0 \leqslant A'' \leqslant N-2\right) \qquad (0 \leqslant A' \leqslant N-1) \end{gathered} \tag{10.5}$$

where $F_{\langle A \rangle}$ and $F_{i\langle A \rangle}$ given in (4.41) are expressed as

$$F_{\langle A \rangle} = \int_{\mathbb{R}^3} \int_0^\infty f c_{\langle A \rangle} I^\alpha \, dI d\mathbf{c}, \quad F_{i\langle A \rangle} = \int_{\mathbb{R}^3} \int_0^\infty f c_i c_{\langle A \rangle} I^\alpha \, dI d\mathbf{c}. \tag{10.6}$$

with

$$c_{\langle A \rangle} = \begin{cases} 1 & \text{for } A = 0 \\ c_{i_1} & \text{for } A = 1 \\ c_{\langle i_1} \cdots c_{i_A \rangle} & \text{for } 1 \leqslant A \leqslant N. \end{cases}$$

The unknown fields (independent variables) are $\{F_{\langle A \rangle}, I_{A''}, G_{llA'}\}$. In order that the binary hierarchy converges to the single hierarchy in the limit $D \to 3$, it is necessary that the hierarchy of the variables $I_{A''}$ vanishes. Therefore we will focus particularly on the evolution of the variables $I_{A''}$.

As an example of the structure of (10.5) let us take ET_{14} (see (6.2)). In this case with $N = 2$, $M = N - 1 = 1$,

$$F_{\langle A\rangle} \equiv \begin{pmatrix} F \\ F_{i_1} \\ F_{\langle i_1 i_2\rangle} \end{pmatrix}, \qquad F_{i\langle A\rangle} \equiv \begin{pmatrix} F_i \\ F_{ii_1} \\ F_{i\langle i_1 i_2\rangle} \end{pmatrix}, \qquad P_{\langle A\rangle} \equiv \begin{pmatrix} 0 \\ 0 \\ P_{\langle i_1 i_2\rangle}, \end{pmatrix},$$

$$I_{A''} \equiv \left(F_{ll} - G_{ll} \right), \qquad I_{iA''} \equiv \left(F_{ill} - G_{ill} \right), \qquad R_{A''} \equiv \left(P_{ll} \right),$$

and

$$G_{llA'} \equiv \begin{pmatrix} G_{ll} \\ G_{i_1 ll} \end{pmatrix}, \qquad G_{illA'} \equiv \begin{pmatrix} G_{ill} \\ G_{ii_1 ll} \end{pmatrix}, \qquad Q_{llA'} \equiv \begin{pmatrix} 0 \\ Q_{i_1 ll} \end{pmatrix}.$$

10.2.1 Closure of the New System

We have seen in Chap. 9 that the distribution function that maximizes the entropy is the one given in (9.10). In the present case where the system is given in the form (10.5), the expression of χ is given by

$$\chi = \lambda_{\langle A\rangle} c_{\langle A\rangle} - \frac{2I}{m} \nu_{A''} c_{A''} + \left(c^2 + \frac{2I}{m} \right) \mu_{A'} c_{A'}, \tag{10.7}$$

where $\lambda_{\langle A\rangle}$, $\nu_{A''}$, and $\mu_{A'}$ are the main field components (Lagrange multipliers). These components are linear combination of the main field components of the system (9.2). By inserting (10.7) into (9.10), (10.6), (10.4), and (10.3), the main field components are evaluated, in principle, in terms of the densities $F_{\langle A\rangle}$, $I_{A''}$, and $G_{llA'}$.

For the same reason for the convergence as in the case of monatomic gas, the distribution function (9.10) is expanded around a local equilibrium state:

$$f \approx f^{(E)} \left\{ 1 - \frac{m}{k_B} \left[\tilde{\lambda}_{\langle A\rangle} c_{\langle A\rangle} - \frac{2I}{m} \tilde{\nu}_{A''} c_{A''} + \left(c^2 + \frac{2I}{m} \right) \tilde{\mu}_{A'} c_{A'} \right] \right\} \tag{10.8}$$

with

$$\tilde{\lambda}_{\langle A\rangle} = \lambda_{\langle A\rangle} - \lambda^E_{\langle A\rangle}, \; \tilde{\nu}_{A''} = \nu_{A''} - \nu^E_{A''}, \; \tilde{\mu}_{A'} = \mu_{A'} - \mu^E_{A'}.$$

The equilibrium distribution function $f^{(E)}$ is given by (7.8).

By inserting (10.8) into $(10.6)_1$, $(10.4)_1$, and $(10.3)_2$, a linear algebraic system is obtained, from which we can evaluate the nonequilibrium main field

$\left(\tilde{\lambda}_{\langle A\rangle}, \tilde{\nu}_{A''}, \tilde{\mu}_{A'}\right)$ in terms of the densities $F_{\langle A\rangle}$, $I_{A''}$, and $G_{llA'}$:

$$\begin{pmatrix} J_{\langle A\rangle\langle B\rangle} & -K_{\langle A\rangle B''} & J_{ii\langle A\rangle B'} + K_{\langle A\rangle B'} \\ -K_{A''\langle B\rangle} & L_{A''B''} & -K_{iiA''B'} - L_{A''B'} \\ J_{iiA'\langle B\rangle} + K_{A'\langle B\rangle} & -K_{iiA'B''} - L_{A'B''} & J_{iijjA'B'} + 2K_{iiA'B'} + L_{A'B'} \end{pmatrix} \begin{pmatrix} \tilde{\lambda}_{\langle B\rangle} \\ \tilde{\nu}_{B''} \\ \tilde{\mu}_{B'} \end{pmatrix} = \begin{pmatrix} F_{\langle A\rangle} - F^E_{\langle A\rangle} \\ \Pi_{A''} \\ G_{llA'} - G^E_{llA'} \end{pmatrix}, \tag{10.9}$$

where

$$\begin{aligned} J_{\langle A\rangle\langle B\rangle} &= -\frac{m^2}{k_B}\int_{\mathbb{R}^3}\int_0^\infty f^{(E)} c_{\langle A\rangle} c_{\langle B\rangle} I^\alpha \, dI d\mathbf{c}, \\ K_{\langle A\rangle B''} &= -\frac{m^2}{k_B}\int_{\mathbb{R}^3}\int_0^\infty \frac{2}{m} f^{(E)} c_{\langle A\rangle} c_{B''} I^{\alpha+1} \, dI d\mathbf{c}, \\ L_{A''B''} &= -\frac{m^2}{k_B}\int_{\mathbb{R}^3}\int_0^\infty \frac{4}{m^2} f^{(E)} c_{A''} c_{B''} I^{\alpha+2} \, dI d\mathbf{c}. \end{aligned} \tag{10.10}$$

Inserting (7.8) into (10.10), we can obtain the following useful relations for $J_{\langle A\rangle\langle B\rangle}$, $K_{\langle A\rangle B''}$, and $L_{A''B''}$ in terms of the parameter α and the corresponding quantities of monatomic gases J^M_{AB} given by

$$J^M_{AB} = -\frac{m^2}{k_B}\int_{\mathbb{R}^3} f^{(M)} c_A c_B \, d\mathbf{c} \tag{10.11}$$

with $f^{(M)}$ being the Maxwellian distribution (1.29):

$$\begin{aligned} J_{\langle A\rangle\langle B\rangle} &= J^M_{\langle A\rangle\langle B\rangle}, \\ K_{\langle A\rangle B'} &= 2c_s^2(1+\alpha) J^M_{\langle A\rangle B'}, \\ L_{A''B'} &= 4c_s^4(1+\alpha)(2+\alpha) J^M_{A''B'}, \end{aligned} \tag{10.12}$$

where $c_s = \sqrt{\frac{k_B}{m}T}$. Using these relations, we can order (10.9) and these are expressed in the matrix form as follows:

$$\begin{pmatrix} J^M_{\langle A\rangle\langle B\rangle} & J^M_{\langle A\rangle B''} & J^M_{ii\langle A\rangle B'} + 2c_s^2(1+\alpha) J^M_{\langle A\rangle B'} \\ 0 & -2c_s^2 J^M_{A''B''} & -4c_s^4(1+\alpha) J^M_{iiA''B'} \\ J^M_{iiA'\langle B\rangle} & J^M_{iiA'B''} + 2c_s^2 J^M_{A'B''} & J^M_{iijjA'B'} + 2c_s^2(1+\alpha)\left(J^M_{iiA'B'} + 2c_s^2 J^M_{A'B'}\right) \end{pmatrix} \begin{pmatrix} \tilde{\lambda}_{\langle B\rangle} \\ -2c_s^2(1+\alpha)\tilde{\nu}_{B''} \\ \tilde{\mu}_{B'} \end{pmatrix}$$

$$= \begin{pmatrix} F_{\langle A\rangle} - F^E_{\langle A\rangle} \\ \Pi_{A''} - 2c_s^2(1+\alpha)\left(F_{A''} - F^E_{A''}\right) \\ G_{llA'} - G^E_{llA'} - 2c_s^2(1+\alpha)\left(F_{A'} - F^E_{A'}\right) \end{pmatrix}. \tag{10.13}$$

From (10.13), the main field $\left(\tilde{\lambda}_{\langle A\rangle}, \tilde{\nu}_{A''}, \tilde{\mu}_{A'}\right)$ can be calculated in terms of the densities $\left(F_{\langle A\rangle}, \Pi_{A''}, G_{llA'}\right)$. Then, since the constitutive functions for fluxes satisfy

$$\begin{pmatrix} F_{i\langle N\rangle} - F^E_{i\langle N\rangle} \\ \Pi_{i\langle N-2\rangle} - 2c_s^2(1+\alpha)\left(F_{i\langle N-2\rangle} - F^E_{i\langle N-2\rangle}\right) \\ G_{lli\langle N-1\rangle} - G^E_{lli\langle N-1\rangle} - 2c_s^2(1+\alpha)\left(F_{i\langle N-1\rangle} - F^E_{i\langle N-1\rangle}\right) \end{pmatrix}$$

$$= \begin{pmatrix} J^M_{i\langle N\rangle\langle B\rangle} & J^M_{i\langle N\rangle B''} & J^M_{ikk\langle N\rangle B'} + 2c_s^2(1+\alpha)J^M_{i\langle N\rangle B'} \\ 0 & -2c_s^2 J^M_{i\langle N-2\rangle B''} & -4c_s^4(1+\alpha)J^M_{ikk\langle N-2\rangle B'} \\ J^M_{ikk\langle N-1\rangle\langle B\rangle} & J^M_{ikk\langle N-1\rangle B''} + 2c_s^2 J^M_{i\langle N-1\rangle B''} & J^M_{ikkll\langle N-1\rangle B'} + 2c_s^2(1+\alpha)\left(J^M_{ikk\langle N-1\rangle B'} + 2c_s^2 J^M_{i\langle N-1\rangle B'}\right) \end{pmatrix}$$

$$\times \begin{pmatrix} \tilde{\lambda}_{\langle B\rangle} \\ -2c_s^2(1+\alpha)\tilde{\nu}_{B''} \\ \tilde{\mu}_{B'} \end{pmatrix}, \tag{10.14}$$

the closed field equations are obtained except for the production terms.

10.2.2 Production Terms in the BGK Model

In order to close the system, it is necessary to express the production terms by using the densities. For this purpose, we need the concrete expression of the collision term $Q(f)$. In the present study, however, we avoid entering into the precise modeling of the collision term (see also the remark in Sect. 10.7). Therefore, let us introduce the simplest model, i.e., the BGK model:

$$Q(f) = -\frac{f - f^{(E)}}{\tau},$$

where τ stands for a relaxation time. Although, in practical problems, there exist several kinds of the relaxation time and the difference of the order of magnitude of relaxation times plays an important role such as studied in [308], we assume here that all relaxation processes have a common relaxation time τ. Then the production terms are obtained as

$$P_{\langle A\rangle} = -\frac{1}{\tau}\left(F_{\langle A\rangle} - F^E_{\langle A\rangle}\right), \quad R_{A''} = -\frac{1}{\tau}\Pi_{A''}, \quad Q_{A'} = -\frac{1}{\tau}\left(G_{llA'} - G^E_{llA'}\right).$$

10.3 Singular Limit of a Polyatomic Gas to a Monatomic Gas

Let us consider the singular limit: (1) we take the limit where D approaches 3 continuously, and (2) we adopt an appropriate initial condition compatible with a property of monatomic gases. In this limit, we will show that the variables $I_{A''}$ converge to zero and the remaining fields $F_{\langle A\rangle}$ and $G_{llA'}$ approach the fields in the single hierarchy (4.40).

10.3.1 The Limit D → 3

As explained above, the main field is evaluated in terms of the densities by using the relation (10.13). It is convenient to introduce new fields:

$$\tilde{v}^*_{B''} \equiv -2c_s^2(1+\alpha)\tilde{v}_{B''},$$

then the relation between the main field and the densities (10.13) is rewritten as

$$\begin{pmatrix} J^M_{\langle A\rangle\langle B\rangle} & J^M_{\langle A\rangle B''} & J^M_{ii\langle A\rangle B'} \\ 0 & -2c_s^2 J^M_{A''B''} & 0 \\ J^M_{iiA'\langle B\rangle} & J^M_{iiA'B''} + 2c_s^2 J^M_{A'B''} & J^M_{iijjA'B'} \end{pmatrix} \lim_{D\to 3} \begin{pmatrix} \tilde{\lambda}_{\langle B\rangle} \\ \tilde{v}^*_{B''} \\ \tilde{\mu}_{B'} \end{pmatrix} = \lim_{D\to 3} \begin{pmatrix} F_{\langle A\rangle} - F^E_{\langle A\rangle} \\ \Pi_{A''} \\ G_{llA'} - G^E_{llA'} \end{pmatrix}. \tag{10.15}$$

From (10.4), $(10.10)_2$, and $(10.12)_2$, we have

$$\lim_{D\to 3} I^E_{A''} = \lim_{D\to 3} \frac{k_B}{m} K_{0A''} = 0, \qquad \lim_{D\to 3} I^E_{iA''} = \lim_{D\to 3} \frac{k_B}{m} K_{1A''} = 0.$$

Therefore, in the limit $D \to 3$, $I_{A''}$ and $I_{iA''}$ have only the nonequilibrium part $\Pi_{A''}$ and $\Pi_{iA''}$.

From $(10.15)_2$, the fields $v^*_{B''}$ are expressed explicitly by $\Pi_{A''}$:[1]

$$\lim_{D\to 3} \tilde{v}^*_{B''} = -\frac{1}{2c_s^2}\left(J^M\right)^{-1}_{A''B''} \Pi_{A''}, \tag{10.16}$$

where the element of the matrix J^M is $J^M_{A''B''}$. Concerning the nonequilibrium fluxes of the dynamical pressure tensors:

$$\Pi_{iA''} = I_{iA''} - I^E_{iA''}$$

[1] We continue, for simplicity, using the same letter $\Pi_{A''}$ for the limit value.

evaluated in the limit of $D \to 3$, from (10.14) and (10.16), we obtain

$$\Pi_{iA''} = -2c_s^2 J^M_{iA''B''} \lim_{D\to 3} \tilde{\nu}^*_{B''} = J^M_{iA''B''} (J)^{-1}_{B''C''} \Pi_{C''}.$$

Then, in the limit $D \to 3$, the field equations of the dynamical pressure tensors $(10.5)_2$ become

$$\partial_t \Pi_{A''} + J^M_{iA''B''} (J^M)^{-1}_{B''C''} \, \partial_i \Pi_{C''} = -\left\{ \frac{1}{\tau} \delta_{A''C''} + \partial_i \left[J^M_{iA''B''} (J^M)^{-1}_{B''C''} \right] \right\} \Pi_{C''}, \tag{10.17}$$

where $\delta_{A''C''}$ is the Kronecker delta. It is remarkable that only $\Pi_{A''}$ are involved in these equations, which are the first-order quasi-linear partial differential equations with respect to $\Pi_{A''}$.

10.3.2 Initial Condition Compatible with Monatomic Gas

In the limit to monatomic gas, the initial condition should be compatible with a property of monatomic gases. Therefore we impose the condition:

$$\Pi_{A''}(\boldsymbol{x}, 0) = 0.$$

Under this condition, assuming the uniqueness of the solution of (10.17), we obtain

$$\Pi_{A''}(\boldsymbol{x}, t) = 0, \qquad \forall t.$$

Then we also find that $\lim_{D\to 3} \tilde{\nu}^*_{A''} = 0$ and $\tilde{\nu}_{A''}$ is undetermined. Therefore $\Pi_{A''}$ $(0 \leqslant A'' \leqslant N-2)$ given in (10.2) have a similar role to that of the dynamic pressure in ET_{14}. In particular, when $A'' = 0$, $\Pi_{A''}$ becomes Π itself except for the factor 3 (see (10.1)). For this reason we may call $\Pi_{A''}$ *dynamical pressure tensor.*

10.3.3 Remaining Field Equations

In the singular limit, the balance equations of non-vanishing fields are expressed as follows:

$$\partial_t F_{\langle A \rangle} + \partial_i F_{i\langle A \rangle} = P_{\langle A \rangle}, \qquad (0 \leqslant A \leqslant N)$$

$$\partial_t G_{llA'} + \partial_i G_{lliA'} = Q_{llA'}, \qquad (0 \leqslant A' \leqslant N-1)$$

and, from (10.15), the main field components corresponding to $F_{\langle A \rangle}$ and $G_{llA'}$ are calculated from

$$\begin{pmatrix} J^M_{\langle A \rangle \langle B \rangle} & J^M_{ii\langle A \rangle B'} \\ J^M_{iiA'\langle B \rangle} & J^M_{iijjA'B'} \end{pmatrix} \lim_{D\to 3} \begin{pmatrix} \tilde{\lambda}_{\langle B \rangle} \\ \tilde{\mu}_{B'} \end{pmatrix} = \lim_{D\to 3} \begin{pmatrix} F_{\langle A \rangle} - F^E_{\langle A \rangle} \\ G_{llA'} - G^E_{llA'} \end{pmatrix}. \tag{10.18}$$

Recalling (10.3), from (10.18), we notice that

$$\lim_{D\to 3} (G_{llA'} - G^E_{llA'}) = \lim_{D\to 3} (F_{llA'} - F^E_{llA'}).$$

Taking into account the definition of $G_{llA'}$ (10.3) and the relations (10.12), we notice

$$\lim_{D\to 3} G^E_{llA'} = \lim_{D\to 3} F^E_{llA'} = F^M_{llA'},$$

where $F^M_{llA'}$ is the moment defined in the case of monatomic gas (10.11) in an equilibrium state. Similar relations hold for fluxes and productions. From these relations, the G-hierarchy coincides with the F-hierarchy in the limit:

$$\lim_{D\to 3} G_{llA'} = \lim_{D\to 3} F_{llA'}, \quad \lim_{D\to 3} G_{lliA'} = \lim_{D\to 3} F_{lliA'}, \quad \lim_{D\to 3} Q_{llA'} = \lim_{D\to 3} P_{llA'}.$$

This concludes that the balance equations of non-vanishing fields have the same hierarchy structure as that of monatomic gases of the $((N+1)^{(1)})$-system. The relation between the main field and the densities (10.18) coincides with the relation of monatomic gases (4.59). Then the constitutive equations have the same form as those of monatomic gases, in other words, the equations of non-vanishing fields coincide with those of monatomic gases.

10.3.4 Singular Limit of Other Systems

The above procedure of the singular limit can be applied to the other systems. For example, the results for some typical cases are summarized as follows: $(N, (N-1)^{(1)})$ and $(N^{(1)}, (N-1)^{(1)})$-systems converge to, respectively, $((N+1)^{(2)})$ and $((N+1)^{(2,1)})$-systems, where $((N+1)^{(2)})$-system adopts an $(N+1)$th order density with double trace in addition to the densities up until Nth order as independent fields,[2] and $((N+1)^{(2,1)})$-system adopts the densities of $((N+1)^{(2)})$-system in which the Nth order tensor has only a trace part. As an example of these systems, the densities of $(4^{(2)})$ and $(4^{(2,1)})$-systems are shown in Table 10.1.

[2] $(N+1)^{(2)}$-system is defined for $N \geqslant 3$.

Table 10.1 Correspondence relation between the systems of monatomic and polyatomic gases for particular N and M. Densities and number of moments are shown. $\lambda^* = \lambda^*(D)$ indicates that the equilibrium characteristic velocities λ^* depend on D or not

Monatomic gas			Polyatomic gas			
(N)	Densities	$n_{(N)}$	(N,M)	Densities	$n_{(N,M)}$	$\lambda^* = \lambda^*(D)$
$(2^{(1)})$	F, F_i, F_{ll}	5	(1, 0)	F, F_i, G_{ll}	5	Yes
			$(2^{(1)}, 0)$	F, F_i, F_{ll}, G_{ll}	6	No
$(3^{(1)})$	F, F_i, F_{ij}, F_{lli}	13	(2,1)	$F, F_i, F_{ij}, G_{ll}, G_{lli}$	14	Yes
			$(3^{(1)},1)$	$F, F_i, F_{ij}, F_{lli}, G_{ll}, G_{lli}$	17	No
$(4^{(2,1)})$	$F, F_i, F_{ij}, F_{lli}, F_{lljj}$	14	$(3^{(1)}, 2^{(1)})$	$F, F_i, F_{ij}, F_{lli}, G_{ll}, G_{lli}, G_{lljj}$	18	Yes
			$(4^{(2,1)},2^{(1)})$	$F, F_i, F_{ij}, F_{lli}, F_{lljj}, G_{ll}, G_{lli}, G_{lljj}$	19	No
$(4^{(2)})$	$F, F_i, F_{ij}, F_{ijk}, F_{lljj}$	21	$(3, 2^{(1)})$	$F, F_i, F_{ij}, F_{ijk}, G_{ll}, G_{lli}, G_{lljj}$	25	Yes
			$(4^{(2)}, 2^{(1)})$	$F, F_i, F_{ij}, F_{ijk}, F_{lljj}, G_{ll}, G_{lli}, G_{lljj}$	26	No
$(4^{(1)})$	$F, F_i, F_{ij}, F_{ijk}, F_{llij}$	26	(3,2)	$F, F_i, F_{ij}, F_{ijk}, G_{ll}, G_{lli}, G_{llij}$	30	Yes
			$(4^{(1)}, 2)$	$F, F_i, F_{ij}, F_{ijk}, F_{llij}, G_{ll}, G_{lli}, G_{llij}$	36	No

Table 10.2 Correspondence relation between the systems of monatomic and polyatomic gases $M < N - 1$.

Monatomic gas	Polyatomic gas	
System	System	$\lambda^* = \lambda^*(D)$
$(N+1)^{(1)}$	$(N, N-1)$	Yes
	$((N+1)^{(1)}, M)$	No
$(N+1)^{(2)}$	$(N, (N-1)^{(1)})$	Yes
	$((N+1)^{(2)}, (M+1)^{(1)})$	No
$(N+1)^{(2,1)}$	$(N^{(1)}, (N-1)^{(1)})$	Yes
	$((N+1)^{(2,1)}, (M+1)^{(1)})$	No
N	(N, M)	No

Up to now, we have shown the singular limit of the physically meaningful systems where equilibrium characteristic velocities depend on D. We can also prove the singular limit of (N, M)-system with $M < N - 1$ that adopts, as independent variables, $F_{\langle A\rangle}$ $(0 \leqslant A \leqslant N)$, $I_{A'}$ $(0 \leqslant A' \leqslant M)$, $G_{llA'}$ $(0 \leqslant A' \leqslant M)$, and $F_{ll\alpha}$ $(M+1 \leqslant \alpha \leqslant N-2)$.[3] In the singular limit, $\Pi_{A'} = 0$ and this system also converges to (N)-system of monatomic gas. Similarly the singular limit of $((N+1)^{(1)}, M)$ system converges to $((N+1)^{(1)})$-system.

We summarize the correspondence relation between the systems of polyatomic and monatomic gases in the singular limit in Table 10.2. The correspondence is shown for some systems with particular N and M in Table 10.1.

An interesting remark is that, from the view point of molecular ET of polyatomic gases, appropriate models of monatomic gases have the trace in the highest tensorial equation, such as $(N^{(1)})$, $(N^{(2)})$ and $(N^{(2,1)})$-systems. This fact implies that (N)-system, which is usually utilized in the literature of the RET theory, is not appropriate as a model of monatomic gases.

10.3.5 Singular Limit for the Intrinsic Fields

In practical problems, instead of the system of moments (10.5), the system of field equations in terms of the velocity-independent variables, such as the mass density, internal energy, dynamic pressure, shear viscosity, and heat flux, is useful. In general, the relation between the moments and the velocity-independent fields (intrinsic fields) is derived by imposing the Galilean invariance on the system of balance equations. In this section, we introduce the intrinsic fields and show the equations of these fields. In particular, the closed system of field equations of the velocity-independent part of $\Pi_{A''}$ is obtained and the singular limit for these variables is also shown.

[3] If $M = N - 2$, $F_{ll\alpha} = 0$.

The intrinsic fields (denoted with hat) are expressed, by using the peculiar velocity, as follows:

$$\hat{F}_A = \int_{\mathbb{R}^3}\int_0^\infty mfC_A I^\alpha \, dI d\mathbf{C}, \quad \hat{G}_{llA'} = \int_{\mathbb{R}^3}\int_0^\infty mf\left(C^2 + \frac{2I}{m}\right)C_{A'}I^\alpha \, dI d\mathbf{C}.$$

By imposing the Galilean invariance on the system of balance equations, the moments F_A and $G_{llA'}$ are expressed by these intrinsic fields as follows:

$$\begin{aligned} F_A &= X_{AB}\hat{F}_B, \\ G_{llA'} &= X_{A'B'}\left(\hat{G}_{llB'} + 2v_l\hat{F}_{lB'} + v^2\hat{F}_{B'}\right). \end{aligned} \tag{10.19}$$

The coefficient X_{AB} is a function only of the velocity, and it makes up an $(N \times N)$-matrix X. It should be noted that this matrix is the same as the one that dictates the velocity dependence of the monatomic (N)-system. Similarly $X_{A'B'}$ is an element of $((N-1)\times(N-1))$-matrix. The details of the Galilean invariance and the features of this matrix are discussed in Sect. 2.7,

According to the general result obtained in Sect. 2.7, the system of field equations can be rewritten in terms of the intrinsic fields using the material derivative (indicated by a dot) as follows:

$$\begin{aligned} &\dot{\hat{F}}_A + \hat{F}_A\partial_i v_i + \partial_i\hat{F}_{iA} + A^r_{AC}\left(\hat{F}_C\dot{v}_r + \hat{F}_{iC}\partial_i v_r\right) = -\frac{1}{\tau}\left(\hat{F}_A - \hat{F}^E_A\right), \\ &\dot{\hat{G}}_{llA'} + \hat{G}_{llA'}\partial_i v_i + \partial_i\hat{G}_{lliA'} + 2\dot{v}_l\hat{F}_{lA'} + 2\hat{F}_{ilA'}\partial_i v_l \\ &\qquad + A^r_{A'C'}\left(\hat{G}_{llC'}\dot{v}_r + \hat{G}_{lliC'}\partial_i v_r\right) = -\frac{1}{\tau}\left(\hat{G}_{llA'} - \hat{G}^E_{llA'}\right) \end{aligned} \tag{10.20}$$

with the conservation law of momentum:

$$\rho\dot{v}_i + \partial_j\left\{(p+\Pi)\delta_{ij} - \sigma_{\langle ij\rangle}\right\} = 0.$$

Similarly, for the tensors $I_{A''}$, the velocity-independent densities are introduced as

$$\hat{I}_{A''} = \hat{F}_{llA''} - \hat{G}_{llA''} = -2\int_{\mathbb{R}^3}\int_0^\infty fC_{A''}I^{\alpha+1}\, dI d\mathbf{C},$$

and are related to the moments as follows:

$$I_{A''} = X_{A''B''}\hat{I}_{B''}.$$

The velocity-independent part of $\Pi_{A''}$ is also defined as

$$\hat{\Pi}_{A''} = \hat{I}_{A''} - \hat{I}^E_{A''}.$$

Then the field equations of $\hat{I}_{A''}$ are obtained as follows:

$$\dot{\hat{I}}_{A''} + \hat{I}_{A''}\partial_i v_i + \partial_i \hat{I}_{iA''} + A^r_{A''C''}\left(\hat{I}_{C''}\dot{v}_r + \hat{I}_{iC''}\partial_i v_r\right) = -\frac{1}{\tau}\Pi_{A''}. \tag{10.21}$$

The field equations of the velocity-independent variables corresponding to the system (10.5) with $I_{A''}$, $G_{llA'}$, and $F_{\langle A\rangle}$ are (10.21), $(10.20)_2$, and the traceless part of the field equations $(10.20)_1$. For the traceless tensor $F_{\langle A\rangle}$ we have the following velocity dependence:

$$F_{\langle A\rangle} = X^{tl}_{AB}\hat{F}_B = X^{tl}_{AB}\hat{F}_{\langle B\rangle}, \tag{10.22}$$

where X^{tl}_{AB} indicates the traceless counterpart of X_{AB}. By imposing $v_i = 0$ on (10.9), the velocity-independent main field components $\hat{\lambda}_{\langle A\rangle}$, $\hat{\nu}_{A''}$, and $\hat{\mu}_{A'}$ are expressed by the velocity-independent densities with the coefficient matrix in (10.9) with quantities evaluated at $v_i = 0$, i.e., quantities in (10.10) in terms of the peculiar velocity C_i instead of c_i.

In particular, in the limit $D \to 3$, the closed system of field equations of $\Pi_{A''}$ is obtained as

$$\begin{aligned}
&\dot{\hat{\Pi}}_{A''} + \hat{J}^M_{iA''B''}\left(\hat{J}^M\right)^{-1}_{B''C''}\partial_i \hat{\Pi}_{C''} \\
&= -\Bigg\{\left(\frac{1}{\tau} + \partial_i v_i\right)\delta_{A''C''} + \partial_i\left[\hat{J}^M_{iA''B''}\left(\hat{J}^M\right)^{-1}_{B''C''}\right] \\
&\quad + A^r_{A''B''}\left[\dot{v}_r\delta_{B''C''} + \hat{J}^M_{iB''D''}\left(\hat{J}^M\right)^{-1}_{D''C''}\partial_i v_r\right]\Bigg\}\hat{\Pi}_{C''},
\end{aligned} \tag{10.23}$$

where

$$\hat{J}^M_{A''B''} = -\frac{m^2}{k_B}\int f^{(M)} C_{A''}C_{B''}\mathrm{d}\boldsymbol{C}. \tag{10.24}$$

The singular limit is also achieved as before, that is, by adopting the initial condition compatible with a property of monatomic gases: $\hat{\Pi}_{A''}(\boldsymbol{x}, 0) = 0$. Then the solution of (10.23), that is, the solution of the first-order quasi-linear partial differential equations of $\hat{\Pi}_{A''}$ is obtained, for any time, as

$$\hat{\Pi}_{A''}(\boldsymbol{x}, t) = 0.$$

10.4 Closure and Singular Limit in the One-Dimensional Case

In the one-dimensional case, it is easy to calculate the explicit expression of the constitutive equations and the field equations because the integral (10.24) can be replaced with the simple expression. For this reason, we study the one-dimensional case along the $x(\equiv x_1)$-axis, and introduce the following notations:

$$F_{p,q} = \int_{\mathbb{R}^3}\int_0^\infty mf\,(c_1)^p\left(c^2\right)^q I^\alpha\, dI d\mathbf{c},$$

$$G_{p',q'} = \int_{\mathbb{R}^3}\int_0^\infty mf\left(c^2+\frac{2I}{m}\right)(c_1)^{p'}\left(c^2\right)^{q'} I^\alpha\, dI d\mathbf{c},$$

$$I_{p'',q''} \equiv F_{p'',q''+1} - G_{p'',q''} = -2\int_{\mathbb{R}^3}\int_0^\infty f\,(c_1)^{p''}\left(c^2\right)^{q''} I^{\alpha+1}\, dI d\mathbf{c}$$

with the dynamical pressure tensors that become in the one dimensional case

$$\Pi_{p'',q''} = I_{p'',q''} - I^E_{p'',q''}.$$

The indexes are the non-negative integers satisfying

$$0 \leqslant p+2q \leqslant N, \quad 0 \leqslant p'+2q' \leqslant N-1, \quad 0 \leqslant p''+2q'' \leqslant N-2. \tag{10.25}$$

The balance equations (10.5) become, in the present case,

$$\partial_t F_{p,0} + \partial_x F_{p+1,0} = P_{p,0},$$

$$\partial_t I_{p'',q''} + \partial_x I_{p''+1,q''} = R_{p'',q''}, \quad \partial_t G_{p',q'} + \partial_x G_{p'+1,q'} = Q_{p',q'},$$

where ∂_x denotes the partial derivative with respect to x. The production terms $P_{p,q}$, $R_{p'',q''}$ and $Q_{p',q'}$ are defined in a similar way as follows:

$$P_{p,q} = \int_{\mathbb{R}^3}\int_0^\infty mQ(f)(c_1)^p\left(c^2\right)^q I^\alpha\, dI d\mathbf{c},$$

$$R_{p'',q''} = -2\int_{\mathbb{R}^3}\int_0^\infty Q(f)\,(c_1)^{p''}\left(c^2\right)^{q''} I^{\alpha+1}\, dI d\mathbf{c},$$

$$Q_{p',q'} = \int_{\mathbb{R}^3}\int_0^\infty mQ(f)\left(c^2+\frac{2I}{m}\right)(c_1)^{p'}\left(c^2\right)^{q'} I^\alpha\, dI d\mathbf{c}.$$

Taking into account that the velocity-independent variables denoted with a hat are evaluated with the peculiar velocity $C_1 = c_1 - v_1$ instead of the velocity of a molecule c_1, the relations (10.19) between the moments and the velocity-independent variables become

$$F_{p,q} = \sum_{r=0}^{N} \sum_{s=0}^{[r/2]} X_{p,q,r-2s,s} \hat{F}_{r-2s,s},$$

$$G_{p',q'} = \sum_{r'=0}^{N-1} \sum_{s'=0}^{[r'/2]} X_{p',q',r'-2s',s'} \left(\hat{G}_{r'-2s',s'} + 2v\hat{F}_{r'-2s'+1,s'} + v^2 \hat{F}_{r'-2s',s'} \right),$$

$$I_{p'',q''} = \sum_{r''=0}^{N-2} \sum_{s''=0}^{[r''/2]} X_{p'',q'',r''-2s'',s''} \hat{I}_{r''-2s'',s''}.$$

The coefficients are defined as follows:

$$X_{p,q,r,s} = \begin{cases} \displaystyle\sum_{j=0}^{q-s} \binom{p}{p+q-r-s-j} \binom{q}{q-s} \binom{q-s}{j} 2^{q-s-j} v^{p+2q-(r+2s)} \\ \qquad\qquad\qquad\qquad \text{if } q \geqslant s \text{ and } p+q \geqslant r+s \\ 0 \qquad\qquad\qquad\qquad \text{otherwise} \end{cases},$$

and also

$$A_{p,q,r,s} = \begin{cases} \displaystyle\sum_{j=0}^{q-s} \binom{p}{p+q-r-s-j} \binom{q}{q-s} \binom{q-s}{j} (p+2q-(r+2s))\, 2^{q-s-j} \delta_{p+2q,r+2s+1} \\ \qquad\qquad\qquad\qquad \text{if } q \geqslant s \text{ and } p+q \geqslant r+s \\ 0 \qquad\qquad\qquad\qquad \text{otherwise} \end{cases},$$

where $v \equiv v_1$.

The closure for the system of balance equations is obtained by the same procedure summarized in Sect. 10.2.1. In particular, we show the closure for the velocity-independent part of the constitutive fluxes. In this case, the velocity-independent and nonequilibrium parts of the main field components $\tilde{\lambda}_{r,s}$, $\tilde{\nu}_{r'',s''}$, $\tilde{\mu}_{r',s'}$ are determined by the following equation corresponding to (10.9):

$$\begin{pmatrix} \hat{J}^M_{p+r,q+s} & -\hat{K}_{p+r'',q+s''} & \hat{J}^M_{p+r',q+s'+1} + \hat{K}_{p+r',q+s'} \\ -\hat{K}_{p''+r,q''+s} & \hat{L}_{p''+r'',q''+s''} & -\hat{K}_{p''+r',q''+s'+1} - \hat{L}_{p''+r',q''+s'} \\ \hat{J}^M_{p'+r,q'+s+1} + \hat{K}_{p'+r,q'+s} & -\hat{K}_{p'+r'',q'+s''+1} - \hat{L}_{p'+r'',q'+s''} & \hat{J}^M_{p'+r',q'+s'+2} + 2\hat{K}_{p'+r',q'+s'+1} + \hat{L}_{p'+r',q'+s'} \end{pmatrix}$$

$$\times \begin{pmatrix} \tilde{\lambda}_{r,s} \\ \tilde{\nu}_{r'',s''} \\ \tilde{\mu}_{r',s'} \end{pmatrix} = \begin{pmatrix} \tilde{F}_{p,q} \\ \tilde{\Pi}_{p'',q''} \\ \tilde{G}_{p',q'} \end{pmatrix}, \tag{10.26}$$

where a quantity with a tilde indicates the velocity-independent nonequilibrium part of the quantity, and the coefficients are expressed as follows [119]:

$$\hat{J}^M_{p,q} = -\frac{m^2}{k_B}\int_{\mathbb{R}^3} f^{(M)}(C_1)^p \left(C^2\right)^q d\mathbf{C} = -\frac{m}{k_B}\rho\left(\frac{k_B T}{m}\right)^{\frac{p}{2}+q}\frac{2^{\frac{p}{2}+q}}{p+1}\Gamma\left(\frac{p+3}{2}+q\right)\frac{1+(-1)^p}{\sqrt{\pi}},$$

$$\hat{K}_{p,q} = 2c_s^2(1+\alpha)\hat{J}^M_{p,q},$$

$$\hat{L}_{p,q} = 4c_s^4(1+\alpha)(2+\alpha)\hat{J}^M_{p,q}.$$

In (10.26), the summations for repeated indexes are taken over all integers which satisfy (10.25). By using $\tilde{\lambda}_{r,s}$, $\tilde{v}_{r'',s''}$, $\tilde{\mu}_{r',s'}$ evaluated by $\tilde{F}_{p,q}$, $\hat{\Pi}_{p'',q''}$, $\tilde{G}_{p',q'}$, the constitutive equations for the fluxes $\tilde{F}_{t+1,u}$, $\hat{\Pi}_{t''+1,u''}$, $\tilde{G}_{t'+1,u'}$ are obtained where the non-negative integers satisfy $t+2u = N$, $t''+2u'' = N-2$ and $t'+2u' = N-1$.

Finally, after the constitutive equations are obtained, in the limit $D \to 3$, the closed system of field equations for the dynamical pressure tensors corresponding to (10.23) becomes as follows:

$$\dot{\hat{\Pi}}_{p'',q''} + \sum_{t''=0}^{N-2}\sum_{u''=0}^{[t''/2]} \hat{H}^{(1)}_{p'',q'',t''-2u'',u''}\partial_x \hat{\Pi}_{t''-2u'',u''}$$
$$= -\sum_{t''=0}^{N-2}\sum_{u''=0}^{[t''/2]}\left\{\left(\frac{1}{\tau}+\partial_x v\right)\delta_{p'',t''-2u''}\delta_{q'',u''} + \partial_x \hat{H}^{(1)}_{p'',q'',t''-2u'',u''} + \hat{H}^{(2)}_{p'',q'',t''-2u'',u''}\right\}\hat{\Pi}_{t''-2u'',u''}. \tag{10.27}$$

Here we define

$$\hat{H}^{(1)}_{p,q,t,u} = \sum_{t=0}^{N-2}\sum_{u=0}^{[t/2]} \hat{J}^M_{p+r-2s+1,q+s}\left(\hat{J}^M\right)^{-1}_{r-2s+t,s+u},$$

$$\hat{H}^{(2)}_{p,q,t,u} = \sum_{t=0}^{N-2}\sum_{u=0}^{[t/2]} A_{p,q,r-2s,s}\left(\dot{v}\delta_{r-2s,t}\delta_{s,u} + (\partial_x v)\hat{H}^{(1)}_{r-2s,s,t,u}\right),$$

where $\hat{J}^M$ is a matrix of which element is $\hat{J}^M_{p,q}$.

10.5 Examples of Particular Systems

The closed systems of field equations and the singular limit for particular cases are studied in this section. Solving (10.26), we can obtain the velocity-independent nonequilibrium part of the main field, and then the constitutive fluxes are closed. To avoid showing the long calculations and expressions of the field equations, we show only the closure equations and the field equations of $\hat{\Pi}_{p'',q''}$ (10.27) in the limit that D approaches 3. As particular examples, we choose the case of $(1,0)$, $(2^{(1)},0)$, $(2,1)$, $(3^{(1)},1)$, $(3^{(1)},2^{(1)})$, $(3,2)$, and $(4,3)$-systems. For simplicity, we study the one-dimensional case.

10.5.1 The 5-Moment System ($N = 1$, $M = 0$)

The independent variables $F_{0,0}$, $F_{1,0}$, $G_{0,0}$ are

$$F_{0,0} = \rho, \quad F_{1,0} = \rho v, \quad G_{0,0} = 2\rho\varepsilon + \rho v^2,$$

where the velocity-independent variables are denoted as $\hat{F}_{0,0} = \rho, \hat{F}_{1,0} = 0, \hat{G}_{0,0} = 2\rho\varepsilon$. The field equations are

$$\begin{aligned}&\partial_t F_{0,0} + \partial_x F_{1,0} = 0,\\ &\partial_t F_{1,0} + \partial_x F_{2,0} = 0,\\ &\qquad\qquad\qquad \partial_t G_{0,0} + \partial_x G_{1,0} = 0,\end{aligned}$$

where the constitutive functions $F_{2,0}$, $G_{1,0}$ are determined as follows:

$$F_{2,0} = p + \rho v^2, \quad G_{1,0} = v(D+2)p + \rho v^3.$$

In this case there is no quantity like $\hat{I}_{A''}$, and therefore the number of field equations does not change in the singular limit.

10.5.2 The 6-Moment System ($N = 2^{(1)}$, $M = 0$)

The independent variables $F_{0,0}$, $F_{1,0}$, $I_{0,0}$, $G_{0,0}$ are

$$F_{0,0} = \rho, \quad F_{1,0} = \rho v, \quad I_{0,0} = -(D-3)p + 3\Pi, \quad G_{0,0} = 2\rho\varepsilon + \rho v^2,$$

where $\Pi = \hat{\Pi}_{0,0}/3$ is the dynamic pressure and the velocity-independent part of $I_{0,0}$ is denoted as $\hat{I}_{0,0} = -(D-3)p + 3\Pi$. The field equations of these variables are

$$\begin{aligned}&\partial_t F_{0,0} + \partial_x F_{1,0} = 0,\\ &\partial_t F_{1,0} + \partial_x F_{2,0} = 0,\\ &\qquad\qquad \partial_t I_{0,0} + \partial_x I_{1,0} = R_{0,0}, \qquad \partial_t G_{0,0} + \partial_x G_{1,0} = 0,\end{aligned}$$

and the constitutive functions $F_{2,0}$, $I_{1,0}$, $G_{1,0}$ are determined as follows:

$$F_{2,0} = p + \Pi + \rho v^2, \quad I_{1,0} = -v\{(D-3)p - 3\Pi\}, \quad G_{1,0} = v\{(D+2)p + 2\Pi\} + \rho v^3.$$

The characteristic variable of polyatomic gases is only Π. Its field equation when D approaches 3 is given by

$$\dot{\Pi} = -\left(\frac{1}{\tau} + \partial_x v\right)\Pi. \tag{10.28}$$

In the singular limit with the null dynamical pressure at initial time, this system converges to the Euler system of monatomic gas [112].

10.5.3 The 14-Moment System ($N = 2$, $M = 1$)

The independent variables $F_{0,0}$, $F_{1,0}$, $F_{2,0}$, $I_{0,0}$, $G_{0,0}$, $G_{1,0}$ are

$$\begin{aligned} &F_{0,0} = \rho, \quad F_{1,0} = \rho v, \quad F_{2,0} = p + \Pi - \sigma + \rho v^2, \quad I_{0,0} = -(D-3)p + 3\Pi, \\ &G_{0,0} = 2\rho\varepsilon + \rho v^2, \quad G_{1,0} = 2q + v\{(D+2)p + 2\Pi - 2\sigma\} + \rho v^3, \end{aligned} \tag{10.29}$$

where $\sigma \equiv \sigma_{\langle 11\rangle}$ and $q \equiv q_1$, and the velocity-independent parts of $F_{2,0}$ and $G_{1,0}$ are denoted as $\hat{F}_{2,0} = p + \Pi - \sigma$ and $\hat{G}_{1,0} = 2q$. The field equations of these variables are

$$\begin{aligned} &\partial_t F_{0,0} + \partial_x F_{1,0} = 0, \\ &\partial_t F_{1,0} + \partial_x F_{2,0} = 0, \\ &\partial_t F_{2,0} + \partial_x F_{3,0} = P_{2,0}, \qquad \partial_t I_{0,0} + \partial_x I_{1,0} = R_{0,0}, \qquad \partial_t G_{0,0} + \partial_x G_{1,0} = 0, \\ &\qquad\qquad\qquad\qquad\qquad\qquad\qquad\qquad\qquad\qquad \partial_t G_{1,0} + \partial_x G_{2,0} = Q_{1,0}. \end{aligned}$$

In the present case, the constitutive functions $F_{3,0}$, $I_{1,0}$, $G_{2,0}$ are determined as follows:

$$\begin{aligned} F_{3,0} &= \frac{6}{D+2}q + 3v(p + \Pi - \sigma) + \rho v^3, \\ I_{1,0} &= -2\frac{D-3}{D+2}q - v[(D-3)p - 3\Pi], \\ G_{2,0} &= \frac{p}{\rho}\{(D+2)p + (D+4)(\Pi - \sigma)\} \\ &\quad + \frac{4(D+5)}{D+2}qv + v^2\{(D+5)p + 5\Pi - 5\sigma\} + \rho v^4. \end{aligned}$$

As is the case with the 6-moment system, the characteristic variable of polyatomic gases is only the dynamic pressure Π. When $D \to 3$, the field equation of Π is the same as (10.28), and therefore the 14-moment system converges to the Grad 13-moment system [112] in the singular limit.

Hereafter, we will show the constitutive equations of the $(3^{(1)}, 1)$, $(3^{(1)}, 2^{(1)})$, $(3, 2)$, and $(4, 3)$-systems, whose first 14 moments $(F_{0,0}, F_{1,0}, F_{2,0}, I_{0,0}, G_{0,0}, G_{1,0})$ are in common with (10.29).

10.5.4 The 17-Moment System ($N = 3^{(1)}$, $M = 1$)

The independent variables are $F_{0,0}, F_{1,0}, F_{2,0}, I_{0,0}, I_{1,0}, G_{0,0}, G_{1,0}$, where $I_{1,0}$ is given by

$$I_{1,0} = \hat{\Pi}_{1,0} - v\left[(D-3)p + 3\Pi\right].$$

The constitutive functions $F_{3,0}, I_{2,0}, G_{2,0}$ are determined as follows:

$$\begin{aligned}
F_{3,0} &= \frac{3}{5}\left(\hat{\Pi}_{1,0} + 2q\right) + 3v(p + \Pi - \sigma) + \rho v^3, \\
I_{2,0} &= -\frac{p}{\rho}\left\{(D-3)p + (D-6)\Pi - (D-3)\sigma\right\} + 2v\hat{\Pi}_{1,0} - v^2\left[(D-3)p - 3\Pi\right], \\
G_{2,0} &= \frac{p}{\rho}\left\{(D+2)p + (D+4)(\Pi - \sigma)\right\} + \frac{6}{5}v\left(\hat{\Pi}_{1,0} + \frac{16}{3}q\right) \\
&\quad + v^2\left\{(D+5)p + 5\Pi - 5\sigma\right\} + \rho v^4.
\end{aligned}$$

The closed system of field equations with these constitutive equations completely coincides with that of Zhdanov's theory [317] except for the production terms. In Zhdanov's theory, the translational part of the heat flux $Q = \tilde{F}_{1,1}/2$ is adopted as the nonequilibrium variable instead of $\hat{\Pi}_{1,0}$. The field equations of the characteristic variables of polyatomic gases Π and $\hat{\Pi}_{1,0}(= 2(Q-q))$ in the limit $D \to 3$ are given as follows:

$$\begin{aligned}
\dot{\Pi} + \frac{1}{3}\partial_x\hat{\Pi}_{1,0} &= -\left(\frac{1}{\tau} + \partial_x v\right)\Pi, \\
\dot{\hat{\Pi}}_{1,0} + 3\frac{k_B}{m}T\partial_x\Pi &= -\left(\frac{1}{\tau} + 2\partial_x v\right)\hat{\Pi}_{1,0} - 3\left(\frac{k_B}{m}\partial_x T + \dot{v}\right)\Pi.
\end{aligned}$$

As mentioned before, the 13-moment system of monatomic gases is achieved as the singular limit of the 14-moment system and also of the 17-moment system.

10.6 Examples of the Convergence of Solution in the Singular Limit

In the paper [125], numerical examples of the convergence in the limit $D \to 3$ for the cases of sound wave and shock waves were shown. The results obtained are in good agreement with the theoretical results. In Fig. 10.1, an example of convergence for phase velocity and attenuation when D approaches 3, while in Fig. 10.2, we depict the wave profiles of the independent variables for $D = 4, 3.5, 3.1, 3$ at $M_0 = 1.5$ where the Mach number in the unperturbed state M_0 is defined as $M_0 = v_0/c_0$. We notice from the figures that the wave profiles approach uniformly the corresponding profiles of monatomic gases with $\Pi = 0$ and $\hat{\Pi}_{1,0} = 0$ $(Q = q)$ in the singular limit.

For more details see the original paper [125]. See also the paper [321] where, in the singular limit, the asymptotic behavior of attenuation per wavelength of sound wave (see Chap. 16) and the behavior of the dynamic structure factor (see Chap. 21) are shown numerically.

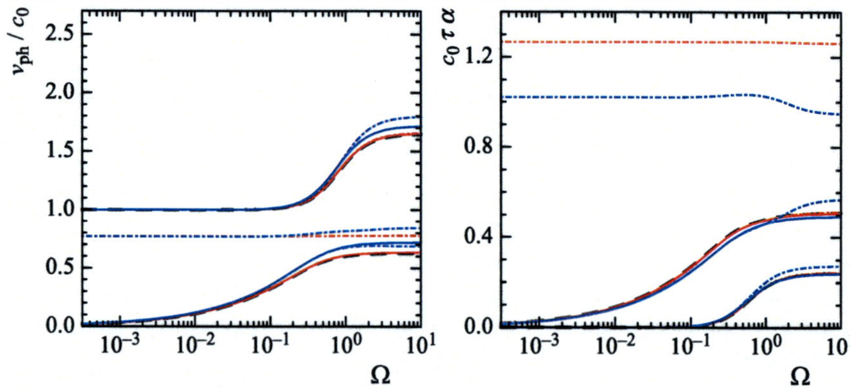

Fig. 10.1 Dependence of the dimensionless phase velocity (left) and attenuation factor (right) on the dimensionless frequency. The dashed line denotes the monatomic $(3^{(1)})$-system. The solid and dot-dashed lines with red $(D = 3.1)$ and blue $(D = 5)$ lines denote, respectively, the polyatomic (2, 1)-system and $(3^{(1)}, 1)$-system

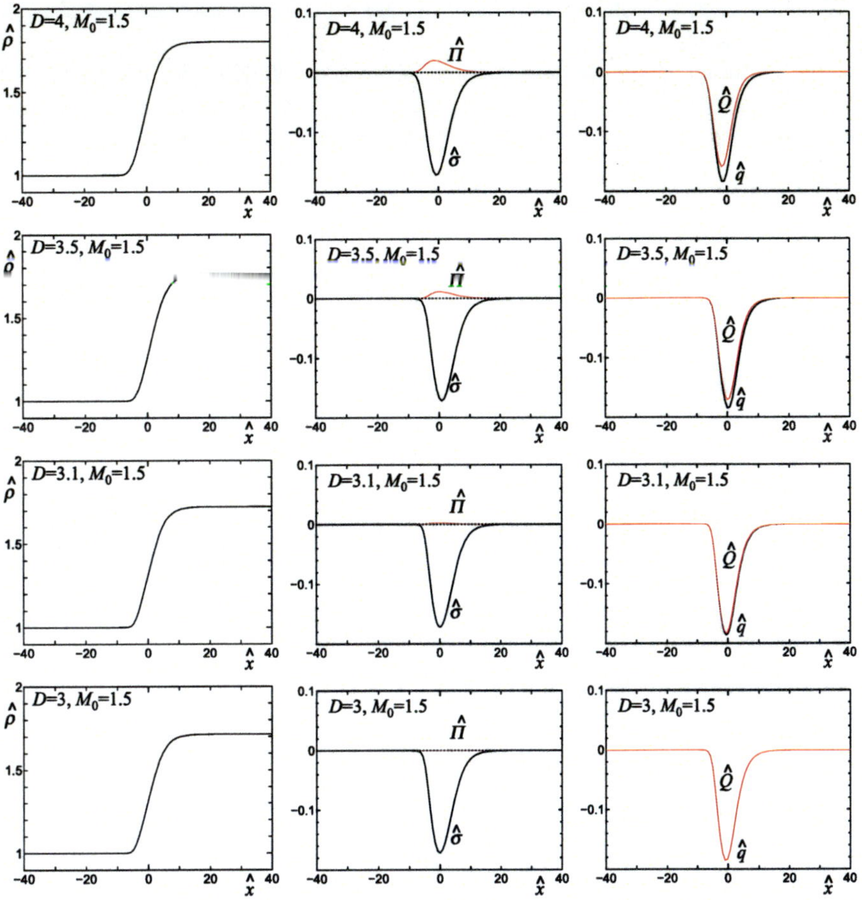

Fig. 10.2 Mass density profiles (left) and the profiles of the dynamic pressure (center, red) and the shear stress (center, black), and the heat flux (right, black) and Q (right, red) for $D = 4, 3.5, 3.1, 3$ at $M_0 = 1.5$. Here $\hat{x}$ is the dimensionless position coordinate defined by $\hat{x} \equiv x/(c_0\tau)$

10.7 Concluding Remarks

The singular limit of the polyatomic $(N, N-1)$-system to the one of monatomic gases in molecular ET has been studied. This limit is achieved by taking the degrees of freedom of a molecule $D \to 3$ continuously under an appropriate initial condition compatible with a property of monatomic gases. In this limit, the characteristic variables of polyatomic gases, i.e., the dynamical pressure tensors defined by the difference between the *momentum like*-hierarchy and the *energy like*-hierarchy converge to zero for any time. The remaining field equations coincide with those of the monatomic $((N+1)^{(1)})$-system. Therefore, in this sense, the molecular

ET theory of polyatomic gases is valid not only for polyatomic gases but also for monatomic gases.

In the present study, we have adopted the simplest BGK model as the collision term. If each velocity-independent nonequilibrium variable has its own relaxation time with no cross effect between the production terms, the present proof is also valid.

Chapter 11
Many Moments with Molecular Rotation and Vibration

Abstract Dealing with the rotational mode and the vibration mode separately, we can study also the RET theories with many moments, which include the ET_{15} theory (Chap. 8) as a special one with 15 moments. In this chapter, we present very briefly the general triple hierarchy with the discussion about the MEP.

11.1 Triple Hierarchy of Moment Equations

In a similar way adopted in Chap. 8, we introduce three kinds of moments F, H^R, and H^V as follows:

$$
\begin{aligned}
F &= \int_{\mathbb{R}^3}\int_0^{\infty}\int_0^{\infty} mf\,\varphi\left(I^R\right)\psi\left(I^V\right)\,dI^R dI^V d\mathbf{c},\\
F_{i_1\dots i_j} &= \int_{\mathbb{R}^3}\int_0^{\infty}\int_0^{\infty} mc_{i_1}\cdots c_{i_j} f\,\varphi\left(I^R\right)\psi\left(I^V\right)\,dI^R dI^V d\mathbf{c},\\
H^R_{ll} &= \int_{\mathbb{R}^3}\int_0^{\infty}\int_0^{\infty} 2I^R f\,\varphi\left(I^R\right)\psi\left(I^V\right)\,dI^R dI^V d\mathbf{c},\\
H^R_{lli_1\dots i_k} &= \int_{\mathbb{R}^3}\int_0^{\infty}\int_0^{\infty} 2I^R c_{i_1}\cdots c_{i_k} f\,\varphi\left(I^R\right)\psi\left(I^V\right)\,dI^R dI^V d\mathbf{c},\\
H^V_{ll} &= \int_{\mathbb{R}^3}\int_0^{\infty}\int_0^{\infty} 2I^V f\,\varphi\left(I^R\right)\psi\left(I^V\right)\,dI^R dI^V d\mathbf{c},\\
H^V_{lli_1\dots i_l} &= \int_{\mathbb{R}^3}\int_0^{\infty}\int_0^{\infty} 2I^V c_{i_1}\cdots c_{i_l} f\,\varphi\left(I^R\right)\psi\left(I^V\right)\,dI^R dI^V d\mathbf{c},
\end{aligned}
$$

where $j, k, l = 1, 2, \cdots$. From the Boltzmann equation (1.20), we obtain three hierarchies (a triple hierarchy) of balance equations, i.e., F, H^R, and H^V-hierarchies in

T. Ruggeri, M. Sugiyama, *Classical and Relativistic Rational Extended Thermodynamics of Gases*, https://doi.org/10.1007/978-3-030-59144-1_11

the following form:

$$
\begin{aligned}
&\partial_t F + \partial_i F_i = 0, \\
&\partial_t F_{i_1} + \partial_i F_{ii_1} = 0, \\
&\partial_t F_{i_1 i_2} + \partial_i F_{ii_1 i_2} = P^K_{i_1 i_2}, \qquad \partial_t H^R_{ll} + \partial_i H^R_{lli} = P^R_{ll}, \qquad \partial_t H^V_{ll} + \partial_i H^V_{lli} = P^V_{ll}, \\
&\partial_t F_{i_1 i_2 i_3} + \partial_i F_{ii_1 i_2 i_3} = P^K_{i_1 i_2 i_3}, \quad \partial_t H^R_{lli_1} + \partial_i H^R_{llii_1} = P^R_{lli_1}, \quad \partial_t H^V_{lli_1} + \partial_i H^V_{llii_1} = P^V_{lli_1}, \\
&\qquad \vdots \qquad\qquad\qquad\qquad \vdots \qquad\qquad\qquad\qquad \vdots
\end{aligned}
$$

where the production terms are related to the collision term as follows:

$$
\begin{aligned}
P^K_{i_1 \dots i_j} &= \int_{\mathbb{R}^3} \int_0^\infty \int_0^\infty m c_{i_1} \cdots c_{i_j} Q(f) \varphi\left(I^R\right) \psi\left(I^V\right) dI^R dI^V d\mathbf{c}, \\
P^R_{lli_1 \dots i_k} &= \int_{\mathbb{R}^3} \int_0^\infty \int_0^\infty 2I^R c_{i_1} \cdots c_{i_k} Q(f) \varphi\left(I^R\right) \psi\left(I^V\right) dI^R dI^V d\mathbf{c}, \\
P^V_{lli_1 \dots i_l} &= \int_{\mathbb{R}^3} \int_0^\infty \int_0^\infty 2I^V c_{i_1} \cdots c_{i_l} Q(f) \varphi\left(I^R\right) \psi\left(I^V\right) dI^R dI^V d\mathbf{c}.
\end{aligned}
$$

We notice that the first and second equations of the F-hierarchy represent the conservation laws of mass and momentum, while the sum of the balance equations of F_{ll}, H^R_{ll} and H^V_{ll} represents the conservation law of energy with

$$
Q_{ll} = P^K_{ll} + P^R_{ll} + P^V_{ll} = 0. \tag{11.1}
$$

In each of the three hierarchies, the flux in one equation appears as the density in the next equation.

Remark 11.1 Equivalently, instead of the one of the three hierarchies, we may adopt the hierarchy of the total energy (G-hierarchy):

$$
\begin{aligned}
&\partial_t G_{ll} + \partial_i G_{lli} = 0, \\
&\partial_t G_{lli_1 \cdots i_m} + \partial_i G_{llii_1 \cdots i_m} = Q_{lli_1 \cdots i_m}, \quad m = 1, 2, \cdots,
\end{aligned}
$$

where G_{ll} is given by (8.2)$_3$ and

$$
\begin{aligned}
G_{lli_1 \cdots i_m} = \int_{\mathbb{R}^3} \int_0^\infty \int_0^\infty & \left(mc^2 + 2I^R + 2I^V\right) \\
& \times c_{i_1} \cdots c_{i_m} f \varphi\left(I^R\right) \psi\left(I^V\right) dI^R dI^V d\mathbf{c},
\end{aligned}
$$

and, for the production terms,

$$Q_{lli_1\cdots i_m} = P^K_{lli_1\cdots i_m} + P^R_{lli_1\cdots i_m} + P^V_{lli_1\cdots i_m} \quad m = 1, 2, \cdots .$$

11.1.1 Truncated System of Balance Equations and Its Closure

To have a finite system of balance equations, we truncate the F, H^R, and H^V-hierarchies at the orders of N, M, and L, respectively. We can express the densities as follows:

$$\begin{aligned}
F_A &= \int_{\mathbb{R}^3}\int_0^\infty\int_0^\infty m c_A f \varphi\left(I^R\right)\psi\left(I^V\right) dI^R dI^V d\mathbf{c}, \\
H^R_{llA'} &= \int_{\mathbb{R}^3}\int_0^\infty\int_0^\infty 2I^R c_{A'} f \varphi\left(I^R\right)\psi\left(I^V\right) dI^R dI^V d\mathbf{c}, \\
H^V_{llA''} &= \int_{\mathbb{R}^3}\int_0^\infty\int_0^\infty 2I^V c_{A''} f \varphi\left(I^R\right)\psi\left(I^V\right) dI^R dI^V d\mathbf{c}.
\end{aligned} \tag{11.2}$$

The fluxes F_{iA}, $H^R_{lliA'}$, $H^V_{lliA''}$ and the productions P^K_A, $P^R_{llA'}$, $P^V_{llA''}$ are also expressed in a similar way.

Then a triple hierarchy of moments truncated at the orders N, M and L ((N, M, L)-*system*) is compactly expressed as

$$\begin{aligned}
&\partial_t F_A + \partial_i F_{iA} = P^K_A, \\
&(0 \leqslant A \leqslant N) \qquad \partial_t H^R_{llA'} + \partial_i H^R_{lliA'} = P^R_{llA'}, \quad \partial_t H^V_{llA''} + \partial_i H^V_{lliA''} = P^V_{llA''} \\
&\qquad\qquad\qquad\qquad \left(0 \leqslant A' \leqslant M\right) \qquad\qquad \left(0 \leqslant A'' \leqslant L\right)
\end{aligned} \tag{11.3}$$

with $P^K = 0$ and $P^K_1 = 0$ and with the condition (11.1) representing the conservation laws of mass, momentum, and energy.

11.1.1.1 Galilean Invariance

Since the velocity-independent variables are the moments in terms of the peculiar velocity C_i instead of c_i, it is possible to express the velocity dependence of the densities $\mathbf{F} = (F_A, H^R_{llA'}, H^V_{llA''})^T$, the non-convective fluxes $\boldsymbol{\Phi} = (F_{iA} - F_A v_i, H^R_{lliA'} - H^R_{llA'}v_i, H^V_{lliA''} - H^V_{llA''}v_i)^T$, and the production terms $\mathbf{P} = (P^K_A, P^R_{llA'}, P^V_{llA''})$ as proved in (2.49):

$$\mathbf{F} = \mathbf{X}(\mathbf{v})\hat{\mathbf{F}}, \quad \boldsymbol{\Phi} = \mathbf{X}(\mathbf{v})\hat{\boldsymbol{\Phi}}, \quad \mathbf{P} = \mathbf{X}(\mathbf{v})\hat{\mathbf{P}},$$

where a hat on a quantity indicates the velocity-independent part of the quantity.

We assume that the constitutive quantities $\hat{F}_{iN}$, $\hat{H}^R_{lliM}$, $\hat{H}^V_{lliL}$, $\hat{P}^K_A$, $\hat{P}^R_{llA'}$, and $\hat{P}^V_{llA''}$, which we express as $\hat{\Psi}$ generically, depend on the densities locally and instantaneously:

$$\hat{\Psi} = \hat{\Psi}(\hat{F}_A, \hat{H}^R_{llA'}, \hat{H}^V_{llA''}). \tag{11.4}$$

Remark 11.2 In principle, the truncated orders N, M and L may be chosen independently. However, if we naturally impose the condition that the (N, M, L)-system can make the G-hierarchy be Galilean invariant, the inequality; $\min(M, L) \leqslant N-1$ should be satisfied because of the relation:

$$G_{lla} = X_{ab}\left(\hat{F}_{llb} + \hat{H}^R_{llb} + \hat{H}^V_{llb} + 2v_l\hat{F}_{lb} + v^2\hat{F}_b\right)$$
$$(0 \leqslant a, b \leqslant \min(M, L)).$$

11.1.1.2 MEP and the Closure of the System

To obtain the constitutive equations (11.4) explicitly, we utilize the MEP. That is, the most suitable distribution function $f_{(N,M,L)}$ is the one that maximizes the functional defined by (we omit the symbol of summation for the repeated indices: A from 0 to N, A' from 0 to M, and A'' from 0 to L)

$$\begin{aligned}
&\mathscr{L}_{(N,M,L)}(f)\\
&= -k_B \int_{\mathbb{R}^3}\int_0^\infty\int_0^\infty f\log f\,\varphi\left(I^R\right)\psi\left(I^V\right)dI^R dI^V d\mathbf{c}\\
&\quad + \lambda_A\left(F_A - \int_{\mathbb{R}^3}\int_0^\infty\int_0^\infty mc_A f\varphi\left(I^R\right)\psi\left(I^V\right)dI^R dI^V d\mathbf{c}\right)\\
&\quad + \mu^R_{A'}\left(H^R_{llA'} - \int_{\mathbb{R}^3}\int_0^\infty\int_0^\infty 2I^R c_{A'} f\varphi\left(I^R\right)\psi\left(I^V\right)dI^R dI^V d\mathbf{c}\right)\\
&\quad + \mu^V_{A''}\left(H^V_{llA''} - \int_{\mathbb{R}^3}\int_0^\infty\int_0^\infty 2I^V c_{A''} f\varphi\left(I^R\right)\psi\left(I^V\right)dI^R dI^V d\mathbf{c}\right),
\end{aligned}$$

where λ_A, $\mu^R_{A'}$ and $\mu^V_{A''}$ are the Lagrange multipliers. As a consequence, we have

$$f_{(N,M,L)} = \exp\left(-1 - \frac{m}{k_B}\chi_{(N,M,L)}\right),$$

$$\chi_{(N,M,L)} = \lambda_A c_A + \frac{2I^R}{m}\mu^R_{A'}c_{A'} + \frac{2I^V}{m}\mu^V_{A''}c_{A''}.$$

Due to the Galilean invariance, the distribution function can be expressed in terms of the velocity-independent quantities:

$$
\begin{aligned}
f_{(N,M,L)} &= \exp\left(-1 - \frac{m}{k_B}\hat{\chi}_{(N,M,L)}\right), \\
\hat{\chi}_{(N,M,L)} &= \hat{\lambda}_A C_A + \frac{2I^R}{m}\hat{\mu}^R_{A'} C_{A'} + \frac{2I^V}{m}\hat{\mu}^V_{A''} C_{A''}.
\end{aligned}
\tag{11.5}
$$

Therefore we obtain the velocity dependence of the Lagrange multipliers $\boldsymbol{\lambda} = (\lambda_A, \mu^R_{A'}, \mu^V_{A''})$ as follows (see (2.55)):

$$
\boldsymbol{\lambda} = \hat{\boldsymbol{\lambda}}\mathbf{X}(-\mathbf{v}). \tag{11.6}
$$

By inserting (11.5) into (11.2), the Lagrange multipliers λ_A, $\mu^R_{A'}$ and $\mu^V_{A''}$ are evaluated in terms of the densities F_A, $H^R_{llA'}$ and $H^V_{llA''}$. And, finally, by plugging (11.5) into the last fluxes and production terms, the system is closed. In this way we obtain the RET theory for the (N, M, L)-system.

Remark 11.3 An alternative approach to achieve the closure (phenomenological closure) of the system makes use of the entropy principle. In this case, it is required that all the solutions of (11.3) satisfy the entropy law (2.8) where h, h_i, and Σ are given by (1.24). According with the general results given in Sect. 9.2.3, the two closure methods give the same closed system of balance equations. Moreover we obtain the following relations:

$$
\begin{aligned}
\mathrm{d}h &= \lambda_A \mathrm{d}F_A + \mu^R_{A'}\mathrm{d}H^R_{llA'} + \mu^V_{A''}\mathrm{d}H^V_{llA''}, \\
\mathrm{d}h_i &= \lambda_A \mathrm{d}F_{iA} + \mu^R_{A'}\mathrm{d}H^R_{lliA'} + \mu^V_{A''}\mathrm{d}H^V_{lliA''}, \\
\Sigma &= \lambda_A P^K_A + \mu^R_{A'} P^R_{llA'} + \mu^V_{A''} P^V_{llA''} \geqslant 0.
\end{aligned}
\tag{11.7}
$$

Part IV
Nonlinear Theories Far from Equilibrium

Chapter 12
Phenomenological Nonlinear RET with 6 Fields

Abstract In this chapter, we present a RET theory of rarefied polyatomic gases with 6 independent fields (ET_6), i.e., the mass density, the velocity, the temperature, and the dynamic pressure, without adopting the near-equilibrium approximation. This model takes into account the dissipation process in a gas only through the dynamic pressure. By ignoring both the shear viscosity and the heat conductivity, it can highlight specifically the role of the dynamic pressure. We prove its compatibility with the universal principles (the Galilean invariance, the entropy principle, and the stability), and obtain the symmetric hyperbolic system with respect to the main field. The correspondence between the ET_6 theory and the Meixner theory of relaxation processes is discussed. The internal variable and the nonequilibrium temperature in the Meixner theory are expressed in terms of the quantities adopted in the ET_6 theory, in particular, the dynamic pressure. We study the monatomic-gas limit where the system of ET_6 converges to the Euler system of a perfect fluid. Lastly the nature of dynamic pressure is discussed in the case of rarefied polyatomic gas.

12.1 Introduction

The ET_{14} theory explained in Chaps. 6 and 7 gives us a satisfactory phenomenological model, but its differential system is rather complex. For this reason a simplified theory with 6 fields (ET_6) [128, 322]: the mass density ρ, the velocity $\mathbf{v}$, the temperature T, and the dynamic pressure Π is constructed. This simplified theory preserves the main physical properties of the more complex RET theory of 14 variables, in particular, when the bulk viscosity plays more important role than the shear viscosity and the heat conductivity. This situation is observed in many gases such as rarefied hydrogen gas and carbon dioxide gas at some temperature ranges (see Chaps. 16 and 17) [17, 106, 109]. ET_6 has another advantage to offer us a more affordable hyperbolic partial differential system. In fact, it is the simplest system that takes into account a dissipation mechanism after the Euler system of perfect fluids.

T. Ruggeri, M. Sugiyama, *Classical and Relativistic Rational Extended Thermodynamics of Gases*, https://doi.org/10.1007/978-3-030-59144-1_12

Although RET can go beyond the local equilibrium assumption and therefore go beyond the Navier-Stokes and Fourier theory, it has been developed only for processes not so far from equilibrium. In the phenomenological ET_{14} theory, for example, we have studied only constitutive equations that are linear with respect to the nonequilibrium variables $(\sigma_{\langle ij \rangle}, \Pi, q_i)$ because of the enormous difficulties in the analysis. While, in the molecular RET approach, we have seen, in Chap. 7 for the case of 14 moments and in Chap. 9 for the case of generic number of moments, that there exists the problematic point about the convergence of the integrals as we discussed in Sect. 4.4.

In this chapter we consider the ET_6 theory in detail. The simplicity of the 6-field model permits us to construct a full nonlinear theory. From a mathematical point of view, the advantage to have a full nonlinear theory is that the hyperbolicity region exists, if an equilibrium state is in it, not only in the neighborhood of an equilibrium state but everywhere provided that the solutions exist and are bounded. For any bounded solutions, there exist upper and lower bounds of the dynamic pressure. The nature of dynamic pressure in the case of rarefied polyatomic gas is also discussed on the basis of the ET_6 theory.

12.2 RET Theory with 6 Fields

We adopt 6 independent field-variables (ρ, v_l, T, Π), and we retain the general structure of the binary hierarchy (F-series and G-series) (9.2) but assume only the $(2^{(1)}, 0)$-system (see Definition 9.2):

$$\begin{aligned}
&\frac{\partial F}{\partial t} + \frac{\partial F_i}{\partial x_i} = 0, \\
&\frac{\partial F_j}{\partial t} + \frac{\partial F_{ji}}{\partial x_i} = 0, \\
&\frac{\partial F_{ll}}{\partial t} + \frac{\partial F_{lli}}{\partial x_i} = P_{ll}, \qquad \frac{\partial G_{ll}}{\partial t} + \frac{\partial G_{lli}}{\partial x_i} = 0,
\end{aligned} \tag{12.1}$$

where $(12.1)_{1,2,4}$ represent the conservation laws of mass, momentum, and energy provided that $F = \rho$, $F_i = \rho v_i$, $F_{ij} = \rho v_i v_j + (p + \Pi)\delta_{ij}$, $G_{ll} = \rho v_l v_l + 2\rho\varepsilon$, and $G_{lli} = (\rho v_l v_l + 2\rho\varepsilon + 2p + 2\Pi) v_i$ with p and ε being, respectively, the pressure and the specific internal energy. The shear stress $\sigma_{\langle ij \rangle}$ and the heat flux q_i are ignored in this theory. While $(12.1)_3$ is the new balance law corresponding to the dynamic pressure Π.

Therefore undetermined functions at the moment are only F_{lli}, P_{ll}, and the functions h, φ^i, and Σ in the entropy law (2.8) with (2.54). These functions will be determined by the universal principles in the following analysis.

12.2.1 Galilean Invariance

In the present case, the Galilean invariance (2.49) implies:

$$F_{lli} = \left(5(p+\Pi) + \rho v^2\right) v_i, \quad P_{ll} = \hat{P}_{ll},$$

and

$$\mathbf{X} = \begin{pmatrix} 1 & 0 & 0 & 0 \\ v_j & \delta_{jk} & 0 & 0 \\ v^2 & 2v_j & 1 & 0 \\ v^2 & 2v_j & 0 & 1 \end{pmatrix}, \tag{12.2}$$

$$\mathbf{A}^r = \left(\frac{\partial \mathbf{X}}{\partial v_r}\right)\bigg|_{v_j=0} = \begin{pmatrix} 0 & 0 & 0 & 0 \\ \delta_{jr} & 0 & 0 & 0 \\ 0 & 2\delta_{jr} & 0 & 0 \\ 0 & 2\delta_{jr} & 0 & 0 \end{pmatrix}.$$

Then, by using the physical variables, the system (12.1) is written as

$$\begin{aligned} &\frac{\partial \rho}{\partial t} + \frac{\partial}{\partial x_i}(\rho v_i) = 0, \\ &\frac{\partial \rho v_j}{\partial t} + \frac{\partial}{\partial x_i}\left[(p+\Pi)\delta_{ij} + \rho v_i v_j\right] = 0, \\ &\frac{\partial}{\partial t}(2\rho\varepsilon + \rho v^2) + \frac{\partial}{\partial x_i}\left\{\left[2(p+\Pi) + 2\rho\varepsilon + \rho v^2\right] v_i\right\} = 0, \\ &\frac{\partial}{\partial t}\left[3(p+\Pi) + \rho v^2\right] + \frac{\partial}{\partial x_i}\left\{\left[5(p+\Pi) + \rho v^2\right] v_i\right\} = \hat{P}_{ll}. \end{aligned} \tag{12.3}$$

We see that, in this exceptional case of ET$_6$, the Galilean invariance fixes completely the structure of the undetermined flux F_{lli}, and we notice that the left-hand side of (12.3) is linear in Π without assuming any approximation. Instead the production term $\hat{P}_{ll}$ can be nonlinear. Therefore the present theory is valid for any nonequilibrium processes as far as the continuum description is valid.

As was observed by Curró and Manganaro [323], Π appears in the left side of (12.3) always together with p. Therefore, by introducing the total nonequilibrium pressure $\mathcal{P} = p + \Pi$, the previous system can be rewritten as

$$\begin{aligned} &\frac{\partial \rho}{\partial t} + \frac{\partial}{\partial x_i}(\rho v_i) = 0, \\ &\frac{\partial \rho v_j}{\partial t} + \frac{\partial}{\partial x_i}\left[\mathcal{P}\delta_{ij} + \rho v_i v_j\right] = 0, \end{aligned} \tag{12.4}$$

$$\frac{\partial}{\partial t}(2\rho\varepsilon + \rho v^2) + \frac{\partial}{\partial x_i}\left\{\left[2\mathscr{P} + 2\rho\varepsilon + \rho v^2\right] v_i\right\} = 0,$$

$$\frac{\partial}{\partial t}\left[3\mathscr{P} + \rho v^2\right] + \frac{\partial}{\partial x_i}\left\{\left[5\mathscr{P} + \rho v^2\right] v_i\right\} = \hat{P}_{ll}.$$

The system (12.4) with independent field-variables $(\rho, v_j, \varepsilon, \mathscr{P})$ looks very unusual because there are 6 equations for 6 unknowns. So that means no constitutive relation is necessary to close the system except for the production term:

$$\hat{P}_{ll} \equiv \hat{P}_{ll}(\rho, \varepsilon, \mathscr{P})$$

that we need to determine.

12.2.2 Entropy Principle

We here study the compatibility with the entropy principle in the case of non-polytropic gas.

Let us rewrite the entropy density h in the form:

$$h = \rho\eta = \rho s + \rho k, \tag{12.5}$$

where $\eta(\rho, \varepsilon, \mathscr{P})$ is the specific entropy density, $s(\rho, \varepsilon)$ is the specific entropy density in equilibrium, and $k(\rho, \varepsilon, \mathscr{P})$ is the nonequilibrium part of the specific entropy density with the condition:

$$k(\rho, \varepsilon, p(\rho, \varepsilon)) = 0. \tag{12.6}$$

Let us denote the components of the main field $\mathbf{u}'$ as

$$\mathbf{u}' \equiv \left(\lambda, \lambda_j, \mu, \zeta\right),$$

then we have, from (2.58),

$$\hat{\lambda}_j = 0, \qquad \hat{h}'^0 = -2(\hat{\mu} + \hat{\zeta})\mathscr{P}. \tag{12.7}$$

From (2.57)$_1$, we obtain

$$\hat{\lambda}d\rho + 2\hat{\mu}d(\rho\varepsilon) + 3\hat{\zeta}d\mathscr{P} = (\eta + \rho\eta_\rho)d\rho + \rho\eta_\varepsilon d\varepsilon + \rho\eta_{\mathscr{P}} d\mathscr{P}, \tag{12.8}$$

where the suffix indicates a partial differentiation with respect to the corresponding variable. Then, from (12.8), we have

$$\hat{\lambda} = \eta + \rho\eta_\rho - \varepsilon\eta_\varepsilon, \qquad \hat{\mu} = \frac{1}{2}\eta_\varepsilon, \qquad \hat{\zeta} = \frac{1}{3}\rho\eta_{\mathscr{P}}. \tag{12.9}$$

From (2.13) and (12.7), we obtain

$$h = \hat{\lambda}\,\rho + 2\hat{\mu}(\rho\,\varepsilon + \mathscr{P}) + 5\hat{\zeta}\,\mathscr{P} = \rho\eta. \tag{12.10}$$

Substituting (12.9) into (12.10), we have a linear first-order partial differential equation for the function $\eta \equiv \eta(\rho, \varepsilon, \mathscr{P})$:

$$\eta_\rho + \frac{\mathscr{P}}{\rho^2}\eta_\varepsilon + \frac{5\mathscr{P}}{3\rho}\eta_{\mathscr{P}} = 0.$$

This equation can be solved easily by using the method of characteristics. The general integral is expressed as

$$\eta = f(X, Y), \tag{12.11}$$

where f is an arbitrary function of the following variables:

$$X = \frac{\mathscr{P}}{\rho^{\frac{5}{3}}}, \quad Y = \varepsilon - \frac{3}{2}\frac{\mathscr{P}}{\rho}. \tag{12.12}$$

Therefore the system admits seemingly infinite number of expressions of the entropy. However, the function f can be determined uniquely by the condition (12.6), which says that, in an equilibrium state with $\mathscr{P} = p(\rho, \varepsilon)$, η should be identical to the equilibrium specific entropy density $s(\rho, \varepsilon)$:

$$f(X_E, Y_E) = s(\rho, \varepsilon); \quad X_E = \frac{p}{\rho^{\frac{5}{3}}}, \quad Y_E = \varepsilon - \frac{3}{2}\frac{p}{\rho}. \tag{12.13}$$

Therefore the function f coincides with the equilibrium specific entropy density s as a function of the variables (X_E, Y_E). Hereafter it is denoted by $s(X_E, Y_E)$.

As is well known, the entropy s is obtained from the Gibbs equation:

$$T ds = d\varepsilon - \frac{p}{\rho^2}d\rho \tag{12.14}$$

with the equations of state:

$$p = p(\rho, T) = \frac{k_B}{m}\rho T, \quad \varepsilon = \varepsilon_E(T). \tag{12.15}$$

Then, from (12.11) and (12.5), the solution k is given by

$$k = s(X, Y) - s(X_E, Y_E), \tag{12.16}$$

where, from (12.14),

$$s = \frac{k_B}{m}\left\{\int_{T_0}^{T} \frac{\hat{c}_v(x)}{x}\,dx - \log\rho\right\} \tag{12.17}$$

with $\hat{c}_v$ and T_0 being the dimensionless specific heat at constant volume (see (6.4)) and a reference temperature. From (12.13) we have

$$X_E = \frac{k_B}{m}\frac{T}{\rho^{\frac{2}{3}}}, \qquad Y_E = \varepsilon_E(T) - \frac{3}{2}\frac{k_B}{m}T.$$

Since T depends only on Y_E, we have

$$T = T(Y_E), \qquad \rho = \left(\frac{k_B}{m}\frac{T(Y_E)}{X_E}\right)^{\frac{3}{2}}. \tag{12.18}$$

Substituting (12.18) into (12.17), we obtain s as the function of (X_E, Y_E):

$$s(X_E, Y_E) = \frac{k_B}{m}\left\{\int_{T_0}^{T(Y_E)} \frac{\hat{c}_v(x)}{x}\,dx - \frac{3}{2}\log\left(\frac{k_B}{m}\frac{T(Y_E)}{X_E}\right)\right\}.$$

Let $\omega(\phi)$ be the inverse function of $R(\omega)$ defined by

$$\phi = R(\omega) = \varepsilon_E(\omega) - \frac{3}{2}\frac{k_B}{m}\omega. \tag{12.19}$$

The inverse function always exists because the inequality

$$\frac{\mathrm{d}R}{\mathrm{d}\omega} = \frac{k_B}{m}\left(\hat{c}_v - \frac{3}{2}\right) > 0$$

holds for any rarefied polyatomic gases, which satisfy $\hat{c}_v(T) > 3/2$, $\forall T$. Let

$$T = \omega(Y_E), \qquad \Theta = \omega(Y). \tag{12.20}$$

Then, by using (12.16), the solution k is expressed as

$$k = -\frac{k_B}{m}\left\{\int_{\Theta}^{T} \frac{\hat{c}_v(x)}{x}\,dx + \frac{3}{2}\log\left(\frac{p}{\mathcal{P}}\frac{\Theta}{T}\right)\right\}. \tag{12.21}$$

Subtracting Y_E, (12.13)$_3$, from Y, (12.12)$_2$, we obtain

$$Y - Y_E = -\frac{3}{2}\frac{\Pi}{\rho},$$

and taking into account that, from (12.19) and (12.20), $Y = R(\Theta)$ and $Y_E = R(T)$, we obtain the following relation between the parameter Θ and the dynamic pressure Π:

$$\frac{\Pi}{p} = \frac{2}{3T}\int_{\Theta}^{T}\left(\hat{c}_v(x) - \frac{3}{2}\right)dx. \tag{12.22}$$

Defining

$$Z = \frac{\Pi}{p},$$

we have the following statement:

Statement 12.1 *For a non-polytropic rarefied gas characterized by (12.15), the nonequilibrium specific entropy k depends only on the two variables T and Z, and has the following parametric expression through the parameter* Θ:

$$\begin{aligned} k(T,\Theta) &= -\frac{k_B}{m}\left\{\int_{\Theta}^{T}\frac{\hat{c}_v(x)}{x}dx - \frac{3}{2}\log\left[\left(1 + \frac{2}{3T}\int_{\Theta}^{T}\left(\hat{c}_v(x) - \frac{3}{2}\right)dx\right)\frac{T}{\Theta}\right]\right\}, \\ Z(T,\Theta) &= \frac{2}{3T}\int_{\Theta}^{T}\left(\hat{c}_v(x) - \frac{3}{2}\right)dx. \end{aligned} \tag{12.23}$$

Because the argument of log in Eq. (12.21) or (12.23)$_1$ must be positive, we have the condition:

$$\frac{\mathscr{P}}{p}\frac{T}{\Theta} > 0, \quad \rightarrow \quad \frac{\mathscr{P}}{\Theta} > 0.$$

And since $\Theta = T > 0$ and $\mathscr{P} = p > 0$ in equilibrium, we obtain, by the continuity argument, the following condition of existence:

$$\mathscr{P} > 0, \qquad \Theta > 0. \tag{12.24}$$

For a fixed value of T, from (12.23), we obtain

$$\begin{aligned}\frac{\partial k}{\partial Z} &= \frac{3}{2}\frac{k_B}{m}\left(-\frac{T}{\Theta}+\frac{1}{1+Z}\right),\\ \frac{\partial \Theta}{\partial Z} &= \frac{3T}{3-2\hat{c}_v(\Theta)}.\end{aligned} \tag{12.25}$$

In an equilibrium state where $\Theta = T$ and $Z = 0$, from $(12.25)_1$, we have

$$\left.\frac{\partial k}{\partial Z}\right|_{Z=0} = 0.$$

Taking into account that $\hat{c}_v(T) > 3/2$, we have

$$\left.\frac{\partial^2 k}{\partial Z^2}\right|_{Z=0} = \frac{k_B}{m}\left(\frac{3\hat{c}_v(T)}{3-2\hat{c}_v(T)}\right) < 0.$$

The equilibrium state is the only one extremum because the condition of extrema $(12.25)_1$ is satisfied if and only if $Z = 0$ and $T = \Theta$ due to the relation $(12.23)_2$.

We conclude as follows:

Statement 12.2 *The equilibrium state with $\Pi = Z = 0$ is a global maximum of k for any T, and the inequality $k < 0$ holds for any T and for any $\Pi \neq 0$ in the interval of existence* (12.24)*. Therefore, from* (12.5)*, the entropy has a maximum at the equilibrium state and is convex. Then the system is symmetric hyperbolic.*

Remark 12.1 We have observed that we need not any constitutive functions for the principal part of the system (12.4) because of the 6 unknowns for 6 equations. Nevertheless, in order to have the explicit expression of entropy density, we need to fix the thermal and caloric equations of state (12.15). This is not surprising. In equilibrium the entropy density s is explicitly determined only when we know the equilibrium constitutive equations due to the Gibbs equation (12.14). And this influences also the nonequilibrium part k of the entropy density (see (12.16)).

12.2.2.1 Main Field

From (12.5), the components of the main field (12.9) can be rewritten as

$$\hat{\lambda} = -\frac{g}{T} + k + \rho k_\rho - \varepsilon k_\varepsilon, \qquad \hat{\mu} = \frac{1}{2T} + \frac{1}{2}k_\varepsilon, \qquad \hat{\zeta} = \frac{1}{3}\rho k_{\mathscr{P}}, \tag{12.26}$$

where $g = \varepsilon + p/\rho - Ts$ is the chemical potential.

To evaluate these components explicitly, it is convenient to rewrite (12.23) in the following equivalent form:

$$k(\Theta, T) = \int_T^\Theta \frac{\varepsilon_E^I(x)}{x^2}\,dx + \frac{3}{2}\frac{k_B}{m}\log\left(\frac{\varepsilon_E(T) - \varepsilon_E^I(\Theta)}{\varepsilon_E^K(T)}\right) + \frac{\varepsilon_E^I(\Theta)}{\Theta} - \frac{\varepsilon_E^I(T)}{T},$$
$$Z(\Theta, T) = \frac{\Pi}{p} = \frac{\varepsilon_E^I(T) - \varepsilon_E^I(\Theta)}{\varepsilon_E^K(T)}, \tag{12.27}$$

where $\varepsilon_E^I(x)$ and $\varepsilon_E^K(x)$ are defined by

$$\varepsilon_E^I(x) = \varepsilon_E(x) - \varepsilon_E^K(x) = \varepsilon_E(x) - \frac{3}{2}\frac{k_B}{m}x. \tag{12.28}$$

From (12.27)$_2$ we have

$$\mathscr{P} = \frac{2}{3}\rho(\varepsilon_E(T) - \varepsilon_E^I(\Theta)). \tag{12.29}$$

Choosing $(\rho, T, \mathscr{P})$ as independent variables, and differentiating (12.29) with respect to $(\rho, T, \mathscr{P})$, we have

$$\Theta_\rho = \frac{3}{2}\frac{\mathscr{P}}{\rho^2 c_v^I(\Theta)}, \quad \Theta_T = \frac{c_v(T)}{c_v^I(\Theta)}, \quad \Theta_{\mathscr{P}} = -\frac{3}{2\rho c_v^I(\Theta)}, \tag{12.30}$$

where $c_v^I(x)$ is the specific heat defined by

$$c_v^I(x) = \frac{d\varepsilon_E^I(x)}{dx}.$$

On the other hand, from (12.27), we have

$$k_\Theta = \frac{c_v^I(\Theta)}{\Theta}\left(1 - \frac{p}{\mathscr{P}}\frac{\Theta}{T}\right), \quad k_T = -\frac{c_v^I(T)}{T} + \frac{3}{2}\frac{k_B}{mT}\left(\frac{T c_v(T)}{\varepsilon_E(T) - \varepsilon_E^I(\Theta)} - 1\right). \tag{12.31}$$

Now we are ready to calculate the derivatives of k with respect to the field variables $(\rho, \varepsilon, \mathscr{P})$. From (12.31) and (12.30), we have

$$\begin{aligned} k_\rho &= k_\Theta \Theta_\rho = \frac{3}{2}\frac{k_B}{m\rho}\left(\frac{T}{\Theta}(1+Z)-1\right), \\ k_\varepsilon &= (k_\Theta \Theta_T + k_T)\, T_\varepsilon = \frac{1}{\Theta} - \frac{1}{T}, \\ k_{\mathscr{P}} &= k_\Theta \Theta_{\mathscr{P}} = \frac{3}{2}\frac{k_B}{mp}\left(1 - \frac{Z}{1+Z} - \frac{T}{\Theta}\right) \end{aligned} \tag{12.32}$$

with k, Z, Θ given in (12.27).

Inserting (12.32) into (12.26), we have the final expressions for the components of the intrinsic main field in the nonlinear non-polytropic case:

$$\begin{aligned} \hat{\lambda} &= -\frac{g}{T} + k(T,\Theta) + \frac{\varepsilon_E(T)}{T} - \frac{\varepsilon_E(\Theta)}{\Theta} \\ &= -\frac{g}{T} + \int_T^\Theta \frac{\varepsilon_E^I(x)}{x^2}\,dx + \frac{3}{2}\frac{k_B}{m}\log\left(\frac{\varepsilon_E(T) - \varepsilon_E^I(\Theta)}{\varepsilon_E^K(T)}\right), \\ \hat{\lambda}_i &= 0, \\ \hat{\mu} &= \frac{1}{2\Theta}, \\ \hat{\zeta} &= \frac{1}{2T}\frac{\varepsilon_E^K(T)}{\varepsilon_E(T) - \varepsilon_E^I(\Theta)} - \frac{1}{2\Theta}. \end{aligned} \tag{12.33}$$

Taking into account (2.55) and (12.2), we obtain the components of the main field as follows:

$$\begin{aligned} \lambda &= \hat{\lambda} + (\hat{\mu} + \hat{\zeta})v^2 = -\frac{g}{T} + k(T,\Theta) + \frac{\varepsilon_E(T)}{T} - \frac{\varepsilon_E(\Theta)}{\Theta} + \frac{v^2}{2T}\frac{\varepsilon_E^K(T)}{\varepsilon_E(T) - \varepsilon_E^I(\Theta)}, \\ \lambda_i &= \hat{\lambda}_i - 2v_i\left(\hat{\mu} + \hat{\zeta}\right) = -\frac{v_i}{T}\frac{\varepsilon_E^K(T)}{\varepsilon_E(T) - \varepsilon_E^I(\Theta)}, \\ \mu &= \hat{\mu} = \frac{1}{2\Theta}, \\ \zeta &= \hat{\zeta} = \frac{1}{2T}\frac{\varepsilon_E^K(T)}{\varepsilon_E(T) - \varepsilon_E^I(\Theta)} - \frac{1}{2\Theta}. \end{aligned} \tag{12.34}$$

The system (12.4) becomes symmetric hyperbolic with respect to the main field given by (12.34) for any values of the fields provided that the inequalities (12.24) are satisfied. This is a unique example of completely nonlinear RET that is symmetric hyperbolic globally.

12.2.2.2 Intrinsic Entropy Flux, Residual Entropy Inequality, and Production Term

Moreover, from (2.57), (2.13), and (2.11)$_2$, we have

$$\varphi^{\prime} = 0, \qquad \Sigma = \zeta P_{ll} > 0. \tag{12.35}$$

From the residual inequality (12.35)$_2$, we have

$$\hat{P}_{ll} = \alpha\zeta, \qquad \alpha > 0. \tag{12.36}$$

As in all phenomenological theory, the residual inequality gives only some inequalities. The determination of α, which in principle depends on all field variables, may be a hard task. Nevertheless, we will prove in the next section that we can estimate α in terms of the bulk viscosity.

12.2.2.3 Alternative Form

Subtracting the equation of the energy (12.4)$_3$ from the equation of the dynamic pressure (12.4)$_4$, we can rewrite the system (12.4) in an equivalent form:

$$\begin{aligned}
&\frac{\partial \rho}{\partial t} + \frac{\partial}{\partial x_i}(\rho v_i) = 0,\\
&\frac{\partial \rho v_j}{\partial t} + \frac{\partial}{\partial x_i}\left[\mathscr{P}\delta_{ij} + \rho v_i v_j\right] = 0,\\
&\frac{\partial}{\partial t}(2\rho\varepsilon + \rho v^2) + \frac{\partial}{\partial x_i}\left\{\left[2\mathscr{P} + 2\rho\varepsilon + \rho v^2\right] v_i\right\} = 0,\\
&\frac{\partial}{\partial t}\left[3\mathscr{P} - 2\rho\varepsilon\right] + \frac{\partial}{\partial x_i}\left\{\left[3\mathscr{P} - 2\rho\varepsilon\right] v_i\right\} = \hat{P}_{ll} = \alpha\zeta.
\end{aligned} \tag{12.37}$$

It is easy to verify, from (2.11)$_1$, that the corresponding main field $(\bar{\lambda}, \bar{\lambda}_i, \bar{\mu}, \bar{\zeta})$ that symmetrizes the system (12.37) has the same components of the main field of the system (12.3) except for the Lagrange multiplier $\bar{\mu}$ of the energy equation (12.37)$_3$. It is given by $\bar{\mu} = \mu + \zeta$. Therefore for this alternative form we have the following main field components:

$$\begin{aligned}
\bar{\lambda} &= -\frac{g}{T} + k(T,\Theta) + \frac{\varepsilon_E(T)}{T} - \frac{\varepsilon_E(\Theta)}{\Theta} + \frac{v^2}{2T}\frac{\varepsilon_E^K(T)}{\varepsilon_E(T)-\varepsilon_E^I(\Theta)},\\
\bar{\lambda}_i &= -\frac{v_i}{T}\frac{\varepsilon_E^K(T)}{\varepsilon_E(T)-\varepsilon_E^I(\Theta)},\\
\bar{\mu} &= \frac{1}{2T}\frac{\varepsilon_E^K(T)}{\varepsilon_E(T)-\varepsilon_E^I(\Theta)},\\
\bar{\zeta} &= \frac{1}{2T}\frac{\varepsilon_E^K(T)}{\varepsilon_E(T)-\varepsilon_E^I(\Theta)} - \frac{1}{2\Theta}.
\end{aligned} \tag{12.38}$$

From (12.27), these can be rewritten in an equivalent form:

$$\begin{aligned}
\bar{\lambda} &= -\frac{g}{T} + k(T,\Theta) + \frac{\varepsilon_E(T)}{T} - \frac{\varepsilon_E(\Theta)}{\Theta} + \frac{v^2}{2T}\frac{1}{1+Z},\\
\bar{\lambda}_i &= -\frac{v_i}{T}\frac{1}{1+Z},\\
\bar{\mu} &= \frac{1}{2T}\frac{1}{1+Z},\\
\bar{\zeta} &= \frac{1}{2T}\frac{1}{1+Z} - \frac{1}{2\Theta}.
\end{aligned} \tag{12.39}$$

By introducing the material time derivative, the system (12.37) can be rewritten as

$$\begin{aligned}
&\dot{\rho} + \rho\frac{\partial v_k}{\partial x_k} = 0,\\
&\rho\dot{v}_i + \frac{\partial}{\partial x_i}\mathscr{P} = 0,\\
&\rho\dot{\varepsilon} + \mathscr{P}\frac{\partial v_k}{\partial x_k} = 0,\\
&\left(\frac{\mathscr{P}}{\rho} - \frac{2}{3}\varepsilon\right)^{\bullet} = \frac{\hat{P}_{ll}}{3\rho} = \frac{\alpha\zeta}{3\rho}.
\end{aligned} \tag{12.40}$$

We observe that any differentiable solution of (12.40) satisfies the entropy law (2.8) that can be rewritten, from (12.5) and (12.35), as

$$\dot{\eta} = \frac{\alpha}{\rho}\zeta^2. \tag{12.41}$$

12.2.2.4 Polytropic Gas

In the case of polytropic gas (7.18), from $(12.27)_2$, we have

$$\frac{\Theta}{T} = 1 - \frac{3}{D-3}\frac{\Pi}{p}, \tag{12.42}$$

and then, from $(12.23)_1$, we obtain

$$k = \frac{k_B}{m} \log \left\{ \left(1 + \frac{\Pi}{p}\right)^{\frac{3}{2}} \left(1 - \frac{3}{D-3}\frac{\Pi}{p}\right)^{\frac{D-3}{2}} \right\}. \tag{12.43}$$

We notice that k depends only on a single variable $Z = \Pi/p$ or $\mathscr{P}/p$. In fact (12.43) can be rewritten as

$$k = \frac{k_B}{m} \log \left\{ \left(\frac{\mathscr{P}}{p}\right)^{\frac{3}{2}} \left[\frac{1}{D-3}\left(D - 3\frac{\mathscr{P}}{p}\right)\right]^{\frac{D-3}{2}} \right\}.$$

For $D > 3$, k exists and is bounded in the domain, which contains an equilibrium state, given by the existence conditions (12.24). In the present case, the conditions are given by

$$-1 < Z < \frac{D-3}{3} \qquad \Leftrightarrow \qquad 0 < \frac{\Theta}{T} < \frac{D}{D-3},$$

in which $k(Z) < 0$, $\forall Z \neq 0$ and k has a global maximum $k(0) = 0$ at the equilibrium state. See Fig. 12.1.

From (12.42), the intrinsic main field (12.33) becomes

$$\begin{aligned}
\bar{\lambda} &= -\frac{g}{T} + k + \frac{v^2}{2T}\frac{1}{1+Z}, \\
\bar{\lambda}_i &= -\frac{v_i}{T}\frac{1}{1+Z}, \\
\bar{\mu} &= \frac{1}{2T}\frac{1}{1+Z}, \\
\bar{\zeta} &= -\frac{1}{2T}\frac{1}{1+Z}\frac{D}{D-3}Z\left(1 + \frac{3}{D-3}Z\right)^{-1}
\end{aligned}$$

with k given in (12.43) and Θ related to Π through (12.42).

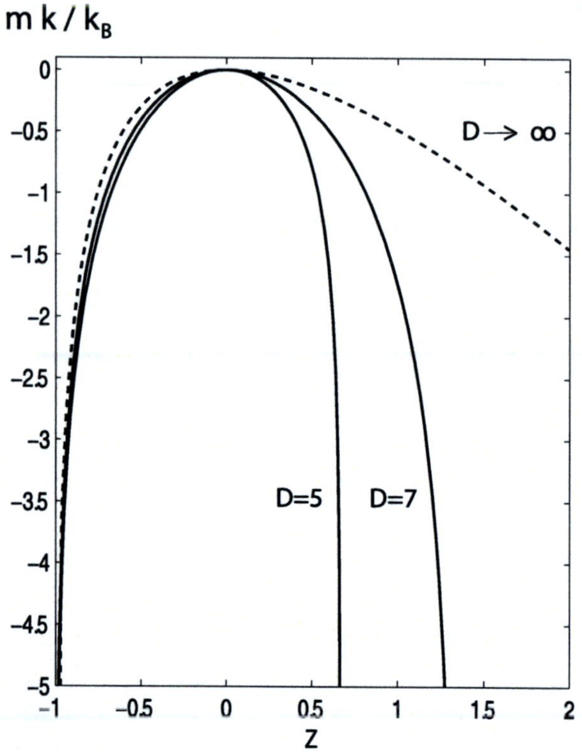

Fig. 12.1 Nonequilibrium part of the entropy density, k, as function of Z for different values of D

12.2.2.5 Theory Near Equilibrium

In a near-equilibrium state, Π is small. Then we can expand (12.27) around an equilibrium state with $\Pi = 0$ and $\Theta = T$ with respect to Z.

Non-polytropic Case

$$\begin{aligned} k(T, Z) &= -\frac{3}{4}\frac{k_B}{m}\frac{c_v^I(T)}{c_v(T)}Z^2, \\ Z &= \frac{\Pi}{p} = \frac{2}{3}\hat{c}_v^I(T)\left(1 - \frac{\Theta}{T}\right) \end{aligned} \tag{12.44}$$

with the main field:

$$
\begin{aligned}
\bar{\lambda} &= -\frac{g}{T} - \frac{3}{2T\hat{c}_v(T)}\left[c_v(T)T - \varepsilon(T)\right] Z + \frac{v^2}{2T}(1-Z), \\
\bar{\lambda}_i &= -\frac{v_i}{T}(1-Z), \\
\bar{\mu} &= \frac{1}{2T}(1-Z), \\
\bar{\zeta} &= -\frac{1}{2T}\frac{\hat{c}_v(T)}{\hat{c}_v^I(T)}Z,
\end{aligned}
$$

where $\hat{c}_v^I = (m/k_B)c_v^I$ is the dimensionless specific heat.

Polytropic Case

$$
\begin{aligned}
k(Z) &= -\frac{3}{4}\frac{k_B}{m}\frac{D-3}{D}Z^2, \\
Z &= \frac{\Pi}{p} = \frac{D-3}{3}\left(1 - \frac{\Theta}{T}\right)
\end{aligned}
$$

with the main field:

$$
\begin{aligned}
\bar{\lambda} &= -\frac{g}{T} + \frac{v^2}{2T}(1-Z), \\
\bar{\lambda}_i &= -\frac{v_i}{T}(1-Z), \\
\bar{\mu} &= \frac{1}{2T}(1-Z), \\
\bar{\zeta} &= -\frac{1}{2T}\frac{D}{D-3}Z.
\end{aligned}
\tag{12.45}
$$

From (12.34)$_4$, the production term (12.36) is given by

$$
\hat{P}_{ll} = -\frac{\alpha_E}{2T}\frac{\hat{c}_v(T)}{\hat{c}_v^I(T)}\frac{\Pi}{p},
$$

where α_E denotes the function α at an equilibrium state. If we introduce the relaxation time τ such that the previous equations can be written in the form:

$$
\hat{P}_{ll} = -3\frac{\Pi}{\tau},
\tag{12.46}
$$

α_E is expressed as

$$\alpha_E = \frac{6pT}{\tau} \frac{\hat{c}_v^I(T)}{\hat{c}_v(T)}. \tag{12.47}$$

By using the Maxwellian iteration, Eq. (12.3)$_4$ becomes the Navier-Stokes constitutive equation for the dynamic pressure [112, 128]:

$$\Pi = -\nu \,\mathrm{div}\, \mathbf{v} \tag{12.48}$$

with the bulk viscosity ν expressed by the relaxation time τ as follows:

$$\nu = \left(\frac{2}{3} - \frac{1}{\hat{c}_v}\right) p\tau = \frac{2}{3}\frac{\hat{c}_v^I}{\hat{c}_v} p\tau. \tag{12.49}$$

From (12.49) and (12.47), α_E is expressed in terms of the bulk viscosity:

$$\alpha_E = \frac{4p^2T}{\nu}\left(\frac{\hat{c}_v^I}{\hat{c}_v}\right)^2.$$

Remark 12.2 The evaluation of α in a far-from-equilibrium case is still an open problem. If we may assume that $\alpha = \alpha_E$, the production term is completely known in terms of the bulk viscosity. In many applications, we can use the linear approximation (12.46) that corresponds to the BGK approximation in the kinetic theory.

12.2.3 Euler Fluid as a Principal Subsystem of the ET$_6$ System and Subcharacteristic Conditions

System of the Euler fluid is a principal subsystem of the ET$_6$ theory according to the definition given by Boillat and Ruggeri in Sect. 2.4. We require $\zeta = 0$ and neglect the last equation in (12.3). We have, from (12.34), $\varepsilon_E(T) = \varepsilon_E(\Theta)$. As $\varepsilon_E(T)$ is a monotonous increasing function with respect to T, we deduce $\Theta = T$, i.e., $\Pi = 0$. Therefore the principal subsystem, i.e., the system of the remaining 5 equations is the Euler system. The remaining 5 components of the main field in (12.34) converge to the symmetrizable variables, which were deduced first by Godunov [179] for the Euler-fluid system:

$$\bar{\lambda} \to -\frac{g}{T} + \frac{v^2}{2T}, \quad \lambda_i \to -\frac{v_i}{T}, \quad \mu \to \frac{1}{2T}. \tag{12.50}$$

According to the result in Sect. 2.4, the convexity of a system implies automatically the so-called subcharacteristic condition, i.e., the spectrum of the characteristic velocities of the principal subsystem is contained in the spectrum of the full system evaluated in the subspace of the subsystem. This implies, in particular, that the maximum characteristic velocity of the subsystem is not greater than the one of the full system. In the present particular case of ET$_6$, the maximum characteristic velocity is given by (see [322])

$$v_i n_i + \sqrt{\frac{5}{3}\frac{p+\Pi}{\rho}} \tag{12.51}$$

with $\mathbf{n} \equiv (n_i)$ being the unit normal to the wave front. While the maximum characteristic velocity of the Euler fluid is given by

$$v_i n_i + \sqrt{\left(\frac{\partial p}{\partial \rho}\right)_s}. \tag{12.52}$$

It is easy to verify the relation:

$$\left(\frac{\partial p}{\partial \rho}\right)_s = \left(\frac{\partial p}{\partial \rho}\right)_T + \frac{T\left(\frac{\partial p}{\partial T}\right)_\rho^2}{\rho^2\left(\frac{\partial \varepsilon}{\partial T}\right)_\rho} = \gamma(T)\frac{k_B}{m}T,$$

where $\gamma(T)$ is the ratio of the specific heats: $\gamma(T) = (1 + \hat{c}_v(T))/\hat{c}_v(T)$. Then, as $1 < \gamma(T) \leqslant 5/3$, the subcharacteristic condition is indeed satisfied, i.e., the maximum characteristic velocity (12.51) evaluated in equilibrium ($\Pi = 0$) is not less than the corresponding one of the Euler fluid (12.52).

12.3 Comparison with the Meixner Theory

In order to describe thermodynamic properties of a material with internal structures, theories with internal variables were developed [7, 324, 325], and have been applied to various phenomena like chemical reaction, sound wave and shock wave propagation in a polyatomic gas, inelastic behavior of a solid, dynamics of a dielectric material, and so on. See, for example, the reference [326]. The Meixner theory [129, 130] may be regarded as the prototype of the theories of internal variables, but it is still frequently applied to various nonequilibrium phenomena. For a recent study, where dynamic degrees of freedom and internal variables are discussed in a uniform way, see, for example, the reference [327].

In this section, we compare the ET$_6$ theory with the Meixner theory, and discuss the physical implications involved.

12.3.1 Far-From-Equilibrium Case

We assume the simplest version of the Meixner theory, that is, the system composed of the Euler equations and only one relaxation equation for an internal variable ξ. In this case the system of field equations is summarized as follows [7, 129, 130]:

$$
\begin{aligned}
&\dot{\rho} + \rho \frac{\partial v_k}{\partial x_k} = 0, \\
&\rho \dot{v}_i + \frac{\partial \mathscr{P}}{\partial x_i} = 0, \\
&\rho \dot{\mathscr{E}} + \mathscr{P} \frac{\partial v_k}{\partial x_k} = 0, \\
&\dot{\xi} = -\beta \mathscr{A},
\end{aligned}
\tag{12.53}
$$

where $\mathscr{P}$, $\mathscr{E}$, and $\mathscr{A}$ are, respectively, the pressure, the specific internal energy, and the affinity of a relaxation process, and β is a positive phenomenological coefficient. We will see below that the pressure $\mathscr{P}$ here is the nonequilibrium pressure defined above (see (12.56)). The generalized Gibbs relation in the Meixner theory is assumed to be

$$
\mathscr{T} \mathrm{d}\mathscr{S} = \mathrm{d}\mathscr{E} - \frac{\mathscr{P}}{\rho^2} \mathrm{d}\rho - \mathscr{A} \mathrm{d}\xi, \tag{12.54}
$$

where $\mathscr{T}$ is the temperature and $\mathscr{S}$ is the specific entropy. Note that the quantities $\mathscr{T}$, $\mathscr{S}$, $\mathscr{P}$, and $\mathscr{A}$ depend not only on the mass density ρ and the specific internal energy $\mathscr{E}$ but also on the internal variable ξ. From (12.53) and (12.54), we obtain

$$
\dot{\mathscr{S}} = \beta \frac{\mathscr{A}^2}{\mathscr{T}}. \tag{12.55}
$$

Comparing the system of the ET$_6$ theory: (12.40) and (12.41) with the system of the Meixner theory: (12.53) and (12.55), we may say that the ET$_6$ theory belongs to the Meixner theory through the following interpretation of the internal variable ξ and the affinity $\mathscr{A}$:

$$
\begin{aligned}
&\xi = \frac{p + \Pi}{\rho} - \frac{2}{3}\varepsilon, \\
&\mathscr{P} = p + \Pi, \\
&\mathscr{E} = \varepsilon, \\
&\mathscr{A} = -3\mathscr{T}\bar{\zeta}, \\
&\mathscr{S} = \eta, \\
&\beta = \frac{\alpha}{9\rho\mathscr{T}},
\end{aligned}
\tag{12.56}
$$

where the Meixner temperature $\mathscr{T}$ is given by

$$\mathscr{T} = T(1+Z) = T\,\frac{\varepsilon_E(T) - \varepsilon_E^I(\Theta)}{\varepsilon_E^K(T)}, \tag{12.57}$$

or

$$\frac{\mathscr{T}}{T} = \frac{\mathscr{P}}{p}. \tag{12.58}$$

From (12.54), it is easy to determine the intrinsic main-field components for the Meixner theory:

$$\bar{\lambda} = -\frac{\mathscr{G}}{\mathscr{T}} + \frac{v^2}{2\mathscr{T}} \qquad \bar{\lambda}_i = -\frac{v_i}{\mathscr{T}},$$
$$\bar{\mu} = \frac{1}{2\mathscr{T}}, \qquad \bar{\zeta} = -\frac{\mathscr{A}}{3\mathscr{T}},$$

where we define the chemical potential $\mathscr{G}$ in the Meixner theory:

$$\mathscr{G} \equiv \mathscr{E} + \frac{\mathscr{P}}{\rho} - \mathscr{T}\mathscr{S} - \mathscr{A}\xi.$$

The chemical potential and the affinity in the Meixner theory are expressed explicitly in terms of the quantities in the ET_6 theory as follows:

$$\mathscr{G} = (1+Z)\left\{g - T\left[\int_T^{\Theta} \frac{\varepsilon_E^I(x)}{x^2}\,dx + \frac{3}{2}\frac{k_B}{m}\log(1+Z)\right]\right\} \tag{12.59}$$

and

$$\mathscr{A} = \frac{3}{2}\left\{(1+Z)\frac{T}{\Theta} - 1\right\}.$$

Remark 12.3 The nonequilibrium temperature $\mathscr{T}$ of the Meixner theory is introduced through the generalized Gibbs relation (12.54). This can be interpreted as the temperature of a state *in equilibrium* with a constraint that the system is kept at a given ξ by applying a suitable external field [328], which may be or may not be virtual. We sometimes call such a field thermodynamic external field or effective field. This technique has been used, for example, in the linear response theory [24, 83]. On the other hand, the temperature T of ET_6 is defined in Chap. 6 as the temperature of an equilibrium state with the same values of the mass density ρ and the specific internal energy ε in a nonequilibrium state under consideration.

This definition has been usually adopted, for example, in the kinetic theory [17, 84], in nonequilibrium statistical mechanics [83]. We will discuss, in a general way, the nonequilibrium temperature (and also chemical potential) in Chap. 15. A survey on nonequilibrium temperatures is given in references [157, 329]. See also [155] for the analysis of the thermodynamic temperature in the RET theory.

Remark 12.4 In the Meixner theory, it is usually difficult to identify a proper internal variable. We notice, however, that, in the case of rarefied polyatomic gas, the explicit dependence of the internal variable ξ on the dynamic pressure Π is given by $(12.56)_1$. We will discuss this point in Sect. 12.5.

12.3.2 Near-Equilibrium Case

From (12.56), (12.58), and (12.44), the correspondence relations between the ET_6 theory and the Meixner theory are the same as those in (12.56) except that the last three equations are replaced by

$$\begin{aligned}
\mathscr{A} &= \frac{3}{2}\frac{\hat{c}_v(T)}{\hat{c}_v^I(T)}Z + O(Z^2),\\
\mathscr{S} &= s - \frac{3}{4}\frac{k_B}{m}\frac{c_v^I(T)}{c_v(T)}Z^2 + O(Z^3),\\
\beta &= \frac{2}{3}\frac{\hat{c}_v^I(T)}{\hat{c}_v(T)}\frac{p}{\rho\tau} + O(Z).
\end{aligned} \tag{12.60}$$

The temperature $\mathscr{T}$ in (12.57) and the chemical potential $\mathscr{G}$ in (12.59) are now given by

$$\mathscr{T} = T(1+Z), \qquad \mathscr{G} = g(T) + Z\left(g(T) + \frac{3\varepsilon_E^I(T)}{\hat{c}_v^I(T)} - \varepsilon_E^K(T)\right) + O(Z^2).$$

Noteworthy points are summarized as follows:

1. We can find explicit expressions of the quantities $\mathscr{P}$, $\mathscr{E}$, $\mathscr{A}$, and $\mathscr{S}$ in terms of the internal variable ξ. In fact, eliminating Π in (12.60), we obtain

$$\begin{aligned}
\mathscr{P} &= p + \rho(\xi - \xi_{eq}),\\
\mathscr{E} &= \varepsilon,\\
\mathscr{A} &= \frac{3\rho}{2p}\frac{c_v(T)}{\hat{c}_v^I(T)}(\xi - \xi_{eq}),\\
\mathscr{S} &= s - \frac{3\rho^2}{4p^2}\frac{k_B}{m}\frac{c_v^I(T)}{c_v(T)}(\xi - \xi_{eq})^2,
\end{aligned} \tag{12.61}$$

where ξ_{eq} is defined by

$$\xi_{eq} \equiv \frac{p}{\rho} - \frac{2}{3}\varepsilon.$$

2. The relaxation time in the Meixner theory is defined by de Groot and Mazur [7]

$$\tau_M \equiv \frac{1}{\beta}\left(\frac{\partial \xi}{\partial \mathscr{A}}\right)_{\rho,\mathscr{E}}. \tag{12.62}$$

From (12.60)$_3$, (12.61)$_{2,3}$, and (12.62), we can prove that this relaxation time coincides with the relaxation time τ in the ET$_6$ theory:

$$\tau_M = \tau. \tag{12.63}$$

3. As seen before, by using the Maxwellian iteration, Eq. (12.3)$_4$ with linear production (12.46) becomes the Navier-Stokes constitutive equation for the dynamic pressure (12.48). The equivalence between the relaxation times (12.63) and the relation (12.49) enables us to evaluate the value of the bulk viscosity from the data on the relaxation time τ_M for the internal degrees of freedom, which are accumulated in literature [330] for many kinds of polyatomic gases.
4. If we apply the Maxwellian iteration to (12.3)$_4$ with nonlinear production (12.36), we have a nonlinear constitutive equation for Π with respect to the velocity gradient, that is, the so-called non-Newtonian constitutive equation.
5. It is interesting to point out that there exist works [331–333] that, in the parabolic Navier-Stokes framework, neglect the shear viscosity but take into account the bulk viscosity expressed by Eq. (12.48). And the qualitative analysis of the system was made.

12.4 Monatomic-Gas Limit

The monatomic-gas limit of ET$_6$ in the case of polytropic gas with linear production term was already studied in Sect. 10.5.2 although it is also possible to study the case of non-polytropic gas.

The first 5 components of the main field (12.45) converge to the ones of the Euler fluid (12.50) while the last component of the main field, the Lagrange multiplier $\bar{\zeta}$ of the vanishing equation in the system of ET$_6$, remains undetermined. Then the convergence of the system of ET$_6$ to the Euler system when $D \to 3$ is fully verified.

Remark 12.5 We observe that, although, for $D > 3$, the two differential symmetric systems (12.3) and (12.37) are equivalent to each other, only the main field (12.39) of (12.37) converges to the main field of the Euler fluid. In fact, as we have noticed, the difference between the main fields (12.34) and (12.39) of the two equivalent systems consists only in the component of the energy equation, i.e., μ and $\bar{\mu}$ with the relation $\mu = \bar{\mu} - \bar{\zeta}$. Now $\bar{\mu}$ converges to the corresponding value of the Euler fluids $1/(2T)$, instead the limit of μ is undetermined due to the indetermination of $\bar{\zeta}$. This indicates that, for the convergence in the monatomic-gas limit, it is preferable to use the system (12.37) instead of the system (12.3).

12.5 Nature of the Dynamic Pressure

In this section, we investigate the ET$_6$ theory constructed above from a different point of view, and thereby try to understand the origin of the dynamic pressure. We emphasize the importance of energy exchange between the molecular translational motion and the internal motion of a molecule as one of fundamental irreversible processes.

12.5.1 *Thermal and Caloric Equations of State, Revisited*

In a rarefied gas, the pressure p is only due to the translational motion of molecules [334], and has the well-known relation:

$$p(\rho, T) = \frac{k_B}{m}\rho T.$$

Instead the internal energy ε can be divided into two parts, that is, ε^K due to the translational motion of molecules and ε^I due to the molecular internal motion such as molecular rotation and vibration (see (12.28)):

$$\varepsilon_E(x) = \varepsilon_E^K(x) + \varepsilon_E^I(x), \qquad \varepsilon_E^K(x) = \frac{3}{2}\frac{k_B}{m}x, \quad \forall x > 0.$$

In the following analysis, we also need the expression of the pressure in terms of ρ and ε^K:

$$p(\rho, \varepsilon^K) = \frac{2}{3}\rho\varepsilon^K. \tag{12.64}$$

12.5.2 *Origin of the Dynamic Pressure*

From (12.27)$_2$, we notice an interesting relation:

$$\Pi = -\frac{p(\rho, T)}{\varepsilon_E^K(T)}\Delta = -\frac{2}{3}\rho\Delta, \tag{12.65}$$

where

$$\Delta = \varepsilon_E^I(\Theta) - \varepsilon_E^I(T).$$

Therefore the relationship between Π and $-\Delta$ is the same as the one between p and ε^K in (12.64). Since the quantity Δ is the energy exchange between the energy of translational motion and the energy of molecular internal motion in a nonequilibrium process, we may say from (12.65) that the cause of the dynamic pressure is such a nonequilibrium energy exchange. This suggests us strongly that we may adopt Δ as an independent field instead of Π in order to understand the nonequilibrium phenomena from the viewpoint of energy exchange. In the next subsection, we will rewrite the system of balance equations in terms of the set of independent fields $\{\rho, v_i, T, \Delta\}$.

The other expression of the dynamic pressure is given by

$$\Pi = p(\rho, \varepsilon_E^K(T) - \Delta) - p(\rho, \varepsilon_E^K(T)). \tag{12.66}$$

This expression shows the physical meaning of the dynamic pressure in a self-evident way. The dynamic pressure Π emerges as the nonequilibrium part of the pressure due to the energy exchange Δ in the kinetic energy of molecular translational motion!

By using the energy exchange Δ, the internal variable ξ in the Meixner theory given in $(12.56)_1$ is expressed as

$$\xi = -\frac{2}{3}(\varepsilon_E^I(T) + \Delta) \tag{12.67}$$

or

$$\xi - \xi_{eq} = -\frac{2}{3}\Delta$$

Therefore, in the case of rarefied polyatomic gas, the internal variable is related to the energy exchange in a nonequilibrium process.

Lastly it is worth noting that, in the context of the kinetic theory for rarefied gases [17, 317, 335], the dynamic pressure is related to the nonequilibrium energy exchange among several kinds of motions of a molecule. See also Chap. 14.

12.5.3 System of Balance Equations in Terms of $\{\rho, v_i, T, \Delta\}$

Instead of the balance equation for the density F_{ll}, it is possible to adopt the balance equation for the density $G_{ll} - F_{ll}$, which is the nonequilibrium internal energy with the energy exchange:

$$G_{ll} - F_{ll} = 2\rho(\varepsilon_E^I(T) + \Delta).$$

Then, we have an alternative system of balance equations:

$$\begin{aligned}
&\frac{\partial F}{\partial t} + \frac{\partial F_i}{\partial x_i} = 0, \\
&\frac{\partial F_j}{\partial t} + \frac{\partial F_{ij}}{\partial x_i} = 0, \\
&\frac{\partial}{\partial t}(G_{ll} - F_{ll}) + \frac{\partial}{\partial x_i}(G_{lli} - F_{lli}) = -P_{ll}, \qquad \frac{\partial G_{ll}}{\partial t} + \frac{\partial G_{lli}}{\partial x_i} = 0.
\end{aligned}$$

This system of balance equations is expressed explicitly in terms of the set of independent fields $\{\rho, v_i, T, \Delta\}$ as follows:

$$\begin{aligned}
&\frac{\partial \rho}{\partial t} + \frac{\partial}{\partial x_i}(\rho v_i) = 0, \\
&\frac{\partial \rho v_j}{\partial t} + \frac{\partial}{\partial x_i}\left[p(\rho, \varepsilon_E^K(T) - \Delta)\delta_{ij} + \rho v_i v_j\right] = 0, \\
&\frac{\partial}{\partial t}(2\rho\varepsilon_E(T) + \rho v^2) + \frac{\partial}{\partial x_i}\left\{\left[2p(\rho, \varepsilon_E^K(T) - \Delta) + 2\rho\varepsilon_E(T) + \rho v^2\right] v_i\right\} = 0, \\
&\frac{\partial}{\partial t}\left[2\rho(\varepsilon_E^I(T) + \Delta)\right] + \frac{\partial}{\partial x_i}\left[2\rho(\varepsilon_E^I(T) + \Delta)v_i\right] = -\hat{P}_{ll}.
\end{aligned}$$

12.5.4 Nonequilibrium Temperatures ϑ and Θ

We introduce nonequilibrium temperatures ϑ and Θ through the following relation:

$$\Delta = \varepsilon_E^I(\Theta) - \varepsilon_E^I(T) = -\left(\varepsilon_E^K(\vartheta) - \varepsilon_E^K(T)\right). \tag{12.68}$$

The temperature T is related to the temperatures ϑ and Θ by this relation. Θ is the nonequilibrium temperature of the internal modes of a molecule induced by Δ. While ϑ is the nonequilibrium temperature of the translational modes of molecules induced by $-\Delta$. Now it is evident that the temperatures ϑ and Θ are nothing but the temperatures θ^K and θ^I introduced in (7.35).

The expression of the dynamic pressure (12.66) is rewritten by using ϑ as follows:

$$\Pi = p(\rho, \vartheta) - p(\rho, T).$$

And, from (12.27), the nonequilibrium specific entropy density k is rewritten as

$$\begin{aligned} k = \eta - s &= \int_T^{\vartheta} \frac{c_v^K(x)}{x}\,dx + \int_T^{\Theta} \frac{c_v^I(x)}{x}\,dx \\ &= \frac{3}{2}\frac{k_B}{m}\log\left(\frac{\vartheta}{T}\right) + \int_T^{\Theta} \frac{c_v^I(x)}{x}\,dx, \end{aligned} \tag{12.69}$$

where $c_v^K = \frac{3}{2}\frac{k_B}{m}$. Then it is easy to prove that η is expressed as

$$\eta = s^K(\rho, \vartheta) + s^I(\Theta),$$

where

$$\begin{aligned} s(\rho, T) &= s^K(\rho, T) + s^I(T), \\ s^K(\rho, T) &= \frac{k_B}{m}\left(\log \frac{T^{3/2}}{\rho}\right) + s_0^K, \\ s^I(T) &= \int_{T_0}^{T} \frac{c_v^I(x)}{x}\,dx + s_0^I \end{aligned}$$

with constants s_0^K, s_0^I, and T_0 at a reference state.

Finally it is interesting to note that, from (12.68), the Meixner temperature $\mathscr{T}$ given in (12.57) is identified as ϑ (or θ^K):

$$\mathscr{T} = \vartheta = \theta^K. \tag{12.70}$$

Therefore, in the case of rarefied polyatomic gas, $\mathscr{T}$ is just the kinetic temperature.

12.5.5 System of Balance Equations in Terms of $\{\rho, v_i, \vartheta, \Theta\}$

The system of balance equations in terms of the set independent fields $\{\rho, v_i, \vartheta, \Theta\}$ is given as follows:

$$\begin{aligned} &\frac{\partial \rho}{\partial t} + \frac{\partial}{\partial x_i}(\rho v_i) = 0, \\ &\frac{\partial \rho v_j}{\partial t} + \frac{\partial}{\partial x_i}\left\{p(\rho, \vartheta)\delta_{ij} + \rho v_i v_j\right\} = 0, \\ &\frac{\partial}{\partial t}\left(2\rho\varepsilon_E^K(\vartheta) + \rho v^2\right) + \frac{\partial}{\partial x_i}\left\{\left[2p(\rho, \vartheta) + 2\rho\varepsilon_E^K(\vartheta) + \rho v^2\right]v_i\right\} = \hat{P}_{ll}, \\ &\frac{\partial}{\partial t}\left[2\rho\varepsilon_E^I(\Theta)\right] + \frac{\partial}{\partial x_i}\left[2\rho\varepsilon_E^I(\Theta)v_i\right] = -\hat{P}_{ll}. \end{aligned} \tag{12.71}$$

12.6 Concluding Remarks

1. We have found the relationship between the RET theory and the Meixner theory within the simplest cases, namely, the ET_6 theory and the Meixner theory composed of the Euler equations with only one relaxation equation. The following question may naturally arise: Is there a possibility that, by establishing some correspondence relations like (12.60), the RET theory with more fields (for example, 14 fields studied in Chaps. 6 and 7, many fields in Chap. 9) has a counterpart in the Meixner theory with many internal variables? This question is interesting because several authors have tried to present models with internal variables to describe nonequilibrium processes and shown its applicability (see, for example, [324, 327] and the review paper [71]). In our opinion the answer is negative except for some special cases. In fact, only in the simplest RET theory like the ET_6 theory, a combination of the balance laws $(12.3)_3$ and $(12.3)_4$ gives us the possibility to have the equation in the form $(12.40)_4$ in which the spatial derivative exists only within the material time derivative, which is typical in the internal variable equation $(12.53)_4$. When the number of independent fields becomes large, however, we have a system of balance laws (2.1) for the fields that is perfectly consistent with the kinetic theory [119, 336]. In our opinion, this is the reason why the RET theory is much more powerful than other theories using internal variables in a similar way to the Meixner theory.
2. In the case of rarefied polyatomic gas, there exists large literature on kinetic models to explain relaxation processes and chemical reacting flows. See for example [337].
3. On the basis of nonlinear ET_6, Currò and Manganaro studied nonlinear wave interactions [338].
4. One of the promising applications of the present nonlinear analysis is the study of shock wave phenomena in a *non-polytropic* gas. This study is highly expected to enrich the previous studies [109, 127] by using the concepts of the internal variable and the nonequilibrium temperature of the Meixner theory, because some authors studied this subject by using these quantities [138]. We will discuss this subject in Chap. 18.

Chapter 13
Nonlinear Molecular ET Theory with 6 Fields

Abstract We establish extended thermodynamics of rarefied polyatomic gases with six independent fields via the maximum entropy principle. The distribution function is not necessarily near equilibrium. The result is in perfect agreement with the phenomenological RET theory explained in Chap. 12. This is the first example of molecular ET with a nonlinear closure. The integrability condition of the moments requires that the dynamic pressure should be bounded from below and above.

13.1 Introduction

We discussed the MEP (maximum entropy principle) procedure in Sect. 1.8.5, and in particular we put in evidence the problem of the convergence of moments in a far-from-equilibrium case in Sect. 4.4. All closures by the MEP procedure studied before including the 14-moment theory of polyatomic gases discussed in Chap. 7 are valid only near equilibrium. One of the issues in these closures is that the hyperbolicity condition is valid only in some small domain of the configuration space near equilibrium (see [25, 250, 261] for monatomic non-degenerate gas case and [262] for Fermi and Bose gases).

The aim of this chapter is to prove that, in the case of ET_6 of rarefied polyatomic gases, a theory can be established with the closure that is valid even far from equilibrium. We will show that this nonlinear closure matches completely the previous result presented in Chap. 12 by using only the macroscopic method [132]. The result of this chapter is presented firstly in the paper [131] in the case of polytropic gas and in [134] for the general case of non-polytropic one.

13.2 Non-polytropic Gas

We study a non-polytropic gas, the equilibrium properties of which are explained in Sect. 7.2. In particular we have deduced the equilibrium distribution function of a polyatomic gas in Theorem 7.1 (see (7.8)).

T. Ruggeri, M. Sugiyama, *Classical and Relativistic Rational Extended Thermodynamics of Gases*, https://doi.org/10.1007/978-3-030-59144-1_13

Now we consider the system (12.1) with 6 moments:

$$\begin{pmatrix} F \\ F_i \\ F_{ll} \end{pmatrix} = \begin{pmatrix} \rho \\ \rho v_i \\ \rho v^2 + 3(p + \Pi) \end{pmatrix} = \int_{\mathbb{R}^3} \int_0^\infty m \begin{pmatrix} 1 \\ c_i \\ c^2 \end{pmatrix} f \varphi(I) \, dI \, d\mathbf{c} \tag{13.1}$$

and

$$G_{ll} = \rho v^2 + 2\rho \varepsilon = \int_{\mathbb{R}^3} \int_0^\infty m(c^2 + 2I/m) f \varphi(I) \, dI \, d\mathbf{c}. \tag{13.2}$$

We prove the following theorem about the nonequilibrium distribution function:

Theorem 13.1 *The distribution function that maximizes the entropy (7.7) under the constraints (13.1) and (13.2) has the form:*

$$f = \frac{\rho}{mA(\Theta)} \left(\frac{m}{2\pi k_B T} \frac{1}{1 + \frac{\Pi}{p}} \right)^{3/2} \exp\left\{ -\frac{1}{k_B T} \left(\frac{1}{2} m C^2 \left(\frac{1}{1 + \frac{\Pi}{p}} \right) + I \frac{T}{\Theta} \right) \right\}, \tag{13.3}$$

where the nonequilibrium temperature Θ is related to the dynamic pressure Π through the relation:

$$\frac{\varepsilon_E^I(T) - \varepsilon_E^I(\Theta)}{\varepsilon_E^K(T)} = \frac{\Pi}{p}, \tag{13.4}$$

and $A(\Theta)$ is given by

$$A(\Theta) = A_0 \exp\left(\frac{m}{k_B} \int_{T_0}^{\Theta} \frac{\varepsilon_E^I(x)}{x^2} dx \right). \tag{13.5}$$

All the moments are convergent and the bounded solutions satisfy the inequalities:

$$-1 < \frac{\Pi}{p} < \frac{\varepsilon_E^I(T)}{\varepsilon_E^K(T)}. \tag{13.6}$$

The distribution function is nonlinear in the dynamic pressure.

Proof The theorem is proved by the method of Lagrange multipliers. Introducing the vector of the multipliers $(\lambda, \lambda_i, \mu, \zeta)$, we define the functional:

$$\begin{aligned} \mathscr{L} = &- \int_{\mathbb{R}^3} \int_0^\infty k_B f \log f \, \varphi(I) \, dI \, d\mathbf{c} + \lambda \left(\rho - \int_{\mathbb{R}^3} \int_0^\infty m f \, \varphi(I) \, dI \, d\mathbf{c} \right) \\ &+ \lambda_i \left(\rho v_i - \int_{\mathbb{R}^3} \int_0^\infty m f c_i \, \varphi(I) \, dI \, d\mathbf{c} \right) \end{aligned}$$

$$+\mu\left(\rho v^2+2\rho\varepsilon-\int_{\mathbb{R}^3}\int_0^\infty m\left(c^2+2\frac{I}{m}\right)f\,\varphi(I)\,dI\,d\mathbf{c}\right)$$
$$+\zeta\left(\rho v^2+3(p+\Pi)-\int_{\mathbb{R}^3}\int_0^\infty mc^2 f\,\varphi(I)\,dI\,d\mathbf{c}\right).$$

Because we want to maximize this functional with respect to f with the given macroscopic quantities, this can be substituted by the following one:

$$\begin{aligned}\mathscr{L}=&-\int_{\mathbb{R}^3}\int_0^\infty k_B f\log f\,\varphi(I)\,dI\,d\mathbf{c}-\lambda\int_{\mathbb{R}^3}\int_0^\infty mf\,\varphi(I)\,dI\,d\mathbf{c}\\&-\lambda_i\int_{\mathbb{R}^3}\int_0^\infty mfc_i\,\varphi(I)\,dI\,d\mathbf{c}\\&-\mu\int_{\mathbb{R}^3}\int_0^\infty m\left(c^2+2\frac{I}{m}\right)f\,\varphi(I)\,dI\,d\mathbf{c}-\zeta\int_{\mathbb{R}^3}\int_0^\infty mc^2 f\,\varphi(I)\,dI\,d\mathbf{c}.\end{aligned}\tag{13.7}$$

Furthermore, since $\mathscr{L}$ is a scalar, it must retain the same value in the case of zero hydrodynamic velocity $\mathbf{v}=\mathbf{0}$ due to the Galilean invariance. Therefore we have

$$\begin{aligned}\mathscr{L}=&-\int_{\mathbb{R}^3}\int_0^\infty k_B f\log f\,\varphi(I)\,dI\,d\mathbf{C}-\hat{\lambda}\int_{\mathbb{R}^3}\int_0^\infty mf\,\varphi(I)\,dI\,d\mathbf{C}\\&-\hat{\lambda}_i\int_{\mathbb{R}^3}\int_0^\infty mfC_i\,\varphi(I)\,dI\,d\mathbf{C}\\&-\hat{\mu}\int_{\mathbb{R}^3}\int_0^\infty m\left(C^2+2\frac{I}{m}\right)f\,\varphi(I)\,dI\,d\mathbf{C}-\hat{\zeta}\int_{\mathbb{R}^3}\int_0^\infty mC^2 f\,\varphi(I)\,dI\,d\mathbf{C}.\end{aligned}\tag{13.8}$$

Comparison between (13.7) and (13.8) yields the relations between the Lagrange multipliers and the intrinsic Lagrange multipliers indicated by hat:

$$\lambda=\hat{\lambda}-\hat{\lambda}_i v_i+(\hat{\mu}+\hat{\zeta})v^2,\quad \lambda_i=\hat{\lambda}_i-2(\hat{\mu}+\hat{\zeta})v_i,\quad \mu=\hat{\mu},\quad \zeta=\hat{\zeta},$$

which dictate the velocity dependence of the Lagrange multipliers. We notice that these relations are in accordance with the general results, valid for a generic system of balance laws, of the Galilean invariance explained in Sect. 2.7.

The Euler-Lagrange equation $\delta\mathscr{L}/\delta f=0$ leads to the following form of the distribution function:

$$f=\exp\left(-1-\frac{m}{k_B}\chi\right),$$

where

$$\chi=\hat{\lambda}+\hat{\lambda}_i C_i+\hat{\mu}\left(C^2+2\frac{I}{m}\right)+\hat{\zeta}C^2.$$

By introducing the following new multipliers:

$$\xi = \frac{m}{k_B}(\hat{\mu} + \hat{\zeta}), \quad \eta_i = \frac{m}{k_B}\hat{\lambda}_i, \quad \omega = \frac{2}{k_B}\hat{\mu}, \quad \Omega = \exp\left(-1 - \frac{m}{k_B}\hat{\lambda}\right),$$

the distribution function can be rewritten as

$$f = \Omega\, e^{-\omega I}\, e^{-\xi C^2 - \eta_i C_i}. \tag{13.9}$$

Inserting (13.9) into (13.1)$_2$ evaluated at zero velocity, we obtain immediately $\eta_i = 0$. Then the remaining equations of (13.1) and (13.2) evaluated at zero velocity become

$$\begin{aligned}
\rho &= \int_{\mathbb{R}^3}\int_0^\infty m f \varphi(I)\, dI\, d\mathbf{C} = m\pi^{3/2}\Omega \frac{A(\omega)}{\xi^{3/2}}, \\
p + \Pi &= \frac{1}{3}\int_{\mathbb{R}^3}\int_0^\infty m f C^2 \varphi(I)\, dI\, d\mathbf{C} = m\pi^{3/2}\Omega \frac{A(\omega)}{2\xi^{5/2}}, \\
\rho\varepsilon &= \int_{\mathbb{R}^3}\int_0^\infty m f \left(\frac{C^2}{2} + \frac{I}{m}\right)\varphi(I)\, dI\, d\mathbf{C} \\
&= m\pi^{3/2}\Omega \frac{A(\omega)}{4\xi^{5/2}}\left(3 - \frac{4}{m}\xi \frac{A'(\omega)}{A(\omega)}\right),
\end{aligned} \tag{13.10}$$

with

$$A(\omega) = \mathscr{L}_u\left[\varphi(I)\right](\omega) = \int_0^\infty e^{-\omega I}\varphi(I) dI.$$

From (13.10), we can invert the Lagrange multipliers (ξ, ω, Ω) in terms of the physical variables (ρ, T, Π) through the parameter Θ:

$$\begin{aligned}
\xi &= \frac{\rho}{2p}\frac{1}{1+\frac{\Pi}{p}}, \\
\omega &= \frac{1}{k_B\Theta}, \\
\Omega &= \frac{\rho}{mA(\Theta)}\left(\frac{m}{2\pi k_B T}\frac{1}{1+\frac{\Pi}{p}}\right)^{3/2},
\end{aligned} \tag{13.11}$$

where the parameter Θ depends on T and Π through

$$\Pi = \frac{2}{3}\rho\left(\varepsilon_E^I(T) - \varepsilon_E^I(\Theta)\right) \tag{13.12}$$

and $A(\Theta)$ is given by (13.5). Equation $(13.11)_1$ is obtained by the ratio between $(13.10)_1$ and $(13.10)_2$. With the relation $(13.11)_2$, Eq. $(13.11)_3$ is obtained by $(13.10)_1$. While from $(13.10)_3$ we have

$$(\ln A(\omega))' = \frac{m}{2\xi}\{3(p+\Pi) - 2\rho\varepsilon\}.$$

Rewriting (7.16) as (13.5), we obtain immediately the expression (13.12). Inserting (13.11) into the distribution function (13.9), we obtain (13.3).

Concerning the inequalities (13.6), we observe that, from the integrability conditions in (13.10), we have

$$\omega > 0, \qquad \xi > 0. \tag{13.13}$$

Taking into account $(13.11)_1$ and (13.13), we obtain the left inequality of (13.6), while from the fact that ε^I is a strictly increasing function and that (13.13) requires $\Theta > 0$, we can show, from (13.12), that for a fixed T the maximum value of Π/p is obtained when $\Theta \to 0$. Since $\varepsilon^I(0) = 0$, we obtain the right part of the inequality (13.6). QED

When $\Pi \to 0$, it follows that $\Theta \to T$ and that (13.3) reduces to the equilibrium distribution function (7.8).

13.2.1 Closure and Field Equations

Substituting (13.3) into the fluxes and also into the production term, we obtain

$$\begin{aligned}
F_{ik} &= \int_{\mathbb{R}^3}\int_0^\infty m c_i c_k f \varphi(I)\, dI\, d\mathbf{c} = \rho v_i v_k + (p+\Pi)\delta_{ik},\\
F_{llk} &= \int_{\mathbb{R}^3}\int_0^\infty m c^2 c_k f \varphi(I)\, dI\, d\mathbf{c} = \left(5(p+\Pi) + \rho v^2\right) v_k,\\
G_{llk} &= \int_{\mathbb{R}^3}\int_0^\infty m\left(c^2 + \frac{2I}{m}\right) c_k f \varphi(I)\, dI\, d\mathbf{c} = (\rho v^2 + 2\rho\varepsilon + 2p + 2\Pi) v_k,
\end{aligned} \tag{13.14}$$

and

$$P_{ll} = \hat{P}_{ll} = \int_{\mathbb{R}^3}\int_0^\infty m C^2 Q(f)\, \varphi(I)\, dI\, d\mathbf{C}. \tag{13.15}$$

From the balance equations of momentum and of energy in continuum mechanics, we know that

$$F_{ik} = \rho v_i v_k - t_{ik}, \qquad G_{llk} = (\rho v^2 + 2\rho\varepsilon) v_k - 2t_{ik} v_i + 2q_k,$$

where

$$t_{ij} = -p\delta_{ij} + \sigma_{ij} = -(p+\Pi)\delta_{ij} + \sigma_{\langle ij \rangle}.$$

Comparing these with the closure relations (13.14)$_{1,3}$, we conclude that, in the 6-moment theory, $\sigma_{\langle ik \rangle} = 0$ and $q_k = 0$. This is the expected result: no shear stress and no heat flux in the 6-moment theory.

For what concerns (13.14)$_2$, taking into account the Galilean invariance, we obtain the intrinsic part of F_{llk}:

$$\hat{F}_{llk} = \int_{\mathbb{R}^3}\int_0^\infty mC^2 C_k f\varphi(I)\, dI\, d\mathbf{C} = 0,$$

Concerning the production term P_{ll}, the main problem is that, in order to have explicit expression of the production, we need a model for the collision term, which is, in general, not easy to obtain in the case of polyatomic gas. However, in the case of BGK model, we obtain $\hat{P}_{ll} = -3\Pi/\tau$, which is the linearized version of the production term.

With (13.14), the differential system of 6 moments (12.1) becomes

$$\begin{aligned}
&\frac{\partial \rho}{\partial t} + \frac{\partial}{\partial x_i}(\rho v_i) = 0,\\
&\frac{\partial(\rho v_j)}{\partial t} + \frac{\partial}{\partial x_i}\left[(p+\Pi)\delta_{ij} + \rho v_i v_j\right] = 0,\\
&\frac{\partial}{\partial t}(2\rho\varepsilon + \rho v^2) + \frac{\partial}{\partial x_i}\left\{\left[2(p+\Pi) + 2\rho\varepsilon + \rho v^2\right] v_i\right\} = 0,\\
&\frac{\partial}{\partial t}\left[3(p+\Pi) + \rho v^2\right] + \frac{\partial}{\partial x_i}\left\{\left[5(p+\Pi) + \rho v^2\right] v_i\right\} = \hat{P}_{ll}.
\end{aligned} \tag{13.16}$$

In order to obtain the monatomic-gas limit, it is convenient to substitute the last equation in (13.16) with the difference between (13.16)$_4$ and (13.16)$_3$. Then we have the system corresponding to the system (12.37) obtained in the macroscopic theory. Therefore this system with the thermal and caloric equations of state (12.15) is a closed system for the 6 unknown variables (ρ, v_i, T, Π) provided that we know the collision term in (13.15). These results are in perfect agreement with the results derived from the phenomenological theory explained in Chap. 12. The differential system is symmetric hyperbolic, and it satisfies the K-condition. Therefore smooth solutions exist for any time provided that the initial data are sufficiently small.

13.2.2 Entropy Density

Let us study the entropy density h with nonequilibrium distribution function:

$$h = -k_B \int_{\mathbb{R}^3}\int_0^\infty f \log f \varphi(I) \mathrm{d}I\, \mathrm{d}\boldsymbol{C}.$$

Inserting (13.9) into this equation, we obtain the nonequilibrium part of the specific entropy k as a function of T and Z in a parametric form with the nonequilibrium temperature Θ:

$$k = \frac{h - \rho s}{\rho} = \int_T^\Theta \frac{\varepsilon_E^I(x)}{x^2}\,dx + \frac{3}{2}\frac{k_B}{m}\log(1+Z) + \frac{\varepsilon_E^I(\Theta)}{\Theta} - \frac{\varepsilon_E^I(T)}{T},$$

$$Z = \frac{\Pi}{p} = \frac{\varepsilon_E^I(T) - \varepsilon_E^I(\Theta)}{\varepsilon_E^K(T)}.$$

These expressions are exactly the same as those in (12.27).

Therefore, in conclusion, we have the perfect match between the present results and those obtained in the phenomenological RET approach presented in Chap. 12.

13.3 Polytropic Gas

In the polytropic case (7.18), the expressions become the ones obtained in Chap. 12. In particular, the nonequilibrium distribution function (13.3) is expressed as

$$f = \frac{\rho}{m\,(k_B T)^{1+\alpha}\Gamma(1+\alpha)}\left(\frac{m}{2\pi k_B T}\,\frac{1}{1+\frac{\Pi}{p}}\right)^{3/2}\left(\frac{1}{1-\frac{3}{2(1+\alpha)}\frac{\Pi}{p}}\right)^{1+\alpha}$$
$$\exp\left\{-\frac{1}{k_B T}\left(\frac{1}{2}mC^2\left(\frac{1}{1+\frac{\Pi}{p}}\right) + I\left(\frac{1}{1-\frac{3}{2(1+\alpha)}\frac{\Pi}{p}}\right)\right)\right\}.$$

13.4 Nonequilibrium Temperatures

As in Chap. 12, we have seen again in (13.12) that we can express the dynamic pressure in terms of the nonequilibrium temperature Θ. Therefore we can interpret the dynamic pressure as the one caused by the energy exchange which puts the internal modes into the nonequilibrium state with a nonequilibrium temperature Θ from the state with the local equilibrium temperature T. For this reason we may say that Θ is the temperature of the internal modes of a molecule.

Moreover by the analogy of $(12.15)_2$ we can define a nonequilibrium temperature ϑ such that the total pressure $\mathscr{P} = p + \Pi$ has analogous expression of the equilibrium one:

$$\mathscr{P} = p + \Pi = \frac{k_B}{m}\rho\vartheta. \tag{13.17}$$

Therefore from (12.15)$_2$ and (13.17) we can express the dynamic pressure in terms of ϑ:

$$\Pi = \frac{k_B}{m}\rho(\vartheta - T) = p(\rho, \vartheta) - p(\rho, T). \tag{13.18}$$

Comparison between (13.18) and (13.12) gives a relation between the two nonequilibrium temperatures (see (12.68)):

$$\varepsilon_E^I(T) - \varepsilon_E^I(\Theta) = \varepsilon_E^K(\vartheta) - \varepsilon_E^K(T).$$

We observe the two possible meanings of the dynamic pressure. In the momentum equation (12.37)$_2$, Π given by (13.18) is the difference between the pressure corresponding to the nonequilibrium temperature ϑ and the equilibrium one. At the same time, Π appearing in the last Eq. (12.37) has, according to (13.12), the physical meaning of difference of energy due to the internal motion between equilibrium and nonequilibrium. In fact, using the expressions (13.12) and $p = \frac{2}{3}\rho\varepsilon_K$, we can rewrite (12.37)$_4$ as (see (12.71)$_4$)

$$\frac{\partial}{\partial t}\left[2\rho\varepsilon_E^I(\Theta)\right] + \frac{\partial}{\partial x_i}\left[2\rho\varepsilon_E^I(\Theta)v_i\right] = -\hat{P}_{ll}. \tag{13.19}$$

In the case of polytropic gas, for which (7.18) holds, above results of non-polytropic gas reduce to the ones obtained in Chap. 12. In particular, since ε^I is given by

$$\varepsilon_E^I(T) = \frac{D-3}{2}\frac{k_B}{m}T,$$

the relation between the nonequilibrium temperatures (13.4) becomes

$$(D-3)(T-\Theta) = 3(\vartheta - T),$$

while Eq. (13.19) becomes

$$\frac{\partial}{\partial t}(\rho\Theta) + \frac{\partial}{\partial x_i}(\rho\Theta v_i) = -\frac{m\hat{P}_{ll}}{k_B(D-3)}$$

with Π given by the simple relation:

$$\Pi = \frac{D-3}{3}\frac{k_B}{m}\rho(T-\Theta).$$

Chapter 14
Nonlinear ET_7 Theory with Molecular Rotational and Vibrational Modes

Abstract In Chap. 8, we have established the ET_{15} theory of rarefied polyatomic gases where molecular rotational and vibrational modes are taken into account individually. Thereby, compared with the ET_{14} theory, we have obtained a more refined version with a wider applicability range. And, in Chaps. 12 and 13, we have constructed the nonlinear ET_6 theory by ignoring the shear viscosity and heat conductivity. The features of the dynamic pressure can be highlighted by using the ET_6 theory. Similar to the relationship between ET_{14} and ET_{15}, in this chapter, we derive the ET_7 theory of rarefied polyatomic gas with molecular rotational and vibrational modes by ignoring the shear viscosity and heat conductivity as a generalization of the ET_6 theory. In this case, as in ET_{15}, we need a triple hierarchy of the moment system. The system of balance equations is closed via the maximum entropy principle (MEP). Three different types of the production term in the system, which are studied in Chap. 8 on the basis of the generalized BGK model for the collision term in the Boltzmann equation, are also adopted. Some characteristic features of the ET_7 theory are discussed.

14.1 RET Theory with Seven Independent Fields: ET_7

The simplest system of (11.3) with a triple hierarchy next to the Euler system is the system with seven independent fields (ET_7):

$$
\begin{aligned}
&\text{mass density:} && F = \rho, \\
&\text{momentum density:} && F_i = \rho v_i, \\
&\text{translational energy density:} && F_{ll} = 2\rho\varepsilon^K + \rho v^2, \\
&\text{rotational energy density:} && H_{ll}^R = 2\rho\varepsilon^R, \\
&\text{vibrational energy density:} && H_{ll}^V = 2\rho\varepsilon^V.
\end{aligned}
\tag{14.1}
$$

T. Ruggeri, M. Sugiyama, *Classical and Relativistic Rational Extended Thermodynamics of Gases*, https://doi.org/10.1007/978-3-030-59144-1_14

By ignoring the dissipation due to the shear stress and heat flux, the ET$_7$ theory focuses on the description of the internal relaxation processes in a molecule.

In this section, by means of the MEP closure, we derive the nonequilibrium distribution function and the closed system of field equations [135]. We assume the generalized BGK model of the collision term in the Boltzmann equation, and derive the explicit expression of the production terms. We use the same notation as in Chap. 8.

14.1.1 System of Balance Equations

From (11.3), the system of balance equations is expressed as follows:

$$\begin{aligned}
&\frac{\partial F}{\partial t}+\frac{\partial F_i}{\partial x_i}=0,\\
&\frac{\partial F_j}{\partial t}+\frac{\partial F_{ij}}{\partial x_i}=0,\\
&\frac{\partial F_{ll}}{\partial t}+\frac{\partial F_{lli}}{\partial x_i}=P^K_{ll}, \qquad \frac{\partial H^R_{ll}}{\partial t}+\frac{\partial H^R_{lli}}{\partial x_i}=P^R_{ll}, \qquad \frac{\partial H^V_{ll}}{\partial t}+\frac{\partial H^V_{lli}}{\partial x_i}=P^V_{ll},
\end{aligned} \tag{14.2}$$

where $(F_{lli}, H^R_{lli}, H^V_{lli})$ and $(P^K_{ll}, P^R_{ll}, P^V_{ll})$ are, respectively, the fluxes and productions of the densities $(F_{ll}, H^R_{ll}, H^V_{ll})$. It is easy to verify that the production terms are velocity independent.

14.1.2 Nonequilibrium Distribution Function

First of all, we start with the following statement:

Theorem 14.1 *The nonequilibrium distribution function $f^{(7)}$ for the truncated system* (14.2) *obtained by using the MEP is given by*

$$f^{(7)}=\frac{\rho}{mA^R(\theta^R)A^V(\theta^V)}\left(\frac{m}{2\pi k_B\theta^K}\right)^{3/2}\exp\left(-\frac{mC^2}{2k_B\theta^K}-\frac{I^R}{k_B\theta^R}-\frac{I^V}{k_B\theta^V}\right), \tag{14.3}$$

where $A^R(\theta^R)$ and $A^V(\theta^V)$ are normalization factors given by (8.5). *Nonequilibrium temperatures θ^K, θ^R, and θ^V of K, R, and V-modes are determined through* *the relations:*

$$\varepsilon^K = \varepsilon_E^K(\theta^K), \quad \varepsilon^R = \varepsilon_E^R(\theta^R), \quad \varepsilon^V = \varepsilon_E^V(\theta^V).$$

Proof Let us introduce the Lagrange multipliers $\{\lambda, \lambda_i, \mu^K(\equiv \lambda_{ll}/3), \mu^R, \mu^V\}$ that correspond to the densities $\{F, F_i, F_{ll}, H_{ll}^R, H_{ll}^V\}$. In the present case, the velocity dependence of the Lagrange multipliers (11.6) is explicitly expressed as follows:

$$\lambda = \hat{\lambda} - \hat{\lambda}_i v_i + \hat{\mu}^K v^2, \quad \lambda_i = \hat{\lambda}_i - 2\hat{\mu}^K v_i,$$
$$\mu^K = \hat{\mu}^K, \quad \mu^R = \hat{\mu}^R, \quad \mu^V = \hat{\mu}^V.$$

From (11.5), it is possible to express the distribution function of the truncated system (14.2) as follows

$$f^{(7)} = \Omega \mathrm{e}^{-\eta_i C_i} \mathrm{e}^{-\beta^K \frac{mC^2}{2}} \mathrm{e}^{-\beta^R I^R} \mathrm{e}^{-\beta^V I^V}, \tag{14.4}$$

where

$$\Omega = \exp\left(-1 - \frac{m}{k_B}\hat{\lambda}\right), \quad \eta_i = \frac{m}{k_B}\hat{\lambda}_i,$$
$$\beta^K = \frac{2}{k_B}\hat{\mu}^K, \quad \beta^R = \frac{2}{k_B}\hat{\mu}^R, \quad \beta^V = \frac{2}{k_B}\hat{\mu}^V.$$

In addition, we introduce the following three parameters θ^K, θ^R, and θ^V through β^K, β^R, and β^V as follows:

$$\theta^K = \frac{1}{k_B\beta^K}, \quad \theta^R = \frac{1}{k_B\beta^R}, \quad \theta^V = \frac{1}{k_B\beta^V}.$$

Recalling (8.5) and substituting (14.4) into (14.1) (see also (8.2) and (8.14)) evaluated at zero velocity, we obtain $\eta_i = 0$ and

$$\rho = m\left(\frac{2\pi k_B \theta^K}{m}\right)^{3/2} A^R(\theta^R) A^V(\theta^V)\Omega,$$
$$\varepsilon_E^K(\theta^K) = \frac{3}{2}\frac{k_B}{m}\theta^K,$$
$$\varepsilon_E^R(\theta^R) = \frac{k_B}{m}\theta^{R2}\frac{\mathrm{d}\log A^R(\theta^R)}{\mathrm{d}\theta^R},$$
$$\varepsilon_E^V(\theta^V) = \frac{k_B}{m}\theta^{V2}\frac{\mathrm{d}\log A^V(\theta^V)}{\mathrm{d}\theta^V}.$$

These indicate that θ^K, θ^R, and θ^V are the nonequilibrium temperatures of K, R, and V-modes, respectively. Then Ω, β^K, β^R and β^V are expressed in terms of ρ, θ^K, θ^R, and θ^V as follows:

$$\Omega = \frac{\rho}{mA^R(\theta^R)A^V(\theta^V)}\left(\frac{m}{2\pi k_B \theta^K}\right)^{3/2},$$

$$\beta^K = \frac{1}{k_B\theta^K}, \quad \beta^R = \frac{1}{k_B\theta^R}, \quad \beta^V = \frac{1}{k_B\theta^V},$$

and we finally obtain the nonequilibrium distribution function (14.3).

The intrinsic Lagrange multipliers are expressed in terms of the independent fields as follows:

$$\hat{\lambda} = -\frac{k_B}{m}(1+\log\Omega), \quad \hat{\mu}^K = \frac{1}{2\theta^K}, \quad \hat{\mu}^R = \frac{1}{2\theta^R}, \quad \hat{\mu}^V = \frac{1}{2\theta^V}.$$

Recalling (8.11), we obtain the following relations:

$$\begin{aligned}
-\frac{k_B}{m} + \frac{k_B}{m}\log\left[\frac{m}{\rho}\left(\frac{2\pi k_B\theta^K}{m}\right)^{3/2}\right] &= -\frac{p(\rho,\theta^K)}{\rho\theta^K} + s^K(\rho,\theta^K) - \frac{\varepsilon_E^K(\theta^K)}{\theta^K} \\
&= -\frac{g^K(\rho,\theta^K)}{\theta^K}, \\
\frac{k_B}{m}\log A^R(\theta^R) = s^R(\theta^R) - \frac{\varepsilon_E^R(\theta^R)}{\theta^R} &= -\frac{g^R(\theta^R)}{\theta^R}, \\
\frac{k_B}{m}\log A^V(\theta^V) = s^R(\theta^V) - \frac{\varepsilon_E^V(\theta^V)}{\theta^V} &= -\frac{g^V(\theta^V)}{\theta^V}.
\end{aligned}$$

Therefore, we obtain the expressions of the Lagrange multipliers in terms of $(\rho, v_i, \theta^K, \theta^R, \theta^V)$:

$$\begin{aligned}
&\lambda = -\frac{g^K(\rho,\theta^K)}{\theta^K} - \frac{g^R(\theta^R)}{\theta^R} - \frac{g^V(\theta^V)}{\theta^V} + \frac{v^2}{2\theta^K}, \quad \lambda_i = -\frac{v_i}{\theta^K}, \\
&\mu^K = \frac{1}{2\theta^K}, \quad \mu^R = \frac{1}{2\theta^R}, \quad \mu^V = \frac{1}{2\theta^V}.
\end{aligned} \tag{14.5}$$

where $g^K(\rho,\theta^K)$, $g^R(\theta^R)$, and $g^V(\theta^V)$ are the nonequilibrium chemical potentials of the three modes.

Remark 14.1 From (14.3), we notice that, within ET$_7$, any nonequilibrium state can be identified by assigning the nonequilibrium temperatures θ^K, θ^R, and θ^V together with ρ and v_i. In other words, ET$_7$ adopts the approximation that K, R, and V-modes are always in equilibrium but, in general, with different temperatures from each other. Therefore ET$_7$ does not take into account the relaxation (i) with the relaxation time τ_K (see Section 8.5.1). See also Fig. 8.1 and Section 14.1.5 below.

Remark 14.2 In the context of RET, the nonequilibrium temperature is defined through the Lagrange multiplier corresponding to the conservation law of energy. Indeed the expression of the Lagrange multiplies of ET$_7$, (14.5), ensures the present definition of the nonequilibrium temperatures θ^K, θ^R and θ^V. For details, see Chap. 15.

14.1.3 Closed System of Field Equations

Using the distribution function (14.3), we obtain the constitutive equations for the fluxes as follows:

$$\begin{aligned}
F_{ij} &= \int_{\mathbb{R}^3} \int_0^\infty \int_0^\infty m c_i c_j f^{(7)} \varphi\left(I^R\right) \psi\left(I^V\right) dI^R dI^V d\mathbf{c} \\
&= p(\rho, \theta^K)\delta_{ij} + \rho v_i v_j, \\
F_{lli} &= \int_{\mathbb{R}^3} \int_0^\infty \int_0^\infty m c^2 c_i f^{(7)} \varphi\left(I^R\right) \psi\left(I^V\right) dI^R dI^V d\mathbf{c} \\
&= \left\{2\rho\varepsilon_E^K(\theta^K) + 2p(\rho, \theta^K) + \rho v^2\right\} v_i, \\
H_{lli}^R &= \int_{\mathbb{R}^3} \int_0^\infty \int_0^\infty 2 c_i I^R f^{(7)} \varphi\left(I^R\right) \psi\left(I^V\right) dI^R dI^V d\mathbf{c} \\
&= 2\rho\varepsilon_E^R(\theta^R) v_i, \\
H_{lli}^V &= \int_{\mathbb{R}^3} \int_0^\infty \int_0^\infty 2 c_i I^V f^{(7)} \varphi\left(I^R\right) \psi\left(I^V\right) dI^R dI^V d\mathbf{c} \\
&= 2\rho\varepsilon_E^V(\theta^V) v_i.
\end{aligned} \tag{14.6}$$

We notice that the velocity-independent parts of F_{lli}, H_{lli}^R and H_{lli}^V vanish.

The trace part of the momentum flux F_{ll} is related to the pressure p and the dynamic pressure Π in continuum mechanics as follows:

$$F_{ll} = 3(p + \Pi) + \rho v^2.$$

Comparing this relation with (14.6)$_2$, we notice that Π is given by

$$\Pi = p(\rho, \theta^K) - p(\rho, T),$$

or, from (8.10), it is given by

$$\Pi = \frac{2}{3}\rho\left(\varepsilon_E^K(\theta^K) - \varepsilon_E^K(T)\right).$$

Therefore, as was explained in the previous chapters, the dynamic pressure is related to the energy exchange among the modes.

Using the constitutive equations (14.6), we obtain the closed system of field equations for the independent seven fields, ρ, v_i, θ^K, θ^R, θ^V (the equations of state are given by (8.10) with $(8.9)_1$ and (8.8)):

$$\begin{aligned}
&\frac{\partial \rho}{\partial t} + \frac{\partial}{\partial x_i}(\rho v_i) = 0,\\
&\frac{\partial \rho v_j}{\partial t} + \frac{\partial}{\partial x_i}\left\{p(\rho,\theta^K)\delta_{ij} + \rho v_i v_j\right\} = 0,\\
&\frac{\partial}{\partial t}\left\{2\rho\varepsilon_E^K(\theta^K) + \rho v^2\right\}\\
&\qquad + \frac{\partial}{\partial x_i}\left\{\left(2\rho\varepsilon_E^K(\theta^K) + \rho v^2 + 2p(\rho,\theta^K)\right) v_i\right\} = P_{ll}^K,\\
&\frac{\partial}{\partial t}\left\{2\rho\varepsilon_E^R(\theta^R)\right\} + \frac{\partial}{\partial x_i}\left\{2\rho\varepsilon_E^R(\theta^R) v_i\right\} = P_{ll}^R,\\
&\frac{\partial}{\partial t}\left\{2\rho\varepsilon_E^V(\theta^V)\right\} + \frac{\partial}{\partial x_i}\left\{2\rho\varepsilon_E^V(\theta^V) v_i\right\} = P_{ll}^V,
\end{aligned} \tag{14.7}$$

where the production terms will be analyzed in Sect. 14.1.5.

By using the material derivative, the system (14.7) is rewritten as

$$\begin{aligned}
&\dot\rho + \rho\frac{\partial v_i}{\partial x_i} = 0,\\
&\rho \dot v_i + \frac{\partial p(\rho,\theta^K)}{\partial x_i} = 0,\\
&\dot\varepsilon_E^K(\theta^K) + \frac{p(\rho,\theta^K)}{\rho}\frac{\partial v_k}{\partial x_k} = \frac{P_{ll}^K}{2\rho},\\
&\dot\varepsilon_E^R(\theta^R) = \frac{P_{ll}^R}{2\rho},\\
&\dot\varepsilon_E^V(\theta^V) = \frac{P_{ll}^V}{2\rho}.
\end{aligned} \tag{14.8}$$

14.1.4 Entropy Density and Production

The nonequilibrium specific entropy density $\eta = h/\rho$ for the truncated system (14.2) is obtained from (1.24) as follows:

$$\eta = s^K(\rho, \theta^K) + s^R(\theta^R) + s^V(\theta^V), \tag{14.9}$$

where $s^K(\rho, \theta^K)$, $s^R(\theta^R)$, and $s^V(\theta^V)$ are expressed as in (8.11). From (14.9) with (8.12), we obtain the extension of the Gibbs relation in nonequilibrium as follows:

$$\mathrm{d}\eta = \frac{1}{\theta^K}\left(\mathrm{d}\varepsilon^K - \frac{p(\rho, \theta^K)}{\rho^2}\mathrm{d}\rho\right) + \frac{1}{\theta^R}\mathrm{d}\varepsilon^R + \frac{1}{\theta^V}\mathrm{d}\varepsilon^V. \tag{14.10}$$

In the present case, the non-convective part of the entropy flux is zero. Therefore we have

$$h_i = hv_i.$$

Then the balance law of the entropy density is written as follows:

$$\rho\dot{\eta} = \Sigma, \tag{14.11}$$

where, from (11.7)$_3$ with (14.5), we obtain the entropy production:

$$\Sigma = \frac{P^K_{ll}}{2\theta^K} + \frac{P^R_{ll}}{2\theta^R} + \frac{P^V_{ll}}{2\theta^V} \geqslant 0.$$

14.1.5 Production Terms in the Generalized BGK Model

Three different types of the production term in the system, which are studied in Chap. 8 on the basis of the generalized BGK model for the collision term in the Boltzmann equation, are also adopted in this section. For details of definition and notation, see Sect. 8.5.

In the present case, the relaxation time τ_K plays no role in the production term (see also the Remark 14.1 above). By using the collision term (8.30) with $f = f_K^{(E)}$, the production terms are given explicitly as follows:

- Process (KR):

$$
\begin{aligned}
P_{ll}^K &= -\frac{2\rho}{\tau_{KR}}\left(\varepsilon_E^K(\theta^K) - \varepsilon_E^K(\theta^{KR})\right) - \frac{2\rho}{\tau}\left(\varepsilon_E^K(\theta^K) - \varepsilon_E^K(T)\right), \\
P_{ll}^R &= -\frac{2\rho}{\tau_{KR}}\left(\varepsilon_E^R(\theta^R) - \varepsilon_E^R(\theta^{KR})\right) - \frac{2\rho}{\tau}\left(\varepsilon_E^R(\theta^R) - \varepsilon_E^R(T)\right), \\
P_{ll}^V &= -\frac{2\rho}{\tau}\left(\varepsilon_E^V(\theta^V) - \varepsilon_E^V(T)\right),
\end{aligned}
\tag{14.12}
$$

where, from (8.48) and (8.35), θ^{KR} is determined by

$$\varepsilon_E^{K+R}(\theta^{KR}) = \varepsilon_E^K(\theta^K) + \varepsilon_E^R(\theta^R),$$

and, from (8.40), T is determined by

$$\varepsilon_E(T) = \varepsilon_E^K(\theta^K) + \varepsilon_E^R(\theta^R) + \varepsilon_E^V(\theta^V). \tag{14.13}$$

- Process (KV):

$$
\begin{aligned}
P_{ll}^K &= -\frac{2\rho}{\tau_{KV}}\left(\varepsilon_E^K(\theta^K) - \varepsilon_E^K(\theta^{KV})\right) - \frac{2\rho}{\tau}\left(\varepsilon_E^K(\theta^K) - \varepsilon_E^K(T)\right), \\
P_{ll}^R &= -\frac{2\rho}{\tau}\left(\varepsilon_E^R(\theta^R) - \varepsilon_E^R(T)\right), \\
P_{ll}^V &= -\frac{2\rho}{\tau_{KV}}\left(\varepsilon_E^V(\theta^V) - \varepsilon_E^V(\theta^{KV})\right) - \frac{2\rho}{\tau}\left(\varepsilon_E^V(\theta^V) - \varepsilon_E^V(T)\right),
\end{aligned}
$$

where, from (8.50) and (8.37), θ^{KV} is determined by

$$\varepsilon_E^{K+V}(\theta^{KV}) = \varepsilon_E^K(\theta^K) + \varepsilon_E^V(\theta^V),$$

and T is determined by (14.13).

- Process (RV):

$$
\begin{aligned}
P_{ll}^K &= -\frac{2\rho}{\tau}\left(\varepsilon_E^K(\theta^K) - \varepsilon_E^K(T)\right), \\
P_{ll}^R &= -\frac{2\rho}{\tau_{RV}}\left(\varepsilon_E^R(\theta^R) - \varepsilon_E^R(\theta^{RV})\right) - \frac{2\rho}{\tau}\left(\varepsilon_E^R(\theta^R) - \varepsilon_E^R(T)\right), \\
P_{ll}^V &= -\frac{2\rho}{\tau_{RV}}\left(\varepsilon_E^V(\theta^V) - \varepsilon_E^V(\theta^{RV})\right) - \frac{2\rho}{\tau}\left(\varepsilon_E^V(\theta^V) - \varepsilon_E^V(T)\right),
\end{aligned}
$$

where, from (8.51) and (8.39), θ^{RV} is determined by

$$\varepsilon_E^{R+V}(\theta^{RV}) = \varepsilon_E^R(\theta^R) + \varepsilon_E^V(\theta^V),$$

and T is determined by (14.13).

14.2 Characteristic Features of ET$_7$

We summarize some features of the ET$_7$ theory.

14.2.1 Comparison with the Meixner Theory

The system of field equations of the Meixner theory [129, 130] with two internal variables $\xi^{(1)}$ and $\xi^{(2)}$ is expressed as follows:

$$\begin{aligned}
&\dot{\rho} + \rho\frac{\partial v_i}{\partial x_i} = 0,\\
&\rho\dot{v}_i + \frac{\partial \mathscr{P}}{\partial x_i} = 0,\\
&\rho\dot{\mathscr{E}} + \mathscr{P}\frac{\partial v_k}{\partial x_k} = 0,\\
&\dot{\xi}^{(1)} = -\beta^{(1)}\mathscr{A}^{(1)},\\
&\dot{\xi}^{(2)} = -\beta^{(2)}\mathscr{A}^{(2)},
\end{aligned} \tag{14.14}$$

where $\mathscr{P}$, $\mathscr{E}$, and $\mathscr{A}^{(a)} (a = 1, 2)$ are, respectively, the pressure, the specific internal energy, and the affinities of the relaxation processes, and $\beta^{(a)}$ are positive phenomenological coefficients. The generalized Gibbs relation in the Meixner theory is assumed to be

$$\mathscr{T}\mathrm{d}\mathscr{S} = \mathrm{d}\mathscr{E} - \frac{\mathscr{P}}{\rho^2}\mathrm{d}\rho - \sum_{a=1}^{2}\mathscr{A}^{(a)}\mathrm{d}\xi^{(a)}, \tag{14.15}$$

where $\mathscr{T}$ is the temperature and $\mathscr{S}$ is the specific entropy. The quantities $\mathscr{T}$, $\mathscr{S}$, $\mathscr{P}$ and $\mathscr{A}$ depend not only on the mass density ρ and the specific internal energy $\mathscr{E}$ but also on the internal variables $\xi^{(1)}$ and $\xi^{(2)}$. From (14.15), with the use of (14.14), we obtain

$$\dot{\mathscr{S}} = \frac{1}{\mathscr{T}}\sum_{a=1}^{2}\beta^{(a)}\mathscr{A}^{(a)2}. \tag{14.16}$$

Comparing the system of the ET$_7$ theory: (14.8), (14.10), and (14.11) with the system of the Meixner theory: (14.14), (14.15), and (14.16), we have the following relationship between the Meixner theory and the ET_7 theory:

$$\xi^{(1)} = \varepsilon_E^R(\theta^R), \quad \xi^{(2)} = \varepsilon_E^V(\theta^V),$$

$$\mathscr{P} = p(\rho, \theta^K), \quad \mathscr{E} = \varepsilon_E(T), \quad \mathscr{S} = \eta(\rho, \theta^K, \theta^R, \theta^V), \quad \mathscr{T} = \theta^K,$$

$$\mathscr{A}^{(1)} = -\theta^K\left(\frac{1}{\theta^R} - \frac{1}{\theta^K}\right), \quad \mathscr{A}^{(2)} = -\theta^K\left(\frac{1}{\theta^V} - \frac{1}{\theta^K}\right),$$

$$\beta^{(1)} = \frac{1}{2\rho\theta^K}\left(\frac{1}{\theta^R} - \frac{1}{\theta^K}\right)^{-1} P_{ll}^R, \quad \beta^{(2)} = \frac{1}{2\rho\theta^K}\left(\frac{1}{\theta^V} - \frac{1}{\theta^K}\right)^{-1} P_{ll}^V.$$

To sum up, we have identified the quantities in the Meixner theory in terms of the more understandable quantities of ET_7. In particular, the nonequilibrium temperature $\mathscr{T}$ is recognized as the temperature of the translational mode θ^K (see (12.70) in the case of ET_6). This is reasonable because, from (14.15), $\mathscr{T}(=\theta^K)$ is the temperature of a state in equilibrium under a constraint that the system is kept at fixed values of $\xi^{(1)}$ and $\xi^{(2)}$, that is, at the fixed values of $\varepsilon_E^R(\theta^R)$ and $\varepsilon_E^V(\theta^V)$.

14.2.2 *Characteristic Velocity, Subcharacteristic Conditions, and Local Exceptionality*

It is well known that the characteristic velocity V associated with a hyperbolic system of equations can be obtained by using the operator chain rule (see (3.27)):

$$\frac{\partial}{\partial t} \to -V\delta, \quad \frac{\partial}{\partial x_i} \to n_i\delta, \quad \mathbf{f} \to 0,$$

where n_i denotes the i-component of the unit normal to the wave front, $\mathbf{f}$ is the production terms and δ is a differential operator.

In the present case, if we choose $\{\rho, v_i, \eta, \theta^R, \theta^V\}$ as independent variables instead of $\{\rho, v_i, \theta^K, \theta^R, \theta^V\}$, and adopt the entropy law (14.11) instead of the energy equation of the K-mode in $(14.8)_3$, we obtain

$$\textbf{ContactWaves}: \quad V = v_n = 0, \tag{14.17}$$

(multiplicity 5)

$$\textbf{SoundWaves}: \quad V = v_n \pm \sqrt{\left(\frac{\partial p(\rho, \theta^K(\rho, \eta, \theta^R, \theta^V))}{\partial \rho}\right)_{\eta, \theta^R, \theta^V}} \tag{14.18}$$

(each of multiplicity 1),

where $v_n = v_j n_j$. We can rewrite the velocity of the sound wave $U = V - v_n$ as follows:

$$U^2 = p_\rho(\rho, \theta^K) + \frac{\theta^K p_{\theta^K}^2(\rho, \theta^K)}{\rho^2 c_v^K(\theta^K)},$$

where a subscript attached to p indicates a partial derivative and c_v^K is the specific heat of the translational mode defined by $c_v^K(T) = \mathrm{d}\varepsilon^K(T)/\mathrm{d}T$. In an equilibrium case, we have

$$U_E^2 = p_\rho(\rho, T) + \frac{T p_T^2(\rho, T)}{\rho^2 c_v^K(T)}.$$

The sound velocity of the Euler fluid is given by

$$U_{Euler}^2 = p_\rho(\rho, T) + \frac{T p_T^2(\rho, T)}{\rho^2 c_v(T)}, \tag{14.19}$$

where c_v is the specific heat defined by $c_v(T) = \mathrm{d}\varepsilon(T)/\mathrm{d}T$ and

$$c_v = c_v^K + c_v^R + c_v^V$$

with the specific heat of the rotational mode c_v^R and the vibrational mode c_v^V: $c_v^R(T) = \mathrm{d}\varepsilon_E^R(T)/\mathrm{d}T$ and $c_v^V(T) = \mathrm{d}\varepsilon_E^V(T)/\mathrm{d}T$. Since the specific heats of the three modes are positive, we notice that the subcharacteristic condition (see Theorem 2.3) [165] is satisfied:

$$U_E > U_{Euler}.$$

A characteristic velocity associated with a wave is classified as (see (3.14), (3.15), and (3.16)): *genuinely non-linear* if $\delta V = \nabla_{\mathbf{u}} V \cdot \delta \mathbf{u} \propto \nabla_{\mathbf{u}} V \cdot \mathbf{r} \neq 0, \ \forall \mathbf{u}$; *linearly degenerate or exceptional* if $\delta V \equiv 0, \ \forall \mathbf{u}$; *locally linearly degenerate or locally exceptional* if $\delta V = 0$, for some $\mathbf{u}$, where $\mathbf{r}$ is the corresponding eigenvector associated with the system (14.2). The contact waves (14.17) are exceptional while the sound waves (14.18) can be locally exceptional if the condition is satisfied. Simple algebra similar to the one in [206] gives the fact that, if the hyper-surface of local exceptionality exists, the following relation is satisfied on it:

$$\delta V = \frac{1}{2\rho^2 U}\left(\frac{\partial \rho^2 U^2}{\partial \rho}\right)_{\eta, \theta^R, \theta^V} = 0.$$

The results obtained here will be useful in the analysis of nonlinear waves such as shock waves.

14.2.3 ET$_6$ Theories as the Principal Subsystems of the ET$_7$ Theory

Let us consider the ($\mathfrak{bc}$)-process (($\mathfrak{bc}$) = (KR), (KV), (RV)) defined in (8.30) again, and assume that the relaxation time τ is of several orders larger than the relaxation time $\tau_{\mathfrak{bc}}$. In such a case, the composite system of $\mathfrak{b}$-mode and $\mathfrak{c}$-mode quickly reaches a state with the common temperature $\theta^{\mathfrak{bc}}$. Therefore, except for the short period of $O(\tau_{\mathfrak{bc}})$ after the initial time, we have the relation:

$$\theta^{\mathfrak{b}} = \theta^{\mathfrak{c}} = \theta^{\mathfrak{bc}}.$$

As the balance equation of the density $\left(\varepsilon_E^{\mathfrak{b}}(\theta^{\mathfrak{b}}) - \varepsilon_E^{\mathfrak{c}}(\theta^{\mathfrak{c}})\right)$ is identically satisfied in the present approximation, the remaining equations are given by

$$\begin{aligned}
&\frac{\partial \rho}{\partial t} + \frac{\partial}{\partial x_i}(\rho v_i) = 0,\\
&\frac{\partial \rho v_j}{\partial t} + \frac{\partial}{\partial x_i}\left\{(p+\Pi)\delta_{ij} + \rho v_i v_j\right\} = 0,\\
&\frac{\partial}{\partial t}\left\{2\rho\varepsilon + \rho v^2\right\} + \frac{\partial}{\partial x_i}\left\{\left(2\rho\varepsilon + 2p + 2\Pi + \rho v^2\right) v_i\right\} = 0,\\
&\frac{\partial(2\rho\mathscr{E})}{\partial t} + \frac{\partial(2\rho\mathscr{E}_i)}{\partial x_i} = P_{\mathscr{E}},
\end{aligned}$$

where $\mathscr{E}$ is the nonequilibrium energy density characterizing the relaxation process, and $\mathscr{E}_i$ and $P_{\mathscr{E}}$ are its flux and production. We may regard this system as the RET theory with 6 fields, which we call ET$_6^{\mathfrak{bc}}$. In Table 14.1, three possible ET$_6$ theories corresponding to the types of the relaxation process are summarized.

The above argument can be rigorously formulated by using the idea of the principal subsystem (see Sect. 2.4) [25]. In the present case, ET$_6$ is the principal subsystem of ET$_7$. The crucial point is that, all the universal principles of continuum thermomechanics—objectivity, entropy, and causality principles—are automatically preserved also in the subsystem.

Table 14.1 Three possible ET$_6$ theories

	Process	$(\mathfrak{a}, \mathfrak{b}, \mathfrak{c})$	$p+\Pi$	$\mathscr{E}$	$\mathscr{E}_i$	$P_{\mathscr{E}}$
ET$_6^{KR}$	(KR)	(V, K, R)	$p(\rho, \theta^{KR})$	$\varepsilon_E^V(\theta^V)$	$\varepsilon_E^V(\theta^V)v_i$	P_{ll}^V
ET$_6^{KV}$	(KV)	(R, K, V)	$p(\rho, \theta^{KV})$	$\varepsilon_E^R(\theta^R)$	$\varepsilon_E^R(\theta^R)v_i$	P_{ll}^R
ET$_6^{RV}$	(RV)	(K, R, V)	$p(\rho, \theta^{K})$	$\varepsilon_E^{RV}(\theta^{RV})$	$\varepsilon_E^{RV}(\theta^{RV})v_i$	$P_{ll}^R + P_{ll}^V$

The characteristic velocity of ET_6^{RV} is obtained as

$$U^{RV^2} = p_\rho(\rho, \theta^K) + \frac{\theta^K p_{\theta^K}^2(\rho, \theta^K)}{\rho^2 c_v^K(\theta^K)},$$

which is the same as the characteristic velocity of ET_7: $U^{RV} = U$. On the other hand, for $ET_6^{\mathfrak{bc}}$ with $(\mathfrak{b}, \mathfrak{c}) = (K, R)$ or (K, V), we obtain

$$U^{\mathfrak{bc}^2} = p_\rho(\rho, \theta^{\mathfrak{bc}}) + \frac{\theta^{\mathfrak{bc}} p_{\theta^{\mathfrak{bc}}}^2(\rho, \theta^{\mathfrak{bc}})}{\rho^2 c_v^{\mathfrak{b}+\mathfrak{c}}(\theta^{\mathfrak{bc}})},$$

where $c_v^{\mathfrak{b}+\mathfrak{c}} = c_v^{\mathfrak{b}} + c_v^{\mathfrak{c}}$. Since $c_v > c_v^{\mathfrak{b}+\mathfrak{c}} > c_v^K$, we have the following relation in equilibrium:

$$U_E > U_E^{\mathfrak{bc}} > U_{Euler}.$$

Remark 14.3 The ET_6 theory studied in Chaps. 12 and 13 directly corresponds to the ET_6^{RV} theory in the present notation. However, it should be noted that the previous ET_6 theory may also correspond to the ET_7 theories with (KR) and (KV)-processes as far as the V-mode is kept in the ground state and has no role in the phenomena under study.

14.2.4 Near Equilibrium Case

In the $(\mathfrak{bc})$-process $((\mathfrak{bc}) = (KR), (KV), (RV))$, energy exchanges among $\mathfrak{a}$, $\mathfrak{b}$, and $\mathfrak{c}$-modes are characterized by the quantities δ and Δ defined in Section 8.9. Proceeding in the same manner as in Section 8.8, we have the expressions of the production terms near equilibrium that are the same as (8.41).

The entropy production is given by

$$\Sigma = \frac{\rho}{T^2}\frac{c_v^{\mathfrak{b}+\mathfrak{c}}}{c_v^{\mathfrak{b}} c_v^{\mathfrak{c}}}\frac{1}{\tau_\delta}\delta^2 + \frac{\rho}{T^2}\frac{c_v}{c_v^{\mathfrak{a}} c_v^{\mathfrak{b}+\mathfrak{c}}}\frac{1}{\tau}\Delta^2.$$

Since $c_v^K > 0$, $c_v^R > 0$ and $c_v^V > 0$, and $\tau_\delta > 0$, $\tau > 0$, the entropy production is non-negative.

The system of field equations (14.8) is rewritten as follows:

$$\begin{aligned}
&\dot{\rho} + \rho \frac{\partial v_i}{\partial x_i} = 0, \\
&\rho \dot{v}_i + \frac{\partial}{\partial x_i}(p + \Pi) = 0, \\
&\rho c_v \dot{T} + (p + \Pi) \frac{\partial v_i}{\partial x_i} = 0, \\
&\dot{\delta} + \frac{p + \Pi}{\rho} \left\{ A_1 + \frac{1}{c_v} \frac{\mathrm{d}}{\mathrm{d}T} \left(\frac{c_v^{\mathfrak{b}}}{c_v^{\mathfrak{b}+\mathfrak{c}}} \right) \Delta \right\} \frac{\partial v_i}{\partial x_i} = -\frac{\delta}{\tau_\delta}, \\
&\dot{\Delta} + \frac{p + \Pi}{\rho} \frac{A_2 - c_v^{\mathfrak{a}}}{c_v} \frac{\partial v_i}{\partial x_i} = -\frac{\Delta}{\tau},
\end{aligned} \tag{14.20}$$

where $p = p(\rho, T)$, and A_1, A_2 and Π are given in Table 8.3. In the limit $\tau_\delta \to 0$, this system reduces to the system of ET$_6^{\mathfrak{bc}}$.

Applying the Maxwellian iteration to $(14.20)_{4,5}$ and retain the first order terms with respect to the relaxation times τ_δ and τ, we obtain the expressions of the bulk viscosity. For details, see [136] and Section 8.8.

14.2.5 Homogeneous Solution and Relaxation of Nonequilibrium Temperatures

In order to focus our attention on the behavior of the internal molecular relaxation processes, we study first a simple case: homogeneous solutions of the system (14.8), i.e., solutions in which the unknowns are independent of space coordinates and depend only on the time t. The system (14.8) reduces now to an ODE system:

$$\begin{aligned}
&\frac{d\rho}{dt} = 0, \\
&\frac{d\mathbf{v}}{dt} = 0, \\
&\frac{d\varepsilon_E^K(\theta^K)}{dt} = \frac{P_{ll}^K}{2\rho}, \\
&\frac{d\varepsilon_E^R(\theta^R)}{dt} = \frac{P_{ll}^R}{2\rho}, \\
&\frac{d\varepsilon_E^V(\theta^V)}{dt} = \frac{P_{ll}^V}{2\rho}.
\end{aligned} \tag{14.21}$$

From the first two equations, we notice that ρ and $\mathbf{v}$ are constant. Due to the Galilean invariance we can assume without any loss of generality that $\mathbf{v} = 0$. Moreover, from (14.13), summing the last three equations of (14.21), and taking into account that the sum of the productions is zero and that $\varepsilon(T)$ is a monotonous function, we conclude that also T is constant. Therefore there remain only the last three equations of (14.21) that govern the relaxation of the nonequilibrium temperatures.

For simplicity, we assume a process near equilibrium and then consider a linearized version. Taking into account (14.12), we obtain the following linear ODE system:

$$\begin{aligned}
\frac{d\bar{\theta}^{\mathrm{a}}}{dt} &= -\frac{1}{\tau}\bar{\theta}^{\mathrm{a}}, \\
\frac{d\bar{\theta}^{\mathrm{b}}}{dt} &= -\frac{1}{\tau}\bar{\theta}^{\mathrm{b}} - \frac{1}{\tau_{\mathrm{bc}}}(\bar{\theta}^{\mathrm{b}} - \bar{\theta}^{\mathrm{bc}}), \\
\frac{d\bar{\theta}^{\mathrm{c}}}{dt} &= -\frac{1}{\tau}\bar{\theta}^{\mathrm{c}} - \frac{1}{\tau_{\mathrm{bc}}}(\bar{\theta}^{\mathrm{c}} - \bar{\theta}^{\mathrm{bc}}),
\end{aligned} \tag{14.22}$$

where $\bar{\theta}^{\mathrm{a}} \equiv \theta^{\mathrm{a}} - T, \bar{\theta}^{\mathrm{b}} \equiv \theta^{\mathrm{b}} - T, \bar{\theta}^{\mathrm{c}} \equiv \theta^{\mathrm{c}} - T$, and

$$\bar{\theta}^{\mathrm{bc}} \equiv \theta^{\mathrm{bc}} - T = \frac{c_v^{\mathrm{b}}\bar{\theta}^{\mathrm{b}} + c_v^{\mathrm{c}}\bar{\theta}^{\mathrm{c}}}{c_v^{\mathrm{b+c}}}. \tag{14.23}$$

The solution with the initial data $\bar{\theta}_0^{\mathrm{a}} = \bar{\theta}^{\mathrm{a}}|_{t=0}$, $\bar{\theta}_0^{\mathrm{b}} = \bar{\theta}^{\mathrm{b}}|_{t=0}$ and $\bar{\theta}_0^{\mathrm{c}} = \bar{\theta}^{\mathrm{c}}|_{t=0}$ is given by

$$\begin{aligned}
\bar{\theta}^{\mathrm{a}} &= \bar{\theta}_0^{\mathrm{a}} \mathrm{e}^{-\hat{t}}, \\
\bar{\theta}^{\mathrm{b}} &= \frac{1}{\hat{c}_v^{\mathrm{b+c}}}(\hat{c}_v^{\mathrm{b}}\bar{\theta}_0^{\mathrm{b}} + \hat{c}_v^{\mathrm{c}}\bar{\theta}_0^{\mathrm{c}})\mathrm{e}^{-\hat{t}} + \frac{\hat{c}_v^{\mathrm{c}}}{\hat{c}_v^{\mathrm{b+c}}}(\bar{\theta}_0^{\mathrm{b}} - \bar{\theta}_0^{\mathrm{c}})\mathrm{e}^{-\hat{t}/\hat{\tau}_\delta}, \\
\bar{\theta}^{\mathrm{c}} &= \frac{1}{\hat{c}_v^{\mathrm{b+c}}}(\hat{c}_v^{\mathrm{b}}\bar{\theta}_0^{\mathrm{b}} + \hat{c}_v^{\mathrm{c}}\bar{\theta}_0^{\mathrm{c}})\mathrm{e}^{-\hat{t}} - \frac{\hat{c}_v^{\mathrm{b}}}{\hat{c}_v^{\mathrm{b+c}}}(\bar{\theta}_0^{\mathrm{b}} - \bar{\theta}_0^{\mathrm{c}})\mathrm{e}^{-\hat{t}/\hat{\tau}_\delta},
\end{aligned} \tag{14.24}$$

where

$$\hat{t} = \frac{t}{\tau}, \quad \hat{\tau}_\delta = \frac{\tau_\delta}{\tau} = \frac{\tau_{\mathrm{bc}}/\tau}{1 + \tau_{\mathrm{bc}}/\tau}.$$

We have also the following relations:

$$\begin{aligned}
\bar{\theta}^{\mathrm{b}} - \bar{\theta}^{\mathrm{bc}} &= \frac{\hat{c}_v^{\mathrm{c}}}{\hat{c}_v^{\mathrm{b+c}}}(\bar{\theta}_0^{\mathrm{b}} - \bar{\theta}_0^{\mathrm{c}})\mathrm{e}^{-\hat{t}/\hat{\tau}_\delta}, \\
\bar{\theta}^{\mathrm{c}} - \bar{\theta}^{\mathrm{bc}} &= -\frac{\hat{c}_v^{\mathrm{b}}}{\hat{c}_v^{\mathrm{b+c}}}(\bar{\theta}_0^{\mathrm{b}} - \bar{\theta}_0^{\mathrm{c}})\mathrm{e}^{-\hat{t}/\hat{\tau}_\delta}.
\end{aligned} \tag{14.25}$$

As is expected, we can clearly see, from (14.24), (14.25), and (14.23), that the temperatures θ^{b} and θ^{c} relax to the temperature θ^{bc} with the relaxation time τ_δ, while the temperatures θ^{bc} and θ^{a} relax to the equilibrium temperature T with the relaxation time τ.

From experimental data on polyatomic gases such as CO_2, Cl_2, Br_2 gases, the (KR)-process is a suitable process [313, 314] (see also the analysis in Sect. 16.7.5). Therefore, as a typical example, we particularly focus on this process and study the relaxation evolved from a nonequilibrium initial state: $\theta^K|_{t=0} = \theta^V|_{t=0} = T_0$, $\theta^R|_{t=0} = \theta_0^R$ ($> T_0$). This initial state may be generated experimentally as follows: we firstly prepare the equilibrium state with the temperature T_0, then we excite only the R-mode from the temperature T_0 to the temperature θ_0^R instantaneously at the initial time. The relaxation is analyzed by solving (14.22) under the initial condition, in which T should be replaced by T_1. From the condition (14.13), T_1 is given by

$$T_1 = T_0 + \frac{c_v^R(\theta_0^R - T_0)}{c_v}.$$

The time-evolution of the relaxation is shown schematically in Fig. 14.1, from which we understand the two-step relaxation, and the energy redistribution from the R-mode to the K and V-modes. We also notice that, after the elapse of a period of time of $O(\tau_\delta)$ from the initial time, the relation $\theta^K = \theta^R = \theta^{KR}$ is approximately satisfied. Therefore the results derived from ET_6^{KR} and ET_7 with (KR)-process are nearly the same with each other. This means that ET_7 can be safely replaced by the simpler theory, ET_6^{KR}.

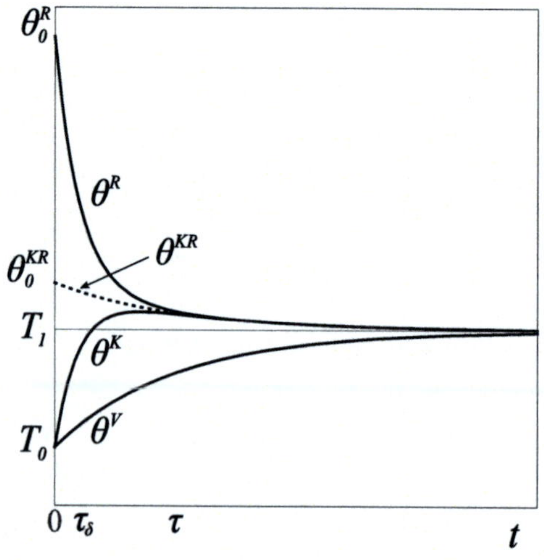

Fig. 14.1 Schematic time-evolution of the relaxation of the nonequilibrium temperatures θ^K, θ^R, θ^V, and θ^{KR} in the (KR)-process. The R-mode is excited from T_0 to θ_0^R instantaneously at the initial time, while K and V-modes are initially at the temperature T_0. The final equilibrium temperature is T_1. Relaxation times τ_δ and τ in this case are also indicated

Chapter 15
Nonequilibrium Temperature and Chemical Potential

Abstract In this chapter, we propose a natural definition of nonequilibrium temperature and chemical potential. The main field, with which the generalized Gibbs equation is expressed in a differential form, is the key quantity in the definition. In the ET_6 theory, in particular, the nonequilibrium temperature and chemical potential coincide exactly with those in the Meixner theory explained in Chap. 12.

15.1 Generalized Gibbs Equation, Nonequilibrium Temperature, and Chemical Potential

As mentioned in Sect. 1.12, one of the most delicate and controversial questions in nonequilibrium thermodynamics is the following one: What is the most appropriate definition of nonequilibrium temperature? This question has been considered by many authors, but we do not intend to discuss here the history and the several attempts to solve this intriguing problem. Some general discussions are summarized in the paper [157]. See also a previous tentative in RET [155, 156]. Instead, we here want to reconsider the idea that was proposed in [293] in a relativistic framework.

For this aim, let us firstly observe the structure of RET given in the case of rarefied monatomic gas by (2.11). For this structure, the temporal part of the differential conditions $(2.57)_1$ derived from the requirement of the entropy principle reads

$$\begin{aligned} dh^0 &= \mathbf{u}' \cdot d\mathbf{F}^0 = \hat{\mathbf{u}}' \cdot d\hat{\mathbf{F}}^0 \\ &= \hat{u}' d\hat{F} + \hat{u}'_{k_1} d\hat{F}_{k_1} + \hat{u}'_{k_1k_2} d\hat{F}_{k_1k_2} + \cdots + \hat{u}'_{k_1k_2\ldots k_n} d\hat{F}_{k_1k_2\ldots k_n} \\ &= \hat{u}' d\hat{F} + \frac{1}{3}\hat{u}'_{ll} d\hat{F}_{kk} + \hat{u}'_{\langle k_1k_2\rangle} d\hat{F}_{\langle k_1k_2\rangle} + \cdots + \hat{u}'_{k_1k_2\ldots k_n} d\hat{F}_{k_1k_2\ldots k_n}, \end{aligned} \tag{15.1}$$

where we write explicitly the components of the main field:

$$\mathbf{u}' \equiv (u', u'_{k_1}, u'_{k_1k_2}, \ldots, u'_{k_1k_2\ldots k_n}).$$

T. Ruggeri, M. Sugiyama, *Classical and Relativistic Rational Extended Thermodynamics of Gases*, https://doi.org/10.1007/978-3-030-59144-1_15

By taking into account the general properties that, in equilibrium, all the components of the main field corresponding to the balance laws vanish (2.45)$_2$ and that the first 5 fields F, F_i, F_{kk} are equilibrium quantities (mass density, momentum, 2 times energy), the relation (15.1) reduces to

$$dh_E^0 = \hat{\mathbf{u}}'_E \cdot d\hat{\mathbf{F}}_E^0 = \hat{u}'_E d\hat{F} + \frac{1}{3}\hat{u}'_{llE} dF_{kk}, \tag{15.2}$$

where suffix E indicates a quantity in equilibrium. This is nothing else the Gibbs equation (1.8) with non-vanishing components of the main field. These components coincide with those of the main field of the Euler equilibrium principal subsystem:

$$\hat{\lambda}_E = \hat{u}'_E = -\frac{g}{T}, \qquad \hat{\mu}_E = \frac{1}{3}\hat{u}'_{llE} = \frac{1}{2T}, \tag{15.3}$$

where we have put $\hat{\lambda} = \hat{u}'$ and $\hat{\mu} = \frac{\hat{u}'_{ll}}{3}$ for simplicity and also for uniformity with the notation in the previous chapters.

Therefore, firstly we may conclude that (15.1) represents the generalized Gibbs equation in nonequilibrium. In this respect, we emphasize the fact that, in RET, we do not adopt the local equilibrium assumption characterized by the equilibrium Gibbs equation (15.2).

Secondly we can understand the physical meaning of the intrinsic Lagrange multipliers in equilibrium corresponding to the mass conservation and the energy conservation owing to the relation (15.3). They are, respectively, the ratios between the chemical potential and the absolute temperature except for the sign, and the coldness (the inverse of the temperature) except for the factor $1/2$. It is therefore natural that, also for a nonequilibrium state, we assume the following relations:

$$\hat{\lambda} = -\frac{\mathcal{G}}{\mathcal{T}}, \qquad \hat{\mu} = \frac{1}{2\mathcal{T}}, \tag{15.4}$$

where $\mathcal{G}$ and $\mathcal{T}$ are interpreted as the nonequilibrium chemical potential and the nonequilibrium temperature. In other words, these nonequilibrium quantities are defined through the relations (15.4).

Finally, a crucial remark is the following: in the old ET [49] or in the EIT [157], the Gibbs equation (15.1) is the *starting point* and therefore is assumed as a hypothesis, while, in RET, (15.1) is not assumed but is deduced. This means that, since we saw and will see in several chapters, we are able to evaluate all the main field components *without any ambiguity*. Therefore we can measure the nonequilibrium temperature and the nonequilibrium chemical potential indirectly through the relations:

$$\mathcal{T} = \frac{1}{2\hat{\mu}}, \qquad \mathcal{G} = -\frac{\hat{\lambda}}{2\hat{\mu}}. \tag{15.5}$$

In the case of rarefied monatomic gas, among other RET theories, the RET theory with 13 fields is the most interesting one in the sense that all fields have concrete physical meanings. For such a theory, the definition (15.5) gives a simple but non-trivial answer. In fact, from the expression of the main field of RET with 13 fields (4.28), the definition (15.5) gives the relations:

$$\mathscr{T} = T, \qquad \mathscr{G} = g. \tag{15.6}$$

In this case, two kinds of the nonequilibrium quantities coincide with each other.

We have to recall that the RET theory with 13 fields is valid only near equilibrium, where we assume that all quantities are linear with respect to the nonequilibrium quantities. Because of the two facts that (1) the scalar quantities $\mathscr{T}$ and $\mathscr{G}$ can be influenced only by nonequilibrium scalars obtained from the tensor F's and (2) we do not have any nonequilibrium scalar in the first order, we have obtained the relation (15.6). Here, the fact that the dynamic pressure vanishes identically in a rarefied monatomic gas is essential.

In the case of RET of monatomic gases with more fields than 13, however, other nonequilibrium scalars may play a role. For example, in the RET theory of monatomic gases with 14 fields, we adopt one more balance equation for the new independent variable F_{llkk}. Then we have [256]

$$\hat{\lambda} = -\frac{g}{T} - \frac{\Delta}{8pT}, \qquad \hat{\mu} = \frac{1}{2T}\left(1 + \frac{\rho}{6p^2}\Delta\right),$$

where Δ is the nonequilibrium part of $\hat{F}_{llkk}$.

Then, in the case of rarefied polyatomic gas, the situation is quite different. The nonequilibrium temperature $\mathscr{T}$ and chemical potential $\mathscr{G}$ are different from the temperature T and chemical potential g due to the existence of the dynamic pressure Π. We will establish, in the following sections, an explicit relationship between ($\mathscr{T}$, $\mathscr{G}$) and (T, g) in the cases of ET$_6$, ET$_{14}$, and RET with arbitrary number of fields.

15.2 Nonequilibrium Temperature and Chemical Potential in RET with the Binary Hierarchy

In the case of RET of polyatomic gases with the binary hierarchy (9.1), one important point concerning the Lagrange multiplier of the energy equation should be clarified in order to define the nonequilibrium temperature and chemical potential properly. We start to consider this point in the case of ET$_6$.

In ET$_6$, we already observed that we can write the system of field equations in two different ways. The one is written in the form of F-series and G-series (12.1), which corresponds to (12.3) and to the Lagrange multipliers $(\lambda, \lambda_i, \mu, \zeta)$ given by (12.34). The other one is written in the form (12.1) except that, instead of the equation for F_{ll}, we adopt the equation:

$$\partial_t (F_{ll} - G_{ll}) + \partial_i (F_{ill} - G_{ill}) = P_{ll}, \tag{15.7}$$

which is obtained by subtracting the energy equation from the equation of the dynamic pressure (12.37). In this case, the Lagrange multipliers (12.38), indicated by a bar, are the same of the former ones except that the Lagrange multiplier of the energy equation is now given by

$$\bar{\mu} = \mu + \zeta$$

As explained in Remark 12.5 in Sect. 12.5, the two possible forms of the system of ET$_6$ are completely equivalent to each other, but only the second form gives the possibility to obtain the singular limit $D \to 3$ of a monatomic gas. In fact, Eq. (15.7) coincides with (10.28) in the limit, and admits the solution $\Pi = 0$. The Lagrange multipliers that converge to those of Euler fluids are the ones of the second form.

Therefore, from the existence condition of the limit of the monatomic case, the correct intrinsic Lagrange multiplier of the energy equation must be $\hat{\bar{\mu}}$. And the previous definition (15.5) for $\mathscr{T}$ and $\mathscr{G}$ should be expressed as

$$\mathscr{T} = \frac{1}{2\hat{\bar{\mu}}}, \qquad \mathscr{G} = -\frac{\hat{\bar{\lambda}}}{2\hat{\bar{\mu}}}.$$

Or equivalently, in terms of the intrinsic Lagrange multipliers of the F- and G-series, we have

$$\mathscr{T} = \frac{1}{2(\hat{\mu} + \hat{\zeta})}, \qquad \mathscr{G} = -\frac{\hat{\lambda}}{2(\hat{\mu} + \hat{\zeta})}, \tag{15.8}$$

where $\hat{\lambda}$, $\hat{\zeta}$, $\hat{\mu}$, are, respectively, the intrinsic Lagrange multipliers with respect to the conservation equations for F, F_{ll}, G_{ll}. Above considerations are valid for RET with any number of fields.

15.3 Nonequilibrium Temperature and Chemical Potential in ET$_6$ and ET$_{14}$

It is easy to see that, in the case of rarefied polyatomic gas, the previous quantities in (15.8) coincide with the Meixner nonequilibrium temperature $\mathscr{T}(=\theta^K)$ given in (12.57) and the nonequilibrium chemical potential $\mathscr{G}$ given in (12.59). In Sect. 18.6, we will see that the behavior of the nonequilibrium temperature $\mathscr{T}$ can be quite different from the behavior of the temperature T in the shock wave structure.

In the case of polytropic gas, $\mathscr{T}$ is the same as the one in non-polytropic case (12.57), while the chemical potential has a simple expression:

$$\mathscr{T} = T\left(1+\frac{\Pi}{p}\right), \qquad \mathscr{G} = (g - Tk)\left(1+\frac{\Pi}{p}\right).$$

We can see that the dynamic pressure plays a dominant role to cause the difference between the temperature T and the nonequilibrium temperature $\mathscr{T}$, and the same is true for the chemical potential.

In the case of ET_{14} within a linear approximation, as the nonequilibrium scalar is only Π, we have no difference of the nonequilibrium temperature and the nonequilibrium chemical potential from those of the ET_6.

For a general N-moment system, the definition of nonequilibrium quantities is given by the general one, i.e., (15.8). As $\mathscr{T}$ and $\mathscr{G}$ are scalars, in order to have a new nonequilibrium contribution to these quantities, it is necessary to consider more moments such that there appear other scalars of the densities. For example, we expect a new contribution in a $(3^{(1)}, 2^{(1)})$-system.

15.4 Conclusion

We have proposed a natural definition of nonequilibrium temperature and chemical potential by comparing the generalized Gibbs equation and the equilibrium one. The components of the main field are the coefficients of the generalized Gibbs equation in a differential form. In equilibrium, only the components that are related to the mass density and the energy density remain to be non-zero. The first one is strictly related to the chemical potential and the second one to the coldness. Therefore we can define the nonequilibrium coldness (or temperature) and chemical potential by assuming that the coefficients do not change their meanings even in nonequilibrium.

We have proved that, in the most interesting cases, that is, ET_6 and ET_{14} theories where all fields have concrete physical meanings, the dynamic pressure is responsible for the difference between the nonequilibrium quantities in RET and the corresponding ones in the Meixner theory. In the limit of monatomic gas, the difference disappears. This means that the internal degrees of freedom in a polyatomic molecule play a crucial role in the definition of temperature. Appropriateness of the definition is supported, in the case of ET_6, by the perfect coincidence of the RET nonequilibrium quantities with those in the Meixner theory. See also the discussions made in Sect. 12.3.

In the case of RET with many moments, all nonequilibrium scalars constructed by the tensors of the density-fields play roles in the nonequilibrium temperature and chemical potential.

Finally, we remark that the same definitions can be adopted also in the case of RET of dense gases (see Chaps. 24 and 25).

Part V
Applications of the RET Theory of Polyatomic Gas

Chapter 16
Linear Sound Wave in a Rarefied Polyatomic Gas

Abstract In this chapter, we study a linear sound wave in a rarefied polyatomic gas in equilibrium. Thereby we can clarify the validity and the features of the ET_{14} and ET_{15} theories established in Chaps. 6–8.

In the first half of this chapter, we derive the dispersion relations on the basis of the ET_{14} theory and of the classical Navier-Stokes and Fourier (NSF) theory. Comparison of these relations with experimental data reveals clearly the superiority of the ET_{14} theory to the NSF theory. We confine our analysis within sound waves in some rarefied diatomic gases (hydrogen, deuterium, and hydrogen deuteride gases) because suitable experimental data are scarce. We also evaluate the relaxation times, and the shear and bulk viscosities, and the heat conductivity of the gases.

In the second half, firstly we emphasize the necessity of the ET_{15} theory. Then using this theory we study the dispersion relation of a rarefied polyatomic gas with molecular rotational- and vibrational-relaxation processes. We study, in particular, its temperature dependence in the cases where the rotational and vibrational modes may or may not be excited. It is shown that the curve of the attenuation per wavelength with respect to the frequency has up to three peaks depending on the temperature and on the relaxation times. The peak in a low frequency region is studied especially in detail because its experimental measurement is relatively easy.

16.1 Introduction

The study of the dispersion and absorption of sound wave in a gas has a long history [7, 129, 130, 311, 312, 339–341]. For polyatomic gases, in particular, an interesting experimental fact is that there exists the absorption due to energy exchanges among the molecular translational, rotational, and vibrational degrees of freedom, which is intimately related to the bulk viscosity. In fact, for gases with large bulk viscosity, for example, hydrogen gas, carbon dioxide gas, such absorption can be observed. The absorption appears in a lower frequency region while the absorption due to shear viscosity and heat conduction appears in a higher region.

T. Ruggeri, M. Sugiyama, *Classical and Relativistic Rational Extended Thermodynamics of Gases*, https://doi.org/10.1007/978-3-030-59144-1_16

Although there are several thermodynamic theories and kinetic theories [312] for understanding such absorption, most of these, including the Navier-Stokes and Fourier (NSF) theory, are valid only within very low frequency region where the local thermodynamic equilibrium assumption can be adopted. It has been highly desired to have a satisfactory theory that is valid in much more wide frequency region.

In the first half of the present chapter (Sects. 16.2–16.4), we derive the dispersion relations on the basis of the ET_{14} theory and of the classical NSF theory [106, 107]. Comparison of these relations with experimental data in some rarefied diatomic gases (hydrogen, deuterium, and hydrogen deuteride gases) is made. We also evaluate the relaxation times, and the shear and bulk viscosities, and the heat conductivity of the gases.

In the latter half (Sects. 16.5–16.7), on the basis of the ET_{15} theory (and also the ET_{7} theory), we study the dispersion relation in the case where the molecular rotational and vibrational modes play important roles simultaneously [135, 136, 342]. We make clear the characteristic features of ET_{15} theory and of its subsystems with 14 fields through studying wave propagation phenomena.

16.2 Basic Equations: Linearized System of ET_{14}

We study a linear sound wave in a rarefied polyatomic gas in equilibrium with the thermal and caloric equations of state given by (12.15). We assume that a nonequilibrium state can be characterized by the 14 independent field variables $\mathbf{u} \equiv (\rho, v_i, T, \Pi, \sigma_{\langle ij\rangle}, q_i)$. Let $\mathbf{u}_0 \equiv (\rho_0, 0, T_0, 0, 0, 0)$ be an equilibrium state, then, from (6.33), the linearized system for the perturbed field $\bar{\mathbf{u}} = (\bar{\rho}, \bar{v}_i, \bar{T}, \bar{\Pi}, \bar{\sigma}_{\langle ij\rangle}, \bar{q}_i)$ in the neighborhood of $\mathbf{u}_0$ is given by

$$
\begin{aligned}
&\frac{\partial \rho}{\partial t} + \rho_0 \frac{\partial v_k}{\partial x_k} = 0,\\
&\rho_0 \frac{\partial v_i}{\partial t} + \frac{k_B}{m} T_0 \frac{\partial \rho}{\partial x_i} + \frac{k_B}{m}\rho_0 \frac{\partial T}{\partial x_i} - \frac{\partial \sigma_{\langle ij\rangle}}{\partial x_j} + \frac{\partial \Pi}{\partial x_i} = 0,\\
&\frac{k_B}{m}\rho_0 \hat{c}_v \frac{\partial T}{\partial t} + \frac{k_B}{m}\rho_0 T_0 \frac{\partial v_k}{\partial x_k} + \frac{\partial q_k}{\partial x_k} = 0,\\
&\frac{\partial \Pi}{\partial t} + \left(\frac{2}{3} - \frac{1}{\hat{c}_v}\right)\frac{k_B}{m}\rho_0 T_0 \frac{\partial v_k}{\partial x_k} + \frac{2\hat{c}_v - 3}{3\hat{c}_v(1+\hat{c}_v)}\frac{\partial q_k}{\partial x_k} = -\frac{1}{\tau_\Pi}\Pi,\\
&\frac{\partial \sigma_{\langle ij\rangle}}{\partial t} - 2\frac{k_B}{m}\rho_0 T_0 \frac{\partial v_{\langle i}}{\partial x_{j\rangle}} - \frac{2}{1+\hat{c}_v}\frac{\partial q_{\langle i}}{\partial x_{j\rangle}} = -\frac{1}{\tau_\sigma}\sigma_{\langle ij\rangle},\\
&\frac{\partial q_i}{\partial t} + (1+\hat{c}_v)\left(\frac{k_B}{m}\right)^2 \rho_0 T_0 \frac{\partial T}{\partial x_i} - \frac{k_B}{m} T_0 \frac{\partial \sigma_{\langle ik\rangle}}{\partial x_k} + \frac{k_B}{m} T_0 \frac{\partial \Pi}{\partial x_i} = -\frac{1}{\tau_q} q_i.
\end{aligned}
\tag{16.1}
$$

Here and hereafter we ignore a bar on a perturbed quantity for simplicity. The dimensionless specific heat $\hat{c}_v$ (see (6.4)) and the relaxation times τ_σ, τ_Π, and τ_q in (16.1) are evaluated at the equilibrium state. The relations between the relaxation times and the shear viscosity μ, the bulk viscosity ν, and the heat conductivity κ are given by (6.36).

Let us confine our study within a one-dimensional problem, that is, a plane longitudinal wave propagating in the positive x-direction. Therefore, by considering the symmetry of the wave, we assume

$$v_i \equiv \begin{pmatrix} v \\ 0 \\ 0 \end{pmatrix}, \quad \sigma_{\langle ij\rangle} \equiv \begin{pmatrix} \sigma & 0 & 0 \\ 0 & -\frac{1}{2}\sigma & 0 \\ 0 & 0 & -\frac{1}{2}\sigma \end{pmatrix}, \quad q_i \equiv \begin{pmatrix} q \\ 0 \\ 0 \end{pmatrix}. \tag{16.2}$$

Then, the linearized basic field equations (16.1) are neatly written as

$$\frac{\partial \mathbf{u}}{\partial t} + \mathbf{A}_0 \frac{\partial \mathbf{u}}{\partial x} = \mathbf{B}_0 \mathbf{u}, \tag{16.3}$$

where $\mathbf{u}$ is now redefined as $\mathbf{u} \equiv (\rho, v, T, \Pi, \sigma, q)$, and $\mathbf{A}_0$ and $\mathbf{B}_0$ are given by

$$\mathbf{A}_0 = \begin{pmatrix} 0 & \rho_0 & 0 & 0 & 0 & 0 \\ \frac{k_B}{m}\frac{T_0}{\rho_0} & 0 & \frac{k_B}{m} & \frac{1}{\rho_0} & -\frac{1}{\rho_0} & 0 \\ 0 & \frac{T_0}{\hat{c}_v} & 0 & 0 & 0 & \frac{m}{k_B \hat{c}_v \rho_0} \\ 0 & \left(\frac{2}{3} - \frac{1}{\hat{c}_v}\right)\frac{k_B}{m}\rho_0 T_0 & 0 & 0 & 0 & \frac{2\hat{c}_v - 3}{3\hat{c}_v(1+\hat{c}_v)} \\ 0 & -\frac{4}{3}\frac{k_B}{m}\rho_0 T_0 & 0 & 0 & 0 & -\frac{4}{3(1+\hat{c}_v)} \\ 0 & 0 & (1+\hat{c}_v)\left(\frac{k_B}{m}\right)^2 \rho_0 T_0 & \frac{k_B}{m} T_0 & -\frac{k_B}{m} T_0 & 0 \end{pmatrix},$$

$$\mathbf{B}_0 = \begin{pmatrix} 0 & 0 & 0 & 0 & 0 & 0 \\ 0 & 0 & 0 & 0 & 0 & 0 \\ 0 & 0 & 0 & 0 & 0 & 0 \\ 0 & 0 & 0 & -\frac{1}{\tau_\Pi} & 0 & 0 \\ 0 & 0 & 0 & 0 & -\frac{1}{\tau_\sigma} & 0 \\ 0 & 0 & 0 & 0 & 0 & -\frac{1}{\tau_q} \end{pmatrix}.$$

16.3 Dispersion Relation for Sound

In this section, using the general theory of linear wave in Sect. 3.1, we derive the dispersion relation, and then obtain the high-frequency limit of the phase velocity and the attenuation factor.

16.3.1 Dispersion Relation, Phase Velocity, and Attenuation Factor

We study a plane harmonic wave with angular frequency ω and (complex) wave number k. From Eq. (16.3), the dispersion relation is expressed by (see (3.3))

$$\det\left(\mathbf{I} - z\mathbf{A}_0 + \frac{i}{\omega}\mathbf{B}_0\right) = 0, \tag{16.4}$$

where $z \equiv k/\omega$ and $\mathbf{I}$ is the unit matrix. Then the phase velocity v_{ph} and the attenuation factor α are calculated as the functions of the frequency ω by using the relation (3.4). In addition, it is useful to introduce the attenuation per wavelength:

$$\alpha_\lambda(\omega) = \alpha\lambda = \frac{2\pi v_{ph}\alpha}{\omega} = -2\pi\frac{\mathscr{I}m(z)}{\mathscr{R}e(z)},$$

where λ is the wavelength.

By introducing the dimensionless parameters defined by

$$\Omega = \tau_\sigma\omega, \quad \tau_q^* = \frac{\tau_q}{\tau_\sigma}, \quad \tau_\Pi^* = \frac{\tau_\Pi}{\tau_\sigma}, \tag{16.5}$$

the dispersion relation (16.4) is expressed explicitly as

$$\begin{aligned}
&\frac{\hat{c}_v(c_0z)^4}{3\Omega^2\left(1+\hat{c}_v\right)^2\tau_\Pi^*}\left\{-3(1+\hat{c}_v) - i\Omega\left(3+7\hat{c}_v+5\hat{c}_v\tau_\Pi^*\right) + 9\Omega^2\hat{c}_v\tau_\Pi^*\right\}\\
&+\frac{(c_0z)^2}{3\Omega^3(1+\hat{c}_v)^2\tau_q^*\tau_\Pi^*}\Big\{-3\mathrm{i}\left(1+\hat{c}_v\right)^2 + \Omega\left(1+\hat{c}_v\right)\left[3+7\hat{c}_v+5\hat{c}_v\tau_\Pi^*+6\left(1+\hat{c}_v\right)\tau_q^*\right]\\
&+\mathrm{i}\Omega^2\left[2\left(3+10\hat{c}_v+5\hat{c}_v^2\right)\tau_q^*+9\hat{c}_v\left(1+\hat{c}_v\right)\tau_\Pi^*+\hat{c}_v\left(13+8\hat{c}_v\right)\tau_q^*\tau_\Pi^*\right]\\
&-3\Omega^3\hat{c}_v\left(7+4\hat{c}_v\right)\tau_\Pi^*\tau_q^*\Big\} + \frac{(\Omega-i)(\tau_\Pi^*\Omega-i)(\tau_q^*\Omega-i)}{\Omega^3\tau_\Pi^*\tau_q^*} = 0
\end{aligned} \tag{16.6}$$

with c_0 being the sound velocity in an equilibrium state:

$$c_0 = \sqrt{\left(\frac{\partial p}{\partial \rho}\right)_0 + \frac{T_0 \left(\frac{\partial p}{\partial T}\right)_0^2}{\rho_0^2 \left(\frac{\partial \varepsilon}{\partial T}\right)_0}} = \sqrt{\frac{k_B}{m} T_0 \left(1 + \frac{1}{\hat{c}_v}\right)_0} = \sqrt{\frac{k_B}{m} T_0 \gamma(T_0)}, \qquad (16.7)$$

where the suffix 0 indicates the values at the equilibrium state and

$$\gamma(T) = (1 + \hat{c}_v(T))/\hat{c}_v(T) \qquad (16.8)$$

is the ratio of the specific heats. Therefore, for given $\hat{c}_v$, τ_q^*, and τ_Π^*, the quantity $c_0 z (= c_0 k/\omega)$ is calculated from (16.6) as the function of Ω $(= \tau_\sigma \omega)$. Hereafter in this chapter, we will confine our study within the *fastest* sound wave because the experiments give us the data on this wave.

16.3.2 High-Frequency Limit of the Phase Velocity and the Attenuation Factor

From (3.7), we have the relations:

$$v_{ph}^{(\infty)} \equiv \lim_{\omega \to \infty} v_{ph}(\omega) = \lambda_0, \qquad \alpha^{(\infty)} \lambda_0 \equiv \lim_{\omega \to \infty} \alpha(\omega) \lambda_0 = -\mathbf{l}_0 \cdot \mathbf{B}_0 \cdot \mathbf{d}_0,$$

where the characteristic velocity λ_0 is the largest eigenvalue of $\mathbf{A}_0$, and $\mathbf{l}_0$ and $\mathbf{d}_0$ are the corresponding left and right eigenvectors of $\mathbf{A}_0$. Then we obtain the limits:

$$v_{ph}^{(\infty)} = \sqrt{\frac{\frac{k_B}{m} T_0 (4\hat{c}_v + 7 + F)}{2(1 + \hat{c}_v)}},$$

$$\alpha^{(\infty)} = \frac{\sqrt{2(1+\hat{c}_v)^3}\left(F\left(4 + \hat{c}_v\right) - 22 - 11\hat{c}_v + 2\hat{c}_v^2\right)}{9\hat{c}_v \tau_\sigma \sqrt{\frac{k_B}{m} T_0} \sqrt{7 + 4\hat{c}_v + F}\left(7 + 4\hat{c}_v - F\right)^2 F} \times \left(4\hat{c}_v + \frac{3\hat{c}_v \left(8 + 2\hat{c}_v - F\right)}{\tau_q^*} + \frac{-3 + 2\hat{c}_v}{\tau_\Pi^*}\right),$$

where F is given by

$$F = \sqrt{37 + 32\hat{c}_v + 4\hat{c}_v^2}.$$

The dependence of the phase velocity $v_{ph}^{(\infty)}$ on $\hat{c}_v$ is shown in Fig. 16.1. In a rarefied monatomic gas with $\hat{c}_v = 3/2$, the phase velocity $v_{ph}^{(\infty)}$ is given

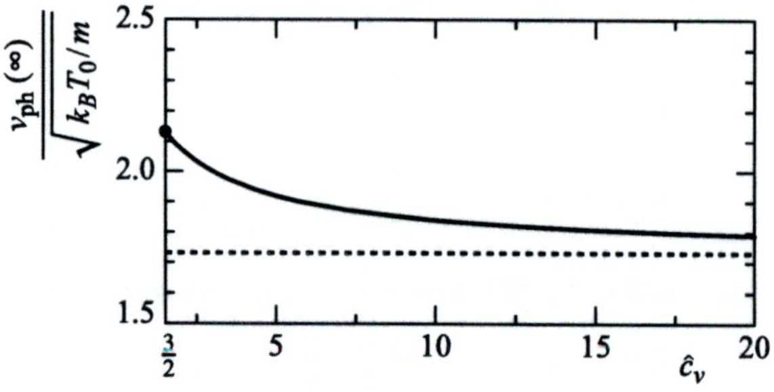

Fig. 16.1 Dependence of the phase velocity in the high-frequency limit $v_{ph}^{(\infty)}$ on the dimensionless specific heat $\hat{c}_v$. Rarefied monatomic gases correspond to the case with $\hat{c}_v = 3/2$. The dotted line is the asymptote

by $2.13051\sqrt{k_B T_0/m}$ [25]. For large $\hat{c}_v$, it approaches $\sqrt{3k_B T_0/m}$. On the other hand, the attenuation factor $\alpha^{(\infty)}$ depends not only on $\hat{c}_v$ but also on the relaxation times. In a rarefied monatomic gas, the attenuation factor $\alpha^{(\infty)}$ is given by $\left(0.0951852 + 0.0931368/\tau_q^*\right)/(\tau_\sigma\sqrt{k_B T_0/m})$ [25]. For large $\hat{c}_v$, it approaches $(1 + 2\tau_\Pi^*)/(9\sqrt{3}\tau_\Pi^*\tau_\sigma\sqrt{k_B T_0/m})$.

16.4 Comparison with Experimental Data

The dispersion relation obtained above, in particular, the phase velocity v_{ph}, the attenuation factor α, and the attenuation per wavelength $\alpha_\lambda$as the functions of the frequency ω are compared with the experimental data on normal hydrogen (n-H_2), para hydrogen (p-H_2), normal deuterium (n-D_2), ortho deuterium (o-D_2), and hydrogen deuteride (HD) gases at temperatures from 77.3 K to 1073.15 K [343–346]. The comparison is also made with the predictions by the classical NSF theory.

Before discussing the subject, we need to make preliminary calculations for determining the values of $\hat{c}_v$, τ_q^*, and τ_Π^* defined in (6.4) and $(16.5)_{2,3}$ at the equilibrium state.

16.4.1 Preliminary Calculations

16.4.1.1 Specific heat

We calculate the specific heat $\hat{c}_v$ of hydrogen, deuterium, and hydrogen deuteride gases on the basis of statistical mechanics [347, 348]. We assume that the translational mode satisfies the equipartition law of energy. Then $\hat{c}_v$ is expressed as

$$\hat{c}_v = \frac{3}{2} + \hat{c}_v^R + \hat{c}_v^V,$$
$$\hat{c}_v^R = \beta^2 \frac{\partial^2 \log Z_{\text{rot}}}{\partial \beta^2}, \qquad \hat{c}_v^V = \beta^2 \frac{\partial^2 \log Z_{\text{vib}}}{\partial \beta^2}, \qquad \left(\beta \equiv \frac{1}{k_B T}\right)$$

where $\hat{c}_v^R$, Z_{rot} and $\hat{c}_v^V$, Z_{vib} are the specific heat and the partition function due to the rotational and vibrational modes, respectively.

For gases composed of heteronuclear diatomic molecules (HD), the partition function of rotational motion is given by

$$Z_{\text{rot}} = \sum_{l=0}^{\infty} (2l+1) \exp[-\beta B l(l+1)],$$

where l is the quantum number of the orbital angular momentum and $B = \hbar^2/2I$ with I and $\hbar$ being the moment of inertia of a molecule and the Planck constant divided by 2π, respectively. While, for gases composed of diatomic homonuclear molecules (H_2 and D_2), the partition function of rotational motion is given by

$$Z_{\text{rot}} = Z_g^{g_g} Z_u^{g_u},$$
$$Z_g = \sum_{l=even} (2l+1) \exp[-\beta B l(l+1)],$$
$$Z_u = \sum_{l=odd} (2l+1) \exp[-\beta B l(l+1)],$$

where g_g and g_u are defined by

$$H_2 \begin{cases} \text{normal} - H_2 : g_u = 3/4, \; g_g = 1/4 \\ \text{para} - H_2 : \quad g_u = 0, \quad g_g = 1 \end{cases}, \quad D_2 \begin{cases} \text{normal} - D_2 : g_u = 1/3, \; g_g = 2/3 \\ \text{ortho} - D_2 : \quad g_u = 0, \quad g_g = 1 \end{cases}.$$

For the vibrational modes, we assume the harmonic oscillator model. Then the partition function of the vibrational modes is obtained as follows:

$$Z_{\text{vib}} = \frac{e^{-\frac{\hbar \omega_v \beta}{2}}}{1 - e^{-\hbar \omega_v \beta}},$$

where ω_v is the characteristic frequency.

Numerically calculated values of $\hat{c}_v$ are shown in Table 16.1 and in Fig. 16.2. The values of B of H_2, D_2, and HD adopted are 12.09×10^{-22} [J], 6.047×10^{-22} [J], and 9.068×10^{-22} [J], respectively, and the values of ω_v are $6332 k_B/\hbar$ [Hz], $4483 k_B/\hbar$ [Hz], and $5486 k_B/\hbar$ [Hz], respectively [349].

Table 16.1 Values of the temperature T_0, dimensionless specific heat $\hat{c}_v$, sound speed in equilibrium c_0, shear viscosity μ [345, 350–353], heat conductivity κ [345, 350–353] and the ratio of the relaxation times of the heat flux and the deviatoric part of the viscous stress τ_q^* adopted in the present analysis. And the values of the parameter ν/μ, bulk viscosity ν, and the ratio of the relaxation times of the bulk viscosity and the deviatoric part of the viscous stress τ_Π^* evaluated by the present analysis

Gas	T_0[K]	$\hat{c}_v$	c_0 [$\frac{\text{m}}{\text{s}}$]	μ [μPa · s]	κ [$\frac{\text{mW}}{\text{m}\cdot\text{K}}$]	τ_q^*	ν/μ	ν [μPa · s]	τ_Π^*
$n-H_2$	77.3	1.57	723	3.50	49.8	1.34	27.2	95.2	960
	273	2.42	1260	8.33	173	1.47	41.9	349	165
	295.15	2.45	1310	8.95	187	1.47	33.1	296	128
	873.15	2.54	2240	18.7	403	1.48	36.8	685	135
	1073.15	2.60	2480	21.0	462	1.49	40.3	846	143
$p-H_2$	77.3	1.76	707	3.50	52.7	1.33	75.0	263	773
	90.2	1.99	748	3.97	63.6	1.30	83.6	332	512
	170	2.96	968	6.10	113	1.14	54.8	334	166
	293	2.61	1290	8.82	192	1.46	28.8	254	101
$n-D_2$	77.3	2.54	472	4.82	45.6	1.30	35.7	172	131
	273.15	2.50	888	11.8	136	1.60	24.7	291	92.6
	295.15	2.50	923	12.6	141	1.55	20.9	264	78.3
	773.15	2.60	1490	24.2	260	1.45	30.9	747	109
	1073.15	2.78	1740	30.4	337	1.42	35.9	1092	117
$o-D_2$	77.3	2.93	463	4.82	49.4	1.26	45.4	219	140
	90.2	2.96	499	5.50	55.6	1.24	33.6	185	102
	293	2.50	920	12.3	131	1.47	22.6	278	84.7
HD	77.3	2.55	544	4.21	51.9	1.26	1.84	7.75	6.72
	293	2.50	1060	10.8	149	1.43	2.27	24.5	8.51

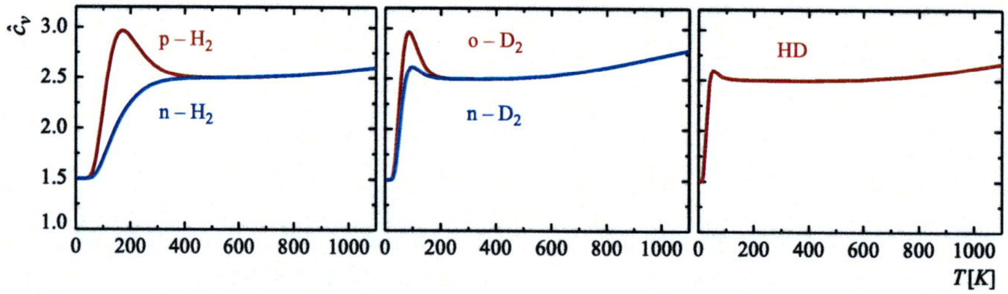

Fig. 16.2 Dependence of the dimensionless specific heat $\hat{c}_v$ for n-H_2 and p-H_2 (left), n-D_2 and o-D_2 (center), and HD (right) on the temperature T

16.4.2 Relaxation Times

From (6.36), we have the following relations for the ratios τ_q^* and τ_Π^*:

$$\tau_q^* = \left(1 + \hat{c}_v\right)^{-1} \frac{\kappa}{\frac{k_B}{m}\mu}, \qquad \tau_\Pi^* = \left(\frac{2}{3} - \frac{1}{\hat{c}_v}\right)^{-1} \frac{\nu}{\mu}.$$

Therefore, in principle, with the help of the experimental data on μ, κ, and ν, we can estimate the values of τ_q^* and τ_Π^*. However, at present, since we have the reliable data only on μ and κ [345, 350–353], we adopt, in the analysis below, an adjustable parameter: ν/μ. We summarize the adopted values of $\hat{c}_v$, c_0, μ, κ, τ_q^*, and the evaluated values of ν/μ and τ_Π^* in Table 16.1, details of which will be discussed in the next subsection.

16.4.3 Experimental Data and Theoretical Prediction for the Dispersion Relation

16.4.3.1 Hydrogen Gases: n-H_2 and p-H_2

For n-H_2, the dimensionless phase velocity, v_{ph}/c_0, the dimensionless attenuation factor, $c_0\tau_\sigma\alpha$, and the attenuation per wavelength, α_λ, are shown as the functions of the dimensionless frequency Ω in Fig. 16.3. We see the experimental data on the phase velocity at T_0 = 295.15K and 296.8K by Winter and Hill [343] and Rhodes [344], on the attenuation factor at T_0 = 293K by Sluijter et al. [345], and on the attenuation per wavelength at T_0 = 295.15K by Winter and Hill [343] accompanied by the theoretical results at T_0 = 295.15K predicted by the ET$_{14}$ theory and the NSF theory.

Noticeable points in Fig. 16.3 are summarized as follows: (1) In the region with small Ω, as is expected, the predictions by the two theories coincide with each other. The value of the parameter ν/μ is determined to be 33.1 as the best fit with the experimental data in this region. This procedure of determining ν/μ will be adopted throughout the present chapter. (2) When we go into the ultrasonic frequency region with larger Ω, the prediction by the ET$_{14}$ theory is evidently superior to that by the NSF theory. The difference between the two theories emerges around $\Omega = \omega\tau_\sigma = 10^{-3}$. We will evaluate τ_σ, which depends on T_0 and p_0, later. (3) The ET$_{14}$ theory seems to be valid at least up to the experimental data with the maximum dimensionless frequency $\Omega = 10^{-1}$. (iv) We have found an interesting fact that $\nu >> \mu$. We will discuss its physical meaning below.

At other temperatures, there exists the experimental data of α at T_0 = 77.3 K by Sluijter et al. [345], those of v_{ph} and α_λ at around 273 K, respectively, by Rhodes [344] (at T_0 = 273.5 K) and Stewart and Stewart [346] (at T_0 = 273.15 K) and those of v_{ph} and α_λ at T_0 = 873.15, 1073.15 K by Winter and Hill [343]. The

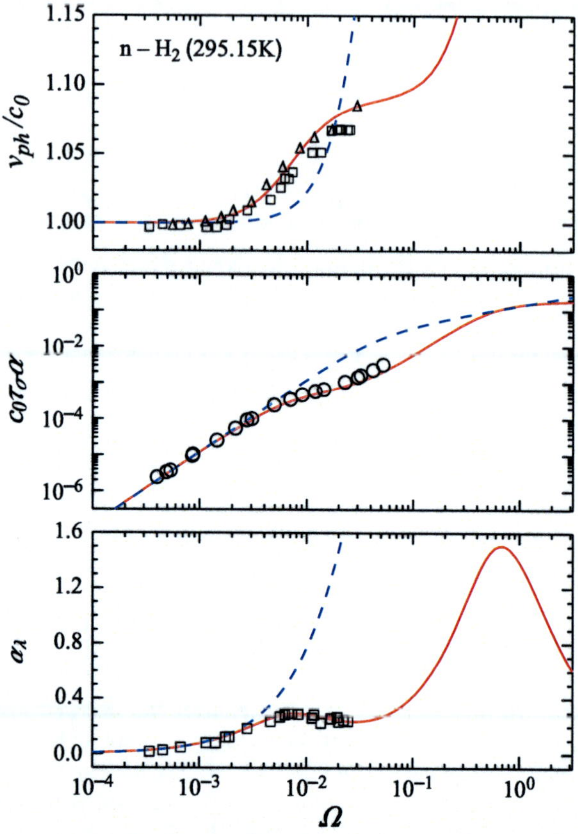

Fig. 16.3 Dependence of the dimensionless phase velocity v_{ph}/c_0, the attenuation factor $c_0\tau_\sigma\alpha$ and the attenuation per wavelength α_λ on the dimensionless frequency Ω for n-H_2. The circles, squares, and triangles in the figures are, respectively, the experimental data at 293 K by Sluijter et al. [345], those at 295.15 K by Winter and Hill [343], and those at 296.8 K by Rhodes [344]. The solid and dashed lines are predictions at 295.15 K by the ET_{14} and NSF theories, respectively. We adopt $\nu/\mu = 33.1$

comparison of the theoretical predictions with these experimental data is shown in Fig. 16.4. We see again that the ET_{14} theory can describe the experimental data very well. The values of the parameter ν/μ are selected to be 27.2, 41.9, 36.8, and 40.3, respectively.

For p-H_2, we compare the theoretical predictions with the experimental data on the phase velocity at $T_0 = 273.8, 298.4\,\text{K}$ by Rhodes [344] and on the attenuation factor at $T_0 = 77.3, 90.2, 170, 293\,\text{K}$ by Sluijter et al. [345, 350].

We have a similar result as shown in Fig. 16.5, where the selected values of the parameter ν/μ are 75.0, 83.6, 54.8, 28.8, respectively, for $T_0 = 77.3, 90.2, 170, 293\,\text{K}$. Remarkable points in this case are qualitatively the same as in the case of n-H_2 above.

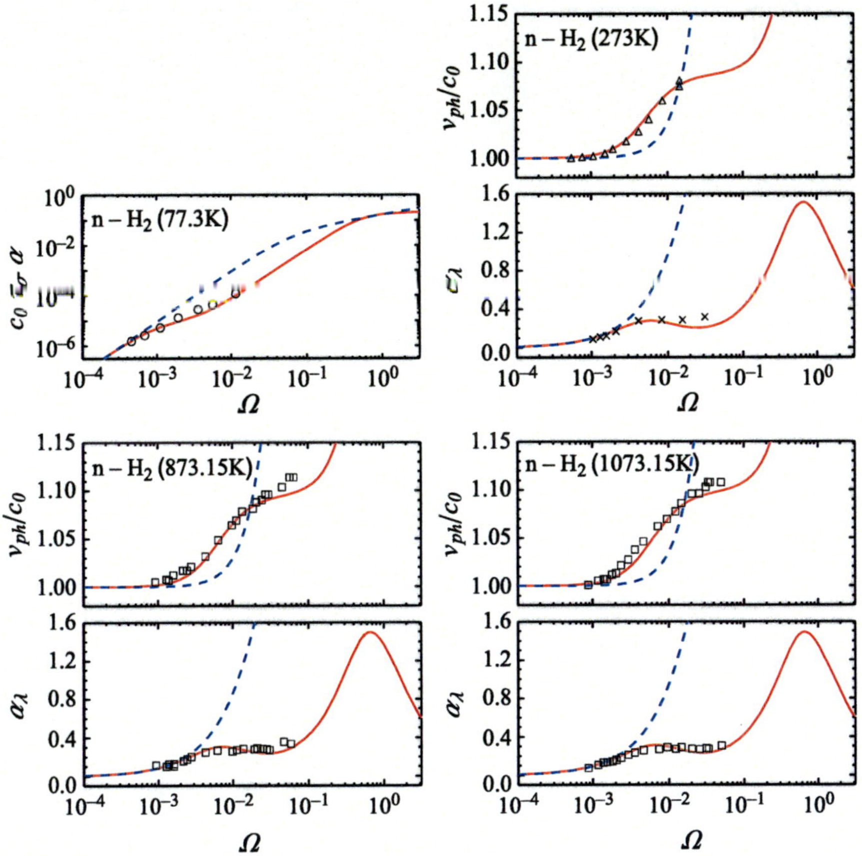

Fig. 16.4 Dependence of the dimensionless phase velocity v_{ph}/c_0, the attenuation factor $c_0\tau_\sigma\alpha$, and the attenuation per wavelength α_λ on the dimensionless frequency Ω for n-H_2. The circles, squares, triangles, and crosses in the figures are, respectively, the experimental data by Sluijter et al. [345] at 77.3 K, Winter and Hill [343] at 873.15, 1073.15 K, Rhodes [344] at 273.5 K, and Stewart and Stewart [346] at 273.15 K. The solid and dashed lines are predictions at $T_0 = 77.3, 273, 873.15, 1073.15$ K by the ET$_{14}$ and NSF theories, respectively. We adopt $\nu/\mu = 27.2, 41.9, 36.8, 40.3$, respectively, for $T_0 = 77.3, 273, 873.15, 1073.15$ K

16.4.3.2 Deuterium Gases: n-D_2 and o-D_2

Comparisons are also made for n-D_2 at $T_0 = 295.15$ K with $\nu/\mu = 20.9$ in Fig. 16.6, and it shows the superiority of the prediction by the ET$_{14}$ theory to that by the NSF theory in high-frequency region. Also comparisons at other temperatures $T_0 = 77.3, 273.15, 773.15$, and 1073.15 K with $\nu/\mu = 35.7, 24.7, 30.9$, and 35.9, respectively, are shown in Fig. 16.7. Comparisons for o-D_2 at $T_0 = 77.3, 90.2$, and 293 K with $\nu/\mu = 45.4, 33.6$, and 22.6 are shown in Fig. 16.8. From these figures, we have qualitatively the same observations as those in the case of hydrogen gases.

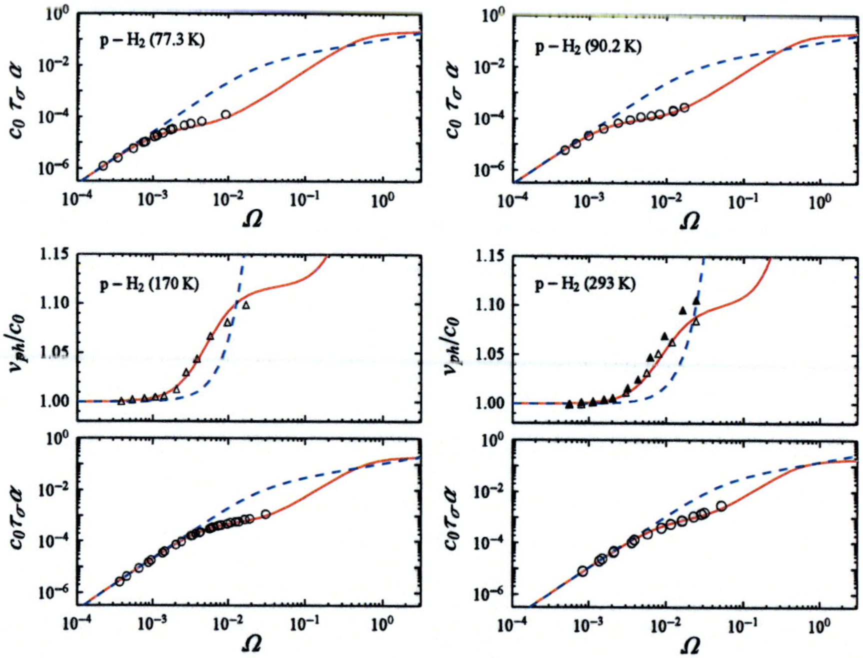

Fig. 16.5 Dependence of the dimensionless phase velocity v_{ph}/c_0 and the attenuation factor $c_0\tau_\sigma\alpha$ on the dimensionless frequency Ω for p-H_2. The circles and empty triangles in the figures are, respectively, the experimental data by Sluijter et al. [345, 350] at $T_0 = 77.3, 90.2, 170, 293$ K and by Rhodes [344] at $T_0 = 298.4$ K. As a reference the experimental data by Rhodes [344] at $T_0 = 273.8$ K are on the figure as the filled triangles. The solid and dashed lines are predictions at $T_0 = 77.3, 90.2, 170, 293$ K by the ET_{14} and NSF theories, respectively. We adopt ν/μ=75.0, 83.6, 54.8, 28.8, respectively, for $T_0 = 77.3, 90.2, 170, 293$ K

16.4.3.3 Hydrogen Deuteride Gases: HD

Lastly we show the results of HD gases at $T_0 = 77.3$ and 293 K in Fig. 16.9. We notice the following points: (1) The difference between the two theories is small and the theoretical predictions are consistent with the experimental data in the range: $\Omega \leqslant 10^{-1}$. This means that the local equilibrium assumption holds well up to $\Omega \fallingdotseq 10^{-1}$, while, for the other gases analyzed above, the assumption holds until $\Omega \fallingdotseq 10^{-3}$. (2) The values of ν/μ adopted here are 1.84 and 2.27. These values are $O(1)$, that is, $\nu \sim \mu$, and are very small compared with those obtained for the other gases discussed above. We will discuss this interesting fact below.

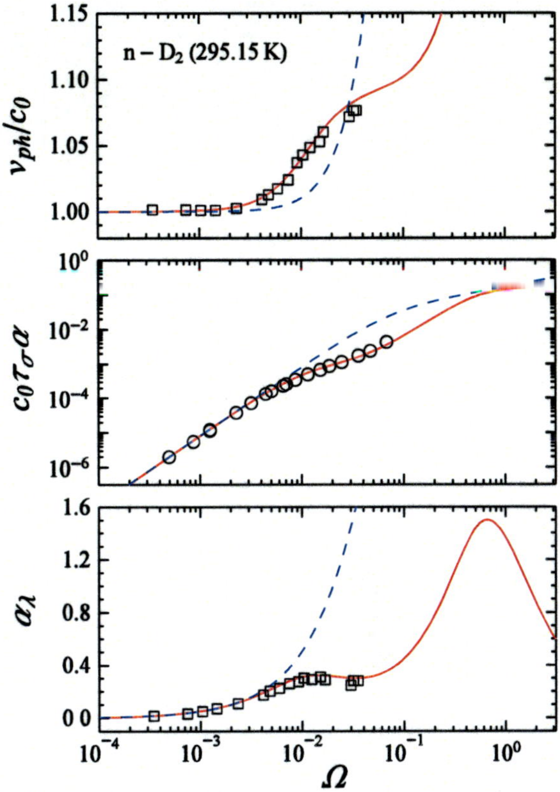

Fig. 16.6 Dependence of the dimensionless phase velocity v_{ph}/c_0, the attenuation factor $c_0\tau_\sigma\alpha$, and the attenuation per wavelength α_λ on the dimensionless frequency Ω for n-D_2. The circles and squares in the figures are, respectively, the experimental data by Sluijter et al. [345] at 293 K and Winter and Hill [343] at 295.15 K. The solid and dashed lines are predictions at 295.15 K by the ET_{14} and NSF theories, respectively. We adopt $\nu/\mu = 20.9$

16.4.4 *Some Remarks*

Remarks (A)–(E) are made:

(A) We have seen clearly that the ET_{14} theory is consistent with the experimental data even in the high-frequency range where the local equilibrium assumption is no longer valid. There are potentially many research fields where the ET_{14} theory may play a crucial role, for example, fields of acoustics [312] and gas dynamics [137].

(B) From the values of the ratios τ_q^* and τ_Π^* in Table 16.1, we have noticed an interesting fact that, except for HD gas, τ_Π is much larger than τ_σ, while τ_σ and τ_q are comparable with each other. This fact was reported also in some kinetic theoretical studies [17, 354]. By using the result summarized in Table 16.2, the relaxation times for given T_0 and p_0 can be estimated. For example, the relaxation times in a n-H_2 gas at $p_0 = 10^3$ [Pa] and $T_0 = 77.3$ [K] can be calculated: $\tau_\sigma = 3.50 \times 10^{-9}$ [s], $\tau_\Pi = 3.36 \times 10^{-6}$ [s], and $\tau_q = 4.70 \times 10^{-9}$ [s].

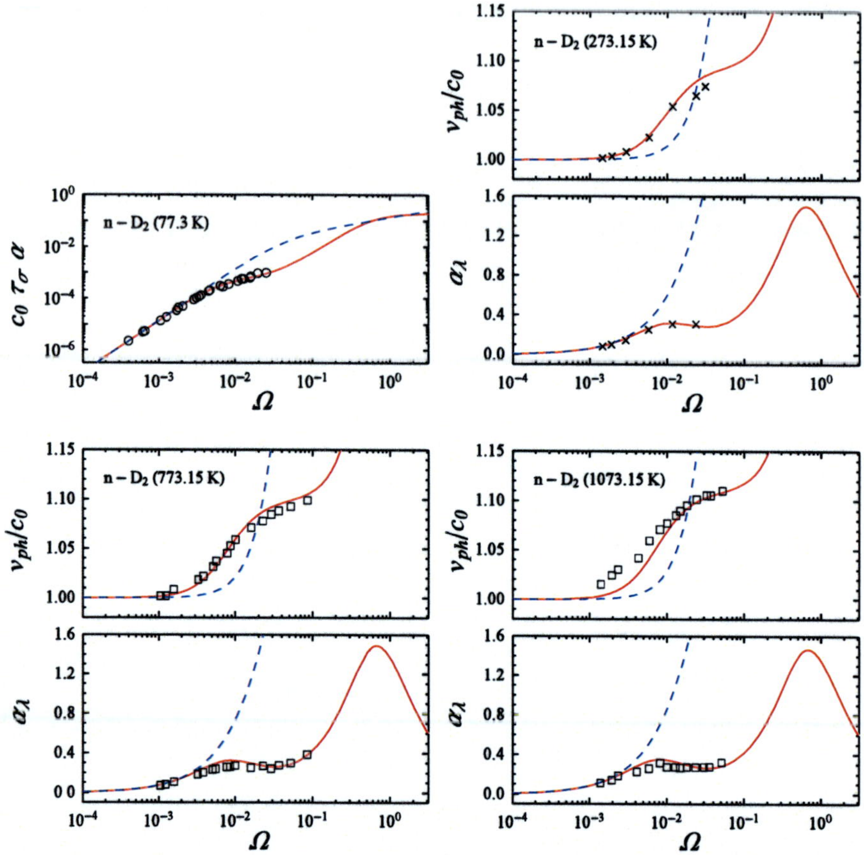

Fig. 16.7 Dependence of the dimensionless phase velocity v_{ph}/c_0, the attenuation factor $c_0\tau_\sigma\alpha$, and the attenuation per wavelength α_λ on the dimensionless frequency Ω for n-D_2. The circles, squares, and crosses in the figures are, respectively, the experimental data by Sluijter et al. [345] at 77.3 K, Winter and Hill [343] at 773.15, 1073.15 K, and Stewart and Stewart [346] at 273.15 K. The solid and dashed lines are predictions by the ET_{14} and NSF theories, respectively. We adopt $\nu/\mu = 35.7, 24.7, 30.9$, and 35.9, respectively for $T_0 = 77.3, 273.15, 773.15$, and 1073.15 K

In the paper [322], it was pointed out that the relaxation time τ_Π is in the same order of magnitude as the relaxation time of the energy exchange between the molecular translational mode and the internal modes. The results obtained above suggest that the sharp temperature change of the specific heat due to the rotational modes $\hat{c}_v^R$ depicted in Fig. 16.2 is somehow related to the emergence of the large value of τ_Π. The detailed study of this subject is, however, beyond the scope of the present phenomenological study, and its statistical-mechanical or kinetic-theoretical study by taking into account the realistic collision processes between the constituent molecules is required.

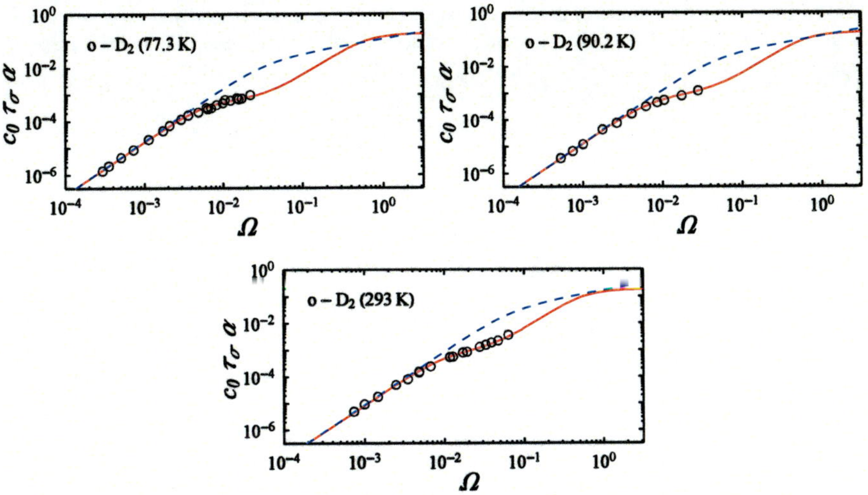

Fig. 16.8 Dependence of the dimensionless attenuation factor $c_0\tau_\sigma\alpha$ on the dimensionless frequency Ω for o-D_2 at $T_0 = 77.3$ K, 90.2 K, and 293 K. The circles are the experimental data by Sluijter et al. [345]. The solid and dashed lines are predictions by the ET_{14} and NSF theories, respectively. We adopt $\nu/\mu = 45.4$, 33.6, and 22.6 from 77.3 K to 293 K

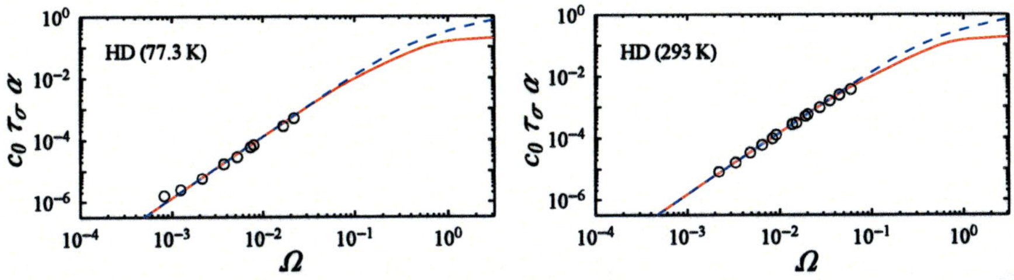

Fig. 16.9 Dependence of the dimensionless attenuation factor $c_0\tau_\sigma\alpha$ on the dimensionless frequency Ω for HD at $T_0 = 77.3$ K and 293 K. The circles are the experimental data by Sluijter et al. [345]. The solid and dashed lines are predictions by the ET_{14} and NSF theories, respectively. We adopt $\nu/\mu = 1.84$ at 77.3 K and 2.27 at 293 K

(C) From the values of ν/μ in Table 16.1, we have also noticed a similar fact that, except for HD gas, the bulk viscosity ν is much larger than the shear viscosity μ. The similarity is natural because there are relations between the viscosities and the relaxation times as shown in (6.36). A point to be emphasized here is that, since the direct experiments to measure the bulk viscosity are usually difficult, the method for the evaluation of the bulk viscosity utilized here through analyzing the dispersion relation on the basis of the ET_{14} theory is quite useful. The values of ν thus evaluated are summarized in Table 16.1. See the paper [107] where the temperature dependence of the bulk viscosity in hydrogen and deuterium gases is obtained. See also the recent studies of the bulk viscosity [355–359].

Table 16.2 Relaxation times of the deviatoric part of the viscous stress τ_σ, dynamic pressure τ_Π and heat flux τ_q multiplied by the pressure p_0 for several values of T_0 in H_2, D_2 and HD gases

Gas	T_0[K]	$\tau_\sigma p_0$[s $\cdot$ μPa]	$\tau_\Pi p_0$[s $\cdot$ μPa]	$\tau_q p_0$[s $\cdot$ μPa]
$n - H_2$	77.3	3.50	3360	4.70
	273	8.33	1380	12.2
	295.15	8.95	1150	13.1
	873.15	18.6	2510	27.6
	1073.15	21.0	3000	31.2
$p - H_2$	77.3	3.50	2710	4.64
	90.2	3.97	2030	5.16
	170	6.10	1010	6.93
	293	8.82	894	12.9
$n - D_2$	77.3	4.82	632	6.25
	273.15	11.8	1090	18.8
	295.15	12.6	990	20.0
	773.15	24.2	2640	35.0
	1073.15	30.4	3560	43.2
$o - D_2$	77.3	4.82	673	6.09
	90.2	5.50	562	6.81
	293	12.3	1040	18.1
HD	77.3	4.21	28.3	5.32
	293	10.8	91.9	15.5

(D) There has been a phenomenological theory of the dispersion relation for sound, the basic equations of which are composed of the relaxation equations for some nonequilibrium parameters and the Euler (or NSF) equations for the conservation laws [7, 360]. One crucial point is that the theory is based on the local equilibrium assumption. In this respect, this may be regarded as a theory in the framework of thermodynamics of irreversible processes [7]. Because of this, we have compared the RET theory only with the NSF theory as a representative one. In Chap. 12, the relationship between the simplified RET theory with 6 fields and the Meixner theory with one relaxation equation was studied in detail.

(E) In order to study the effect of the large value of the relaxation time τ_Π on various nonequilibrium phenomena such as shock wave phenomena, it seems to be reasonable to adopt a simpler model than the one adopted here. The ET$_6$ and ET$_7$ theories explained in Chaps. 12–14 play an important role in such studies. See Sect. 16.7, and also Chap. 18.

16.5 Necessity of the ET_{15} Theory

It is shown in the above that the ET_{14} theory can make much better predictions in the ultrasonic wave propagation than the NSF theory. However, as discussed in Chap. 8, we know that there is a limitation in the applicability of the ET_{14} theory, because the theory treats the molecular rotational and vibrational degrees of freedom as a unit.

In the first half of this chapter, the comparison of the theoretical predictions with experimental data on hydrogen, deuterium, and hydrogen deuteride gases was made in low temperature range where the molecular vibrational modes are not so much excited as seen in Table 16.1 and in Fig. 16.2. In such cases, the vibrational modes do not play an important role, and therefore the ET_{14} theory works well.

In the second half of the present Chapter, using the ET_{15} theory explained in Chap. 8, we revisit the dispersion relation of a gas in a wider temperature range where molecular rotational and vibrational modes play significant roles individually [342]. In particular, we study the cross effect among the molecular relaxations, shear viscosity, and heat conduction on the dispersion and absorption of sound wave. In what follows, after summarizing the linearized system of field equations in ET_{15}, we derive the dispersion relation of sound wave. The temperature and frequency dependences of the phase velocity and the attenuation per wavelength in the case with typical relaxation times are discussed. A peak of the attenuation per wavelength in a low-frequency region, which will be useful in the analysis of experimental data, is predicted. The ET_7 theory is also utilized for the study in such a low-frequency region [135] because the difference between the results derived from the ET_{15} and ET_7 theories is useful for understanding the molecular relaxation phenomena. Theoretical prediction of the attenuation is compared with the experimental data for CO_2, Cl_2, and Br_2 gases.

16.6 Linearized ET_{15} System of Field Equations

We summarize the linearized ET_{15} system in terms of the 15 independent field variables $\mathbf{u} \equiv (\rho, v_i, T, \Delta^R, \Delta^V, \sigma_{\langle ij\rangle}, q_i)$. For details of the definitions of quantities, see Chap. 8. Let $\mathbf{u}_0 \equiv (\rho_0, 0, T_0, 0, 0, 0, 0)$ be a reference equilibrium state, then we have the linearized system in the neighborhood of $\mathbf{u}_0$ for the perturbed fields $\bar{\mathbf{u}} = (\bar{\rho}, \bar{v}_i, \bar{T}, \bar{\Delta}^R, \bar{\Delta}^V, \bar{\sigma}_{\langle ij\rangle}, \bar{q}_i)$ as follows:

$$\frac{\partial \rho}{\partial t} + \rho_0 \frac{\partial v_k}{\partial x_k} = 0,$$

$$\rho_0 \frac{\partial v_i}{\partial t} + \frac{k_B}{m} T_0 \frac{\partial \rho}{\partial x_i} + \frac{k_B}{m} \rho_0 \frac{\partial T}{\partial x_i} - \frac{2}{3} \rho_0 \frac{\partial \Delta^R}{\partial x_i} - \frac{2}{3} \rho_0 \frac{\partial \Delta^V}{\partial x_i} - \frac{\partial \sigma_{\langle ij\rangle}}{\partial x_j} = 0,$$

$$\frac{k_B}{m}\rho_0 \hat{c}_v \frac{\partial T}{\partial t} + p_0 \frac{\partial v_k}{\partial x_k} + \frac{\partial q_k}{\partial x_k} = 0,$$

$$\frac{\partial \Delta^R}{\partial t} - \frac{k_B}{m} T_0 \frac{\hat{c}^R_{v0}}{\hat{c}_{v0}} \frac{\partial v_k}{\partial x_k} - \frac{\hat{c}^R_{v0}}{\rho_0 \hat{c}_{v0}(\hat{c}_{v0}+1)} \frac{\partial q_k}{\partial x_k} = P^R,$$

$$\frac{\partial \Delta^V}{\partial t} - \frac{k_B}{m} T_0 \frac{\hat{c}^V_{v0}}{\hat{c}_{v0}} \frac{\partial v_k}{\partial x_k} - \frac{\hat{c}^V_{v0}}{\rho_0 \hat{c}_{v0}(\hat{c}_{v0}+1)} \frac{\partial q_k}{\partial x_k} = P^V,$$

$$\frac{\partial \sigma_{\langle ij \rangle}}{\partial t} - 2p_0 \frac{\partial v_{\langle i}}{\partial x_{j\rangle}} - \frac{2}{\hat{c}_{v0}+1} \frac{\partial q_{\langle i}}{\partial x_{j\rangle}} = P_{\langle ij \rangle},$$

$$\frac{\partial q_i}{\partial t} + \left(\hat{c}_{v0}+1\right) \frac{k_B}{m} p_0 \frac{\partial T}{\partial x_i} - \frac{2}{3} p_0 \frac{\partial \Delta^R}{\partial x_i} - \frac{2}{3} p_0 \frac{\partial \Delta^V}{\partial x_i} - \frac{p_0}{\rho_0} \frac{\partial \sigma_{\langle ik \rangle}}{\partial x_k} = Q_i,$$

where $\hat{c}_v = mc_v/k_B$ is the dimensionless specific heat, a quantity with the index 0 indicates the one evaluated at the reference equilibrium state, and P^R, P^V, $P_{\langle ij \rangle}$, and Q_i are the velocity independent (intrinsic) part of the production terms of Δ^R, Δ^V, $\sigma_{\langle ij \rangle}$, and q_i, respectively. Near equilibrium, we have the following expressions:

$$P^R = \begin{cases} -\dfrac{1}{\tau_\delta}\Delta^R - \dfrac{\hat{c}^R_v}{\hat{c}^{K+R}_v}\left(\dfrac{1}{\tau_\delta} - \dfrac{1}{\tau}\right)\Delta^V & \text{for (KR)-process,} \\ -\dfrac{1}{\tau}\Delta^R & \text{for (KV)-process,} \\ -\dfrac{1}{\hat{c}^{R+V}_v}\left(\dfrac{\hat{c}^R_v}{\tau} + \dfrac{\hat{c}^V_v}{\tau_\delta}\right)\Delta^R + \dfrac{\hat{c}^R_v}{\hat{c}^{R+V}_v}\left(\dfrac{1}{\tau_\delta} - \dfrac{1}{\tau}\right)\Delta^V & \text{for (RV)-process,} \end{cases}$$

$$P^V = \begin{cases} -\dfrac{1}{\tau}\Delta^V & \text{for (KR)-process,} \\ -\dfrac{\hat{c}^V_v}{\hat{c}^{K+V}_v}\left(\dfrac{1}{\tau_\delta} - \dfrac{1}{\tau}\right)\Delta^R - \dfrac{1}{\tau_\delta}\Delta^V & \text{for (KV)-process,} \\ \dfrac{\hat{c}^V_v}{\hat{c}^{R+V}_v}\left(\dfrac{1}{\tau_\delta} - \dfrac{1}{\tau}\right)\Delta^R - \dfrac{1}{\hat{c}^{R+V}_v}\left(\dfrac{\hat{c}^V_v}{\tau} + \dfrac{\hat{c}^R_v}{\tau_\delta}\right)\Delta^V & \text{for (RV)-process,} \end{cases}$$

and

$$P_{\langle ij \rangle} = -\frac{1}{\tau_\sigma}\sigma_{\langle ij \rangle}, \quad Q_i = -\frac{1}{\tau_q} q_i.$$

In usual cases, the order of magnitude of the relaxation times is: $O(\tau_\sigma) \sim O(\tau_q) \sim O(\tau_K)$. Here and hereafter we ignore a bar on a perturbed quantity for simplicity.

We remark that, from (8.22), it is possible to adopt the linearized field equation of Δ^K instead of the field equation of Δ^R or Δ^V, which is expressed as follows:

$$\frac{\partial \Delta^K}{\partial t} + \frac{k_B}{m} T_0 \frac{\hat{c}_{v0} - \hat{c}_{v0}^K}{\hat{c}_{v0}} \frac{\partial v_k}{\partial x_k} + \frac{\hat{c}_{v0} - \hat{c}_{v0}^K}{\rho_0 \hat{c}_{v0}(\hat{c}_{v0} + 1)} \frac{\partial q_k}{\partial x_k} = P^K.$$

Here the production term P^K satisfies the following relation due to the energy conservation law:

$$P^K + P^R + P^V = 0.$$

16.7 Dispersion Relation for Sound: Revisited

Dependence of the phase velocity and the attenuation per wavelength on the frequency is studied. In particular, we study the attenuation due to the relaxation processes, and clarify the characteristic features of ET_{15} by comparing to ET_{14}. Temperature dependence of a peak of attenuation per wavelength in a lower frequency region, which is relatively easy to be measured in experiments, is discussed particularly in detail.

16.7.1 Phase Velocity and Attenuation Factor

As in the first half of this chapter, let us confine our study within one-dimensional problem. Taking into account the expressions (16.2), we study a harmonic plane wave for the fields $\mathbf{u} = (\rho, v, T, \Delta^R, \Delta^V, \sigma, q)$ with the angular frequency ω and the complex wave number k. Then the dispersion relation $\omega = \omega(k)$, phase velocity v_{ph}, and attenuation factor α are obtained in a similar way as above. For simplicity, we omit their explicit expressions but show only final results by using the figures below. We note that τ_δ does not appear explicitly in the dispersion relation of ET_{15} with (RV)-process (ET_{15}^{RV}), and the dispersion relation coincides with the one derived from ET_{14}.

16.7.2 Dimensionless Variables and the Order of Magnitude of the Ratio of Relaxation Times

Let us introduce the following dimensionless parameters:

$$\Omega = \tau_\sigma \omega = \frac{\mu\omega}{p}, \quad \tau_q^* = \frac{\tau_q}{\tau_\sigma} = \frac{1}{\frac{k_B}{m}\left(\hat{c}_v + 1\right)} \frac{\kappa}{\mu}, \quad \tau^* = \frac{\tau}{\tau_\sigma}, \quad \tau_\delta^* = \frac{\tau_\delta}{\tau_\sigma}.$$

Then the dispersion relation depends on these parameters with dimensionless specific heats, $\hat{c}_v^K$, $\hat{c}_v^R$ and $\hat{c}_v^V$. We emphasize that $\omega = \omega(k)$ does not depend on ρ, and that its temperature dependence is determined through the dimensionless specific heats that can be evaluated from statistical mechanics or experimental data. The ratio τ_q^* is evaluated by the experimental data on shear viscosity and heat conductivity as shown in (8.43). As mentioned above, usually $O(\tau_q^*) = 1$. Since we have not enough experimental data on the bulk viscosity ν, we may adopt τ^* and τ_δ^* as adjustable parameters. We just remark that, from (8.43), these two parameters are related to the ratio of the shear and bulk viscosities as follows:

$$\frac{\nu}{\mu} = \begin{cases} \dfrac{\hat{c}_v^{\mathfrak{c}}}{\hat{c}_v^{\mathfrak{b}}\hat{c}_v^{\mathfrak{b}+\mathfrak{c}}}\tau_\delta^* + \dfrac{\hat{c}_v - \hat{c}_v^{\mathfrak{b}+\mathfrak{c}}}{\hat{c}_v^{\mathfrak{b}+\mathfrak{c}}\hat{c}_v}\tau^* & \text{for } (\mathfrak{bc})\text{-process } ((\mathfrak{bc}) = (KR) \text{ or } (KV)) \\ \dfrac{\hat{c}_v^{R+V}}{\hat{c}_v^K \hat{c}_v}\tau^* & \text{for (RV)-process} \end{cases}.$$

Its value varies greatly depending on the gas under study. For example, $O(\nu/\mu) > 10^3$ in Cl_2 and CO_2 gases, while $O(\nu/\mu) \sim O(1)$ in N_2 and CO gases [126].

16.7.3 *Frequency Dependence of Phase Velocity and Attenuation per Wavelength*

Since the dispersion relation depends on the temperature through the specific heats, we firstly remark on the specific heats. We estimate the specific heats on the basis of the statistical-mechanical considerations in the same way as in 16.4.1.1. That is, the specific heats c_v^R and c_v^V are evaluated by using the rotational and vibrational partition functions Z^R and Z^V of a molecule. In the case of diatomic molecule, a typical temperature dependence of the specific heats c_v, $c_v^{K+R} \equiv c_v^K + c_v^R$, and $c_v^{K+V} \equiv c_v^K + c_v^V$ is shown in Fig. 16.10 where Θ_R and Θ_V are, respectively, the characteristic rotational temperature and vibrational temperature. For details, see [342].

To grasp the feature of the phase velocity v_{ph} and the attenuation per wavelength α_λ, we study the (KR)-process, as an example, with the relaxation times $\tau_q^* = 1$, $\tau^* = 10^4$, and $\tau_\delta^* = 10^2$ or 10^1 in the five typical cases (A)–(E) in Table 16.3. The temperature of the reference equilibrium state $\mathbf{u}_0$ increases from the case (A) to the case (E) as seen in Fig. 16.10. Then we have Fig. 16.11.

From the figure, we notice the following points:

1. Around $\Omega \sim \tau_q^{*-1}(= 1)$, α_λ has a large peak due to the dissipation by the shear viscosity and heat conduction. The previous approaches based on the NSF theory [7], however, can not capture this peak because this theory only predicts a curve which increases monotonically with the increase of the frequency and approaches a limit value. This indicates that the attenuation due to the dissipation in such a

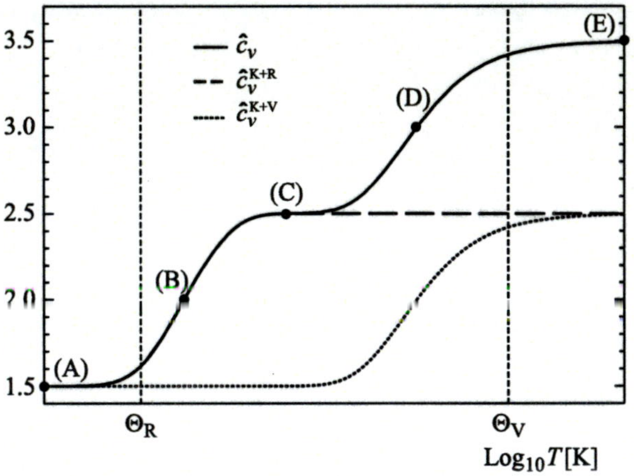

Fig. 16.10 Typical temperature dependence of the dimensionless specific heats; $\hat{c}_v$, $\hat{c}_v^{K+R}$, and $\hat{c}_v^{K+V}$. The five cases (A)–(E) listed in Table 16.3 are also indicated. Θ_R and Θ_V are, respectively, the characteristic rotational temperature and vibrational temperature

Table 16.3 Five typical cases, in which the translational mode is fully excited

Case	Specific heats	Rotational mode	Vibrational mode
(A)	$\hat{c}_v^K = 3/2, \hat{c}_v^R = 0, \hat{c}_v^V = 0$	Ground state	Ground state
(B)	$\hat{c}_v^K = 3/2, \hat{c}_v^R = 1/2, \hat{c}_v^V = 0$	Partly excited	Ground state
(C)	$\hat{c}_v^K = 3/2, \hat{c}_v^R = 1, \hat{c}_v^V = 0$	Fully excited	Ground state
(D)	$\hat{c}_v^K = 3/2, \hat{c}_v^R = 1, \hat{c}_v^V = 1/2$	Fully excited	Partly excited
(E)	$\hat{c}_v^K = 3/2, \hat{c}_v^R = 1, \hat{c}_v^V = 1$	Fully excited	Fully excited

high frequency region is out of the applicability range of the NSF theory. For v_{ph}, there appears a steep change. Such peak and steep change have been predicted also by ET_{14}.

2. In the case with $\tau_\delta^* = 10^2$, around $\Omega \sim \tau_\delta^{*-1}(= 10^{-2})$, we can observe clearly a peak of α_λ and a steep change of v_{ph}. Since this is due to the relaxation of rotational mode in the (KR)-process, such peak and steep change appear except for the case (A) in Table 16.3. While, in the case with $\tau_\delta^* = 10^1$, no peak and steep change around $\Omega \sim \tau_\delta^{*-1}(= 10^{-1})$ are observed because the small peak (small steep change) is absorbed to the big peak (big steep change). Therefore, when we have experimental data on the curve around such a big peak, it is necessary to pick out the information of the rotational relaxation from the curve. Such a separation can be carried out by using the ET_{15} and ET_7 theories simultaneously.
3. Around $\Omega \sim \tau^{*-1}(= 10^{-4})$, due to the contribution of the slow relaxation with τ, a small peak of α_λ and a steep change of v_{ph} appear. Since this relaxation is due to the vibrational mode in the (KR)-process, such peak and steep change disappear in the cases (A), (B), and (C) where there appear up to two peaks.

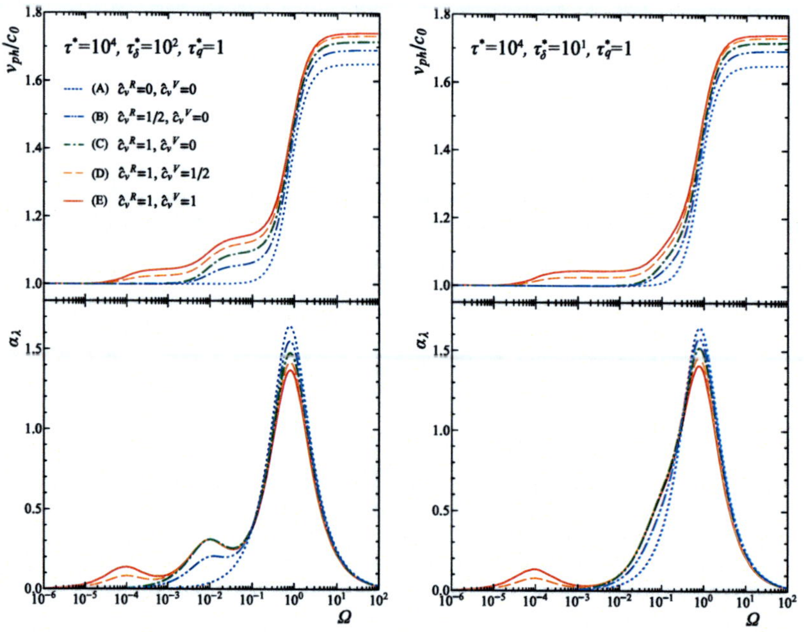

Fig. 16.11 Typical dependence of the dimensionless phase velocity v_{ph}/c_0 and the attenuation per wavelength α_λ on the dimensionless frequency Ω predicted by the ET_{15}^{KR} theory in five cases (A)–(E) listed in Table 16.3. The left and right figures correspond to the cases with $\tau_\delta^* = 10^2$ and 10^1, respectively

In such cases, by choosing a suitable relaxation time, ET_{14} becomes a good approximation of ET_{15}. We remark that the curve is similar to the one derived from a simple relaxation equation of the energy exchange [312]. However, the applicability of such an approach is limited only to a low-frequency region.

For (KV)-process, we have similar temperature dependences of v_{ph} and α_λ although the rapid relaxation with τ_δ is due to the vibrational mode and the slow relaxation with τ is due to the rotational mode. For (RV)-process, as was mentioned above, the contribution of the rapid relaxation does not appear explicitly. There appear at most two steep changes in v_{ph} and two peaks in α_λ.

We remark that, in a lower temperature range where the vibrational mode is not excited, all ET_{15} theories give the same prediction, and ET_{14} is a good approximation of ET_{15} if we choose a suitable relaxation time. On the other hand, in a higher temperature range where both rotational and vibrational modes are excited, ET_{14} is no longer valid.

Finally we note that the dispersion relation of strongly nonequilibrium gases was studied on the basis of the state-to-state kinetic model that takes into account the vibrational level populations [361]. In particular, the influence of vibrational excitations on the phase velocity and the attenuation coefficient in a wide temperature range was elucidated in the paper. Comparison between these results and the present results must be an interesting future work.

16.7.4 Peak of α_λ Corresponding to the the Slow Relaxation with τ

We study in detail the peak of α_λ due to the the slow relaxation with τ, i.e., the left peak in Fig. 16.11. Since such a peak value in a low-frequency region is relatively easy to be measured experimentally, comparison between the present result and experimental data on α_λ can afford a suitable method (selection method) to identify the relaxation process among (KR), (KV), and (RV) in a gas under study. It is also worthy of note that, if we focus our attention only on the peak in a low-frequency region, the simple ET$_7$ theory can give us almost the same results as those given by ET$_{15}$ under the condition that $\tau^* >> 1$.

As an example, let us study a homonuclear diatomic gas of which specific heats are given, for example, in [334, 362], and we assume, as a typical example, the rotational and vibrational characteristic temperatures: $\Theta_R = 10$[K] and $\Theta_V = 800$[K]. Under the condition $\tau_\delta^* = 10$, we study the effect of τ^* on the peak value α_λ^{peak} by adopting three cases with $\tau^* = 10^2, 10^3, 10^4$. As mentioned above, by using the ET$_7$ theory, the peak value can be estimated well by the value at the frequency given by

$$\Omega^{peak} = \frac{1}{\tau^* \hat{U}_E^{\mathfrak{bc}}} \qquad ((\mathfrak{bc}) = (\mathrm{KR}), (\mathrm{KV}), (\mathrm{RV})),$$

where

$$\hat{U}_E^{\mathfrak{bc}} = \sqrt{\frac{\hat{c}_v}{\hat{c}_v^{\mathfrak{b}+\mathfrak{c}}} \frac{1+\hat{c}_v^{\mathfrak{b}+\mathfrak{c}}}{1+\hat{c}_v}} \qquad ((\mathfrak{bc}) = (\mathrm{KR}), (\mathrm{KV}), (\mathrm{RV})).$$

One remark is that α_λ^{peak} does not necessarily mean that there exists a peak because there may occur a merger of two peaks as seen above.

The temperature and τ^* dependence of α_λ^{peak} is shown in Fig. 16.12.

When τ^* becomes large, the curve approaches the one expressed by

$$\alpha_\lambda^{peak} = 2\pi \frac{\hat{U}_E^{\mathfrak{bc}} - 1}{\hat{U}_E^{\mathfrak{bc}} + 1} \qquad ((\mathfrak{bc}) = (\mathrm{KR}), (\mathrm{KV}), (\mathrm{RV})),$$

which was obtained also by using ET$_7$. On the other hand, when τ^* becomes small, the contributions by the rapid relaxations play role. The value in a temperature range smaller than Θ_R reflects the contribution from the dissipation, i.e., the peak

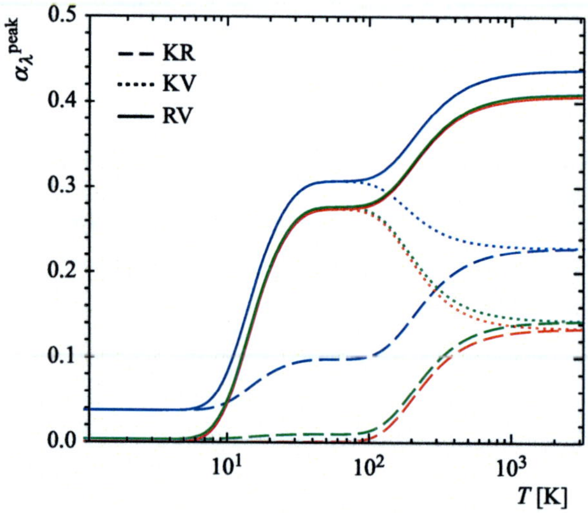

Fig. 16.12 Temperature and τ^* dependence of α_λ^{peak} in ET_{15} for (KR), (KV), and (RV)-processes. The red, green, and blue curves correspond, respectively, to the cases with $\tau^* = 10^4, 10^3, 10^2$

related to τ_q^*. In a temperature range between Θ_R and Θ_V, the relaxation of the rotational mode contributes. In the temperature range larger than Θ_V, the vibrational relaxation also contributes.

Remark 16.1 We have discussed the general features of the dispersion relation by studying some typical cases. Therefore, for given experimental data, we can determine the most suitable relaxation process among possible (KR), (KV), (RV)-processes. As is discussed above, this is just a selection method for the most suitable relaxation process.

Remark 16.2 We point out that the general features of the dispersion relation discussed above can be found not only diatomic gases but also in polyatomic gases because such features come mainly from the global dependence of the specific heats on the temperature.

16.7.5 Comparison with Experimental Data

We compare the theoretical prediction of α_λ by ET_7 [135] with the experimental data of CO_2 [363], Cl_2, and Br_2 gases [364].

We evaluate the specific heats of CO_2, Cl_2, and Br_2 gases by the statistical-mechanical method. In these gases, the characteristic rotational temperature Θ_R is very low. In fact, from the data on the rotational constant at the ground state [349], it is estimated as 0.56 K for CO_2, 0.35 K for Cl_2, and 0.12 K for Br_2. Therefore,

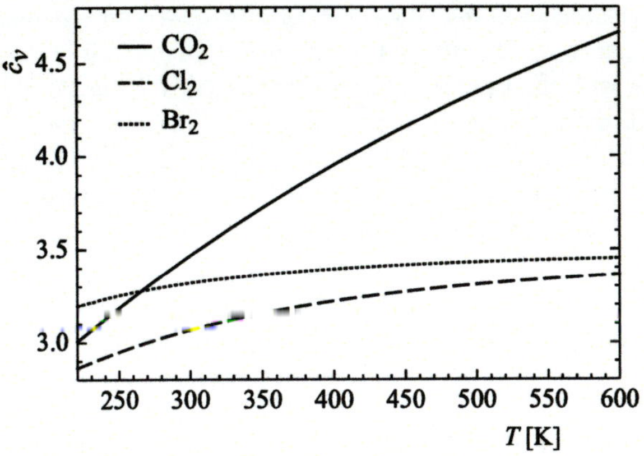

Fig. 16.13 Dependence of $\hat{c}_v$ on T

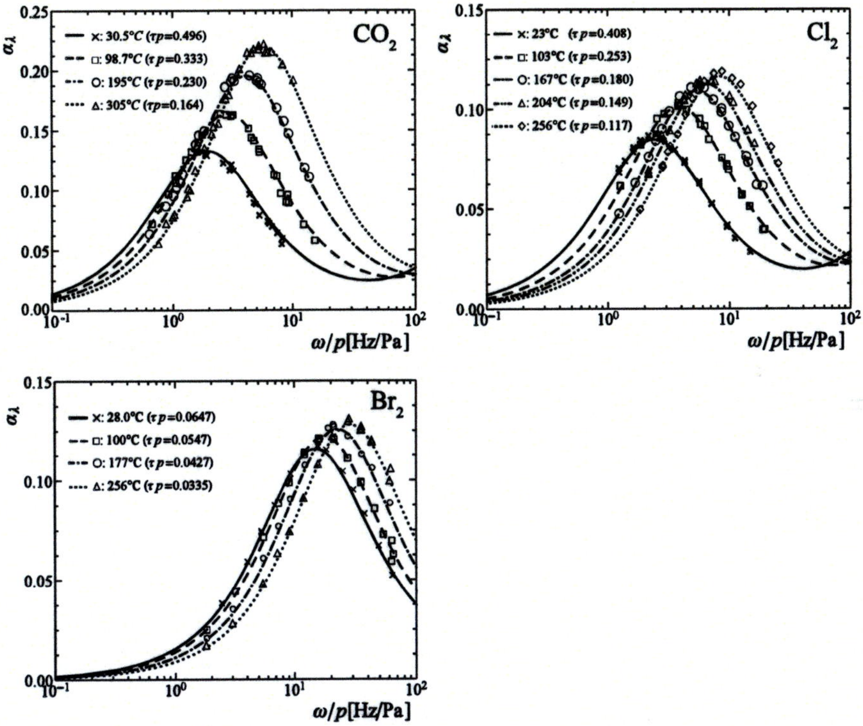

Fig. 16.14 Dependence of α_λ on ω/p [Hz/Pa] for several temperatures with $\hat{c}_v^{K+R} = 5/2$, $\hat{\tau}_\delta = 10^{-3}$ in rarefied CO_2, Cl_2, and Br_2 gases [363, 364]. A parameter $p\tau$ is chosen to fit the experimental data by the least square method

in the temperature range higher than the room temperature, the rotational degrees of freedom of these gases are in a fully excited state with $\hat{c}_v^{K+R} = 5/2$. While the temperature dependence of the vibrational specific heat is approximately calculated by using the partition function for molecular harmonic-vibrational modes. For CO_2 molecule with the number of vibrational modes $N = 4$, the characteristic vibrational temperatures are given by $\Theta_{V_1} = \Theta_{V_2} = 960\,\mathrm{K}$, $\Theta_{V_3} = 1997\,\mathrm{K}$, and $\Theta_{V_4} = 3380\,\mathrm{K}$ [349]. For Cl_2 and Br_2 molecules with $N = 1$, the characteristic vibrational temperatures are, respectively, $\Theta_V = 805\,\mathrm{K}$ and $\Theta_V = 468\,\mathrm{K}$ [349]. The temperature dependence of $\hat{c}_v$ is shown in Fig. 16.13.

Applying the selection method discussed above to the experimental data on α_λ [363, 364], we conclude that these gases have the (KR)-process and the relaxation time τ is several orders larger than the relaxation time τ_δ. Therefore, since the present comparison is made only in the low frequency region, we may safely assume $\hat{\tau}_\delta = 10^{-3}$.

Because the experimental data were summarized as the relationship between α_λ and f/p [Hz/Pa] $(f = \omega/2\pi)$ [363, 364], we use the quantity ω/p instead of the frequency ω and adopt the quantity τp as a fitting parameter determined by the least square method.

The comparison is made in Fig. 16.14. These figures show the excellent agreement between the theoretical prediction of ET$_7$ (and also ET$_{15}$) and the experimental data. The selected parameter τp and the bulk viscosity ν, which is estimated by using $(8.43)_1$ in the case that $\tau_\delta \ll \tau$ and $\hat{c}_v^a$ has a value of $O(1)$, are summarized in Table 16.4. We also emphasize the importance of the dynamic pressure in the wave propagation phenomena. This is because the bulk viscosity coefficients of CO_2, Cl_2, and Br_2 gases are much larger than the shear viscosity coefficients that are estimated as 1.49×10^{-5}[Pa·s] for CO_2, 1.363×10^{-5}[Pa·s] for Cl_2, and 9.42×10^{-4}[Pa·s] for Br_2 at $T = 298\,\mathrm{K}$ and $p = 1$ atm [365].

Remark 16.3 Many studies of the dispersion relation of sound in polyatomic gases have been made based on nonequilibrium thermodynamics and/or the kinetic theory [339, 366, 367]. Except for different definitions of the relaxation times,

Table 16.4 The parameter τp and the bulk viscosity

Gas	T [°C]	τp [Pa · s]	ν [Pa · s]
CO_2	30.5	4.96×10^{-1}	5.61×10^{-2}
	98.7	3.33×10^{-1}	4.62×10^{-2}
	195	2.30×10^{-1}	3.75×10^{-2}
	305	1.64×10^{-1}	2.99×10^{-2}
Cl_2	23	4.08×10^{-1}	2.98×10^{-2}
	103	2.53×10^{-1}	2.19×10^{-2}
	167	1.80×10^{-1}	1.68×10^{-2}
	204	1.49×10^{-1}	1.43×10^{-2}
	256	1.17×10^{-1}	1.16×10^{-2}
Br_2	28.0	6.47×10^{-2}	6.40×10^{-3}
	100	5.47×10^{-2}	5.69×10^{-3}
	177	4.27×10^{-2}	4.57×10^{-3}
	256	3.35×10^{-2}	3.65×10^{-3}

these theories equally describe well the absorption of sound due to the energy exchange among the degrees of freedom of a molecule up to some limited frequency [311, 312] (see also [368] for the classification of the previous studies). In particular, the Meixner theory with the relaxation processes of the molecular internal energies [129, 130] has been used to describe the attenuation of sound phenomenologically. By using the correspondence relationship between the Meixner theory and the ET_7 theory discussed in Sect. 14.2.1, the Meixner theory seems to be valid also for phenomena out of local equilibrium to which ET_7 is applicable. However, in the high frequency region where shear viscosity and heat conduction play roles, the ET_{15} theory becomes indispensable because there exists no such correspondence relationship.

Chapter 17
Shock Wave in a Polyatomic Gas Analyzed by ET_{14}

Abstract In this chapter, we study shock wave structure in a rarefied polyatomic gas by using the ET_{14} theory. We show how the ET_{14} theory can overcome the difficulties encountered in the previous approaches: Bethe–Teller approach and Gilbarg–Paolucci approach.

Firstly, the predictions derived from the ET_{14} theory are shown and compared with the results from the Navier–Stokes and Fourier (NSF) theory. Secondly, the Bethe–Teller theory is reexamined in the light of the ET_{14} theory. Lastly, comparison between the theoretical predictions derived from the ET_{14} theory and the experimental data is made, where we show a very good agreement. We are able to explain in a unified manner the three different shock wave profiles Types A, B, and C (Fig. 1.2) for increasing Mach number.

17.1 Introduction

In Sect. 1.9.7.2, characteristic features of shock wave structure in a rarefied polyatomic gas have been briefly explained. In order to understand such structure, two different approaches by Bethe–Teller and by Gilbarg–Paolucci were proposed many years ago and are still in activity.

1. Bethe–Teller approach [145]: In their celebrated theory, at the very beginning, the internal degrees of freedom of a molecule are assumed to be classified into two parts; one part is composed of the "active" degrees of freedom that relax instantaneously, and the other part is composed of the "inert" degrees of freedom that relax slowly with a finite relaxation time. Except for a hydrogen gas, the translational and rotational modes are regarded as the "active" degrees of freedom but the vibrational modes are considered as "inert" degrees of freedom. In order to analyze the thin layer Δ shown in Fig. 1.2, the system of Euler equations is adopted. The Rankine–Hugoniot relations (see (3.29)) of the system for the jumps of the physical quantities at Δ are derived under the following assumption: The internal energy due to the "inert" degrees of freedom

T. Ruggeri, M. Sugiyama, *Classical and Relativistic Rational Extended Thermodynamics of Gases*, https://doi.org/10.1007/978-3-030-59144-1_17

is unchanged in the thin layer Δ because only "active" degrees of freedom are able to adjust to such an instantaneous change. Therefore, the thin layer in this theory is just the jump discontinuity with no thickness. While, in order to analyze the relaxation process in the thick layer Ψ in Fig. 1.2, a variant of the Euler system with an additional linear relaxation equation for the internal vibrational modes is adopted.

This approach can describe the shock wave structure of Type C. The jumps of the physical quantities at the thin layer Δ can be calculated without using any adjustable parameters. The agreement between the theoretical predictions and the existing experimental data seems to be good.

It should be emphasized, however, that the basis of the Bethe–Teller theory is not clear enough. The assumption of the classification of the internal degrees of freedom should be regarded as a rough approximation even though it seems to be plausible intuitively. Because the classification is not so much clear-cut in reality, it may introduce some arbitrariness into the theory. And furthermore two different systems of equations are adopted in their theory for analyzing the thin and thick layers separately. The compatibility between the two systems is, however, unclear from both mathematical and physical points of view. It is highly preferable, of course, to have one unified system of equations from which all Types A, B, and C can be derived in a fully consistent way.

2. Gilbarg–Paolucci approach [146]: This is based on the system of the Navier–Stokes and Fourier (NSF) equations. They studied, as a typical example, the shock wave structure in a rarefied carbon dioxide (CO_2) gas by adopting a very large value of the bulk viscosity. Although they could predict a thick shock wave structure, the shock profiles are always symmetric (Type A). No asymmetric shock wave structure (Type B) nor thin layer (Type C) could be explained by this theory.

   One crucial point to be noted is that, because the NSF theory is constructed with the assumption of the local equilibrium (see Sect. 1.3.1), the theory is, in general, unsatisfactory for analyzing highly nonequilibrium phenomena such as shock wave phenomena.

Fortunately we have alternative theories as discussed below:

For rarefied monatomic gases, there already exist theories which can describe the phenomena out of local equilibrium, that is, the kinetic theory with the use of the Boltzmann equation (the Chapman–Enskog method [17] and the moment method [84]), and the theories of RET [25] and of molecular ET with the closure by the maximum entropy principle [56, 88]. These theories can indeed describe the structure of strong shock waves in a rarefied monatomic gas [25, 76]. Numerical techniques for solving the Boltzmann equation, such as the Direct Simulation Monte Carlo (DSMC) method [369], have also been developed, and their usefulness has been confirmed through the comparison between their predictions and experimental data.

For rarefied polyatomic gases, the kinetic theory (the Chapman–Enskog method [17] and the moment method [304, 317, 318, 370]) has been developed. Numerical methods for solving the Boltzmann equation have also been developed [369]. However, because the appropriate modeling of the collision term in the Boltzmann equation between two polyatomic molecules is very complicated, some simplifications are usually introduced into the modeling. It is therefore not self-evident that the numerical results thus obtained is compatible with the second law of thermodynamics (the entropy principle).

However, we know well that there is still another theory, that is, the ET_{14} theory of rarefied polyatomic gases explained in Chap. 6. As will be seen below, it is remarkable that the RET theory is totally free from the difficulties mentioned just above [109].

17.2 Basic Equations

In this section, we summarize the basic equations for the present analysis.

17.2.1 Equations of State, Internal Energy, and Sound Velocity

We study a shock wave in a rarefied polyatomic gas with thermal and caloric equations of state given by (12.15). The functional form of the specific internal energy $\varepsilon_E(T)$ is determined by (6.3) through the specific heat at constant volume c_v. The velocity of sound is expressed in (16.7).

17.2.2 Balance Equations

We analyze one-dimensional (plane) shock waves propagating along the x-axis. The vectorial and tensorial quantities are expressed in the form (16.2). Then, from (6.33), the system of equations in the present problem is given by

$$\frac{\partial \rho}{\partial t} + \frac{\partial}{\partial x}(\rho v) = 0,$$

$$\frac{\partial \rho v}{\partial t} + \frac{\partial}{\partial x}(p + \Pi - \sigma + \rho v^2) = 0,$$

$$
\begin{aligned}
&\frac{\partial}{\partial t}(2\rho\varepsilon + \rho v^2)+ \\
&\qquad + \frac{\partial}{\partial x}\left\{2\rho\varepsilon v + 2(p+\Pi-\sigma)v + \rho v^3 + 2q\right\} = 0, \\
&\frac{\partial}{\partial t}\left\{3(p+\Pi) + \rho v^2\right\}+ \\
&\qquad + \frac{\partial}{\partial x}\left\{(5p+5\Pi-2\sigma)v + \rho v^3 + \frac{5}{1+\hat{c}_v}q\right\} = -\frac{3\Pi}{\tau_\Pi}, \\
&\frac{\partial}{\partial t}(p+\Pi-\sigma+\rho v^2)+ \\
&\qquad + \frac{\partial}{\partial x}\left\{3(p+\Pi-\sigma)v + \rho v^3 + \frac{3}{1+\hat{c}_v}q\right\} = \frac{\sigma}{\tau_\sigma} - \frac{\Pi}{\tau_\Pi}, \\
&\frac{\partial}{\partial t}\left\{2\rho\varepsilon v + 2(p+\Pi-\sigma)v + \rho v^3 + 2q\right\}+ \\
&\qquad + \frac{\partial}{\partial x}\left\{2\rho\varepsilon v^2 + 5(p+\Pi-\sigma)v^2 + \rho v^4 + \right. \\
&\qquad\quad \left. + 2\left(\varepsilon + \frac{k_B}{m}T\right)p + 2\left(\varepsilon + 2\frac{k_B}{m}T\right)(\Pi-\sigma) + \frac{10+4\hat{c}_v}{1+\hat{c}_v}qv\right\} \\
&\qquad\quad = -2\left\{\frac{q}{\tau_q} + \left(\frac{\Pi}{\tau_\Pi} - \frac{\sigma}{\tau_\sigma}\right)v\right\}.
\end{aligned}
\tag{17.1}
$$

From (2.5), characteristic velocities λ of the hyperbolic system (17.1) evaluated in an equilibrium state are given by

$$
\begin{aligned}
\frac{\lambda}{c} = 0,\quad 0,\quad &\pm\sqrt{\frac{\hat{c}_v\left(7+4\hat{c}_v-\sqrt{37+32\hat{c}_v+4\hat{c}_v^2}\right)}{2(1+\hat{c}_v)^2}}, \\
&\pm\sqrt{\frac{\hat{c}_v\left(7+4\hat{c}_v+\sqrt{37+32\hat{c}_v+4\hat{c}_v^2}\right)}{2(1+\hat{c}_v)^2}}
\end{aligned}
\tag{17.2}
$$

with $c = \sqrt{\frac{k_B}{m}\gamma(T)T}$.

As explained in Sect. 3.3, these velocities play an essential role in the study of shock wave propagation. The RET theory gives a differential system of hyperbolic type and, as a consequence, it predicts the shock wave structure with discontinuous part when the Mach number becomes large. According to the Theorem 3.1 (subshock formation) of Boillat and Ruggeri in Sect. 3.4, a subshock emerges when the shock velocity s exceeds the maximum characteristic velocity λ_0^{max} of the hyperbolic system in the unperturbed state.

17.3 Setting of the Problem

In this section, several conditions that we adopt for the present analysis are summarized. The parameters are fixed and the numerical method for the computation is explained.

Because the differential system is Galilean invariant, we can consider, without loss of generality, that the shock wave is stationary using the coordinate system moving with the shock wave (co-moving coordinate system). Both *the unperturbed state* (the state at $x = -\infty$ before and far from a shock wave) and *the perturbed state* (the state at $x = \infty$ after and far from a shock wave) are assumed to be in thermal equilibrium.

17.3.1 Dimensionless Form of the Field Equations

For convenience we introduce the following dimensionless quantities:

$$\begin{aligned}
&\hat{\rho} \equiv \frac{\rho}{\rho_0}, \quad \hat{v} \equiv \frac{v}{c_0}, \quad \hat{T} \equiv \frac{T}{T_0}, \\
&\hat{\sigma} \equiv \frac{\sigma}{\rho_0 \frac{k_B}{m} T_0}, \quad \hat{\Pi} \equiv \frac{\Pi}{\rho_0 \frac{k_B}{m} T_0}, \quad \hat{q} \equiv \frac{q}{\rho_0 \frac{k_B}{m} T_0 c_0}, \\
&\hat{x} \equiv \frac{x}{\tau_\Pi(\rho_0, T_0) c_0}, \quad \hat{t} \equiv \frac{t}{\tau_\Pi(\rho_0, T_0)}, \\
&\hat{\tau}_\Pi \equiv \frac{\tau_\Pi(\rho, T)}{\tau_\Pi(\rho_0, T_0)}, \quad \hat{\tau}_\sigma \equiv \frac{\tau_\sigma(\rho, T)}{\tau_\Pi(\rho_0, T_0)}, \quad \hat{\tau}_q \equiv \frac{\tau_q(\rho, T)}{\tau_\Pi(\rho_0, T_0)},
\end{aligned} \tag{17.3}$$

where the quantities with subscript 0 represent the quantities in the unperturbed state. The balance equations (17.1) are now rewritten in terms of the dimensionless quantities as follows:

$$\begin{aligned}
&\frac{\mathrm{d}}{\mathrm{d}\hat{x}} (\hat{\rho}\hat{v}) = 0, \\
&\frac{\mathrm{d}}{\mathrm{d}\hat{x}} \left\{ \frac{1}{\gamma_0} \left(\hat{\rho}\hat{T} + \hat{\Pi} - \hat{\sigma} \right) + \hat{\rho}\hat{v}^2 \right\} = 0, \\
&\frac{\mathrm{d}}{\mathrm{d}\hat{x}} \left\{ \frac{2}{\gamma_0} \left(\hat{\rho}\hat{v} \frac{1}{T_0} \int_{T_R}^{T} \hat{c}_v(\xi)\, d\xi + (\hat{\rho}\hat{T} + \hat{\Pi} - \hat{\sigma})\hat{v} + \hat{q} \right) + \hat{\rho}\hat{v}^3 \right\} = 0,
\end{aligned}$$

$$\frac{d}{d\hat{x}}\left\{\frac{1}{\gamma_0}\left((5\hat{\rho}\hat{T}+5\hat{\Pi}-2\hat{\sigma})\hat{v}+\frac{5}{\hat{c}_v(T)+1}\hat{q}\right)+\hat{\rho}\hat{v}^3\right\}=-\frac{3}{\gamma_0}\frac{\hat{\Pi}}{\hat{\tau}_\Pi},$$

$$\frac{d}{d\hat{x}}\left\{\frac{3}{\gamma_0}\left((\hat{\rho}\hat{T}+\hat{\Pi}-\hat{\sigma})\hat{v}+\frac{1}{\hat{c}_v(T)+1}\hat{q}\right)+\hat{\rho}\hat{v}^3\right\}=\frac{1}{\gamma_0}\left(\frac{\hat{\sigma}}{\hat{\tau}_\sigma}-\frac{\hat{\Pi}}{\hat{\tau}_\Pi}\right),$$

$$\frac{d}{d\hat{x}}\left\{\frac{1}{\gamma_0}\left\{\left(\frac{1}{\gamma_0}\left(\hat{\rho}\hat{T}+\hat{\Pi}-\sigma\right)+\hat{\rho}\hat{v}^2\right)\frac{2}{T_0}\int_{T_R}^{T}\hat{c}_v(\xi)\,d\xi+\right.\right.$$
$$\left.\left.+\frac{2}{\gamma_0}\hat{T}\left(\hat{\rho}\hat{T}+2\hat{\Pi}-2\hat{\sigma}\right)+5\left(\hat{\rho}\hat{T}+\hat{\Pi}-\hat{\sigma}\right)\hat{v}^2+\frac{4\hat{c}_v(T)+10}{\hat{c}_v(T)+1}\hat{q}\hat{v}\right\}+\hat{\rho}\hat{v}^4\right\}$$
$$=-\frac{2}{\gamma_0}\left\{\frac{\hat{q}}{\hat{\tau}_q}+\left(\frac{\hat{\Pi}}{\hat{\tau}_\Pi}-\frac{\hat{\sigma}}{\hat{\tau}_\sigma}\right)\hat{v}\right\},\tag{17.4}$$

where $\gamma_0 \equiv \gamma(T_0)$. Since the conservation laws $(17.4)_{1-3}$ can be easily integrated, we can simplify the balance equations as follows:

$$\hat{\rho}=\frac{M_0}{\hat{v}},$$

$$\frac{1}{\gamma_0}\left(\frac{M_0\hat{T}}{\hat{v}}+\hat{\Pi}-\hat{\sigma}\right)+M_0\hat{v}=\frac{1}{\gamma_0}+M_0^2,$$

$$\frac{2}{\gamma_0}\left(\frac{M_0}{T_0}\int_{T_0}^{T}\hat{c}_v(\xi)\,d\xi+M_0\hat{T}+\hat{\Pi}\hat{v}-\hat{\sigma}\hat{v}+\hat{q}\right)+M_0\hat{v}^2=\frac{2}{\gamma_0}M_0+M_0^3,$$

$$\frac{d}{d\hat{x}}\left\{\frac{1}{\gamma_0}\left(5M_0\hat{T}+5\hat{\Pi}\hat{v}-2\hat{\sigma}\hat{v}+\frac{5}{\hat{c}_v(T)+1}\hat{q}\right)+M_0\hat{v}^2\right\}=-\frac{3}{\gamma_0}\frac{\hat{\Pi}}{\hat{\tau}_\Pi},$$

$$\frac{d}{d\hat{x}}\left\{\frac{3}{\gamma_0}\left(\left(M_0\hat{T}+\hat{\Pi}\hat{v}-\hat{\sigma}\hat{v}\right)+\frac{1}{\hat{c}_v(T)+1}\hat{q}\right)+M_0\hat{v}^2\right\}=\frac{1}{\gamma_0}\left(\frac{\hat{\sigma}}{\hat{\tau}_\sigma}-\frac{\hat{\Pi}}{\hat{\tau}_\Pi}\right),$$

$$\frac{d}{d\hat{x}}\left\{\frac{1}{\gamma_0}\left\{\left(\frac{1}{\gamma_0}+M_0^2\right)\frac{2}{T_0}\int_{T_R}^{T}\hat{c}_v(\xi)\,d\xi+\frac{2}{\gamma_0}\hat{T}\left(\frac{M_0\hat{T}}{\hat{v}}+2\hat{\Pi}-2\hat{\sigma}\right)+\right.\right.$$
$$\left.\left.+5\left(M_0\hat{T}+\hat{\Pi}\hat{v}-\hat{\sigma}\hat{v}\right)\hat{v}+\frac{4\hat{c}_v(T)+10}{\hat{c}_v(T)+1}\hat{q}\hat{v}\right\}+M_0\hat{v}^3\right\}=-\frac{2}{\gamma_0}\left\{\frac{\hat{q}}{\hat{\tau}_q}+\left(\frac{\hat{\Pi}}{\hat{\tau}_\Pi}-\frac{\hat{\sigma}}{\hat{\tau}_\sigma}\right)\hat{v}\right\},\tag{17.5}$$

where M_0 represents the Mach number in the unperturbed state:

$$M_0 \equiv \frac{v_0}{c_0}.$$

17.3.2 Boundary Conditions: Rankine–Hugoniot Conditions for the System of Euler Equations

The boundary conditions for the basic system of equations expressed above are determined as follows: Inserting $\hat{\Pi} = 0$, $\hat{\sigma} = 0$ and $\hat{q} = 0$ into $(17.5)_{1-3}$, we obtain the expressions for the quantities in the perturbed state:

$$\begin{aligned}
&\rho_1 = \frac{M_0}{\hat{v}_1},\\
&\frac{1}{\gamma_0}\left(\frac{M_0\hat{T}_1}{\hat{v}_1} - 1\right) + M_0\left(\hat{v}_1 - M_0\right) = 0,\\
&\frac{2}{\gamma_0}M_0\left(\hat{T}_1 + \frac{1}{T_0}\int_{T_0}^{T_1}\hat{c}_v(\xi)\,d\xi - 1\right) + M_0\left(\hat{v}_1^2 - M_0^2\right) = 0,
\end{aligned} \tag{17.6}$$

where the quantities with subscript 1 are those in the perturbed state. These relations express the Rankine–Hugoniot (RH) conditions for the system of Euler equations.

17.3.3 Parameters

In order to compare the theoretical predictions with experimental data, we will focus our study on the experimental data for the shock wave structure in a rarefied CO_2 gas at $T = 295$ K and $p = 69$ mmHg in the unperturbed state [143].

We determine the dependence of the specific heat (6.3) on the temperature, which is shown in Fig. 16.13, by inserting the data on the temperature dependence of the sound velocity [365] into (16.7). The values of the dimensionless specific heat $\hat{c}_v$, sound velocity c_0, heat conductivity κ [371], and shear viscosity μ [371] in the unperturbed state are summarized in Table 17.1. We note that the rotational modes are completely excited and the vibrational modes are partially excited at this temperature. From (17.2), the maximum characteristic velocity at $T = 295$ K is estimated as $\lambda_0^{max}/c \approx 1.74$. Therefore we recognize that, from the Theorem 3.1 (subshock formation), the shock wave structure predicted by the present analysis is continuous up to $M_0 \approx 1.74$.

Table 17.1 Experimental values of the dimensionless specific heat $\hat{c}_v$ [365], sound velocity c_0 [365], heat conductivity κ [371], and shear viscosity μ [371] of a rarefied CO_2 gas at $T = 295$ K and $p = 69$ mmHg

	$\hat{c}_v$	c_0 [m/s]	κ [W/(m · K)]	μ [Pa · s]
CO_2	3.45	269	1.68×10^{-2}	1.5×10^{-5}

For a rarefied CO_2 gas, the temperature dependence of the phenomenological coefficients was already estimated by both the kinetic theoretical considerations and the experimental data [146] as follows:

$$\mu \propto T^{n_1}, \qquad \nu \propto T^{n_2}, \qquad \kappa \propto T^{n_1}\hat{c}_v(T), \tag{17.7}$$

where the exponents n_1 and n_2 were estimated as $n_1 = 0.935$ and $n_2 = -1.3$ [126, 146]. We adopt the same temperature dependence in the present analysis. We have confirmed that the shock wave structure studied below depends weakly on the values of the exponents n_1 and n_2.

Substituting (17.7) into the relations (6.36), we have the following dependence of the relaxation times on the mass density and the temperature:

$$\begin{aligned}
\hat{\tau}_\Pi &= \frac{1}{\hat{\rho}\hat{T}^{1-n_2}}\frac{5-3\gamma_0}{5-3\gamma(T)}, \\
\hat{\tau}_\sigma &= \frac{\tau_\sigma(\rho_0, T_0)}{\tau_\Pi(\rho_0, T_0)}\frac{1}{\hat{\rho}\hat{T}^{1-n_1}}, \\
\hat{\tau}_q &= \frac{\tau_q(\rho_0, T_0)}{\tau_\Pi(\rho_0, T_0)}\frac{1}{\hat{\rho}\hat{T}^{1-n_1}}\frac{\gamma_0}{\gamma(T)}.
\end{aligned} \tag{17.8}$$

Inserting the values of the phenomenological coefficients in Table 17.1 into the relations (6.36), we obtain the values of the relaxation times τ_σ and τ_q for the shear stress and the heat flux in the unperturbed state as shown in Table 17.2. The remaining undetermined parameter is only the relaxation time τ_Π for the dynamic pressure, which is proportional to the bulk viscosity ν. Because of the lack of knowledge of reliable data on ν due to the difficulty in its experimental measurements, as was already done in Chap. 16, we will use τ_Π in the unperturbed state as a fitting parameter. As will be explained below, the value of τ_Π in the unperturbed state is determined by the comparison of the theoretical prediction with the experimental data for $M_0 = 1.47$. See also Figs. 17.4 and 17.5 below. It is noticeable that the value of τ_Π is much larger, with different order of magnitude, than the other two relaxation times.

Table 17.2 Relaxation times for a CO_2 gas at $T = 295$ K and $p = 69$ mmHg. The relaxation times τ_σ and τ_q are obtained from the experimental data shown in Table 17.1. Only the relaxation time for the dynamic pressure τ_Π is remained as a fitting parameter

	$\tau_\sigma(\rho_0, T_0)$[s]	$\tau_q(\rho_0, T_0)$[s]	$\tau_\Pi(\rho_0, T_0)$[s]
CO_2	1.6×10^{-9}	2.2×10^{-9}	3.3×10^{-5}

17.3.4 Numerical Methods

We solve numerically the system of balance equations (17.5) under the boundary conditions (17.6) by adopting the methods proposed by Weiss [25, 228].

We introduce the $N+1$ grid points such that the range $[-\hat{L}/2, \hat{L}/2]$ in the $\hat{x}$−axis is discretized with constant intervals $\Delta\hat{x} = \hat{L}/N$ as follows:

$$\hat{x}^i = -\frac{\hat{L}}{2} + \frac{\hat{L}}{N}i \quad \text{for } i = 0, 1, \cdots, N,$$

where superscript i represents the number of the grid point.

Because the mass density $\hat{\rho}$ is already expressed by other variables in $(17.5)_1$, we need to solve the system $(17.5)_{2-6}$ for $\mathbf{u} = (\hat{v}, \hat{T}, \hat{\Pi}, \hat{\sigma}, \hat{q})$. The boundary conditions (17.6) give

$$\begin{aligned} \mathbf{u}^0 &= \mathbf{u}_0, \\ \mathbf{u}^N &= \mathbf{u}_1, \end{aligned} \tag{17.9}$$

where $\mathbf{u}^i$ represents $\mathbf{u}|_{x=x^i}$, $\mathbf{u}_0 = (M_0, 1, 0, 0, 0)$, and $\mathbf{u}_1 = (\hat{v}_1, \hat{T}_1, 0, 0, 0)$. For the conservation laws $(17.5)_{2,3}$ expressed as $\mathbf{F}(\mathbf{u}) = \mathbf{F}(\mathbf{u}_0)$ with $\mathbf{F}$ being the general flux, we have

$$\mathbf{F}(\mathbf{u}^i) = \mathbf{F}(\mathbf{u}_0) \quad \text{for } i = 1, 2, \cdots, N-1. \tag{17.10}$$

Replacing the differentiation in the balance equations $(17.5)_{4-6}$, which we express as $\mathrm{d}\mathbf{F}(\mathbf{u})/d\hat{x} = \mathbf{P}(\mathbf{u})$ briefly with $\mathbf{P}$ being the general production, by the central difference, we get

$$\frac{\mathbf{F}(\mathbf{u}^{i+1}) - \mathbf{F}(\mathbf{u}^{i-1})}{2\Delta\hat{x}} = \mathbf{P}(\mathbf{u}^i) \quad \text{for } i = 1, 2, \cdots, N-1. \tag{17.11}$$

The nonlinear algebraic equations (17.10) and (17.11) with the condition (17.9) may be solved with the help of numerical solvers equipped with softwares for numerical computations. In the present analysis, we have constructed numerical codes by adopting the numerical solver implemented in the Mathematica based on the Newton's method. The computation starts from an appropriate initial guess, e.g.,

$$\mathbf{u}^i = \begin{cases} \mathbf{u}_0 & \text{for } i = 0, 1, \cdots, \frac{N}{2}, \\ \mathbf{u}_1 & \text{for } i = \frac{N}{2}+1, \frac{N}{2}+2, \cdots, N, \end{cases}$$

and the iterative calculations are repeated until the numerical solution converges to the one that satisfies the system (17.5) and the boundary conditions (17.6) within the appropriate accuracy we have set; eight digits of the precision in the present analysis.

We have chosen $\Delta\hat{x}$ small enough and have confirmed that the dependence of the profiles in Figs. 17.1, 17.2, 17.3, 17.4, and 17.5 on $\Delta\hat{x}$ is negligibly small.

17.4 Navier–Stokes and Fourier Theory

Because both results obtained by the ET_{14} theory and by the NSF theory are compared with each other in the next section, we here summarize also the NSF system of equations. It is obtained as the first approximation of the ET_{14} system (6.33) by using the Maxwellian iteration (see Sect. 6.3.11):

$$
\begin{aligned}
&\frac{\partial \rho}{\partial t} + \frac{\partial}{\partial x}(\rho v) = 0,\\
&\frac{\partial \rho v}{\partial t} + \frac{\partial}{\partial x}\left(p + \Pi - \sigma + \rho v^2\right) = 0,\\
&\frac{\partial}{\partial t}\left(2\rho\varepsilon + \rho v^2\right) + \frac{\partial}{\partial x}\left\{2\rho\varepsilon v + 2(p + \Pi - \sigma)v + \rho v^3 + 2q\right\} = 0,\\
&\Pi = -\left(\frac{2}{3} - \frac{1}{\hat{c}_v}\right) p\tau_\Pi \frac{\partial v}{\partial x},\\
&\sigma = \frac{4}{3} p\tau_\sigma \frac{\partial v}{\partial x},\\
&q = -\left(1 + \hat{c}_v\right)\frac{k_B}{m} p\tau_q \frac{\partial T}{\partial x}.
\end{aligned}
\tag{17.12}
$$

From the Eqs. $(17.12)_{4-6}$, we obtain the relationship between relaxation times and the phenomenological coefficients (6.36). Note that the system (17.12) is of parabolic type although the original RET system (6.33) is hyperbolic.

The dimensionless form of the conservation laws are the same as $(17.5)_{1-3}$. While the dimensionless constitutive relations are expressed by

$$
\begin{aligned}
\hat{\Pi} &= -\left(\frac{2}{3} - \frac{1}{\hat{c}_v(T)}\right)\frac{M_0\hat{T}}{\hat{v}}\hat{\tau}_\Pi \frac{\mathrm{d}\hat{v}}{\mathrm{d}\hat{x}},\\
\hat{\sigma} &= \frac{4}{3}\frac{M_0\hat{T}}{\hat{v}}\hat{\tau}_\sigma \frac{\mathrm{d}\hat{v}}{\mathrm{d}\hat{x}},\\
\hat{q} &= -\left(1 + \hat{c}_v(T)\right)\frac{1}{\gamma_0}\frac{M_0\hat{T}}{\hat{v}}\hat{\tau}_q \frac{\mathrm{d}\hat{T}}{\mathrm{d}\hat{x}}.
\end{aligned}
$$

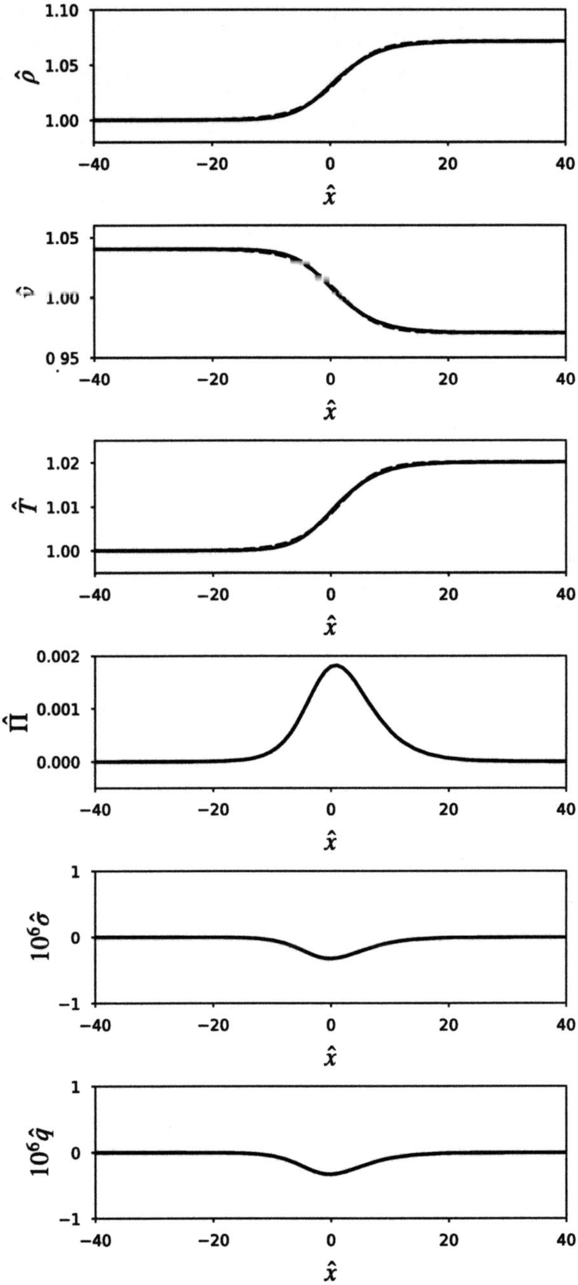

Fig. 17.1 Type A: profiles of the dimensionless mass density, velocity, temperature, dynamic pressure, shear stress, and heat flux predicted by the ET_{14} theory (solid curves). Profiles of the dimensionless mass density, velocity, and temperature predicted by the NSF theory (dashed curves) are also shown. $M_0 = 1.04$. The conditions for the numerical calculations are $\hat{L} = 100$, and $N = 100$

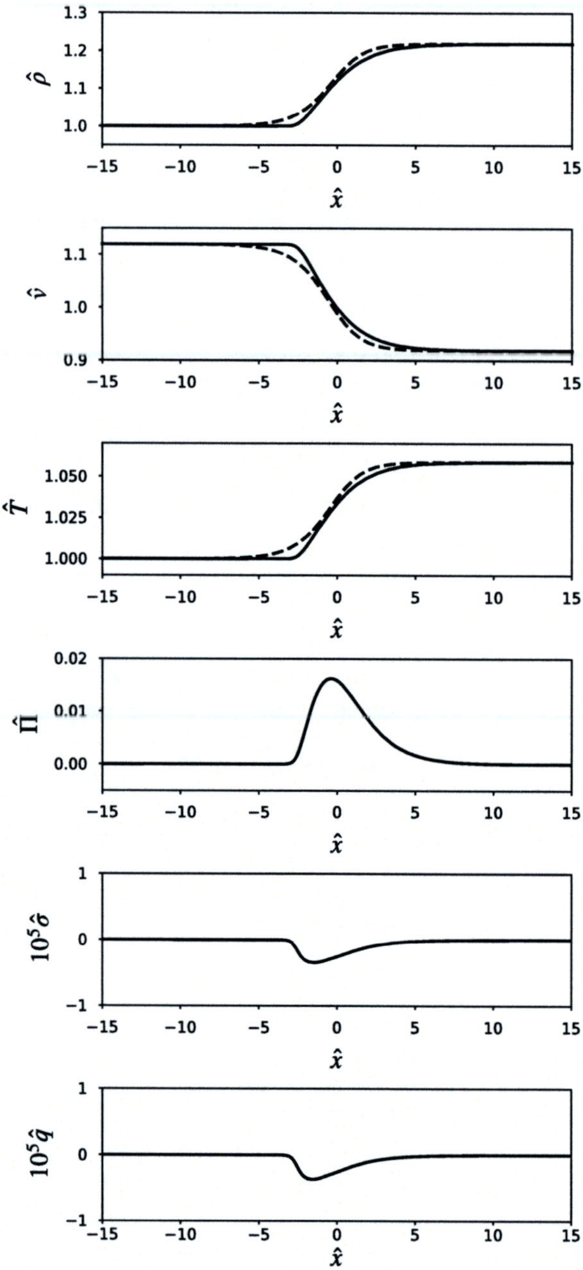

Fig. 17.2 Type B: shock wave structure predicted by the ET_{14} theory (solid curves) and by the NSF theory(dashed curves). $M_0 = 1.12$. The numerical conditions are $\hat{L} = 50$, $N = 5000$ for the ET_{14} theory, and $N = 100$ for the NSF theory

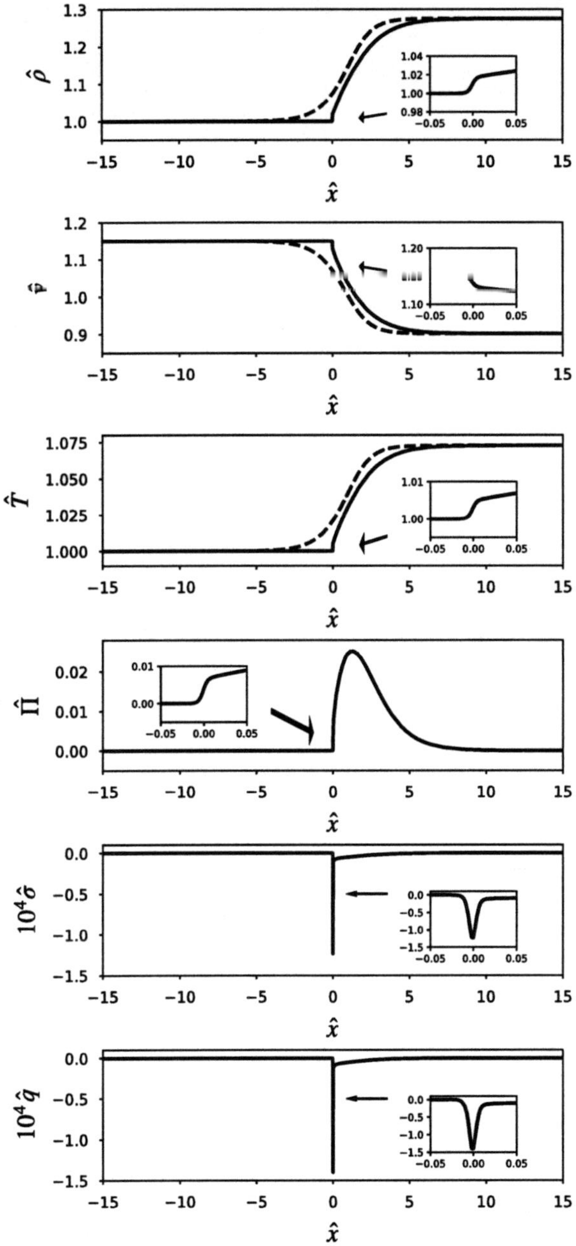

Fig. 17.3 Type C: shock wave structure predicted by the ET_{14} theory (solid curves) and by the NSF theory (dashed curves). $M_0 = 1.15$. The numerical conditions are $\hat{L} = 40$, $N = 20{,}000$ for the ET_{14} theory and $N = 100$ for the NSF theory

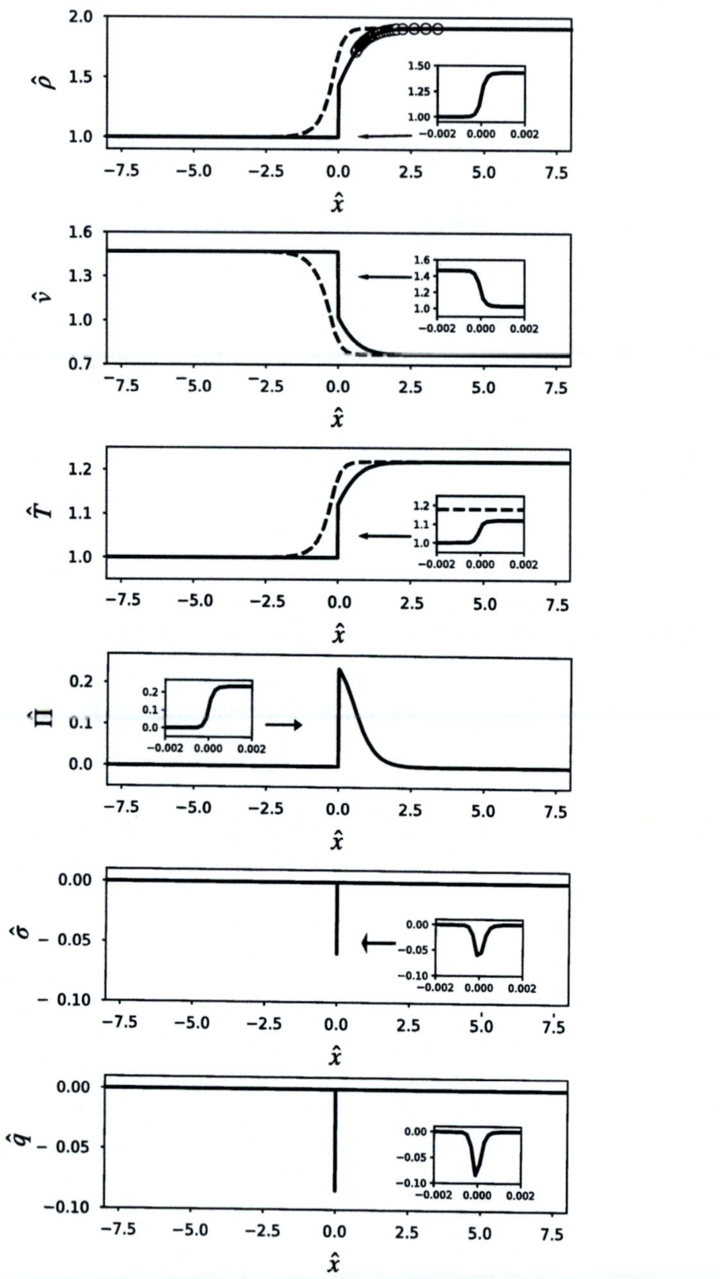

Fig. 17.4 Type C: shock wave structure predicted by the ET theory (solid curves) and by the NSF theory (dashed curves). The experimental data [143] in the thick layer are also shown by circles. $M_0 = 1.47$. The numerical conditions are $\hat{L} = 20$, $N = 100{,}000$ for the ET$_{14}$ theory and $N = 100$ for the NSF theory

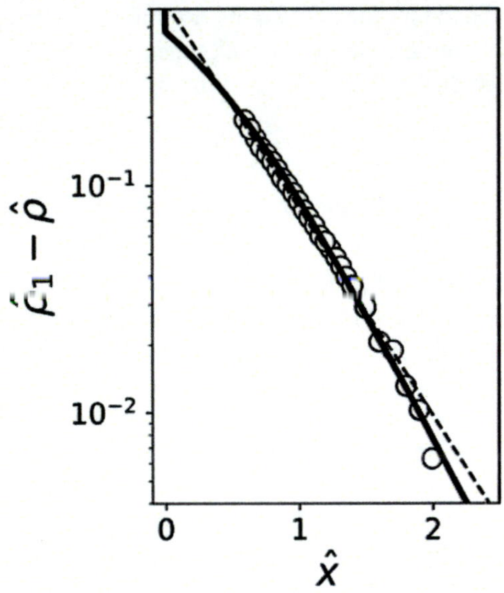

Fig. 17.5 Profile of the mass density difference $\hat{\rho}_1 - \hat{\rho}$ predicted by the RET theory (solid curve) and the experimental data (circles) [143]. The dotted line shows the exponential decay

17.5 Shock Wave Structure

In this section, we show that all of the three types of the shock wave structure, Types A, B, and C, can be described naturally within the ET$_{14}$ theory.

17.5.1 *Type A: Nearly Symmetric Shock Wave Structure*

The nearly symmetric shock wave structure appears in a small Mach number region just above unity. The typical example of the shock wave structure of Type A is obtained at $M_0 = 1.04$ as shown in Fig. 17.1. We have depicted the profiles of all independent variables; mass density, velocity, temperature, dynamic pressure, shear stress, and heat flux. We can confirm that the shock wave structure is indeed nearly the same as the one predicted by the NSF theory.

We notice that the thickness of a shock wave is very large even at several centimeters order because of the large characteristic length estimated as $\tau_\Pi(\rho_0, T_0)c_0 = 0.60$ cm. And the dimensionless dynamic pressure is also several orders larger than the dimensionless shear stress and heat flux. These features are, of course, due to the fact that the relaxation time for the dynamic pressure τ_Π, which is proportional to the bulk viscosity ν, is much larger than the other two relaxation times τ_σ and τ_q that are, respectively, proportional to the shear viscosity μ and the heat conductivity κ.

Because of the large thickness and the small Mach number, i.e., small gradients of physical quantities, a shock wave is not so much far from local equilibrium.

Therefore the predictions from the ET_{14} theory and the predictions from the NSF theory are similar to each other. It is this Type A that Gilbarg and Paolucci [146] studied.

17.5.2 Type B: Asymmetric Shock Wave Structure

When the Mach number increases further, the gradient of the physical quantities in the shock wave structure near the unperturbed state becomes much steeper than the gradient near the perturbed state. The shock wave structure now becomes evidently asymmetric. The NSF theory cannot describe such asymmetric profiles. Typical shock wave structure of Type B is shown in Fig. 17.2 where $M_0 = 1.12$.

From Fig. 17.2, we can see that the dimensionless shear stress and heat flux are still several orders smaller than the dimensionless dynamic pressure in the whole range of the shock wave structure. Therefore we may conclude that the dynamic pressure plays much more important role in the global structure of a shock wave of Types A and B than the shear stress and the heat flux.

17.5.3 Type C: Shock Wave Structure Composed of Thin and Thick Layers

When the Mach number increases further more, the shock wave structure changes from a single-layer asymmetric structure (Type B) to a structure composed of thick and thin layers (Type C). Typical examples of Type C are shown in Fig. 17.3 with $M_0 = 1.15$ and Fig. 17.4 with $M_0 = 1.47$. It is this Type C that Bethe and Teller mainly studied. The NSF theory again cannot describe such shock wave structures with two layers.

We notice from Figs. 17.3 and 17.4 clearly that the thickness of the thin layer is finite although it is still much smaller than that of the thick layer, the thickness of which is at several centimeters order. Therefore we can analyze the detailed structure in the thin layer, which is impossible to be addressed by the Bethe–Teller theory. For example, as shown in Figs. 17.3 and 17.4, we understand the detailed profiles of the dissipative quantities in the thin layer.

We see that the shear stress and the heat flux are negligibly small everywhere except for the thin-layer region. On the other hand, the dynamic pressure is large in both the thick and thin layers. Therefore we may say that, in the thin layer with finite thickness, all dissipative quantities together play a crucial role, while, in the thick layer, only the dynamic pressure seems to be essential.

Within the present theory, as is pointed out above, the continuous shock wave structure is obtained until $M_0 \approx 1.74$. If we want to study the shock wave structure at larger Mach numbers than 1.74, we need the RET theory with more independent variables, which is discussed in Chap. 9.

17.5.4 *Critical Mach Numbers for the Transitions Between the Types A–B and B–C*

We have estimated numerically the critical Mach numbers for the transition of the type of shock wave structure: The asymmetric character of Type B becomes evident when the Mach number is around 1.08. And, for the transition between Type B and Type C, we have the critical Mach number $M_0 \approx 1.14$. Note that these values of the Mach number are merely rough indications because the boundary between two different types cannot be clearly defined.

In the Bethe–Teller theory, from the stability analysis of the discontinuous part of the shock wave structure (nowadays known as the Lax condition (see Sect. 3.3.2.1)), the critical Mach number between Type B and Type C was estimated as $M_0 \approx 1.04$. This value does not agree with the value mentioned above, but is not so far from it.

17.5.5 *Reexamination of the Bethe–Teller Theory*

We summarize the features of the Bethe–Teller theory in the light of the ET_{14} theory.

(A) As explained above, the Bethe–Teller theory describes the shock wave structure of Type C by adopting the two systems of equations under the assumption that the internal degrees of freedom of a molecule can be divided into two parts, that is, "inert" part and "active" part. One system is applied to analyze a thin layer and the other system to a thick layer. The compatibility of the two system of equations is, however, not self-evident. In the ET_{14} theory, on the other hand, a single system of equations can describe all Types A, B, and C without any ambiguity. There is no compatibility problem.

(B) The thin layer is a jump discontinuity with zero thickness. While, in the ET_{14} theory, the thin layer has a structure with finite thickness.

(C) The thick layer is described essentially by a relaxation equation with a finite relaxation time. If necessary, the theory may be generalized so as to have several relaxation equations with different relaxation times such as ET_{15} theory. Usually the relaxation equation is assumed to be linear. The RET theory includes, in a natural way, the relaxation mechanism of the internal degrees of freedom. In this respect, see also Chap. 12.

(D) The critical Mach number between Type B and Type C can be estimated by the stability analysis. Its predicted values by the Bethe–Teller theory and by the ET_{14} theory are not far from each other.

(E) There is a qualitative difference between the Bethe–Teller theory and the ET_{14} theory in the temperature profile. The temperature just after the discontinuous jump derived from the Bethe–Teller theory may have the possibility to be larger than the temperature in the perturbed state (i.e., so-called temperature overshoot), while the temperature profile derived from the RET theory is always smaller than the perturbed temperature.

Experiments to observe the temperature profile, however, seem to be extremely difficult because a shock wave has a very steep and rapid change in space and time. There is another difficulty from a theoretical point of view. We should be careful about the definition of the temperature in nonequilibrium. We will discuss this point later in Sect. 18.6 (see also Chap. 15).

17.6 Comparison with Experimental Data

The experiments of the shock wave structure in a CO_2 gas at the room temperature and the atmospheric pressure indicate that the shock profiles with no thin layer are obtained at least in the range $1 < M_0 < 1.04$ [144]. The present result is consistent with this.

The experimental results at $M_0 = 1.134$ and $M_0 = 1.16$ are available [141, 142]. We have confirmed that our theoretical predictions shown in Fig. 17.3 are qualitatively the same as the shock profiles obtained by the experiments. Quantitative comparison is, however, impossible because only the interferograms are shown in the papers.

The experimental data on the mass density profile [143] and the theoretical mass density profile derived from the ET_{14} theory at $M_0 = 1.47$ are shown in Fig. 17.4. Note that only the experimental data in the thick layer are reported in the paper, in which the authors said that the accurate measurement in the region near the thin layer was impossible because the change of physical quantities is so steep. In order to study in more detail, Fig. 17.4 is shown in a different way: the single logarithmic plot of the profile of the mass density difference $\hat{\rho}_1 - \hat{\rho}$ as shown in Fig. 17.5.

We can see that the agreement between the theoretical prediction and the experimental data is excellent. It is also remarkable that the ET_{14} theory seems to explain the deviation of the experimental data $\hat{\rho}_1 - \hat{\rho}$ from the the dotted line in Fig. 17.5, i.e., from the purely exponential decay.

Unfortunately, only the experimental data of the mass density profile at $M_0 = 1.47$ are available at present. More detailed experimental studies of the shock wave structure are highly expected.

17.7 Concluding Remarks

Some concluding remarks are made below:

1. We have shown that the ET_{14} theory can describe the shock wave structure in a rarefied polyatomic gas in a consistent way. We have found the fact that the dynamic pressure Π is essentially important in the shock wave structure but the shear stress and the heat flux are not so important everywhere except for the inside of a thin layer. Therefore it is natural to expect that, by neglecting all dissipative fluxes but the dynamic pressure, we can study the shock wave structure properly and more easily. In fact, in the next Chap. 18, we will show that the RET theory with 6 independent fields (ET_6) can describe the shock wave structure of Types A, B, and C reasonably well.
2. Comparison of predictions about the shock wave structure by using the RET theory with the numerical results by the kinetic theory will be discussed in the next Chap. 18.
3. The study of shock wave structure in the framework of the ET_{15} theory must be interesting to understand it in more detail. However, we postpone the study as a future work.
4. The analyses of contact waves and rarefaction waves, in addition to shock waves, are also remained as one of the future studies. In this respect, in the case of monatomic gas, the shock tube problem was analyzed by using RET in [372].

Chapter 18
Shock Wave and Subshock Formation Analyzed by ET_6

Abstract In this chapter, we show the usefulness of both linear and nonlinear ET_6 theories for the analysis of shock wave structure in a rarefied polyatomic gas. Firstly we compare the theoretical predictions derived from the linear ET_6 theory with those from the ET_{14} theory. We see, in particular, that the thin layer in Type C with finite thickness described by the ET_{14} theory is replaced by a discontinuous jump, which is a subshock. The strength and the stability of a subshock is also discussed. Secondly, by using the nonlinear ET_6 theory, we show that the linear ET_6 is reliable at least up to the Mach number $M_0 = 5$. The temperature overshoot at a subshock in terms of Meixner's temperature (kinetic temperature) defined in Sect. 12.3 is discovered. Lastly, we comment on the numerical results by using the kinetic theory briefly.

18.1 Introduction

In Chap. 17, the shock wave structure in a rarefied polyatomic gas was studied on the basis of the ET_{14} theory. We found that the ET_{14} theory can describe three types of shock wave structure, Types A, B, and C, in a rarefied polyatomic gas in a unified and consistent way. We also found that the dynamic pressure Π plays an essential role in the formation of the shock wave structure but the shear stress and the heat flux are negligibly small everywhere except for the inside of a thin layer. This is due to the fact that the relaxation time of dynamic pressure is much larger than the other relaxation times of shear stress and heat flux in a rarefied polyatomic gas, such as H_2 gas, CO_2 gas. Therefore, it is natural to expect that we may study the shock wave structure properly on the basis of the ET_6 theory.

In this chapter, we study again the one-dimensional problem in Chap. 17, but with the use of the ET_6 theory [110, 127, 133, 373] . (For the study of a spherical shock wave by ET_6, see [374].) Firstly we study shock waves by the ET_6 theory with linear production term (linear ET_6) in order to compare the results with those in the previous chapter. Secondly we study shock waves by using the ET_6 theory with nonlinear production term (nonlinear ET_6) to elucidate its nonlinear effects.

T. Ruggeri, M. Sugiyama, *Classical and Relativistic Rational Extended Thermodynamics of Gases*, https://doi.org/10.1007/978-3-030-59144-1_18

Lastly we mention briefly the recent interesting results obtained by using the kinetic theory.

18.2 Basis of the Analysis by the Linear ET_6 Theory

18.2.1 Characteristic Velocities

Let us consider the problem of one space dimension. The characteristic velocities in equilibrium λ of the ET_6 system have the following expression for any gas (see Sect. 20.1 for the proof):

$$\lambda = 0, \ \ 0, \ \ \pm\sqrt{\frac{5}{3}\frac{k_B}{m}T}.$$

Let us consider a non-polytropic gas. In this case it is convenient to write the ratio with the sound velocity c given in (16.7) :

$$\frac{\lambda}{c} = 0, \ \ 0, \ \ \pm\sqrt{\frac{5}{3}\frac{1}{\gamma(T)}},$$

where $\gamma(T)$ is the ratio of specific heat given in (16.8).

As in Chap. 17, we study the shock wave structure in a CO_2 gas at $T = 295$ [K] and $p = 69$ [mmHg] in the unperturbed state. In this case, the maximum characteristic velocity in the unperturbed state, λ_0^{max}, is estimated as $\lambda_0^{max}/c_0 \approx 1.137$. Therefore, from Theorem 3.1, the continuous shock wave structure exists in the Mach number region: $1 < M_0 < 1.137$. It is interesting to notice that, in the case of the ET_{14} system studied in Chap. 17, $\lambda_0^{max}/c_0 \approx 1.74$.

18.2.2 Parameters

The specific heat of rarefied CO_2 gas is given as shown in Fig. 16.13. The temperature and mass density dependence of the relaxation time of the dynamic pressure is expressed by (see $(17.8)_1$)

$$\frac{\tau_\Pi(\rho, T)}{\tau_\Pi(\rho_0, T_0)} = \frac{5 - 3\gamma_0}{5 - 3\gamma(T)}\frac{\rho_0}{\rho}\left(\frac{T_0}{T}\right)^{1-n},$$

where T_0 is the temperature in the unperturbed state and $\gamma_0 \equiv \gamma(T_0)$. The exponent n is, as before, given by $n = -1.3$ [126]. The values of the specific heat, the sound velocity, and the relaxation time of the dynamic pressure are summarized in Tables 17.1 and 17.2.

Remark 18.1 The ET_6 theory is also applicable to some other gases like rarefied hydrogen and deuterium gases. In fact, the analysis of ultrasonic sounds in Chap. 16 reveals that τ_Π of these gases is much larger than τ_σ and τ_q. However, since there is no suitable experimental datum on the shock wave structure in these gases, we take only CO_2 gas as a typical example in the present chapter.

18.2.3 Dimensionless form of the Balance Equations

We hereafter study the structure of a stationary plane shock wave. For convenience we again introduce the dimensionless quantities (17.3). Then the balance equations (12.3) are rewritten as

$$\begin{aligned}
&\frac{\mathrm{d}}{\mathrm{d}\hat{x}}(\hat{\rho}\hat{v}) = 0,\\
&\frac{\mathrm{d}}{\mathrm{d}\hat{x}}\left\{\frac{1}{\gamma_0}\left(\hat{\rho}\hat{T} + \hat{\Pi}\right) + \hat{\rho}\hat{v}^2\right\} = 0,\\
&\frac{\mathrm{d}}{\mathrm{d}\hat{x}}\left\{\frac{2}{\gamma_0}\left((\hat{\rho}\hat{T} + \hat{\Pi})\hat{v} + \hat{\rho}\hat{v}\frac{1}{T_0}\int_{T_R}^{T}\hat{c}_v(\xi)\,d\xi\right) + \hat{\rho}\hat{v}^3\right\} = 0,\\
&\frac{\mathrm{d}}{\mathrm{d}\hat{x}}\left\{\frac{5}{\gamma_0}(\hat{\rho}\hat{T} + \hat{\Pi})\hat{v} + \hat{\rho}\hat{v}^3\right\} = -\frac{3}{\gamma_0}\frac{\hat{\Pi}}{\hat{\tau}_\Pi}.
\end{aligned} \tag{18.1}$$

Here we have assumed the linear relation (12.46) for $\hat{P}_{ll}$ with the relaxation time τ_Π so that we can compare the results of ET_6 with those of ET_{14} in Chap. 17.

By integrating the conservation laws $(18.1)_{1-3}$, the balance equations (18.1) become

$$\begin{aligned}
&\hat{\rho} = \frac{M_0}{\hat{v}},\\
&\frac{1}{\gamma_0}\left(\frac{M_0\hat{T}}{\hat{v}} + \hat{\Pi}\right) + M_0\hat{v} = \frac{1}{\gamma_0} + M_0^2,\\
&\frac{2}{\gamma_0}\left(M_0\hat{T} + \hat{\Pi}\hat{v} + \frac{M_0}{T_0}\int_{T_0}^{T}\hat{c}_v(\xi)\,d\xi\right) + M_0\hat{v}^2 = \frac{2}{\gamma_0}M_0 + M_0^3,\\
&\frac{\mathrm{d}}{\mathrm{d}\hat{x}}\left\{\frac{5}{\gamma_0}\left(M_0\hat{T} + \hat{\Pi}\hat{v}\right) + M_0\hat{v}^2\right\} = -\frac{3}{\gamma_0}\frac{\hat{\Pi}}{\hat{\tau}_\Pi}.
\end{aligned} \tag{18.2}$$

18.2.4 Boundary Conditions

The quantities in the perturbed state are derived from the Rankine–Hugoniot (RH) conditions (17.6).

18.2.5 RH Conditions for a Subshock in Type C

From the general conditions (3.29), we obtain the RH conditions for a subshock in Type C:

$$\begin{aligned}
&\hat{\rho}_* = \frac{M_0}{\hat{v}_*},\\
&\frac{1}{\gamma_0}\left(\frac{M_0\hat{T}_*}{\hat{v}_*} - 1 + \hat{\Pi}_*\right) + M_0\left(\hat{v}_* - M_0\right) = 0,\\
&\frac{2}{\gamma_0}\left(M_0\hat{T}_* + \hat{\Pi}_*\hat{v}_* + \frac{M_0}{T_0}\int_{T_0}^{T_*}\hat{c}_v(\xi)\,d\xi - M_0\right) + M_0\left(\hat{v}_*^2 - M_0^2\right) = 0,\\
&\frac{5}{\gamma_0}\left(M_0\hat{T}_* + \hat{\Pi}_*\hat{v}_* - M_0\right) + M_0\left(\hat{v}_*^2 - M_0^2\right) = 0.
\end{aligned} \tag{18.3}$$

Quantities with the subscript $*$ mean the quantities in the state just after a subshock. Note that the state $*$ in Type C is different from the perturbed state 1. See the shock profile of Type C in Fig. 1.2.

If we subtract (18.3)$_3$ from (18.3)$_4$, we obtain:

$$\frac{\hat{\Pi}_*}{\hat{\rho}_*} = \frac{2}{3}\frac{1}{T_0}\int_{T_0}^{T_*}\left(\hat{c}_v(x) - \frac{3}{2}\right)dx$$

that gives the value of $\hat{\Pi}_*$. If the gas is monatomic, we notice that $\hat{c}_v = 3/2$ and then Π vanishes everywhere. Furthermore, from (18.3)$_2$ and (18.3)$_4$, we have

$$4M_0\hat{v}_*^2 - 5\left(M_0 + \frac{1}{\gamma_0}\right)\hat{v}_* + \frac{5}{\gamma_0}M_0 + M_0^3 = 0.$$

Then we have

$$\hat{v}_* = M_0 \quad \text{and} \quad \frac{1}{4M_0}\left(M_0^2 + \frac{5}{\gamma_0}\right).$$

The first solution is a trivial one, and the second one corresponds to the solution for a subshock.

18.2.6 *Numerical Methods*

We solve numerically the system (18.2) under the boundary conditions by adopting the numerical methods proposed by Weiss [25, 228]. Because the mass density is already expressed by the other variables $(18.2)_1$, we need to solve the system $(18.2)_{2-4}$ for $\mathbf{u} = (\hat{v}, T, \hat{\Pi})$.

18.2.7 *Case 1: $M_0 < \lambda_0^{max}/c_0$*

In this case, no subshock appears. Therefore the numerical method explained in Sect. 17.3.4 can be adopted also in this case.

18.2.8 *Case 2: $M_0 > \lambda_0^{max}/c_0$*

Since a subshock appears in this case, the boundary condition is replaced by the RH conditions for a subshock (18.3). The subshock is assumed to be at $x = 0$. We introduce the $N+1$ grid points such that the range $[0, L]$ in the x-axis is discretized with constant intervals $\Delta x = L/N$ as follows:

$$x^i = \frac{L}{N} i \quad \text{for} \quad i = 0, 1, \cdots, N. \tag{18.4}$$

The boundary conditions are given by

$$\mathbf{u}^0 = \mathbf{u}_*,$$
$$\mathbf{u}^N = \mathbf{u}_1,$$

where $\mathbf{u}^i$ represents $\mathbf{u}|_{x=x^i}$, $\mathbf{u}_0 = (M_0, T_0, 0)$ and $\mathbf{u}_* = (\hat{v}_*, T_*, \hat{\Pi}_*)$. We solve numerically the basic equations in the same way as before.

18.3 Shock Wave Structure with and Without a Subshock

18.3.1 Shock Wave Structure Without a Subshock

Let us analyze the continuous shock wave structure in the Mach number region $1 < M_0 < \lambda_0^{max}/c_0$. The profiles of mass density, velocity, temperature, and dynamic pressure at $M_0 = 1.04$ are shown in Fig. 18.1. This is a typical example of the shock

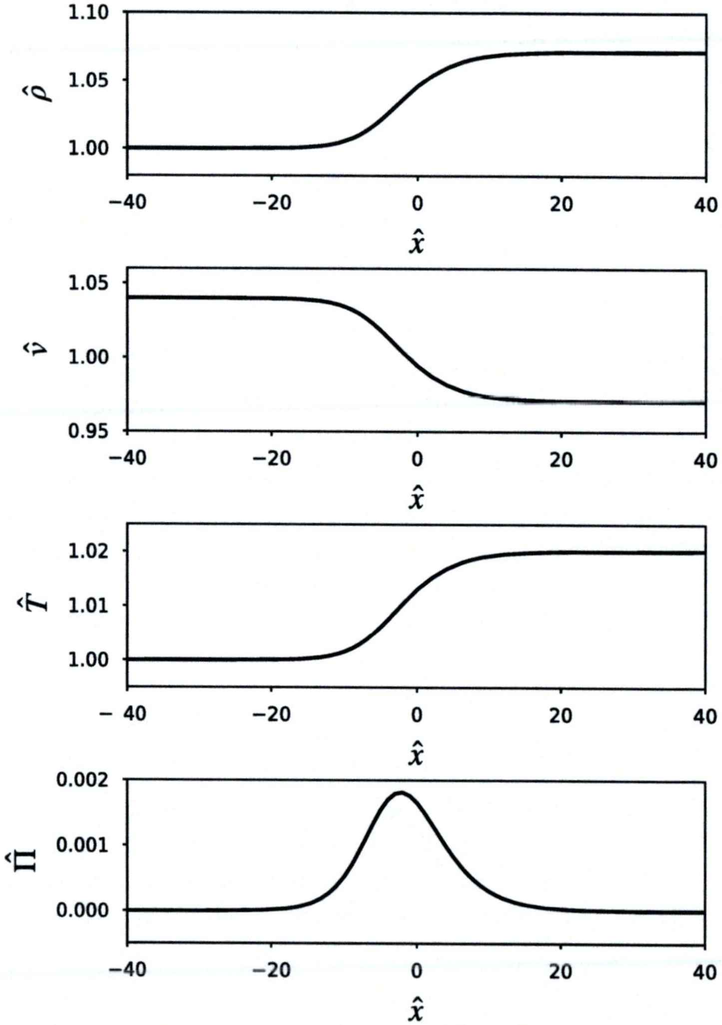

Fig. 18.1 Type A: profiles of the dimensionless mass density, velocity, temperature, and dynamic pressure predicted by the ET_6 theory (black thick curves) and by the ET_{14} theory (red thin curves). $M_0 = 1.04$. The black and red curves coincide with each other within the thickness of the black curve

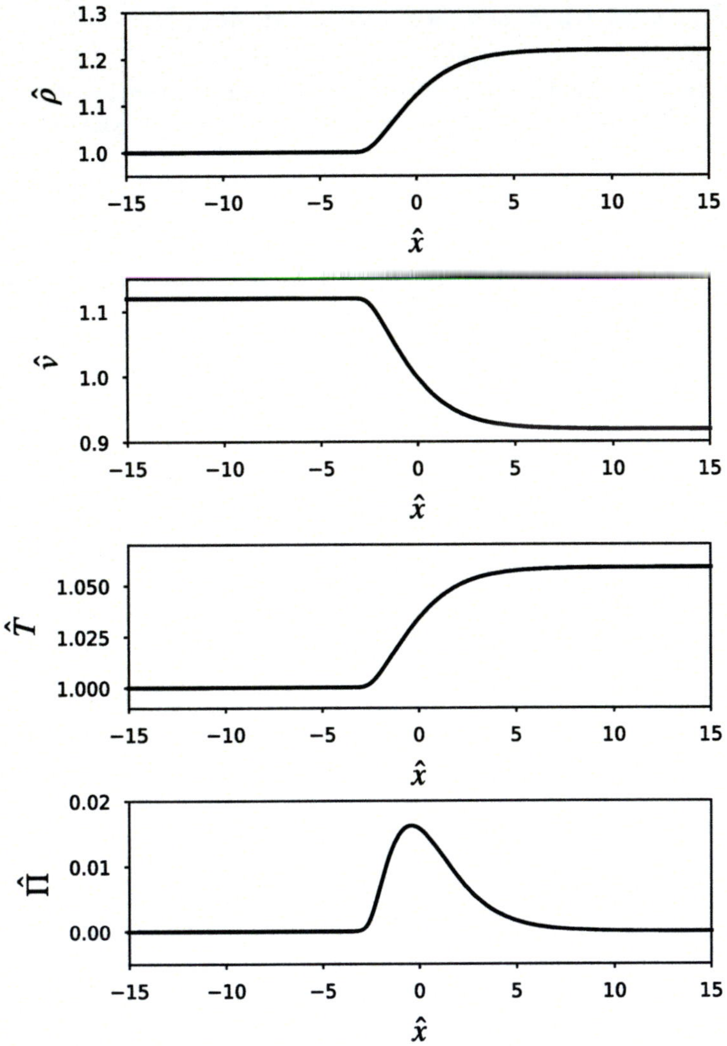

Fig. 18.2 Type B: shock wave structures predicted by the ET_6 theory (black thick curves) and by the ET_{14} theory (red thin curves). $M_0 = 1.12$. The black and red curves coincide with each other within the thickness of the black curve

wave structure of Type A. We also depict the shock wave structure at $M_0 = 1.12$ as an example of Type B in Fig. 18.2.

From these figures, we understand clearly that the ET_6 theory can reproduce nearly the same shock waves structures of Types A and B as those by the ET_{14} theory. Note that this conclusion is true only when the relaxation time of the dynamic pressure is much larger than the other two relaxation times.

18.3.2 Shock Wave Structure with a Subshock

Figures 18.3 and 18.4 show the predictions of the shock wave structure at $M_0 = 1.15$ and $M_0 = 1.47$ by the ET_6 and ET_{14} theories. These shock wave structures belong to Type C.

We notice from Figs. 18.3 and 18.4 that the ET_6 theory can describe the shock wave structure quite well. Only the difference between ET_6 and ET_{14} can be detected in the thin layer. See the magnifications shown in the sub-figures of Figs. 18.3 and 18.4. The thin layer in the ET_6 theory is represented by a subshock,

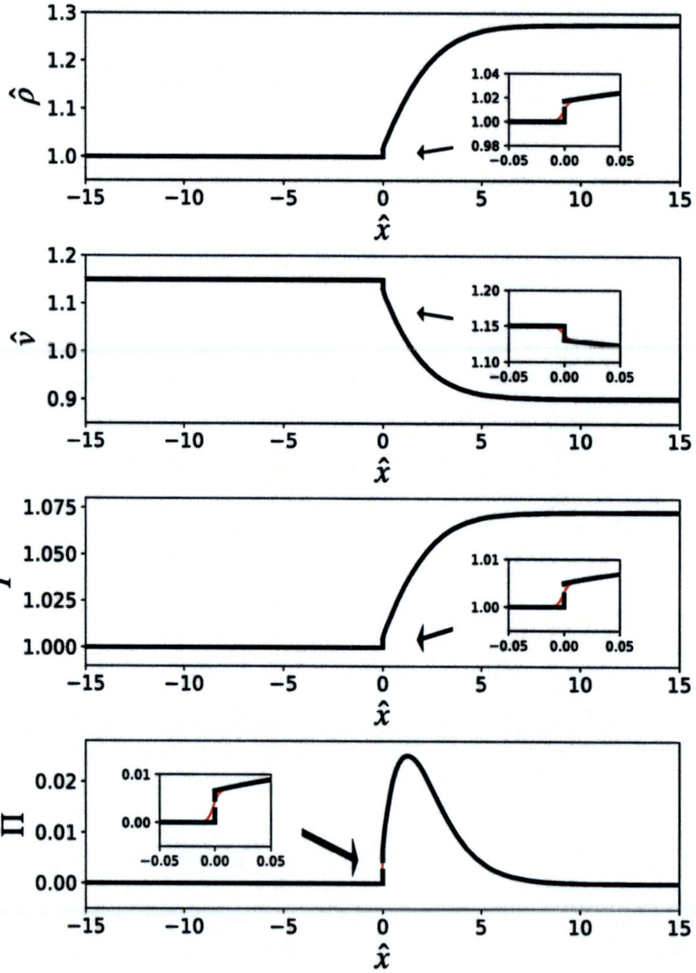

Fig. 18.3 Type C at $M_0 = 1.15$: shock wave structures predicted by the ET_6 theory (black thick curves) and by the ET_{14} theory (red thin curves)

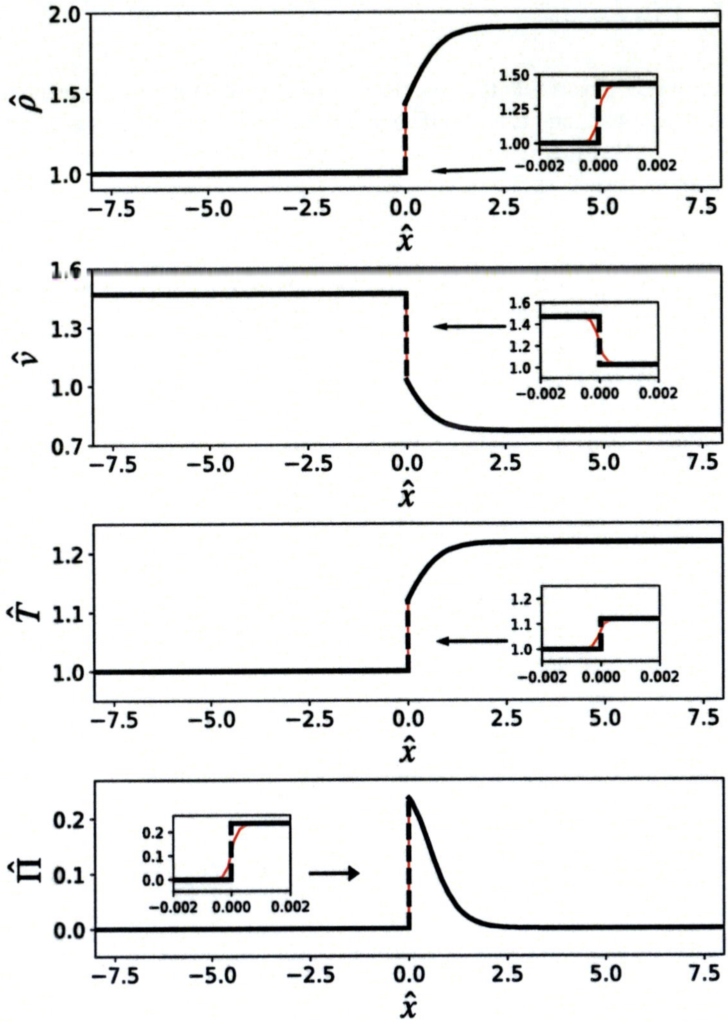

Fig. 18.4 Type C at $M_0 = 1.47$: shock wave structures predicted by the ET_6 theory (black thick curves) and by the ET_{14} theory (red thin curves)

but, in the ET_{14} theory, it is represented by a thin layer with finite thickness. In other words, the ET_6 theory can describe the thin layer only as a discontinuous surface because of the limited resolution inherent in the theory. On the other hand, owing to its finer resolution, the ET_{14} theory can describe the fine structure of the thin layer. In this respect, it is also interesting to note that the Bethe–Teller theory [145] also regards the thin layer as a discontinuous surface.

In conclusion, the ET_6 theory can describe also the shock wave structure of Type C very well if we are not interested in the fine structure of the thin layer.

18.3.3 Discussions

1. We have seen above that the ET_6 theory can describe the shock wave structures of all Types A, B, and C in a unified and consistent way within its resolution. We may use the ET_6 theory for analyzing the shock wave structure even at the Mach number larger than λ_0^{max}/c_0.
2. The interpretation in (1) is consistent with the fact that the maximum characteristic velocity monotonically increases with the increase of the number of the independent variables in hyperbolic systems (see Sect. 4.3). The RET theory can describe more and more fine shock-wave-structure as the number of the independent variables increases.
3. As pointed out before, in the ET_{14} theory, the continuous shock wave structure is obtained up to the Mach number $M_0 \approx 1.74$. Above this Mach number, the ET_{14} theory describes the thin layer as the layer with a subshock. If we want to analyze the continuous shock wave structure at a larger Mach number than 1.74, we need to adopt more independent variables.
4. We can define, without ambiguity, the critical Mach number at which the transition between Types B and C occurs by using the characteristic velocity of ET_6 theory. We obtain $M_0 \approx 1.137$, which is consistent with the experimental data [139–144] although, at present, the experimental data are too few to determine the definite value of the critical Mach number.

18.4 Strength and Stability of a Subshock

18.4.1 Mach Number Dependence of the Strength of a Subshock

We depict the Mach number dependence of mass density, velocity, temperature, and dynamic pressure in the state $*$, that is, the state just after a subshock in Fig. 18.5. We can see that the mass density, the temperature, and the dynamic pressure (the velocity) increase (decreases) with the increase of the Mach number. When $M_0 = \lambda_0^{max}/c_0$, the values of the quantities in the state $*$ coincide with the values in the unperturbed state 0. It is interesting to note that the values of the mass density and the temperature (the velocity) in the state $*$ are always smaller (larger) than the values of those in the perturbed state 1. This means that there is no overshoot of these quantities in a subshock. See also Sect. 18.6.

18.4.2 Stability of a Subshock

All solutions of the RH conditions for a subshock (18.3) are not necessarily stable (admissible). In order to select a stable solution, we can utilize the Lax condition

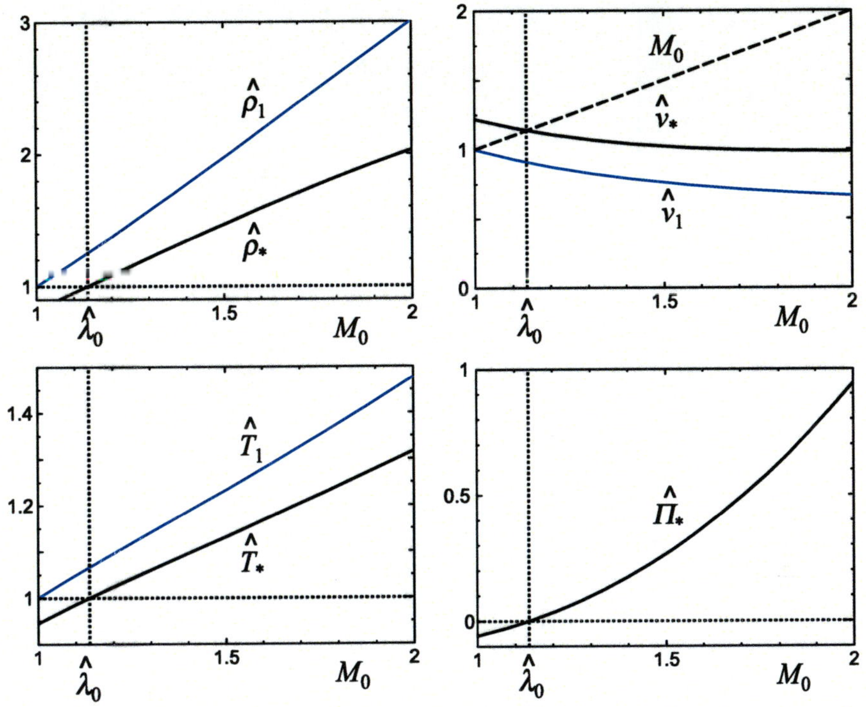

Fig. 18.5 Mach number dependence of mass density, velocity, temperature, and dynamic pressure in the state indicated by an asterisk $*$ just after a subshock (black curves). Similar dependence of the quantities in the perturbed state 1 (blue curves) is also shown. Here $\hat{\lambda}_0$ is defined by $\hat{\lambda}_0 \equiv \lambda_0^{max}/c_0$

(see Sect. 3.3.2.1), which, in the present case, is written by

$$\hat{\lambda}_0 < M_0 < \hat{\lambda}_*$$

with $\hat{\lambda}_0$ and $\hat{\lambda}_*$ being

$$\hat{\lambda}_0 = \sqrt{\frac{5}{3}\frac{1}{\gamma_0}} \quad \text{and} \quad \hat{\lambda}_* = \sqrt{\frac{5}{3}\frac{\hat{\rho}_*\hat{T}_* + \hat{\Pi}_*}{\gamma_0\hat{\rho}_*}} + M_0 - \hat{v}_*.$$

The dependence of $\hat{\lambda}_0$ and $\hat{\lambda}_*$ on the Mach number is shown in Fig. 18.6. We can see that the Lax condition is satisfied when the Mach number is larger than the dimensionless maximum characteristic velocity in the unperturbed state, that is, $M_0 > \hat{\lambda}_0 \equiv \lambda_0^{max}/c_0$. This is consistent with the theorem about the formation of a subshock explained in Sect. 3.4. We conclude that, if a subshock is *compressive*, it is admissible.

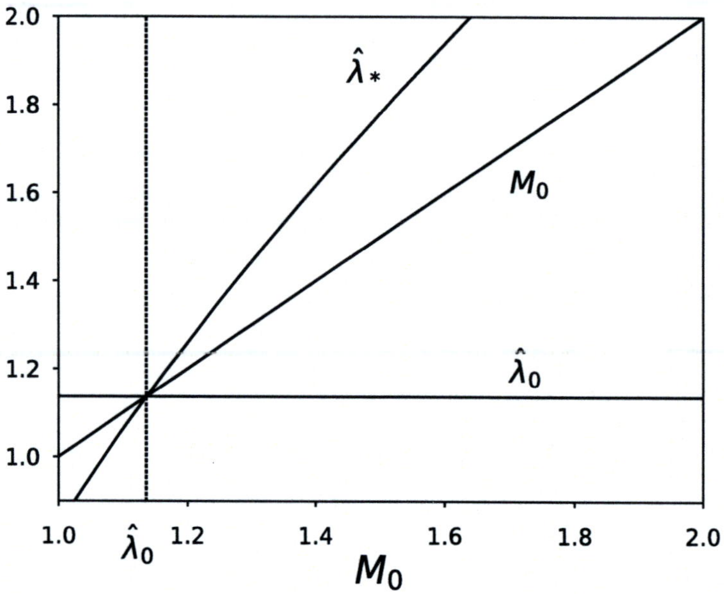

Fig. 18.6 Dependence of the characteristic velocities $\hat{\lambda}_0$ and $\hat{\lambda}_*$ on the Mach number for a subshock

18.5 Nonlinear Effect Analyzed by Nonlinear ET$_6$

In this section, we study the nonlinear effect by comparing the predictions based on the linear and nonlinear ET$_6$ theories. We again consider a plane shock wave propagating along the x-axis with the velocity $\mathbf{v} \equiv (v, 0, 0)$ in a rarefied carbon dioxide gas with $T_0 = 295$ K and $p_0 = 69$ mmHg in the unperturbed state.

In the present case, the balance equations (12.37) are rewritten as [133]

$$
\begin{aligned}
&\frac{\mathrm{d}}{\mathrm{d}\hat{x}}(\hat{\rho}\hat{v}) = 0,\\
&\frac{\mathrm{d}}{\mathrm{d}\hat{x}}\left\{\frac{1}{\gamma_0}\left(\hat{\rho}\hat{T} + \hat{\Pi}\right) + \hat{\rho}\hat{v}^2\right\} = 0,\\
&\frac{\mathrm{d}}{\mathrm{d}\hat{x}}\left\{\frac{2}{\gamma_0}\left((\hat{\rho}\hat{T} + \hat{\Pi})\hat{v} + \hat{\rho}\hat{v}\frac{1}{T_0}\int_{T_R}^{T} \hat{c}_v(\xi)\, d\xi\right) + \hat{\rho}\hat{v}^3\right\} = 0,\\
&\frac{\mathrm{d}}{\mathrm{d}\hat{x}}\left\{\frac{5}{\gamma_0}(\hat{\rho}\hat{T} + \hat{\Pi})\hat{v} + \hat{\rho}\hat{v}^3\right\} = -\frac{3}{\hat{\tau}_\Pi}\left(\frac{2}{3} - \frac{1}{\hat{c}_v(T)}\right)\frac{\hat{\rho}\hat{T}^2}{\hat{\Theta}}\frac{3\int_\Theta^T \hat{c}_v(\xi)\, d\xi}{2\int_\Theta^T \hat{c}_v(\xi)\, d\xi + 3\Theta},
\end{aligned}
\tag{18.5}
$$

where Θ is related with Π/p by (12.22), and $\hat{\Theta}$ is defined by

$$\hat{\Theta} \equiv \frac{\Theta}{T_0}.$$

If we take the linear approximation in the production term of $(18.5)_4$, the above system reduces to the linear case (18.1).

The profiles of the dimensionless mass density $\hat{\rho}$, the velocity $\hat{v}$, the temperature $\hat{T}$, and the dynamic pressure $\hat{\Pi}$ based on the nonlinear ET$_6$ theory (18.5) are numerically obtained as shown in Figs. 18.7, 18.8, and 18.9 with the Mach numbers $M_0 = 1.3,\ 3,\ 5$, respectively. The numerical procedure is the same as in the study based on the linear ET$_6$ theory above. For comparison, the shock wave structure obtained by the linear ET$_6$ theory in the same condition is also shown.

We observe that the difference between the predictions based on the nonlinear and linear ET$_6$ theories is negligible for moderate Mach numbers; $M_0 = 1.3$ and $M_0 = 3$. The difference becomes evident only for strong shock waves with large Mach numbers such as $M_0 = 5$.

We understand the reason why the difference is not so drastic even at a very high Mach number as follows: If a shock wave is enough strong, the shock wave structure is composed of thin and thick layers (shock wave structure of Type C). Within the resolution of the ET$_6$ theory, the thin layer is obtained as a subshock. The strength of subshock is determined by the Rankine–Hugoniot (RH) conditions that are independent of the production term. On the other hand, the difference between the systems of the linear and nonlinear ET$_6$ theories resides only in the functional form of the production term. Therefore the nonlinear and linear ET$_6$ theories predict the same strength of a subshock. The difference appears only in the thick layer. However, because the gradient of the physical quantities is not large in the thick layer, the difference is not so remarkable at least up to $M_0 = 5$. Such a small difference implies that the analysis based on the linear ET$_6$ is reliable at least up to $M_0 = 5$.

18.6 Shock Wave Structure in Terms of Meixner's Variables: Temperature Overshoot

We have discussed in Chap. 12 that we can also describe nonequilibrium phenomena in terms of the quantities in the Meixner theory. Therefore, by using the correspondence relation (12.56), the shock wave structure described by the nonlinear ET$_6$ theory can be converted into the structure from the viewpoint of the Meixner theory. Comparison of the nonequilibrium temperatures in both theories is especially interesting. The following fact is also worth noting: In Chap. 15, we have proposed a well-defined nonequilibrium temperature and chemical potential, which, within the ET$_6$ theory, coincide with the corresponding quantities in the Meixner theory.

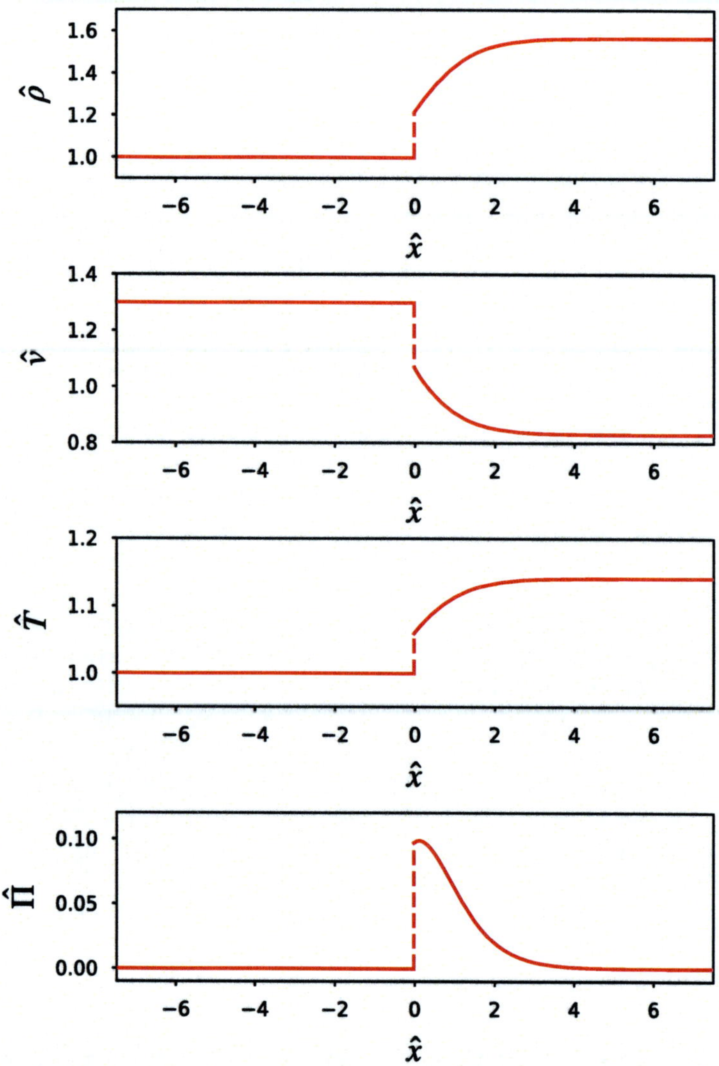

Fig. 18.7 Theoretical predictions of the profiles of the dimensionless mass density, velocity, temperature, and dynamic pressure based on the nonlinear ET$_6$ theory (red thick curves). The theoretical predictions obtained by the linear ET$_6$ theory (black thin curves) are also shown. $M_0 = 1.3$

In addition to (17.3), we introduce the following dimensionless quantities in the Meixner theory:

$$\hat{\mathscr{T}} \equiv \frac{\mathscr{T}}{T_0}, \quad \hat{\mathscr{P}} \equiv \frac{\mathscr{P}}{\rho_0 \frac{k_B}{m} T_0}, \quad \hat{\xi} \equiv \frac{\xi}{\frac{k_B}{m} T_0}.$$

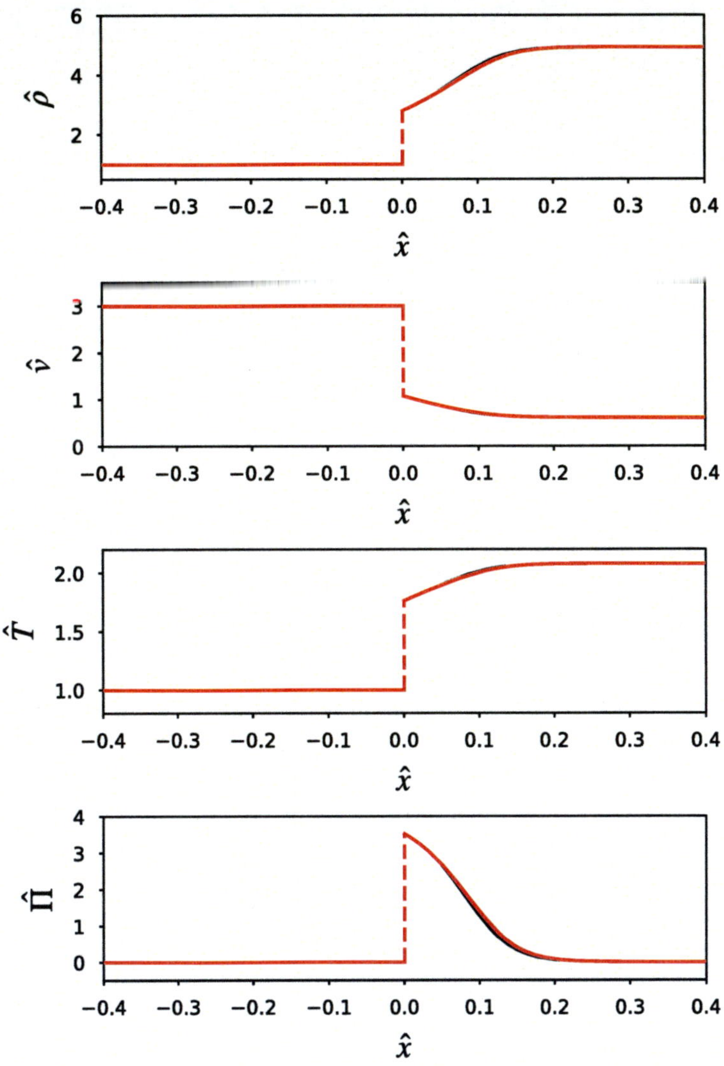

Fig. 18.8 Shock wave structure based on the nonlinear ET_6 theory (red thick curves) and the linear ET_6 theory (black thin curves). $M_0 = 3$

18.6.1 Shock Wave Structure of Types A, B, and C

When the Mach number is near but above unity, the shock wave structure of Type A appears. Figure 18.10 shows the typical profiles of the dimensionless mass density $\hat{\rho}$, velocity $\hat{v}$, temperature $\hat{\mathscr{T}}$, pressure $\hat{\mathscr{P}}$, and internal variable $\hat{\xi}$ at $M_0 = 1.04$ in terms of Meixner's variables. For comparison, the profile of the temperature T in the ET_6 theory is also shown.

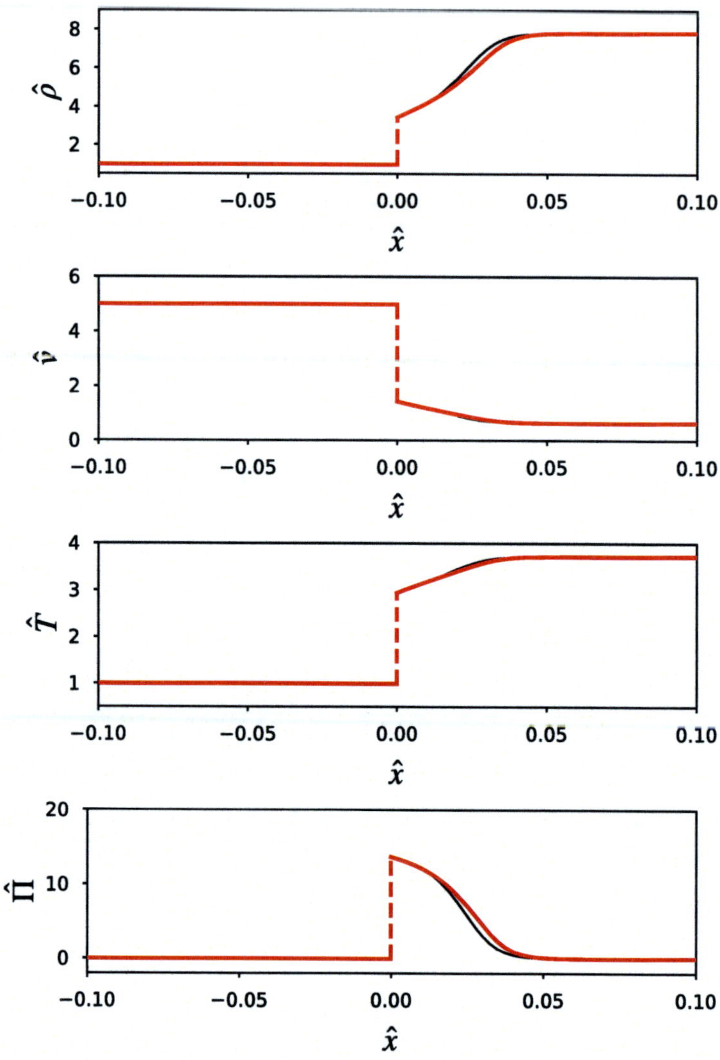

Fig. 18.9 Shock wave structure based on the nonlinear ET_6 theory (red thick curves) and the linear ET_6 theory (black thin curves). $M_0 = 5$

As the Mach number increases, the asymmetric character of the shock wave structure (Type B) emerges. Figure 18.11 shows a typical example at $M_0 = 1.12$. We notice clearly the difference between the two temperatures $\mathcal{T}$ and T. Note also that $\mathcal{T} = \theta^K$ (see (12.70)).

When the velocity of a shock wave is larger than the maximum characteristic velocity in the unperturbed state, a subshock appears. As the Mach number increases more, the overshoot of the temperature $\mathcal{T}$ emerges, while there appears no overshoot of T in the ET_6 theory. Here the overshoot is defined by the temperature

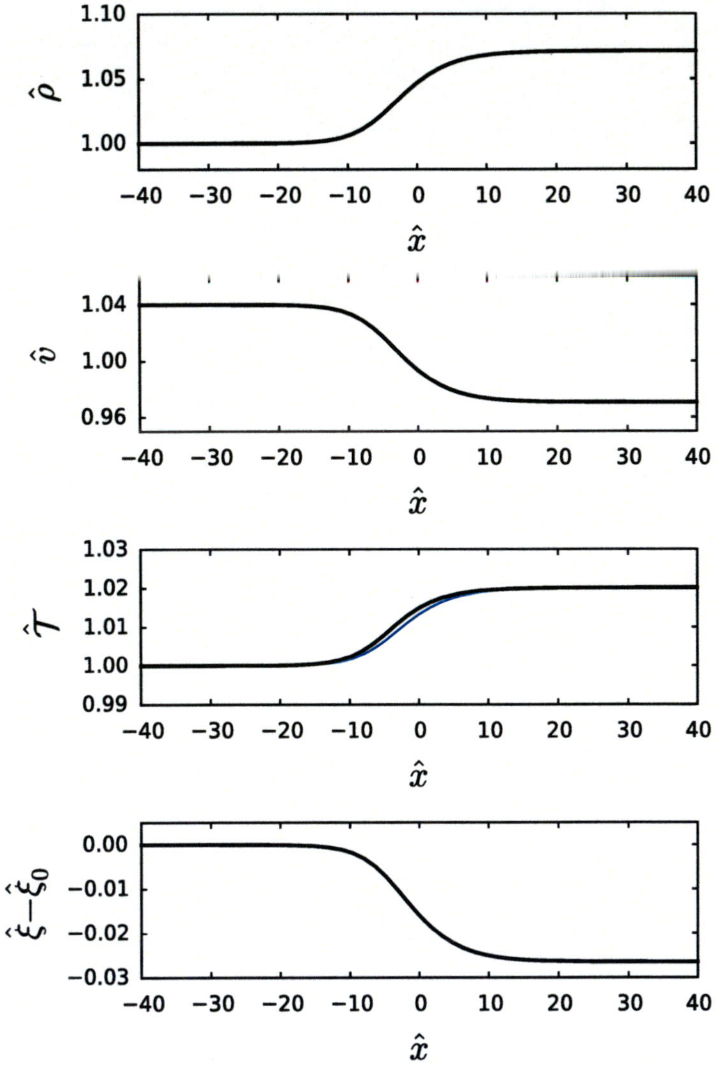

Fig. 18.10 Type A at $M_0 = 1.04$: profiles of the dimensionless mass density $\hat{\rho}$, velocity $\hat{v}$, temperature $\hat{\mathcal{T}}$, and internal variable $\hat{\xi}$ in the Meixner theory (black thick curves). The temperature $\hat{T}$ in the ET_6 theory (blue thin curve) is also shown

profile that has a part being larger than the perturbed temperature. There are three sub-Types C1–C3 in Type C as seen in Figs. 18.12, 18.13, and 18.14:

Type C1 ($1.137 < M_0 \lesssim 1.17$): Monotonic profiles without any overshoot.

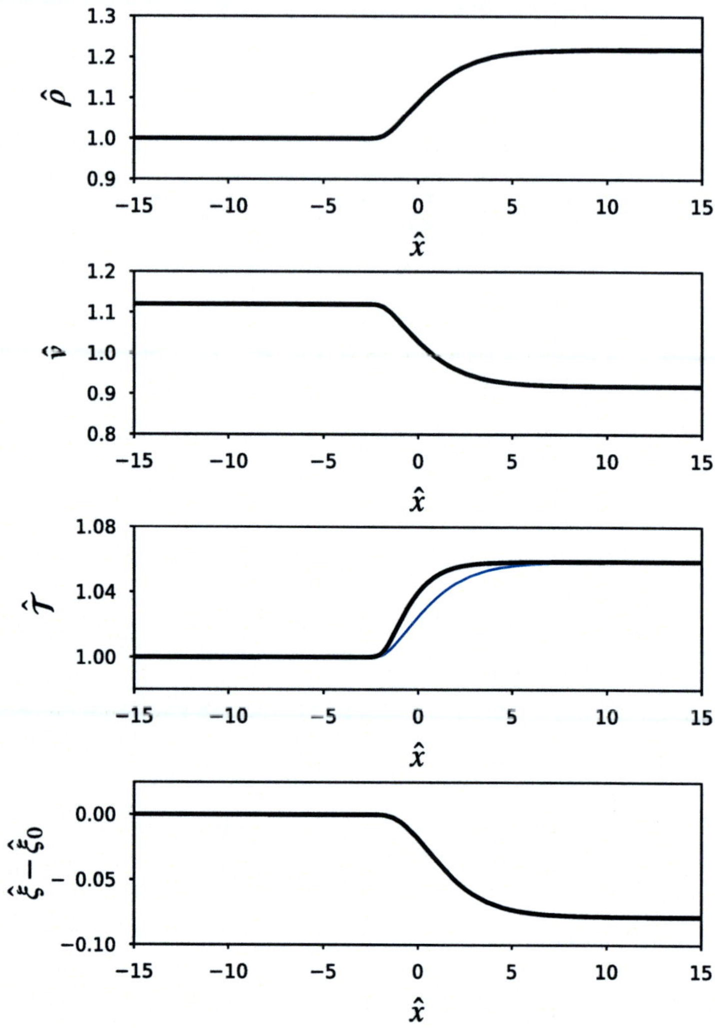

Fig. 18.11 Type B at $M_0 = 1.12$: shock wave structure described by the Meixner theory (Black thick curves). The temperature $\hat{T}$ in the ET_6 theory (blue thin curve) is also shown

Type C2 ($1.17 \lesssim M_0 < 1.30$): Non-monotonic change in the thick layer and the overshoot exists. The temperature just after a subshock $\mathcal{T}_*$ is smaller than the one in the perturbed state T_1.

Type C3 ($1.30 < M_0$): The overshoot exists. The temperature $\mathcal{T}_*$ is larger than T_1.

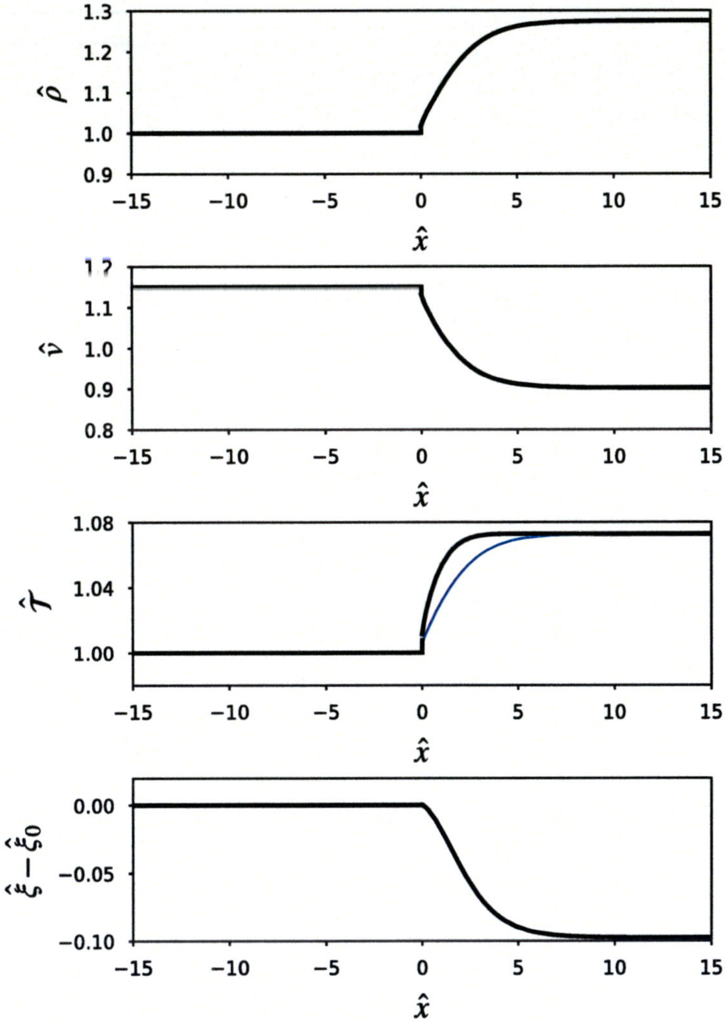

Fig. 18.12 Type C1: the shock wave structure described in the Meixner theory (Black thick curves). The temperature profile in the ET_6 theory (blue thin curve) is also shown. $M_0 = 1.15$

18.6.2 Rankine–Hugoniot Conditions for a Subshock

The Mach number dependence of the strength of a subshock in the Meixner theory is also obtained from the Rankine–Hugoniot (RH) conditions for a subshock.

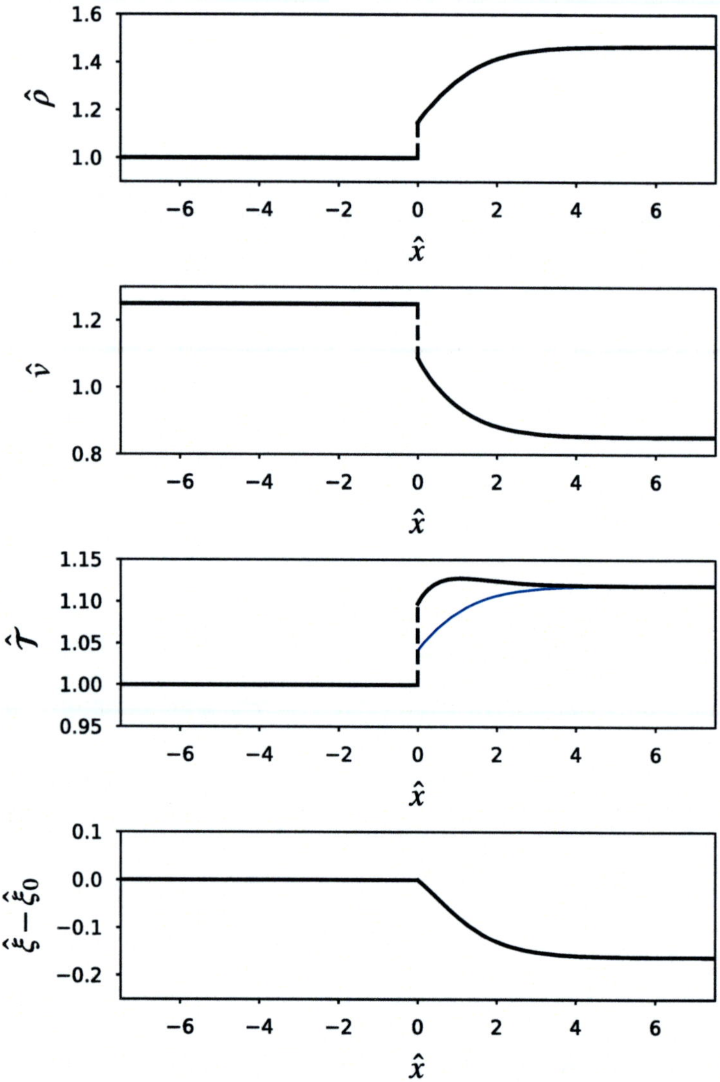

Fig. 18.13 Type C2: the shock wave structure in the Meixner theory (Black thick curves). The temperature profile in the ET$_6$ theory (blue thin curve) is also shown. $M_0 = 1.25$

The results are summarized in Fig. 18.15. Remarkable points to be noticed are as follows:

1. When the Mach number is larger than 1.30, the temperature $\mathcal{T}_*$ is larger than $\mathcal{T}_1$.
2. The internal variable ξ is continuous across a subshock. It can be proved that this is true for any thermal and caloric equations of state.

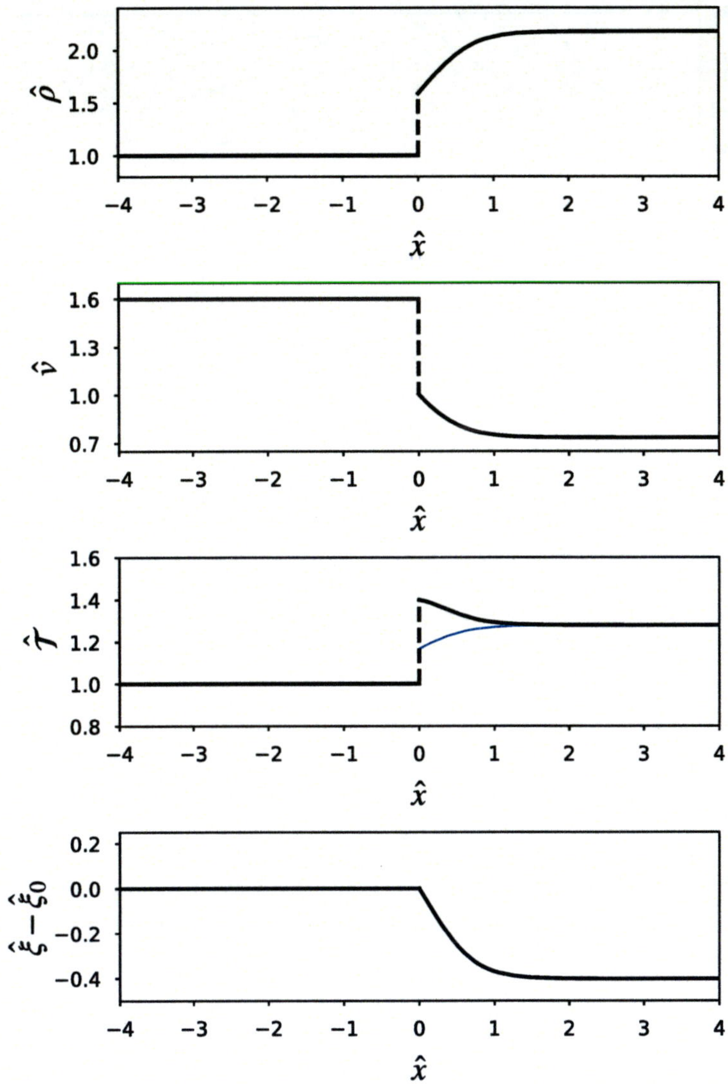

Fig. 18.14 Type C3: the shock wave structure in the Meixner theory (Black thick curves). The temperature profile in the ET_6 theory (blue thin curve) is also shown. $M_0 = 1.6$

18.6.3 Discussions

(A) From the analysis above, we see that we should be careful not to confuse the nonequilibrium temperatures T and $\mathcal{T}(= \theta^K)$. The difference between these temperatures is, in general, not negligible, and even more these may possibly be qualitatively different from each other. Therefore, in order to compare

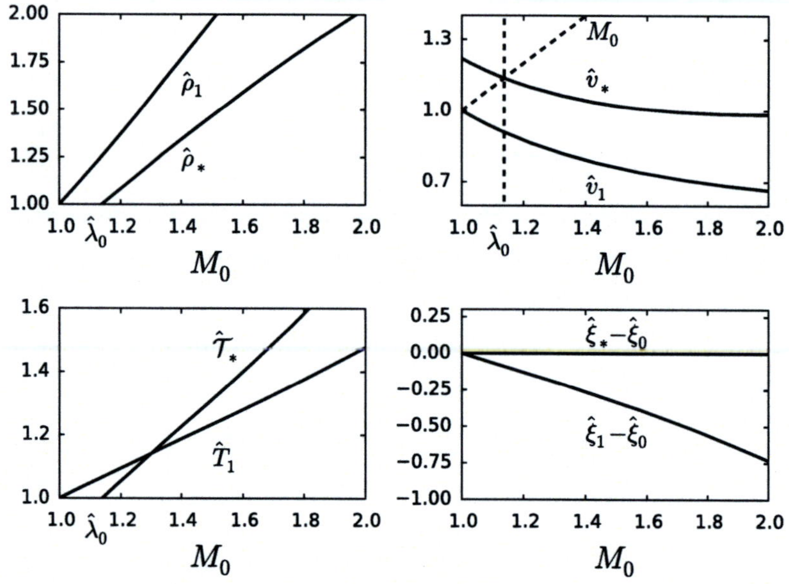

Fig. 18.15 Mach number dependence of the strength of a subshock. Here $\hat{\lambda}_0 \equiv \lambda_0^{max}/c_0$ represents the dimensionless maximum characteristic velocity in the unperturbed state

the theoretical and/or experimental results with each other, we should use a nonequilibrium temperature in a consistent manner. For example, the non-monotonic profile of the temperature $\mathscr{T}$ is qualitatively different from the previous studies assuming the exponential decay for the thick layer in the Type C shock wave structure [140, 142, 145]. Re-examination of the experimental and theoretical results from this point of view seems to be necessary.

(B) Inserting the equations of state (12.15) into the correspondence relation (12.56), we have (see (12.67))

$$\xi = -\frac{2}{3}\varepsilon^I. \tag{18.6}$$

In the Bethe–Teller theory [145], by making the separation between the "inert" and "active" degrees of freedom, it is assumed that the internal energy due to the "inert" degrees of freedom is unchanged in the thin layer in the Type C shock wave structure. Such separation, however, seems to be artificial and obscure. While in the ET$_6$ theory, from the relation (18.6), we notice that the continuous profile of ξ implies that the specific internal energy due to all internal modes ε^I does not change across a subshock. It should be emphasized that this fact is derived without any ad hoc assumptions.

18.7 Analysis of Shock Wave Structure by the Kinetic Theory

Shock wave structure was also analyzed on the basis of the kinetic theory [147–149], where the ellipsoidal statistical model of the Boltzmann equation for a polyatomic gas with constant specific heats proposed by Andries et al. [375] and its extended model with temperature-dependent specific heats [149] were adopted.

In the numerical analysis, the shock wave structures of Type A, B, and C in a CO_2 gas were also found, and the profiles of mass density, velocity, temperature, and dynamic pressure are in excellent agreement with the results obtained above and in [373].

Interesting studies of shock wave structure in CO_2 gas were also made in the recent papers [376, 377].

Evidently, further cooperation between the communities working in the kinetic theory and RET would be highly desired for deeper understanding of the shock wave phenomena.

Chapter 19
Steady Flow of a Polyatomic Gas

Abstract As seen in the previous chapters, the role of the dynamic pressure in nonequilibrium phenomena can be highlighted in the ET_6 and ET_7 theories by ignoring other dissipative processes. These theories are particularly important for gases with the large relaxation time of the dynamic pressure. In this chapter, taking the role into special account, we study a steady flow of a rarefied polyatomic gas by the ET_7 theory. It is shown that, in contrast to the Euler theory of perfect fluid, the ET_7 theory predicts some remarkable effects on a flow due to the relaxation processes of the rotational and vibrational modes of a molecule. And thereby characteristic features of the ET_7 theory itself can be clarified.

19.1 Basic Equations

In the ET_7 theory of rarefied polyatomic gases, a nonequilibrium state is specified by the quantities $\{\rho, v_i, \theta^K, \theta^R, \theta^V\}$. For details about the quantities, see Chap. 14. Then, from (14.7), a steady flow is governed by the following system of equations:

$$
\begin{aligned}
&\frac{\partial}{\partial x_i}(\rho v_i) = 0,\\
&\frac{\partial p(\rho, \theta^K)}{\partial x_j} + \rho v_i \frac{\partial v_j}{\partial x_i} = 0,\\
&v_i \frac{\partial}{\partial x_i}\left(2\frac{p(\rho, \theta^K)}{\rho} + 2(\varepsilon^K_E(\theta^K) + \varepsilon^R_E(\theta^R) + \varepsilon^V_E(\theta^V)) + v^2\right) = 0,\\
&2\rho v_i \frac{\partial \varepsilon^R_E(\theta^R)}{\partial x_i} = \hat{P}^R_{ll},\\
&2\rho v_i \frac{\partial \varepsilon^V_E(\theta^V)}{\partial x_i} = \hat{P}^V_{ll}.
\end{aligned}
\tag{19.1}
$$

T. Ruggeri, M. Sugiyama, *Classical and Relativistic Rational Extended Thermodynamics of Gases*, https://doi.org/10.1007/978-3-030-59144-1_19

The entropy balance equation (14.11) now reduces to

$$\rho v_i \frac{\partial \eta}{\partial x_i} = \Sigma \geqslant 0, \tag{19.2}$$

where the nonequilibrium specific entropy density η is given by (14.9).

Let us pay attention to the variation of quantities *along a streamline*. From $(19.1)_{2,3,4}$, we have [378]

$$\begin{aligned}
&\delta p(\rho, \theta^K) + \rho v \delta v = 0, \\
&v\delta \left(2\frac{p(\rho,\theta^K)}{\rho} + 2\varepsilon_E^K(\theta^K) + 2\varepsilon_E^R(\theta^R) + 2\varepsilon_E^V(\theta^V) + v^2 \right) = 0, \\
&2\rho v \delta \varepsilon_E^R(\theta^R) = \hat{P}_{ll}^R, \\
&2\rho v \delta \varepsilon_E^V(\theta^V) = \hat{P}_{ll}^V,
\end{aligned} \tag{19.3}$$

where

$$\delta \equiv \frac{v_i}{v}\frac{\partial}{\partial x_i} \qquad (v \equiv \sqrt{v_i v_i})$$

is the directional derivative along a streamline. And, from (19.2), we have

$$\rho v \delta \eta = \Sigma. \tag{19.4}$$

Equations (19.3) are rewritten in terms of $\{\delta v, \delta \rho, \delta\theta^K, \delta\theta^R, \delta\theta^V\}$ as follows:

$$\begin{aligned}
&\rho v \delta v + p_\rho(\rho,\theta^K)\delta\rho + p_{\theta^K}(\rho,\theta^K)\delta\theta^K = 0, \\
&v\delta v + \left(-\frac{p(\rho,\theta^K)}{\rho^2} + \frac{p_\rho(\rho,\theta^K)}{\rho} \right)\delta\rho \\
&\quad + \left(\frac{p_{\theta^K}(\rho,\theta^K)}{\rho} + c_v^K(\theta^K) \right)\delta\theta^K + c_v^R(\theta^R)\delta\theta^R + c_v^V(\theta^V)\delta\theta^V = 0, \\
&2\rho v c_v^R(\theta^R)\delta\theta^R = \hat{P}_{ll}^R, \\
&2\rho v c_v^V(\theta^V)\delta\theta^V = \hat{P}_{ll}^V,
\end{aligned} \tag{19.5}$$

where the suffix indicates a partial derivative, for example, $p_\rho(\rho,\theta) \equiv \partial p(\rho,\theta)/\partial\rho$, and c_v^K, c_v^R, and c_v^V are the specific heats of the subsystems defined by

$$c_v^K(\theta^K) = \frac{d\varepsilon_E^K(\theta^K)}{d\theta^K}, \quad c_v^R(\theta^R) = \frac{d\varepsilon_E^R(\theta^R)}{d\theta^R}, \quad c_v^V(\theta^V) = \frac{d\varepsilon_E^V(\theta^V)}{d\theta^V}.$$

These specific heats are positive, and the total specific heat is given by $c_v = c_v^K + c_v^R + c_v^V$.

19.2 Discussions

The following qualitative analysis is valid for any production terms $\hat{P}_{ll}^R$ and $\hat{P}_{ll}^V$ provided that it is compatible with the entropy principle ($\Sigma \geq 0$).

From (19.5)$_{1,2}$, the variation of the mass flux $\delta(\rho v)$ is expressed in terms of δv, $\delta\theta^R$, and $\delta\theta^V$ as follows:

$$\delta(\rho v) = \rho\left(1 - \frac{v^2}{U^2}\right)\delta v + \frac{v}{U^2}\frac{p_{\theta^K}(\rho,\theta^K)}{c_v^K(\theta^K)}(c_v^R(\theta^R)\delta\theta^R + c_v^V(\theta^V)\delta\theta^V), \tag{19.6}$$

where U is the characteristic velocity given by (see Sect. 14.2.2)

$$U^2 = p_\rho(\rho,\theta^K) + \frac{\theta^K p_{\theta^K}^2(\rho,\theta^K)}{\rho^2 c_v^K(\theta^K)}.$$

In particular, the velocity U in an equilibrium case with the condition $\theta^K = \theta^R = \theta^V = T$, is given by

$$U_E^2 = p_\rho(\rho,T) + \frac{T p_T^2(\rho,T)}{\rho^2 c_v^K(T)}.$$

It is interesting to remind that the counterpart of equation (19.6) in the Euler theory is given by

$$\delta(\rho v) = \rho\left(1 - \frac{v^2}{U_{Euler}^2}\right)\delta v, \tag{19.7}$$

where U_{Euler}^2 is the sound velocity given by (see Sect. 14.2.2)

$$U_{Euler}^2 = p_\rho(\rho,T) + \frac{T p_T^2(\rho,T)}{\rho^2 c_v(T)}.$$

Because the system of ET$_7$ includes the Euler system as a principal subsystem, we have the following *subcharacteristic condition* (see Theorem 2.3):

$${U_E}^2 > U_{Euler}^2.$$

Other interesting relations along a streamline are as follows: From (19.5)$_{1,2}$, we have

$$\delta\varepsilon = \frac{p(\rho, \theta^K)}{\rho^2}\delta\rho \tag{19.8}$$

and, from (14.9), (19.8), and (19.4), we have

$$\delta\eta = c_v^R(\theta^R)\left(\frac{1}{\theta^R} - \frac{1}{\theta^K}\right)\delta\theta^R + c_v^V(\theta^V)\left(\frac{1}{\theta^V} - \frac{1}{\theta^K}\right)\delta\theta^V = \frac{\Sigma}{\rho v}. \tag{19.9}$$

Some noticeable points from the relations above are summarized as follows:

1. In the Euler theory, as seen from (19.7), we notice the well-known result that the mass flux ρv reaches the maximum value at the place where the velocity v is equal to the local velocity of sound U_{Euler} [8, 379].

   While, the ET$_7$ theory predicts that, from (19.6), mass flux attains the maximum value at the place where the following condition is satisfied:

$$\rho\left(1 - \frac{v^2}{U^2}\right)\delta v = -\frac{v}{U^2}\frac{p_{\theta^K}(\rho, \theta^K)}{c_v^K(\theta^K)}(c_v^R(\theta^R)\delta\theta^R + c_v^V(\theta^V)\delta\theta^V).$$

2. When we take the monatomic gas limit by letting ε^R and ε^V tend to zero, it is easy to see that the relation (19.6) reduces to (19.7).
3. For simplicity, let us study, in particular, the low-temperature region where $\varepsilon^V = 0$. From (19.6), we have

$$\delta(\rho v) = \rho\left(1 - \frac{v^2}{U^2}\right)\delta v + \frac{v}{U^2}\frac{p_{\theta^K}(\rho, \theta^K)}{c_v^K(\theta^K)}c_v^R(\theta^R)\delta\theta^R,$$

   At the place where the condition $v = U$ is fulfilled, we have the relation:

$$\delta(\rho v) = \frac{1}{U}\frac{p_\theta(\rho, \theta^K)c_v^R(\theta^R)}{c_v^K(\theta^K)}\delta\theta^R = \frac{1}{U^2}\frac{p_{\theta^K}(\rho, \theta^K)}{\rho c_v^K(\theta^K)}\frac{\Sigma}{\left(\frac{1}{\theta^R} - \frac{1}{\theta^K}\right)}.$$

   Therefore, if $p_{\theta^K} > 0$, the quantities $\delta(\rho v)$, $\delta\theta^R$, and $\theta^K - \theta^R$ have the same sign. From the relation (19.9), we notice that the quantities $\delta\theta^R$ and $\theta^K - \theta^R$ have the same sign at all places on a streamline.

Remark 19.1 In a quantitative analysis, for example numerical analysis, the choice of explicit expressions for $\hat{P}_{ll}^K$, $\hat{P}_{ll}^R$, and $\hat{P}_{ll}^V$ is required. Their expressions derived from the generalized BGK model have been given in Sect. 14.1.5.

19.3 Nozzle Flow

In this section, we study the nozzle flow assuming, for simplicity, the case with the temperature range where $\varepsilon^V = 0$. The generalization with finite ε^V is straightforward.

19.3.1 Basic Equations

Let us study a steady flow in a nozzle parallel to the x-direction. We adopt the approximation of a quasi-one-dimensional flow. Then, in addition to the system of equations (19.3), we have an equation that comes from the conservation law of mass:

$$\delta(\rho v A) = 0, \tag{19.10}$$

where A is the cross section of a nozzle, and δ in this case is

$$\delta \equiv \frac{d}{dx}.$$

From (19.6) and (19.10), we have the relation:

$$\begin{aligned} \frac{1}{v}\left(1 - \frac{v^2}{U^2}\right)\delta v &= -\frac{1}{A}\delta A - \frac{1}{U^2}\frac{p_{\theta^K}(\rho,\theta^K)}{\rho c_v^K(\theta^K)} c_v^R(\theta^R)\delta\theta^R \\ &= -\frac{1}{A}\delta A - \frac{1}{U^2}\frac{p_{\theta^K}(\rho,\theta^K)}{\rho^2 c_v^K(\theta^K)} \frac{\Sigma}{\left(\frac{1}{\theta^R} - \frac{1}{\theta^K}\right)}. \end{aligned} \tag{19.11}$$

While, in the Euler theory, we have the well-known relation from (19.7) and (19.10):

$$\frac{1}{v}\left(1 - \frac{v^2}{U^2_{Euler}}\right)\delta v = -\frac{1}{A}\delta A.$$

From the relation (19.11), we understand that the internal molecular relaxation process gives a new effect on a flow in addition to the effect of the variation of the cross section.

A numerical simulation of the nozzle flow is now being prepared and its results will be shown elsewhere.

Chapter 20
Acceleration Wave, K-condition, and Global Existence in ET_6

Abstract We verify the K-condition for the nonlinear ET_6 model and show, for any gas, the existence of global smooth solutions provided that initial data are sufficiently small. As an example, in the case of polyatomic gas, we study acceleration waves. We evaluate the Bernoulli equation for the amplitude of the wave. If the initial amplitude of an acceleration wave is sufficiently small compared with the critical amplitude, the acceleration wave exists for all time and decays to zero as time t becomes large. Vice versa, for large initial amplitude, there exists a critical time at which we have the blow up of the solution and the formation of a shock wave. We show the peculiarity of this model, that is, the velocity of a disturbance and the critical time are universal: these are independent of the degrees of freedom D of a constituent molecule.

20.1 Characteristic Velocities and the K-condition

We discussed, in Sects. 2.6.2 and 2.6.3, the role of the K-condition (2.48) [189]. Together with the entropy convexity, it is a sufficient condition for the existence of global smooth solutions. Lou and Ruggeri [268] noticed a connection between the K-condition and the global existence of acceleration waves (see Sect. 3.2) and they rewrote the condition (2.48) in the form (3.28). In this chapter, we study acceleration waves and prove that the K-condition is satisfied for the ET_6 model for any gas with a convex entropy.

T. Ruggeri, M. Sugiyama, *Classical and Relativistic Rational Extended Thermodynamics of Gases*, https://doi.org/10.1007/978-3-030-59144-1_20

The system (12.40) in the BGK approximation is given by

$$\begin{aligned}
&\dot{\rho} + \rho\frac{\partial v_k}{\partial x_k} = 0,\\
&\rho\dot{v}_i + \frac{\partial}{\partial x_i}(p+\Pi) = 0,\\
&\rho\dot{\varepsilon} + (p+\Pi)\frac{\partial v_k}{\partial x_k} = 0,\\
&\left(\frac{p+\Pi}{\rho} - \frac{2}{3}\varepsilon\right)^{\bullet} = -\frac{\Pi}{\rho\tau}.
\end{aligned} \tag{20.1}$$

Using the chain rule (3.27) we obtain, from the differential system, characteristic eigenvalues and right eigenvectors. In particular, for the material time derivative, the chain rule is expressed by

$$\bullet \to -V\delta, \quad V = \lambda - v_n, \quad v_n = v_i n_i.$$

In the present case, from the system (20.1), we obtain

$$\begin{aligned}
&-V\delta\rho + \rho\delta v_n = 0,\\
&-\rho V\delta\mathbf{v} + \mathbf{n}\delta(p+\Pi) = 0,\\
&-\rho V\delta\varepsilon + (p+\Pi)\delta v_n = 0,\\
&-V\delta\left(\frac{p+\Pi}{\rho} - \frac{2}{3}\varepsilon\right) = 0.
\end{aligned}$$

Taking into account the equations of state (12.15)$_1$ and (7.18), we have:

$$1) \qquad V = 0 \qquad \longleftrightarrow \qquad \lambda = v_n, \qquad \text{contact waves,}$$

with $\delta\rho$, $\delta\mathbf{v}_T$, δp arbitrary (multiplicity 4), and $\delta v_n = 0$, $\delta\Pi = -\delta p$ ($\mathbf{v}_T$ denotes the tangential velocity);

$$2) \quad V = \pm\sqrt{\frac{5}{3}\frac{p+\Pi}{\rho}} \quad \longleftrightarrow \quad \lambda = v_n \pm \sqrt{\frac{5}{3}\frac{p+\Pi}{\rho}}, \quad \text{sound waves,} \tag{20.2}$$

with $\delta\rho$ arbitrary,

$$\delta\mathbf{v} = \mathbf{n}V\frac{\delta\rho}{\rho}, \quad \delta\varepsilon = \frac{\delta\rho}{\rho}\left(\frac{p+\Pi}{\rho}\right), \quad \delta\Pi = \frac{\delta\rho}{\rho}\Gamma, \tag{20.3}$$

where Γ is given by

$$\Gamma = (p + \Pi)\left(\frac{5}{3} - \frac{p_\varepsilon}{\rho}\right) - \rho p_\rho.$$

Since only the last component of the production term $\mathbf{f}$ of the generic system (2.1) is non-zero (see (20.1)), the K-condition (2.48) is satisfied if $\delta\Pi \neq 0$ in equilibrium ($\Pi = 0$). This is true for both contact waves and for sound waves because of (20.3) and the inequality $\Gamma_E > 0$ due to the convexity condition, where Γ_E is the value of Γ at equilibrium. Therefore the K-condition is satisfied for any gas and, together with the convexity of the entropy, we can conclude that, according to the general theorems of Sect. 2.6.3, the ET$_6$ system has global smooth solutions for all time, and the solution converges to the equilibrium one provided that the initial data are sufficiently smooth. This proof was given in [218].

We notice that the sound velocity in (20.2) is independent of the degrees of freedom D of a molecule and that, in equilibrium, it coincides with the sound velocity of a monatomic gas. In the case of polytropic polyatomic gas this result is in agreement with the Theorem 9.5. In fact, for polytropic polyatomic gases, the ET$_6$ system is a $(2^{(1)}, 0)$ system and belongs to the class for which the characteristic velocities are independent of D (see Sect. 9.4.2).

20.2 Time-Evolution of Amplitude of an Acceleration Wave and the Critical Time

Let us consider the system (20.1) in the one-dimensional case for the field $\mathbf{u} \equiv (\rho, v, \varepsilon, \Pi)^T$ with $\mathbf{v} \equiv (v, 0, 0)$. We consider, for simplicity, a polytropic fluid. We study time-evolution of the amplitude of an acceleration wave along the characteristic line of the transport equation. We focus on the fastest wave propagating in an equilibrium state at rest: $\mathbf{u}_0 \equiv (\rho_0, 0, \varepsilon_0, 0)$.

In the present case, the amplitude evolves according to the Eq. (3.23) with a and b given by (3.22). The system (20.1) can be rewritten as

$$\dot{\mathbf{u}} + \mathbf{A}\mathbf{u}_x = \mathbf{f}$$

with

$$\mathbf{u} \equiv \begin{pmatrix} \rho \\ v \\ \varepsilon \\ \Pi \end{pmatrix}, \quad \mathbf{A} \equiv \begin{pmatrix} 0 & \rho & 0 & 0 \\ \frac{3c^2}{5\rho} & 0 & \frac{2}{D} & \frac{1}{\rho} \\ 0 & \frac{3c^2}{5} + \frac{\Pi}{\rho} & 0 & 0 \\ 0 & \frac{2c^2(D-3)\rho}{5D} + \frac{5D-6}{3D}\Pi & 0 & 0 \end{pmatrix}, \quad \mathbf{f} \equiv \begin{pmatrix} 0 \\ 0 \\ 0 \\ -\frac{\Pi}{\tau} \end{pmatrix},$$

where we have put

$$c = \sqrt{\frac{5}{3}\frac{p}{\rho}}. \tag{20.4}$$

The eigenvalues of $\mathbf{A}$ are given by

$$V = \left\{0, 0, -\sqrt{\frac{5}{3}\frac{p+\Pi}{\rho}}, \sqrt{\frac{5}{3}\frac{p+\Pi}{\rho}}\right\}.$$

Therefore the fastest velocity is expressed in terms of the field $\mathbf{u}$ as

$$\lambda = v + \sqrt{10\varepsilon/(3D) + 5\Pi/(3\rho)}.$$

The gradient evaluated at an equilibrium state (we omit the index E) is given by

$$\mathrm{grad}\,\lambda = \left(\partial_\rho\lambda, \partial_v\lambda, \partial_\varepsilon\lambda, \partial_\Pi\lambda\right) = \left(0, 1, \frac{5}{3cD}, \frac{5}{6c\rho}\right).$$

The orthonormal right and left eigenvectors in equilibrium are obtained as

$$\mathbf{d}_1 = \left(-\frac{5}{3c^2}, 0, 0, 1\right)^T, \quad \mathbf{d}_2 \equiv \left(-\frac{10\rho}{3c^2D}, 0, 1, 0\right)^T,$$

$$\mathbf{d}_3 \equiv \left(\frac{5D}{2c^2(D-3)}, -\frac{5D}{2c\rho(D-3)}, \frac{3D}{2(D-3)\rho}, 1\right)^T,$$

$$\mathbf{d} = \mathbf{d}_4 \equiv \left(\frac{5D}{2c^2(D-3)}, \frac{5D}{2c\rho(D-3)}, \frac{3D}{2(D-3)\rho}, 1\right)^T,$$

$$\mathbf{l}_1 \equiv \left(-\frac{6c^2(D-3)}{25D}, 0, -\frac{4(D-3)\rho}{5D^2}, \frac{3(2+D)}{5D}\right),$$

$$\mathbf{l}_2 \equiv \left(\frac{3c^2(6-5D)}{25D\rho} + \frac{6c^2(D-3)}{25D\rho}, 0, -\frac{6-5D}{5D}, -\frac{3}{5\rho}\right),$$

$$\mathbf{l}_3 \equiv \left(\frac{3c^2(D-3)}{25D}, -\frac{c(D-3)\rho}{5D}, \frac{2(D-3)\rho}{5D^2}, \frac{-3+D}{5D}\right),$$

$$\mathbf{l} = \mathbf{l}_4 \equiv \left(\frac{3c^2(-3+d)}{25D}, \frac{c(D-3)\rho}{5D}, \frac{2(D-3)\rho}{5D^2}, \frac{-3+D}{5D}\right).$$

Therefore, from (3.22), we have

$$a = \frac{10D}{3c(D-3)\rho}, \quad b = \frac{1}{\tau}.$$

The condition (3.10) with (3.11) becomes, in the present case,

$$\begin{pmatrix} \left[\frac{[\partial\rho]}{\partial x}\right] \\ \left[\frac{[\partial v]}{\partial x}\right] \\ \left[\frac{[\partial\varepsilon]}{\partial x}\right] \\ \left[\frac{[\partial\Pi]}{\partial x}\right] \end{pmatrix} = \mathscr{A} \begin{pmatrix} \frac{5D}{2c^2(D-3)} \\ \frac{5D}{2c\rho(D-3)} \\ \frac{3D}{2(D-3)\rho} \\ 1 \end{pmatrix}. \tag{20.5}$$

Introducing the acceleration jump:

$$G = \left[\frac{\partial v}{\partial t}\right]$$

and taking into account both the Hadamard condition and the second component of (20.5), we obtain

$$\mathscr{A} = -\frac{2\rho(D-3)}{5D}G.$$

Therefore the transport Bernoulli equation (3.12) becomes very simple:

$$\frac{dG}{dt} - \frac{4}{3c}G^2 + \frac{G}{\tau} = 0. \tag{20.6}$$

An interesting remark is that this equation is independent of D as is the sound velocity (20.4)! The solution of (20.6) (see (3.23)) is

$$G(t) = \frac{G(0)e^{-\frac{t}{\tau}}}{1 + G(0)\frac{4\tau}{3c}\left(e^{-\frac{t}{\tau}} - 1\right)}. \tag{20.7}$$

Therefore if the initial amplitude of the acceleration jump satisfies the condition:

$$G(0) < G_{cr} = \frac{3}{4}\frac{c}{\tau}, \tag{20.8}$$

we have no critical time. The acceleration jump decays in agreement with the global existence and the K-condition. Instead, if $G(0)$ is greater than the critical value G_{cr}, we have the critical time (see (3.25)):

$$t_{cr} = -\tau \log\left(1 - \frac{3c}{4G(0)\tau}\right) \tag{20.9}$$

and blow-up arises with the formation of a shock wave.

It is interesting to observe that the Bernoulli equation for Euler monatomic fluids is the same as (20.6) if the last term is absent. In the case of Euler fluids, any compressive wave $G(0) > 0$ has the critical time that, in the case of monatomic gas, is given by [218]

$$t_{cr}^{5M} = \frac{3c}{4G(0)}. \tag{20.10}$$

We can rewrite (20.9) by using (20.10):

$$\frac{t_{cr}}{\tau} = -\log\left(1 - \frac{t_{cr}^{5M}}{\tau}\right).$$

If $w = t_{cr}^{5M}/\tau \ll 1$, then

$$\frac{t_{cr}}{\tau} = \frac{t_{cr}^{5M}}{\tau} + O(w^2).$$

20.3 Conclusion

We have evaluated the time-evolution of an acceleration wave for the ET_6 model. And we have seen such peculiarity of this simple model that the velocity of disturbance and the critical time are universal in the sense that these are independent of the degrees of freedom D of a constituent molecule. This is due to the fact that the model is oversimplified, and a more realistic model like the ET_{14} model gives more realistic results.

Nevertheless, even though the model may be oversimplified, it has the advantage that the system is the simplest dissipative system that corrects the classical Euler fluid. The dissipation is enough to produce a competition with hyperbolicity. Therefore, since the K-condition is satisfied and the entropy is convex, smooth solutions exist for all time and they converge to the equilibrium state.

The acceleration wave is a good example to understand this situation. In fact, if the initial acceleration amplitude is sufficiently small compared with the critical amplitude (20.8), the acceleration wave exists for all time and decays to zero as the time t becomes large as seen in (20.7).

Chapter 21
Light Scattering

Abstract Experiments of light scattering in a gas afford us with precise information about irreversible processes in a gas even out of local equilibrium. These are a good test for checking the validity of a nonequilibrium thermodynamic theory. In this chapter, we study light scattering on the basis of the ET_{14} theory.

21.1 Introduction

In this chapter, light scattering in a rarefied polyatomic gas is studied by the ET_{14} theory [108]. The results obtained are compared with the results derived from the Navier–Stokes and Fourier (NSF) theory and also with some experimental data.

Light scattering occurs due to the fluctuation in the mass density ρ through the dielectric constant $\epsilon(\rho)$. The intensity of scattered light is directly related to the dynamic structure factor [25]:

$$S(\boldsymbol{q}, \omega) = \frac{1}{\pi}\left(\frac{\partial \epsilon}{\partial \rho}\right)^2 \Re\langle \delta\rho^*(\boldsymbol{q}, 0)\delta\hat{\rho}(\boldsymbol{q}, s)\rangle_{s=i\omega},$$

where $\boldsymbol{q}$ is the scattering vector, magnitude of which is $|\boldsymbol{q}| = (4\pi/\Lambda)\sin(\theta/2)$ with Λ and θ being the wavelength of the incident light and the scattering angle, ω is the shift in angular frequency, and $\langle \delta\rho^*(\boldsymbol{q}, 0)\delta\hat{\rho}(\boldsymbol{q}, s)\rangle$ is the Laplace transform of the autocorrelation of the density fluctuation $\langle \delta\rho^*(\boldsymbol{q}, 0)\delta\rho(\boldsymbol{q}, t)\rangle$ where $\langle\ \rangle$ stands for the thermal average and $\delta\rho(\boldsymbol{q}, t)$ is the Fourier transform of the mass density fluctuation $\delta\rho(\boldsymbol{x}, t)$.

Usually the dynamic structure factor has been studied theoretically by means of two approaches: the hydrodynamic approach [380–382] based on the NSF theory, and the approach by using the kinetic theory of gases based on a special model [383–389].

The analyses of light scattering in monatomic gases on the basis of the RET theory were also made in detail [25, 390]. The RET theory is shown to be a

T. Ruggeri, M. Sugiyama, *Classical and Relativistic Rational Extended Thermodynamics of Gases*, https://doi.org/10.1007/978-3-030-59144-1_21

better theory to describe light scattering than the NSF theory. The analysis of the light scattering in polyatomic gases by using the RET theory is, however, not well developed until now [108]. The result presented in this chapter should be seen as a preliminary one.

21.2 Basic Equations

21.2.1 ET_{14} Theory

Basic equations for the present analysis are given by the linearized system of equations (16.1) with the thermal and caloric equations of state given by (6.8) and (6.3). This linear system can be decomposed into two uncoupled systems with independent variables:

system-L $\rho,\ \psi = \frac{\partial v_i}{\partial x_i},\ T,\ \sigma = \frac{\partial^2 \sigma_{\langle ij \rangle}}{\partial x_i \partial x_j},\ \Pi,\ Q = \frac{\partial q_i}{\partial x_i},$

system-T rot $\boldsymbol{v},\ \epsilon_{ijk} \frac{\partial^2 \sigma_{\langle jn \rangle}}{\partial x_k \partial x_n},$ rot $\boldsymbol{q},$

where ϵ_{ijk} is the Levi-Civita symbol. Here and hereafter, we denote the fluctuations $\delta\rho$ and δT simply as ρ and T.

For the present purpose, we adopt the system-L because only this system affects the density fluctuations. It is expressed as follows:

$$
\begin{aligned}
&\frac{\partial \rho}{\partial t} + \rho_0 \psi = 0, \\
&\frac{\partial \psi}{\partial t} + \frac{p_0}{\rho_0^2} \Delta \rho + \frac{k_B}{m} \Delta T - \frac{1}{\rho_0} \sigma + \frac{1}{\rho_0} \Delta \Pi = 0, \\
&\frac{\partial T}{\partial t} + \frac{T_0}{\hat{c}_v} \psi + \frac{T_0}{p_0 \hat{c}_v} Q = 0, \\
&\frac{\partial \Pi}{\partial t} + \left(\frac{2}{3} - \frac{1}{\hat{c}_v} \right) p_0 \psi + \frac{2\hat{c}_v - 3}{3\hat{c}_v (1 + \hat{c}_v)} Q = -\frac{1}{\tau_\Pi} \Pi, \\
&\frac{\partial \sigma}{\partial t} - \frac{4}{3} p_0 \Delta \psi - \frac{4}{3(1 + \hat{c}_v)} \Delta Q = -\frac{1}{\tau_\sigma} \sigma, \\
&\frac{\partial Q}{\partial t} + (1 + \hat{c}_v) \frac{k_B}{m} p_0 \Delta T - \frac{p_0}{\rho_0} \sigma + \frac{p_0}{\rho_0} \Delta \Pi = -\frac{1}{\tau_q} q_i,
\end{aligned}
$$

where Δ is the Laplacian. The Fourier (space) and Laplace (time) transformed basic equations provide an algebraic system:

$$\begin{pmatrix} s & \rho_0 & 0 & 0 & 0 & 0 \\ -q^2\frac{p_0}{\rho_0^2} & s & -q^2\frac{k_B}{m} & -\frac{q^2}{\rho_0} & -\frac{1}{\rho_0} & 0 \\ 0 & \frac{T_0}{\hat{c}_v} & s & 0 & 0 & \frac{T_0}{p_0\hat{c}_v} \\ 0 & \left(\frac{2}{3}-\frac{1}{\hat{c}_v}\right)p_0 & 0 & s+\frac{1}{\tau_{\Pi}} & 0 & \frac{2\hat{c}_v-3}{3\hat{c}_v(1+\hat{c}_v)} \\ 0 & \frac{4}{3}q^2 p_0 & 0 & 0 & s+\frac{1}{\tau_\sigma} & \frac{4q^2}{3(1+\hat{c}_v)} \\ 0 & 0 & -q^2(1+\hat{c}_v)\frac{k_B}{m}p_0 & -q^2\frac{p_0}{\rho_0} & -\frac{p_0}{\rho_0} & s+\frac{1}{\tau_q} \end{pmatrix} \cdot \begin{pmatrix} \hat{\rho}(\boldsymbol{q},s) \\ \hat{\psi}(\boldsymbol{q},s) \\ \hat{T}(\boldsymbol{q},s) \\ \hat{\sigma}(\boldsymbol{q},s) \\ \hat{\Pi}(\boldsymbol{q},s) \\ \hat{Q}(\boldsymbol{q},s) \end{pmatrix} = \begin{pmatrix} \rho(\boldsymbol{q},0) \\ \psi(\boldsymbol{q},0) \\ T(\boldsymbol{q},0) \\ \sigma(\boldsymbol{q},0) \\ \Pi(\boldsymbol{q},0) \\ Q(\boldsymbol{q},0) \end{pmatrix}, \tag{21.1}$$

where $\hat{X}(\boldsymbol{q},s)$ denotes the Fourier and Laplace transform of a generic quantity $X(\boldsymbol{x},t)$, and $X(\boldsymbol{q},0)$ denotes the initial value of spatial Fourier transform of $X(\boldsymbol{x},t)$.

Denoting the coefficient matrix of (21.1) as $\boldsymbol{A}$, we obtain the density as

$$\begin{aligned}\hat{\rho}(\boldsymbol{q},s) =&(\boldsymbol{A}^{-1})_{11}\rho(\boldsymbol{q},0)+(\boldsymbol{A}^{-1})_{12}\psi(\boldsymbol{q},0)+(\boldsymbol{A}^{-1})_{13}T(\boldsymbol{q},0)\\ &+(\boldsymbol{A}^{-1})_{14}\Pi(\boldsymbol{q},0)+(\boldsymbol{A}^{-1})_{15}\sigma(\boldsymbol{q},0)+(\boldsymbol{A}^{-1})_{16}Q(\boldsymbol{q},0).\end{aligned}$$

When we consider the adiabatic system with volume V_G which contains the scattering volume V, we obtain

$$\langle\rho^*(\boldsymbol{q},0)\hat{\rho}(\boldsymbol{q},s)\rangle = (\boldsymbol{A}^{-1})_{11}\rho^2 k_B T \kappa_T V Q_0,$$

where κ_T is the isothermal compressibility and the function Q_0 is defined by

$$Q_0 = 1-\frac{1}{VV_G}\left(\int_V e^{-i\boldsymbol{q}\cdot\boldsymbol{x}}d\boldsymbol{x}\right)^2.$$

It is useful to introduce the following dimensionless quantities:

$$x=\frac{\omega}{v_0 q},\quad y=\frac{p_0}{\mu q v_0},$$

where $v_0 = \sqrt{2k_B T_0/m}$. Usually the case of large y is referred to as the hydrodynamic region and the case of small y as the kinetic region. By using x and y, the relative intensity of $S(\boldsymbol{q},\omega)$ can be expressed as a function of x, y, $\hat{c}_v$, τ_q^*, and τ_Π^*, where $\tau_q^* = \tau_q/\tau_\sigma$ and $\tau_\Pi^* = \tau_\Pi/\tau_\sigma$. In fact, the essential part of $S(\boldsymbol{q},\omega)$ is obtained as follows:

$$\frac{S(\boldsymbol{q},\omega)}{\rho^2 k_B T \kappa_T V Q_0 \frac{1}{\pi}\left(\frac{\partial\epsilon}{\partial\rho}\right)^2} = \Re\left(\frac{N}{D}\right),$$

where

$$\begin{aligned}
N(x,y) =& -12ix^5\hat{c}_v(\hat{c}_v+1)\tau_\Pi^*\tau_q^* - 12x^4 y\hat{c}_v(\hat{c}_v+1)(\tau_\Pi^*\tau_q^* + \tau_\Pi^* + \tau_q^*) \\
&+ x^3\left\{18i\hat{c}_v(\hat{c}_v+2)\tau_\Pi^*\tau_q^* + 12iy^2\hat{c}_v(\hat{c}_v+1)(\tau_\Pi^* + \tau_q^* + 1)\right\} \\
&+ x^2\left\{12y^3\hat{c}_v(\hat{c}_v+1) + y\left[2(\hat{c}_v+2)\tau_q^*(5\hat{c}_v\tau_\Pi^* + 7\hat{c}_v + 3) + 12\hat{c}_v(\hat{c}_v+1)\tau_\Pi^*\right]\right\} \\
&+ x\left\{-2iy^2(\hat{c}_v+1)\left[3(\hat{c}_v+2)\tau_q^* + 2\hat{c}_v\tau_\Pi^* + 4\hat{c}_v + 3\right] + 3i(1-2\hat{c}_v)\hat{c}_v\tau_\Pi^*\tau_q^*\right\} \\
&- 6y^3(\hat{c}_v+1) - y\hat{c}_v\tau_q^*\left[(2\hat{c}_v-3)\tau_\Pi^* + 4\hat{c}_v\right], \\
D(x,y) =& 12x^6\hat{c}_v(\hat{c}_v+1)\tau_\Pi^*\tau_q^* - 12ix^5 y\hat{c}_v(\hat{c}_v+1)(\tau_\Pi^*\tau_q^* + \tau_\Pi^* + \tau_q^*) \\
&+ x^4\left\{-6\hat{c}_v(4\hat{c}_v+7)\tau_\Pi^*\tau_q^* - 12y^2\hat{c}_v(\hat{c}_v+1)(\tau_\Pi^* + \tau_q^* + 1)\right\} \\
&+ x^3\left\{12y^3 i\hat{c}_v(\hat{c}_v+1) + 2iy\left[\tau_q^*(\hat{c}_v(8\hat{c}_v+13)\tau_\Pi^* + 10\hat{c}_v(\hat{c}_v+2) + 6) + 9\hat{c}_v(\hat{c}_v+1)\tau_\Pi^*\right]\right\} \\
&+ x^2\left\{2y^2(\hat{c}_v+1)\left[6(\hat{c}_v+1)\tau_q^* + 5\hat{c}_v\tau_\Pi^* + 7\hat{c}_v + 3\right] + 9\hat{c}_v(\hat{c}_v+1)\tau_\Pi^*\tau_q^*\right\} \\
&+ x\left\{-6iy^3(\hat{c}_v+1)^2 - iy(\hat{c}_v+1)\tau_q^*(5\hat{c}_v\tau_\Pi^* + 7\hat{c}_v + 3)\right\} - 3y^2(\hat{c}_v+1)^2\tau_q^*.
\end{aligned}$$

21.2.2 Navier–Stokes and Fourier Theory

In the case of the NSF theory, the Fourier-Laplace transformed field equations are as follows:

$$\begin{pmatrix} s & \rho_0 & 0 \\ -\frac{k_B}{m}\frac{q^2 T_0}{\rho_0} & s + \frac{1}{\rho_0}\left(\frac{4\mu}{3} + \nu\right)q^2 & -\frac{k_B}{m}q^2 \\ 0 & \frac{T_0}{\hat{c}_v} & s + \frac{m}{k_B}\frac{\kappa q^2}{\hat{c}_v\rho_0} \end{pmatrix} \cdot \begin{pmatrix} \hat{\rho}(\boldsymbol{q},s) \\ \hat{\psi}(\boldsymbol{q},s) \\ \hat{T}(\boldsymbol{q},s) \end{pmatrix} = \begin{pmatrix} \rho(\boldsymbol{q},0) \\ \psi(\boldsymbol{q},0) \\ T(\boldsymbol{q},0) \end{pmatrix}.$$

The dynamic structure factor can be derived from this equation in a similar way.

21.3 Comparison with Experimental Data for CO_2

For polyatomic gases (both rarefied and dense), there are many experimens of Rayleigh-Brillouin Scattering based on the traditional (called spontaneous) [391–398] and new (called coherent) scattering techniques [399–401]

Here we compare the theoretical predition of ET_{14} with the laser scattering experiment of Greytak and Benedek [391, 392] in a carbon dioxide gas. The experiments were achieved under an approximately fixed pressure $p_0(= \frac{k_B}{m}\rho_0 T_0) =$ 1[atm], and the parameter y is set as $y = 11.1$ (hydrodynamic region) and $y = 1.01$ (kinetic region) by changing the detecting angle. The wavelength of the laser is $\Lambda = 6328$[Å]. As thermodynamic parameters, we use $\mu = 1.46 \times 10^{-5}$ [kg m^{-1} s^{-1}] and $\kappa = 1.31$ [W K^{-1} m^{-1}] which are adopted in [400] with Euken relation, therefore $\tau_q^* = 1.36$.

In the present comparison, the instrumental function $f_{inst}(\omega)$ is approximated by the Lorentzian with a full-width at half-maximum of about 28 MHz for the case of $y = 11.1$ and 210 MHz for the case of $y = 1.01$, and is convolved with the theoretical predictions:

$$S_{convolved}(\omega) = \int_{-\infty}^{\infty} S(q, \omega') f_{inst}(\omega' - \omega) d\omega.$$

As indicated in [392, 393, 400, 401], since the frequencies in light scattering experiments are much larger than the characteristic frequency of the vibrational modes, the modes seem to freeze. Therefore, in contrast with the study of shock waves in Chap. 17, the effect of the relaxation time of the dynamic pressure (or the bulk viscosity) is not considerably larger than that of the shear stress and the heat flux. This consideration indicates that we need to adopt the specific heat without the vibrational modes, i.e., $\hat{c}_v = 2.5$. Here we adopt these assumptions as a working hypothesis, although further study of these is evidently necessary.

In Fig. 21.1, we show the comparioson of $S(\boldsymbol{q}, \omega)$ derived from ET_{14} and NSF theories with experimental data. The theoretical predictions are regularized to be 1 at $x = 0$.

In the hydrodynamic region, the experiment was made at scattering angle $\theta =$ 10.6 [°] (therefore $y = 11.1$), T=297.9 [K] and $p_0 = 770$ [mmHg]. In this region, the predictions by ET_{14} and NSF give almost no difference from each other and show good agreement with experimental data. The ratio of the relaxation times is estimated as $\tau_\Pi^* = 0.8$ in such a way that the difference between the experiment and theories becomes minimum. Then the relaxation time of the dynamic pressure is 1.9×10^{-10} [s] and the bulk viscosity in this case is 5.2×10^{-6} [kg m^{-1} s^{-1}].

For the kinetic region, the experiment was made at scattering angle $\theta = 169.4$ [°] (therefore $y = 1.01$), T=298.1 [K] and $p_0 = 750$ [mmHg]. The relaxation time of the dynamic pressure is same as the one estimated in hydrodynamic region. From

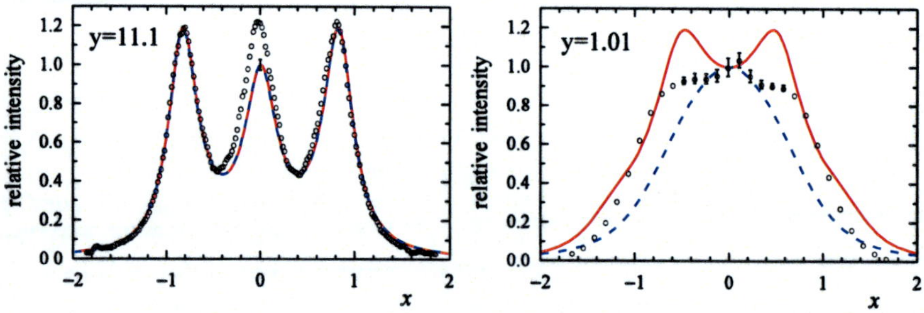

Fig. 21.1 Comparison of the theoretical predictions of relative intensity of the dynamic structure factor derived from RET (solid line) and NSF (dashed line) with the experimental data (circle) [391, 392] in a CO_2 gas. Left and right figures show the cases with $y = 11.1$ and 1.01, respectively

Fig. 21.1, also for the kinetic region, the ET_{14} theory shows better agreement with experimental data than NSF except for small x. To study more details of the kinetic region, as is expected, the theory with more moments [119] (see Chap. 9) will be useful.

Chapter 22
Heat Conduction

Check for updates

Abstract We study stationary heat conduction in a rarefied polyatomic gas at rest confined in a bounded domain in planar or radial (cylindrical and spherical) geometry by using the ET_{14} theory. We are particularly interested in the effect of the dynamic pressure Π on heat conduction phenomena because such an effect cannot be observed in a monatomic gas.

22.1 Introduction

In this chapter, the effect of the dynamic pressure Π on stationary heat conduction in a rarefied polyatomic gas at rest confined in a bounded domain in planar or radial (cylindrical and spherical) geometry is studied by the ET_{14} theory [402]. The effect is observable only in a polyatomic gas because the dynamic pressure vanishes identically in a rarefied monatomic gas and is intrinsically related to the internal degrees of freedom of a polyatomic molecule. The effect in the case of para-hydrogen gas (p-H_2) is explained briefly as a typical example. As discussed in Sect. 4.6.1, we know that, for rarefied monatomic gases, the RET and Navier–Stokes and Fourier (NSF) theories carry mutually different results in such a heat transfer problem [265, 403–407].

The present analysis is also a typical example in the problem of bounded domain discussed in Sect. 4.6.4.

22.2 Basis of the Present Analysis

The basic system of field equations is given by (6.33) with the thermal and caloric equations of state (6.8) and (6.3). We study the one-dimensional heat conduction problem in the planar, radially symmetric cylindrical, and spherical cases. All the quantities will be described in the *physical components* [408].

T. Ruggeri, M. Sugiyama, *Classical and Relativistic Rational Extended Thermodynamics of Gases*, https://doi.org/10.1007/978-3-030-59144-1_22

22.2.1 Basic System of Equations

For convenience, we use the following dimensionless quantities:

$$
\begin{aligned}
&\hat{x} = \frac{x}{L}, \quad \hat{c}_v' = T_0 \frac{\mathrm{d}\hat{c}_v}{\mathrm{d}T}, \quad \hat{T} = \frac{T}{T_0}, \quad \hat{p} = \frac{p}{p_0}, \\
&\hat{\sigma}_{\langle ij \rangle} = \frac{\sigma_{\langle ij \rangle}}{p_0}, \quad \hat{\Pi} = \frac{\Pi}{p_0}, \quad \hat{q} = \frac{q_1}{p_0\sqrt{k_B T_0/m}}, \\
&\hat{\tau}_\sigma = \frac{\tau_\sigma \sqrt{k_B T_0/m}}{L}, \quad \hat{\tau}_\Pi = \frac{\tau_\Pi \sqrt{k_B T_0/m}}{L}, \quad \hat{\tau}_q = \frac{\tau_q \sqrt{k_B T_0/m}}{L},
\end{aligned}
$$

where x represents the position along the axis normal to the plates in the planar case or the radius in the radial geometry, L is the distance between the two boundaries, and the index 0 denotes the value at a reference state.

The basic system of equations can be rewritten as follows [402], in which the mass conservation law is identically satisfied:

$$
\begin{aligned}
&\frac{\mathrm{d}}{\mathrm{d}\hat{x}}(\hat{p} + \hat{\Pi} - \hat{\sigma}_{\langle 11 \rangle}) - \frac{j(j+1)}{2\hat{x}}\hat{\sigma}_{\langle 11 \rangle} - \frac{j(j-2)}{\hat{x}}\hat{\sigma}_{\langle 22 \rangle} = 0, \\
&\frac{\mathrm{d}\hat{q}}{\mathrm{d}\hat{x}} + \frac{j}{\hat{x}}\hat{q} = 0, \\
&\frac{2}{1+\hat{c}_v}\frac{\mathrm{d}\hat{q}}{\mathrm{d}\hat{x}} - \frac{4}{3}\frac{\hat{c}_v'}{(1+\hat{c}_v)^2}\hat{q}\frac{\mathrm{d}\hat{T}}{\mathrm{d}\hat{x}} = \frac{\hat{\sigma}_{\langle 11 \rangle}}{\hat{\tau}_\sigma}, \\
&\frac{2(1-\delta_{j0})}{1+\hat{c}_v}\frac{\hat{q}}{\hat{x}} + \frac{2}{3}\frac{\hat{c}_v'}{(1+\hat{c}_v)^2}\hat{q}\frac{\mathrm{d}\hat{T}}{\mathrm{d}\hat{x}} = \frac{\hat{\sigma}_{\langle 22 \rangle}}{\hat{\tau}_\sigma}, \\
&\frac{5}{3}\frac{\hat{c}_v'}{(1+\hat{c}_v)^2}\hat{q}\frac{\mathrm{d}\hat{T}}{\mathrm{d}\hat{x}} = \frac{\hat{\Pi}}{\hat{\tau}_\Pi}, \\
&\left\{(1+\hat{c}_v)\hat{p} + (2+\hat{c}_v)(\hat{\Pi} - \hat{\sigma}_{\langle 11 \rangle})\right\}\frac{\mathrm{d}\hat{T}}{\mathrm{d}\hat{x}} - \hat{T}\frac{\mathrm{d}\hat{p}}{\mathrm{d}\hat{x}} = -\frac{\hat{q}}{\hat{\tau}_q},
\end{aligned}
\tag{22.1}
$$

where the index j is 0, 1, and 2 for the planar case, the cylindrical case, and the spherical case, respectively. δ_{ij} denotes the Kronecker symbol.

22.2.2 Reduced Basic System of Equations

From the basic system (22.1), several relations are obtained:

$$
\begin{aligned}
\hat{q} &= \frac{Q_0}{\hat{x}^j}, \\
\hat{\sigma}_{\langle 11 \rangle} &= -\frac{2j}{1+\hat{c}_v}\frac{Q_0 \hat{\tau}_\sigma}{\hat{x}^{j+1}} - \frac{4}{5}\frac{\hat{\tau}_\sigma}{\hat{\tau}_\Pi}\hat{\Pi}, \\
\hat{\sigma}_{\langle 22 \rangle} &= \frac{j+\delta_{j1}}{1+\hat{c}_v}\frac{Q_0 \hat{\tau}_\sigma}{\hat{x}^{j+1}} + \frac{2}{5}\frac{\hat{\tau}_\sigma}{\hat{\tau}_\Pi}\hat{\Pi},
\end{aligned}
\tag{22.2}
$$

where Q_0 is a constant determined by the boundary condition. We note that, in the planar and spherical cases, $\sigma_{\langle 22 \rangle} = \sigma_{\langle 33 \rangle}$.

Hereafter, in order to set apart the effect of the dynamic pressure, we neglect the T dependence of the relaxation times assuming that these have constant values determined by the temperature range under consideration. Essential features in the present analysis are not changed by this assumption, which was also adopted in Sect. 4.6.1. Then, the following relation is obtained from $(22.1)_1$ and (22.2):

$$
\hat{p} + \left(1 + \frac{4}{5}\frac{\hat{\tau}_\sigma}{\hat{\tau}_\Pi}\right)\hat{\Pi} = P_0, \tag{22.3}
$$

where P_0 is a constant determined by the boundary condition. The quantities $\hat{\Pi}$, $\hat{\sigma}_{\langle 11 \rangle}$, $\hat{\sigma}_{\langle 22 \rangle}$, and $\hat{q}$ are determined by the solution of the differential equations for $\hat{T}$ and $\hat{p}$:

$$
\begin{aligned}
(A - \hat{p})\frac{\mathrm{d}\hat{T}}{\mathrm{d}\hat{x}} - \hat{T}\frac{\mathrm{d}\hat{p}}{\mathrm{d}\hat{x}} &= -\frac{Q_0}{\hat{\tau}_q \hat{x}^j}, \\
\left(1 + \frac{4}{5}\frac{\hat{\tau}_\sigma}{\hat{\tau}_\Pi}\right)\hat{c}_v' \frac{\mathrm{d}\hat{T}}{\mathrm{d}\hat{x}} &= B\left(P_0 - \hat{p}\right),
\end{aligned}
\tag{22.4}
$$

where

$$
A = (2 + \hat{c}_v)\left(P_0 + \frac{2j}{1+\hat{c}_v}\frac{Q_0 \hat{\tau}_\sigma}{\hat{x}^{j+1}}\right), \qquad B = \frac{3}{5}(1+\hat{c}_v)^2 \frac{\hat{x}^j}{Q_0 \hat{\tau}_\Pi}.
$$

22.2.3 Navier–Stokes and Fourier Theory

For reference, we write down the NSF basic equation:

$$(1+\hat{c}_v)\hat{p}\frac{\mathrm{d}\hat{T}}{\mathrm{d}\hat{x}} = -\frac{\hat{q}}{\hat{\tau}_q} \tag{22.5}$$

with $\hat{p} = P_0$ and $\hat{q} = Q_0/\hat{x}^j$. The viscous stress $\sigma_{\langle ij\rangle}$ and the dynamic pressure Π vanish identically.

22.3 Boundary Condition

We assume, for simplicity, a permeable plate at $\hat{x} = 0$ and an impermeable one at $\hat{x} = 1$. At the permeable plate, we can impose both the density ρ_0 and the temperature T_0. While, at the impermeable plate, a fixed heat flux Q_0 is applied.

Although these boundary data are sufficient to determine completely the solution of the NSF system, we need one more condition for the RET system of equations. This is the so-called "non-controllable boundary data problem" and several approaches have been introduced to overcome it. See the discussions in Sect. 4.6.4.

In the present study we have determined it as follows: We tried to assign different values of $\sigma_{\langle 11\rangle}$ at one plate. We found that the corresponding solutions of $\sigma_{\langle 11\rangle}$ exhibit, in general, thin boundary layers near the plate. We then adopt, as an appropriate boundary value, the one for which no boundary layer is observed. A similar phenomenon was also observed in a different context [409]. We remark that different boundary values for $\sigma_{\langle 11\rangle}$ create minimally different solutions for T and Π, but create significant differences for q and for p.

22.4 Effect of the Dynamic Pressure

From (22.1)$_5$, we notice that

$$\hat{c}'_v = 0 \qquad \longrightarrow \qquad \hat{\Pi} = 0, \tag{22.6}$$

therefore the polyatomic effect ($\hat{\Pi} \neq 0$) is observable only in the case with $\hat{c}'_v \neq 0$, which is the case that we can not see in rarefied monatomic gases. From (22.3), the pressure $\hat{p}$ is no longer constant but depends on $\hat{x}$.

Let us divide this case into the following two cases:

(a) Planar case: It is evident that, even in the planar case, the RET theory predicts the existence of the polyatomic effect.

(b) Cylindrical and spherical cases: From the numerical analysis [402], we see that the radial geometry enhances the polyatomic effect significantly.

Remark 22.1 Two quantities $\hat{c}_v'$ and $\hat{\Pi}$ have qualitatively different characters from each other: $\hat{c}_v'$ is, by definition, an equilibrium quantity, while $\hat{\Pi}$ is a nonequilibrium one. However, both quantities are intrinsically related to the internal degrees of freedom of a polyatomic molecule. In stationary heat conduction in a rarefied polyatomic gas, we have found that both quantities play an important role simultaneously.

Remark 22.2 In the NSF theory (22.5), the dynamic pressure Π vanishes identically even in the case with $\hat{c}_v' \neq 0$. From a viewpoint of RET, this fact reveals an inconsistency in the NSF theory of a rarefied polyatomic gas.

Before closing this section, we discuss the case with $\hat{c}_v' = 0$ for the sake of completeness. When the specific heat $\hat{c}_v$ is constant, we have

$$\Pi = 0, \quad \hat{p} = P_0$$

in all geometries. There is no polyatomic effect. Then, $(22.4)_2$ is identically satisfied, and the temperature profile is governed by

$$\frac{\mathrm{d}\hat{T}}{\mathrm{d}\hat{x}} = -\frac{Q_0}{(A - P_0)\hat{\tau}_q \hat{x}^j}. \tag{22.7}$$

We discuss this case further as follows:

(i) $\hat{c}_v = 3/2$: This case corresponds to a monatomic gas or a polyatomic gas in a low temperature range where no internal degrees of freedom of a molecule are excited. Equation (22.7) is exactly the same as that studied in Sect. 4.6.1, where it was shown that the temperature profiles in the planar case predicted by the RET theory and the NSF theory coincide with each other, and that unphysical singularities of the temperature on the axis of the cylinder and at the center of the sphere predicted by the NSF theory can be removed by the RET theory.
(ii) $\hat{c}_v = constant > 3/2$: This case corresponds to a polyatomic gas with excited internal degrees of freedom. However, except for the value of $\hat{c}_v$, the analytical expression of the temperature profile is the same as that of the case (i) above.

22.5 An Example: Polyatomic Effect in a Para-Hydrogen Gas

By taking the importance of the dynamic pressure $\hat{\Pi}$ and the relationship (22.6) into consideration, the numerical study of the polyatomic effect in a para-hydrogen (p-H_2) gas was made as a typical example. See its temperature dependence of the specific heat in Fig. 16.2, where $\hat{c}_v' \neq 0$ except for one special point.

It is shown that, both in the planar case and in the radial case, the polyatomic effect due to the presence of the dynamic pressure Π appears. The presence of Π affects both the heat flux and the mass density, which may be detected by the experimental techniques described in [410]. The radial geometry enhances significantly the difference between the RET and NSF predictions although the difference of the temperature profile is still small. For details of the numerical results, see [402].

Chapter 23
Fluctuating Hydrodynamics

Abstract In this chapter, we present the theory of fluctuating hydrodynamics based on the ET_{14} theory of rarefied polyatomic gases. And we discuss the link between the two levels of description of fluctuating hydrodynamics, that is, the ET_{14} theory and the Landau and Lifshitz theory of fluctuating hydrodynamics.

23.1 Introduction

Landau and Lifshitz developed the theory of fluctuating hydrodynamics for viscous, heat-conducting fluids with constitutive equations of Navier–Stokes and Fourier type [8, 334, 348] based on thermodynamics of irreversible processes (TIP). They introduced additional stochastic flux terms (generalized random forces) into the constitutive equations of viscous stress and heat flux by applying the fluctuation-dissipation theorem [411–413]. Therefore, in their theory, the random force and the constitutive equations are combined in an indistinct way. See also review articles on fluctuating hydrodynamics [414–416].

In recent years, the Landau-Lifshitz (LL) theory has been applied to, in particular, nano-technology [417, 418] and molecular biology [419, 420]. Numerical analyses of the fluctuations by using the theory have been made extensively [421–427]. The fluctuating-hydrodynamic approach can also contribute to the study of fluctuations in nonequilibrium states [416, 428, 429]. However, as TIP rests essentially on the local equilibrium assumption, it is probable that TIP may no longer be valid for highly nonequilibrium cases such as the cases where nanoflows are involved, or the cases where rarefied gases play a role.

The purpose of the present chapter is to summarize briefly the theory of fluctuating hydrodynamics based on RET through the study of the ET_{14} theory of rarefied polyatomic gases as a representative case. See also [430, 431] for fluctuating hydrodynamics based on ET_{13} theory of rarefied monatomic gases.

T. Ruggeri, M. Sugiyama, *Classical and Relativistic Rational Extended Thermodynamics of Gases*, https://doi.org/10.1007/978-3-030-59144-1_23

23.2 Theory of Fluctuating Hydrodynamics Based on RET

Basic equations in the present study are the linearized equations of ET_{14} for a rarefied polyatomic gas (16.1) but, for simplicity, in a polytropic case with $\hat{c}_v = D/2$.

Let us now try to introduce the random forces into RET. Following the general theory [347], we can have the expressions for the productions P_{jj}, $P_{\langle ij\rangle}$, and Q_i in terms of Π, $\sigma_{\langle ij\rangle}$, q_i and the Gaussian white random forces $\mathfrak{r}$, $\mathfrak{r}_{\langle ij\rangle}$, $\mathfrak{s}_i$ (see (6.29)):

$$
\begin{aligned}
P_{jj} &= -\frac{3}{\tau_\Pi}\Pi + \mathfrak{r}, \\
P_{\langle ij\rangle} &= \frac{1}{\tau_\sigma}\sigma_{\langle ij\rangle} + \mathfrak{r}_{\langle ij\rangle}, \\
Q_i &= -\frac{2}{\tau_q}q_i + \mathfrak{s}_i.
\end{aligned}
\tag{23.1}
$$

The means of the random forces $\mathfrak{r}$, $\mathfrak{r}_{\langle ij\rangle}$, and $\mathfrak{s}_i$ vanish. And their correlations are given by

$$
\begin{aligned}
\langle \mathfrak{r}(\boldsymbol{x},t)\mathfrak{r}(\boldsymbol{x}',t')\rangle &= k_B \frac{12\frac{D-3}{D}\frac{k_B}{m}\rho_0 T_0^2}{\tau_\Pi}\delta(\boldsymbol{x}-\boldsymbol{x}')\delta(t-t'), \\
\langle \mathfrak{r}_{\langle ij\rangle}(\boldsymbol{x},t)\mathfrak{r}_{\langle mn\rangle}(\boldsymbol{x}',t')\rangle &= k_B \frac{2\frac{k_B}{m}\rho_0 T_0^2}{\tau_\sigma} \\
&\quad\times(\delta_{im}\delta_{jn}+\delta_{in}\delta_{jm}-\frac{2}{3}\delta_{ij}\delta_{mn})\delta(\boldsymbol{x}-\boldsymbol{x}')\delta(t-t'), \\
\langle \mathfrak{s}_i(\boldsymbol{x},t)\mathfrak{s}_j(\boldsymbol{x}',t')\rangle &= k_B \frac{4(D+2)\left(\frac{k_B}{m}\right)^2\rho_0 T_0^3}{\tau_q}\delta_{ij}\delta(\boldsymbol{x}-\boldsymbol{x}')\delta(t-t'), \\
\langle \mathfrak{r}(\boldsymbol{x},t)\mathfrak{r}_{\langle ij\rangle}(\boldsymbol{x}',t')\rangle &= \langle \mathfrak{r}(\boldsymbol{x},t)\mathfrak{s}_m(\boldsymbol{x}',t')\rangle = \langle \mathfrak{r}_{\langle ij\rangle}(\boldsymbol{x},t)\mathfrak{s}_m(\boldsymbol{x}',t')\rangle = 0,
\end{aligned}
\tag{23.2}
$$

where brackets $\langle\ \ \rangle$ in the left-hand side stand for the statistical average at the reference equilibrium state. Note that, as seen in (23.1), the random force appears as a part of the production term in contrast to the LL theory.

Field equations (16.1) with (23.1) and (23.2) constitute *the basic system of equations for fluctuating hydrodynamics based on* ET_{14}. The relaxation times can be evaluated by experiments or kinetic-theoretical analysis.

23.3 Two Subsystems of the Stochastic Field Equations

The system of equations obtained above may be decomposed into two uncoupled subsystems, that is, the subsystem composed of longitudinal modes (System-L) and the subsystem of transverse modes (System-T). (See also Sect. 21.2.1.)

System-L: The relevant quantities of the system are given by

$$\rho,\ T,\ \psi\left(\equiv \frac{\partial v_i}{\partial x_i}\right),\ \Pi,\ \tau\left(\equiv \frac{\partial^2 \sigma_{\langle ij\rangle}}{\partial x_i \partial x_j}\right),\ \varphi\left(\equiv \frac{\partial q_i}{\partial x_i}\right),$$
$$\mathfrak{u}\left(\equiv \frac{1}{3}\mathfrak{t}\right),\ \mathfrak{v}\left(\equiv -\frac{\partial^2 \mathfrak{t}_{\langle ij\rangle}}{\partial x_i \partial x_j}\right),\ \text{and}\ \mathfrak{w}\left(\equiv \frac{1}{2}\frac{\partial \mathfrak{s}_i}{\partial x_i}\right).$$

The spatial Fourier transform of the system is the system of the rate-type differential equations in the space of the wave number $\boldsymbol{k}$ and time t ($\boldsymbol{k}t$-representation) as follows:

$$\frac{\partial \rho(\boldsymbol{k},t)}{\partial t} + \rho_0 \psi(\boldsymbol{k},t) = 0,$$

$$\frac{\partial \psi(\boldsymbol{k},t)}{\partial t} - \frac{\frac{k_B}{m}T_0}{\rho_0}k^2\rho(\boldsymbol{k},t) - \frac{k_B}{m}k^2 T(\boldsymbol{k},t) - \frac{1}{\rho_0}k^2\Pi(\boldsymbol{k},t) - \frac{1}{\rho_0}\tau(\boldsymbol{k},t) = 0,$$

$$\frac{\partial T(\boldsymbol{k},t)}{\partial t} + \frac{2}{D}T_0\psi(\boldsymbol{k},t) + \frac{m}{k_B}\frac{2}{D\rho_0}\varphi(\boldsymbol{k},t) = 0,$$

$$\frac{\partial \Pi(\boldsymbol{k},t)}{\partial t} + \frac{4(D-3)}{3D(D+2)}\varphi(\boldsymbol{k},t) + \frac{2(D-3)}{3D}\frac{k_B}{m}\rho_0 T_0 \psi(\boldsymbol{k},t) = -\frac{1}{\tau_\Pi}\Pi(\boldsymbol{k},t) + \mathfrak{u}(\boldsymbol{k},t),$$

$$\frac{\partial \tau(\boldsymbol{k},t)}{\partial t} + \frac{8}{3(D+2)}k^2\varphi(\boldsymbol{k},t) + \frac{4}{3}\frac{k_B}{m}\rho_0 T_0 k^2 \psi(\boldsymbol{k},t) = -\frac{1}{\tau_\sigma}\tau(\boldsymbol{k},t) + \mathfrak{v}(\boldsymbol{k},t),$$

$$\begin{aligned}\frac{\partial \varphi(\boldsymbol{k},t)}{\partial t} - \frac{k_B}{m}T_0 k^2 \Pi(\boldsymbol{k},t) - \frac{k_B}{m}T_0\tau(\boldsymbol{k},t) - \frac{D+2}{2}\left(\frac{k_B}{m}\right)^2 \rho_0 T_0 k^2 T(\boldsymbol{k},t) \\ = -\frac{1}{\tau_q}\varphi(\boldsymbol{k},t) + \mathfrak{w}(\boldsymbol{k},t),\end{aligned}$$

where $\rho(\boldsymbol{k},t)$ is the spatial Fourier transform of $\rho(\boldsymbol{x},t)$ and so on.

From (23.2), the quantities $\mathfrak{v}(\boldsymbol{k},t)$ and $\mathfrak{w}(\boldsymbol{k},t)$ are the Gaussian white random forces with null means and correlations:

$$\langle \mathfrak{u}(\boldsymbol{k},t)\mathfrak{u}(\boldsymbol{k}',t')\rangle = k_B \frac{\frac{D-3}{D}\frac{k_B}{m}\rho_0 T_0^2}{6\pi^3 \tau_\Pi}\delta(\boldsymbol{k}+\boldsymbol{k}')\delta(t-t'),$$

$$\langle \mathfrak{v}(\boldsymbol{k},t)\mathfrak{v}(\boldsymbol{k}',t')\rangle = k_B \frac{\frac{k_B}{m}\rho_0 T_0^2}{3\pi^3 \tau_\sigma}k^4\delta(\boldsymbol{k}+\boldsymbol{k}')\delta(t-t'),$$

$$\langle \mathfrak{w}(\boldsymbol{k},t)\mathfrak{w}(\boldsymbol{k}',t')\rangle = k_B \frac{(D+2)\left(\frac{k_B}{m}\right)^2 \rho_0 T_0^3}{8\pi^3 \tau_q} k^2 \delta(\boldsymbol{k}+\boldsymbol{k}')\delta(t-t'),$$

$$\langle \mathfrak{u}(\boldsymbol{k},t)\mathfrak{v}(\boldsymbol{k}',t')\rangle = \langle \mathfrak{u}(\boldsymbol{k},t)\mathfrak{w}(\boldsymbol{k}',t')\rangle = \langle \mathfrak{v}(\boldsymbol{k},t)\mathfrak{w}(\boldsymbol{k}',t')\rangle = 0.$$

System-T: The relevant quantities of the system are given by

$$\omega_i \ (\equiv (\text{rot}\boldsymbol{v})_i), \quad \sigma_i \left(\equiv \epsilon_{ijk}\frac{\partial^2 \sigma_{\langle kn\rangle}}{\partial x_j \partial x_n}\right), \quad \pi_i (\equiv (\text{rot}\boldsymbol{q})_i),$$

$$\mathfrak{x}_i \left(\equiv -\epsilon_{ijk}\frac{\partial^2 \mathfrak{r}_{\langle kn\rangle}}{\partial x_j \partial x_n}\right), \quad \text{and} \quad \mathfrak{y}_i \left(\equiv \frac{1}{2}(\text{rot}\mathfrak{s})_i\right).$$

The field equations in the $\boldsymbol{k}t$-representation are as follows:

$$\frac{\partial \omega_i(\boldsymbol{k},t)}{\partial t} - \frac{1}{\rho_0}\sigma_i(\boldsymbol{k},t) = 0,$$

$$\frac{\partial \sigma_i(\boldsymbol{k},t)}{\partial t} + \frac{2}{D+2}k^2\pi_i(\boldsymbol{k},t) + \frac{k_B}{m}\rho_0 T_0 k^2 \omega_i(\boldsymbol{k},t) = -\frac{1}{\tau_\sigma}\sigma_i(\boldsymbol{k},t) + \mathfrak{x}_i(\boldsymbol{k},t),$$

$$\frac{\partial \pi_i(\boldsymbol{k},t)}{\partial t} - \frac{k_B}{m}T_0\sigma_i(\boldsymbol{k},t) = -\frac{1}{\tau_q}\pi_i(\boldsymbol{k},t) + \mathfrak{y}_i(\boldsymbol{k},t).$$

Note that, for given $\mathfrak{x}_i$ and $\mathfrak{y}_i$, the equations for the set of variables $(\omega_i, \sigma_i, \pi_i)$ with the same suffix i can be solved separately from those with the different suffix $j (\neq i)$. In view of (23.2), $\mathfrak{x}_i$ and $\mathfrak{y}_i$ are the Gaussian white random forces with null means and correlations:

$$\langle \mathfrak{x}_i(\boldsymbol{k},t)\mathfrak{x}_m(\boldsymbol{k}',t')\rangle = k_B \frac{\frac{k_B}{m}\rho_0 T_0^2}{4\pi^3 \tau_\sigma} k^4 \left(\delta_{im} - \frac{k_i k_m}{k^2}\right) \times \delta(\boldsymbol{k}+\boldsymbol{k}')\delta(t-t'),$$

$$\langle \mathfrak{y}_i(\boldsymbol{k},t)\mathfrak{y}_m(\boldsymbol{k}',t')\rangle = k_B \frac{(D+2)\left(\frac{k_B}{m}\right)^2 \rho_0 T_0^3}{8\pi^3 \tau_q} k^2 \left(\delta_{im} - \frac{k_i k_m}{k^2}\right) \times \delta(\boldsymbol{k}+\boldsymbol{k}')\delta(t-t'),$$

$$\langle \mathfrak{x}_i(\boldsymbol{k},t)\mathfrak{y}_m(\boldsymbol{k}',t')\rangle = 0.$$

23.4 Relationship to the Landau-Lifshitz Theory

In what follows, we adopt the coarse-graining approximation where the fast modes are eliminated [430], and show explicitly the coarse-grained solutions for the System-L and System-T. We will see that these solutions are just the ones in the LL theory.

System-L: We have the following relation up to the leading term with respect to τ_Π, τ_σ, and τ_q:

$$\begin{bmatrix} \Pi(\boldsymbol{k},t) \\ \tau(\boldsymbol{k},t) \\ \varphi(\boldsymbol{k},t) \end{bmatrix} = \begin{bmatrix} -\frac{2(D-3)}{3D}\frac{k_B}{m}\rho_0 T_0 \tau_\Pi \psi(\boldsymbol{k},t) + \mathfrak{f}(\boldsymbol{k},t) \\ -\frac{4}{3}\frac{k_B}{m}\rho_0 T_0 \tau_\sigma k^2 \psi(\boldsymbol{k},t) + \mathfrak{g}(\boldsymbol{k},t) \\ \frac{D+2}{2}\left(\frac{k_B}{m}\right)^2 \rho_0 T_0 \tau_q k^2 T(\boldsymbol{k},t) + \mathfrak{h}(\boldsymbol{k},t) \end{bmatrix}. \tag{23.3}$$

The Gaussian white random forces $\mathfrak{f}$, $\mathfrak{g}$, and $\mathfrak{h}$ have null means and correlations:

$$\begin{aligned} \langle \mathfrak{f}(\boldsymbol{k},t)\mathfrak{f}(\boldsymbol{k}',t')\rangle &= \frac{1}{6\pi^3}\frac{D-3}{D}k_B\frac{k_B}{m}\rho_0 T_0^2 \tau_\Pi \delta(\boldsymbol{k}+\boldsymbol{k}')\delta(t-t'), \\ \langle \mathfrak{g}(\boldsymbol{k},t)\mathfrak{g}(\boldsymbol{k}',t')\rangle &= \frac{1}{3\pi^3}k_B\frac{k_B}{m}\rho_0 T_0^2 \tau_\sigma k^4 \delta(\boldsymbol{k}+\boldsymbol{k}')\delta(t-t'), \\ \langle \mathfrak{h}(\boldsymbol{k},t)\mathfrak{h}(\boldsymbol{k}',t')\rangle &= \frac{D+2}{8\pi^3}k_B\left(\frac{k_B}{m}\right)^2 \rho_0 T_0^3 \tau_q k^2 \delta(\boldsymbol{k}+\boldsymbol{k}')\delta(t-t'), \\ \langle \mathfrak{f}(\boldsymbol{k},t)\mathfrak{g}(\boldsymbol{k}',t')\rangle &= \langle \mathfrak{f}(\boldsymbol{k},t)\mathfrak{h}(\boldsymbol{k}',t')\rangle = \langle \mathfrak{g}(\boldsymbol{k},t)\mathfrak{h}(\boldsymbol{k}',t')\rangle = 0. \end{aligned} \tag{23.4}$$

System-T: We obtain the following relations in a similar way as above:

$$\begin{bmatrix} \sigma_i(\boldsymbol{k},t) \\ \pi_i(\boldsymbol{k},t) \end{bmatrix} = \begin{bmatrix} -\frac{k_B}{m}\rho_0 T_0 \tau_\sigma k^2 \omega_i(\boldsymbol{k},t) + \mathfrak{k}_i(\boldsymbol{k},t) \\ \mathfrak{l}_i(\boldsymbol{k},t) \end{bmatrix}. \tag{23.5}$$

Note that there is no deterministic part in $\pi_i(\boldsymbol{k},t)$, therefore, only the random force plays a role. The correlations between the zero-mean Gaussian white random forces $\mathfrak{k}_i$ and $\mathfrak{l}_i$ are given by

$$\begin{aligned} \langle \mathfrak{k}_i(\boldsymbol{k},t)\mathfrak{k}_m(\boldsymbol{k}',t')\rangle &= \frac{1}{4\pi^3}k_B\frac{k_B}{m}\rho_0 T_0^2 \tau_\sigma k^4 \left(\delta_{im} - \frac{k_i k_m}{k^2}\right) \\ &\quad \times \delta(\boldsymbol{k}+\boldsymbol{k}')\delta(t-t'), \\ \langle \mathfrak{l}_i(\boldsymbol{k},t)\mathfrak{l}_m(\boldsymbol{k}',t')\rangle &= \frac{D+2}{8\pi^3}k_B\left(\frac{k_B}{m}\right)^2 \rho_0 T_0^3 \tau_q k^2 \left(\delta_{im} - \frac{k_i k_m}{k^2}\right) \\ &\quad \times \delta(\boldsymbol{k}+\boldsymbol{k}')\delta(t-t'), \\ \langle \mathfrak{k}_i(\boldsymbol{k},t)\mathfrak{l}_m(\boldsymbol{k}',t')\rangle &= 0. \end{aligned} \tag{23.6}$$

The Relationship Between the Present Theory and the LL Theory: We can now confirm that the expressions in (23.3), (23.4), (23.5), and (23.6) are exactly the same as those derived from the LL theory where viscosities ν, μ, and heat conductivity κ are identified by the relations (6.39). Thus we have proved that the LL theory can be derived from the ET_{14} theory by using the coarse-graining approximation, and that the LL theory is included in the present theory as a limiting case.

The ET_{14} theory and the LL theory belong to the two different levels of description of fluctuating hydrodynamics (see Sect. 1.2). As we analyzed above, the rapidly changing deterministic modes (fast modes) in RET have been consistently re-normalized into the random forces in the LL theory. Therefore, from a physical point of view, the delta functions appeared in the correlations have their own validity range depending on the spatio-temporal resolution of their description level.

23.5 Discussion

In the present chapter, we have summarized the theory of fluctuating hydrodynamics based on RET. And we have made clear the link between the two levels of description of fluctuating hydrodynamics, that is, the ET_{14} theory and the LL theory.

Generally speaking, there are many such levels. As explained in Sect. 2.4, Boillat and Ruggeri found the hierarchy structure in RET, and important concept called principal subsystem of field equations. Each principal subsystem gives us one level of description with different resolution from each other. And, in a similar way as above, we can develop the corresponding fluctuating hydrodynamics based on a given principal subsystem. Detailed discussions are omitted here for simplicity.

Part VI
Polyatomic Dense Gas

Chapter 24
RET of Dense Polyatomic Gas with Six Fields

Abstract A RET theory of dissipative dense gases is presented. In this chapter, we study, in particular, the RET theory with six fields, where we ignore the shear viscosity and heat conductivity and we treat the internal (rotational and vibrational) motion of a molecule as a unit. We postulate a principle of duality between rarefied gas and dense gas. This principle is based on the microscopic analysis of the energy exchange between different modes of the molecular motion. The basic system of field equations satisfies all principles of RET, that is, Galilean invariance and objectivity, entropy principle, and thermodynamic stability (entropy convexity), and, as in the RET theory of rarefied gases, the constitutive equations are completely determined by the thermal and caloric equations of state. The present theory includes the RET theory of rarefied polyatomic gases with six fields (ET_6) explained in Chap. 12 as a special case in the rarefied-gas limit. The system is the simplest one after the Euler system, but, in contrast to the Euler system, we may have a global smooth solution due to the fact that the system is dissipative symmetric hyperbolic and satisfies the K-condition. Similar to the ET_6 theory, there emerge two nonequilibrium temperatures. Furthermore we evaluate the characteristic velocities associated with the hyperbolic system, and address the fluctuation-dissipation relation of the bulk viscosity. As a typical example, we analyze van der Waals fluids by using the present theory.

24.1 Introduction

RET of *dense gases* is still at an early stage in its development. The RET theories of *rarefied gases* have been developed by enjoying the benefit of the mathematical structure of the balance laws that are dictated by the moment theory associated with the Boltzmann equation. In the case of dense gas, there also exist kinetic theories and statistical-mechanical theories (see for example [83, 432–435]). And many studies on macroscopic theories of dense gases and liquids have been made with the help of such microscopic theories. However, we still do not have a tractable counterpart in kinetic theories of dense gases corresponding to the moment theory

T. Ruggeri, M. Sugiyama, *Classical and Relativistic Rational Extended Thermodynamics of Gases*, https://doi.org/10.1007/978-3-030-59144-1_24

for rarefied gases. Therefore, in this chapter, we adopt *the phenomenological RET approach*. That is, we aim to construct a RET theory of dense gases in a purely phenomenological way preserving all the general principles of RET, i.e., Galilean invariance and objectivity, entropy principle, and thermodynamical stability and convexity of entropy. One of the keys to construct such a theory that includes a RET theory of rarefied gases as a special case in the rarefied-gas limit is the duality principle, which indicates a possible way of passage from a theory of rarefied gases to a theory of *moderately dense* gases.

One of the main difficulties encountered in the construction of a RET theory of dense gases comes from the convexity condition relating to the dynamic pressure. While, as seen in the previous chapters, the role of the dynamic pressure can be highlighted by using the RET theory with 6 fields (ET_6): mass density, velocity, specific internal energy, and dynamic pressure. Therefore it seems to be reasonable to study ET_6 of dense gases ($\mathrm{ET}_6^{\mathrm{D}}$) as a first step toward an unexplored fertile research field.

A first tentative to go beyond the previous ET_6 theory was made in [436], where the binary hierarchy of balance equations no longer has the feature: the flux in one equation becomes the density in the next equation. The authors of this paper adopted the following type of binary hierarchy:

$$
\begin{aligned}
&\frac{\partial F}{\partial t} + \frac{\partial F_k}{\partial x_k} = 0, \\
&\frac{\partial F_i}{\partial t} + \frac{\partial F_{ik}}{\partial x_k} = 0, \\
&\frac{\partial \bar{F}_{ii}}{\partial t} + \frac{\partial F_{iik}}{\partial x_k} = P_{ii}, \qquad \frac{\partial G_{ii}}{\partial t} + \frac{\partial G_{iik}}{\partial x_k} = 0.
\end{aligned}
$$

Note that, in general, $\bar{F}_{ii} \neq F_{ii}$. Here, since the balance equations of the densities F, F_i, and G_{ii} represent, respectively, the conservation laws of mass, momentum, and energy without shear stress and heat flux [127, 128, 322], we have the following expressions of the densities and fluxes in terms of the commonly used macroscopic variables:

$$
\begin{aligned}
&F = \rho, \qquad F_i = \rho v_i, \qquad F_{ij} = (p + \Pi)\delta_{ij} + \rho v_i v_j, \\
&G_{ii} = 2\rho\varepsilon + \rho v^2, \qquad G_{iik} = 2(\rho\varepsilon + p + \Pi)v_k + \rho v^2 v_k,
\end{aligned}
$$

where ρ, v_i, p, ε, and Π are the mass density, the velocity, the pressure, the specific internal energy density, and the dynamic pressure, respectively. The problem remained is the determination of the quantities: density $\bar{F}_{ii}$, flux F_{iik}, and production P_{ii}. According to the Galilean invariance, it was proved [436] that the above system can be rewritten in the form:

$$
\begin{aligned}
&\frac{\partial \rho}{\partial t}+\frac{\partial}{\partial x_i}(\rho v_i)=0,\\
&\frac{\partial \rho v_j}{\partial t}+\frac{\partial}{\partial x_i}\left\{\rho v_i v_j+(p+\Pi)\delta_{ij}\right\}=0,\\
&\frac{\partial}{\partial t}(2\rho\varepsilon+\rho v^2)+\frac{\partial}{\partial x_i}\left\{\left(2\rho\varepsilon+\rho v^2+2(p+\Pi)\right)v_i\right\}=0,\\
&\frac{\partial}{\partial t}\left\{3(\bar{p}+\bar{\Pi})+\rho v^2\right\}+\\
&\qquad+\frac{\partial}{\partial x_i}\left\{(\rho v^2+3(\bar{p}+\bar{\Pi})+2(p+\Pi))v_i\right\}=\hat{P}_{ll},
\end{aligned}
\tag{24.1}
$$

where $\hat{P}_{ll}$ is the velocity-independent production term. We have now two undetermined quantities $\bar{p}$ and $\bar{\Pi}$, which, in the case of rarefied-gas limit, approach p and Π, respectively. Unfortunately it was proved [436] that the entropy principle is unable to specify the new functions in a unique way.

As is explained in the previous chapters, in the study of rarefied gases, it is shown that $G_{ii}-\bar{F}_{ii}$ is a characteristic variable for polyatomic gases that vanishes in the monatomic-gas limit. Therefore it seems to be reasonable to substitute the difference between $(24.1)_3$ and $(24.1)_4$ for the last equation of (24.1):

$$
\begin{aligned}
&\frac{\partial \rho}{\partial t}+\frac{\partial}{\partial x_i}(\rho v_i)=0,\\
&\frac{\partial \rho v_j}{\partial t}+\frac{\partial}{\partial x_i}\left\{\rho v_i v_j+(p+\Pi)\delta_{ij}\right\}=0,\\
&\frac{\partial}{\partial t}(2\rho\varepsilon+\rho v^2)+\frac{\partial}{\partial x_i}\left\{\left(2\rho\varepsilon+\rho v^2+2(p+\Pi)\right)v_i\right\}=0,\\
&\frac{\partial}{\partial t}\left(2\rho\varepsilon-3\bar{p}-3\bar{\Pi}\right)+\frac{\partial}{\partial x_i}\left\{(2\rho\varepsilon-3\bar{p}-3\bar{\Pi})v_i\right\}=-\hat{P}_{ll}.
\end{aligned}
\tag{24.2}
$$

In this chapter, we study an ET_6 theory of dense gases ($\mathrm{ET}_6^{\mathrm{D}}$) that is free from the difficulties mentioned above by adopting a simple principle that bridges the gap between rarefied gas and dense gas [150]. Some characteristic features of this theory are also discussed.

24.2 Equations of State

In this section, as a preliminary discussion for understanding the duality principle explained in the next section, we write down the equations of state of rarefied gas and of dense gas in a parallel way.

Rarefied Gas From a microscopic point of view, the Hamiltonian of a rarefied polyatomic gas $\mathscr{H}$ is given in the form:

$$\mathscr{H} = \mathscr{H}^K + \mathscr{H}^I,$$

where $\mathscr{H}^K$ is the *kinetic energy* of molecular translational motion and $\mathscr{H}^I$ is the *energy of the internal motion* around the center-of-mass positions of molecules such as molecular rotation and vibration.

We study non-polytropic gases within a temperature range where the classical (i.e., non-degenerate) equations of state are valid. Therefore the pressure p and the specific internal energy density ε are expressed by the mass density ρ and the temperature T as follows:

$$p = p(\rho, T) \equiv \frac{k_B}{m}\rho T, \quad \varepsilon = \varepsilon_E(T).$$

Note that ε has, in general, nonlinear dependence on T.

The pressure p can be divided into two parts [334]: the pressure p^K due to $\mathscr{H}^K$ and the pressure p^I due to $\mathscr{H}^I$. However, since $p^I = 0$, we have

$$p = p^K; \quad p^K = \frac{k_B}{m}\rho T.$$

Similarly we have

$$\varepsilon = \varepsilon^K + \varepsilon^I; \quad \varepsilon^K = \frac{3}{2}\frac{k_B}{m}T, \quad \varepsilon^I = \varepsilon_E^I(T).$$

Dense Gas The Hamiltonian of a dense polyatomic gas is given in the form:

$$\mathscr{H} = \mathscr{H}^{K+U} + \mathscr{H}^I,$$

where $\mathscr{H}^{K+U}$ is the sum of the *kinetic energy* K of molecular translational motion and the *potential energy* U among molecules, and $\mathscr{H}^I$ is the energy of the internal motion of molecules. The thermal and caloric equations of state in terms of ρ and T are expressed as

$$p = p(\rho, T), \quad \varepsilon = \varepsilon_E(\rho, T).$$

Note that ε depends also on ρ due to the potential energy.

We assume that $\mathscr{H}^{K+U}$ is the function only of center-of-mass positions of molecules and their conjugate momenta. Then, the pressure p can be divided into two parts: the pressure p^{K+U} due to $\mathscr{H}^{K+U}$ and the pressure p^I due to $\mathscr{H}^I$. However, since $p^I = 0$ under this assumption, we have

$$p = p^{K+U}; \quad p^{K+U} = p(\rho, T).$$

Similarly we have

$$\varepsilon = \varepsilon^{K+U} + \varepsilon^{I}; \quad \varepsilon^{K+U} = \varepsilon_E^{K+U}(\rho, T), \quad \varepsilon^{I} = \varepsilon_E^{I}(T).$$

Note that ε^I depends only on T.

Remark 24.1 Configuration of one polyatomic molecule can be specified by both the position vector for the center of mass of the molecule and the relative (internal) positions with respect to the center-of-mass position. Therefore, the potential energy U for the system of molecules is, in general, a function of the configurations of all molecules. In other words, U depends not only on the center-of-mass positions but also on the relative positions. However, for *moderately dense* gases, we may assume that U can be expressed in terms of *effective inter-molecular potential* such as the Lennard-Jones potential, which takes into account the effect due to the relative motion inside a molecule partially. Then, U depends only on the center-of-mass positions of molecules. Therefore, $\mathscr{H}^{K+U}$ is the function only of center-of-mass positions of molecules and their conjugate momenta. In the above discussion, we have adopted such an assumption. In Chaps. 24 and 25, we will study solely this case.

24.3 Nonequilibrium Temperatures and Duality Principle

Since $\varepsilon_E^I(T) \neq 0$ for polyatomic gases, we may define two *positive* parameters ϑ and Θ through the corresponding quantities Π and $\bar{\Pi}$:

$$\Pi = p(\rho, \vartheta) - p(\rho, T), \tag{24.3}$$

$$\bar{\Pi} = \frac{2}{3}\rho\left(\varepsilon_E^I(T) - \varepsilon_E^I(\Theta)\right). \tag{24.4}$$

As will be discussed below, ϑ and Θ can be regarded as nonequilibrium temperatures.

24.3.1 *Rarefied Gas*

In the case of rarefied gases, we know that $\bar{p} = p$ and $\bar{\Pi} = \Pi$ [436]. Therefore, from (24.3) and (24.4), we have the following relations between the two parameters:

$$\varepsilon_E^I(T) - \varepsilon_E^I(\Theta) = \varepsilon_E^K(\vartheta) - \varepsilon_E^K(T) \equiv \Delta, \tag{24.5}$$

and

$$p = p(\rho, T), \quad \bar{p} = \frac{2}{3}\rho\varepsilon_E^K(T). \tag{24.6}$$

Moreover, as seen in Sect. 12.5.4, we know that the nonequilibrium part of specific entropy density k is given by (12.69).

Remark 24.2 We observe that $\bar{p}$ and $\bar{\Pi}$ appear in the last equation of (24.2). Therefore we can regard this equation as the equation that describes the time-evolution of the energy exchange Δ. Then $\bar{p}$ and $\bar{\Pi}$ are regarded as the quantities that are closely related to energy more than pressure. While p and Π appear in the momentum flux, and therefore these are necessarily related to pressure. This justifies the expressions (24.3), (24.4), and (24.6), where we have introduced, respectively, pressure and energy in accordance with this rule.

24.3.2 Dense Gas

In order to construct the ET$_6^{\mathrm{D}}$ theory of *dense* gases, we need to find a suitable bridge between rarefied gas and dense gas. For this purpose, it seems natural to adopt the duality principle mentioned in Sect. 1.9.

According to this principle we have, from (24.6),

$$p = p^{K+U} = p(\rho, T), \quad \bar{p} = \frac{2}{3}\rho\varepsilon_E^{K+U}(\rho, T),$$

while ϑ and Θ are now related through, instead of (24.5), the following relation due to the duality principle:

$$\varepsilon_E^I(\Theta) - \varepsilon_E^I(T) = \varepsilon_E^{K+U}(\rho, T) - \varepsilon_E^{K+U}(\rho, \vartheta) \equiv \Delta. \tag{24.7}$$

Then the expressions (24.3) and (24.4) are explicitly expressed, also for dense gases, in terms of the nonequilibrium temperatures. Remarkable point is that $\bar{\Pi} \neq \Pi$ and $\bar{p} \neq p$ as we expect for dense gases.

Remark 24.3 What we have done in the above based on the duality principle is, physically speaking, that we have assumed that the dynamic pressure Π is caused by the energy exchange Δ between H^{K+U} and H^I. In polyatomic gases, it seems that this nonequilibrium process is the most dominant mechanism for the emergence of the dynamic pressure Π. In reality, however, there exist several mechanisms. Because we may conceive different kinds of energy exchange, the above one is probably one of them. In fact, in the monatomic-gas limit, the present mechanism predicts that the dynamic pressure Π disappears not only in a rarefied gas case but also in a dense gas case, then the bulk viscosity is always zero. This is, of course, not acceptable. It is, therefore, evident that we need to introduce other mechanisms to explain the dynamic pressure in a *dense monatomic* gas. For example, in the paper [437], Hirai and Eyring pointed out the two possible mechanisms. The first one is essentially the same as the present one. The second one is due to the so-called structural relaxation mechanism: the change from a structure to other one takes time and the lag is the cause of the dynamic pressure. The hierarchy structure of the system of field equations, therefore, will be changed. Therefore, although the applicability range of the present model seems to be rather wide, it is fair to say that the model is appropriate only for a gas where the mechanism adopted here is overwhelming and other mechanisms can be safely neglected.

24.4 RET Model of Dense Polyatomic Gas: $\mathrm{ET}_6^{\mathrm{D}}$

We here construct the $\mathrm{ET}_6^{\mathrm{D}}$ theory step by step.

24.4.1 System of Field Equations

When we choose (ρ, v_j, T, Θ) as independent variables, according to the above discussions, we may propose the differential system for dense gases as follows:

$$\begin{aligned}
&\frac{\partial \rho}{\partial t}+\frac{\partial}{\partial x_i}(\rho v_i)=0,\\
&\frac{\partial \rho v_j}{\partial t}+\frac{\partial}{\partial x_i}\left\{\rho v_i v_j+(p(\rho,T)+\Pi)\delta_{ij}\right\}=0,\\
&\frac{\partial}{\partial t}(2\rho\varepsilon_E(\rho,T)+\rho v^2)+\frac{\partial}{\partial x_i}\left\{\Big(2\rho\varepsilon_E(\rho,T)+\rho v^2+2(p(\rho,T)+\Pi)\Big)v_i\right\}=0,\\
&\frac{\partial}{\partial t}\left[2\rho\varepsilon_E^I(\Theta)\right]+\frac{\partial}{\partial x_i}\left[2\rho\varepsilon_E^I(\Theta)v_i\right]=-\hat{P}_{ll},
\end{aligned} \tag{24.8}$$

where Π is expressed in terms of Θ in the parametric form with the parameter ϑ (see (24.3) and (24.7)).

With the material derivative, the system can be rewritten as follows:

$$\begin{aligned}
&\dot{\rho}+\rho\frac{\partial v_k}{\partial x_k}=0,\\
&\rho\dot{v}_i+\frac{\partial}{\partial x_i}(p(\rho,T)+\Pi)=0,\\
&\rho c_v(\rho,T)\dot{T}+\left(p(\rho,T)-\rho^2\frac{\partial\varepsilon_E(\rho,T)}{\partial\rho}+\Pi\right)\frac{\partial v_k}{\partial x_k}=0,\\
&\dot{\varepsilon}_E^I(\Theta)=-\frac{P_{ll}}{2\rho},
\end{aligned}$$

where $c_v(\rho,T)\equiv\frac{\partial\varepsilon_E(\rho,T)}{\partial T}$ is the specific heat.

24.4.2 Galilean Invariance and the Entropy Principle

We study the consequence from the Galilean invariance and the entropy principle following the general discussion made in Chap. 2. In the present case ($N = 6$), we have

$$\mathbf{F} \equiv \left(\rho, \rho v_j, 2\rho\varepsilon + \rho v^2, 2\rho\varepsilon^I\right)^T,$$
$$\boldsymbol{\Phi}^i \equiv \left(0, (p+\Pi)\delta_{ij}, 2(p+\Pi)v_i, 0\right)^T,$$
$$\mathbf{f} \equiv \left(0, 0_j, 0, -P_{ll}\right)^T,$$

where 0_i denotes the zero raw $\mathbb{R}^3$ vector. Then we have

$$\mathbf{X}(\mathbf{v}) = \begin{pmatrix} 1 & 0_k & 0 & 0 \\ v^k & \delta^j_k & 0 & 0 \\ v^2 & 2v_k & 1 & 0 \\ 0 & 0_k & 0 & 1 \end{pmatrix}, \tag{24.9}$$

while, from (2.51), the matrices $\mathbf{A}^r$ are given by

$$\mathbf{A}^r = \left.\frac{\partial \mathbf{X}}{\partial v_r}\right|_{\mathbf{v}=0} = \begin{pmatrix} 0 & 0_k & 0 & 0 \\ \delta^{kr} & 0^j_k & 0 & 0 \\ 0 & 2\delta^r_k & 0 & 0 \\ 0 & 0_k & 0 & 0 \end{pmatrix},$$

where 0^j_k denotes the null 3×3 matrix.

Let us express the main field $\mathbf{u}'$ in the component form:

$$\mathbf{u}' \equiv (\lambda, \lambda_i, \mu, \zeta), \tag{24.10}$$

then we have, from $(2.58)_1$, $\hat{\lambda}_i = 0$, and, from $(2.57)_1$, we have

$$\begin{aligned} \hat{\lambda} &= \eta + \rho\eta_\rho - \frac{1}{c_v(\rho, T)}(\varepsilon + \rho\varepsilon_\rho)\eta_T - \frac{\varepsilon^I_E(\Theta)}{c^I_v(\Theta)}\eta_\Theta, \\ \hat{\mu} &= \frac{1}{2c_v(\rho, T)}\eta_T, \\ \hat{\zeta} &= \frac{1}{2c^I_v(\Theta)}\eta_\Theta, \end{aligned} \tag{24.11}$$

where a subscript attached to η or ε indicates a partial derivative. Equation (2.57)$_2$ is automatically satisfied and gives

$$\hat{h}^i = 0, \quad \rightarrow \quad h^i = hv^i.$$

While, inserting (24.11) into (2.58)$_2$, we obtain the following partial differential equation for η:

$$\eta_\rho + \frac{1}{c_v}\left(\frac{p+\Pi}{\rho^2} - \varepsilon_\rho\right)\eta_T = 0. \tag{24.12}$$

The solution η is expressed compactly by using the nonequilibrium temperatures ϑ and Θ, if we divide the equilibrium specific entropy density s into two parts: s^{K+U} due to H^{K+U} and s^I due to H^I. Then we can prove, according with the duality principle (see also (12.69)), that a solution of (24.12) is given by

$$\eta = s^{K+U}(\rho, \vartheta) + s^I(\Theta) \tag{24.13}$$

for a given temperature T. The proof of (24.13) can be done directly by noticing the following relations:

$$d\eta = \frac{1}{\vartheta}\left\{\left(\frac{\partial \varepsilon_E(\rho, T)}{\partial \rho} - \frac{p(\rho, \vartheta)}{\rho^2}\right) d\rho + c_v(\rho, T)dT\right\} + \left(\frac{1}{\Theta} - \frac{1}{\vartheta}\right) c_v^I(\Theta)d\Theta, \tag{24.14}$$

and

$$\begin{aligned}
\eta_\rho &= \frac{1}{\vartheta}\left(\frac{\partial \varepsilon_E(\rho, T)}{\partial \rho} - \frac{p(\rho, \vartheta)}{\rho^2}\right), \\
\eta_T &= \frac{1}{\vartheta} c_v(\rho, T), \\
\eta_\Theta &= \left(\frac{1}{\Theta} - \frac{1}{\vartheta}\right) c_v^I(\Theta).
\end{aligned} \tag{24.15}$$

Using (24.15), we can rewrite the components of $\hat{\mathbf{u}}'$ in (24.11) as follows:

$$\begin{aligned}
\hat{\lambda} &= -\frac{g^{K+U}(\rho, \vartheta)}{\vartheta} - \frac{g^I(\Theta)}{\Theta}, \\
\hat{\mu} &= \frac{1}{2\vartheta}, \\
\hat{\zeta} &= \frac{1}{2}\left(\frac{1}{\Theta} - \frac{1}{\vartheta}\right),
\end{aligned} \tag{24.16}$$

where we have defined the chemical potentials of $(K+U)$-part and I-part:

$$g^{K+U}(\rho,\vartheta) = \varepsilon_E^{K+U}(\rho,\vartheta) + \frac{p(\rho,\vartheta)}{\rho} - \vartheta s^{K+U}(\rho,\vartheta),$$
$$g^I(\Theta) = \varepsilon_E^I(\Theta) - \Theta s^I(\Theta).$$

Therefore the nonequilibrium temperature $\mathscr{T}$ defined by (15.5) is as follows:

$$\mathscr{T} = \vartheta. \tag{24.17}$$

Note that, in a rarefied-gas case, we have the relation: $\mathscr{T} = \theta^K$. From (24.9), (24.10), (24.16), and (2.55), we deduce the main field components:

$$\begin{aligned}
\lambda &= -\frac{g^{K+U}(\rho,\vartheta)}{\vartheta} - \frac{g^I(\Theta)}{\Theta} + \frac{v^2}{2\vartheta},\\
\lambda_i &= \frac{v_i}{\vartheta},\\
\mu &= \frac{1}{2\vartheta},\\
\zeta &= \frac{1}{2}\left(\frac{1}{\Theta} - \frac{1}{\vartheta}\right).
\end{aligned} \tag{24.18}$$

The residual inequality of the entropy principle $(2.57)_3$ is expressed as

$$\Sigma = \frac{1}{3}\eta_\Theta \hat{P}_{ll} \geqslant 0. \tag{24.19}$$

From this inequality, we have

$$\hat{P}_{ll} = \alpha\eta_\Theta, \qquad \alpha \geqslant 0.$$

In the simplest case, we can assume that α is an equilibrium quantity, i.e., it is independent of Θ: $\alpha \equiv \alpha(\rho, T)$. See also the discussion in Sect. 12.2.2.2 in the case of rarefied gas.

24.4.3 Convexity Principle

The thermodynamic stability condition requires that the entropy must be convex with respect to the densities, i.e., (2.10). From $(2.57)_1$, this condition corresponds to the following negative quadratic form:

$$Q = \delta\mathbf{u}' \cdot \delta\mathbf{F} < 0.$$

Taking into account (2.49)$_1$ and (2.55), we have [167]:

$$Q = \bar{Q} - 2\hat{\mathbf{u}}'\mathbf{A}^r\delta\mathbf{F}\delta v_r - g^{rs}\delta v_r \delta v_s, \tag{24.20}$$

where

$$\bar{Q} = \delta\hat{\mathbf{u}}' \cdot \delta\hat{\mathbf{F}}, \qquad g^{rs} = \hat{\mathbf{u}}'\mathbf{A}^r\mathbf{A}^s\hat{\mathbf{F}}.$$

In the present case, since the second term in the right hand side of (24.20) vanishes and $g^{rs} = (\rho/\vartheta)\delta_{rs}$, we have

$$Q = \bar{Q} - \frac{\rho}{\vartheta}||\delta\mathbf{v}||^2$$

$$\text{with} \quad \bar{Q} = \delta\rho\delta\hat{\lambda} + 2\delta(\rho\varepsilon)\delta\hat{\mu} + 2\delta(\rho\varepsilon_E^I(\Theta))\delta\hat{\zeta}$$
$$= -\frac{p_\rho(\rho,\vartheta)}{\rho\vartheta}(\delta\rho)^2 - \frac{\rho c_v^{K+U}(\rho,\vartheta)}{\vartheta^2}(\delta\vartheta)^2 - \frac{\rho c_v^I(\Theta)}{\Theta^2}(\delta\Theta)^2 < 0.$$

Then the state (ρ, T, Θ) is stable if and only if the following inequalities are satisfied:

$$\begin{aligned} &\left(\frac{\partial p}{\partial \rho}\right)_\vartheta (\rho, \vartheta(\rho, T, \Theta)) > 0, \\ &\left(\frac{\partial \varepsilon_E^{K+U}}{\partial \vartheta}\right)_\rho (\rho, \vartheta(\rho, T, \Theta)) > 0, \quad \frac{d\varepsilon_E^I}{d\Theta}(\Theta) > 0. \end{aligned} \tag{24.21}$$

Therefore under these conditions, the system of the present theory can be put in a symmetric form by choosing the main field given in (24.18) as independent fields. In particular, an equilibrium state (ρ, T, T) is stable if

$$\begin{aligned} &\left(\frac{\partial p}{\partial \rho}\right)_T (\rho, T) > 0, \\ &\left(\frac{\partial \varepsilon_E^{K+U}}{\partial T}\right)_\rho (\rho, T) > 0, \quad \frac{d\varepsilon_E^I}{dT}(T) > 0. \end{aligned}$$

24.4.4 *Upper and Lower Bounds for Nonequilibrium Temperatures*

Differentiating (24.7) with respect to ϑ keeping ρ and T fixed, we easily obtain the relation:

$$\frac{d\Theta}{d\vartheta} = -\frac{c_v^{K+U}(\rho,\vartheta)}{c_v^I(\Theta)} < 0. \tag{24.22}$$

If ϑ increases (decreases) then Θ decreases (increases), and, since both Θ and ϑ are equal to T in equilibrium, if $\vartheta \gtrless T$ then $\Theta \lessgtr T$.

Moreover, in the definition of the nonequilibrium temperatures, we have assumed that these are positive quantities. This is also in consistency with the fact that the entropy must be an increasing function of the energy. Therefore all solutions must satisfy the condition $\vartheta > 0, \Theta > 0$. Taking into account (24.22) and (24.7), we obtain immediately the upper bound for these quantities:

$$\begin{aligned} &0 < \Theta < \Theta_{\max} \quad \text{with} \quad \varepsilon_E^I(\Theta_{\max}) = \varepsilon_E(\rho, T) - \varepsilon_E^{K+U}(\rho, 0), \\ &0 < \vartheta < \vartheta_{\max} \quad \text{with} \quad \varepsilon_E^{K+U}(\vartheta_{\max}) = \varepsilon_E(\rho, T) - \varepsilon_E^I(0). \end{aligned} \tag{24.23}$$

24.4.5 *Characteristic Velocity, Subcharacteristic Conditions, and Local Exceptionality*

The characteristic velocities V are given as follows:

$$\textbf{Contact Waves:} \quad V = v_n = 0, \tag{24.24}$$

(multiplicity 4)

$$\textbf{Sound Waves:} \quad V = v_n \pm \sqrt{\left(\frac{\partial p(\rho, \vartheta(\rho, \eta, \Theta))}{\partial \rho}\right)_{\eta, \Theta}}, \tag{24.25}$$

(each of multiplicity 1)

where $v_n = v_j n_j$ with n_j being the j-component of the unit normal to the wave front. Taking into account the relation (24.14), we can rewrite the velocity of the sound wave as follows:

$$\begin{aligned} U^2 &= p_\rho(\rho, \vartheta) + \frac{p_\vartheta(\rho, \vartheta)}{c_v^{K+U}(\rho, \vartheta)} \left\{ \frac{p^{K+U}(\rho, \vartheta)}{\rho^2} - \varepsilon_{E\rho}^{K+U}(\rho, \vartheta) \right\} \\ &= p_\rho(\rho, \vartheta) + \frac{\vartheta p_\vartheta^2(\rho, \vartheta)}{\rho^2 c_v^{K+U}(\rho, \vartheta)}, \end{aligned} \tag{24.26}$$

where $U = V - v_n$. In particular, in an equilibrium case, we have

$$U_E^2 = p_\rho(\rho, T) + \frac{T p_T^2(\rho, T)}{\rho^2 c_v^{K+U}(\rho, T)}. \tag{24.27}$$

Since the system of ET_6^D includes the Euler system as a principal subsystem, we have the *subcharacteristic condition* (see Theorem 2.3): $U_E > U_{Euler}$, where U_{Euler} is given in (14.19).

The contact waves (24.24) are exceptional, while the sound waves (24.25) can be locally exceptional if the condition:

$$\delta V = 0, \quad \text{for some} \quad \mathbf{u}$$

is satisfied (see Sect. 3.2). Simple algebra similar to the one in [206] gives that, if the hyper-surface of local exceptionality exists, the following relation is satisfied on it:

$$\frac{\partial}{\partial \rho}\left\{\rho^2\left(\frac{\partial p(\rho,\vartheta)}{\partial \rho}\right)_{\eta,\Theta}\right\}_{\eta,\Theta} = \frac{\partial}{\partial \rho}\left\{\rho^2 U^2\right\}_{\eta,\Theta} = 0. \tag{24.28}$$

24.4.6 K-Condition

As is well known that the Euler fluid cannot have global smooth solutions for all time because there appears shock formation or blowup. However, the present system with the dynamical pressure belongs to the group of dissipative hyperbolic systems due to the presence of the production term in the last equation in (24.8). In this case, global smooth solutions can exist due to the interrelationship between the conservation laws and the remaining dissipative one. In fact, for generic hyperbolic systems composed of conservation laws and balance laws, endowed with a convex entropy law, the so-called Shizuta-Kawashima condition (K-condition: see 2.6.2) becomes a sufficient condition for the existence of global smooth solutions, provided that the initial data are sufficiently small [169–172]. It was proved in [173] that the K-condition corresponds to:

$$\delta \mathbf{f}|_E \neq 0.$$

In the present case, from (24.8), the above condition is expressed as

$$\delta \hat{P}_{ll}|_E \neq 0. \tag{24.29}$$

Taking into account that $\hat{P}_{ll}|_E = 0$ and that only the linear part of the nonequilibrium quantities enters into the expression (24.29), we have, from (24.19),

$$\hat{P}_{ll} = \alpha\eta_\Theta = \alpha\left(\eta_{\Theta\Theta}|_E(\Theta - T) + O(2)\right),$$

where $O(2)$ indicates terms of second order or more with respect to the nonequilibrium quantity $\Theta - T$. Therefore we have:

$$\delta \hat{P}_{ll}|_E = \alpha \left(\eta_{\Theta\Theta}(\delta\Theta - \delta T)\right)|_E. \tag{24.30}$$

But, for sonic waves, from (24.25), we have $\delta\Theta = \delta\eta = 0$, and from (24.14) we obtain:

$$\delta T|_E = \frac{1}{c_v(\rho, T)}\left(\frac{\partial \varepsilon_E(\rho, T)}{\partial \rho} - \frac{p(\rho, T)}{\rho^2}\right)\delta\rho = \frac{T p_T}{\rho^2 c_v}\delta\rho.$$

Inserting the last expression into (24.30) we obtain

$$\delta \hat{P}_{ll}|_E = -\alpha\left(\eta_{\Theta\Theta}|_E \frac{T p_T}{\rho^2 c_v}\right)\delta\rho.$$

If $p_T \neq 0$ as usual, the term in the right-hand side is always different from zero due to the convexity argument of η. For contact waves, we can also prove that the K-condition holds. Therefore, in contrast with the Euler system, the present system that expresses the simplest dissipative system after the Euler system has global smooth solution for sufficiently small initial data. This is an interesting result that reinforces the physical meaning of the present model!

24.4.7 Comparison with the Meixner Theory

Since the balance equations (24.8) have the same mathematical structure as those of rarefied polyatomic gases, the correspondence relations between the present theory and the Meixner theory with one internal variable are the same as those in the case of rarefied polyatomic gas in Chap. 12. Therefore $\varepsilon^I_E(\Theta)$ plays the role of the internal variable and the nonequilibrium temperature ϑ is identified as the Meixner temperature (see also (24.17)).

24.4.8 Alternative Form of the System of Balance Equations

When the initial and boundary conditions are given in terms of the nonequilibrium temperatures Θ and ϑ and/or when we want to explicitly trace the evolution of these temperatures, it is more convenient to adopt the independent fields $\{\rho, v_i, \Theta, \vartheta\}$ instead of $\{\rho, v_i, T, \Theta\}$. Therefore, for completeness, we write down the system of balance equations of the fields $\{\rho, v_i, \Theta, \vartheta\}$:

$$
\begin{aligned}
&\frac{\partial \rho}{\partial t}+\frac{\partial}{\partial x_i}(\rho v_i)=0, \\
&\frac{\partial \rho v_j}{\partial t}+\frac{\partial}{\partial x_i}\left\{p(\rho, \vartheta)\delta_{ij}+\rho v_i v_j\right\}=0, \\
&\frac{\partial}{\partial t}\left(2\rho\varepsilon_E^{K+U}(\rho, \vartheta)+\rho v^2\right) \\
&\qquad+\frac{\partial}{\partial x_i}\left\{\left[2p(\rho, \vartheta)+2\rho\varepsilon_E^{K+U}(\rho, \vartheta)+\rho v^2\right] v_i\right\}=\hat{P}_{ll}, \\
&\frac{\partial}{\partial t}\left[2\rho\varepsilon_E^I(\Theta)\right]+\frac{\partial}{\partial x_i}\left[2\rho\varepsilon_E^I(\Theta) v_i\right]=-\hat{P}_{ll}.
\end{aligned}
\tag{24.31}
$$

24.5 Near-Equilibrium Case and the Bulk Viscosity

The simplest way to obtain the linear approximation of the system around an equilibrium state (ρ, T) is to expand the system with respect to the energy exchange Δ up to the first order. The other quantities are expressed by using Δ. In fact, we have, from (24.7),

$$
\Theta - T = \frac{\Delta}{c_v^I(T)}, \qquad \vartheta - T = -\frac{\Delta}{c_v^{K+U}(\rho, T)},
$$

and, from (24.3),

$$
\Pi = p_T(\rho, T)(\vartheta - T) = -\frac{p_T(\rho, T)}{c_v^{K+U}(\rho, T)}\Delta. \tag{24.32}
$$

The production term is given by

$$
\hat{P}_{ll} = \frac{2\rho}{\tau_\Delta}\Delta,
$$

where $\tau_\Delta(\rho, T)$ is the relaxation time for Δ, which is positive by the entropy principle. In particular, from (24.8), the linear equation of Δ is obtained as

$$
\dot{\Delta} - \frac{p_T c_v^I}{\rho c_v}\left(T - \frac{\Delta}{c_v^{K+U}}\right)\frac{\partial v_k}{\partial x_k} = -\frac{1}{\tau_\Delta}\Delta. \tag{24.33}
$$

24.5.1 Maxwellian Iteration

Let us derive the relationship between the relaxation time τ_Δ and the bulk viscosity. When the relaxation time is small, we can apply the Maxwellian iteration to (24.33). The first iterate of Δ is obtained by putting $\Delta = 0$ in the left hand side of (24.33):

$$\Delta^{(1)} = \frac{p_T c_v^I}{\rho c_v} T \tau_\Delta \frac{\partial v_k}{\partial x_k}. \tag{24.34}$$

Inserting (24.34) into (24.32), we have the relation between Π and $\partial v_k/\partial x_k$. Then we obtain the relationship between τ_Δ and the bulk viscosity ν:

$$\nu = \frac{T p_T^2 c_v^I}{\rho c_v c_v^{K+U}} \tau_\Delta. \tag{24.35}$$

In the rarefied-gas limit, this expression is consistent with the previous results obtained by different approaches [17, 335]. If the experimental data on the bulk viscosity are available, the relaxation time τ_Δ can be estimated by using this relationship.

24.5.2 Dispersion Relation

Let us study a linear plane harmonic wave. From the linearized system of field equations, we can obtain the dispersion relation between frequency ω and complex wave number k, $\omega = \omega(k)$, as the function of the dimensionless frequency $\Omega = \tau_\Delta \omega$ and the dimensionless characteristic velocity $\hat{U}_E = U_E/U_{Euler} (> 1)$ (see Sect. 3.1):

$$\frac{k}{\omega} = \frac{1}{U_{Euler}} \sqrt{\frac{1 + \mathrm{i}\Omega}{1 + \mathrm{i}\hat{U}_E^2 \Omega}} = \frac{1}{U_{Euler}} \sqrt{w_2 + \frac{\mathrm{i}}{\Omega}(w_1 - w_2)}, \tag{24.36}$$

where

$$w_1 = \frac{1 + \Omega^2}{1 + \hat{U}_E^4 \Omega^2}, \qquad w_2 = \frac{1 + \hat{U}_E^2 \Omega^2}{1 + \hat{U}_E^4 \Omega^2}.$$

It is interesting to note that the dispersion relation (24.36) is essentially determined by the ratio of the characteristic velocity of $\mathrm{ET}_6^{\mathrm{D}}$ to the characteristic velocity of Euler system.

From the dispersion relation, the phase velocity v_{ph} and the attenuation factor α are obtained as follows:

$$
\begin{aligned}
v_{\text{ph}} &= \frac{\omega}{\Re(k)} = \pm U_{Euler} \frac{\sqrt{2}}{\sqrt{\sqrt{w_1}+w_2}}, \\
\alpha &= -\Im(k) = \pm \frac{\Omega}{\sqrt{2} U_{Euler} \tau_\Delta} \sqrt{\sqrt{w_1} - w_2}.
\end{aligned}
\tag{24.37}
$$

In the high-frequency limit $\Omega \to \infty$, we have

$$
\lim_{\Omega\to\infty} v_{\text{ph}} = \pm U_E,
$$

$$
\lim_{\Omega\to\infty} \alpha = \pm \frac{1}{2U_{Euler}\tau_\Delta} \frac{\hat{U}_E^2 - 1}{\hat{U}_E^3}.
$$

In experimental studies of sound waves, the attenuation per wavelength α_λ defined by

$$
\alpha_\lambda = \frac{2\pi v_{\text{ph}} \alpha}{\omega},
$$

and the peak value of α_λ are measured. The dependence of α_λ on Ω is shown in Fig. 24.1 in the cases of $\hat{U}_E = 1.1, 1.2, 1.3$. From (24.37) we notice that α_λ has the maximum value:

$$
\alpha_\lambda^{max} = 2\pi \frac{\hat{U}_E - 1}{\hat{U}_E + 1} \quad \text{at } \Omega = \frac{1}{\hat{U}_E}.
$$

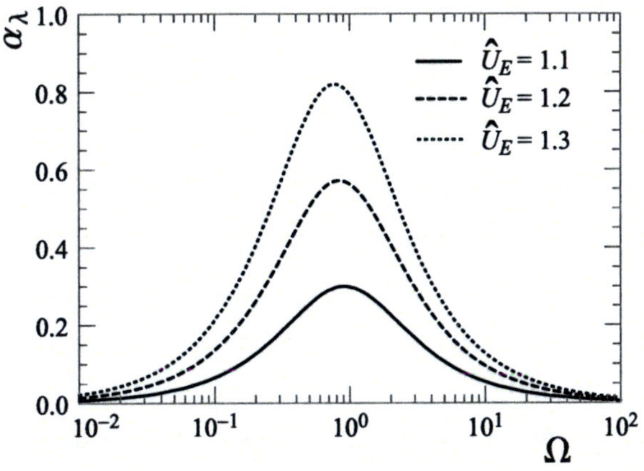

Fig. 24.1 Dependence of attenuation per wavelength α_λ on dimensionless frequency Ω with $\hat{U}_E = 1.1, 1.2, 1.3$

From the experimental data on α_λ^{max}, we can estimate the bulk viscosity as was done by the previous theory of ultrasonic waves [312].

24.5.3 Fluctuation-Dissipation Relation

The entropy density η is also expanded with respect to Δ around an equilibrium state up to the second order:

$$\eta = s(\rho, T) - \frac{1}{2\Psi}\Delta^2,$$
$$\Psi = \left[\frac{1}{T^2}\left(\frac{1}{c_v^{K+U}} + \frac{1}{c_v^I}\right)\right]^{-1} = T^2\frac{c_v^{K+U}c_v^I}{c_v}. \tag{24.38}$$

Since $c_v^{K+U} > 0$ and $c_v^I > 0$, the convexity condition is always satisfied. Moreover, in the case of rarefied gas, the result (24.38) reduces to the one obtained in (12.44).

We can estimate the fluctuation of the energy exchange Δ, which obeys the Gaussian distribution functional $f(\Delta(\boldsymbol{x}))$. Except for the normalization factor, the distribution functional is given by

$$f(\Delta(\boldsymbol{x})) \sim \exp\left(\int_{\mathcal{V}} \frac{\rho\eta}{k_B} d\boldsymbol{x}\right)$$

with η being given by (24.38). Here the integration is taken over the whole system $\mathcal{V}$. Therefore we have the estimation given as a simultaneous spatial correlation:

$$\langle\Delta(\boldsymbol{x})\Delta(\boldsymbol{x}')\rangle = \frac{k_B T^2}{\rho}\frac{c_v^{K+U}c_v^I}{c_v}\delta(\boldsymbol{x}-\boldsymbol{x}'),$$

where $\langle\ \rangle$ indicates the thermal average at an equilibrium state. From (24.32), the fluctuation of the dynamic pressure Π is also estimated as

$$\langle\Pi(\boldsymbol{x})\Pi(\boldsymbol{x}')\rangle = \frac{k_B T^2 p_T^2 c_v^I}{\rho c_v c_v^{K+U}}\delta(\boldsymbol{x}-\boldsymbol{x}').$$

Then the bulk viscosity (24.35) is rewritten as follows:

$$\nu = \frac{\tau_\Delta}{k_B T\,|\mathcal{V}|}\iint_{\mathcal{V}} \langle\Pi(\boldsymbol{x})\Pi(\boldsymbol{x}')\rangle d\boldsymbol{x}d\boldsymbol{x}',$$

where $|\mathcal{V}|$ is the volume of the system. This is a fluctuation-dissipation relation. It is also interesting to notice the following relation:

$$\nu = \rho\tau_\Delta(U_E^2 - U_{Euler}^2).$$

24.6 An Example: $\mathrm{ET}_6^{\mathrm{D}}$ Theory of van der Waals Gases

24.6.1 *Equations of State, Nonequilibrium Temperatures, and Dynamic Pressure*

Let us study, as a typical example, a polytropic van der Waals gas of which thermal and caloric equations of state are given by

$$p = \frac{k_B}{m}\frac{T\rho}{1 - b\rho} - a\rho^2, \quad \varepsilon = \frac{D}{2}\frac{k_B}{m}T - a\rho, \tag{24.39}$$

where D is related to the degrees of freedom of a molecule and the material-dependent constants a and b represent, respectively, a measure of the strength of the attraction between constituent molecules and the effective volume (or exclusion volume) of a molecule. In this case, we have

$$\varepsilon_E^{K+U}(\rho, T) = \frac{3}{2}\frac{k_B}{m}T - a\rho, \quad \varepsilon_E^I(T) = \frac{D-3}{2}\frac{k_B}{m}T,$$

and, from (24.3), (24.4) and (24.7), we have

$$\Pi = \frac{k_B}{m}\frac{\rho}{1 - b\rho}(\vartheta - T), \quad \bar{\Pi} = \frac{k_B}{m}\rho(\vartheta - T),$$
$$\frac{\Theta - T}{\vartheta - T} = -\frac{3}{D-3}.$$

Moreover the relation between the bulk viscosity and the relaxation time (24.35) is explicitly obtained as follows:

$$\nu = \frac{k_B}{m}\frac{2(D-3)}{3D}\frac{T\rho}{(1 - b\rho)}\tau_\Delta.$$

24.6.2 *Nonequilibrium Entropy and Bounded Domain of* Θ

From equilibrium thermodynamics, the equilibrium entropy density s is given by

$$s = s^{K+U} + s^I,$$
$$s^{K+U} = \frac{k_B}{m}\log\left(T^{\frac{3}{2}}\frac{1 - b\rho}{\rho}\right) + s_0^{K+U},$$
$$s^I = \frac{k_B}{m}\log T^{\frac{D-3}{2}} + s_0^I,$$

where s_0^{K+U} and s_0^I are constants at a reference state. From (24.13) the nonequilibrium entropy density η is obtained as follows:

$$\eta = s^{K+U}(\rho, \vartheta) + s^I(\Theta) = \frac{k_B}{m}\left\{\log\left(\vartheta^{\frac{3}{2}}\frac{1-b\rho}{\rho}\right) + \log\Theta^{\frac{D-3}{2}}\right\} + s_0, \tag{24.40}$$

where $s_0 = s_0^{K+U} + s_0^I$.

From (24.23), we have the condition for the nonequilibrium temperature ϑ:

$$0 < \vartheta < \frac{D}{3}T. \tag{24.41}$$

This condition corresponds to the condition of Θ:

$$0 < \Theta < \frac{D}{D-3}T.$$

24.6.3 Convexity

Since the conditions $(24.21)_{2,3}$ are identically satisfied, the convexity condition of $h = \rho\eta$ with η given by (24.40) comes only from $(24.21)_1$ and is expressed as

$$\hat{p}(\hat{\rho}, \hat{\vartheta}) > (3 - 2\hat{\rho})\hat{\rho}^2, \tag{24.42}$$

where we have introduced the following dimensionless variables:

$$\hat{p} = \frac{p}{p_{cr}}, \quad \hat{\rho} = \frac{\rho}{\rho_{cr}}, \quad \hat{T} = \frac{T}{T_{cr}}, \quad \hat{\vartheta} = \frac{\vartheta}{T_{cr}}, \tag{24.43}$$

with $p_{cr} = a/(27b^2)$, $\rho_{cr} = 1/(3b)$ and $T_{cr} = 8a/(27\frac{k_B}{m}b)$ being the quantities at the critical point. We notice that the boundary (spinodal curve) is independent of $\hat{\vartheta}$ as in the case of the Euler system in which the spinodal curve for equilibrium pressure is independent of the temperature. From (24.41), $\hat{\vartheta}$ is bounded as follows:

$$0 < \hat{\vartheta} < \frac{D}{3}\hat{T}.$$

The critical point is the point on the spinodal curve with the condition that the second derivative also vanishes:

$$\left(\frac{\partial^2 p(\rho, \vartheta)}{\partial \rho^2}\right)_{\vartheta} = 0.$$

With (24.42), we have, similar to the Euler system,

$$\hat{\rho} = 1, \quad \hat{\vartheta} = 1, \quad \hat{p}(\hat{\rho}, \hat{\vartheta}) = 1.$$

It should be noted that, in the space (ρ, p, Θ), we can observe now a spinodal surface instead of the spinodal curve and a critical line instead of the critical point. Using the present RET model, we may develop critical dynamics [438, 439].

24.6.4 *Characteristic Velocities*

From (24.26), (24.27), (14.19), and the equations of state (24.39), the velocities U, U_E, and U_{Euler} in a van der Waals gas are obtained as follows:

$$\begin{aligned} U^2 &= \frac{5}{3}\frac{k_B}{m}\frac{\vartheta}{(1-b\rho)^2} - 2a\rho \\ &= \frac{5}{3}\frac{k_B}{m}\frac{T}{(1-b\rho)^2} - 2a\rho + \frac{5(D-3)}{9}\frac{k_B}{m}\frac{T-\Theta}{(1-b\rho)^2}. \end{aligned} \tag{24.44}$$

In an equilibrium case with $\Theta = T$, we have

$$U_E^2 = \frac{5}{3}\frac{k_B}{m}\frac{T}{(1-b\rho)^2} - 2a\rho.$$

It is remarkable that if $\Theta \lesseqgtr T$ then $U \gtreqless U_E$. For Eulerian fluids, we have

$$U_{Euler}^2 = \frac{D+2}{D}\frac{k_B}{m}\frac{T}{(1-b\rho)^2} - 2a\rho < U_E^2$$

for any $D > 3$. This inequality represents the subcharacteristic condition. Lastly, the dimensionless characteristic velocity, which determines the dispersion relation, is obtained as follows:

$$\hat{U}_E = \sqrt{\frac{\frac{20}{3}\hat{T} - (\hat{\rho}-3)^2\hat{\rho}}{4\frac{D+2}{D}\hat{T} - (\hat{\rho}-3)^2\hat{\rho}}}. \tag{24.45}$$

From (24.45) with (24.42), we can prove $\hat{U}_E \geqslant \hat{U}_E^{rarefied}$ where $\hat{U}_E^{rarefied} = \lim_{\hat{\rho}\to 0} \hat{U}_E = \sqrt{\frac{5}{3}\frac{D}{D+2}}$ is the dimensionless characteristic velocity in rarefied gases. Moreover, we notice that $\hat{U}_E$ has an extremum at $\hat{\rho} = 1, 3$ and $\lim_{\hat{\rho}\to 3} \hat{U}_E = \hat{U}_E^{rarefied}$. The dependence of $\hat{U}_E$ on $\hat{\rho}$ is shown in Fig. 24.2 for $\hat{T} = 1.0, 1.5, 2.0$ with $D = 5$.

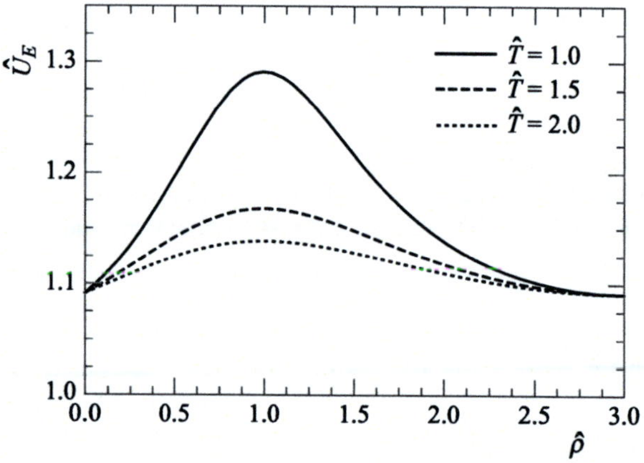

Fig. 24.2 Dependence of $\hat{U}_E$ on dimensionless mass density for $\hat{T} = 1.0, 1.5, 2.0$ with $D = 5$

Remark 24.4 Inequalities $\hat{U}_E^{rarefied} \leqslant \hat{U}_E \leqslant \hat{U}_E^{spinodal}$ were recently proved for a van der Waals fluid in [440], where $\hat{U}_E^{spinodal}$ is the dimensionless characteristic velocity evaluated on the spinodal curve. These inequalities are valid not only for polytropic gas but also for non-polytropic gas.

24.6.5 Critical Derivative

Solving $p = p(\rho, \vartheta)$ given by (24.39) with respect to ϑ and substituting it into (24.44), we obtain immediately the result that, from (24.28), the locus of the local exceptionality in the dimensionless form with (24.43) is expressed as

$$\hat{p}(\hat{\rho}, \hat{\vartheta}) = \hat{\rho}^2 \left\{ \frac{9}{20}(3 - \hat{\rho})^2 - 3 \right\}. \tag{24.46}$$

It is interesting to observe that the locus is independent of the nonequilibrium variables and coincides with the curve for a monatomic Euler fluid. As is well known among fluid dynamics researchers, this curve has been called the critical derivative curve (for details and references see [206]). Comparing this curve with the spinodal curve (24.42), we notice that the curve of local exceptionality always resides in the unstable region. Therefore, as far as the van der Waals equation of state is concerned, the rarefaction shock does not appear in the present RET theory in contrast to the Euler system [206]. It is necessary to clarify the admissibility of the rarefaction shock by the more realistic RET theory with shear viscosity and heat conductivity. On the other hand, from an experimental point of view, it is well known that the existence of the rarefaction shock is still controversial. We hope that our theoretical prediction may give an insight into this longstanding problem. In Fig. 24.3, the locus of the local exceptionality is depicted with the spinodal curve in $(\hat{\rho}, \hat{p}(\hat{\rho}, \hat{\theta}))$-plane.

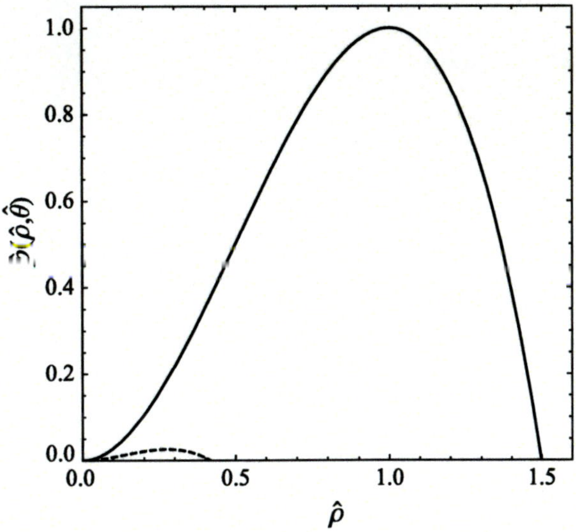

Fig. 24.3 Spinodal curve, that is the boundary of (24.42) (solid line) and locus of the local exceptionality (24.46) (dashed line) in $(\hat{\rho}, \hat{p}(\hat{\rho}, \hat{\theta}))$-plane

Chapter 25
RET of Dense Polyatomic Gas with Seven Fields

Abstract In this chapter, we study a RET theory of dense polyatomic gases taking into account the experimental evidence that the relaxation times of molecular rotation and that of molecular vibration are quite different from each other. For simplicity, as in Chap. 24, we focus on the bulk viscosity but ignore the shear viscosity and the heat conductivity. The present theory includes the previous RET theory of dense gases with six fields as a principal subsystem, and it also includes the RET theory of rarefied polyatomic gases with seven fields (ET_7 in Chap. 14) in the rarefied-gas limit. The closure is carried out by using the universal principles. In addition, the duality principle connecting rarefied gas to dense gas is adopted. A discussion is devoted to the expression of the production terms in the system of balance equations. As typical examples, we study a gas with virial equations of state and a van der Waals gas. Lastly the dispersion relation of a linear harmonic wave is derived, and its comparison with experimental data is made briefly.

25.1 Introduction

In Chap. 24, we have constructed the RET theory with 6 fields (ET_6^D). However, in a case with a high temperature where both rotational and vibrational modes exist, more than one molecular relaxation processes should be taken into account to have a more realistic RET theory. The purpose of the present chapter is to explain such a RET model of dense gases [151], and thereby to extend the applicability range of the ET_6^D theory. The present theory includes the ET_6^D theory as a principal subsystem, and it also includes the RET theory of rarefied polyatomic gases with seven fields (ET_7) explained in Chap. 14 as a special case in the rarefied-gas limit. The approach to the model-construction adopted in this chapter is more systematic than that in Chap. 24.

T. Ruggeri, M. Sugiyama, *Classical and Relativistic Rational Extended Thermodynamics of Gases*, https://doi.org/10.1007/978-3-030-59144-1_25

25.2 RET Model of Dense Polyatomic Gases: ET$_7^{\mathrm{D}}$

In this section, we establish a RET theory in which rotational mode R and vibrational mode V are treated individually. Therefore we divide the Hamiltonian of molecular internal motion $\mathscr{H}^I$ into two:

$$\mathscr{H}^I = \mathscr{H}^R + \mathscr{H}^V,$$

where $\mathscr{H}^R$ and $\mathscr{H}^V$ are, respectively, Hamiltonians of molecular rotation and vibration.

25.2.1 *System of Field Equations*

The ET$_7^{\mathrm{D}}$ theory has the following 7 independent fields:

mass density:	$F = \rho,$
momentum density:	$F_i = \rho v_i,$
translational and potential energy density:	$H_{ll}^{K+U} = \rho v^2 + 2\rho \varepsilon^{K+U},$
rotational energy density:	$H_{ll}^R = 2\rho\varepsilon^R,$
vibrational energy density:	$H_{ll}^V = 2\rho\varepsilon^V,$

where v_i is the velocity, $v^2 = v_i v_i$, and ε^{K+U}, ε^R, and ε^V are the specific internal energies of kinetic and potential mode, rotational mode, and vibrational mode of a molecule, respectively.

We adopt the following system of balance equations:

$$\begin{aligned}
&\frac{\partial F}{\partial t} + \frac{\partial F_i}{\partial x_i} = 0,\\
&\frac{\partial F_i}{\partial t} + \frac{\partial F_{ij}}{\partial x_i} = 0,\\
&\frac{\partial H_{ll}^{K+U}}{\partial t} + \frac{\partial H_{lli}^{K+U}}{\partial x_i} = P_{ll}^{K+U}, \quad \frac{\partial H_{ll}^R}{\partial t} + \frac{\partial H_{lli}^R}{\partial x_i} = P_{ll}^R, \quad \frac{\partial H_{ll}^V}{\partial t} + \frac{\partial H_{lli}^V}{\partial x_i} = P_{ll}^V,
\end{aligned} \tag{25.1}$$

where F_{ij}, H_{lli}^{K+U}, H_{lli}^R, and H_{lli}^V are the fluxes of F_i, H_{ll}^{K+U}, H_{ll}^R, and H_{ll}^V, respectively, and P_{ll}^{K+U}, P_{ll}^R, and P_{ll}^V are the productions of H_{ll}^{K+U}, H_{ll}^R, and H_{ll}^V, respectively. The first two equations express the conservation laws of mass and momentum. Equations (25.1)$_{3,4,5}$ are, respectively, the balance equations of the kinetic and potential energy, rotational energy, and vibrational energy. The sum of (25.1)$_{3,4,5}$ represents the conservation law of energy for the specific total internal energy ε:

$$\varepsilon = \varepsilon^{K+U} + \varepsilon^R + \varepsilon^V. \tag{25.2}$$

Therefore, we have the following condition for the production terms:

$$P_{ll}^{K+U} + P_{ll}^R + P_{ll}^V = 0.$$

We remark that $F_{ll} \neq H_{ll}^{K+U}$ in contrast to the case of rarefied gases.

25.2.2 Galilean Invariance

The system (25.1) is in a balance form:

$$\frac{\partial \mathbf{F}}{\partial t} + \frac{\partial \mathbf{F}^i}{\partial x_i} = \mathbf{f},$$

with density vector $\mathbf{F}$, non-convective flux $\boldsymbol{\Phi}^i = \mathbf{F}^i - \mathbf{F}v_i$, and production $\mathbf{f}$ as

$$\begin{aligned}
\mathbf{F} &\equiv \left(\rho, \rho v^j, 2\rho\varepsilon^{K+U} + \rho v^2, 2\rho\varepsilon^R, 2\rho\varepsilon^V\right)^T, \\
\boldsymbol{\Phi}^i &\equiv \left(0, F_{ij} - \rho v_i v_j, H_{lli}^{K+U} - H_{ll}^{K+U} v_i, H_{lli}^R - H_{ll}^R v_i, H_{lli}^V - H_{ll}^V v_i\right)^T, \\
\mathbf{f} &\equiv \left(0, 0_j, P_{ll}^{K+U}, P_{ll}^R, P_{ll}^V\right)^T.
\end{aligned}$$

Following the general theory explained in Sect. 2.7, the system is Galilean invariant if and only if there exists a matrix $\mathbf{X}(\mathbf{v})$ such that:

$$\mathbf{F} = \mathbf{X}(\mathbf{v})\hat{\mathbf{F}}, \quad \boldsymbol{\Phi}^i = \mathbf{X}(\mathbf{v})\hat{\boldsymbol{\Phi}}^i, \quad \mathbf{f} = \mathbf{X}(\mathbf{v})\hat{\mathbf{f}},$$

where $\hat{\mathbf{F}}$, $\hat{\boldsymbol{\Phi}}_i$, and $\hat{\mathbf{f}}$ are the intrinsic (velocity-independent) parts of the density, flux, and production, respectively. The components of these intrinsic fields can be expressed in terms of the conventional fields as follows:

$$\begin{aligned}
\hat{\mathbf{F}} &\equiv \left(\rho, 0, 2\rho\varepsilon^{K+U}, 2\rho\varepsilon^R, 2\rho\varepsilon^V\right)^T, \\
\hat{\boldsymbol{\Phi}}^i &\equiv \left(0, -t_{ij}, q_i^{K+U}, q_i^R, q_i^V\right)^T, \\
\hat{\mathbf{f}} &\equiv \left(0, 0_j, \hat{P}_{ll}^{K+U}, \hat{P}_{ll}^R, \hat{P}_{ll}^V\right)^T,
\end{aligned}$$

where 0_i denotes the zero raw $\mathbb{R}^3$ vector, $t_{ij} = -\mathscr{P}\delta_{ij} + \sigma_{\langle ij\rangle}$ is the stress tensor with the nonequilibrium pressure $\mathscr{P}$ and the shear stress $\sigma_{\langle ij\rangle}$ (traceless part of the viscous stress σ_{ij}), and q_i^{K+U}, q_i^R, q_i^V are the heat fluxes of the three modes. The nonequilibrium pressure is divided into the equilibrium pressure p and the dynamic pressure Π such that $\mathscr{P} = p + \Pi$ where $\Pi = -\sigma_{ll}/3$.

In the present case, $\mathbf{X}(\mathbf{v})$ is a polynomial matrix of degree 2 given by (2.51) and $\mathbf{A}^r$ are nilpotent matrices of index 3:

$$\mathbf{X}(\mathbf{v}) = \begin{pmatrix} 1 & 0_k & 0\,0\,0 \\ v^k & \delta_k^j & 0\,0\,0 \\ v^2 & 2v_k & 1\,0\,0 \\ 0 & 0_k & 0\,1\,0 \\ 0 & 0_k & 0\,0\,1 \end{pmatrix}, \qquad \mathbf{A}^r = \left.\frac{\partial \mathbf{X}}{\partial v_r}\right|_{\mathbf{v}=0} = \begin{pmatrix} 0 & 0_k & 0\,0\,0 \\ \delta^{kr} & 0_k^j & 0\,0\,0 \\ 0 & 2\delta_k^r & 0\,0\,0 \\ 0 & 0_k & 0\,0\,0 \\ 0 & 0_k & 0\,0\,0 \end{pmatrix},$$

where 0_k^j denotes the null 3×3 matrix. Then we obtain the following expressions:

$$\begin{aligned}
F_{ij} &= \rho v_i v_j + \mathscr{P}\delta_{ij} - \sigma_{\langle ij\rangle}, \\
H_{lli}^{K+U} &= \rho v^2 + 2(\rho\varepsilon^{K+U} + \mathscr{P})v_i + q_i^{K+U}, \\
H_{lli}^{R} &= 2\rho\varepsilon^R v_i + q_i^R, \\
H_{lli}^{V} &= 2\rho\varepsilon^V v_i + q_i^V, \\
P_{ll}^{K+U} &= \hat{P}_{ll}^{K+U}, \quad P_{ll}^R = \hat{P}_{ll}^R, \quad P_{ll}^V = \hat{P}_{ll}^V.
\end{aligned}$$

Since the intrinsic parts of the independent fields are $\{\rho, v_i, \varepsilon^{K+U}, \varepsilon^R, \varepsilon^V\}$, quantities $\sigma_{\langle ij\rangle}$, q_i^{K+U}, q_i^R, q_i^V vanish identically due to the representation theorem. Therefore the system of the balance equations is rewritten in the following form:

$$\begin{aligned}
&\frac{\partial \rho}{\partial t} + \frac{\partial}{\partial x_i}(\rho v_i) = 0, \\
&\frac{\partial \rho v_j}{\partial t} + \frac{\partial}{\partial x_i}\left(\mathscr{P}\delta_{ij} + \rho v_i v_j\right) = 0, \\
&\frac{\partial}{\partial t}\left(2\rho\varepsilon^{K+U} + \rho v^2\right) + \frac{\partial}{\partial x_i}\left\{\left(2\rho\varepsilon^{K+U} + 2\mathscr{P} + \rho v^2\right) v_i\right\} = P_{ll}^{K+U}, \\
&\frac{\partial}{\partial t}\left(2\rho\varepsilon^R\right) + \frac{\partial}{\partial x_i}\left(2\rho\varepsilon^R v_i\right) = P_{ll}^R, \\
&\frac{\partial}{\partial t}\left(2\rho\varepsilon^V\right) + \frac{\partial}{\partial x_i}\left(2\rho\varepsilon^V v_i\right) = P_{ll}^V.
\end{aligned} \tag{25.3}$$

25.2.3 Entropy Principle and Nonequilibrium Pressure

Let us express the main field $\mathbf{u}'$ in the component form:

$$\mathbf{u}' \equiv \left(\lambda, \lambda_i, \mu^{K+U}, \mu^R, \mu^V\right).$$

Then we have, from (2.58)$_1$, $\hat{\lambda}_l = 0$. And, from (2.58)$_2$, we have

$$\eta = \hat{\lambda} + 2\left(\varepsilon^{K+U} + \frac{\mathscr{P}}{\rho}\right)\hat{\mu}^{K+U} + 2\varepsilon^R\hat{\mu}^R + 2\varepsilon^V\hat{\mu}^V. \tag{25.4}$$

From (2.57)$_1$, we have

$$\begin{aligned} \hat{\lambda} &= \eta + \rho\eta_\rho - 2\hat{\mu}^{K+U}\varepsilon^{K+U} - 2\hat{\mu}^R\varepsilon^R - 2\hat{\mu}^V\varepsilon^V, \\ \hat{\mu}^{K+U} &= \frac{1}{2}\eta_{\varepsilon^{K+U}}, \quad \hat{\mu}^R = \frac{1}{2}\eta_{\varepsilon^R}, \quad \hat{\mu}^V = \frac{1}{2}\eta_{\varepsilon^V}, \end{aligned} \tag{25.5}$$

where a subscript attached to η stands for a partial derivative with respect to the corresponding subscript. In what follows we will use a similar notational convention.

With (25.4), we obtain the relation between η and $\mathscr{P}$ as follows:

$$\eta_\rho + \frac{\mathscr{P}}{\rho^2}\eta_{\varepsilon^{K+U}} = 0. \tag{25.6}$$

From (25.6), changing the variables of $\mathscr{P}$ from $\{\rho, \varepsilon^{K+U}, \varepsilon^R, \varepsilon^V\}$ to $\{\rho, \eta, \varepsilon^R, \varepsilon^V\}$, we have the expression of the total pressure as follows:

$$\mathscr{P} = \rho^2\left(\frac{\partial\varepsilon^{K+U}}{\partial\rho}\right)_{\rho,\eta,\varepsilon^R,\varepsilon^V}, \tag{25.7}$$

which has formally the same form as in equilibrium thermodynamics [441].

25.2.4 Nonequilibrium Temperatures in Terms of the Main Field

In RET, the nonequilibrium temperature and the chemical potential are introduced through the components of the main field that appear in the generalized Gibbs equation (see Chap. 15). To find these nonequilibrium variables in the present case, we adopt ε as an independent field instead of ε^{K+U} and let $(\lambda', \lambda_i', \mu', \mu^{R'}, \mu^{V'})$ be the component of the main field corresponding to the independent fields $(\rho, \rho v_i, 2\rho\varepsilon +$

$\rho v^2, 2\rho\varepsilon^R, 2\rho\varepsilon^V$). In equilibrium, the generalized Gibbs equation (2.57)$_1$ reduces to the Gibbs equation. Then we have the relations:

$$\hat{\lambda}'_E = -\frac{g}{T}, \quad \hat{\mu}'_E = \frac{1}{2T}, \quad \hat{\mu}^{R\prime}_E = 0, \quad \hat{\mu}^{V\prime}_E = 0,$$

where a quantity with the suffix E is the one evaluated at equilibrium, and g ($= \varepsilon + p/\rho - Ts$) is the chemical potential. The nonequilibrium temperature $\mathcal{T}$ and the chemical potential $\mathcal{G}$ are introduced in terms of $\hat{\mu}'$ and $\hat{\lambda}'$ as follows:

$$\mathcal{T} = \frac{1}{2\hat{\mu}'}, \quad \mathcal{G} = -\frac{\hat{\lambda}'}{2\hat{\mu}'}. \tag{25.8}$$

The relationship between $(\lambda, \lambda_i, \mu^{K+U}, \mu^R, \mu^V)$ and $(\lambda', \lambda'_i, \mu', \mu^{R\prime}, \mu^{V\prime})$ is determined from (2.57)$_1$ as follows:

$$\begin{aligned} dh &= \hat{\lambda}d\rho + \hat{\mu}^{K+U}d\left(2\rho\varepsilon^{K+U}\right) + \hat{\mu}^R d\left(2\rho\varepsilon^R\right) + \hat{\mu}^V d\left(2\rho\varepsilon^V\right) \\ &= \hat{\lambda}'d\rho + \hat{\mu}'d\left(2\rho\varepsilon\right) + \hat{\mu}^{R\prime}d\left(2\rho\varepsilon^R\right) + \hat{\mu}^{V\prime}d\left(2\rho\varepsilon^V\right) \end{aligned}$$

with $\hat{\lambda}_i = \hat{\lambda}'_i = 0$. Then we obtain

$$\hat{\lambda} = \hat{\lambda}', \quad \hat{\mu}^{K+U} = \hat{\mu}', \quad \hat{\mu}^R = \hat{\mu}' + \hat{\mu}^{R\prime}, \quad \hat{\mu}^V = \hat{\mu}' + \hat{\mu}^{V\prime}.$$

Therefore, in the case of the independent fields $(\rho, \rho v_i, 2\rho\varepsilon^{K+U} + \rho v^2, 2\rho\varepsilon^R, 2\rho\varepsilon^V)$, the components of the main field in equilibrium are given by

$$\hat{\lambda}_E = -\frac{g}{T}, \quad \hat{\mu}_E^{K+U} = \frac{1}{2T}, \quad \hat{\mu}_E^R = \frac{1}{2T}, \quad \hat{\mu}_E^V = \frac{1}{2T}.$$

This suggests that, instead of (25.8), we may define three nonequilibrium temperatures and a chemical potential as follows:

$$\mathcal{T}^{K+U} = \frac{1}{2\hat{\mu}^{K+U}}, \quad \mathcal{T}^R = \frac{1}{2\hat{\mu}^R}, \quad \mathcal{T}^V = \frac{1}{2\hat{\mu}^V}, \quad \mathcal{G} = -\frac{\hat{\lambda}}{2\hat{\mu}^{K+U}}, \tag{25.9}$$

where $\mathcal{T}^{K+U} = \mathcal{T}$. From (25.5), we obtain

$$\frac{1}{\mathcal{T}^{K+U}} = \frac{\partial\eta}{\partial\varepsilon^{K+U}}, \quad \frac{1}{\mathcal{T}^R} = \frac{\partial\eta}{\partial\varepsilon^R}, \quad \frac{1}{\mathcal{T}^V} = \frac{\partial\eta}{\partial\varepsilon^V}. \tag{25.10}$$

With (25.6) and (25.10), we have

$$d\eta = -\frac{\mathcal{P}}{\rho^2\mathcal{T}^{K+U}}d\rho + \frac{1}{\mathcal{T}^{K+U}}d\varepsilon^{K+U} + \frac{1}{\mathcal{T}^R}d\varepsilon^R + \frac{1}{\mathcal{T}^V}d\varepsilon^V. \tag{25.11}$$

25.2.5 *Assumption of the Entropy Density*

In equilibrium, the energies ε^{K+U}, ε^{R}, and ε^{V} are represented by the caloric equations of state in the following form:

$$\varepsilon^{K+U}=\varepsilon_E^{K+U}(\rho,T),\quad \varepsilon^{R}=\varepsilon_E^{R}(T),\quad \varepsilon^{V}=\varepsilon_E^{V}(T).$$

In nonequilibrium, we may introduce parameters with the dimension of temperature, θ^{K+U}, θ^{R}, and θ^{V} through the caloric equations of state as follows:

$$\varepsilon^{K+U}=\varepsilon_E^{K+U}\left(\rho,\theta^{K+U}\right),\quad \varepsilon^{R}=\varepsilon_E^{R}\left(\theta^{R}\right),\quad \varepsilon^{V}=\varepsilon_E^{V}\left(\theta^{V}\right).$$

The temperature T of the whole system is evaluated by these parameters since the total specific internal energy ε $(=\varepsilon_E(\rho,T))$ is the sum of the energies of the three modes (25.2), i.e.,

$$\varepsilon_E(\rho,T)=\varepsilon_E^{K+U}\left(\rho,\theta^{K+U}\right)+\varepsilon_E^{R}\left(\theta^{R}\right)+\varepsilon_E^{V}\left(\theta^{V}\right).$$

According with the duality principle (1.36), we assume, from the expression of the nonequilibrium entropy density for a *rarefied gas* explained in Chap. 14, that the nonequilibrium specific entropy density η is the sum of the specific entropies of the three modes as follows:

$$\eta=s^{K+U}\left(\rho,\theta^{K+U}\right)+s^{R}\left(\theta^{R}\right)+s^{V}\left(\theta^{V}\right),\tag{25.12}$$

where the corresponding Gibbs relations of $K+U$, R, and V-parts hold individually as follows:

$$\begin{aligned}\theta^{K+U}ds^{K+U}&=d\varepsilon^{K+U}-\frac{p^{K+U}}{\rho^2}d\rho,\\ \theta^{R}ds^{R}&=d\varepsilon^{R},\\ \theta^{V}ds^{V}&=d\varepsilon^{V}.\end{aligned}\tag{25.13}$$

From (25.11), it is easy to verify that the three parameters θ^{K+U}, θ^{R}, and θ^{V} coincide with the nonequilibrium temperatures defined above, that is,

$$\mathcal{T}^{K+U}=\theta^{K+U},\quad \mathcal{T}^{R}=\theta^{R},\quad \mathcal{T}^{V}=\theta^{V}.$$

From (25.7), the entropy density (25.12) also provides the expression of the total pressure:

$$\mathcal{P}=p\left(\rho,\theta^{K+U}\right).\tag{25.14}$$

Using (25.5), we can rewrite the components of the main field as follows:

$$\hat{\lambda} = -\frac{\mathscr{G}^{K+U}\left(\rho, \theta^{K+U}\right)}{\theta^{K+U}} - \frac{\mathscr{G}^{R}\left(\theta^{R}\right)}{\theta^{R}} - \frac{\mathscr{G}^{V}\left(\theta^{V}\right)}{\theta^{V}},$$

$$\hat{\mu}^{K+U} = \frac{1}{2\theta^{K+U}}, \quad \hat{\mu}^{R} = \frac{1}{2\theta^{R}}, \quad \hat{\mu}^{V} = \frac{1}{2\theta^{V}},$$

where we have defined nonequilibrium chemical potentials of $K + U$, R, and V-parts as follows:

$$\mathscr{G}^{K+U}\left(\rho, \theta^{K+U}\right) = \varepsilon_E^{K+U}\left(\rho, \theta^{K+U}\right) + \frac{p\left(\rho, \theta^{K+U}\right)}{\rho} - \theta^{K+U} s^{K+U}\left(\rho, \theta^{K+U}\right),$$

$$\mathscr{G}^{R}\left(\theta^{R}\right) = \varepsilon_E^{R}\left(\theta^{R}\right) - \theta^{R} s^{R}\left(\theta^{R}\right), \qquad \mathscr{G}^{V}\left(\theta^{V}\right) = \varepsilon_E^{V}\left(\theta^{V}\right) - \theta^{V} s^{V}\left(\theta^{V}\right).$$

From (25.9), the nonequilibrium chemical potential is expressed as

$$\mathscr{G} = \mathscr{G}^{K+U}\left(\rho, \theta^{K+U}\right) + \frac{\theta^{K+U}}{\theta^{R}}\mathscr{G}^{R}\left(\theta^{R}\right) + \frac{\theta^{K+U}}{\theta^{V}}\mathscr{G}^{V}\left(\theta^{V}\right).$$

25.2.5.1 System of Field Equations in Terms of $\{\rho, v_i, \theta^{K+U}, \theta^R, \theta^V\}$

By using the material derivative denoted by a dot on a quantity, the system (25.3) is rewritten in terms of the independent fields $\{\rho, v_i, \theta^{K+U}, \theta^R, \theta^V\}$ as follows:

$$\begin{aligned}
&\dot{\rho} + \rho\frac{\partial v_i}{\partial x_i} = 0,\\
&\rho\dot{v}_i + \frac{\partial p\left(\rho, \theta^{K+U}\right)}{\partial x_i} = 0,\\
&c_v^{K+U}\left(\rho, \theta^{K+U}\right)\dot{\theta}^{K+U} + \frac{1}{\rho}\left\{p\left(\rho, \theta^{K+U}\right) - \rho^2\frac{\partial \varepsilon_E^{K+U}(\rho, \theta^{K+U})}{\partial \rho}\right\}\frac{\partial v_k}{\partial x_k}\\
&\quad = \frac{P_{ll}^{K+U}}{2\rho},\\
&c_v^{R}\left(\theta^{R}\right)\dot{\theta}^{R} = \frac{P_{ll}^{R}}{2\rho},\\
&c_v^{V}\left(\theta^{V}\right)\dot{\theta}^{V} = \frac{P_{ll}^{V}}{2\rho},
\end{aligned} \tag{25.15}$$

where

$$c_v^{K+U}\left(\rho, \theta^{K+U}\right) = \frac{\partial \varepsilon^{K+U}}{\partial \theta^{K+U}}, \quad c_v^R\left(\theta^R\right) = \frac{\partial \varepsilon^R}{\partial \theta^R}, \quad c_v^V\left(\theta^V\right) = \frac{\partial \varepsilon^V}{\partial \theta^V}$$

are the specific heats of the three modes evaluated at the nonequilibrium temperatures. It is explicitly shown that the above system indeed satisfies the duality principle (1.36).

25.2.6 *Conditions on the Entropy Density, Flux, and Production*

From $(2.57)_1$, the stability condition (2.10) corresponds to the following negative definite form:

$$Q = \delta \mathbf{u}' \cdot \delta \mathbf{F} < 0. \tag{25.16}$$

Taking into account $(2.49)_1$ and (2.55), we have (see for details [167]):

$$Q = \bar{Q} - 2\hat{\mathbf{u}}'\mathbf{A}^r \delta\hat{\mathbf{F}} \delta v_r - g^{rs} \delta v_r \delta v_s, \tag{25.17}$$

where

$$\bar{Q} = \delta\hat{\mathbf{u}}' \cdot \delta\hat{\mathbf{F}}, \qquad g^{rs} = \hat{\mathbf{u}}'\mathbf{A}^r\mathbf{A}^s\hat{\mathbf{F}}.$$

In the present case, the second term in the right hand side of (25.17) vanishes and $g^{rs} = (\rho/\theta^{K+U})\delta_{rs}$. Adopting $\{\rho, v_i, \theta^{K+U}, \theta^R, \theta^V\}$ as independent fields, we have

$$Q = \bar{Q} - \frac{\rho}{\theta^{K+U}} ||\delta \mathbf{v}||^2$$

with

$$\begin{aligned}
\bar{Q} = &- \frac{\mathscr{P}_\rho}{\rho \theta^{K+U}} (\delta\rho)^2 \\
&- \rho \frac{c_v^{K+U}(\rho, \theta^{K+U})}{\theta^{K+U^2}} \left(\delta\theta^{K+U}\right)^2 - \rho \frac{c_v^R(\theta^R)}{\theta^{R^2}} \left(\delta\theta^R\right)^2 \\
&- \rho \frac{c_v^V(\theta^V)}{\theta^{V^2}} \left(\delta\theta^V\right)^2 < 0.
\end{aligned}$$

Then the state $(\rho, \theta^{K+U}, \theta^R, \theta^V)$ is stable if and only if the following inequalities are satisfied:

$$\mathscr{P}_\rho > 0, \quad c_v^{K+U}\left(\rho, \theta^{K+U}\right) > 0, \quad c_v^R\left(\theta^R\right) > 0, \quad c_v^V\left(\theta^V\right) > 0, \tag{25.18}$$

Therefore, under these conditions, the system of the present theory can be put in a symmetric form by choosing the main field as independent fields.

In particular, an equilibrium state is stable if

$$\left(\frac{\partial p(\rho, T)}{\partial \rho}\right)_T > 0, \quad c_v^{K+U}(\rho, T) > 0, \quad c_v^R(T) > 0, \quad c_v^V(T) > 0.$$

The expressions of the entropy flux and the production are determined by the constitutive equations. The entropy flux is given by the representation theorem as follows:

$$\hat{h}^i = 0 \quad \rightarrow \quad h^i = hv^i. \tag{25.19}$$

Then the equation (2.57)$_2$ is automatically satisfied. The entropy production is deduced from (2.57)$_3$ as follows:

$$\begin{aligned} \Sigma &= \hat{\mu}^{K+U}\hat{P}^{K+U} + \hat{\mu}^R\hat{P}^R + \hat{\mu}^V\hat{P}^V \\ &= \frac{\hat{P}_{ll}^{K+U}}{2\theta^{K+U}} + \frac{\hat{P}_{ll}^R}{2\theta^R} + \frac{\hat{P}_{ll}^V}{2\theta^V}. \end{aligned} \tag{25.20}$$

In Sect. 25.3 we will determine the explicit expressions of the production terms that satisfy the entropy inequality (25.20).

25.2.7 *Characteristic Velocity, Subcharacteristic Condition, and Local Exceptionality*

The characteristic velocity V is obtained in a similar way as in Sect. 24.4.5. We have the following results in terms of the independent variables $\{\rho, \eta, \theta^R, \theta^V\}$:

$$\textbf{Contact Waves}: \quad V = v_n = 0 \quad \text{(multiplicity 5)}, \tag{25.21}$$

$$\textbf{Sound Waves}: \quad V = v_n \pm \sqrt{\left(\frac{\partial p\left(\rho, \theta^{K+U}\left(\rho, \eta, \theta^R, \theta^V\right)\right)}{\partial \rho}\right)_{\eta, \theta^R, \theta^V}} \tag{25.22}$$

(each of multiplicity 1),

where $v_n = v_j n_j$ with n_j being the j-component of the unit normal to the wave front. With the fields $\{\rho, \theta^{K+U}, \theta^R, \theta^V\}$, we can rewrite the velocity of the sound wave $U = V - v_n$ as follows:

$$U^2 = p_\rho\left(\rho, \theta^{K+U}\right) + \frac{\theta^{K+U} p^2_{\theta^{K+U}}\left(\rho, \theta^{K+U}\right)}{\rho^2 c_v^{K+U}\left(\rho, \theta^{K+U}\right)}. \tag{25.23}$$

In an equilibrium case, we have the expression (24.27) for U_E. While, for Eulerian fluids, we have the expression (14.19) for U_{Euler}. Since the system of ET_7^D includes the Euler system as a principal subsystem, we have the *subcharacteristic condition* (see Theorem 2.3): $U_E > U_{Euler}$.

The contact waves (25.21) are exceptional while the sound waves (25.22) can be locally exceptional. The hyper-surface of the local exceptionality exists when the following relation is satisfied:

$$\delta V = \frac{1}{2\rho^2 U}\left(\frac{\partial \rho^2 U^2}{\partial \rho}\right)_{\eta, \theta^R, \theta^V} = 0.$$

25.2.8 Rarefied-Gas Limit

In the rarefied-gas limit $\rho \ll \rho_{cr}$ with ρ_{cr} being the mass density at the critical point, which implies $\varepsilon^U \to 0$, we obtain ET_7 for rarefied polyatomic gases explained in Chap. 14. In this sense, as a matter of course, the present theory ET_7^D includes the previous RET theory of rarefied gases, ET_7, as a special case.

25.3 Energy Exchange Processes and Production Terms

In this section, we propose explicit expressions of the production terms.

25.3.1 Production Terms with Relaxation Processes

As is evident from (25.3), the system expresses the relaxation of energy exchange among $K+U$, R, and V-modes. In general, this energy-exchange process depends on a gas, and is complicated. For rarefied gases, however, as the simplest model of the process, the generalized BGK model has been proposed in Chap. 14. It is natural and is also suitable to the spirit of the duality principle to extend this model to the case of dense gases. We therefore propose, from a phenomenological point of view, a possible model for describing the relaxation process with three stages (i)–(iii) as follows:

(i) The translational and potential mode relaxes most rapidly to its partial equilibrium state characterized by a temperature θ^{K+U}. Since this relaxation process affects only on the production terms of the shear stress and total heat flux, it plays no role explicitly in the present model.

(ii) After stage (i), there are three possibilities, of which a gas takes one:

- (K+U+R)-process: The $(K + U)$ and R modes relax with relaxation time τ_{KUR} to their partial equilibrium state where these modes can be treated as a unit. We characterize the state by a temperature θ^{K+U+R} which is defined through the following relation:

$$\varepsilon_E^{K+U}\left(\rho,\theta^{K+U}\right)+\varepsilon_E^{R}\left(\theta^{R}\right)=\varepsilon_E^{K+U}\left(\rho,\theta^{K+U+R}\right)+\varepsilon_E^{R}\left(\theta^{K+U+R}\right). \tag{25.24}$$

- (K+U+V)-process: The $(K + U)$ and V modes relax with relaxation time τ_{KUV} to their partial equilibrium state that is characterized by a temperature θ^{K+U+V} defined by

$$\varepsilon_E^{K+U}\left(\rho,\theta^{K+U}\right)+\varepsilon_E^{V}\left(\theta^{V}\right)=\varepsilon_E^{K+U}\left(\rho,\theta^{K+U+V}\right)+\varepsilon_E^{V}\left(\theta^{K+U+V}\right).$$

- (R+V)-process: The R and V modes relax with relaxation time τ_{RV} to their partial equilibrium state with a temperature θ^{R+V}:

$$\varepsilon_E^{R}\left(\theta^{R}\right)+\varepsilon_E^{V}\left(\theta^{V}\right)=\varepsilon_E^{R}\left(\theta^{R+V}\right)+\varepsilon_E^{V}\left(\theta^{R+V}\right).$$

(iii) All modes relax with a relaxation time τ to a state in which all modes are characterized by only one temperature T:

$$\varepsilon_E^{K+U}\left(\rho,\theta^{K+U}\right)+\varepsilon_E^{R}\left(\theta^{R}\right)+\varepsilon_E^{V}\left(\theta^{V}\right)=\varepsilon_E^{K+U}\left(\rho,T\right)+\varepsilon_E^{R}\left(T\right)+\varepsilon_E^{V}\left(T\right). \tag{25.25}$$

In each stage of the relaxation process, we assume the production terms in expansion forms around a corresponding equilibrium state up to the first order, where the expansion coefficients are the relaxation times. Depending on the three possibilities in the stage (ii), we have the following production terms:

- For (K+U+R)-process:

$$\begin{aligned}\hat{P}_{ll}^{K+U}=&-\frac{2\rho}{\tau_{KUR}}\left(\varepsilon_E^{K+U}\left(\rho,\theta^{K+U}\right)-\varepsilon_E^{K+U}\left(\rho,\theta^{K+U+R}\right)\right)\\&-\frac{2\rho}{\tau}\left(\varepsilon_E^{K+U}\left(\rho,\theta^{K+U}\right)-\varepsilon_E^{K+U}\left(\rho,T\right)\right),\end{aligned}$$

$$\hat{P}_{ll}^{R} = -\frac{2\rho}{\tau_{KUR}}\left(\varepsilon_E^R\left(\theta^R\right) - \varepsilon_E^R\left(\theta^{K+U+R}\right)\right) - \frac{2\rho}{\tau}\left(\varepsilon_E^R\left(\theta^R\right) - \varepsilon_E^R(T)\right),$$

$$\hat{P}_{ll}^{V} = -\frac{2\rho}{\tau}\left(\varepsilon_E^V\left(\theta^V\right) - \varepsilon_E^V(T)\right).$$

- For (K+U+V)-process:

$$\begin{aligned}\hat{P}_{ll}^{K+U} = &-\frac{2\rho}{\tau_{KUV}}\left(\varepsilon_E^{K+U}\left(\rho, \theta^{K+U}\right) - \varepsilon_E^{K+U}\left(\rho, \theta^{K+U+V}\right)\right)\\ &- \frac{2\rho}{\tau}\left(\varepsilon_E^{K+U}\left(\rho, \theta^{K+U}\right) - \varepsilon_E^{K+U}(\rho, T)\right),\end{aligned}$$

$$\hat{P}_{ll}^{R} = -\frac{2\rho}{\tau}\left(\varepsilon_E^R\left(\theta^R\right) - \varepsilon_E^R(T)\right),$$

$$\hat{P}_{ll}^{V} = -\frac{2\rho}{\tau_{KUV}}\left(\varepsilon_E^V\left(\theta^V\right) - \varepsilon_E^V\left(\theta^{K+U+V}\right)\right) - \frac{2\rho}{\tau}\left(\varepsilon_E^V\left(\theta^V\right) - \varepsilon_E^V(T)\right).$$

- For (R+V)-process:

$$\hat{P}_{ll}^{K+U} = -\frac{2\rho}{\tau}\left(\varepsilon_E^{K+U}\left(\rho, \theta^{K+U}\right) - \varepsilon_E^{K+U}(\rho, T)\right),$$

$$\hat{P}_{ll}^{R} = -\frac{2\rho}{\tau_{RV}}\left(\varepsilon_E^R\left(\theta^R\right) - \varepsilon_E^R\left(\theta^{R+V}\right)\right) - \frac{2\rho}{\tau}\left(\varepsilon_E^R\left(\theta^R\right) - \varepsilon_E^R(T)\right),$$

$$\hat{P}_{ll}^{V} = \frac{2\rho}{\tau_{RV}}\left(\varepsilon_E^V\left(\theta^V\right) - \varepsilon^V\left(\theta_E^{R+V}\right)\right) - \frac{2\rho}{\tau}\left(\varepsilon_E^V\left(\theta^V\right) - \varepsilon_E^V(T)\right).$$

With these production terms, we obtain the entropy production (25.20) for $(\mathfrak{b}+\mathfrak{c})$-process $((\mathfrak{a}, \mathfrak{b}, \mathfrak{c})$ =(V,K+U,R), (R,K+U,V), (K+U,R,V)) expressed as follows:

$$\begin{aligned}\Sigma = &-\frac{\rho}{\tau_{\mathfrak{bc}}}\Bigg\{\left(\varepsilon_E^{\mathfrak{b}}\left(\rho, \theta^{\mathfrak{b}}\right) - \varepsilon_E^{\mathfrak{b}}\left(\rho, \theta^{\mathfrak{b}+\mathfrak{c}}\right)\right)\left(\frac{1}{\theta^{\mathfrak{b}}} - \frac{1}{\theta^{\mathfrak{b}+\mathfrak{c}}}\right)\\ &\qquad + \left(\varepsilon_E^{\mathfrak{c}}\left(\rho, \theta^{\mathfrak{c}}\right) - \varepsilon_E^{\mathfrak{c}}\left(\rho, \theta^{\mathfrak{b}+\mathfrak{c}}\right)\right)\left(\frac{1}{\theta^{\mathfrak{c}}} - \frac{1}{\theta^{\mathfrak{b}+\mathfrak{c}}}\right)\Bigg\}\\ &-\frac{\rho}{\tau}\Bigg\{\left(\varepsilon_E^{\mathfrak{b}}\left(\rho, \theta^{\mathfrak{b}}\right) - \varepsilon_E^{\mathfrak{b}}(\rho, T)\right)\left(\frac{1}{\theta^{\mathfrak{b}}} - \frac{1}{T}\right)\\ &\qquad + \left(\varepsilon_E^{\mathfrak{c}}\left(\rho, \theta^{\mathfrak{c}}\right) - \varepsilon_E^{\mathfrak{c}}(\rho, T)\right)\left(\frac{1}{\theta^{\mathfrak{c}}} - \frac{1}{T}\right)\\ &\qquad + \left(\varepsilon_E^{\mathfrak{a}}\left(\rho, \theta^{\mathfrak{a}}\right) - \varepsilon_E^{\mathfrak{a}}(\rho, T)\right)\left(\frac{1}{\theta^{\mathfrak{a}}} - \frac{1}{T}\right)\Bigg\}.\end{aligned}$$

Recalling the monotonicity of $\varepsilon^{\mathfrak{a},\mathfrak{b},\mathfrak{c}}$ with respect to the temperature, i.e., the positivity of the specific heats, the non-negativity of the entropy production

$$\Sigma \geqslant 0$$

is certainly fulfilled. From (25.24) and (25.25), the monotonicity also gives the relation among the temperatures as follows:

$$\begin{aligned}
&\theta^{\mathfrak{b}} < \theta^{\mathfrak{b}+\mathfrak{c}} < \theta^{\mathfrak{c}} \quad \text{or} \quad \theta^{\mathfrak{c}} < \theta^{\mathfrak{b}+\mathfrak{c}} < \theta^{\mathfrak{b}}, \\
&\theta^{\mathfrak{a}} < T < \theta^{\mathfrak{b}+\mathfrak{c}} \quad \text{or} \quad \theta^{\mathfrak{b}+\mathfrak{c}} < T < \theta^{\mathfrak{a}}.
\end{aligned}$$

25.3.2 Linearized Constitutive Equations and Maxwellian Iteration

As explained above, in the present model, there are rapid and slow relaxation processes. In the $(\mathfrak{b}+\mathfrak{c})$-process $((\mathfrak{a}, \mathfrak{b}, \mathfrak{c})$ = (V,K+U,R), (R,K+U,V), (K+U,R,V)), these processes may be characterized by the following quantities:

$$\begin{aligned}
\delta &\equiv \varepsilon_E^{\mathfrak{b}}\left(\rho, \theta^{\mathfrak{b}}\right) - \varepsilon_E^{\mathfrak{b}}\left(\rho, \theta^{\mathfrak{b}+\mathfrak{c}}\right) = -\varepsilon_E^{\mathfrak{c}}\left(\rho, \theta^{\mathfrak{c}}\right) + \varepsilon_E^{\mathfrak{c}}\left(\rho, \theta^{\mathfrak{b}+\mathfrak{c}}\right), \\
\Delta &\equiv \varepsilon_E^{\mathfrak{a}}\left(\rho, \theta^{\mathfrak{a}}\right) - \varepsilon_E^{\mathfrak{a}}\left(\rho, T\right) = -\varepsilon_E^{\mathfrak{b}+\mathfrak{c}}\left(\rho, \theta^{\mathfrak{b}+\mathfrak{c}}\right) + \varepsilon_E^{\mathfrak{b}+\mathfrak{c}}\left(\rho, T\right).
\end{aligned}$$

Note that, for the (R+V)-process, there is the minus-sign difference between the definition of Δ here and the previous one given by (24.7).

By expanding the nonequilibrium energies of the three modes with respect to the nonequilibrium temperatures around a local equilibrium temperature T up to the first order, we obtain

$$\begin{aligned}
\delta &= c_v^{\mathfrak{b}}\left(\theta^{\mathfrak{b}} - \theta^{\mathfrak{b}+\mathfrak{c}}\right) = -c_v^{\mathfrak{c}}\left(\theta^{\mathfrak{c}} - \theta^{\mathfrak{b}+\mathfrak{c}}\right), \\
\Delta &= c_v^{\mathfrak{a}}\left(\theta^{\mathfrak{a}} - T\right) = -c_v^{\mathfrak{b}+\mathfrak{c}}\left(\theta^{\mathfrak{b}+\mathfrak{c}} - T\right),
\end{aligned}$$

where $c_v^{\mathfrak{b}+\mathfrak{c}} = c_v^{\mathfrak{b}} + c_v^{\mathfrak{c}}$. Here and hereafter we use the notation $c_v^{\mathfrak{a}}$ instead of $c_v^{\mathfrak{a}}(\rho, T)$ and so on for simplicity. Inversely, the nonequilibrium temperatures are expressed as follows:

$$\begin{aligned}
&\theta^{\mathfrak{a}} - T = \frac{\Delta}{c_v^{\mathfrak{a}}}, \quad \theta^{\mathfrak{b}+\mathfrak{c}} - T = -\frac{\Delta}{c_v^{\mathfrak{b}+\mathfrak{c}}}, \\
&\theta^{\mathfrak{b}} - T = \frac{\delta}{c_v^{\mathfrak{b}}} - \frac{\Delta}{c_v^{\mathfrak{b}+\mathfrak{c}}}, \quad \theta^{\mathfrak{c}} - T = -\frac{\delta}{c_v^{\mathfrak{c}}} - \frac{\Delta}{c_v^{\mathfrak{b}+\mathfrak{c}}}.
\end{aligned}$$

Then the dynamic pressure near an equilibrium state:

$$\Pi = \mathscr{P} - p = p_T \left(\theta^{K+U} - T\right) \tag{25.26}$$

is expressed in terms of δ and Δ. We summarize the expression of the dynamic pressure in Table 25.1.

The production terms are now given by

$$P_{ll}^{\mathfrak{a}} = -2\rho\frac{\Delta}{\tau},$$

$$P_{ll}^{\mathfrak{b}} = -2\rho\frac{\delta}{\tau_\delta} + 2\rho\frac{c_v^{\mathfrak{b}}}{c_v^{\mathfrak{b}+\mathfrak{c}}}\frac{\Delta}{\tau},$$

$$P_{ll}^{\mathfrak{c}} = 2\rho\frac{\delta}{\tau_\delta} + 2\rho\frac{c_v^{\mathfrak{c}}}{c_v^{\mathfrak{b}+\mathfrak{c}}}\frac{\Delta}{\tau},$$

where τ_δ is defined as

$$\frac{1}{\tau_\delta} \equiv \frac{1}{\tau_{\mathfrak{bc}}} + \frac{1}{\tau}.$$

Then the entropy production is given by

$$\Sigma = \frac{\rho}{T^2}\frac{c_v^{\mathfrak{b}+\mathfrak{c}}}{c_v^{\mathfrak{b}}c_v^{\mathfrak{c}}}\frac{1}{\tau_\delta}\delta^2 + \frac{\rho}{T^2}\frac{c_v}{c_v^{\mathfrak{a}}c_v^{\mathfrak{b}+\mathfrak{c}}}\frac{1}{\tau}\Delta^2.$$

Since $c_v^{K+U} > 0,\ c_v^R > 0,\ c_v^V > 0$ and $\tau_\delta > 0,\ \tau > 0$, the entropy production is non-negative.

The system of field equations (25.15) is rewritten in terms of $\{\rho, v_i, T, \Delta, \delta\}$ as follows:

$$\begin{aligned}
&\dot{\rho} + \rho\frac{\partial v_i}{\partial x_i} = 0,\\
&\rho\dot{v}_i + \frac{\partial}{\partial x_i}(p + \Pi) = 0,\\
&\rho c_v\dot{T} + (Tp_T + \Pi)\frac{\partial v_i}{\partial x_i} = 0,\\
&\dot{\Delta} + \frac{Tp_T + \Pi}{\rho}\frac{A_2 - c_v^{\mathfrak{a}}}{c_v}\frac{\partial v_i}{\partial x_i} = -\frac{\Delta}{\tau},\\
&\dot{\delta} + \frac{Tp_T + \Pi}{\rho}\left\{A_1 + \frac{1}{c_v}\frac{\mathrm{d}}{\mathrm{d}T}\left(\frac{c_v^{\mathfrak{b}}}{c_v^{\mathfrak{b}+\mathfrak{c}}}\right)\Delta\right\}\frac{\partial v_i}{\partial x_i} = -\frac{\delta}{\tau_\delta},
\end{aligned} \tag{25.27}$$

where $p = p(\rho, T)$, and A_1, A_2, and Π are also given in Table 25.1.

Table 25.1 Explicit expressions of A_1, A_2 and Π

$(\mathfrak{b}, \mathfrak{c})$	A_1	A_2	Π
$(K+U, R)$ or $(K+U, V)$	$\frac{c_v^{\mathfrak{c}}}{c_v^{\mathfrak{b}+\mathfrak{c}}}$	0	$\frac{p_T}{c_v^{\mathfrak{b}}}\delta - \frac{p_T}{c_v^{\mathfrak{b}+\mathfrak{c}}}\Delta$
(R, V)	0	c_v	$\frac{p_T}{c_v^{K+U}}\Delta$

When we apply the Maxwellian iteration to (25.27)$_{4,5}$ and retain the first order terms with respect to the relaxation times τ_δ and τ, we obtain the following approximations:

$$\delta = -\tau_\delta \frac{T p_T}{\rho} A_1 \frac{\partial v_i}{\partial x_i}, \qquad \Delta = -\tau \frac{T p_T}{\rho} \frac{A_2 - c_v^{\mathfrak{a}}}{c_v} \frac{\partial v_i}{\partial x_i}.$$

For $(\mathfrak{b}+\mathfrak{c})$-process ($(\mathfrak{a}, \mathfrak{b}, \mathfrak{c})$ = (V,K+U,R) or (R,K+U,V)), from (25.26), we have

$$\Pi = \Pi^{\mathfrak{b}+\mathfrak{c}} + \Pi^{\mathfrak{a}} \qquad \text{with}$$

$$\Pi^{\mathfrak{b}+\mathfrak{c}} = p\left(\rho, \theta^{K+U}\right) - p\left(\rho, \theta^{\mathfrak{b}+\mathfrak{c}}\right) = \frac{p_T}{c_v^{K+U}}\delta = -\tau_\delta \frac{T p_T^2}{\rho} \frac{c_v^{\mathfrak{c}}}{c_v^{\mathfrak{b}} c_v^{\mathfrak{b}+\mathfrak{c}}} \frac{\partial v_i}{\partial x_i},$$

$$\Pi^{\mathfrak{a}} = p\left(\rho, \theta^{\mathfrak{b}+\mathfrak{c}}\right) - p(\rho, T) = -\frac{p_T}{c_v^{\mathfrak{b}+\mathfrak{c}}}\Delta = -\tau \frac{T p_T^2}{\rho} \frac{c_v^{\mathfrak{a}}}{c_v^{\mathfrak{b}+\mathfrak{c}} c_v} \frac{\partial v_i}{\partial x_i}.$$

Therefore we have the expression of the bulk viscosity ν as follows:

$$\nu = \tau_\delta \frac{T p_T^2}{\rho} \frac{c_v^{\mathfrak{c}}}{c_v^{\mathfrak{b}} c_v^{\mathfrak{b}+\mathfrak{c}}} + \tau \frac{T p_T^2}{\rho} \frac{c_v^{\mathfrak{a}}}{c_v^{\mathfrak{b}+\mathfrak{c}} c_v}. \tag{25.28}$$

For (R+V)-process, we have

$$\delta = 0, \qquad \Pi = -\tau \frac{T p_T^2}{\rho} \frac{c_v^{R+V}}{c_v^{K+U} c_v} \frac{\partial v_i}{\partial x_i},$$

and the bulk viscosity is evaluated as (see (24.35))

$$\nu = \tau \frac{T p_T^2}{\rho} \frac{c_v^{R+V}}{c_v^{K+U} c_v}. \tag{25.29}$$

25.4 Coarse Graining of Rapid Relaxation Process, and ET_6^D Theories as Principal Subsystems of ET_7^D Theory

Let us study the case where the relaxation time τ is of several orders larger than the relaxation time $\tau_{\mathfrak{bc}}$ (($\mathfrak{a}$, $\mathfrak{b}$, $\mathfrak{c}$) = (V,K+U,R), (R,K+U,V), (K+U,R,V)). In such a case, the composite system of $\mathfrak{b}$-mode and $\mathfrak{c}$-mode quickly reaches a state with the common temperature $\theta^{\mathfrak{b}+\mathfrak{c}}$. If fact, in the limit $\tau_\delta \to 0$, the equation of δ disappears and we obtain a six field theory, namely, ET_6^D. Therefore, except for the short period of $O(\tau_{\mathfrak{bc}})$ after the initial time, we have the relation:

$$\theta^{\mathfrak{b}} = \theta^{\mathfrak{c}} = \theta^{\mathfrak{b}+\mathfrak{c}}.$$

The above argument can be rigorously formulated by using the idea of the principal subsystem explained in Sect. 2.4. Let us adopt $\{\rho,\ \rho v_i,\ 2\rho\varepsilon_E^{\mathfrak{a}}(\rho,\theta^{\mathfrak{a}}),\ 2\rho\{\varepsilon_E^{\mathfrak{b}}(\rho,\theta^{\mathfrak{b}}) + \varepsilon_E^{\mathfrak{c}}(\rho,\theta^{\mathfrak{c}})\},\ 2\rho\{\varepsilon_E^{\mathfrak{b}}(\rho,\theta^{\mathfrak{b}}) - \varepsilon_E^{\mathfrak{c}}(\rho,\theta^{\mathfrak{c}})\}\}$ as independent fields, then the velocity-independent part of the corresponding main field $\{\hat{\lambda}'', \hat{\mu}''^{\mathfrak{a}}, \hat{\mu}''^{\mathfrak{b}+\mathfrak{c}}, \hat{\mu}''^{\mathfrak{b}-\mathfrak{c}}\}$ is given by

$$\hat{\lambda}'' = \hat{\lambda}, \quad \hat{\mu}''^{\mathfrak{a}} = \frac{1}{2\theta^{\mathfrak{a}}}, \quad \hat{\mu}''^{\mathfrak{b}+\mathfrak{c}} = \frac{1}{4}\left(\frac{1}{\theta^{\mathfrak{b}}} + \frac{1}{\theta^{\mathfrak{c}}}\right), \quad \hat{\mu}''^{\mathfrak{b}-\mathfrak{c}} = \frac{1}{4}\left(\frac{1}{\theta^{\mathfrak{b}}} - \frac{1}{\theta^{\mathfrak{c}}}\right).$$

As a principal subsystem of ET_7^D, let $\hat{\mu}''^{\mathfrak{b}-\mathfrak{c}} = 0$, then the remaining equations are given by

$$\begin{aligned}
&\frac{\partial \rho}{\partial t} + \frac{\partial}{\partial x_i}(\rho v_i) = 0,\\
&\frac{\partial \rho v_j}{\partial t} + \frac{\partial}{\partial x_i}\left(\mathscr{P}\delta_{ij} + \rho v_i v_j\right) = 0,\\
&\frac{\partial}{\partial t}\left(2\rho\varepsilon + \rho v^2\right) + \frac{\partial}{\partial x_i}\left\{\left(2\rho\varepsilon + 2\mathscr{P} + \rho v^2\right) v_i\right\} = 0,\\
&\frac{\partial(2\rho\mathscr{E})}{\partial t} + \frac{\partial(2\rho\mathscr{E}v_i)}{\partial x_i} = P_{\mathscr{E}},
\end{aligned}$$

where $\mathscr{E}$ is the nonequilibrium energy density characterizing the relaxation process and $P_{\mathscr{E}}$ is its production. We may regard this system as the RET theory with 6 fields, which we call $ET_6^{D,\mathfrak{b}+\mathfrak{c}}$. In Table 25.2, three possible ET_6^D theories are summarized. All the universal principles of continuum thermo-mechanics—Galilean invariance and objectivity, entropy principle, and causality principle—are automatically preserved also in these subsystems. We notice that $ET_6^{D,R+V}$ is the same as the previous theory ET_6^D explained in Chap. 24.

Table 25.2 Three possible $\mathrm{ET}_6^{\mathrm{D}}$ theories

	Process	$(\mathfrak{a}, \mathfrak{b}, \mathfrak{c})$	$\mathscr{P}$	ε	P_ε
$\mathrm{ET}_6^{\mathrm{D},K+U+R}$	$(K+U+R)$	$(V, K+U, R)$	$p\left(\rho, \theta^{K+U+R}\right)$	$\varepsilon_E^V\left(\theta^V\right)$	P_{ll}^V
$\mathrm{ET}_6^{\mathrm{D},K+U+V}$	$(K+U+V)$	$(R, K+U, V)$	$p\left(\rho, \theta^{K+U+V}\right)$	$\varepsilon_E^R\left(\theta^R\right)$	P_{ll}^R
$\mathrm{ET}_6^{\mathrm{D},R+V}$	$(R+V)$	$(K+U, R, V)$	$p\left(\rho, \theta^{K+U}\right)$	$\varepsilon_E^{R+V}\left(\theta^{R+V}\right)$	$P_{ll}^R + P_{ll}^V$

The characteristic velocity of $\mathrm{ET}_6^{\mathrm{D},R+V}$ is obtained as

$$U^{R+V^2} = p_\rho\left(\rho, \theta^{K+U}\right) + \frac{\theta^{K+U} p^2_{\theta^{K+U}}\left(\rho, \theta^{K+U}\right)}{\rho^2 c_v^{K+U}\left(\rho, \theta^{K+U}\right)},$$

which is the same as the characteristic velocity of $\mathrm{ET}_7^{\mathrm{D}}$: $U^{R+V} = U$ (see (25.23)). On the other hand, for $\mathrm{ET}_6^{\mathrm{D},\mathfrak{b}+\mathfrak{c}}$ with $(\mathfrak{b}, \mathfrak{c}) = (K+U, R)$ or $(K+U, V)$, we obtain

$$U^{\mathfrak{b}+\mathfrak{c}^2} = p_\rho\left(\rho, \theta^{\mathfrak{b}+\mathfrak{c}}\right) + \frac{\theta^{\mathfrak{b}+\mathfrak{c}} p^2_{\theta^{\mathfrak{b}+\mathfrak{c}}}\left(\rho, \theta^{\mathfrak{b}+\mathfrak{c}}\right)}{\rho^2 c_v^{\mathfrak{b}+\mathfrak{c}}\left(\rho, \theta^{\mathfrak{b}+\mathfrak{c}}\right)}.$$

It is required that the principal subsystems satisfy the subcharacteristic condition (Theorem 2.3). In fact, the condition in the present case:

$$U_E = U_E^{R+V} \geqq U_E^{K+U+a} \geqq U_E^{K+U+b} \geqq U_{Euler} \qquad \begin{cases} a = R,\ b = V \text{ for } c_v^V \geqq c_v^R \\ a = V,\ b = R \text{ for } c_v^R \geqq c_v^V \end{cases},$$

is satisfied since $c_v^{K+U} < c_v^{K+U+a} < c_v^{K+U+b} < c_v$.

The bulk viscosity in $\mathrm{ET}_6^{\mathrm{D},R+V}$ is the same as the one of ET_7 for (R+V)-process (25.29). On the other hand, the bulk viscosity in $\mathrm{ET}_6^{\mathrm{D},\mathfrak{b}+\mathfrak{c}}$ with $(\mathfrak{b}, \mathfrak{c}) = (K+U, R)$ or $(K+U, V)$ is given by

$$\nu = \tau \frac{T p_T^2}{\rho} \frac{c_v^{\mathfrak{a}}}{c_v^{\mathfrak{b}+\mathfrak{c}} c_v} = \left({U^{\mathfrak{b}+\mathfrak{c}}}_E^2 - U_{Euler}^2\right) \tau\rho. \tag{25.30}$$

25.5 $\mathrm{ET}_7^{\mathrm{D}}$ Theory for a Specific Dense Gas

We study $\mathrm{ET}_7^{\mathrm{D}}$ with specific equations of state: equations in the form of virial expansion and equations of van der Waals. The total pressure and the entropy density are explicitly expressed. In particular, for a van der Waals gas, we study also the characteristic velocity and the local exceptionality.

25.5.1 Gas Characterized by the Virial Expansion

The thermal equation of state is given by

$$p = \frac{k_B}{m}\rho T\left(1 + B_2(T)\rho + O(\rho^2)\right), \tag{25.31}$$

where $B_2(T)$ is the second virial coefficient. The caloric equation of state is obtained by integrating the integrability condition of the entropy s^{K+U}, that is, $\rho^2 \varepsilon_\rho^{K+U} = p - Tp_T$ (see (25.13)). The equation for $(K+U)$ mode is expressed as follows:

$$\varepsilon^{K+U} = \frac{3}{2}\frac{k_B}{m}T - \frac{k_B}{m}\rho T^2 B_2'(T) + O(\rho^2),$$

where a prime means a derivative with respect to the temperature T. From (25.14) with (25.31), the total pressure is given by

$$\mathscr{P} = \frac{k_B}{m}\rho\theta^{K+U}\left(1 + B_2(\theta^{K+U})\rho + O(\rho^2)\right).$$

The nonequilibrium entropy (25.12) is given by

$$\begin{aligned}\eta =& \frac{k_B}{m}\left\{\log\left(\frac{\left(\theta^{K+U}\right)^{3/2}}{\rho}\right) - B_2(\theta^{K+U})\rho - \theta^{K+U}B_2'(\theta^{K+U})\rho + O(\rho^2)\right\} \\ &+ \int^{\theta^R}\frac{c_v^R(x)}{x}\,dx + \int^{\theta^V}\frac{c_v^V(x)}{x}\,dx + s_0,\end{aligned}$$

where s_0 is the entropy at a reference state.

The characteristic velocity is expressed as

$$U^2 = \frac{k_B}{m}T\left\{\frac{5}{3} + \frac{2}{9}\rho\left(15B_2(\theta^{K+U}) + 10TB_2'(\theta^{K+U}) + 2T^2B_2''(\theta^{K+U})\right)\right\} + O(\rho^2).$$

The bulk viscosity of the system (25.28) for (K+U+R) or (K+U+V)-process $((\mathfrak{a}, \mathfrak{b}, \mathfrak{c}) = (V, K+U, R)$ or $(R, K+U, V))$ is expressed as

$$\begin{aligned}\nu =& \frac{k_B}{m}\rho T\tau_\delta\frac{\hat{c}_v^{\mathfrak{c}}}{\hat{c}_v^K\hat{c}_v^{K+\mathfrak{c}}}\left\{1 + \left[2B_2(T) + 2TB_2'(T) + \frac{\hat{c}_v^K + \hat{c}_v^{K+\mathfrak{c}}}{\hat{c}_v^K\hat{c}_v^{K+\mathfrak{c}}}\left(2TB_2'(T)\right.\right.\right. \\ &\left.\left.\left.+T^2B_2''(T)\right)\right]\rho\right\} \\ &+ \frac{k_B}{m}\rho T\tau\frac{\hat{c}_v^{\mathfrak{a}}}{\hat{c}_v^{K+\mathfrak{c}}\hat{c}_v^{K+R+V}}\left\{1 + \left[2B_2(T) + 2TB_2'(T)\right.\right.\end{aligned}$$

$$+\frac{\hat{c}_v^{K+c}+\hat{c}_v^{K+R+V}}{\hat{c}_v^{K+c}\hat{c}_v^{K+R+V}}\left(2TB_2'(T)+T^2B_2''(T)\right)\Big]\rho\Big\}+O(\rho^3),$$

where $\hat{c}_v^{a,b,c} = mc_v^{a,b,c}/k_B$ and $c_v^{K+R+V} = c_v^K(T) + c_v^R(T) + c_v^V(T)$ with the specific heat of the translational mode $c_v^K = \frac{3}{2}\frac{k_B}{m}$. For $(R + V)$-process, from (25.29), the bulk viscosity is expressed as

$$\nu = \frac{k_B}{m}\rho T\tau\frac{\hat{c}_v^{R+V}}{\hat{c}_v^K\hat{c}_v^{K+R+V}}\Big\{1+\Big[2B_2(T)+2TB_2'(T)$$
$$+\frac{\hat{c}_v^{K}+\hat{c}_v^{K+R+V}}{\hat{c}_v^{K}\hat{c}_v^{K+R+V}}\left(2TB_2'(T)+T^2B_2''(T)\right)\Big]\rho\Big\}+O(\rho^3).$$

25.5.2 *van der Waals Gas*

Let us study a van der Waals (vdW) gas of which thermal and caloric equations of state are given by

$$p = \frac{k_B}{m}\frac{T\rho}{1-b\rho} - a\rho^2, \quad \varepsilon^{K+U} = \frac{3}{2}\frac{k_B}{m}T - a\rho, \quad \varepsilon^R = \varepsilon_E^R(T), \quad \varepsilon^V = \varepsilon_E^V(T),$$

where a and b are constants. The specific potential energy is given by $\varepsilon^U = -a\rho$ and its specific heat vanishes. Therefore we have

$$c_v^{K+U}(T) = c_v^K(T) = \frac{3}{2}\frac{k_B}{m}T.$$

This indicates that, in the energy exchange process, the intermolecular potential has no role as in a rarefied gas. On the other hand, the total pressure has the effect from the intermolecular potential:

$$\mathcal{P} = \frac{k_B}{m}\frac{\rho}{1-b\rho}\theta^{K+U} - a\rho^2.$$

Therefore, we may say that the equations of state for the vdW gas are peculiar from the viewpoint of the $\mathrm{ET}_7^{\mathrm{D}}$ theory. More realistic equations of state may be necessary in the analysis by using $\mathrm{ET}_7^{\mathrm{D}}$.

From (25.12), the nonequilibrium entropy density η is obtained as

$$\eta = \frac{k_B}{m}\log\left(\left(\theta^{K+U}\right)^{3/2}\frac{1-b\rho}{\rho}\right) + \int^{\theta^R}\frac{c_v^R(x)}{x}\mathrm{d}x + \int^{\theta^V}\frac{c_v^V(x)}{x}\mathrm{d}x + s_0,$$

where s_0 is the entropy at a reference state.

The relationship between the bulk viscosity and the relaxation time (25.28) for (K+U+R) or (K+U+V)-process $((\mathfrak{a}, \mathfrak{b}, \mathfrak{c}) = (V, K + U, R)$ or $(R, K + U, V))$ is explicitly obtained as follows:

$$\nu = \frac{k_B}{m} \frac{T\rho}{(1 - b\rho)^2} \left(\frac{\hat{c}_v^{\mathfrak{a}}}{\hat{c}_v \hat{c}_v^{\mathfrak{b}+\mathfrak{c}}} \tau + \frac{\hat{c}_v^{\mathfrak{c}}}{\hat{c}_v^{\mathfrak{b}} \hat{c}_v^{\mathfrak{b}+\mathfrak{c}}} \tau_\delta \right).$$

For $(R + V)$-process, from (25.29) the bulk viscosity is expressed as

$$\nu = \frac{k_B}{m} \frac{T\rho}{(1 - b\rho)^2} \frac{\hat{c}_v^{R+V}}{\hat{c}_v^{K+U} \hat{c}_v} \tau.$$

As the conditions $(25.18)_{2,3,4}$ are identically satisfied, the convexity condition of the entropy density comes only from $(25.18)_1$. Then we have the condition:

$$\hat{p}\left(\hat{\rho}, \hat{\theta}^{K+U}\right) > (3 - 2\hat{\rho})\, \hat{\rho}^2, \tag{25.32}$$

where we have introduced the following dimensionless variables:

$$\hat{p} = \frac{p}{p_{cr}}, \quad \hat{\rho} = \frac{\rho}{\rho_{cr}}, \quad \hat{\theta}^{K+U} = \frac{\theta^{K+U}}{T_{cr}}, \tag{25.33}$$

with $p_{cr} = a/(27b^2)$, $\rho_{cr} = 1/(3b)$, and $T_{cr} = 8a/(27\frac{k_B}{m}b)$ at the critical point. In the $(\hat{\rho}, \hat{p}(\hat{\rho}, \hat{\theta}^{K+U}))$-plane, the boundary (spinodal curve) is independent of $\hat{\theta}^{K+U}$ as is similar to the case of Euler system in which the spinodal curve for equilibrium pressure is independent of the temperature.

The critical point is the one on the spinodal curve with the condition that the second derivative also vanishes:

$$\left(\frac{\partial^2 p(\rho, \theta^{K+U})}{\partial \rho^2} \right)_{\theta^R, \theta^V} = 0.$$

With (25.32), we have, as for the Euler system,

$$\hat{\rho} = 1, \quad \hat{\theta}^{K+U} = 1, \quad \hat{p}(\hat{\rho}, \hat{\theta}^{K+U}) = 1.$$

In the present case, the characteristic velocity U is obtained as follows:

$$U^2 = \frac{5}{3} \frac{k_B}{m} \frac{\theta^{K+U}}{(1 - b\rho)^2} - 2a\rho.$$

The locus of the local exceptionality, called critical derivative curve (see [206]), is expressed, in dimensionless form with (25.33), as

$$\hat{p}(\hat{\rho}, \hat{\theta}^{K+U}) = \hat{\rho}^2 \left(\frac{9}{20}(3 - \hat{\rho})^2 - 3 \right).$$

In comparison with the spinodal curve (25.32), we notice that the locus of the local exceptionality always resides in the unstable region. It was reported in [442] that the ET_6^D theories for (K+U+R) and (K+U+V) processes predict the locus in the stable region.

In the paper [440], on the basis of ET_7^D of van der Waals gas, the dense-gas effect on the characteristic velocities was studied. It was shown that the lower bound of the characteristic velocity is the same as the one for rarefied gases, while the upper bound is the one evaluated on the spinodal curve.

25.6 Dispersion Relation of Harmonic Wave

We study the dispersion relation of a plane harmonic wave. In particular, we focus on the dispersion relation in the low frequency region where the relations of ET_7^D and ET_6^D coincide with each other. The comparison of the theoretical predictions with experimental data in the case of liquefied CO_2 [443] in the low-frequency region is carried out by using ET_6^D theories.

25.6.1 Dispersion Relation

From the linearized system of field equations (25.27), we obtain the dispersion relation as follows (see Sect. 3.1):

$$\left(\frac{\omega}{k\, U_{Euler}} \right)^2 = 1 + \left(\hat{U}_E^2 - \hat{U}_E^{\mathfrak{b}+\mathfrak{c}^2} \right) \frac{\mathrm{i}\Omega \hat{\tau}_\delta}{1 + \mathrm{i}\Omega \hat{\tau}_\delta} + \left(\hat{U}_E^{\mathfrak{b}+\mathfrak{c}^2} - 1 \right) \frac{\mathrm{i}\Omega}{1 + \mathrm{i}\Omega}, \tag{25.34}$$

where ω is the angular frequency, and k is the complex wave number, $\Omega \equiv \omega\tau$, $\hat{\tau}_\delta \equiv \tau_\delta/\tau$, and the dimensionless characteristic velocities: $\hat{U}_E \equiv U_E/U_{Euler}$ and $\hat{U}_E^{\mathfrak{b}+\mathfrak{c}} \equiv U_E^{\mathfrak{b}+\mathfrak{c}}/U_{Euler}$. For (R+V)-process, δ does not play any role in the dispersion relation as seen from the linearized equations of (25.27).

From the dispersion relation, the phase velocity v_{ph} and the attenuation factor α are derived as follows:

$$
\begin{aligned}
v_{ph} &= \frac{\omega}{\Re(k)} \\
&= \sqrt{2}U_{Euler}\sqrt{\frac{(1+w_1\Omega\hat{\tau}_\delta + w_2\Omega)^2 + (w_1+w_2)^2}{\sqrt{(1+w_1\Omega\hat{\tau}_\delta + w_2\Omega)^2 + (w_1+w_2)^2} + 1 + w_1\Omega\hat{\tau}_\delta + w_2\Omega}}, \\
\alpha &= -\Im(k) \\
&= \frac{\Omega}{\sqrt{2}U_{Euler}\tau}\sqrt{\frac{\sqrt{(1+w_1\Omega\hat{\tau}_\delta + w_2\Omega)^2 + (w_1+w_2)^2} - (1+w_1\Omega\hat{\tau}_\delta + w_2\Omega)}{(1+w_1\Omega\hat{\tau}_\delta + w_2\Omega)^2 + (w_1+w_2)^2}},
\end{aligned}
\tag{25.35}
$$

where

$$
w_1 = \left(\hat{U}_E^2 - \hat{U}_E^{\mathfrak{b}+\mathfrak{c}^2}\right)\frac{\Omega\hat{\tau}_\delta}{1+\Omega^2\hat{\tau}_\delta^2}, \quad w_2 = \left(\hat{U}_E^{\mathfrak{b}+\mathfrak{c}^2} - 1\right)\frac{\Omega}{1+\Omega^2}.
$$

In the high-frequency limit $\Omega \to \infty$, we have

$$
\begin{aligned}
v_{ph,\infty} &\equiv \lim_{\omega\to\infty} v_{ph} = \pm U_E, \\
\alpha_\infty &\equiv \lim_{\omega\to\infty} \alpha = \pm\frac{1}{2U_{Euler}\tau}\frac{\hat{U}_E^2 - \hat{U}_E^{\mathfrak{b}+\mathfrak{c}^2} + \hat{\tau}_\delta(\hat{U}_E^{\mathfrak{b}+\mathfrak{c}^2} - 1)}{\hat{U}_E^3\hat{\tau}_\delta}.
\end{aligned}
$$

25.6.2 Linear Wave in Low-Frequency Region

In a similar way, we can also derive the dispersion relations on the basis of the ET_6^D theories explained in Sect. 25.4. The dispersion relation of $ET_6^{D,\mathfrak{b}+\mathfrak{c}}$ ($(\mathfrak{b},\mathfrak{c})$=(K+U,R) or (K+U,V)) is also derived from the dispersion relation of ET_7^D with $(\mathfrak{b}+\mathfrak{c})$-process by taking the limit $\tau_\delta \to 0$ in (25.34). While the dispersion relation of $ET_6^{D,R+V}$ coincides with the dispersion relation of ET_7^D with (R+V)-process.

In experimental studies of sound waves (see e.g. [311, 312]), the attenuation per wavelength $\alpha_\lambda = 2\pi v_{ph}\alpha/\omega$, in particular, its peak value have been measured. From (25.35) with the condition $\tau_\delta \to 0$, we notice that α_λ has the maximum value:

$$
\alpha_\lambda^{max} = 2\pi\frac{\hat{U}_E^{\mathfrak{b}+\mathfrak{c}} - 1}{\hat{U}_E^{\mathfrak{b}+\mathfrak{c}} + 1} \quad \text{at} \quad \Omega = \frac{1}{\hat{U}_E^{\mathfrak{b}+\mathfrak{c}}}.
$$

From the experimental data on α_λ, it is possible to determine the relaxation process among the three processes that a gas actually takes, and also to determine the bulk viscosity in a similar way to the case of rarefied polyatomic gas in [135] and in [311, 312] with the classical theory for ultrasonic waves .

In the paper [440], the dense-gas effect on the dispersion relation in a van der Waals gas was also studied. In a low frequency region, the upper and lower bounds of the phase velocity and the attenuation per wavelength are explicitly derived.

25.6.3 Comparison with Experimental Data

The attenuation factor per squared frequency, α/f^2, with $f = \omega/(2\pi)$, which can be estimated theoretically by (25.35)$_2$, is compared with the experimental data on liquefied CO_2 [443]. Since the experiments were carried out in the low frequency region where a single relaxation process occurs, we compare the experimental data with the predictions by ET_6^{D}. We adopt, as a trial, the equations of state of a van der Waals (vdW) gas and a Peng-Robinson (PR) equation of state [444]. The PR equation of state is one of the modified vdW equations to predict the liquid density more accurately. In this model, the temperature dependence of the term due to the attractive part of the intermolecular potential is taken into account:

$$p = \frac{k_B}{m}\frac{T\rho}{1-b'\rho} - \frac{a'\mathscr{A}\rho^2}{1+2b'\rho - b'^2\rho^2}$$

with

$$a' = 0.45724\left(\frac{k_B}{m}\right)^2\frac{T_{cr}^2}{p_{cr}}, \quad b' = 0.07780\frac{k_B}{m}\frac{T_{cr}}{p_{cr}},$$
$$\mathscr{A} = \left(1 + k\left(1 - \sqrt{T/T_{cr}}\right)\right)^2, \quad k = 0.37464 + 1.54226w - 0.26992w^2,$$

where w is the acentric factor for the species. From the integrability condition of the entropy, the caloric equation of state for (K+U)-mode is given by

$$\varepsilon_E^{K+U}(\rho, T) = \frac{3}{2}\frac{k_B}{m}T + \int^{\rho}\left(\frac{p(\rho', T) - Tp_T(\rho', T)}{\rho'^2}\right)\mathrm{d}\rho'.$$

The comparison is also made by using the classical Navier–Stokes and Fourier (NSF) theory with PR equation of state. The bulk viscosity of the NSF theory is evaluated from (25.30) for (K+U+R) and (K+U+V) processes and from (25.29) for (RV) process. Since the bulk viscosity depends on the specific heat of the slow relaxation mode $\hat{c}_v^a$, the theoretical prediction by NSF depends on the relaxation process. In the comparison, the relaxation time τ is adopted as an adjustable parameter. On the other hand, we ignore the shear viscosity and heat conductivity.

The values at the critical point for CO_2 are $T_{cr} = 304.15\,\text{K}$ and $p_{cr} = 73\,\text{atm}$. Then the material constants of a vdW gas are obtained from these quantities. The acentric factor of PR equation is $w = 0.239$ [445]. The comparison is made in the case at $T = 273\,\text{K}$ and $p = 37.5\,\text{atm}$, where the specific heat of the vibrational

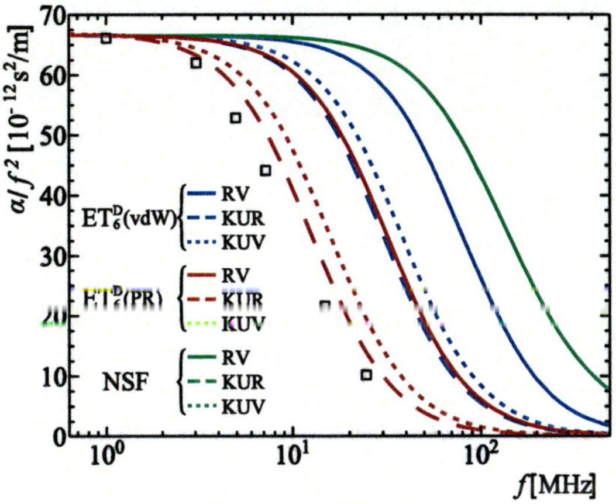

Fig. 25.1 Dependence of the attenuation factor per squared frequency α/f^2 on the frequency f. The squares are the experimental data [443]. The blue, red, and green lines are, respectively, predictions by $\mathrm{ET}_6^{\mathrm{D}}$ theories with the vdW equation, $\mathrm{ET}_6^{\mathrm{D}}$ theories with the PR equation, and NSF theories with the PR equation. The solid, dashed, and dotted lines indicate the (K+U+R), (K+U+V) and (R+V) processes, respectively

Table 25.3 Values of the relaxation time τ

Processes	τ[ns] ($\mathrm{ET}_6^{\mathrm{D}}$: vdW equation)	τ[ns] ($\mathrm{ET}_6^{\mathrm{D}}$: PR equation)	τ[ns] (NSF: PR equation)
(K+U+R)-process	4.49	11.4	11.4
(K+U+V)-process	3.40	8.91	8.88
(R+V)-process	1.21	3.85	3.85

mode is estimated as $\hat{c}_v^V = 0.818$ by the statistical-mechanical consideration with the harmonic oscillator model [135, 342]. Since the characteristic rotational temperature of CO_2 is very low (0.35 K) [349], we have adopted $\hat{c}_v^R = 1$.

In Fig. 25.1, the comparison of the theoretical predictions with experimental data is shown. When the frequency approaches zero, all theoretical predictions tend to coincide with each other. For this reason, we determine the value of τ so as to fit the experimental data of the lowest frequency, and its value is summarized in Table 25.3. Noticeable points in Fig. 25.1 are as follows: (1) The predictions by $\mathrm{ET}_6^{\mathrm{D}}$ theories are evidently superior to those by the NSF theory. In particular, the predictions by $\mathrm{ET}_6^{\mathrm{D}}$ theories with PR equation are in good agreement with the experimental data compared to the predictions by $\mathrm{ET}_6^{\mathrm{D}}$ with the vdW equation. This indicates that a more accurate model of equations of state is necessary to explain the experimental data more precisely. (2) The frequency dependences of α/f^2 predicted by the $\mathrm{ET}_6^{\mathrm{D}}$ theories depend on the relaxation process. In the present case, the (K+U+R)-process is suitable to describe the experimental data on CO_2 as in the case of rarefied gas explained in Chap. 16 (see also [342]). On the other hand, the predictions by the NSF theory seem to be independent of the relaxation process.

The results of the comparison with other experimental conditions given in [443] are essentially the same as those of the present comparison.

25.7 Remarks

Some promising future studies are remarked:

- In the present chapter, we compared the theoretical predictions by ET_6^D theories with experimental data [443], and found that the RET theory with the (K+U+R)-process is an appropriate model for CO_2 in the present temperature range. Comprehensive study of sound waves in a wider temperature range and also in a wider frequency range is highly expected.
- The ET_7^D theory is a simplified model for phenomena in a gas with dissipation. Therefore it is evident that, as a next study, we should construct a RET theory that takes into account also shear viscosity and heat flux. The ET_7^D theory constructed here must be a good starting point for such an enterprise.
- In monatomic-gas limit, the ET_7^D theory reduces to the Euler theory of perfect gas. This indicates that there exists at least one more mesoscopic mechanism that causes the dynamic pressure in a gas. See the papers [126, 446]. See also Remark 24.3 in Sect. 24.3.2. Clarification of such a mechanism is another future subject to be studied.

Part VII
Relativistic Polyatomic Gas

Chapter 26
Relativistic Polyatomic Gas

Abstract The goal of this chapter is to present the relativistic extended thermodynamics (relativistic ET) theory of rarefied polyatomic gases with 14 fields. This is achieved by adopting the closure procedure for the generalized moments of a distribution function that, as in the classical case, depends on an additional continuous variable representing the energy of the internal modes of a molecule. This permits the theory to take into account the energy exchange between translational modes and internal modes of a molecule.

First, we consider the non-dissipative Eulerian fluids and we obtain, via MEP, the equilibrium distribution function that is the natural generalization of the Jüttner distribution function. And we evaluate the thermal and caloric equations of state. We prove that the thermal equation of state is the same as that of monatomic gas, while the caloric equation of state is a generalization of the Synge one. Second, we consider a rarefied polyatomic gas with dissipation. We present the theory given recently by Pennisi and Ruggeri (Ann Phys 377:414, 2017). The theory includes the relativistic ET theory of monatomic gases as a singular limit, and it converges to the corresponding RET theory of polyatomic gases in the classical limit.

In contrast to a monatomic gas, the dynamic pressure in a polyatomic gas is not small due to the internal motion of a molecule as is the case in the classical limit. Therefore the present theory might be particularly useful in cosmology to describe some aspects of the post-recombination era.

26.1 Introduction

In Chap. 5, both the relativistic theory of Eulerian fluids and the dissipative 14-moment hyperbolic theory (LMR theory) of monatomic gases are explained.

The aim of this chapter is to summarize the recent papers of Pennisi and Ruggeri [153, 447], in which the authors constructed a relativistic theory of rarefied polyatomic gases following the ideas of the classical counterpart explained in the previous chapters. We consider the theory of non-dissipative polyatomic Eulerian fluids, and then we present the causal dissipative theory with 14 moments. The latter theory, in the classical limit, coincides with the theory of 14 moments (ET_{14})

T. Ruggeri, M. Sugiyama, *Classical and Relativistic Rational Extended Thermodynamics of Gases*, https://doi.org/10.1007/978-3-030-59144-1_26

described in Chaps. 6 and 7, and, in the monatomic-gas limit, it reduces to the LMR theory.

The advantage of this new theory is twofold: (1) We can describe a more realistic physical situations beyond the monatomic gas at least in the regime in which the temperature is large enough to allow the raise of relativistic effects without affecting the formation and stability of (some) polyatomic molecules. (2) More importantly, we can understand the crucial role of the dynamic pressure, which is expected to be small in a monatomic gas. The relevant dynamic pressure and the bulk viscosity are very important in the field of relativity and cosmology because, following the ideas of [448] and [449], the dynamic pressure might affect the evolution of the Universe.

26.2 Eulerian Rarefied Polyatomic Gas

First of all we consider an Eulerian fluid, for which the balance laws reduce to the conservation laws of particle number and energy-momentum tensor (5.1) with V^α and $T^{\alpha\beta}$ given by (5.2). In order to calculate the thermal and caloric equations of state for a polyatomic rarefied gas, we now adopt the so-called *molecular approach* in which the moment-tensors appearing in (5.1) are related to the extended distribution function $f \equiv f(x^\alpha, p^\beta, I)$. Then, instead of (5.19), we assume, by analogy with the classical case, the following moments:

$$\begin{aligned} V^\alpha &= mc \int_{\mathbb{R}^3} \int_0^{+\infty} f p^\alpha \phi(I)\, d\,I\, d\mathbf{P}, \\ T^{\alpha\beta} &= \frac{1}{mc} \int_{\mathbb{R}^3} \int_0^{+\infty} f \left(mc^2 + I\right) p^\alpha p^\beta\, \phi(I)\, d\,I\, d\mathbf{P}. \end{aligned} \tag{26.1}$$

The meaning of $(26.1)_2$ is that the energy and the momentum in relativity are components of a single energy-momentum tensor. Then we notice that, in addition to the rest mass energy mc^2, there is another contribution I due to the internal degrees of freedom of a molecule. Here $\phi(I)$ is the relativistic counterpart of the weighting function in the classical RET theory of polyatomic gases.

26.2.1 Equilibrium Distribution Function, and Thermal and Caloric Equations of State

First of all we deduce the equilibrium distribution function $f^{(E)}$ of a polyatomic gas that generalizes the Jüttner one of a monatomic gas. Let us search for the distribution function $f \equiv f(x^\alpha\,,\; p^\alpha\,,\; I)$ that maximizes the entropy:

$$\rho s = h = h^\alpha U_\alpha = -k_B c\, U_\alpha \int_{\mathbb{R}^3} \int_0^{+\infty} f \ln f p^\alpha \phi(I)\, d\,I\, d\mathbf{P}$$

under the constraints that the temporal parts $V^\alpha U_\alpha$ and $T^{\alpha\beta}U_\beta$ of (26.1)$_{1,2}$ are prescribed. This means that we have to consider the functional with the Lagrange multipliers λ and λ_β:

$$\begin{aligned}\mathscr{L} =U_\alpha \Bigg\{ & -k_B c \int_{\mathbb{R}^3}\int_0^{+\infty} f \ln f p^\alpha \phi(I)\, d\, I\, d\mathbf{P} \\ & + \lambda \left[V^\alpha - mc \int_{\mathbb{R}^3}\int_0^{+\infty} f p^\alpha \phi(I)\, d\, I\, d\mathbf{P}\right] \\ & + \lambda_\beta \left[T^{\alpha\beta} - \frac{1}{mc}\int_{\mathbb{R}^3}\int_0^{+\infty} f\left(mc^2 + I\right) p^\alpha p^\beta \phi(I)\, d\, I\, d\mathbf{P}\right]\Bigg\}.\end{aligned}$$

This functional $\mathscr{L}$ has to be maximized with respect to the distribution function. So we have the condition: $\frac{\delta \mathscr{L}}{\delta f} = 0$, that is,

$$\int_{\mathbb{R}^3}\int_0^{+\infty}\left[-\frac{k_B}{c}(\ln f + 1) - \lambda\frac{m}{c} - \frac{1}{mc^3}\lambda_\beta p^\beta\left(mc^2 + I\right)\right] U_\alpha p^\alpha \phi(I)\, d\, I\, d\mathbf{P} = 0.$$

Then we have the equilibrium distribution function:

$$f^{(E)} = \exp\left(-1 - \frac{\chi}{k_B}\right), \tag{26.2}$$

where

$$\chi = m\lambda + \left(1 + \frac{I}{mc^2}\right)\lambda_\beta p^\beta. \tag{26.3}$$

We insert (26.2) into (26.1)$_{1,2}$. Then, taking into account (26.3), we have

$$\frac{1}{c^2}V^\alpha U_\alpha = nm = \frac{m}{c}\int_{\mathbb{R}^3}\int_0^{+\infty} e^{-1-\frac{1}{k_B}\left[m\lambda + \left(1+\frac{I}{mc^2}\right)\lambda^\beta p_\beta\right]}(U_\alpha p^\alpha)\phi(I)\, d\, I\, d\mathbf{P}. \tag{26.4}$$

On the other hand, we know that the Lagrange multipliers coincide with the *main field* given in (5.4):

$$\lambda = -\frac{g_r}{T}, \qquad \lambda_\beta = \frac{U_\beta}{T}. \tag{26.5}$$

To perform the integrations, we consider the variables in the reference frame where U^α has only the zeroth component. And we use the following change of integration variables:

$$
\begin{aligned}
&p^1 = mc\sinh s \sin\vartheta \cos\varphi, \quad p^2 = mc\sinh s \sin\vartheta \sin\varphi, \\
&p^3 = mc\sinh s \cos\vartheta, \quad p^0 = mc\cosh s, \\
&s \in [0, +\infty, \quad \vartheta \in [0, \pi[\ \varphi \in [0, 2\pi[
\end{aligned} \tag{26.6}
$$

whose transformation has the Jacobian $m^3c^3\sinh^2 s\cosh s\sin\vartheta$. In this way, (26.4) becomes

$$
n = 4\pi m^3 c^3 e^{-1-\frac{m}{k}\lambda}\int_0^{+\infty} J_{2,1}\left(\gamma + \frac{I}{k_B T}\right)\phi(I)\, d\, I, \tag{26.7}
$$

where we have introduced the functions:

$$
J_{m,n}(\gamma) = \int_0^{+\infty} \sinh^m s\cosh^n s \quad e^{-\gamma\cosh s}\, d\, s \tag{26.8}
$$

satisfying the recurrence relation:

$$
J_{m+2,n} = J_{m,n+2} - J_{m,n} \quad , \quad J_{m,n} = \sum_{h=0}^{r}\binom{r}{h}(-1)^h J_{m-2r,n+2r-2h} \quad \text{for} \forall r \in \aleph .
$$

The second relation can be proved, starting from the first one, with the iterative procedure.

Moreover, $\cosh^n s$ can be written as a linear combination of $\cosh ns$. Therefore all the coefficients can be written in terms of the modified Bessel functions $K_n(\gamma)$:

$$
K_n(\gamma) = \int_0^{+\infty} \cosh ns \quad e^{-\gamma\cosh s}\, d\, s,
$$

which, in turn, can be written in terms of K_2 and K_3 by means of the recurrence relation:

$$
K_{n+1}(\gamma) = K_{n-1}(\gamma) + 2\frac{n}{\gamma}K_n(\gamma). \tag{26.9}
$$

For details see [55].

To relate the equilibrium distribution function of a polyatomic gas with the Jüttner distribution function of a monatomic gas, we proceed in the following way. We observe that the number particle density is unconcerned with the difference between monatomic gas and polyatomic gas. Therefore we rewrite n using the monatomic theory:

$$n = \frac{1}{c} \int_{\mathbb{R}^3} f_J \, U_\alpha p^\alpha \, d\mathbf{P} = \frac{m}{c} \int_{\mathbb{R}^3} e^{-1-\frac{1}{k_B}\left[m\hat{\lambda}+\hat{\lambda}^\beta p_\beta\right]} U_\alpha p^\alpha \, d\mathbf{P}, \tag{26.10}$$

where f_J is the Jüttner distribution function (5.20):

$$f_J = e^{-1-\frac{1}{k_B}\left[m\hat{\lambda}+\hat{\lambda}^\beta p_\beta\right]} \tag{26.11}$$

with

$$\hat{\lambda} = -\frac{g_{rM}}{T}, \qquad \hat{\lambda}_\beta = \frac{U_\beta}{T} \tag{26.12}$$

and g_{rM} is the relativistic chemical potential for a monatomic gas being different from g_r of a polyatomic gas. Using the same arguments as before we can rewrite (26.10) in the form:

$$n = 4\pi m^3 c^3 e^{-1-\frac{m}{k}\hat{\lambda}} J_{2,1}(\gamma). \tag{26.13}$$

Comparing (26.7) with (26.13), we define:

$$A(\gamma) = \exp \frac{m}{k_B}(\lambda - \hat{\lambda}) = \frac{1}{J_{2,1}(\gamma)} \int_0^{+\infty} J_{2,1}\left(\gamma^*\right) \phi(I) \, d\,I, \tag{26.14}$$

where

$$\gamma^* = \gamma + \frac{I}{k_B T}.$$

Taking into account the following recurrence relations [55]:

$$k_B \gamma \, I_{m+2,n} = n \, J_{m,n-1} - (n+m+1) \, J_{m,n+1}, \qquad I_{m,n} = -\frac{1}{k_B} J_{m,n},$$

we have

$$\begin{aligned} J_{2,1}(\gamma) &= \frac{1}{\gamma}[2J_{0,2}(\gamma) - J_{0,0}(\gamma)] = \int_0^{+\infty} \left(2\cosh^2 s - 1\right) e^{-\gamma \cosh s} \, d\,s \\ &= \frac{1}{\gamma} \int_0^{+\infty} \cosh(2s) e^{-\gamma \cosh s} \, d\,s = \frac{1}{\gamma} K_2(\gamma)\,. \end{aligned} \tag{26.15}$$

The expression (26.14) can be rewritten as

$$A(\gamma) = \frac{\gamma}{K_2(\gamma)} \int_0^{+\infty} J_{2,1}\left(\gamma^*\right) \phi(I)\, d\, I\,,$$

or, equivalently, (26.16)

$$A(\gamma) = \frac{\gamma}{K_2(\gamma)} \int_0^{+\infty} \frac{1}{\gamma^*} K_2\left(\gamma^*\right) \phi(I)\, d\, I.$$

Taking the ratio between $f^{(E)}$ given in (26.2) with (26.3) and f_J given in (26.11), and taking into account the definition of $A(\gamma)$, (26.14), we obtain the following relation between the equilibrium distribution functions of polyatomic and monatomic gases:

$$f^{(E)} = \frac{1}{A(\gamma)} f_J \exp\left(-\frac{I}{mc^2} \frac{U^\beta p_\beta}{k_B T}\right). \tag{26.17}$$

Evaluating $\hat{\lambda}$ in terms of n from (26.13) and (26.15), and inserting it into (26.11), we can rewrite the Jüttner distribution function as

$$f_J = \frac{n\gamma}{K_2(\gamma)} \frac{1}{4\pi m^3 c^3} e^{-\frac{1}{k_B T} U_\beta p^\beta}. \tag{26.18}$$

Then inserting (26.18) into (26.17), we obtain the following:

Theorem 26.1 *The equilibrium distribution function for a rarefied polyatomic gas that maximizes the entropy has the following expression:*

$$f^{(E)} = \frac{n\gamma}{A(\gamma) K_2(\gamma)} \frac{1}{4\pi m^3 c^3} e^{-\frac{1}{k_B T}\left[\left(1+\frac{I}{mc^2}\right) U_\beta p^\beta\right]} \tag{26.19}$$

with $A(\gamma)$ being given in (26.16)*. This is the natural generalization of the Jüttner distribution function to the distribution function of a rarefied polyatomic gas.*

Now since we have the distribution function, we can calculate the thermal and caloric equations of state for a non-degenerate rarefied polyatomic gas. In [153], the following theorem was proved:

Theorem 26.2 *The pressure and the energy for a polyatomic gas compatible with the distribution function* (26.19) *are given by*

$$p = \frac{nmc^2}{\gamma} = \frac{k_B}{m} \rho T\,,$$

(26.20)

$$e = \frac{nmc^2}{A(\gamma) K_2(\gamma)} \int_0^{+\infty} \left[K_3\left(\gamma^*\right) - \frac{1}{\gamma^*} K_2\left(\gamma^*\right)\right] \phi(I)\, d\, I.$$

In particular, (26.20)$_2$ *is the generalization of the Synge energy to the case of polyatomic gas.*

Proof Inserting (26.19) into (26.29)$_2$, we have

$$
\begin{aligned}
&p h^{\alpha\beta} + \frac{e}{c^2} U^\alpha U^\beta \\
&= \frac{1}{mc} \int_{\mathbb{R}^3} \int_0^{+\infty} \frac{n\gamma}{A(\gamma)K_2(\gamma)} \frac{1}{4\pi m^3 c^3} e^{-\frac{\gamma}{mc^2}\left[\left(1+\frac{I}{mc^2}\right) U_\mu p^\mu\right]} p^\alpha p^\beta (mc^2 + I)\phi(I)\, d\, I\, d\mathbf{P}.
\end{aligned}
$$

Contracting this with $h_{\alpha\beta}$ and taking into account the change of variables (26.6), we obtain

$$
\begin{aligned}
3p &= \frac{nmc^2}{A(\gamma)K_2(\gamma)} \int_0^{+\infty} \int_0^{+\infty} e^{-\gamma^* \cosh s} \gamma^* \sinh^4 s\, \phi(I)\, ds\, d\, I \\
&= \frac{nmc^2}{A(\gamma)K_2(\gamma)} \int_0^{+\infty} J_{4,0}\left(\gamma^*\right) \phi(I)\, d\, I\, .
\end{aligned}
$$

But since $J_{4,0}(\gamma) = \frac{3}{\gamma^2} K_2(\gamma)$ [55], we have

$$
p = \frac{nmc^2}{A(\gamma)K_2(\gamma)} \int_0^{+\infty} \frac{1}{\gamma^*} K_2\left(\gamma^*\right) \phi(I)\, d\, I\, .
$$

Substituting $A(\gamma)$ in (26.16) into this equation, we obtain

$$
p = \frac{nmc^2}{\gamma} = \rho \frac{k_B}{m} T\, .
$$

Therefore the thermal equation of state is the same as that of a monatomic gas, and remains in the same expression in the classical limit.

Concerning the energy, we have

$$
\begin{aligned}
e &= T^{\alpha\beta} U_\alpha U_\beta \frac{1}{c^2} \\
&= \frac{nmc^2}{A(\gamma)K_2(\gamma)} \int_0^{+\infty} \int_0^{+\infty} e^{-\gamma^* \cosh s} \gamma^* \sinh^2 s \cosh^2 s\, \phi(I)\, ds\, d\, I \\
&= \frac{nmc^2}{A(\gamma)K_2(\gamma)} \int_0^{+\infty} J_{2,2}\left(\gamma^*\right) \gamma^* \phi(I)\, d\, I\, .
\end{aligned}
$$

But since $J_{2,2}(\gamma) = \frac{1}{\gamma} K_3(\gamma) - \frac{1}{\gamma^2} K_2(\gamma)$, we have (26.20)$_2$. The proof is completed.

From (26.14), (26.5), and (26.12), we obtain the chemical potential of a polyatomic gas g_r in terms of the chemical potential of a monatomic gas g_{rM}:

$$g_r = g_{rM} - \frac{k_B T}{m} \log A(\gamma)$$

with A being given by (26.16). Moreover, from the definition of chemical potential (5.6) and (5.8), we can calculate the entropy S of a polyatomic gas in terms of the entropy of a monatomic gas S_M:

$$S = S_M + \frac{1}{\rho T}(e - e_M - \rho(g_r - g_{rM})),$$

where e denotes the energy of a polyatomic gas $(26.20)_2$, while e_M denotes the Synge energy given in (5.21).

Until now the measure $\phi(I)$ is not prescribed explicitly. To determine it we consider the classical limit of the above results.

26.2.2 *Classical Limit of Relativistic Polyatomic Euler Gas*

It is well known that, for large c, the Jüttner distribution function converges to the Maxwellian distribution function except for the factor $1/m^3$:

$$\lim_{c \to \infty} f_J = \frac{1}{m^3} f^{(M)} = \frac{1}{m^3} \frac{\rho}{m} \left(\frac{m}{2\pi k_B T} \right)^{\frac{3}{2}} \exp\left\{ - \frac{mC^2}{2k_B T} \right\}. \tag{26.21}$$

Thanks to the property:

$$\lim_{\gamma \to +\infty} \sqrt{\frac{2}{\pi} \gamma} \, e^{\gamma} K_2(\gamma) = 1 \,,$$

we have

$$\lim_{\gamma \to +\infty} A(\gamma) = A(T) = \int_0^\infty \exp\left\{ - \frac{I}{k_B T} \right\} \varphi(I) \, dI. \tag{26.22}$$

Inserting (26.21) and (26.22) into (26.17), we obtain

$$\lim_{\gamma \to +\infty} f^{(E)} = \frac{1}{m^3} f_C^{(E)},$$

where

$$f_C^{(E)} = \frac{\rho}{mA(T)} \left(\frac{m}{2\pi k_B T}\right)^{3/2} \exp\left(-\frac{1}{k_B T}\left(\frac{1}{2}mC^2 + I\right)\right).$$

This is the distribution function deduced in the classical case (7.8).

Now we study the classical limit of energy. We observe that a part of energy is due to the presence of the rest mass, that is, nmc^2. Then the residual part is the internal energy:

$$\varepsilon = \frac{e - \rho c^2}{\rho}. \tag{26.23}$$

From $(26.20)_2$, it follows that

$$\varepsilon = \frac{c^2}{A(\gamma)K_2(\gamma)} \int_0^{+\infty} \left[K_3\left(\gamma^*\right) - \frac{1}{\gamma^*} K_2\left(\gamma^*\right)\right] \phi(I)\, d\,I - c^2. \tag{26.24}$$

By taking into account the expansion of the Bessel functions (5.22), after some calculations, it is possible to prove that this converges to the classical internal energy:

$$\lim_{c\to\infty} \varepsilon = \varepsilon^C = \frac{3}{2}\frac{k_B}{m} T + \frac{1}{m}\frac{B(T)}{A(T)}, \tag{26.25}$$

where $A(T)$ is given in (26.22) and

$$B(T) = \int_0^{\infty} I \exp\left\{\frac{-I}{k_B T}\right\} \varphi(I)\, dI,$$

i.e., the classical non-polytropic internal energy $\varepsilon^K(T) + \varepsilon^I(T)$ in equilibrium with ε^I given by (7.15). In a polytropic gas, choosing

$$\varphi(I) = I^a, \qquad a = \frac{D-5}{2},$$

we obtain, from (26.25), the usual internal energy for a polytropic polyatomic gas (1.32). Therefore, in the classical limit, we know, for a polytropic gas, that

$$\lim_{\gamma\to\infty} \phi(I) = \varphi(I) = I^\alpha. \tag{26.26}$$

Some considerations concerning the MEP procedure seem to indicate that the measure $\phi(I)$ must be independent of γ. Therefore a possible natural measure in the relativistic regime seems to be the same as in the classical limit (26.26). The integrals in (26.20) are convergent provided that $\alpha > -1$ [450].

26.2.3 *Ultra-Relativistic Limit of a Relativistic Polyatomic Euler Gas*

In the ultra-relativistic limit ($\gamma \to 0$), as is well-known, the Synge energy for a monatomic gas (5.21) converges to

$$\lim_{\gamma \to 0} e_{\text{Singe}} = 3nk_B T = 3p.$$

This result is obtained from (5.21) by taking into account that, for $\gamma \to 0$, we have

$$K_3(\gamma) \sim \frac{8}{\gamma^3}, \quad K_2(\gamma) \sim \frac{2}{\gamma^2}.$$

Now we analyze the ultra-relativistic limit for the energy of a polyatomic gas $(26.20)_2$. It is possible to prove the following:

Theorem 26.3 *The energy of a gas with structure* $(26.20)_2$ *in the ultra-relativistic limit converges to*

$$\lim_{\gamma \to 0} e = 3nk_B T = 3p \quad \textit{for} \quad -1 < a \leqslant 2,$$

$$\lim_{\gamma \to 0} e = (a+1)\, nk_B T = (a+1)\, p \quad \textit{for} \quad a > 2.$$

The proof is given in [447]. This theorem is interesting because it states that, when the gas has small degrees of freedom of a molecule, any relativistic polyatomic gas has the same behavior as of a monatomic gas in the ultra-relativistic case. While, for a being larger, the energy is higher. This latter result is difficult to understand, and maybe this implies that the present model or the choice of $\phi(I) = I^a$ is valid only for $a \leqslant 2$.

26.2.4 *Relativistic Energy for Diatomic Gas*

The generalization of the Synge energy given in $(26.20)_2$ is difficult to evaluate in an explicit way because the integral has not an analytical expression in general. Nevertheless, in [284], it was observed that, in the case of diatomic gas for which $a = 0$ and $\phi = 1$, the integral is expressed in terms of other Bessel functions. In fact it is easy to verify that, if $a = 0$, we have

$$e = p\left(\gamma \frac{K_0(\gamma)}{K_1(\gamma)} + 3\right)$$

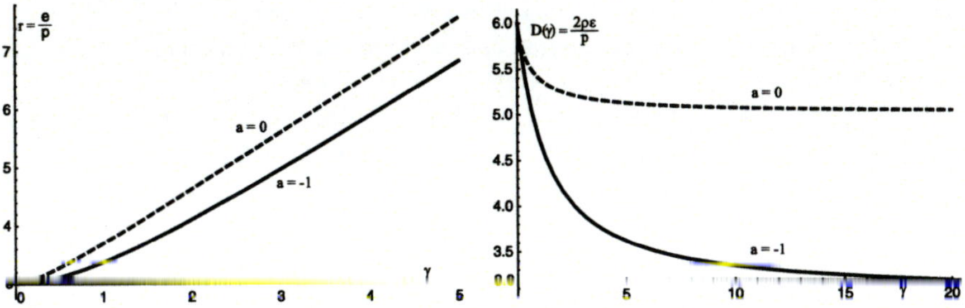

Fig. 26.1 $r = e/p$ (left) and $D(\gamma) = 2\rho\varepsilon/p$ (right) as functions of γ for diatomic gas ($a = 0$) and for monatomic gas ($a = -1$)

having an expression of the same type of the Synge energy (5.21). In the left side of Fig. 26.1, there is the plot of the ratio $r = e/p$, which depends only on γ, in the cases: $a = 0$ and $a = -1$. According with Theorem 26.3, for $\gamma \to 0$, both graphs tend to the same value 3. And the ratio increases with γ. In the right side, we plot the ratio:

$$D(\gamma) = 2\frac{\rho\varepsilon}{p} = 2(r(\gamma) - \gamma) \tag{26.27}$$

in the cases: $a = 0$ and $a = -1$. In relativity the internal energy ε (26.23) due to the Synge energy assumes the form (26.24) and therefore is a nonlinear function of the temperature together with $D(\gamma)$ given by (26.27). Therefore all relativistic gases compatible with the kinetic theory are non-polytropic. According with the results stated in Sect. 26.2.2, ε converges, in the classical limit, to the one in polytropic expression (1.32). As seen in the left side of Fig. 26.1, $D(\gamma)$ tends to a constant when $\gamma \to \infty$, that is, it tends to 3 for a monatomic gas and 5 for a diatomic gas. In the ultra-relativistic regime, according with Theorem 26.3, $D(0)$ converges to 6 for both monatomic and diatomic gases.

26.3 Relativistic Dissipative Rarefied Polyatomic Gas with 14 Fields

For a relativistic dissipative rarefied polyatomic gas, Pennisi and Ruggeri [153] proposed the following set of balance equations:

$$\partial_\alpha V^\alpha = 0 \quad , \quad \partial_\alpha T^{\alpha\beta} = 0 \quad , \quad \partial_\alpha A^{\alpha\langle\beta\gamma\rangle} = I^{\langle\beta\gamma\rangle} \, , \tag{26.28}$$

where $\langle\cdots\rangle$ denotes the traceless part of a tensor. The difference between (5.25) of a monatomic gas and (26.28) is apparently minimal, i.e., we have abandoned only the trace condition (5.26). The motivation came, also in this relativistic case, from the kinetic counterpart as we will see in the next section. Therefore the LMR model is more restrictive than the present model due to the trace condition (5.26). We can use the results given in [55] until we impose the trace condition.

Important step to close the system in LMR approach is to require the entropy principle, i.e., we require the existence of the entropy 4-vector h^α specified by a local function of the fields:

$$h^\alpha \equiv h^\alpha \left(V^\alpha, T^{\alpha\beta}\right)$$

that satisfies, for thermodynamic processes,

$$\partial_\alpha h^\alpha = \Sigma \geqslant 0.$$

Using the symmetrization technique given by Ruggeri and Strumia [164] (see Theorem 2.1), we can show that there exists a 4-vector potential h'^α and the *main field* $\mathbf{u}' \equiv (\lambda, \lambda_\beta, \Sigma_{\beta\gamma})$ ($\Sigma_{\beta\gamma}$ is symmetric and deviatoric tensor) such that

$$\begin{aligned}
&V^\alpha = \frac{\partial h'^\alpha}{\partial \lambda}, \quad T^{\alpha\beta} = \frac{\partial h'^\alpha}{\partial \lambda_\beta}, \quad A^{\alpha\langle\beta\gamma\rangle} = \frac{\partial h'^\alpha}{\partial \Sigma_{\beta\gamma}} - \frac{1}{4} g^{\beta\gamma} g_{\mu\nu} \frac{\partial h'^\alpha}{\partial \Sigma_{\mu\nu}},\\
&h'^\alpha = \lambda V^\alpha + \lambda_\beta T^{\alpha\beta} + \Sigma_{\beta\gamma} A^{\alpha\beta\gamma} - h^\alpha,\\
&\Sigma = \Sigma_{\beta\gamma} I^{\beta\gamma} \geqslant 0.
\end{aligned}$$

By the representation theorem, we have the expression (A.2) of [55], which, up to the second order around an equilibrium state, is expressed as

$$h'^\alpha = \sum_{A=0}^{2} \gamma_A \lambda^{(A)\alpha}.$$

Imposing the symmetry of $T^{\alpha\beta}$ and $A^{\alpha\beta\gamma}$, we obtain the expression (A.3) of [55], which, up to the second order, is given by

$$\begin{aligned}
&\gamma_2 = \Gamma_2,\\
&\gamma_1 = \Gamma_1 + \frac{\partial \Gamma_2}{\partial G_0} G_1,\\
&\gamma_0 = \Gamma_0 + \frac{\partial \Gamma_1}{\partial G_0} G_1 + \frac{1}{2}\frac{\partial^2 \Gamma_2}{\partial G_0^2} G_1^2 + \frac{\partial \Gamma_2}{\partial G_0} G_2 + \frac{1}{4} \Gamma_2 \Sigma^{\mu\nu} \Sigma_{\mu\nu},
\end{aligned}$$

where Γ_0, Γ_1, and Γ_2 are arbitrary functions of λ and $G_0 = \lambda^\alpha \lambda_\alpha$, while $G_1 = \Sigma_{\alpha\beta}\lambda^\alpha\lambda^\beta$ and $G_2 = \Sigma^2_{\alpha\beta}\lambda^\alpha\lambda^\beta$.

Consequently, at every order around equilibrium, one arbitrary function appears, that is Γ_0 at the order zero, Γ_1 at the order 1, and Γ_2 at the order 2. After that, the knowledge of V^α, $T_E^{\alpha\beta}$, $A_E^{\alpha\beta\gamma}$ allows us to determine Γ_0 and Γ_1 except for integration constants, which can be assumed to be zero in harmony with the present molecular approach. Γ_2 remains arbitrary, while it is completely determined by the molecular approach.

26.3.1 Molecular Approach

In the present case we have

$$\begin{aligned}
V^\alpha &= mc \int_{\mathbb{R}^3} \int_0^{+\infty} f p^\alpha \phi(I)\, d\,I\, d\mathbf{P}, \\
T^{\alpha\beta} &= \frac{1}{mc} \int_{\mathbb{R}^3} \int_0^{+\infty} f \left(mc^2 + I\right) p^\alpha p^\beta\, \phi(I)\, d\,I\, d\mathbf{P}, \\
A^{\alpha\beta\gamma} &= \frac{1}{m^2 c} \int_{\mathbb{R}^3} \int_0^{+\infty} f \left(mc^2 + 2I\right) p^\alpha p^\beta p^\gamma \phi(I)\, d\,I\, d\mathbf{P}, \\
I^{\beta\gamma} &= \frac{c}{m} \int_{\mathbb{R}^3} \int_0^{+\infty} Q\, p^\beta p^\gamma\, \phi(I)\, d\,I\, d\mathbf{P}.
\end{aligned} \tag{26.29}$$

One remark is that the coefficient of I in $(26.29)_3$ is different from the coefficient of I in $(26.29)_2$. This comes from the requirement that the non-relativistic limit of the present system must coincide with (1.34) (see paper [153] for details).

26.3.2 Triple Tensor in Equilibrium

Finally, let us calculate $A^{\alpha\beta\gamma}$ at equilibrium. Inserting (26.19) into $(26.29)_3$, we obtain

$$\begin{aligned}
A_E^{\alpha\beta\gamma} &= A_1^0 U^\alpha U^\beta U^\gamma + 3A_{11}^0 h^{(\alpha\beta} U^{\gamma)} \\
&= \frac{1}{4\pi m^5 c^4} \frac{n\gamma}{A(\gamma) K_2(\gamma)} \int_{\mathbb{R}^3} \int_0^{+\infty} e^{-\gamma^* \cosh s} p^\alpha p^\beta p^\gamma (mc^2 + 2I) \phi(I)\, d\,I\, d\mathbf{P},
\end{aligned} \tag{26.30}$$

where the first equality is suggested by the representation theorem. Contracting (26.30) with $U_\alpha U_\beta U_\gamma$, we obtain

$$A_1^0 = \frac{m\, n\gamma}{A(\gamma) K_2(\gamma)} \int_0^{+\infty} J_{2,3}\left(\gamma^*\right) \left(1 + \frac{2I}{mc^2}\right) \phi(I)\, d\,I\,.$$

Contracting (26.30) with $h_{\alpha\beta}U_{\gamma}$, we obtain

$$A^0_{11} = \frac{1}{3}\frac{m\,n\gamma c^2}{A(\gamma)K_2(\gamma)}\int_0^{+\infty} J_{4,1}\left(\gamma^*\right)\left(1+\frac{2I}{mc^2}\right)\phi(I)\,d\,I\,,$$

and by the same considerations between the formula (26.8) and (26.9) these can be written in terms of the Bessel functions:

$$A^0_1 = \frac{m\,n\gamma}{A(\gamma)K_2(\gamma)}\int_0^{+\infty}\left[3\frac{K_3\left(\gamma^*\right)}{\gamma^{*2}} + \frac{K_2(\gamma^*)}{\gamma^*}\right]\left(1+\frac{2I}{mc^2}\right)\phi(I)\,d\,I\,,$$

$$A^0_{11} = \frac{c^2 m\,n\gamma}{A(\gamma)K_2(\gamma)}\int_0^{+\infty}\frac{K_3\left(\gamma^*\right)}{\gamma^{*2}}\left(1+\frac{2I}{mc^2}\right)\phi(I)\,d\,I\,.$$

26.4 Nonequilibrium Distribution Function and the Closure

Let us search for the nonequilibrium distribution function $f \equiv f(x^\alpha, p^\alpha, I)$ that maximizes the entropy under the constraints (26.29). Then we introduce the functional with the Lagrange multipliers λ, λ_β, $\Sigma_{\beta\gamma}$:

$$\begin{aligned}
\mathscr{L} = U_\alpha \Bigg\{ &-k_B c\int_{\mathbb{R}^3}\int_0^{+\infty} f\ln f p^\alpha \phi(I)\,d\,I\,d\mathbf{P} \\
&+\lambda\left[V^\alpha - mc\int_{\mathbb{R}^3}\int_0^{+\infty} f p^\alpha\phi(I)\,d\,I\,d\mathbf{P}\right] \\
&+\lambda_\beta\left[T^{\alpha\beta} - \frac{1}{mc}\int_{\mathbb{R}^3}\int_0^{+\infty} f\left(mc^2+I\right)p^\alpha p^\beta\,\phi(I)\,d\,I\,d\mathbf{P}\right] \\
&+\Sigma_{\beta\gamma}\left[A^{\alpha\beta\gamma} - \frac{1}{m^2c}\int_{\mathbb{R}^3}\int_0^{+\infty} f\left(mc^2+2I\right)p^\alpha p^\beta p^\gamma\phi(I)\,d\,I\,d\mathbf{P}\right]\Bigg\}\,.
\end{aligned}$$

Obviously, the multiplier $\Sigma_{\beta\gamma}$ is symmetric and traceless. This functional $\mathscr{L}$ has to be maximized with respect to the distribution function. So we have $\frac{\delta\mathscr{L}}{\delta f} = 0$, that is,

$$\begin{aligned}
\int_{\mathbb{R}^3}\int_0^{+\infty}\Bigg[&-\frac{k_B}{c}(\ln f+1) - \lambda\frac{m}{c} - \frac{1}{mc^3}\lambda_\beta p^\beta\left(mc^2+I\right) \\
&-\frac{1}{m^2c^3}\Sigma_{\beta\gamma}p^\beta p^\gamma\left(mc^2+2I\right)\Bigg]U_\alpha p^\alpha\phi(I)\,d\,I\,d\mathbf{P} = 0\,.
\end{aligned}$$

It follows that we have

$$f = \exp\left(-1-\frac{\chi}{k_B}\right), \tag{26.31}$$

where

$$\chi = m\lambda + \left(1 + \frac{I}{mc^2}\right)\lambda_\beta p^\beta + \frac{1}{m}\left(1 + \frac{2I}{mc^2}\right)\Sigma_{\beta\gamma}p^\beta p^\gamma .$$

In equilibrium (see (26.2), (26.3), and (26.5)) we have

$$\lambda - \lambda_E = -\frac{g}{T} \qquad \lambda^\beta = \lambda_E^\beta = \frac{U^\beta}{T}, \qquad \Sigma_E^{\beta\gamma} = 0$$

and the distribution function (26.31) becomes $f^{(E)}$ given in (26.19).

If we confine the theory within not so far from equilibrium in such a way that we can invert the Lagrange multipliers in terms of the physical field variables, we can expand, as usual in RET, the distribution function (26.31) in the neighborhood of equilibrium:

$$f - f^{(E)} = -\frac{f^{(E)}}{k_B}\Delta\chi, \tag{26.32}$$

where

$$\Delta\chi = m(\lambda - \lambda_E) + \left(1 + \frac{I}{mc^2}\right)\left(\lambda_\beta - \frac{U_\beta}{T}\right)p^\beta + \frac{1}{m}\left(1 + \frac{2I}{mc^2}\right)\Sigma_{\beta\gamma}p^\beta p^\gamma .$$

26.4.1 Inversion between Lagrange Multipliers and Field Variables

We prove that, in a theory not far from equilibrium (linear with respect the nonequilibrium variables), the closure using MEP can be done.

Inserting (26.32) into (26.29) and taking into account (27.2) and the equilibrium values of V^α and $T^{\alpha\beta}$ given in (5.2), we obtain the following linear algebraic system:

$$\begin{aligned}
&V_E^\alpha(\lambda - \lambda_E) + T_E^{\alpha\mu}\left(\lambda_\mu - \frac{U_\mu}{T}\right) + A_E^{\alpha\mu\nu}\Sigma_{\mu\nu} = 0\,,\\
&T_E^{\alpha\beta}(\lambda - \lambda_E) + A_{11}^{\alpha\beta\mu}\left(\lambda_\mu - \frac{U_\mu}{T}\right) + A_{12}^{\alpha\beta\mu\nu}\Sigma_{\mu\nu}\\
&\qquad = -\frac{k_B}{m}\left(t^{\langle\alpha\beta\rangle_3} + \Pi h^{\alpha\beta} + \frac{2}{c^2}U^{(\alpha}q^{\beta)}\right),\\
&A_E^{\alpha\beta\gamma}(\lambda - \lambda_E) + A_{12}^{\alpha\beta\gamma\mu}\left(\lambda_\mu - \frac{U_\mu}{T}\right) + A_{22}^{\alpha\beta\gamma\mu\nu}\Sigma_{\mu\nu} = -\frac{k_B}{m}(A^{\alpha\beta\gamma} - A_E^{\alpha\beta\gamma}),
\end{aligned} \tag{26.33}$$

where the new tensors are given by

$$A_{11}^{\alpha\beta\mu} = \frac{1}{m^3c^3}\int_{\mathbb{R}^3}\int_0^{+\infty} f^{(E)} p^\alpha p^\beta p^\mu \left(mc^2 + I\right)^2 \phi(I)\, d\,I\, d\mathbf{P} = -\frac{k_B}{m}\frac{\partial}{\partial \lambda_\mu^E} T_E^{\alpha\beta},$$

$$A_{12}^{\alpha\beta\mu\nu} = \frac{1}{m^4c^3}\int_{\mathbb{R}^3}\int_0^{+\infty} f^{(E)} p^\alpha p^\beta p^\mu p^\nu \left(mc^2 + I\right)\left(mc^2 + 2I\right) \phi(I)\, d\,I\, d\mathbf{P} = -\frac{k_B}{m}\frac{\partial}{\partial \lambda_\nu^E} A_E^{\alpha\beta\mu},$$

$$A_{22}^{\alpha\beta\gamma\mu\nu} = \frac{1}{m^5c^3}\int_{\mathbb{R}^3}\int_0^{+\infty} f^{(E)} p^\alpha p^\beta p^\gamma p^\mu p^\nu \left(mc^2 + 2I\right)^2 \phi(I)\, d\,I\, d\mathbf{P}.$$

After obtaining $(\lambda - \lambda_E)$, $\left(\lambda_\mu - \frac{U_\mu}{T}\right)$, and $\Sigma_{\mu\nu}$ from $(26.33)_{1,2}$, we substitute them into $(26.33)_3$ to obtain the requested closure, that is, the expression of $A^{\alpha\beta\gamma} - A_E^{\alpha\beta\gamma}$. It is true that, in effect, we need only the expression of $A^{\alpha\langle\beta\gamma\rangle} - A_E^{\alpha\langle\beta\gamma\rangle}$, but there is no problem if we deduce firstly all the tensor $A^{\alpha\beta\gamma} - A_E^{\alpha\beta\gamma}$ and, subsequently, take only its traceless part. In this way calculations become easier.

It is possible to prove the following:

Theorem 26.4 *The closure near equilibrium via MEP gives the following triple tensor $A^{\alpha\beta\gamma}$ as a function of the* 14 *physical variables associated with V^α and $T^{\alpha\beta}$:*

$$A^{\alpha\beta\gamma} - A_E^{\alpha\beta\gamma} = \frac{-3}{c^2}\frac{N_1^\pi}{D_1^\pi}\Pi U^\alpha U^\beta U^\gamma - 3\frac{N_{11}^\pi}{D_1^\pi}\Pi h^{(\alpha\beta}U^{\gamma)} + \frac{3}{c^2}\frac{N_3}{D_3} q^{(\alpha}U^\beta U^{\gamma)} + \\ + \frac{3}{5}\frac{N_{31}}{D_3} h^{(\alpha\beta}q^{\gamma)} + 3C_5 t^{(\langle\alpha\beta\rangle}U^{\gamma)}. \tag{26.34}$$

The proof is given in [153] together with the explicit expression of the scalar coefficients in (26.34).

26.5 Production Term in Relativistic Polyatomic Gas

To have the full closure, we need the explicit expression of the production term $I^{\beta\gamma}$ appearing in (26.28). From $(26.29)_4$, we need to know an explicit expression of the collision term Q. But it is not simple to evaluate Q even in the classical limit. Since we are now considering a theory not far from equilibrium, the expression of the tensor must be in the following form:

$$I^{\langle\beta\gamma\rangle} = B_1\Pi\left(g^{\beta\gamma} - \frac{4}{c^2}U^\beta U^\gamma\right) + B_3\, t^{\langle\beta\gamma\rangle} + B_4^q\left(U^\beta q^\gamma + U^\gamma q^\beta\right). \tag{26.35}$$

Recently Carrisi, Pennisi, and Ruggeri [451] was able to evaluate the coefficients in (26.35) using a new relativistic BGK model for the collision term Q proposed in [452].

26.5.1 A New Relativistic BGK Model

In the relativistic framework, the most important generalization of the BGK approximation was made by Marle [453] and successively by Anderson and Witting [454]. The Marle model is an extension of the non-relativistic BGK model in the Eckart frame:

$$Q = -\frac{m}{\tau}\left(f - f^{(E)}\right),$$

where τ is the relaxation time in the rest frame where the momentum of particles is zero. The Anderson–Witting model provides another expression of the Q, which has been widely used. It is given in the Landau-Lifshitz frame as follows:

$$Q = -\frac{U_{L\mu} p^{\mu}}{c^2\tau}\left(f - f^{(E)}\right),$$

where $U_{L\mu}$ indicates the four-velocity according with the Landau-Lifshitz definition.

Both proposals have some weak points. Marle's one satisfies the conservation law of particle number and energy momentum in a particular local equilibrium state. Another weak point of Marle model is that the relaxation time becomes unbounded in the case of particle with zero rest mass [280]. The weak point of the Anderson–Witting model is that it is described in the Landau-Lifshitz frame that is less used in literature because of the complexity in the conservation of number of particle [281]. A comparison between the two BGK models and also a study on the macroscopic Marle-Grad 14 moments were made in [455]. In particular, the Cauchy problem has been studied for the linearized kinetic equations with the Marle and Anderson–Witting models, and compared the resulting dispersion relations with the 14-moment theory.

Starting from these considerations, Pennisi and Ruggeri [452] proposed a variant of Anderson–Witting model in the Eckart frame for both relativistic monatomic and polyatomic gases. They proved that the conservation laws of particle number and energy-momentum are satisfied and the H-theorem holds. For details, in particular, for a gas without structure, see [452]. Here we consider only the case of gas with structure.

The proposal of Pennisi and Ruggeri for relativistic gas with structure is as follows:

$$Q = \frac{U_\alpha p^\alpha}{c^2 \tau} \left(f^{(E)} - f - f^{(E)}\, p^\mu q_\mu \frac{1 + \frac{I}{mc^2}}{bmc^2} \right)$$

with

$$\begin{aligned} b =& \frac{mnc^2}{3} \frac{\int_0^{+\infty} J_{4,1}(\gamma^*) \left(1 + \frac{I}{mc^2}\right)^2 \phi(I)\, d\, I}{\int_0^{+\infty} J_{2,1}(\gamma^*)\, \phi(I)\, d\, I} \\ =& \frac{nk_B T}{A(\gamma) K_2(\gamma)} \int_0^{+\infty} K_3(\gamma^*) \phi(I)\, d\, I. \end{aligned}$$

The coefficient b is obtained from the identity:

$$\begin{aligned} \psi^{\alpha\beta\mu} &= \frac{c}{m} \int_{\mathbb{R}^3} \int_0^{+\infty} f^{(E)} p^\alpha p^\beta p^\mu \left(1 + \frac{I}{mc^2}\right)^2 \phi(I) d\mathbf{P}\, d\, I \\ &= aU^\alpha U^\beta U^\mu + b \left(h^{\alpha\beta} U^\mu + h^{\alpha\mu} U^\beta + h^{\beta\mu} U^\alpha\right) \end{aligned}$$

with

$$a = \frac{\rho}{\gamma A(\gamma) K_2(\gamma)} \int_0^{+\infty} \left[3K_3(\gamma^*) + \gamma^* K_2(\gamma^*)\right] \phi(I)\, d\, I.$$

26.5.2 *Production Tensor $I^{\beta\gamma}$, Entropy Inequality, and Convexity*

As mentioned before, by using the previous BGK model in [451], the authors gave explicit expressions for the coefficients in (26.35). They were also able to give the necessary conditions coming from the entropy inequality and from the requirement that the system is symmetric.

26.6 Space-Time Decomposition and the Classical Limit

We now rewrite the closed system of the relativistic theory expressed by the equations (26.28) with (27.2) and (26.34) in the space-time form. For this purpose we recall

$$U^\alpha = \frac{dx^\alpha}{d\tau} = \frac{dx^\alpha}{dt}\frac{dt}{d\tau} \equiv \bar{\Gamma}(c, v^i),$$

$\bar{\Gamma} = \dfrac{1}{\sqrt{1-\frac{v^2}{c^2}}}$ is the macroscopic Lorentz factor,

τ is the proper time: $\dfrac{dt}{d\tau} = \bar{\Gamma}$,

v^i is the velocity components, $v^2 = -v^i v_i$,

$e = nm(c^2 + \varepsilon)$, ε is the internal energy,

$$h^{00} = \bar{\Gamma}^2 - 1, \quad h^{0i} = \bar{\Gamma}^2 \frac{v^i}{c}, \quad h^{ij} = \bar{\Gamma}^2 \frac{v^i v^j}{c^2} + \delta^{ij},$$

from the orthogonality $U_\alpha q^\alpha = 0$, we have $q^0 = \dfrac{1}{c} q_i v^i$,

from the orthogonality $U_\alpha t^{\langle\alpha\beta\rangle_3} = 0$, we have $t^{\langle 0i\rangle} = \dfrac{1}{c} t^{\langle ij\rangle} v^j$,

$$t^{\langle 00\rangle} = \frac{1}{c^2} t^{\langle ij\rangle} v_i v_j .$$

By the previous definitions, (27.2), (26.30), and (26.34) become

$$V^0 = mn\bar{\Gamma}c, \quad V^i = mn\bar{\Gamma}v^i,$$

$$T^{00} = \frac{1}{c^2} t^{\langle ij\rangle} v_i v_j + (p+\Pi)(\bar{\Gamma}^2 - 1) + \frac{2}{c^2}\bar{\Gamma} q_i v^i + \rho c^2 \bar{\Gamma}^2 + \rho\varepsilon\, \bar{\Gamma}^2, \tag{26.36}$$

$$T^{0i} = \frac{1}{c} t^{\langle ij\rangle} v_j + \frac{1}{c}(p+\Pi)\bar{\Gamma}^2 v^i + \frac{1}{c^2}\bar{\Gamma}\left(cq^i + \frac{1}{c} v^i q^j v_j\right) + \rho\left(\frac{\varepsilon}{c^2} + 1\right)\bar{\Gamma}^2 c v^i,$$

$$T^{ij} = t^{\langle ij\rangle} + (p+\Pi)\left(\frac{1}{c^2}\bar{\Gamma}^2 v^i v^j + \delta^{ij}\right) + \frac{1}{c^2}\bar{\Gamma}\left(v^j q^i + v^i q^j\right) + \rho\left(\frac{\varepsilon}{c^2} + 1\right)\bar{\Gamma}^2 v^i v^j .$$

$$
\begin{aligned}
A^{ijk} =& C_5 \bar{\Gamma} \left(t^{\langle ij \rangle} v^k + t^{\langle ik \rangle} v^j + t^{\langle jk \rangle} v^i \right) + \bar{\Gamma}^3 v^i v^j v^k \left\{ A_1^0 + \frac{3}{c^2} \left(A_{11}^0 - \frac{N_1^\pi + N_{11}^\pi}{D_1^\pi} \Pi \right) \right\} \\
&+ \bar{\Gamma}^2 \frac{N_3 + \frac{N_{31}}{5}}{c^2 D_3} \left(v^i v^j q^k + v^i v^k q^j + v^j v^k q^i \right) + \bar{\Gamma} \left(A_{11}^0 - \frac{N_{11}^\pi}{D_1^\pi} \Pi \right) \left(\delta^{ij} v^k + \delta^{ik} v^j + \delta^{jk} v^i \right) \\
&+ \frac{N_{31}}{5 D_3} \left(\delta^{ij} q^k + \delta^{ik} q^j + \delta^{jk} q^i \right), \\
A^{0ij} =& C_5 \frac{\bar{\Gamma}}{c} \left(c^2 t^{\langle ij \rangle} - t^{\langle ik \rangle} v_k v^j - t^{\langle jk \rangle} v_k v^i \right) + \bar{\Gamma}^3 c v^i v^j \left\{ A_1^0 + \frac{3}{c^2} \left(A_{11}^0 - \frac{N_1^\pi + N_{11}^\pi}{D_1^\pi} \Pi \right) \right\} \\
&- \bar{\Gamma}^2 \frac{N_3 + \frac{N_{31}}{5}}{c^3 D_3} q^k v_k v^i v^j + \bar{\Gamma} c \left(A_{11}^0 - \frac{N_{11}^\pi}{D_1^\pi} \Pi \right) \delta^{ij} + \bar{\Gamma}^2 \frac{N_3 + \frac{N_{31}}{5}}{c D_3} (q^i v^j + q^j v^i) \\
&- \frac{N_{31}}{5 D_3} \frac{1}{c} q^k v_k \delta^{ij}, \\
A^{00i} =& C_5 \bar{\Gamma} \left(\frac{1}{c^2} t^{\langle hk \rangle} v_h v_k v^i - 2 t^{\langle ik \rangle} v_k \right) + \bar{\Gamma} v^i \left\{ A_1^0 \bar{\Gamma}^2 c^2 + (3 \bar{\Gamma}^2 - 1) \left(A_{11}^0 - \frac{N_{11}^\pi}{D_1^\pi} \Pi \right) \right. \\
&\left. - 3 \frac{N_1^\pi}{D_1^\pi} \bar{\Gamma}^2 \Pi \right\} + \bar{\Gamma}^2 \frac{N_3 + \frac{N_{31}}{5}}{D_3} \left(q^i - \frac{2}{c^2} v^i q^k v_k \right) - \frac{N_{31}}{5 D_3} q^i .
\end{aligned}
$$

In view of the classical limit for $c \to \infty$, the system of balance laws (26.28) can be rewritten as

$$
\begin{aligned}
&\partial_\alpha V^\alpha = 0 &&\leftrightarrows\quad \partial_\alpha V^\alpha = 0 \\
&\partial_\alpha T^{\alpha\beta} = 0 &&\leftrightarrows\quad \begin{cases} \partial_\alpha T^{\alpha i} = 0 \\ 2\partial_\alpha \left(c T^{\alpha 0} - c^2 V^\alpha \right) = 0 \end{cases} \\
&\partial_\alpha A^{\alpha \langle \beta\gamma \rangle} = I^{\langle \beta\gamma \rangle} &&\leftrightarrows\quad \begin{cases} \partial_\alpha B^{\alpha \langle ij \rangle_3} = I^{\langle ij \rangle_3} \\ \partial_\alpha \left(4 B^{\alpha ij} g_{ij} + 6 c T^{\alpha 0} - 3 c^2 V^\alpha \right) = 4 I^{rs} g_{rs} \\ 2 \partial_\alpha \left(c B^{\alpha 0i} - c^2 T^{\alpha i} \right) = 2 c I^{0i} \end{cases}
\end{aligned}
$$

where we have put

$$
B^{\alpha\beta\gamma} = A^{\alpha \langle \beta\gamma \rangle} = A^{\alpha\beta\gamma} - \frac{1}{4} A^{\alpha\mu\nu} g_{\mu\nu} g^{\beta\gamma},
$$

and $\langle \dots \rangle_3$ indicates the deviatoric part in the 3-dimensional space components. Now it is possible to prove the following convergence when $c \to \infty$:

$$\partial_\alpha V^\alpha = 0 \qquad \to \qquad \partial_t F + \partial_k F_k = 0,$$

$$\partial_\alpha T^{\alpha i} = 0 \qquad \to \qquad \partial_t F_i + \partial_k F_{ki} = 0,$$

$$2\,\partial_\alpha \left(c\, T^{\alpha 0} - c^2\, V^\alpha\right) = 0 \qquad \to \qquad \partial_t G_{ll} + \partial_k G_{llk} = 0,$$

$$\partial_\alpha B^{\alpha \langle ij \rangle_3} = I^{\langle ij \rangle_3} \qquad \to \qquad \partial_t F_{\langle ij \rangle} + \partial_k F_{k\langle ij \rangle} = P_{\langle ij \rangle},$$

$$\partial_\alpha \left(-4\, B^{\alpha ij}\, g_{ij} - 6\, c\, T^{\alpha 0} + 3c^2 V^\alpha\right) = -4 I^{rs} g_{rs} \qquad \to \qquad \partial_t F_{ll} + \partial_k F_{kll} g_{rs} = P_{ll},$$

$$2\,\partial_\alpha \left(c\, B^{\alpha\, 0i} - c^2\, T^{\alpha i}\right) = 2c I^{0i} \qquad \to \qquad \partial_t G_{lli} + \partial_k G_{llik} = Q_{lli}.$$

The equations in the left side correspond to the equations of classical ET$_{14}$: (6.2). Concerning the coefficients of the closure, taking into account the expansion of the integrals $J_{m,n}(\gamma)$, (26.26) and also (26.36), after very straightforward but tedious calculations, we have

$$\lim_{\gamma\to\infty} A_1^0 = \rho, \qquad \lim_{\gamma\to\infty} A_{11}^0 = p, \qquad \lim_{\gamma\to\infty} \frac{N_{11}^\pi}{D_1^\pi} = -1, \qquad \lim_{\gamma\to\infty} \frac{N_1^\pi}{D_1^\pi} = 0,$$

$$\lim_{\gamma\to\infty} C_5 = 1, \qquad \lim_{\gamma\to\infty} \frac{N_{31}}{D_3} = \frac{10}{2a+7}, \qquad \lim_{\gamma\to\infty} \frac{N_3}{D_3} = 2\,,$$

$$\lim_{\gamma\to\infty} (A_1^0 - \rho)\gamma = \rho(2a+5), \qquad \lim_{\gamma\to\infty} (A_{11}^0 - p)\gamma = \frac{p}{2}(2a+7)\,,$$

$$\lim_{\gamma\to\infty} (C_5 - 1)\gamma = \frac{1}{2}(2a+9), \qquad \lim_{\gamma\to\infty} -\left(\frac{N_{11}^\pi}{D_1^\pi} + 1\right)\gamma = \frac{3}{2}(2a+9)\,,$$

$$\lim_{\gamma\to\infty} \tilde{D}_1^\pi = \frac{2}{3}\frac{p^4}{\rho}(a+1).$$

Then, when $c \to \infty$, we obtain the following theorem:

Theorem 26.5 *The relativistic system* (26.28) *converges, for* $c \to \infty$, *to the system for classical extended thermodynamics of polytropic polyatomic gases* (1.33).

The ultra-relativistic limit of the present theory was studied in [456], while the monatomic limit was the object of the paper [154].

Chapter 27
Many-Moment RET of Relativistic Polyatomic Gas and Classical Optimal Limit

Abstract In this chapter, we explain briefly the relativistic RET of a polyatomic gas with many moments associated with the relativistic Boltzmann-Chernikov kinetic equation truncated at the tensorial index N. We consider the classical limit when $c \to \infty$, and we find a natural order and new hierarchies for classical moments that are completely fixed for a given N. We show the main theorem proved by Pennisi and Ruggeri (J Stat Phys 179:231–246, 2020), which is the generalization of Theorem 5.1 for a monatomic gas. Several examples are also given.

27.1 Introduction

In Sect. 5.6, the classical limit of relativistic moments for a monatomic gas is studied, and then it is proved that there exists a unique choice of the classical moments.

A similar study can be done in the case of polyatomic gas as shown by Pennisi and Ruggeri [296]. We assume that, also in the present case of polyatomic gas, the system of the moments associated with the Boltzmann-Chernikov equation (5.18) has the same structure as the one in the monatomic-gas case (5.27):

$$\partial_\alpha A^{\alpha\alpha_1\cdots\alpha_n} = I^{\alpha_1\cdots\alpha_n} \quad \text{with} \quad n = 0, \cdots, N, \tag{27.1}$$

but with the following new moments defined by

$$\begin{aligned} A^{\alpha\alpha_1\cdots\alpha_n} &= \frac{c}{m^{n-1}} \int_{\mathbb{R}^3} \int_0^{+\infty} f\, p^\alpha p^{\alpha_1} \cdots p^{\alpha_n} \left(1 + a_n \frac{I}{m\,c^2}\right) \phi(I)\, d\,I\, d\mathbf{P}, \\ I^{\alpha_1\cdots\alpha_n} &= \frac{c}{m^{n-1}} \int_{\mathbb{R}^3} \int_0^{+\infty} Q\, p^\alpha p^{\alpha_1} \cdots p^{\alpha_n} \left(1 + a_n \frac{I}{m\,c^2}\right) \phi(I)\; dI\, d\mathbf{P}, \end{aligned} \tag{27.2}$$

where a_n $(n = 0, \cdots, N)$ are constant coefficients.

T. Ruggeri, M. Sugiyama, *Classical and Relativistic Rational Extended Thermodynamics of Gases*, https://doi.org/10.1007/978-3-030-59144-1_27

We will prove below that the necessary and sufficient condition to recover, in the classical limit, the F- and G-hierarchies studied in Chap. 9 is given by the relation for the coefficients: $a_n = n$ $(n = 0, \cdots, N)$. In the case of $N = 2$ we obtain a 15-field model, whose principal subsystem with 14 fields is the system established in Chap. 26.

27.2 Classical Limit of Relativistic Moment

Now we consider the classical limit of the moments (27.2) when $c \to \infty$. Then we have the following theorem:

Theorem 27.1 *For a prescribed truncation index N, if the coefficients $a_n = n$, then the relativistic moment system for a polyatomic gas (27.1) with (27.2) converges, when $c \to \infty$, to a system of classical moments with $N + 1$ hierarchies in the form:*

$$\partial_t H_{s,A} + \partial_i H_{s,iA} = J_{s,A}, \tag{27.3}$$

where s is the integer ($0 \leqslant s \leqslant N$), A is the multi-index ($0 \leqslant A \leqslant N - s$), and $H_{s,A}$, $H_{s,iA}$, and $J_{s,A}$ are given by

$$\begin{aligned}
H_{s,A} &= s\, G_{s,A} - (s-1) F_{s,A} = m \int_{\mathbb{R}^3} \int_0^{+\infty} f^C\, \xi_A\, \xi^{2(s-1)} \left(\xi^2 + 2s\frac{I}{m}\right) \phi(I)\, dI\, d\boldsymbol{\xi}\,, \\
H_{s,iA} &= s\, G_{s,iA} - (s-1) F_{s,iA} = m \int_{\mathbb{R}^3} \int_0^{+\infty} f^C\, \xi_i\, \xi_A\, \xi^{2(s-1)} \left(\xi^2 + 2s\frac{I}{m}\right) \phi(I)\, dI\, d\boldsymbol{\xi}\,, \\
J_{s,A} &= s\, Q_{s,A} - (s-1) P_{s,A} = m \int_{\mathbb{R}^3} \int_0^{+\infty} Q^C\, \xi_A\, \xi^{2(s-1)} \left(\xi^2 + 2s\frac{I}{m}\right) \phi(I)\, dI\, d\boldsymbol{\xi}\,.
\end{aligned} \tag{27.4}$$

Here f^C and Q^C are the classical distribution function and the classical collision term, respectively. Moreover

$$\xi_A = 1 \quad \textit{if} \quad A = 0, \qquad \xi_A = \xi_{i_1} \ldots \xi_{i_A} \quad \textit{if} \quad A > 0,$$

and

$$F_{s,A} = F_{j_1 j_1 j_2 j_2 \dots j_s j_s A} = m \int_{\mathbb{R}^3} \int_0^{+\infty} f^C \xi_A \xi^{2s} \phi(I)\, dI\, d\boldsymbol{\xi} \,,$$

$$F_{s,iA} = F_{i j_1 j_1 j_2 j_2 \dots j_s j_s A} = m \int_{\mathbb{R}^3} \int_0^{+\infty} f^C \xi_i \xi_A \xi^{2s} \phi(I)\, dI\, d\boldsymbol{\xi} \,,$$

$$P_{s,A} = P_{j_1 j_1 j_2 j_2 \dots j_s j_s A} = m \int_{\mathbb{R}^3} \int_0^{+\infty} Q^C \xi_A \xi^{2s} \phi(I)\, dI\, d\boldsymbol{\xi}$$

$$G_{s,A} = G_{ll j_2 j_2 \dots j_s j_s A} = m \int_{\mathbb{R}^3} \int_0^{+\infty} f^C \xi_A \xi^{2(s-1)} \left(\xi^2 + 2\frac{I}{m}\right) \phi(I)\, dI\, d\boldsymbol{\xi} \,,$$

$$G_{s,iA} = G_{lli j_2 j_2 \dots j_s j_s A} = m \int_{\mathbb{R}^3} \int_0^{+\infty} f^C \xi_i \xi_A \xi^{2(s-1)} \left(\xi^2 + 2\frac{I}{m}\right) \phi(I)\, dI\, d\boldsymbol{\xi} \,,$$

$$Q_{s,A} = Q_{ll j_2 j_2 \dots j_s j_s A} = m \int_{\mathbb{R}^3} \int_0^{+\infty} Q^C \xi_A \xi^{2(s-1)} \left(\xi^2 + 2\frac{I}{m}\right) \phi(I)\, dI\, d\boldsymbol{\xi} \,.$$

In particular, for $s = 0$ we have the F-hierarchy (9.1) *(left block), for $s = 1$ we have the G-hierarchy* (9.1) *(right block) with $M = N + 1$, and for s ($2 \leqslant s \leqslant N$) we have new mixed hybrid hierarchies* (27.3) *with* (27.4)*.*

For details of the proof see the paper [296]. In the next section, several concrete examples are given to show remarkable consequences of Theorem 27.1.

Remark 27.1 The requirement that $a_n = n$ is a necessary and sufficient condition such that, in the classical limit, we have the F- and G-hierarchies, which are obtained for $s = 0$ and $s = 1$, respectively, and are postulated for the classical RET theory of rarefied polyatomic gases in the previous chapters. Nevertheless the classical limit gives an unexpected and surprising result that, for polyatomic gases, we have in reality a more complex structure with new hybrid hierarchies as we will see in the next examples.

Remark 27.2 It is, however, important to emphasize that the system composed of the F- and G-hierarchies is a principal subsystem of the system with more complex structure. Therefore, as is always the case with any principal subsystems, it represents a simpler model than the original system but it is still valid and useful for analyzing nonequilibrium phenomena according with the nesting theory described in Chap. 9 (see also Remark 4.3 in Chap. 4).

27.3 Examples of Moments in the Classical Limit in the Case of Polyatomic Gas

- $N = 1$: Relativistic hierarchy of moments (Relativistic Euler equation): The system of Eqs. (26.28) reduces to the system:

$$\partial_\alpha A^\alpha = 0, \quad \partial_\alpha A^{\alpha\alpha_1} = 0,$$

and this converges to the classical system of moments (27.3) with 2 hierarchies, which, in the present case, reads

$$\begin{aligned}
(s=0): &\quad \partial_t F + \partial_i F_i = 0, \qquad \partial_t F_i + \partial_i F_{ij} = 0,\\
(s=1): &\quad \partial_t G_{ll} + \partial_i G_{lli} = 0,
\end{aligned}$$

i.e., classical Euler ET_5.

- $N = 2$: Relativistic hierarchy of moments (PR-model): The system:

$$\partial_\alpha A^\alpha = 0, \quad \partial_\alpha A^{\alpha\alpha_1} = 0, \quad \partial_\alpha A^{\alpha\alpha_1\alpha_2} = I^{\alpha_1\alpha_2}$$

converges to the classical system of moments with 3 hierarchies:

$$\begin{aligned}
(s=0): &\quad \partial_t F + \partial_i F_i = 0, \qquad \partial_t F_{i_1} + \partial_i F_{ii_1} = 0, \qquad \partial_t F_{i_1i_2} + \partial_i F_{ii_1i_2} = P_{i_1i_2},\\
(s=1): &\quad \partial_t G_{ll} + \partial_i G_{lli} = 0, \quad \partial_t G_{lli_1} + \partial_i G_{llii_1} = Q_{i_1},\\
(s=2): &\quad \partial_t \left(2G_{llj_2j_2} - F_{j_1j_1j_2j_2}\right) + \partial_i \left(2G_{llij_1j_1} - F_{j_1j_1j_2j_2}\right) = 2Q_{llj_1j_1} - P_{j_1j_1j_2j_2}.
\end{aligned} \tag{27.5}$$

In this case, we have the system of ET_{15}, which is composed of not only the equations of classical ET_{14} but also the 15-th hybrid equation, i.e., the last equation in (27.5).

- $N = 3$: Relativistic hierarchy of moments: The system:

$$\partial_\alpha A^\alpha = 0, \quad \partial_\alpha A^{\alpha\alpha_1} = 0, \quad \partial_\alpha A^{\alpha\alpha_1\alpha_2} = I^{\alpha_1\alpha_2}, \quad \partial_\alpha A^{\alpha\alpha_1\alpha_2\alpha_3} = I^{\alpha_1\alpha_2\alpha_3}$$

converges to the classical system of moments with 4 hierarchies:

$$\begin{aligned}
(s=0): &\quad \partial_t F + \partial_i F_i = 0, \qquad \partial_t F_{i_1} + \partial_i F_{ii_1} = 0,\\
&\quad \partial_t F_{i_1i_2} + \partial_i F_{ii_1i_2} = P_{i_1i_2}, \quad \partial_t F_{i_1i_2i_3} + \partial_i F_{ii_1i_2i_3} = P_{i_1i_2i_3},\\
(s=1): &\quad \partial_t G_{ll} + \partial_i G_{lli} = 0, \quad \partial_t G_{lli_1} + \partial_i G_{llii_1} = Q_{i_1},\\
&\quad \partial_t G_{lli_1i_2} + \partial_i G_{llii_1i_2} = Q_{i_1i_2},\\
(s=2): &\quad \partial_t \left(2G_{llj_2j_2} - F_{j_1j_1j_2j_2}\right) + \partial_i \left(2G_{llij_2j_2} - F_{ij_1j_1j_2j_2}\right) = 2Q_{llj_2j_2} - P_{j_1j_1j_2j_2},\\
&\quad \partial_t \left(2G_{llj_2j_2i_1} - F_{j_1j_1j_2j_2i_1}\right) + \partial_i \left(2G_{llij_2j_2i_1} - F_{ij_1j_1j_2j_2i_1}\right) = 2Q_{llj_2j_2i_1} - P_{j_1j_1j_2j_2i_1},\\
(s=3): &\quad \partial_t \left(3G_{llj_2j_2j_3j_3} - 2F_{j_1j_1j_2j_2j_3j_3}\right) + \partial_i \left(3G_{llij_2j_2j_3j_3} - 2F_{ij_1j_1j_2j_2j_3j_3}\right)\\
&\qquad = 3Q_{llj_2j_2j_3j_3} - 2P_{j_1j_1j_2j_2j_3j_3}.
\end{aligned}$$

This corresponds to ET_{35}.

27.4 Properties of the Moments in the Classical Limit

From the theorem and also from the previous examples, we confirm that the hierarchies satisfy all the requirements for the moments:

1. The F-hierarchy obtained when $s = 0$ has the maximum index of truncation N, while the G-hierarchy obtained when $s = 1$ has the truncation index $M = N - 1$. As a consequence, the classical system is Galilean invariant, and the characteristic velocities depend on the degrees of freedom of a polyatomic molecule [119].
2. The maximum number of truncation index in the classical limit $\bar{N} = 2N$ is obtained when $s = N$ and is even. Therefore all moments obtained by the MEP are integrable.
3. Any classical limit system is a principal subsystem of the previous one. This is automatically true because the set of limit moments of relativistic truncation order N is contained in the set of limit moments of the successive order $N + 1$.

Remark 27.3 We can verify that the number of equations of relativistic moment system is the same as the number of equations in the classical limit. In fact, we recall that the number of the components of a symmetric tensor with respect all indices with rank r and in dimension d is given by

$$\binom{d+r-1}{r} = \frac{(d+r-1)!}{r!\,(d-1)!}.$$

Then we take all tensors of increasing order $0 \leqslant r \leqslant n$. The total number of components is given by

$$\sum_{r=0}^{n} \frac{(d+r-1)!}{r!(d-1)!} = \frac{(n+d)!}{n!\, d!}. \tag{27.6}$$

Therefore the total number of moments $\mathscr{N}$ of (26.28) is

$$\mathscr{N} = \frac{(N+4)!}{24\, N!}. \tag{27.7}$$

Concerning the classical limit, for any value of s there are $N - s$ free indexes. Therefore, from (27.6) with $n = N - s$ and $d = 3$, the total number of moments in (27.3) is

$$\sum_{s=0}^{N} \frac{(N-s+3)!}{6\,(N-s)!} = \frac{(N+4)!}{24\, N!}.$$

The classical hierarchy is completely fixed for a prescribed N. We have a number of moments $\mathcal{N}$ given by (27.7). For example, we write down $\mathcal{N}$ for the first 10 integers N:

$$\begin{pmatrix} N \\ \mathcal{N} \end{pmatrix} = \begin{pmatrix} 1 & 2 & 3 & 4 & 5 & 6 & 7 & 8 & 9 & 10 \\ 5 & 15 & 35 & 70 & 126 & 210 & 330 & 495 & 715 & 1001 \end{pmatrix}.$$

Therefore it is interesting to notice that only particular number of moments are admissible in the limiting case of relativistic theory: ET_5, ET_{15}, ET_{35}, ET_{70}, etc.

The theory of a monatomic gas can be regarded as a limit of the theory of a polyatomic gas, where the tensors A's and I's are given by (5.28). In this case, the G-moments coincide with the F-moments. Therefore Theorem 27.1 reduces to Theorem 5.1.

Part VIII
Classical and Relativistic Mixture of Gases

Chapter 28
Multi-Temperature Mixture of Fluids

Abstract We present a survey on the results concerning some different models of mixture of compressible fluids. In particular, we discuss the most realistic case of mixture where each component has its own temperature (*MT*). We first compare the solutions of this model to the one with unique common temperature (*ST*). In the case of Eulerian fluids, it is shown that the corresponding *ST* differential system is a *principal subsystem* of the *MT* system. Global behavior of smooth solutions for large time for both systems is also discussed through the application of the Shizuta-Kawashima K-condition.

Then we introduce the concept of the average temperature of a mixture based on the consideration that the internal energy of the mixture is the same as that in the case of single-temperature mixture. As a consequence, it is shown that the entropy of the mixture reaches a local maximum in equilibrium. Through the procedure of the Maxwellian iteration, a new constitutive equation for nonequilibrium temperatures of components is obtained in a classical limit, together with the Fick law for the diffusion flux.

In order to justify the Maxwellian iteration, we present, for dissipative fluids, a possible approach to a classical theory of mixtures with the multi-temperature. We prove that the differences of temperatures between the components imply the existence of a new *dynamic pressure* even if fluids have zero bulk viscosities.

28.1 Introduction

Modeling and analysis of mixtures is a challenging and stimulating problem. In the case of gaseous mixture, it can be successfully studied by using not only the method of the kinetic theory of gases but also the method of the continuum theory of fluids. In either case, appropriate macroscopic equations can be derived in order to explain irreversible phenomena like diffusion, heat transfer, and chemical reactions. However, since there still remain many open problems, the study of mixtures is one of the fields of active research.

T. Ruggeri, M. Sugiyama, *Classical and Relativistic Rational Extended Thermodynamics of Gases*, https://doi.org/10.1007/978-3-030-59144-1_28

In the classical theory of diffusion, although different concentrations of the components in a mixture are taken into account, only one common global velocity and one temperature are considered; the velocity of each component is obtained through the constitutive equation, that is, the classical Fick law.

Then this classical model is followed by more sophisticated models that are constructed on the basis of either the continuum theory or the kinetic theory of gases. Among these, there appeared two main approaches with two constitutive theories. They can be classified by the answer to the question: Should the components of a mixture have a common temperature or not? Both of these approaches, i.e., single-temperature (ST) one and multi-temperature (MT) one, gained considerable successes in modeling the behavior of mixture.

The MT approach is naturally embedded into Maxwell's kinetic theory of mixtures [17, 457]. This theory comes on its own especially in gases where atomic masses of the constituents are different, for example, in plasmas where the constituents are electrons, ions, and neutral atoms. The relevance of the MT model is thus put in evidence and can be further supported by the analysis of plasmas at high temperatures. See for details the references by Kannappan and Bose [458, 459] and Bose and Seeniraj [460]. Influence of electron and phonon temperatures on the efficiency of thermoelectric conversion is the subject of a recent paper of Sellitto, Cimmelli, and Jou [461].

The idea of multiple temperatures thus reflects the physically justified intention to get a deeper insight into nonequilibrium processes in mixtures, but this concept seems to be mostly overlooked in the context of macroscopic theories. Nevertheless, it was appreciated and naturally embedded in the kinetic theory of gases, which is perfectly designed to monitor the processes far from equilibrium. For example, it appeared as an efficient tool in nonequilibrium flow computations [462].

Apart from the physical reasons, these two theories are completely different from each other from a mathematical point of view: In general, the MT system does not admit the solution with single temperature $T_1 = T_2 = \ldots = T$ even if we pose this condition initially.

The theory of homogeneous mixtures was developed within the framework of rational thermodynamics by Truesdell [57] under the assumption that each component obeys the same balance laws as a single fluid. A huge amount of literature appeared after that in the context of continuum approach, see, for example, [5, 25, 463–467].

Part of the survey on these results can be seen in [158–160], and also in review papers [468–470].

28.2 Mixtures in Rational Thermodynamics

In the context of rational thermodynamics, the description of a homogeneous mixture of n components is based on the postulate that each component obeys to the same balance laws as those to which a single fluid obeys [5, 25, 57]. The laws express the balance equations of masses, momenta, and energies:

$$\begin{cases} \dfrac{\partial \rho_\alpha}{\partial t} + \operatorname{div}(\rho_\alpha \mathbf{v}_\alpha) = \tau_\alpha, \\[2ex] \dfrac{\partial(\rho_\alpha \mathbf{v}_\alpha)}{\partial t} + \operatorname{div}(\rho_\alpha \mathbf{v}_\alpha \otimes \mathbf{v}_\alpha - \mathbf{t}_\alpha) = \mathbf{m}_\alpha, \qquad (\alpha = 1, 2, \ldots, n) \\[2ex] \dfrac{\partial\left(\frac{1}{2}\rho_\alpha v_\alpha^2 + \rho_\alpha \varepsilon_\alpha\right)}{\partial t} + \operatorname{div}\left\{\left(\frac{1}{2}\rho_\alpha v_\alpha^2 + \rho_\alpha \varepsilon_\alpha\right)\mathbf{v}_\alpha - \mathbf{t}_\alpha \mathbf{v}_\alpha + \mathbf{q}_\alpha\right\} = e_\alpha. \end{cases} \tag{28.1}$$

On the left-hand side, ρ_α is the density, $\mathbf{v}_\alpha$ is the velocity, ε_α is the specific internal energy, $\mathbf{q}_\alpha$ is the heat flux, and $\mathbf{t}_\alpha$ is the stress tensor of the component α. The stress tensor $\mathbf{t}_\alpha$ can be decomposed into a pressure part $-p_\alpha \mathbf{I}$ and a viscous part $\boldsymbol{\sigma}_\alpha$ as

$$\mathbf{t}_\alpha = -p_\alpha \mathbf{I} + \boldsymbol{\sigma}_\alpha.$$

On the right-hand sides, τ_α, $\mathbf{m}_\alpha$, and e_α represent the production terms related to the interactions between components. Due to the total conservation of mass, momentum, and energy of the mixture, the sum of the production terms over all components must vanish:

$$\sum_{\alpha=1}^{n} \tau_\alpha = 0, \quad \sum_{\alpha=1}^{n} \mathbf{m}_\alpha = \mathbf{0}, \quad \sum_{\alpha=1}^{n} e_\alpha = 0. \tag{28.2}$$

Global mixture quantities ρ, $\mathbf{v}$, ε, $\mathbf{t}$ and $\mathbf{q}$ are defined as

$$\begin{aligned} &\rho = \sum_{\alpha=1}^{n} \rho_\alpha && \text{total mass density,} \\ &\mathbf{v} = \frac{1}{\rho}\sum_{\alpha=1}^{n} \rho_\alpha \mathbf{v}_\alpha && \text{mixture velocity,} \\ &\varepsilon = \varepsilon_I + \frac{1}{2\rho}\sum_{\alpha=1}^{n} \rho_\alpha u_\alpha^2 && \text{internal energy,} \\ &\mathbf{t} = -p\mathbf{I} + \boldsymbol{\sigma}_I - \sum_{\alpha=1}^{n} (\rho_\alpha \mathbf{u}_\alpha \otimes \mathbf{u}_\alpha) && \text{stress tensor,} \\ &\mathbf{q} = \mathbf{q}_I + \sum_{\alpha=1}^{n} \rho_\alpha \left(\varepsilon_\alpha + \frac{p_\alpha}{\rho_\alpha} + \frac{1}{2}u_\alpha^2\right)\mathbf{u}_\alpha && \text{flux of internal energy,} \end{aligned} \tag{28.3}$$

where

$$\mathbf{u}_\alpha = \mathbf{v}_\alpha - \mathbf{v} \qquad \left(\sum_{\alpha=1}^{n} \rho_\alpha \mathbf{u}_\alpha = 0\right) \tag{28.4}$$

is the diffusion velocity of the component α,

$$p = \sum_{\alpha=1}^{n} p_\alpha$$

is the total pressure, and

$$\varepsilon_I = \frac{1}{\rho} \sum_{\alpha=1}^{n} \rho_\alpha \varepsilon_\alpha, \quad \mathbf{q}_I = \sum_{\alpha=1}^{n} \mathbf{q}_\alpha, \quad \boldsymbol{\sigma}_I = \sum_{\alpha=1}^{n} \boldsymbol{\sigma}_\alpha$$

are, respectively, the total intrinsic internal energy, heat flux, and shear stress.

Summing up all the Eqs. (28.1) and taking (28.2) into account, we obtain

$$\begin{cases} \dfrac{\partial \rho}{\partial t} + \operatorname{div}(\rho \mathbf{v}) = 0, \\[2ex] \dfrac{\partial (\rho \mathbf{v})}{\partial t} + \operatorname{div}(\rho \mathbf{v} \otimes \mathbf{v} - \mathbf{t}) = \mathbf{0}, \\[2ex] \dfrac{\partial \left(\frac{1}{2}\rho v^2 + \rho \varepsilon\right)}{\partial t} + \operatorname{div}\left\{\left(\dfrac{1}{2}\rho v^2 + \rho \varepsilon\right)\mathbf{v} - \mathbf{t}\mathbf{v} + \mathbf{q}\right\} = 0, \end{cases}$$

which are the conservation laws of mass, momentum, and energy of the mixture. They are in the same form as for a single fluid.

In order to compare the balance equations of mixture and of single fluid, we rewrite (28.1), taking into account that the production terms are not independent (see (28.2)), in the following equivalent form:

$$
\begin{cases}
\dfrac{\partial \rho}{\partial t} + \operatorname{div}(\rho \mathbf{v}) = 0, \\[2ex]
\dfrac{\partial (\rho \mathbf{v})}{\partial t} + \operatorname{div}(\rho \mathbf{v} \otimes \mathbf{v} - \mathbf{t}) = \mathbf{0}, \\[2ex]
\dfrac{\partial \left(\frac{1}{2}\rho v^2 + \rho \varepsilon\right)}{\partial t} + \operatorname{div}\left\{\left(\dfrac{1}{2}\rho v^2 + \rho \varepsilon\right)\mathbf{v} - \mathbf{t}\mathbf{v} + \mathbf{q}\right\} = 0, \\[2ex]
\dfrac{\partial \rho_b}{\partial t} + \operatorname{div}(\rho_b \mathbf{v}_b) = \tau_b, \qquad (b = 1, \dots, n-1) \\[2ex]
\dfrac{\partial (\rho_b \mathbf{v}_b)}{\partial t} + \operatorname{div}(\rho_b \mathbf{v}_b \otimes \mathbf{v}_b - \mathbf{t}_b) = \mathbf{m}_b, \\[2ex]
\dfrac{\partial \left(\frac{1}{2}\rho_b v_b^2 + \rho_b \varepsilon_b\right)}{\partial t} + \operatorname{div}\left\{\left(\dfrac{1}{2}\rho_b v_b^2 + \rho_b \varepsilon_b\right)\mathbf{v}_b - \mathbf{t}_b\mathbf{v}_b + \mathbf{q}_b\right\} = e_b,
\end{cases}
\tag{28.5}
$$

where the index b runs from 1 to $n-1$.

In this multi-temperature model (*MT*), used in particular in plasma physics [471], we have $5n$ independent field variables ρ_α, $\mathbf{v}_\alpha$, and T_α ($\alpha = 1, 2, \dots, n$), where T_α is the temperature of the component α. To close the system (28.5) of the field equations of the mixture, we must write the constitutive equations for the quantities $p_\alpha, \varepsilon_\alpha, \mathbf{q}_\alpha, \boldsymbol{\sigma}_\alpha$ ($\alpha = 1, 2, \dots, n$) and τ_b, $\mathbf{m}_b$, e_b ($b = 1, \dots, n-1$) in terms of the field variables ρ_α, $\mathbf{v}_\alpha$, and T_α ($\alpha = 1, 2, \dots, n$).

28.2.1 Galilean Invariance of Field Equations

The system (28.5) is a particular case of the balance laws (2.1) with (2.50):

$$
\mathbf{F}^0 = \begin{pmatrix} \rho \\ \rho v^j \\ \frac{1}{2}\rho v^2 + \rho \varepsilon \\ \rho c_b \\ \rho c_b (u_b^j + v^j) \\ \frac{1}{2}\rho c_b (\mathbf{u}_b + \mathbf{v})^2 + \rho c_b \varepsilon_b \end{pmatrix}, \quad
\hat{\mathbf{F}}^0 = \begin{pmatrix} \rho \\ 0^k \\ \rho \varepsilon \\ \rho c_b \\ \rho c_b u_b^k \\ \frac{1}{2}\rho c_b u_b^2 + \rho c_b \varepsilon_b \end{pmatrix},
$$

$$\boldsymbol{\Phi}^i = \begin{pmatrix} 0^i \\ -t^{ij} \\ -t^{ik}v_k + q^i \\ \rho c_b u_b^i \\ \rho c_b u_b^i(u_b^j + v^j) - t_b^{ij} \\ \begin{cases} \left(\frac{1}{2}\rho c_b(\mathbf{u}_b + \mathbf{v})^2 + \rho c_b \varepsilon_b\right) u_b^i \\ -t_b^{ik}(u_b^k + v_k) + q_b^i \end{cases} \end{pmatrix}, \quad \hat{\boldsymbol{\Phi}}^i = \begin{pmatrix} 0^i \\ -t^{ik} \\ q^i \\ \rho c_b u_b^i \\ \rho c_b u_b^i u_b^k - t_b^{ik} \\ \begin{cases} \left(\frac{1}{2}\rho c_b u_b^2 + \rho c_b \varepsilon_b\right) u_b^i \\ -t_b^{ik} u_b^k + q_b^i \end{cases} \end{pmatrix},$$

$$\mathbf{f} = \begin{pmatrix} 0 \\ 0^j \\ 0 \\ \tau_b \\ m_b^j \\ e_b \end{pmatrix}, \quad \hat{\mathbf{f}} = \begin{pmatrix} 0 \\ 0^k \\ 0 \\ \hat{\tau}_b \\ \hat{m}_b^k \\ \hat{e}_b \end{pmatrix}. \tag{28.6}$$

Theorem 28.1 *The linear operator* $\mathbf{X}(\mathbf{v})$ *(see (2.61) and (2.49)), which assures the Galilean invariance of the field Eqs. (28.5), has the form:*

$$\mathbf{X}(\mathbf{v}) = \begin{pmatrix} 1 & 0_k & 0 & 0 & 0_k & 0 \\ v^j & \delta^{jk} & 0^j & 0^j & 0^{jk} & 0^j \\ \frac{v^2}{2} & v_k & 1 & 0 & 0_k & 0 \\ 0 & 0_k & 0 & 1 & 0_k & 0 \\ 0^j & 0^{jk} & 0^j & v^j & \delta^{jk} & 0^j \\ 0 & 0_k & 0 & \frac{v^2}{2} & v_k & 1 \end{pmatrix}. \tag{28.7}$$

As a consequence, the following relations between the production terms and their internal (intrinsic) counterparts are obtained:

$$\begin{aligned} \tau_b &= \hat{\tau}_b, \\ m_b^j &= \hat{\tau}_b v^j + \hat{m}_b^j, \\ e_b &= \hat{\tau}_b \frac{v^2}{2} + \hat{m}_b^k v_k + \hat{e}_b. \end{aligned} \tag{28.8}$$

In (28.7), 0^i, 0_i, and 0^{ik} indicate, respectively, the zero column vector (3×1), the zero row vector (1×3), and the (3×3) null matrix. δ^{jk} is the Kronecker delta. And, only in this case, we have difference between v_k and v^j that indicates the velocity components in row and in column, respectively.

The first part of the theorem can be proved by the direct application of relations $(2.49)_{1,2}$. Once the operator $\mathbf{X}(\mathbf{v})$ is determined, the second part of the statement can be derived from $(2.49)_3$.

28.3 Coarse-Grained Theories: Single Temperature Model and Classical Mixture

Due to the difficulty to measure the temperature of each component, a common practice among engineers and physicists is to consider only one temperature for the mixture. When we use a single temperature (*ST*), $(28.5)_6$ disappears and we get a unique global conservation law of the total energy in the form $(28.5)_3$ (see for example [23]):

$$
\begin{cases}
\dfrac{\partial \rho}{\partial t} + \operatorname{div}(\rho \mathbf{v}) = 0, \\[2ex]
\dfrac{\partial (\rho \mathbf{v})}{\partial t} + \operatorname{div}(\rho \mathbf{v} \otimes \mathbf{v} - \mathbf{t}) = \mathbf{0}, \\[2ex]
\dfrac{\partial \left(\frac{1}{2}\rho v^2 + \rho \varepsilon\right)}{\partial t} + \operatorname{div}\left\{\left(\dfrac{1}{2}\rho v^2 + \rho \varepsilon\right)\mathbf{v} - \mathbf{t}\mathbf{v} + \mathbf{q}\right\} = 0, \\[2ex]
\dfrac{\partial \rho_b}{\partial t} + \operatorname{div}(\rho_b \mathbf{v}_b) = \tau_b, \qquad (b = 1, \ldots, n-1) \\[2ex]
\dfrac{\partial (\rho_b \mathbf{v}_b)}{\partial t} + \operatorname{div}(\rho_b \mathbf{v}_b \otimes \mathbf{v}_b - \mathbf{t}_b) = \mathbf{m}_b,
\end{cases}
\tag{28.9}
$$

A further step to a coarse-grained theory is the classical approach of mixtures (*CT*), in which the independent field variables are the density, the mixture velocity, the single temperature of the mixture, and the concentrations of components. In this case, the last equation in (28.9) also disappears and the system reduces to the system of equations:

$$
\begin{cases}
\dfrac{d\rho}{dt} + \rho \operatorname{div} \mathbf{v} = 0, \\[2ex]
\rho \dfrac{d\mathbf{v}}{dt} - \operatorname{div}\mathbf{t} = 0, \\[2ex]
\rho \dfrac{d\varepsilon}{dt} - \mathbf{t} \operatorname{grad} \mathbf{v} + \operatorname{div} \mathbf{q} = 0, \\[2ex]
\rho \dfrac{dc_b}{dt} + \operatorname{div} \mathbf{J}_b = 0, \quad (b = 1, \cdots n-1),
\end{cases}
\tag{28.10}
$$

where

$$
\frac{d}{dt} = \frac{\partial}{\partial t} + \mathbf{v} \cdot \frac{\partial}{\partial \mathbf{x}}
$$

represents the material derivative of the mixture motion,

$$c_\alpha = \frac{\rho_\alpha}{\rho}, \qquad \left(\sum_{\alpha=1}^{n} c_\alpha = 1\right)$$

are the concentrations of the components, and

$$\mathbf{J}_\alpha = \rho_\alpha \mathbf{u}_\alpha = \rho_\alpha \left(\mathbf{v}_\alpha - \mathbf{v}\right) \qquad \left(\sum_{\alpha=1}^{n} \mathbf{J}_\alpha = 0\right) \tag{28.11}$$

are the diffusion fluxes of the components.

In the classical approach, the stress tensor—as in a single fluid—splits into the pressure (isotropic part) and the viscous stress tensor (for Stokesian fluids this is a deviatoric tensor)

$$\mathbf{t} = -p\mathbf{I} + \boldsymbol{\sigma}.$$

The system (28.10) determines the field variables ρ, T, $\mathbf{v}$ and c_b $(b = 1, \cdots n-1)$. Consequently, we need constitutive relations for ε, $\boldsymbol{\sigma}$, $\mathbf{q}$, and $\mathbf{J}_b$ $(b = 1, \cdots n-1)$. The pressure $p(\rho, T, c_b)$, the internal energy $\varepsilon(\rho, T, c_b)$, and the chemical potentials $g_\alpha(\rho, T, c_b)$ $(\alpha = 1, \cdots, n)$ are assumed to be given by the equilibrium equations of state. The Gibbs equation in the case of mixture is given by

$$T ds = d\varepsilon - \frac{p}{\rho^2}\, d\rho - \sum_{b=1}^{n-1} \left(g_b - g_n\right)\, dc_b, \tag{28.12}$$

where s is the entropy density of a mixture [5, 7]. Compare this relation to (1.8) for a single-component fluid.

The entropy balance law is a consequence of Eq. (28.12) and system (28.10). For dissipative fluids, by using similar arguments of thermodynamics of irreversible processes *(TIP)* presented in Chap. 1 (see also [472] and [25]), we obtain the classical constitutive equations of mixtures:

$$\begin{aligned}
\boldsymbol{\sigma} &= \nu\, (\operatorname{div}\mathbf{v})\, \mathbf{I} + 2\,\mu\, \mathbf{D}^D, \\
\mathbf{q} &= L \operatorname{grad}\left(\frac{1}{T}\right) + \sum_{b=1}^{n-1} L_b \operatorname{grad}\left(\frac{g_b - g_n}{T}\right), \\
\mathbf{J}_a &= \tilde{L}_a \operatorname{grad}\left(\frac{1}{T}\right) - \sum_{b=1}^{n-1} L_{ab} \operatorname{grad}\left(\frac{g_b - g_n}{T}\right),
\end{aligned} \tag{28.13}$$

where $\mathbf{D}^D$ denotes the deviatoric part of the strain velocity tensor $\mathbf{D} = \frac{1}{2}\left(\nabla\mathbf{v} + (\nabla\mathbf{v})^T\right)$. The *phenomenological coefficients* $L, L_b, \tilde{L}_a$, and L_{ab} $(a, b = 1, \cdots, n-1)$ are the transport coefficients of heat conduction and diffusion.

Note that relation $(28.13)_1$ is the classical Navier–Stokes equation of a Newtonian fluid $(1.11)_{1,2}$, while $(28.13)_{2,3}$ are the generalizations of the original phenomenological laws of Fourier $(1.11)_3$ and Fick according to which the heat flux and the diffusion flux depend on the gradients of temperature and concentrations, respectively. While *TIP* permits the temperature gradient to influence the diffusion fluxes, and also permits the concentration gradients to influence the heat flux; both effects (cross effects) are indeed observed and they are called, respectively, thermo-diffusion effect and diffusion-thermo effect or Soret effect. Furthermore, the Onsager reciprocal theorem yields the following symmetries of coefficients [473]:

$$L_{ab} = L_{ba}, \quad \tilde{L}_b = L_b, \quad (a, b = 1, \cdots, n-1)$$

and the following inequalities must be satisfied:

$$\begin{bmatrix} L & L_b \\ \tilde{L}_a & L_{ab} \end{bmatrix} \quad \text{is a positive definite matrix}$$
$$\text{and} \quad \mu, \nu \geqslant 0, \tag{28.14}$$

such that the entropy inequality can be satisfied.

28.4 Mixture of Euler Fluids

We return to the general case of mixture with multi-temperature (28.5). We observe that, up to now, it was not necessary to introduce the constitutive equations in order to close the system (28.5). Now, let us introduce the assumption that all the components of the mixture are Eulerian fluids, i.e., neither viscous nor heat-conducting:

$$\mathbf{t}_\alpha = -p_\alpha \mathbf{I}, \quad \mathbf{q}_\alpha = \mathbf{0} \qquad (\alpha = 1, \ldots, n).$$

As a consequence of this assumption the stress tensor and the heat flux (28.3) are reduced to

$$\mathbf{t} = -p\mathbf{I} + \sum_{\alpha=1}^{n} (\rho_\alpha \mathbf{u}_\alpha \otimes \mathbf{u}_\alpha),$$
$$\mathbf{q} = \sum_{\alpha=1}^{n} \left\{ \rho_\alpha \left(\varepsilon_\alpha + \frac{1}{2} u_\alpha^2 \right) + p_\alpha \right\} \mathbf{u}_\alpha,$$

where p_α is the partial pressure of the α-component and

$$p = \sum_{\alpha=1}^{n} p_\alpha$$

is the total pressure.

28.4.1 Entropy Principle and Its Restrictions

The existence of the linear operator (28.7) confirms the Galilean invariance of field Eqs. (28.5) and determines the velocity dependence of the production terms (28.8). Another important restriction comes from the entropy inequality (2.8).

In the present case,

$$h^0 = \rho S = \sum_{\alpha=1}^{n} \rho_\alpha s_\alpha, \tag{28.15}$$

where S is the total entropy density and s_α are the entropy densities of the components.

Statement 28.1 *The entropy density h^0 (28.15) of the mixture is a convex function with respect to the densities $\mathbf{u} \equiv (\rho_\alpha, \rho_\alpha \mathbf{v}_\alpha, \frac{1}{2}\rho_\alpha v_\alpha^2 + \rho_\alpha \varepsilon_\alpha)$.*

The proof of the statement is almost trivial: since for every α the entropy density $\rho_\alpha s_\alpha$ is a convex function of the densities of the corresponding α-component fluid, the entropy density of the mixture, being the sum of convex functions, is also a convex function of the whole densities $\mathbf{u}$ ($\alpha = 1, \dots n$).

Let us recall that the main field components have to satisfy the relation $(2.11)_1$. For the balance-law system (28.1) in the case of Euler fluids, it reads

$$dh^0 = d(\rho S) = \sum_{\alpha=1}^{n} \left\{ \tilde{\Lambda}^{\rho_\alpha} d\rho_\alpha + \tilde{\boldsymbol{\Lambda}}^{\mathbf{v}_\alpha} d(\rho_\alpha \mathbf{v}_\alpha) + \tilde{\Lambda}^{\varepsilon_\alpha} d\left(\frac{1}{2}\rho_\alpha v_\alpha^2 + \rho_\alpha \varepsilon_\alpha\right)\right\}, \tag{28.16}$$

where

$$\tilde{\mathbf{u}}' = (\tilde{\Lambda}^{\rho_\alpha}, \tilde{\boldsymbol{\Lambda}}^{\mathbf{v}_\alpha}, \tilde{\Lambda}^{\varepsilon_\alpha}), \quad (\alpha = 1, \dots, n)$$

is the vector of the main field associated to the system (28.1).

Statement 28.2 *Components of the main field for the mixture of Euler fluids described by the system (28.1) have the form:*

$$\tilde{\Lambda}^{\rho_\alpha} = \frac{-g_\alpha + \frac{1}{2}v_\alpha^2}{T_\alpha}; \quad \tilde{\mathbf{\Lambda}}^{\mathbf{v}_\alpha} = -\frac{\mathbf{v}_\alpha}{T_\alpha}; \quad \tilde{\Lambda}^{\varepsilon_\alpha} = \frac{1}{T_\alpha}, \quad (\alpha = 1, \dots, n), \tag{28.17}$$

where

$$g_\alpha = \varepsilon_\alpha - T_\alpha S_\alpha + \frac{p_\alpha}{\rho_\alpha}$$

are the chemical potentials of the mixture components.

This statement is a consequence of the fact that the system (28.1), for what concerns the main parts of the differential operators, is constituted by uncoupled systems of single fluid equations. Consequently, the Gibbs relation holds for each component and the main field components (28.17) coincide, for each mixture component, with the ones of single fluid (2.29).

The main field components of the system (28.1) are used for calculation of the main field of the equivalent system (28.5). Let us denote this main field as follows:

$$\mathbf{u}' = \left(\Lambda^\rho, \mathbf{\Lambda}^\mathbf{v}, \Lambda^\varepsilon, \Lambda^{\rho_b}, \mathbf{\Lambda}^{\mathbf{v}_b}, \Lambda^{\varepsilon_b}\right) \quad (b = 1, \dots, n-1).$$

Equation $(2.11)_1$ written in new variables reads

$$dh^0 = d(\rho S) = \Lambda^\rho d\rho + \mathbf{\Lambda}^\mathbf{v} d(\rho\mathbf{v}) + \Lambda^\varepsilon d\left(\frac{1}{2}\rho v^2 + \rho\varepsilon\right)$$
$$+ \sum_{b=1}^{n-1}\left\{\Lambda^{\rho_b} d(\rho_b) + \mathbf{\Lambda}^{\mathbf{v}_b} d(\rho_b \mathbf{v}_b) + \Lambda^{\varepsilon_b} d\left(\frac{1}{2}\rho_b v_b^2 + \rho_b\varepsilon_b\right)\right\}. \tag{28.18}$$

The expressions (28.16) and (28.18) should be equivalent to each other. Then we have the relation between the main fields $\tilde{\mathbf{u}}'$ and $\mathbf{u}'$.

Statement 28.3 *The main field components for the mixture of Euler fluids described by the system (28.5) have the form:*

$$\Lambda^\rho = \tilde{\Lambda}^{\rho_n}, \quad \mathbf{\Lambda}^\mathbf{v} = \tilde{\mathbf{\Lambda}}^{\mathbf{v}_n}, \quad \Lambda^\varepsilon = \tilde{\Lambda}^{\varepsilon_n},$$

$$\Lambda^{\rho_b} = \tilde{\Lambda}^{\rho_b} - \tilde{\Lambda}^{\rho_n}, \quad \mathbf{\Lambda}^{\mathbf{v}_b} = \tilde{\mathbf{\Lambda}}^{\mathbf{v}_b} - \tilde{\mathbf{\Lambda}}^{\mathbf{v}_n}, \quad \Lambda^{\varepsilon_b} = \tilde{\Lambda}^{\varepsilon_b} - \tilde{\Lambda}^{\varepsilon_n}$$

for $b = 1, \ldots, n-1$, *i.e.*,

$$
\begin{aligned}
\Lambda^{\rho} &= -\frac{1}{T_n}\left(g_n - \frac{1}{2}\left(\mathbf{u}_n + \mathbf{v}\right)^2\right), \\
\Lambda^{\mathbf{v}} &= -\frac{1}{T_n}\left(\mathbf{u}_n + \mathbf{v}\right), \\
\Lambda^{\varepsilon} &= \frac{1}{T_n}, \\
\Lambda^{\rho_b} &= -\frac{1}{T_b}\left(g_b - \frac{1}{2}\left(\mathbf{u}_b + \mathbf{v}\right)^2\right) + \frac{1}{T_n}\left(g_n - \frac{1}{2}\left(\mathbf{u}_n + \mathbf{v}\right)^2\right), \\
\Lambda^{\mathbf{v}_b} &= -\frac{\mathbf{u}_b}{T_b} + \frac{\mathbf{u}_n}{T_n} + \left(\frac{1}{T_b} - \frac{1}{T_n}\right)\mathbf{v}, \\
\Lambda^{\varepsilon_b} &= -\frac{1}{T_n} + \frac{1}{T_b}.
\end{aligned}
\tag{28.19}
$$

The main field permits to determine the production terms through the application of the residual inequality (2.11)$_2$. i.e.,

$$
\Sigma = \mathbf{u}' \cdot \mathbf{f} = \hat{\mathbf{u}}' \cdot \hat{\mathbf{f}} = \sum_{b=1}^{n-1}\left(\hat{\Lambda}^{\rho_b}\hat{\tau}_b + \hat{\boldsymbol{\Lambda}}^{\mathbf{v}_b} \cdot \hat{\mathbf{m}}_b + \hat{\Lambda}^{\varepsilon_b}\hat{e}_b\right) \geqslant 0,
$$

or explicitly

$$
\Sigma = \sum_{b=1}^{n-1}\left(-\frac{g_b - \frac{1}{2}u_b^2}{T_b} + \frac{g_n - \frac{1}{2}u_n^2}{T_n}\right)\hat{\tau}_b + \left(\frac{\mathbf{u}_n}{T_n} - \frac{\mathbf{u}_b}{T_b}\right)\cdot\hat{\mathbf{m}}_b + \left(\frac{1}{T_b} - \frac{1}{T_n}\right)\hat{e}_b \geqslant 0.
\tag{28.20}
$$

This inequality allows us to obtain the following structure of production terms.

Statement 28.4 *The internal (intrinsic) parts of the production terms (28.6)$_2$ are chosen in such a way that the residual inequality (28.20) is actually a quadratic form:*

$$
\begin{aligned}
\hat{\tau}_b &= \sum_{c=1}^{n-1}\varphi_{bc}\left(\frac{g_n - \frac{1}{2}u_n^2}{T_n} - \frac{g_c - \frac{1}{2}u_c^2}{T_c}\right) + \sum_{c=1}^{n-1}\beta_{bc}\left(\frac{1}{T_c} - \frac{1}{T_n}\right), \\
\hat{\mathbf{m}}_b &= \sum_{c=1}^{n-1}\psi_{bc}\left(\frac{\mathbf{u}_n}{T_n} - \frac{\mathbf{u}_c}{T_c}\right), \\
\hat{e}_b &= \sum_{c=1}^{n-1}\theta_{bc}\left(\frac{1}{T_c} - \frac{1}{T_n}\right) + \sum_{c=1}^{n-1}\beta_{bc}\left(\frac{g_n - \frac{1}{2}u_n^2}{T_n} - \frac{g_c - \frac{1}{2}u_c^2}{T_c}\right),
\end{aligned}
\tag{28.21}
$$

where

$$\begin{bmatrix} \varphi_{bc} & \beta_{bc} \\ \beta_{bc} & \theta_{bc} \end{bmatrix}, \quad \psi_{bc}$$

are phenomenological symmetric positive definite matrices ($b, c = 1, \ldots, n-1$).

For processes not far from equilibrium, the previous matrices depend only on the equilibrium variables ρ_α and T. In the sequel, our analysis will be restricted to a model of non-reacting mixtures, for which $\tau_b = 0$. The matrices ψ_{bc} and θ_{bc} in the linearized case can be deduced via kinetic theory [471, 474].

28.4.2 Symmetric Hyperbolic System and Principal Subsystems

The main field components for the mixture of Euler fluids (28.19) symmetrize the system (28.5) according to Theorem 2.1.

Concerning the principal subsystems (see Sect. 2.4), taking into account (28.19) and (28.5), we can recognize the following principal subsystems:

Case 1: *The single-temperature model is a principal subsystem of the multi-temperature*. Let us suppose that $\Lambda^{\varepsilon_b} = 0$ for $b = 1, \ldots, n-1$, then

$$T_1 = \ldots = T_n = T.$$

This principal subsystem contains only the energy conservation equation for the mixture, while energy balance equations for the components are dropped. Thus, one may conclude that single-temperature model naturally appears as a principal subsystem of the multi-temperature system.

Case 2: *The equilibrium subsystem*. If we set

$$\Lambda^{\varepsilon_b} = \mathbf{\Lambda}^{\mathbf{v}_b} = \mathbf{\Lambda}^{\rho_b} = 0 \quad \forall\, b = 1, \ldots, n-1,$$

i.e.,

$$T_b = T, \quad \mathbf{u}_b = 0, \quad g_b = g \quad \forall\, b = 1, \ldots, n-1,$$

we have the equilibrium Euler subsystem (a single fluid system) with concentrations c_b being solutions of $g_1 = g_2 = \ldots = g_n$.

28.4.3 Characteristic Velocities and their Upper Bound in the ST Model

The characteristic velocities for the *MT* model are simple to evaluate. Since, for each component, they are the same as the ones of a single fluid, i.e.,

$$\lambda_\alpha^{(1)} = v_{\alpha n} - c_{s\alpha}; \quad \lambda_\alpha^{(2,3,4)} = v_{\alpha n}; \quad \lambda_\alpha^{(5)} = v_{\alpha n} + c_{s\alpha}, \tag{28.22}$$

where $v_{\alpha n} = \mathbf{v}_\alpha \cdot \mathbf{n}$ are the normal component of the velocities at the wave front and

$$c_{s\alpha} = \sqrt{\left(\frac{\partial p_\alpha}{\partial \rho_\alpha}\right)_{s_\alpha}}$$

are the sound velocities. For an ideal gas, for example, we have

$$p_\alpha = \frac{k_B}{m_\alpha}\rho_\alpha T_\alpha, \quad \varepsilon_\alpha = c_v^{(\alpha)} T_\alpha, \quad c_v^{(\alpha)} = \frac{k_B}{m_\alpha(\gamma_\alpha - 1)} \tag{28.23}$$

and

$$c_{s\alpha} = \sqrt{\frac{k_B}{m_\alpha}\gamma_\alpha T_\alpha}. \tag{28.24}$$

Instead, in the case of the *ST* model, the evaluation of the velocities is very difficult even in an equilibrium state due to the fact that the characteristic polynomial is, in general, irreducible (see e.g. [25, 475]). But thanks to the subcharacteristic property (2.37) of principal subsystems, we are able to establish the following lower and upper bounds for the characteristic velocities of the ST model:

$$\min_\alpha (v_{\alpha n} - c_{s\alpha}^*) \leqslant \lambda_{\min}^{ST}; \qquad \max_\alpha (v_{\alpha n} + c_{s\alpha}^*) \geqslant \lambda_{\max}^{ST},$$

where

$$c_{s\alpha}^* = \sqrt{\frac{k_B \gamma_\alpha}{m_\alpha} T}\,.$$

28.4.4 Qualitative Analysis and K-condition in Mixture Theories

We made, in Sect. 2.6, the qualitative analysis for a system of hyperbolic type that is composed of two groups, as is usual in RET: one group is formed by conservation laws and the other group by balance laws. In this case, the coupling K-condition given in Sect. 2.6.2 plays an important role.

For *ST* theory without chemical reactions, it was proved [476, 477] that the K-condition is violated for some genuinely nonlinear eigenvalues. Therefore, from the results in [173], global smooth solutions can not exist even though initial data are small enough. Instead, for a *MT* system, it is possible to verify that the K-condition is satisfied for all eigenvalues. Therefore we can conclude [158]:

Statement 28.5 *If the initial data of the MT model are perturbations of equilibrium state, smooth solutions exist for all time and tend to an equilibrium constant state.*

This statement also shows clearly that the *MT* model is more realistic than the *ST* model.

28.5 Average Temperature

The *MT*-mixture theory explained above is the most realistic theory, and is consistent with the kinetic theory [104]. It is also a necessary theory in physics, in particular, in plasma physics [471]. Nevertheless, from a theoretical point of view, a serious problem still remains in it: How to measure the temperature of each component?

In this section, instead of studying this difficult problem directly, we study the problem about the macroscopic average temperature of a mixture. Let us consider the definition of the average temperature firstly proposed by Ruggeri and co-workers in [159, 160, 268, 478]. The main idea is to investigate the definition of internal energy so as to introduce the (average) temperature T as a state variable of the mixture. Then the intrinsic internal energy ε_I (see (28.3)$_3$) of the *MT* mixture resembles the structure of the intrinsic internal energy of a *ST* mixture.

Therefore, the following implicit definition of an average temperature is adopted:

Definition 28.1 *The average temperature T is the one that corresponds to the barycentric intrinsic internal energy, i.e., it is defined through the relation*:

$$\rho\varepsilon_I(\rho_1, \ldots, \rho_n, T) = \sum_{\alpha=1}^{n} \rho_\alpha \varepsilon_\alpha(\rho_\alpha, T) = \sum_{\alpha=1}^{n} \rho_\alpha \varepsilon_\alpha(\rho_\alpha, T_\alpha). \tag{28.25}$$

By expanding this relation in the neighborhood of the average temperature we have

$$T = \frac{\sum_{\alpha=1}^{n} \rho_\alpha c_v^{(\alpha)} T_\alpha}{\sum_{\alpha=1}^{n} \rho_\alpha c_v^{(\alpha)}}, \tag{28.26}$$

where

$$c_v^{(\alpha)} = \left.\frac{\partial \varepsilon_\alpha(\rho_\alpha, T_\alpha)}{\partial T_\alpha}\right|_{T_\alpha = T}$$

is the specific heat at constant volume of component α. We observe that (28.26) gives the exact value of the average temperature in the case of mixture of ideal gases for which $c_v^{(\alpha)}$ are constant.

This definition of the average temperature has some advantages over usual ones used frequently in the literature. Firstly, as a consequence of the definition, the conservation law of energy of a mixture $(28.5)_3$ becomes an evolution equation for the average temperature T as in the cases of *ST* and *CT* mixtures. So, in the case of spatial homogeneous solution of the differential system (28.5) (solutions depend only on time), the nonequilibrium temperature of each component T_α converges to T for large time, as shown below.

The second advantage is related to the entropy of the whole mixture that reaches its maximum value when $T_\alpha = T$, thanks to the introduction of this average temperature. In fact, Ruggeri and Simić introduced the *diffusion temperature flux*

$$\Theta_\alpha = T_\alpha - T, \tag{28.27}$$

and they proved in [160], using the Gibbs equation for each component, that the entropy density near equilibrium becomes a negative definite quadratic form with respect to the nonequilibrium variables Θ_α:

$$\rho S = \sum_{\alpha=1}^{n} \rho_\alpha s_\alpha(\rho_\alpha, T_\alpha) = \sum_{\alpha=1}^{n} \rho_\alpha s_\alpha(\rho_\alpha, T) - \frac{1}{2T^2}\sum_{\alpha=1}^{n} \rho_\alpha c_v^{(\alpha)} \Theta_\alpha^2 + O(\Theta_\alpha^3). \tag{28.28}$$

Therefore, in the present case, the nonequilibrium entropy k (4.15) is given by

$$k = -\frac{1}{2\rho T^2}\sum_{\alpha=1}^{n} \rho_\alpha c_v^{(\alpha)} \Theta_\alpha^2,$$

and the entropy density S has a maximum at equilibrium.

28.6 Examples of Spatially Homogeneous Mixtures

In this section, two simple examples are provided in order to support the previous theoretical considerations, and to stress the main features of the multi-temperature approach and the role of the average temperature.

28.6.1 Solution of a Spatially Homogeneous Mixture

First we consider a non-reacting mixture of gases in spatially homogeneous fields, i.e., a mixture with field variables depending solely on time [160, 479]. The governing equations (28.5) with (28.8) can be written in the following form:

$$
\begin{aligned}
&\frac{d\rho}{dt} = 0, \qquad \frac{d\rho\mathbf{v}}{dt} = 0, \qquad \frac{d}{dt}\left(\rho\varepsilon + \frac{1}{2}\rho v^2\right) = 0, \quad t > 0, \\
&\frac{d\rho_b}{dt} = 0, \qquad \frac{d\rho_b\mathbf{v}_b}{dt} = -\sum_{c=1}^{n-1}\psi_{bc}\left(\frac{\mathbf{u}_c}{T_c} - \frac{\mathbf{u}_n}{T_n}\right), \quad b, c = 1, \cdots, n-1, \\
&\frac{d}{dt}\left(\rho_b\varepsilon_b + \frac{1}{2}\rho_b v_b^2\right) = -\sum_{c=1}^{n-1}\theta_{bc}\left(\frac{1}{T_n} - \frac{1}{T_c}\right) - \mathbf{v}\cdot\sum_{c=1}^{n-1}\psi_{bc}\left(\frac{\mathbf{u}_c}{T_c} - \frac{\mathbf{u}_n}{T_n}\right),
\end{aligned}
\tag{28.29}
$$

where the interaction matrices (ψ_{bc}) and (θ_{bc}) are assumed to be *symmetric* and *positive definite*. From (28.29), we notice that the total density ρ, the component density ρ_b, and the center of mass velocity $\mathbf{v}$ are constant. And, since the system is Galilean invariant, we can choose, without loss of generality, the rest frame with $\mathbf{v} = 0$. Then, in this frame, (28.29) reads:

$$
\begin{aligned}
&\frac{d\varepsilon}{dt} = 0 \\
&\rho_b\frac{d\mathbf{v}_b}{dt} = -\sum_{c=1}^{n-1}\psi_{bc}\left(\frac{\mathbf{v}_c}{T_c} - \frac{\mathbf{v}_n}{T_n}\right), \quad b, c = 1, \cdots, n-1, \\
&\rho_b\frac{d}{dt}\left(\varepsilon_b + \frac{1}{2}v_b^2\right) = -\sum_{c=1}^{n-1}\theta_{bc}\left(\frac{1}{T_n} - \frac{1}{T_c}\right).
\end{aligned}
\tag{28.30}
$$

We will discuss this system in Chap. 30 when we compare a mixture of gases with a flocking system. For the moment, we study the linearized version of (28.30) in a neighborhood of an equilibrium state. In this case, from the first equation, we notice that the average temperature $T = T_0$ is constant. While two remaining equations become

$$
\begin{aligned}
&\rho_b\frac{d\mathbf{v}_b}{dt} = -\sum_{c=1}^{n-1}\frac{\psi_{bc}^0}{T_0}(\mathbf{v}_c - \mathbf{v}_n), \\
&\rho_b c_v^{(b)}\frac{dT_b}{dt} = -\sum_{c=1}^{n-1}\frac{\theta_{bc}^0}{T_0^2}(T_c - T_n),
\end{aligned}
\tag{28.31}
$$

where ψ^0_{bc} and θ^0_{bc} are positive definite matrices evaluated in equilibrium. In a particular case of binary mixture, we have

$$\begin{aligned} &\rho_1\mathbf{v}_1 + \rho_2\mathbf{v}_2 = \rho\mathbf{v} = \mathbf{0}, \\ &\rho_1 c_v^{(1)} T_1 + \rho_2 c_v^{(2)} T_2 = (\rho_1 c_v^{(1)} + \rho_2 c_v^{(2)}) T_0. \end{aligned} \tag{28.32}$$

The solution of (28.31) and (28.32) is obtained as

$$\mathbf{v}_1(t) = \mathbf{v}_1(0)e^{-\frac{t}{\tau_v}}, \qquad \mathbf{v}_2(t) = \mathbf{v}_2(0)e^{-\frac{t}{\tau_v}},$$

$$T_1(t) = T_0 + (T_1(0) - T_0)e^{-\frac{t}{\tau_T}}, \qquad T_2(t) = T_0 + (T_2(0) - T_0)e^{-\frac{t}{\tau_T}},$$

where τ_v and τ_T represent the relaxation times, which, for an ideal polytropic gas, assume the following expressions:

$$\tau_v = \frac{\rho_1\rho_2 T_0}{\psi^0_{11}\rho}, \qquad \tau_T = \frac{k_B \rho_1\rho_2 T_0^2}{\theta^0_{11}(\rho_1 m_2(\gamma_2 - 1) + \rho_2 m_1(\gamma_1 - 1))}. \tag{28.33}$$

It is obvious that, due to the dissipative character of the system, *all the nonequilibrium variables exponentially decay and converge to their equilibrium values*. In order to compare the values of τ_v and τ_T for ideal gases, and also to compute the actual values of variables in numerical examples, the following relations from the kinetic theory have to be recalled [471]:

$$\theta^0_{11} = \frac{3m_1m_2}{(m_1+m_2)^2} k_B T_0^2 \Gamma_{12}; \qquad \psi^0_{11} = \frac{2m_1m_2}{m_1+m_2} T_0 \Gamma_{12}, \tag{28.34}$$

where Γ_{12} represents volumetric collision frequency. Then we have the following estimate:

$$\frac{\tau_T}{\tau_v} = \frac{2}{3}\frac{\rho(m_1+m_2)}{\rho_1 m_2(\gamma_2 - 1) + \rho_2 m_1(\gamma_1 - 1)} > \frac{2}{3(\gamma_{\max} - 1)} \geqslant 1, \tag{28.35}$$

($\gamma_{\max} = \max\{\gamma_1, \gamma_2\} \leqslant 5/3$). In Fig. 28.1, we present the graphs of normalized velocities and diffusion temperature fluxes [160]. It can be observed that, due to inequality (28.35), *the mechanical diffusion vanishes more rapidly than the thermal one*. This is in sharp contrast with the widely adopted approach that ignores the influence of the temperature of each component of the mixture.

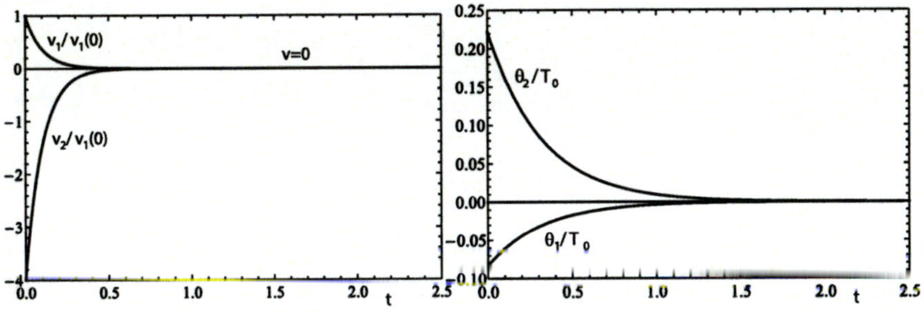

Fig. 28.1 Dimensionless velocities and diffusion temperature fluxes of the components versus time. $\theta_a = T_a - T_0$, $(a = 1, 2)$

28.6.2 *Solution of Static Heat Conduction*

Another simple example is the one-dimensional mixture of gases at rest ($\mathbf{v}_\alpha = 0$) without chemical reactions ($\tau_\alpha = 0$) between two walls $0 \leqslant x \leqslant L$, maintained at two different temperatures $T(0) = T_0$ and $T(L) = T_L$ [268].

In both *CT* and *ST* models, the static field equation reduces to the global energy equation $(28.10)_3$ that reads div $\mathbf{q} = 0$. In the one-dimensional case, this equation, combined with the Fourier law with constant heat conductivity, yields the classical result of a linear temperature profile as for a single fluid:

$$T'' = 0 \iff T = (T_L - T_0)\xi + T_0$$

where $\xi = x/L$ and $'$ denotes $d/d\xi$. For what concerns the densities they are obtained by the conditions that the pressure of each component must be constant due to the momentum equations.

In the *MT* model, the situation is quite different. Let us consider the simple case of a binary mixture ($n = 2$). In the linear case, by taking into account (28.21), system (28.1) reduces to

$$\begin{cases} \dfrac{dp_1}{dx} = 0, \quad \dfrac{dp_2}{dx} = 0, \\ \dfrac{dq_1}{dx} = \beta\,(T_2 - T_1)\,, \\ \dfrac{dq_2}{dx} = \beta\,(T_1 - T_2)\,, \end{cases} \tag{28.36}$$

where $\beta = \theta_{11}/T_0^2$. By using the Fourier law, $(28.36)_{2,3}$ can be rewritten as

$$\begin{cases} T_1'' = \nu_1(T_1 - T_2), \\ T_2'' = \nu_2(T_2 - T_1), \end{cases} \tag{28.37}$$

where we assume that the dimensionless quantities,

$$\nu_1 = \frac{\beta L^2}{\chi_1}, \quad \nu_2 = \frac{\beta L^2}{\chi_2}, \tag{28.38}$$

are constant. The system (28.37) is equivalent to

$$\widehat{T}'' = 0, \qquad \Theta'' - \omega^2\, \Theta = 0$$

with $\widehat{T} = \nu T_1 + (1-\nu)T_2$, $\Theta = T_2 - T_1$, and

$$\nu = \frac{\nu_2}{\nu_1 + \nu_2} = \frac{\chi_1}{\chi_1 + \chi_2}, \qquad \omega = \sqrt{\nu_1 + \nu_2}.$$

Consequently, we get the solution in the form:

$$T_1 = \hat{T} - (1-\nu)\Theta, \quad T_2 = \hat{T} + \nu\Theta \tag{28.39}$$

where

$$\hat{T} = A\,\xi + B, \qquad \Theta = \frac{1}{\sinh(\omega)} \left\{\Theta_L \sinh(\omega\,\xi) + \Theta_0\, \sinh(\omega(1-\xi))\right\}, \tag{28.40}$$

and A, B, Θ_0, Θ_L are constants of integration. In the case of ideal gases, Eqs. $(28.36)_1$ and (28.23) yield the constant internal energy density of each component:

$$\rho_\alpha T_\alpha c_v^{(\alpha)} = P_\alpha = Const., \quad (\alpha = 1, 2). \tag{28.41}$$

And (28.26) yields the average temperature:

$$\frac{1}{T} = \frac{\pi}{T_1} + \frac{1-\pi}{T_2}, \text{ with } \pi = \frac{P_1}{P_1 + P_2}. \tag{28.42}$$

The constant π belongs to $[0, 1]$. It is interesting to observe that the *coldness* $1/T$ (inverse of the average temperature) belongs to the convex envelope of the component coldnesses $1/T_1$ and $1/T_2$. Equations (28.39), (28.40), and (28.42) give the explicit solution of T_1, T_2, and T as the function of ξ and five constants of integration: $(A, B, \Theta_0, \Theta_L, \pi)$. We observe the behavior of T is not a straight line in contrast to the classical case of *CT* or *ST* theory; the multi-temperature effect is that the temperature is not a linear function of x (see Fig. 28.2). Due to (28.36)–(28.38), when $\epsilon = 1/\beta$ tends to zero, the solution of (28.39), (28.40), and (28.42) converges to the classical solution $T_1 = T_2 = T = \hat{T}$ for any $\xi \in]0, 1[$. This result is true also at the boundary when Θ_0 and Θ_1 are of same order as ϵ.

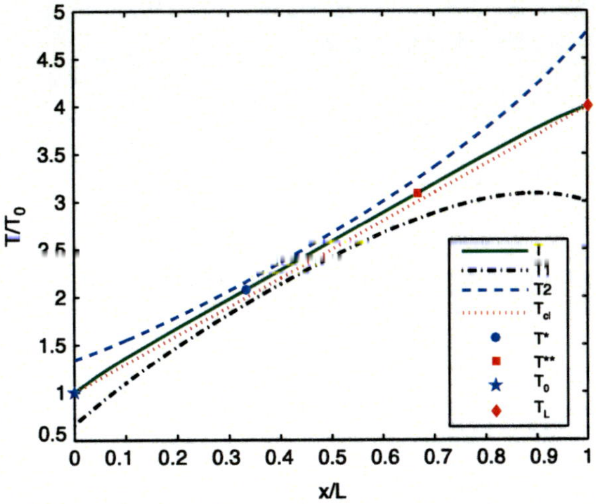

Fig. 28.2 The average temperature T and component temperatures T_1 and T_2 in terms of the dimensionless distance x/L. T_{cl} represents the classical straight line solution. T_0 is the temperature unit. The average temperatures measured are T_0, T_L, T^* and T^{**}

Let us introduce the concentration $c = \rho_1/(\rho_1 + \rho_2)$ with $c(0) = c_0$. Then (28.41), (28.39), and (28.40) imply

$$c = \frac{c_0}{c_0 + \Omega(1 - c_0)} \quad \text{with} \quad \Omega = \frac{T_1}{T_2}\frac{T_{20}}{T_{10}}$$

and

$$T_{10} = B - (1 - \nu)\Theta_0, \quad T_{20} = B + \nu\Theta_0.$$

The concentration is a function of the position x, whereas, in the classical case, $\Omega = 1$ and $c = c_0$. Ruggeri and Lou [268] studied the method how to determine in a unique way the constants of integration. They proved that, for a mixture of n components, the measurement of the average temperature at $2(n - 1)$ points allows to determine the temperature of each component in all points.

28.7 Maxwellian Iteration

It is convenient, in the following analysis, to rewrite the system (28.5) using the material derivatives. And we neglect quadratic terms in the nonequilibrium variables. In particular, in the global energy equation $(28.3)_3$, we neglect the diffusion term and therefore $\varepsilon = \varepsilon_I$. The material derivatives are defined by

$$\frac{d}{dt} = \frac{\partial}{\partial t} + \mathbf{v} \cdot \nabla, \quad \frac{d_b}{dt} = \frac{\partial}{\partial t} + \mathbf{v}_b \cdot \nabla$$

for $b = 1, \cdots n-1$. Taking into account the definition of the average temperature given by (28.25), we have

$$\begin{cases} \dfrac{d\rho}{dt} + \rho \operatorname{div} \mathbf{v} = 0, \\[2ex] \rho \dfrac{d\mathbf{v}}{dt} - \operatorname{div}\mathbf{t} = 0, \\[2ex] \rho \dfrac{\partial \varepsilon_I}{\partial T}\dfrac{dT}{dt} = \rho^2 \dfrac{\partial \varepsilon_I}{\partial \rho} \operatorname{div} \mathbf{v} + \displaystyle\sum_{b=1}^{n-1} \dfrac{\partial \varepsilon_I}{\partial c_b} \operatorname{div} \mathbf{J}_b + \mathbf{t} \operatorname{grad} \mathbf{v} - \operatorname{div} \mathbf{q}, \\[2ex] \dfrac{d_b \rho_b}{dt} + \rho_b \operatorname{div} \mathbf{v}_b = 0, \\[2ex] \rho_b \dfrac{d_b \mathbf{v}_b}{dt} - \operatorname{div}\mathbf{t}_b = \hat{\mathbf{m}}_b, \\[2ex] \rho_b \dfrac{d_b \varepsilon_b}{dt} - \mathbf{t}_b \cdot \nabla \mathbf{v}_b + \operatorname{div} \mathbf{q}_b = \hat{e}_b. \end{cases} \tag{28.43}$$

The differential system of equations governs the evolution of ρ, $\mathbf{v}$, T, ρ_b, $\mathbf{J}_b$, and Θ_b, provided that we assign the constitutive equations of p_α, ε_α, and, for dissipative fluids, the heat fluxes $\mathbf{q}_\alpha$ and the viscous stress tensors $\boldsymbol{\sigma}_\alpha$.

In order to understand a connection among the extended models of *MT*, *ST*, and the classical model *CT*, we use the *Maxwellian iteration* (for more detail see Chap. 33, Sects. 33.2.1 and 33.3). In the present case, the Maxwellian iteration is carried out as follows: put the zeroth iterates, i.e., the values of quantities evaluated in an equilibrium state into the left-hand side of the system $(28.43)_{5,6}$, then we obtain the first iterates from the right-hand side of the system.

Taking into account the fact that in zeroth iteration $\mathbf{v}_\alpha^{(0)} = \mathbf{v}$ and consequently

$$\frac{d_b^{(0)}}{dt} = \frac{d}{dt}, \qquad \mathbf{J}_b^{(0)} = \mathbf{u}_b^{(0)} = \mathbf{0}$$

and moreover

$$T_\alpha^{(0)} = T, \quad \mathbf{q}^{(0)} = \mathbf{q}_b^{(0)} = 0, \quad \mathbf{t}^{(0)} = -p^{(0)}\mathbf{I} = -p_0\mathbf{I}, \quad t_b^{(0)} = -p_b^{(0)}\mathbf{I},$$

we obtain

$$\begin{aligned} &\rho_b \left(\frac{d\mathbf{v}}{dt}\right)^{(0)} + \operatorname{grad} p_b^{(0)} = \hat{\mathbf{m}}_b^{(1)}, \\ &\rho_b \left\{ \left(\frac{\partial \varepsilon_b}{\partial \rho_b}\right)^{(0)} \left(\frac{d\rho_b}{dt}\right)^{(0)} + \left(\frac{\partial \varepsilon_b}{\partial T_b}\right)^{(0)} \left(\frac{dT}{dt}\right)^{(0)} \right\} + p_b^{(0)} \operatorname{div}\mathbf{v} = \hat{e}_b^{(1)}. \end{aligned} \tag{28.44}$$

On the other hand, from the zeroth order of (28.43)$_{2,3,4}$, we have

$$\rho \left(\frac{d\mathbf{v}}{dt} \right)^{(0)} = -\mathrm{grad} p_0,$$
$$\rho \frac{\partial \varepsilon_I}{\partial T} \left(\frac{dT}{dt} \right)^{(0)} = \left(\rho^2 \frac{\partial \varepsilon_I}{\partial \rho} - p_0 \right) \mathrm{div}\mathbf{v}, \tag{28.45}$$
$$\left(\frac{d\rho_b}{dt} \right)^{(0)} = -\rho_b \mathrm{div}\mathbf{v},$$

and therefore, inserting (28.45) into (28.44), we obtain

$$-\frac{\rho_b}{\rho} \mathrm{grad} p_0 + \mathrm{grad} p_b^{(0)} = \hat{\mathbf{m}}_b^{(1)}, \tag{28.46}$$

$$\Omega_b \,\mathrm{div}\,\mathbf{v} = \hat{e}_b^{(1)}, \tag{28.47}$$

where

$$\Omega_b = p_b^{(0)} + \rho_b \left\{ -\rho_b \left(\frac{\partial \varepsilon_b}{\partial \rho_b} \right)^{(0)} + \left(\rho^2 \frac{\partial \varepsilon_I}{\partial \rho} - p_0 \right) \frac{\left(\dfrac{\partial \varepsilon_b}{\partial T_b} \right)^{(0)}}{\rho \dfrac{\partial \varepsilon_I}{\partial T}} \right\}. \tag{28.48}$$

Taking into account the expressions of the productions (28.21) and the definitions of the diffusion flux $\mathbf{J}_a$ (28.11) and of the thermal diffusion Θ_a (28.27), after some arrangement (an interested reader can consult the details in the original paper [160]), we obtain the fact that the approximation of the momentum equation of each component (28.46) gives the Fick law (28.13)$_3$:

$$\mathbf{J}_a^{(1)} = \tilde{L}_a \,\mathrm{grad} \left(\frac{1}{T} \right) - \sum_{b=1}^{n-1} L_{ab} \,\mathrm{grad} \left(\frac{g_b - g_n}{T} \right), \tag{28.49}$$

while for what concerns the approximation of energy equations (28.47) we obtain new constitutive equations:

$$\Theta_a^{(1)} = -k_a \,\mathrm{div}\,\mathbf{v}, \tag{28.50}$$

where k_a is a linear combination of Ω_b given in (28.48).

The equation (28.50), *obtained by means of the Maxwellian iteration, gives the temperature of each component as a constitutive equation. This is similar to the Fick law that gives the velocity of each component.*

It is possible to prove [160] that for a mixture of ideal gases

$$\Omega_b = \rho_b T c_v^{(b)} \frac{\sum_{\alpha=1}^{n} \rho_\alpha c_v^{(\alpha)} (\gamma_b - \gamma_\alpha)}{\sum_{\alpha=1}^{n} \rho_\alpha c_v^{(\alpha)}}.$$

In the next section we will deduce and justify (28.50) using a classical TIP approach.

Equation (28.50) cannot be obtained in the classical theory, and it is interesting to observe that this is probably due to the fact that, if the gas has constituents with the same degrees of freedom, i.e., the same γ, then $\Omega_b = 0$, and consequently $k_a = 0$. And, from (28.50), $\Theta_b^{(1)} = 0$, and then all the temperatures are equal in the first Maxwellian iteration. In this last case, in order to reveal the temperature difference between components, it is necessary to go to the second order Maxwellian iteration. This is the subject of a paper by Ruggeri and Simić [480].

In conclusion, we have proved the following important result in this section:

Statement 28.6 *By using the Maxwellian iteration, the Fick law (28.49) is obtained as the first approximation of the momentum balance equation of each component (28.43)$_5$ and the equivalent thermal diffusion law (28.50) is the first approximation of the energy balance equation of each component (28.43)$_6$.*

28.8 A Classical Approach to Multi-Temperature Mixtures

As seen above, by using the Maxwellian iteration, the *ST* model reduces to the classical model, but when we start from the *MT* model we obtain new constitutive equations (28.50) that do not exist in the classical theory of mixtures.

To justify the results of the Maxwellian iteration, Gouin and Ruggeri [159] constructed a classical theory of mixture with multi-temperature. The idea is to use the usual equations in the classical approach (28.10), but now we suppose that each component has its own temperature. In this approach, the role of the average temperature defined (28.25) is fundamental. In fact, the multi-temperature effect appears through the pressure. Near equilibrium, we have (28.26). We have

$$p = \sum_{\alpha=1}^{n} p_\alpha(\rho_\alpha, T_\alpha) = p_0 + \Pi_\theta,$$

where

$$p_0 = \sum_{\alpha=1}^{n} p_\alpha(\rho_\alpha, T), \quad \Pi_\theta = \sum_{b=1}^{n-1} r_b\, \Theta_b$$

and

$$r_b = \frac{1}{\rho_n c_v^{(n)}} \left\{ \rho_n c_v^{(n)} \frac{\partial p_b}{\partial T_b} (\rho_b, T) - \rho_b c_v^{(b)} \frac{\partial p_n}{\partial T_n} (\rho_n, T) \right\}. \tag{28.51}$$

Therefore, the total pressure p of a mixture is the sum of the equilibrium part p_0 depending on ρ_α and T and a new dynamic pressure part (as a nonequilibrium term) Π_θ due to the difference of temperatures between the components.

We assume the internal energy $\varepsilon(\rho, T, c_b)$ and the equilibrium pressure $p_0(\rho, T, c_b)$ satisfy the Gibbs equation:

$$T dS = d\varepsilon - \frac{p_0}{\rho^2} d\rho - \sum_{b=1}^{n-1} (g_b - g_n) \, dc_b \,. \tag{28.52}$$

The difference between (28.12) and (28.52) consists in the fact that, in (28.52), T means the average temperature when each component α has its own temperature T_α and p_0 takes the place of p. Let us consider first the case of a Stokesian fluid $\operatorname{tr} \boldsymbol{\sigma} = 0$. Taking into account that, from (28.28), ρS depends only on the average temperature up to the first order expansion and using (28.10) to eliminate the time derivatives, we obtain from (28.52) the following entropy balance equation [159]:

$$\begin{aligned} &\rho \frac{dS}{dt} + \operatorname{div} \left\{ \frac{1}{T} \left(\mathbf{q} - \sum_{b=1}^{n-1} (g_b - g_n) \, \mathbf{J}_b \right) \right\} \\ &\quad = \mathbf{q} \cdot \operatorname{grad} \left(\frac{1}{T} \right) - \sum_{b=1}^{n-1} \mathbf{J}_b \cdot \operatorname{grad} \left(\frac{g_b - g_n}{T} \right) + \frac{1}{T} \operatorname{tr} \left(\mathbf{J}_{mech} \, \mathbf{D} \right), \end{aligned} \tag{28.53}$$

where the mechanical flux is given by

$$\mathbf{J}_{mech} = \boldsymbol{\sigma} - \Pi_\theta \, \mathbf{I}. \tag{28.54}$$

Equation (28.53) can be interpreted as a balance of entropy, if we regard

$$\boldsymbol{\Phi} = \frac{1}{T} \left(\mathbf{q} - \sum_{b=1}^{n-1} (g_b - g_n) \, \mathbf{J}_b \right)$$

and

$$\Sigma = \mathbf{q} \cdot \operatorname{grad} \left(\frac{1}{T} \right) - \sum_{b=1}^{n-1} \mathbf{J}_b \cdot \operatorname{grad} \left(\frac{g_b - g_n}{T} \right) + \frac{1}{T} \operatorname{tr} \left(\mathbf{J}_{mech} \, \mathbf{D} \right) \tag{28.55}$$

as the entropy flux and the entropy production, respectively.

We observe that the entropy production is the sum of the products of the following quantities:

Dissipative fluxes	**Thermodynamic forces**
heat flux $\mathbf{q}$	temperature gradient grad $\left(\frac{1}{T}\right)$,
diffusion fluxes $\mathbf{J}_b$	chemical potential gradients grad $\left(\frac{g_b - g_n}{T}\right)$,
mechanical flux $\mathbf{J}_{mech}$	velocity gradient $\mathbf{D}$.

In accordance with the case of a single fluid model [5, 25], which we have discussed in Sect. 1.3, the fluxes depend linearly on the associated forces as in *TIP* near equilibrium (see also for the general methodology of *TIP* [7, 70, 472, 473]):

- For the heat flux and the diffusion fluxes, we obtain the constitutive equations in the form of (28.13)$_{2,3}$.
- For Stokesian fluids, by taking into account (28.54), the last term of (28.55), corresponding to the mechanical production of the entropy, can be written in a separated form:

$$\frac{1}{T}\,\mathrm{tr}\,(\mathbf{J}_{mech}\,\mathbf{D}) = \frac{1}{T}\,\mathrm{tr}\left(\boldsymbol{\sigma}\,\mathbf{D}^D\right) - \frac{1}{T}\,\Pi_\theta\,\mathrm{div}\,\mathbf{v}.$$

  We obtain the constitutive equation of the viscous stress tensor in the form of (28.13)$_1$ (with $\nu = 0$), and the dynamic pressure part due to the difference of temperatures yields

$$\Pi_\theta = \sum_{b=1}^{n-1} r_b\,\Theta_b = -L_\pi\,\mathrm{div}\,\mathbf{v}, \tag{28.56}$$

  where L_π is a scalar coefficient.

  As the production of entropy must be non-negative, (28.55) and therefore the *phenomenological coefficients* must satisfy the inequalities (28.14) and

$$L_\pi \geqslant 0.$$

Taking into account that terms r_b given by (28.51) depend on (ρ_b, T), we deduce, from (28.56), that constitutive quantities Θ_a (depending a priori on $\nabla\mathbf{v}$) must be proportional to div $\mathbf{v}$:

$$\Theta_a = -k_a\,\mathrm{div}\,\mathbf{v} \quad (a = 1, \cdots, n-1).$$

This is in perfect agreement with (28.50) by the Maxwellian iteration procedure presented in the previous section.

Let $\|M_{ab}\|$ be the matrix such that $k_a = \sum_{b=1}^{n-1} M_{ab}\, r_b$, we have

$$\Theta_a = -\sum_{b=1}^{n-1} M_{ab}\, r_b \,\mathrm{div}\,\mathbf{v} \quad (a = 1, \cdots, n-1). \tag{28.57}$$

Introducing expression (28.57) into (28.56), we obtain

$$L_\pi = \sum_{a,b=1}^{n-1} M_{ab}\, r_a r_b \geqslant 0.$$

And assuming the Onsager symmetry property, $M_{ab} = M_{ba}$ $(a, b = 1, \cdots, n-1)$, we deduce that coefficients M_{ab} are *associated with a positive definite quadratic form.*

In conclusion, the results are the same as in the classical theory, but moreover we have obtained new constitutive equations (28.57) for the difference of temperatures.

We have considered the simple case of Stokes fluids. If the fluid is non-Stokesian, the Navier–Stokes stress tensor of viscosity is given by $(28.13)_1$ where ν is the bulk viscosity. The stress tensor $\mathbf{t}$ becomes

$$\mathbf{t} = -(p_0 + \Pi_\theta)\,\mathbf{I} + \boldsymbol{\sigma} = -p\,\mathbf{I} + 2\,\mu\,\mathbf{D}^D,$$

with

$$p = p_0 + \Pi_\theta + \Pi_\sigma.$$

The nonequilibrium part of the pressure, $p - p_0$, is separated into two contributions. The first one $\Pi_\sigma = -\nu\,\mathrm{div}\,\mathbf{v}$ is related to the bulk viscosity and the second one $\Pi_\theta = -L_\pi\,\mathrm{div}\,\mathbf{v}$ is related to the multi-temperature effect between components.

Due to the non-zero dynamic pressure even for Stokes fluids, we may conclude that the model of multi-temperature mixture of fluids is of a great importance. Perhaps such a model may be useful to analyze the evolution of the early universe in which the dynamic pressure seems to be essential [448, 449].

In the present survey, we have not mentioned a chemically reactive mixture, which is also interesting. We here only recall that thermodynamic properties of a multi-component reactive mixture of gases have been widely investigated by Giovangigli and Massot in [481] (see also [482–484]).

Chapter 29
Shock Structure in a Macroscopic Model of Binary Mixtures

Abstract In this chapter, we study the shock wave structure in a mixture on the basis of the model of multi-temperature (MT) mixtures explained in Chap. 28. For simplicity, the study is restricted to shocks in a binary mixture of ideal gases (component-1 and 2) that have no viscosity and no heat conductivity.

In the first part, we focus our attention on subshock formation. We prove that there exist, for a prescribed value of the mass ratio of the constituent molecules, four regions in the plane of the concentration and Mach number: no subshock region, the region where a subshock appears only in the component-1 gas, the region where a subshock appears only in the component-2 gas, and the region where variables of both component-1 and 2 gases have jumps. This is one of few examples for which subshock arises also for the shock velocity less than the maximum characteristic eigenvalue evaluated at the unperturbed state.

The model predicts also the existence of the temperature overshoot of the component gas with heavier constituent molecules. This fact was predicted also by other sophisticated approaches. The overshoot is a consequence of weak energy exchange between the gas components due to either large molecular mass difference or large rarefaction of the mixture. In the range of small Mach number, it is also shown that the shock thickness (or equivalently, the inverse of the Knudsen number) decreases with the increase of the Mach number: a behavior similar to that of a single fluid.

29.1 Introduction

Even though the model of a mixture of Euler fluids seems to be too much simplified since the component fluids have no viscosity and no heat conductivity, the results of the shock structure in a Helium-Argon mixture [485] obtained by using such a hyperbolic multi-temperature (MT) model are in good agreement with available experimental data [486]. In fact, the macroscopic MT model is at the same level of accuracy as more sophisticated models based on the kinetic theory [487] and on DSMC [488].

T. Ruggeri, M. Sugiyama, *Classical and Relativistic Rational Extended Thermodynamics of Gases*, https://doi.org/10.1007/978-3-030-59144-1_29

The aim of this chapter is twofold: Firstly, we make a systematic analysis of the shock structure problem in a binary MT mixture of Euler fluids (component-1 and 2). We clarify the condition under which shock structure may have a subshock inside. In particular, we improve the previous results of Conforto, Mentrelli, and Ruggeri [243] showing that, for a prescribed value of the mass ratio of the constituent molecules, there exist four regions in the plane (c_0, M_0) (c_0 and M_0 are, respectively, the unperturbed concentration and the Mach number): no subshock region, the region where a subshock appears only in the component-1 gas, the region where a subshock appears only in the component-2 gas, and the region where variables of both component-1 and 2 gases have jumps. This is one of few examples for which subshock arises also for shock velocity less than the maximum characteristic eigenvalue evaluated in the unperturbed state as was mentioned in Sect. 3.6.1.

Secondly, focusing on the case with moderate Mach numbers where a continuous solution exists, we present a brief survey of the paper by Madjarevićet al. [489], in which they proved the existence of temperature overshoot of a component with heavier constituent molecules—a region within a shock profile where the temperature is higher than the maximum equilibrium temperature of the mixture. This is consistent with some previous studies [147, 226, 490].

29.2 Shock Structure in a Binary Mixture

We study a binary mixture of ideal gases, which have no viscosity and no heat conductivity. We assume, for simplicity, that constituent molecules of both gases have the same degrees of freedom. The characteristic velocities are given in (28.22) with the sound velocity (28.24) and $\gamma_1 = \gamma_2 = \gamma$. The case in which the components are polyatomic gases with different γ's is only a little more complicated, but gives no conceptual difference from the present discussion (see for more details [243]).

The average temperature of the MT mixture (28.25) becomes, in the present case,

$$(\rho_1 c_{V_1} + \rho_2 c_{V_2})T = \rho_1 c_{V_1} T_1 + \rho_2 c_{V_2} T_2.$$

The total pressure and the intrinsic part of the internal energy are in the same form as in the case of single-component gas:

$$p = p_1 + p_2 = \rho \frac{k_B}{m} T, \qquad \rho\varepsilon_I = \rho_1\varepsilon_1 + \rho_2\varepsilon_2 = \rho \frac{k_B}{(\gamma - 1)m} T,$$

provided that we introduce the average molecular mass $m = m(c)$ and the average temperature T of the mixture in the following form:

$$\frac{1}{m(c)} = \frac{c}{m_1} + \frac{1-c}{m_2}, \qquad T = c\frac{m(c)}{m_1} T_1 + (1-c)\frac{m(c)}{m_2} T_2,$$

where c is the concentration related to the mass densities:

$$\rho_1 = \rho c, \quad \rho_2 = \rho(1-c).$$

As in the previous Chap. 28, we introduce the difference of the temperatures $\Theta = T_2 - T_1$, so-called diffusion temperature, and the ratio of the molecular masses:

$$\mu = \frac{m_1}{m_2}, \quad 0 < \mu \leqslant 1,$$

where, without loss of generality, we have assumed $m_1 \leqslant m_2$. The temperatures of the component gases can now be expressed in terms of the new variables T, Θ, and c, using μ as a parameter:

$$T_1 = T - f(c)\Theta, \quad T_2 = T + (1 - f(c))\Theta,$$

where auxiliary function $f(c)$ has the following form:

$$f(c) = \frac{\mu(1-c)}{c + \mu(1-c)}.$$

Finally, we also use the diffusion flux $\mathbf{J}$ of the component-1, instead of its velocity $\mathbf{v}_1$. Since relative velocities obey the relation $\rho_1\mathbf{u}_1 + \rho_2\mathbf{u}_2 = 0$, due to (28.4), the diffusion flux of the component-1 is defined as

$$\mathbf{J} = \rho_1\mathbf{u}_1 = -\rho_2\mathbf{u}_2.$$

Therefore, we take the change of variables:

$$(\rho, \mathbf{v}, T, \rho_1, \mathbf{v}_1, T_1) \rightarrow (\rho, \mathbf{v}, T, c, \mathbf{J}, \Theta),$$

and rewrite the system of governing equations as if it were a single fluid with extended fields. This is the idea of Ruggeri when he tried to unify the second-sound phenomena in gases and crystals [33]. We will see its explicit form in the next section where we write the shock structure equations.

The source terms satisfy the Galilean invariance (28.8):

$$\mathbf{m}_1 = \hat{\mathbf{m}}_1, \quad e_1 = \hat{e}_1 + \hat{\mathbf{m}}_1 \cdot \mathbf{v},$$

and (28.21) gives

$$\hat{\mathbf{m}}_1 = -\psi_{11}\left(\frac{\mathbf{u}_1}{T_1} - \frac{\mathbf{u}_2}{T_2}\right), \quad \hat{e}_1 = -\theta_{11}\left(-\frac{1}{T_1} + \frac{1}{T_2}\right),$$

where ψ_{11} and θ_{11} are positive phenomenological coefficients. They can be related to state variables and relaxation times for diffusion τ_v and temperature τ_T (28.33) and (28.34). For a monatomic gas, the ratio between relaxation times (28.35) becomes

$$\frac{\tau_T}{\tau_v} = \frac{m_1 + m_2}{c\, m_2 + (1 - c)\, m_1} > 1.$$

29.3 Shock Structure Problem

The shock structure, which we discussed in Sect. 3.4, is a continuous solution with steep gradients of state variables in the neighborhood of singular surface—the shock wave—which is diffused due to dissipative mechanisms taken into account. Our attention will be restricted to the shock structure related to plane shocks, moving at constant speed s in the direction orthogonal to the singular surface. Consequently, one space variable, say x, is enough for the problem, and the shock structure is described as a traveling wave solution, depending on a single variable $\varphi = x - st$, which asymptotically connects equilibrium states in front and behind the shock wave. These assumptions transform the model into a set of ordinary differential equations where the velocity, the diffusion flux, and the source term (momentum exchange) are described by a single component, i.e., $\mathbf{v} = (v, 0, 0)$, $\mathbf{J} = (J, 0, 0)$, and $\hat{\mathbf{m}}_1 = (\hat{m}_1, 0, 0)$, in Cartesian coordinates.

29.3.1 Dimensionless Shock Structure Equations

The problem of the shock structure is studied in a dimensionless form. For this purpose, we introduce the dimensionless variables by scaling the state variables and independent variable φ with appropriate upstream (unperturbed) equilibrium values, indicated by the subscript 0:

$$\hat{\rho} = \frac{\rho}{\rho_0},\ \hat{u} = \frac{u}{a_0},\ \hat{T} = \frac{T}{T_0},\ \hat{J} = \frac{J}{\rho_0 a_0},\ \hat{\Theta} = \frac{\Theta}{T_0},\ \hat{x} = \frac{\varphi}{l_0},\ M_0 = \frac{s - v_0}{a_0} = -\frac{u_0}{a_0}, \tag{29.1}$$

where $u = v - s$ is the relative velocity of a mixture with respect to the shock wave, l_0 is the upstream reference length, $a_0 = \{\gamma (k_B/m_0) T_0\}^{1/2}$ is the upstream sound with $m_0 = m(c_0)$ being the equilibrium average molecular mass, and M_0 is the upstream Mach number. For the sake of simplicity, hat will be dropped in the sequel.

The upstream reference length l_0 is usually taken as the mean free path of the molecules. The average mean free path in the mixture will be expressed in terms of other more primitive properties of the constituent molecules[369]:

$$l_0 = \frac{n_1}{n}(l_1)_0 + \frac{n_2}{n}(l_2)_0,$$

$$(l_\alpha)_0 = \frac{1}{\pi d_{12}^2}\left[n_1\left(1 + \frac{m_\alpha}{m_1}\right)^{1/2} + n_2\left(1 + \frac{m_\alpha}{m_2}\right)^{1/2}\right]^{-1},$$

where n_α are number densities of the molecules, and $n = n_1 + n_2$ is the mixture number density.

Using the scaled variables (29.1), we obtain the following set of dimensionless equations [489]:

$$\begin{aligned}
&\frac{d}{d\varphi}(\rho u) = 0,\\
&\frac{d}{d\varphi}\left(\rho u^2 + \frac{1}{\gamma}\frac{m_0}{m}\rho T + \frac{J^2}{\rho c(1-c)}\right) = 0,\\
&\frac{d}{d\varphi}\left\{\left(\frac{1}{2}\rho u^2 + \frac{1}{\gamma - 1}\frac{m_0}{m}\rho T + \frac{J^2}{2\rho c(1-c)}\right)u + \left(\frac{uJ}{\rho c(1-c)} + \frac{1}{\beta}\right)J\right\} = 0,\\
&\frac{d}{d\varphi}(\rho c u + J) = 0, \qquad (29.2)\\
&\frac{d}{d\varphi}\left\{\rho c u^2 + \frac{J^2}{\rho c} + 2uJ + \frac{1}{\gamma}\frac{m_0}{m_1}\rho c\,(T - f(c)\Theta)\right\} = -\frac{l_0}{\tau_D a_0} m_\mu(T, c, \Theta) J,\\
&\frac{d}{d\varphi}\left\{\left(\frac{1}{2}\rho c\left(u + \frac{J}{\rho c}\right)^2 + \frac{1}{\gamma - 1}\frac{m_0}{m_1}\rho c\,(T - f(c)\Theta)\right)\left(u + \frac{J}{\rho c}\right)\right\}\\
&\qquad = -\frac{l_0}{\tau_D a_0} m_\mu(T, c, \Theta) J u + \frac{l_0}{\tau_T a_0} e_\mu(\rho, T, c, \Theta)\Theta.
\end{aligned}$$

Equations $(29.2)_{1-3}$ represent the conservation laws of mass, momentum, and energy of the mixture, while $(29.2)_{4-6}$ are the balance laws of mass, momentum, and energy of the component-1. Auxiliary functions in the source terms read

$$m_\mu(T, c, \Theta) = \frac{T + [1 - c - f(c)]\Theta}{[T - f(c)\Theta][T + (1 - f(c))\Theta]} T,$$

$$e_\mu(\rho, T, c, \Theta) = \frac{1}{\gamma(\gamma - 1)}\frac{m_0}{m_1}\frac{m}{m_2}\frac{\rho c(1-c)T^2}{[T - f(c)\Theta][T + (1 - f(c))\Theta]}$$

and ratios of molecular masses can be expressed as

$$\frac{m_0}{m} = \frac{c + \mu(1-c)}{c_0 + \mu(1-c_0)}, \quad \frac{m_0}{m_1} = \frac{1}{c_0 + \mu(1-c_0)}, \quad \frac{m}{m_2} = \frac{\mu}{c + \mu(1-c)}.$$

In (29.2) we have also used abbreviation $1/\beta$ for the following expression:

$$\frac{1}{\beta} = \frac{1}{\gamma - 1}\left[\frac{m_0}{m_1}(1-\mu)T - \frac{m_0}{m_1}\frac{m}{m_2}\Theta\right] + \frac{J^2}{2\rho^2}\left[\frac{1}{c^2} - \frac{1}{(1-c)^2}\right],$$

where β can be interpreted as thermal inertia [158].

29.3.2 Boundary Conditions and Numerical Procedure

By using the procedure described in Sect. 3.4, the shock structure problem can be approached with the use of dynamical-system theory. The system (29.2) can be seen formally in the form (3.33), where $\mathbf{u} = (\rho, u, T, c, J, \Theta)^T$ is the column vector of state variables.

Mere observation of the source terms in (29.2) yields that the diffusion flux and the diffusion temperature vanish in both upstream and downstream stationary points, i.e., $J_0 = J_1 = 0$, $\Theta_0 = \Theta_1 = 0$. Thus, for a given upstream equilibrium state $\mathbf{u}_0$, one may determine the downstream equilibrium state $\mathbf{u}_1$ by integrating the conservative part of the system $(29.2)_{1-4}$.

From now on, for simplicity, we consider only the case of $\gamma = 5/3$ corresponding to a mixture of two monatomic gases. The analysis can be done for a generic γ [243]. Moreover we can choose, without loss of generality, the unperturbed velocity $v_0 = 0$. Taking into account the conservation laws, we have the following boundary conditions:

$$\mathbf{u}_0 = \begin{bmatrix} \rho_0 \\ u_0 \\ T_0 \\ c_0 \\ J_0 \\ \Theta_0 \end{bmatrix} = \begin{bmatrix} 1 \\ -M_0 \\ 1 \\ c_0 \\ 0 \\ 0 \end{bmatrix}, \quad \mathbf{u}_1 = \begin{bmatrix} \rho_1 \\ u_1 \\ T_1 \\ c_1 \\ J_1 \\ \Theta_1 \end{bmatrix} = \begin{bmatrix} \frac{4M_0^2}{3+M_0^2} \\ -\frac{3+M_0^2}{4M_0} \\ \frac{1}{16}\left(14 - \frac{3}{M_0^2} + 5M_0^2\right) \\ c_0 \\ 0 \\ 0 \end{bmatrix}. \tag{29.3}$$

In (29.3), we use the indexes 0 and 1 to indicate the two equilibrium state: the unperturbed and perturbed one, respectively. Note that relations between mixture state variables, ρ_1, u_1, T_1 and ρ_0, u_0, T_0, correspond to the solution of the usual Rankine-Hugoniot relations between the state variables at the shock wave for a single fluid (equilibrium subsystem). On the other hand, component-related state variables, c, J, and Θ, have the same equilibrium values in front and behind the shock. Since the diffusion flux J and the diffusion temperature Θ vanish in equilibrium, they can be regarded as genuine nonequilibrium variables. Also, concentration is the same in both equilibrium states, $c_1 = c_0$, and in the sequel it will be denoted as equilibrium concentration, without special regard to upstream or downstream state. Finally, downstream equilibrium can be regarded as a one-parameter family of states parametrized by the Mach number (i.e., shock speed: see (3.41)), $\mathbf{u}_1 = \mathbf{u}_1(\mathbf{u}_0, M_0)$. In this way, the third parameter—Mach number—is naturally introduced in the model through the boundary conditions.

29.3.3 Profile of Shock Structure and Analysis of Subshock Regions

The procedure described above can be applied if the solution of the shock structure equations exists. To analyze the existence of solution in a binary mixture, we write down the characteristic velocities (28.22) of the full system in equilibrium, where both components have common velocity v and temperature T. They read (in increasing order)

$$\lambda^{(1)} = v - \sqrt{\frac{5}{3}\frac{k_B}{m_1}T}, \quad \lambda^{(2)} = v - \sqrt{\frac{5}{3}\frac{k_B}{m_2}T},$$

$$\lambda^{(3)} = \lambda^{(4)} = v,$$

$$\lambda^{(5)} = v + \sqrt{\frac{5}{3}\frac{k_B}{m_2}T}, \quad \lambda^{(6)} = v + \sqrt{\frac{5}{3}\frac{k_B}{m_1}T},$$

while characteristic speeds of equilibrium subsystem are (in increasing order)

$$\mu^{(1)} = v - \sqrt{\frac{5}{3}\frac{k_B}{m_0}T}, \quad \mu^{(2)} = v, \quad \mu^{(3)} = v + \sqrt{\frac{5}{3}\frac{k_B}{m_0}T}.$$

From Theorem 2.3 [90], the subcharacteristic conditions hold

$$\lambda^{(1)} < \mu^{(1)}, \quad \mu^{(3)} < \lambda^{(6)}.$$

From now on, we focus only on positive velocities with respect to the frame moving with the mixture. The negative ones have the same properties.

We study the shock family that bifurcates from the trivial zero shock solution when $s = \mu_0^{(3)}$. In this case, the admissible equilibrium states (29.3) are determined by the Lax condition $\mu_0^{(3)} < s < \mu_1^{(3)}$, which can be equivalently expressed as $M_0 > 1$. Non-existence of the smooth shock structure according with Theorem 3.1 is related to the condition $s > \lambda_0^{(6)}$:

$$s > \lambda_0^{(6)} \quad \Leftrightarrow \quad M_0 > M_0^{(6)} = \frac{\lambda_0^{(6)}}{a_0} = \sqrt{\frac{m_0}{m_1}} = \sqrt{\frac{1}{c_0 + (1 - c_0)\mu}}.$$

For a fixed value of the ratio μ, the graph of $M_0^{(6)}$ is given in Fig. 29.1 ($\mu \leqslant \frac{1}{5}$) and Fig. 29.2 ($\frac{1}{5} < \mu < 1$). In the regions III and IV in the figures, there exists a subshock for the variables of the component-1 gas, which we call S_1^A with superscript A indicating that a subshock emerges after the shock speed becomes larger than the maximum eigenvalue in equilibrium.

In order to check whether there can exist other type of subshock, we observe that, according to the method by Ruggeri [227], a subshock can emerge when s meets an eigenvalue λ evaluated along the shock structure (see (3.34)). In the present case, the only possible positive eigenvalue that can meet s is $\lambda^{(5)}$, the value of which in the state $\mathbf{u}_0$, i.e., $\lambda_0^{(5)}$, is smaller than $\mu_0^{(3)}$ and therefore is smaller than s. Taking into account that $\lambda_1^{(5)} = \lambda^{(5)}(\mathbf{u}_1)$ is an increasing function of s, we notice that the necessary condition for the existence of a value of the field $\mathbf{u}^*$ such that $\lambda^{(5)}(\mathbf{u}*) = s$ is expressed as $\lambda_1^{(5)} > s$. In term of the Mach number, the condition is equivalent to $M_0 > M_0^*$ where M_0^* is the solution of

$$M_1^{(5)}(M_0^*, c, \mu) = M_0^*, \tag{29.4}$$

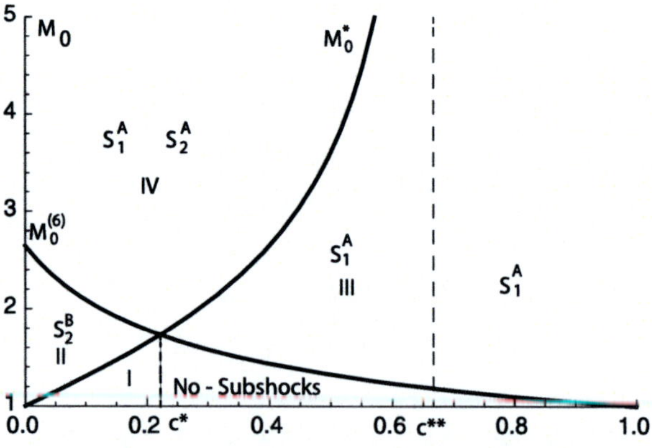

Fig. 29.1 Four regions in the plane (c_0, M_0) for possible subshocks when $\mu = \frac{1}{7} < \frac{1}{5}$. Here $c^* = \frac{2}{9}$ and $c^{**} = \frac{2}{3}$

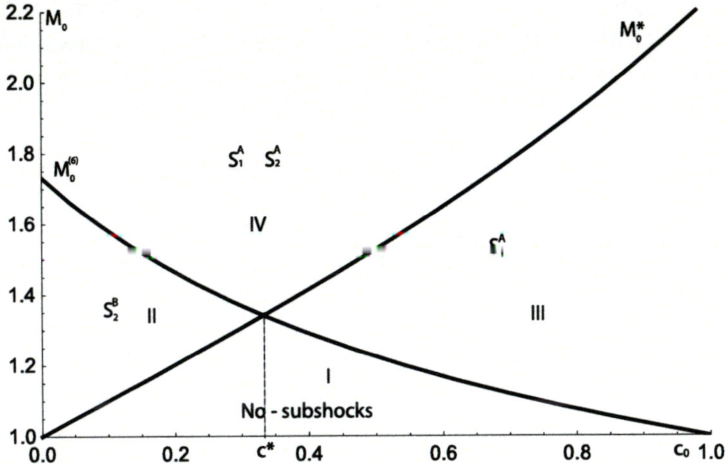

Fig. 29.2 Four regions in the plane (c_0, M_0) for possible subshocks when $\frac{1}{5} < \frac{1}{3} = \mu < 1$. Here $c^* = \frac{1}{3}$

where $M_1^{(5)} = \lambda_1^{(5)}/a_0$. From the relation:

$$M_1^{(5)} = \frac{3}{4}\frac{M_0^2 - 1}{M_0} + \frac{1}{4M_0}\sqrt{\frac{(5M_0^2 - 1)(3 + M_0^2)\mu}{\mu + c_0(1-\mu)}},$$

we have the solution of (29.4):

$$M_0^* = \sqrt{\frac{4\mu + 3(1-\mu)c_0}{4\mu - (1-\mu)c_0}}.$$

The existence of a real value of M_0^* requires

$$c_0 < c^{**} = \frac{4\mu}{1-\mu}.$$

Now we need to distinguish two cases: if $\mu \leqslant 1/5$ (i.e., $c^{**} \leqslant 1$), we have Fig. 29.1. While, if $\mu > 1/5$ (i.e., $c^{**} > 1$), we have Fig. 29.2. The plot of M_0^* as the function of c_0 for a fixed value of μ is given in Figs. 29.1 and 29.2, where the concentration c^* satisfying the condition $M_0^* = M_0^{(6)}$ is given by

$$c^* = \frac{1 + 7\mu - \sqrt{1 + \mu(\mu + 62)}}{6(\mu - 1)}, \quad c^* \in \left[0, \frac{1}{2}\right], \quad \text{for} \quad \mu \in [0, 1].$$

We notice four distinct regions in the plane (c_0, M_0): In the region I, no subshock can exist and the shock structure is regular. In the region II, we can have a subshock S_2^B for the variables of the component-2 gas (as confirmed by the numerical analysis below). This subshock is particularly interesting because this belongs to the type B (before the maximum eigenvalue). In the region III, there exists a subshock S_1^A of type A for the variables of the component-1 gas. Finally, in the region IV, both subshocks S_1^A and S_2^A of type A exist.

The velocity and temperature profiles of the shock structure solutions numerically obtained are given in Figs. 29.4 and 29.5. In particular, Fig. 29.4 shows us an example of the region II in which we have a subshock for variables of component-1 and not for the component-2. In Fig. 29.3, we see an example of the region III in which we have a subshock for variables of component-2 and not for the component-1. Finally, in Fig. 29.5, we observe that both component gases have subshocks in the region IV.

In Figs. 29.4 and 29.5, an ingenious trick [243] has been used to determine whether we have a subshock or not. For any value of the profile, we can construct a hypothetical point solution of the Rankine-Hugoniot of the full system. In this way we obtain a blue curve, denoted as $\mathscr{S}$–curve. Since the vertical gap between $\mathscr{S}$–curve and the profile represents the jump of a virtual shock, $\mathscr{S}$–curve should cease to exist where a shock actually occurs, namely in the location for which the steepness of the solution becomes infinite (therefore, the $\mathscr{S}$–curve should join the profile curve at a point with infinite steepness). Since, in the numerical profiles, the shock steepness is clearly not infinite, numerical errors of the profile and also $\mathscr{S}$–curve increase as a shock is approached. Therefore we know only an approximated localization of the shock. The states $\mathbf{u}_R$ and $\mathbf{u}_L$ delimiting the shock that can be presumed by the analysis of the numerical profiles and the $\mathscr{S}$–curve are indicated in the figures by grey–filled circles.

Remark 29.1 The above finding that a subshock exists before the maximum eigenvalue is not surprising since the maximum speed between $\lambda^{(5)}$ and $\lambda^{(6)}$ depends only on the molecular masses and not on the values of the field variables. Furthermore, as already observed in Sect. 3.6.1, the structure of the differential systems of mixtures is quite particular in the sense that the principal part of the operator has the same structure as that of a simple fluid and the interactions between the components of the mixture occur only through the production terms.

29.4 Regular Shock Structure and Temperature Overshoot

Numerical analysis in the region I, where the shock structure is regular, was given in [489]. The authors found for the first time that there can occur the temperature overshoot in this simple model. This fact clearly shows one of the peculiarities of the shock structure in mixtures whose components have disparate masses. It manifests through the existence of the region of non-zero width where the temperature of one

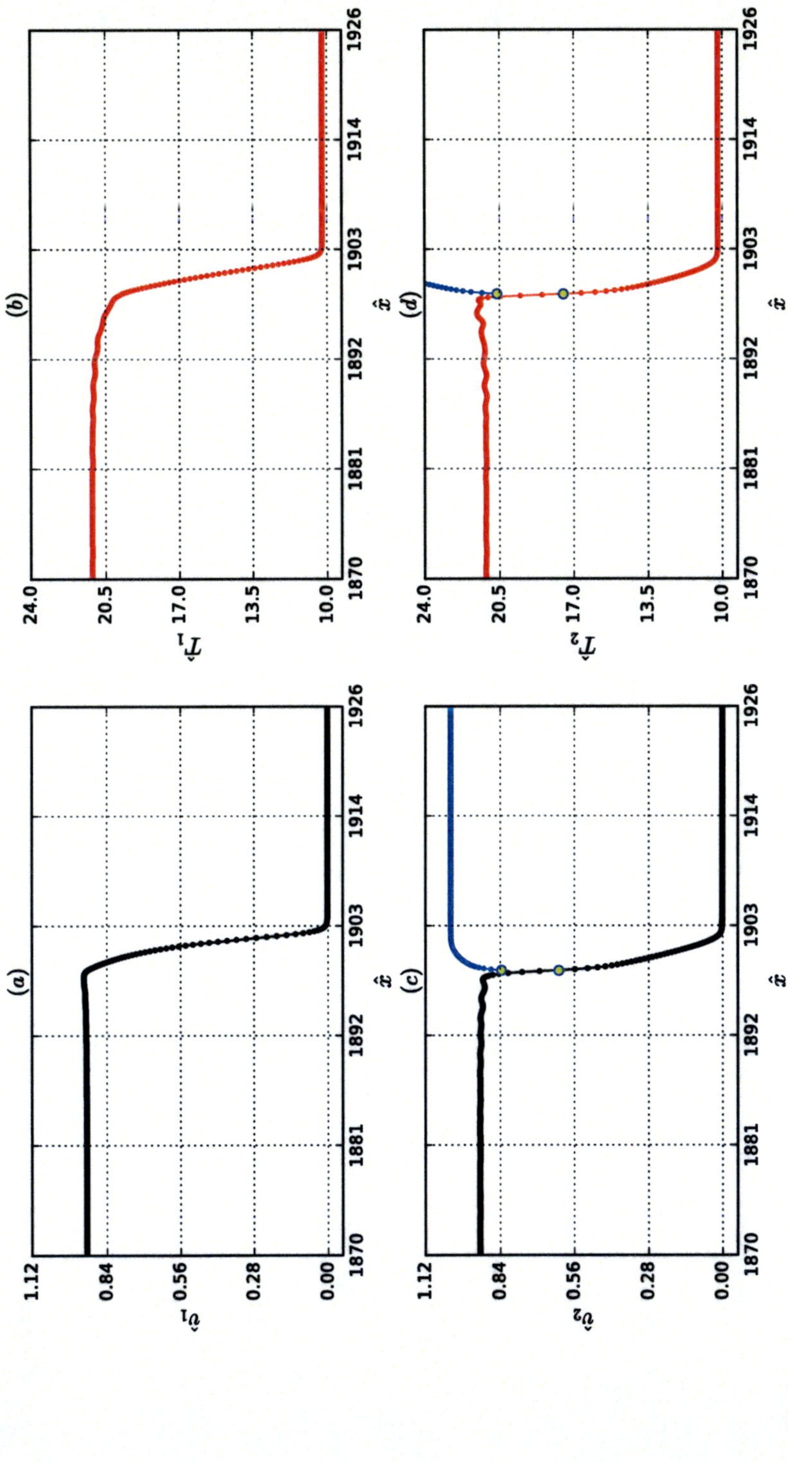

Fig. 29.3 Shock structure profiles of dimensionless velocity and temperature of component 1 (*a*, *b*) and component 2 (*c*, *d*) in the region II

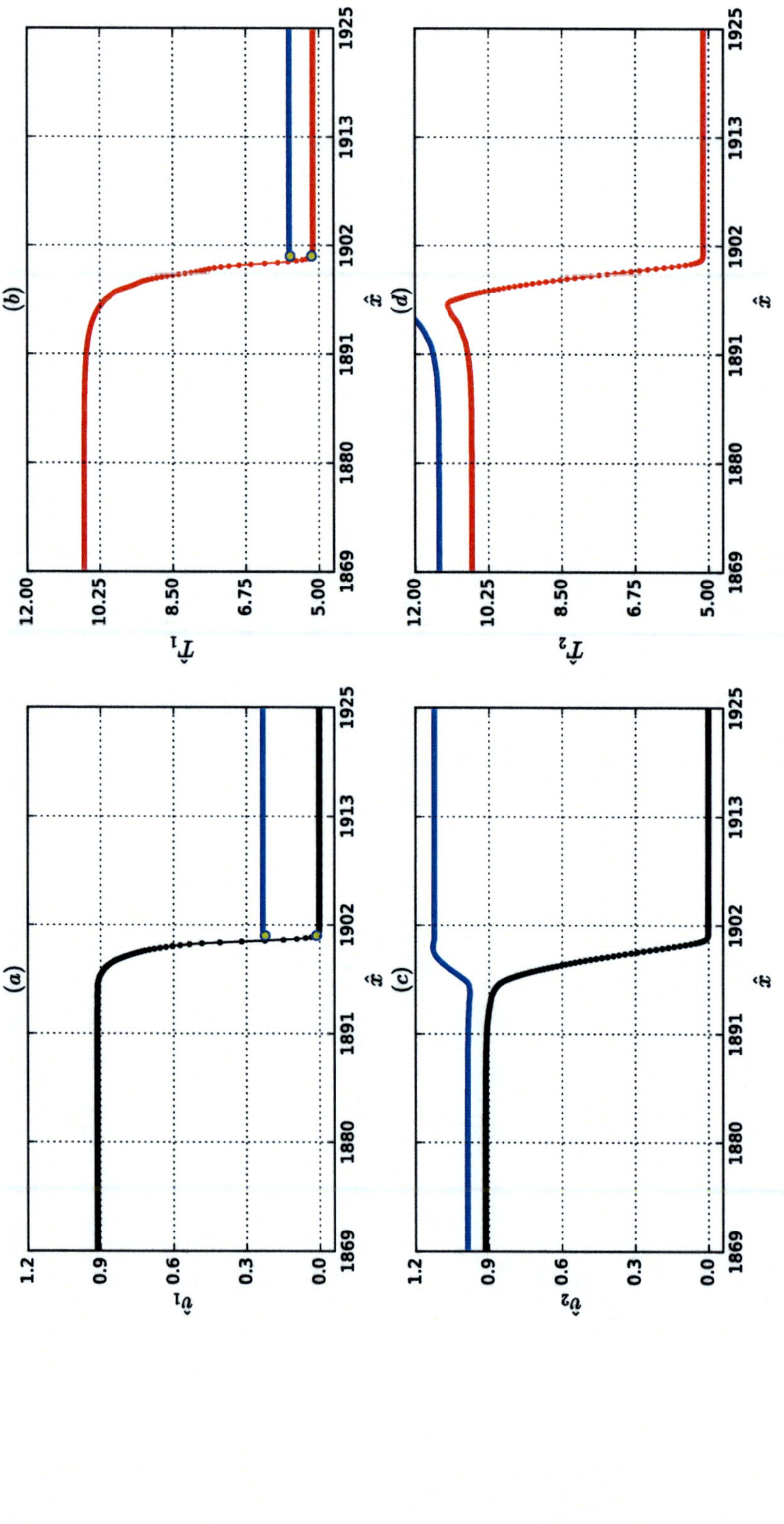

Fig. 29.4 Shock structure profiles of dimensionless velocity and temperature of component 1 (*a*, *b*) and component 2 (*c*, *d*) in the region III

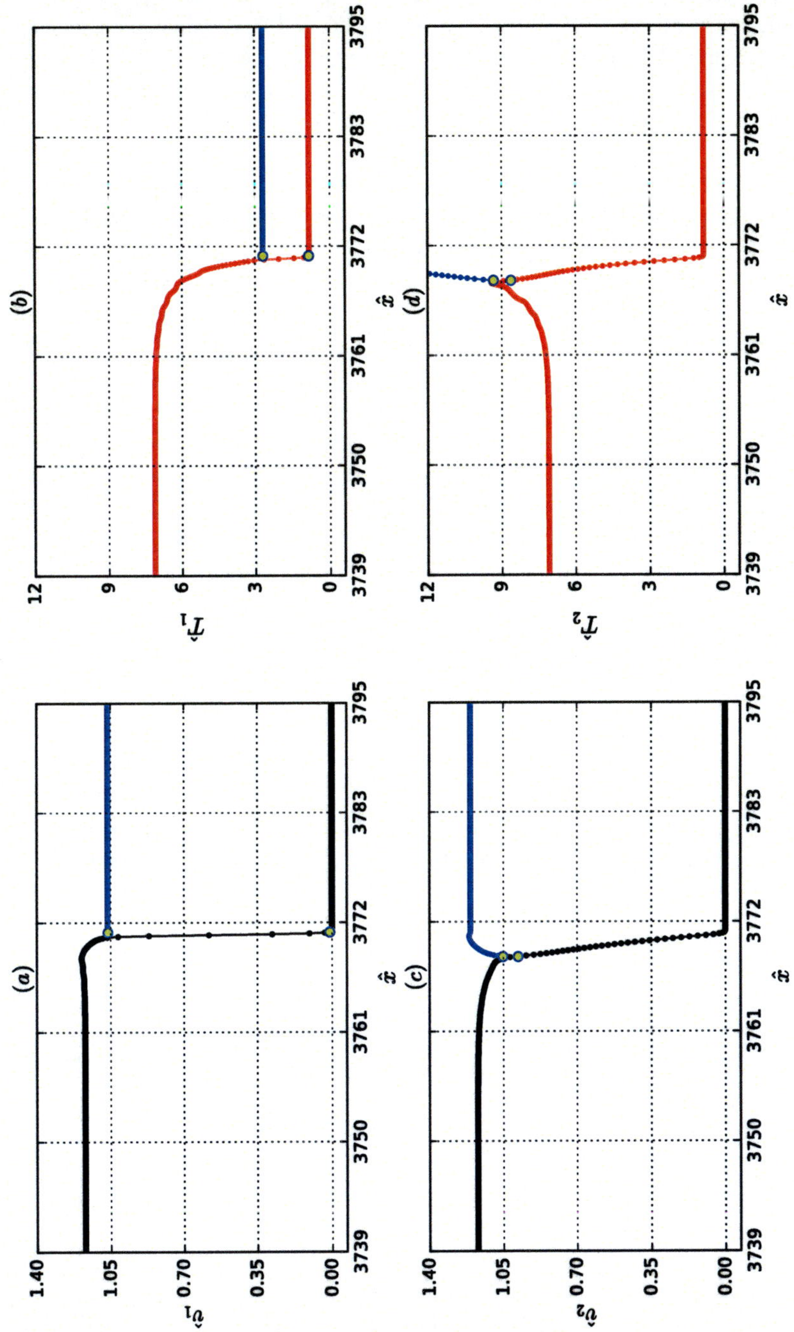

Fig. 29.5 Shock structure profiles of dimensionless velocity and temperature of component 1 (*a*, *b*) and component 2 (*c*, *d*) in the region IV

component raises above the downstream equilibrium temperature of the mixture. This phenomenon was observed in numerical calculations based on Boltzmann equations for mixtures [147, 490, 491] and DSMC [488]. Available experimental data do not provide enough evidence to support numerical simulations, although Harnett and Muntz [486] regard the overshoot of the parallel temperature of Argon as the onset of the overshoot of its mean temperature.

In the studies mentioned above, it was emphasized that the temperature overshoot is the most significant in the case of small molar fraction of heavier component. Abe and Oguchi [487] offered a physical explanation of this phenomenon. They stated that, in the case of vanishingly small mole fraction of heavier component, the main structure of the shock wave is determined by the lighter one. This causes the deceleration of heavier component and, at the same time, conversion of kinetic energy into thermal energy. However, dissipation through conduction is a slow process which cannot diffuse thermal energy gained by deceleration. As a consequence, the internal energy (temperature) of heavier component is raised above the terminal one. In our model, momentum and energy transfer through viscosity and heat conduction are neglected. We are focused on dissipation caused by mutual exchange of momentum and energy between the components, where the most prominent role is played by their molecular mass ratio μ. Thus, we examine the temperature overshoot from this perspective, analyzing its dependence on the mass ratio μ, as well as upstream Mach number M_0, and equilibrium concentration c_0.

The numerical solution of the present model predicts the temperature overshoot. Figure 29.6 shows an example of this effect [489]. It has an outstanding feature, not reported in previous studies, that the temperature overshoot varies non-monotonically with mass ratio. Namely, there exists a value μ^* of the mass ratio which determines the local minimum of temperature overshoot. Since other studies were based on limited number of numerical simulations, which provided information on certain particular cases only, this phenomenon remained unobserved thus far.

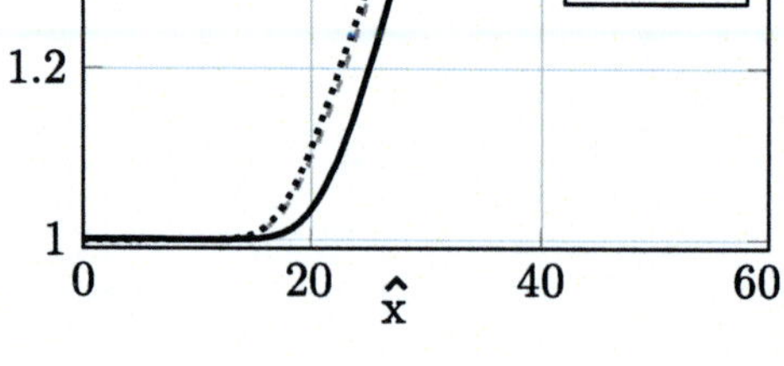

Fig. 29.6 Temperature profiles in the shock structure (T—average temperature of the mixture, T_1—temperature of the lighter component, T_2—temperature of the heavier component): $M_0 = 1.6$, $c_0 = 0.21$, $\mu = 0.1$ [489]

The outstanding feature of non-monotonic behavior of the temperature overshoot can be understood as follows: (1) For $\mu < \mu^*$, the temperature overshoot is increased due to large mass difference and low Knudsen number (Kn). The flow is between hydrodynamic and slip flow regime, but the mass ratio is too small to yield sufficient exchange of energy between the components (2) For $\mu > \mu^*$, Kn is increased, which puts the flow into transition regime. Although the masses of the molecules become comparable, the exchange of energy is prevented by rarefaction of the mixture, i.e., small number of cross-collisions which could cause it. Consequently, the temperature of heavier component cannot be attenuated, and temperature overshoot is increased. A similar overshoot can be noticed in the subplot d of Fig. 29.4.

Therefore, in a simplified model of MT mixtures, where viscosity and heat conductivity are neglected, small mutual exchange of energy between the components can be pointed out as main physical reason for the increase of temperature overshoot. It can occur for two reasons: (a) large mass discrepancy between the molecules (small μ), and (b) more rarefaction of a mixture.

29.5 Shock Thickness and the Knudsen Number

One of the parameters which describe the shock structure globally is the shock thickness. It is usually defined as follows:

$$\delta = \left| \frac{v_1 - v_0}{(dv/d\varphi)_{\max}} \right| .$$

It is important to notice that the dimensionless shock thickness is equal to the reciprocal of the Knudsen number:

$$\tilde{\delta} = \frac{\delta}{l_0} = \frac{1}{\mathrm{Kn}}, \tag{29.5}$$

whose value helps to distinguish between different flow regimes. In view of (29.5), shock thickness will carry also the information about the flow regime. In this section we analyze the shock thickness (and Kn) in terms of the mass ratio, Mach number, and upstream concentration.

The dependence of Kn on the Mach number is monotonous for fixed mass ratio and upstream concentration. It increases with the increase of the Mach number, which amounts to a decrease of the shock thickness. This is rather expected result which is similar to the behavior of a single-component gas. However, experimental facts about shock structure in a single fluid, as well as comparative study based on Navies-Stokes and Fourier model, reveals that this tendency seems to be opposite for larger Mach numbers (see [25]). Since our calculations are confined to small

Mach number flows, at most $M_0 \leqslant 2.0$, the results obtained here are in agreement with the single fluid model in this range.

In this chapter, we have studied only a mixture of non-reacting gases. There are several studies on shock waves in reacting gas-mixture. See, for example, [492].

We finish this chapter quoting the paper of Hantke et al. [493] in which the Riemann problem for two-phase flow model is studied, and the paper of Hantke and Thein [494] making a numerical analysis of two-phase flows with phase transition.

Chapter 30
Flocking and Thermodynamical Cucker-Smale Model

Abstract We present a thermodynamically consistent particle (TCP) model motivated by the theory of multi-temperature mixture of fluids in the spatially homogeneous case. The model incorporates the Cucker-Smale (C-S) type flocking model as its isothermal approximation. It is more complex than the C-S model, because the mutual interactions are not only "*mechanical*" but also "*thermal*". That is, the interactions are also affected by the "*temperature effect*" because individual particles may exhibit distinct internal energies. We develop a framework for asymptotic weak and strong flocking in the context of the proposed model.

30.1 Introduction

Emergent phenomena of interacting particle systems are ubiquitous in biological and chemical complex systems, e.g., the aggregation of bacteria, flocking of birds, swarming of fish, herding of sheep, synchronous flashing of fireflies, synchronization of coupled cells, etc. [495–500], and have been extensively studied with an eye for possible applications to mobile and sensor networks, controls of robots, and unmanned aerial vehicles [501–503]. After the pioneering works of Winfree and Kuramoto, several decades ago, many phenomenological agent-based models have been proposed and studied analytically and numerically. These models typically regard the agents as particles moving under the laws of mechanics, whereas, in this chapter, we consider emergent flocking phenomena in a thermodynamically consistent particle (TCP) model, in which internal energy plays a key role in the dynamics.

Before formulating the TCP model, we recall its natural predecessor, the standard C-S model that mimics Newton's equations for an interacting many-body system.

T. Ruggeri, M. Sugiyama, *Classical and Relativistic Rational Extended Thermodynamics of Gases*, https://doi.org/10.1007/978-3-030-59144-1_30

Let $(\mathbf{x}_\alpha, \mathbf{v}_\alpha) \in \mathbb{R}^{2d}$ be the phase-space coordinate of the α-th particle. The C-S model is governed by the system:

$$\begin{aligned} \frac{d\mathbf{x}_\alpha}{dt} &= \mathbf{v}_\alpha, \quad t > 0, \quad \alpha = 1, \cdots, n, \\ \frac{d\mathbf{v}_\alpha}{dt} &= \frac{1}{n}\sum_{\beta=1}^{n} \omega(\mathbf{x}_\alpha, \mathbf{x}_\beta)(\mathbf{v}_\beta - \mathbf{v}_\alpha), \end{aligned} \tag{30.1}$$

where the positive interaction weight function $\omega = \omega(\mathbf{x}, \mathbf{y})$ is Lipschitz continuous. The question of finding conditions on ω inducing a global flocking, or mono-cluster flocking for solutions of (30.1) at large time was first addressed by Cucker and Smale in their seminal work [162]. They used

$$\omega_{cs}(\mathbf{x}, \mathbf{y}) := \frac{1}{(1 + |\mathbf{x} - \mathbf{y}|^2)^{\frac{\beta}{2}}}, \quad \beta \geqslant 0.$$

Following the work of Cucker and Smale, flocking of solutions to (30.1) for general interaction weight function ω has been established in [504–512]. Although the interactions in (30.1) are of mean-field type, asymptotic flocking in the sense of Definition 30.1 depends on the nature of the interaction weight function ω and on the initial data.

One of obvious limitations of the C-S model is that no internal mode is employed in the interaction (communication) mechanism, because the system (30.1) only involves mechanical properties of particles. However, the states of biological and chemical organisms are not solely determined by their position and velocity, but also by internal modes. Thus, for a better description of the dynamics of such active particle systems, one needs to take into account internal modes in the interaction mechanism. For example, cyanobacteria are photosynthetic microorganisms that sense and respond to changes in the intensity of light and its direction. Under certain conditions depending on internal modes, these bacteria initiate a motion towards a light source [513, 514]. Thus, to simulate such behavior, an internal mode representing excitation level has been employed in the mathematical modelings [515–517]. Systematic study in this new direction of modeling is highly expected.

Ha and Ruggeri [161] noticed that the CS equations (30.1) are the same as those of an isothermal mixture of gases in a spatially homogeneous case. Therefore they had the idea to generalize in a natural way the CS model incorporating not only the momentum equation but also the energy equation. They wrote down the system (28.29), and, for simplicity of presentation, they assumed that the constant densities and constant specific heats are equal to unity:

$$\rho_\alpha = 1, \quad \varepsilon_\alpha = T_\alpha, \quad \forall \alpha = 1, \cdots, n.$$

Thus, the system (28.30) becomes (with the normalization changing $\psi_{bc} \to \psi_{bc}/n$ and also changing similarly for θ_{bc})

$$\begin{aligned}
&\frac{d\mathbf{v}_b}{dt} = -\frac{1}{n}\sum_{c=1}^{n-1}\psi_{bc}\Big(\frac{\mathbf{v}_c}{T_c} - \frac{\mathbf{v}_n}{T_n}\Big), \quad t > 0, \quad b = 1, 2, \dots n-1,\\
&\frac{d}{dt}\left(T_b + \frac{1}{2}v_b^2\right) = -\frac{1}{n}\sum_{c=1}^{n-1}\theta_{bc}\Big(\frac{1}{T_n} - \frac{1}{T_c}\Big), \qquad (30.2)\\
&\mathbf{v}_n = -\sum_{b=1}^{n-1}\mathbf{v}_b, \qquad T_n = n\bar{T}^0 + \frac{1}{2}\sum_{\alpha=1}^{n}(v_\alpha^0)^2 - \sum_{i=1}^{n-1}T_i - \frac{1}{2}\sum_{\alpha=1}^{n}v_\alpha^2.
\end{aligned}$$

The last two equations in (30.2) are consequences of $\mathbf{v} = 0$ and the constancy of total energy ε, where $\bar{T}^0$ is the constant average temperature (28.26).

Now we switch to the original variables $(\mathbf{x}_\alpha, \mathbf{v}_\alpha, T_\alpha)$ by transforming $(n-1) \times (n-1)$ matrices (ψ_{ij}) and (θ_{ij}) to $n \times n$ matrices $(\phi_{\alpha\beta})$ and $(\zeta_{\alpha\beta})$ as follows.

$$\begin{aligned}
&\phi_{ij} := -\psi_{ij}, \quad \forall i \neq j \in \{1, 2, \dots n-1\},\\
&\phi_{in} := \phi_{ni} = \sum_{1 \leqslant j \neq i \leqslant n-1}\psi_{ij}, \quad \forall i = 1, 2, \dots n-1 \qquad (30.3)\\
&\phi_{\alpha\alpha} := \text{arbitrary value } \forall\, \alpha = 1, 2, \dots n,
\end{aligned}$$

and similarly for $(\zeta_{\alpha\beta})$. Then, system (30.2) can be put in the equivalent form:

$$\begin{aligned}
&\frac{d\mathbf{x}_\alpha}{dt} = \mathbf{v}_\alpha, \quad t > 0, \quad \alpha = 1, \cdots, n,\\
&\frac{d\mathbf{v}_\alpha}{dt} = \frac{1}{n}\sum_{\beta=1}^{n}\phi_{\alpha\beta}\Big(\frac{\mathbf{v}_\beta}{T_\beta} - \frac{\mathbf{v}_\alpha}{T_\alpha}\Big), \qquad (30.4)\\
&\frac{d}{dt}\left(T_\alpha + \frac{1}{2}v_\alpha^2\right) = \frac{1}{n}\sum_{\beta=1}^{n}\zeta_{\alpha\beta}\Big(\frac{1}{T_\alpha} - \frac{1}{T_\beta}\Big), \quad v_\alpha^2 = |\mathbf{v}_\alpha|^2,
\end{aligned}$$

where $(\phi_{\alpha\beta})$ and $(\zeta_{\alpha\beta})$ are constant symmetric interaction matrices.

Note that, in the case of constant temperature $T_\alpha = 1$, the first two equations in (30.4) reduce to (30.1) with constant agent-dependent communication weights. Furthermore,

$$\mathbf{x}_\alpha(t) = \mathbf{x}_\alpha^0 + t\mathbf{v}^\infty, \quad \mathbf{v}_\alpha(t) = \mathbf{v}^\infty, \quad T_\alpha(t) = T^\infty, \quad t \geqslant 0, \quad 1 \leqslant \alpha \leqslant n$$

clearly satisfies (30.4).

Conversely, one may return from (30.4) to (30.2) through the inverse transformation:

$$\psi_{ij} = -\phi_{ij}, \quad \forall i \neq j = 1, 2, \ldots n-1;$$
$$\psi_{ii} = \sum_{\beta \neq i=1}^{n} \phi_{i\beta}, \quad \forall i = 1, 2, \ldots n-1 \text{ (no summation with respect to i).} \tag{30.5}$$

Lemma 30.1 *Suppose that all entries of the interaction matrix* $\boldsymbol{\Phi} \equiv (\phi_{\alpha\beta})$ *are positive. Then, the associated matrix* $\boldsymbol{\Psi} \equiv (\psi_{ij})$ *defined by the relation* (30.5) *is positive definite.*

The proofs is given in the paper [161].

Recall that positive and negative signs of the entry $\phi_{\alpha\beta}$ denote the attractive and repulsive interactions between the α-agent and the β-agent, respectively. It may be interesting to study the case where $\boldsymbol{\Psi}$ is positive definite, while the matrix $\boldsymbol{\Phi}$ may contain negative entries.

30.2 Asymptotic Weak Flocking in the TCP Model

We characterize asymptotic weak and strong flockings as follows:

Definition 30.1 Let $\mathscr{X} = \{(\mathbf{x}_\alpha, \mathbf{v}_\alpha, T_\alpha)\}$ be a global solution to (30.4).

1. $\mathscr{X}$ exhibits asymptotic *weak flocking* if

$$\lim_{t\to\infty} (|\mathbf{v}_\alpha(t) - \mathbf{v}_\beta(t)| + |T_\alpha(t) - T_\beta(t)|) = 0, \quad \forall\, 1 \leqslant \alpha, \beta \leqslant n. \tag{30.6}$$

2. $\mathscr{X}$ exhibits asymptotic *strong flocking* if, in addition to (30.6),

$$\sup_{0\leqslant t<\infty} |\mathbf{x}_\alpha(t) - \mathbf{x}_\beta(t)| < \infty, \quad \forall\, 1 \leqslant \alpha, \beta \leqslant n.$$

Note that our characterization of flocking does not require the convergence of $\mathbf{v}_\alpha$ and T_α to constant values as $t \to \infty$. If the solution $\mathscr{X}$ satisfies

$$\mathbf{v}_1 = \cdots = \mathbf{v}_n, \quad T_1 = \cdots = T_n,$$

we call $\mathscr{X}$ a flocking solution.

The main results in the paper [161] are as follows:

- For well prepared small initial data and suitable interaction matrices, using an ℓ^2- energy method, the following stability estimate is derived:

$$\sum_{\alpha=1}^{n}\Big(|\mathbf{v}_\alpha(t)-\mathbf{v}^\infty|^2+|T_\alpha(t)-T^\infty|^2\Big)$$
$$+C\int_0^t\Big(\sum_{\alpha=1}^{n}|\mathbf{v}_\alpha(s)-\mathbf{v}^\infty|^2+\sum_{\alpha,\beta=1}^{n}\zeta_{\alpha\beta}|T_\alpha(s)-T_\beta(s)|^2\Big)ds$$
$$\lesssim\sum_{\alpha=1}^{n}\Big(|\mathbf{v}_\alpha(0)-\mathbf{v}^\infty|^2+|T_\alpha(0)-T^\infty|^2\Big),$$

where $\mathbf{v}_\infty$ and T^∞ are asymptotic values determined by the initial data through conservation laws. This together with uniform bounds on $|\dot{\mathbf{v}}_\alpha|$ and $|\dot{T}_\alpha|$ yields the asymptotic weak flocking estimate:

$$\lim_{t\to\infty}\Big(|\mathbf{v}_\alpha(t)-\mathbf{v}_\beta(t)|+|T_\alpha(t)-T_\beta(t)|\Big)=0,\quad 1\leqslant\alpha,\beta\leqslant n.$$

We do not have a decay rate for the relative velocity that would induce a uniform bound for $|\mathbf{x}_\alpha-\mathbf{x}_\beta|$.
- With the assumption that the situation in which diffusion velocities are sufficiently small (30.4) can be effectively approximated by the system:

$$\frac{d\mathbf{x}_\alpha}{dt}=\mathbf{v}_\alpha,\quad t>0,\quad \alpha=1,\cdots,n,$$
$$\frac{d\mathbf{v}_\alpha}{dt}=\frac{1}{n}\sum_{\beta=1}^{n}\phi_{\alpha\beta}\Big(\frac{\mathbf{v}_\beta}{T_\beta}-\frac{\mathbf{v}_\alpha}{T_\alpha}\Big),$$
$$\frac{dT_\alpha}{dt}=\frac{1}{n}\sum_{\beta=1}^{n}\zeta_{\alpha\beta}\Big(\frac{1}{T_\alpha}-\frac{1}{T_\beta}\Big).$$

For this simplified model, under suitable conditions on the interaction matrices, we establish exponential flocking:

$$\sup_{0\leqslant t<\infty}|\mathbf{x}_\alpha(t)-\mathbf{x}_\beta(t)|<\infty,$$
$$|\mathbf{v}_\alpha(t)-\mathbf{v}_\beta(t)|+|T_\alpha(t)-T_\beta(t)|\leqslant Ce^{-\Lambda t},\quad \text{as } t\to\infty,$$

where C and Λ are positive constants.

Several new results about this TCP model have been obtained. In the paper [518], the TCP model was extended to incorporate the communications among the agents that depend on the positions of the particle. Then asymptotic dynamics, such as formation and nonexistence of mono-cluster flockings in terms of an initial configuration, coupling strength, and ansatz of communication weights, was studied. In [519], a hydrodynamic model for the ensemble of TCP was obtained and its global-in-time well-posedness in Sobolev space was proved.

In the paper [520], a uniform-in-time stability and uniform mean-field limit of a thermodynamic Cucker-Smale model with small diffusion velocity were studied. Emergent dynamics for this model and its coupled models with fluids has been treated in a series of other papers [521–523]. The relativistic version of TCP will be presented in Chap. 32.

Recently by using the TCP a thermodynamical variant of the famous Kuramoto model was proposed [524].

Chapter 31
System of Balance Laws of Mixture Type: Mixture of Dissipative Polyatomic Gases

Abstract *System of balance laws of mixture type* is a system of partial differential equations that has a structure similar to the system of a mixture of continuous media, i.e., its main part of the operator is composed of the main parts of several components and the interactions among the components are caused only through the production terms (see Definition 31.1 below). We study firstly the general properties obtained by imposing the principle of Galilean invariance and the entropy principle. These properties permit to construct, from a single component theory, the corresponding theory of mixtures in an easy way. Secondly, as an illustrative example of the general theory, we write down the hyperbolic system of balance laws of a mixture in which each component is prescribed by RET with 6 fields (ET_6). This is the simplest model after the Eulerian mixture model. Global existence of smooth solutions for small initial data is also proved.

31.1 Introduction

In the previous chapters, we have studied a mixture for which each component is an Euler gas (non-dissipative). In this chapter, we study a mixture of gases in which each component gas is described by the RET theory with 6 fields (ET_6). That is, we take into account the dynamic pressure but we neglect the heat conductivity and the shear viscosity for each component. As seen before, this is the case where the bulk viscosity is much larger than the shear viscosity and heat conductivity. Before studying the main subject, we summarize the results obtained in [525] so as to introduce the general mathematical structure of a mixture of dissipative gases.

The system in continuum mechanics represents balance laws, and under a regularity assumption it is expressed in the divergence form in space-time:

$$\frac{\partial \mathbf{F}_0}{\partial t} + \frac{\partial \mathbf{F}_i}{\partial x_i} = \mathbf{P}; \qquad \mathbf{F}_i = \mathbf{F}_0 v_i + \mathbf{G}_i, \tag{31.1}$$

where the density vector $\mathbf{F}_0$, the flux vector $\mathbf{F}_i$, the non-convective flux vector $\mathbf{G}_i$, and the production vector $\mathbf{P}$ are vectors of $\mathbb{R}^N$ and are expressed through constitutive

T. Ruggeri, M. Sugiyama, *Classical and Relativistic Rational Extended Thermodynamics of Gases*, https://doi.org/10.1007/978-3-030-59144-1_31

equations in terms of a field $\mathbf{U} \equiv \mathbf{U}(\mathbf{x}, t)$. The vector $\mathbf{v}$ is the flow velocity. For the moment we consider generic constitutive equations, i.e., the above quantities may depend on $\mathbf{U}$ either locally or non-locally in space and/or in time. Usually this is a system of balance laws for a single-component material like a single fluid.

In the case of simple mixture, following Truesdell [57], we adopt the assumption, which is also motivated by the kinetic theory, that the form of the system of field equations for any component is the same as that for a single fluid except that there appear new production terms due to the interchange of mass, momentum, energy, and other quantities between components (see Sect. 28.2). In analogy with such a mixture, we have the following:

Definition 31.1 (System of Balance Laws of Mixture Type) A system of balance laws of mixture type is a system of the form:

$$\frac{\partial \mathbf{F}_0^\alpha}{\partial t} + \frac{\partial \mathbf{F}_i^\alpha}{\partial x_i} = \mathbf{P}^\alpha + \mathbf{f}^\alpha; \qquad \mathbf{F}_i^\alpha = \mathbf{F}_0^\alpha v_i^\alpha + \mathbf{G}_\alpha^i, \qquad (\alpha = 1, 2, \ldots, n) \tag{31.2}$$

where the density vector $\mathbf{F}_0^\alpha$, the flux vector $\mathbf{F}_i^\alpha$, and the production vector $\mathbf{P}^\alpha$ have, respectively, the same forms as $\mathbf{F}_0$, $\mathbf{F}_i$, and $\mathbf{P}$ of the single component, and are obtained by the formal substitution of $\mathbf{U}^\alpha$ instead of $\mathbf{U}$. The new production vector $\mathbf{f}^\alpha$ emerges due to the interchange of the densities between the components. The number of components is n.

The production term $\mathbf{P}^\alpha$ indicates the source of the dissipative fluxes that appears also in the single-component fluid. On the other hand, $\mathbf{f}^\alpha$ is the production originated from the effect of mixture, such as chemical reaction and diffusion. In other words, it represents the interchange of mass, momentum, energy, and other fields between the components. The system treated as a whole, which takes all the components of (31.2) into account, must appear as a system of a single material. It is, however, more complex than (31.1) because diffusion contributes to the objective-variables:

$$\frac{\partial \mathscr{F}}{\partial t} + \frac{\partial \mathscr{F}_i}{\partial x_i} = \mathscr{P}, \qquad \mathscr{F}_i = \mathscr{F} v_i + \mathscr{G}_i, \tag{31.3}$$

$$\mathscr{F} = \sum_{\alpha=1}^{n} \mathbf{F}^\alpha, \quad \mathscr{F}_i = \sum_{\alpha=1}^{n} \mathbf{F}_i^\alpha, \quad \mathscr{P} = \sum_{\alpha=1}^{n} \mathbf{P}^\alpha, \quad \sum_{\alpha=1}^{n} \mathbf{f}^\alpha = 0. \tag{31.4}$$

As a consequence of $(31.2)_2$, we have

$$\mathscr{G}_i = \sum_{\alpha=1}^{n} \left(\mathbf{F}^\alpha u_i^\alpha + \mathbf{G}_i^\alpha\right)$$

with the diffusion velocity u_i^α defined by

$$u_i^\alpha = v_i^\alpha - v_i \qquad \left(\sum_{\alpha=1}^{n} \rho^\alpha u_i^\alpha = 0\right).$$

Since the production vectors $\mathbf{f}^\alpha$ are not independent of each other due to the last condition (31.4), it is sometimes convenient to consider another equivalent system that is composed of global balance laws and the component equations with indexes $b = 1, 2, \ldots n-1$:

$$\begin{aligned} &\frac{\partial \mathscr{F}}{\partial t} + \frac{\partial \mathscr{F}_i}{\partial x_i} = \mathscr{P}, \\ &\frac{\partial \mathbf{F}^b}{\partial t} + \frac{\partial \mathbf{F}_i^b}{\partial x_i} = \mathbf{P}^b + \mathbf{f}^b; \qquad \mathbf{F}_i^b = \mathbf{F}^b v_i^b + \mathbf{G}_i^b. \end{aligned} \tag{31.5}$$

As shown in [525], the system (31.5) is the consequence from the Galilean invariance principle and the entropy principle.

The previous study of mixtures made in Chap. 28 is within this framework. In this case, the single fluid system (1.3) corresponds to the system (31.1), while the system for the mixture (28.1) is a particular case of (31.2), and (28.5) belongs to the system (31.5).

In particular, it is possible to prove a theorem about Galilean invariace that permits us to obtain immediately the explicit expressions of the global quantities:

Theorem 31.1 *The global system (31.3) is also Galilean invariant with the same matrix* $\mathbf{X}(\mathbf{v})$ *of the single model (31.1):*

$$\begin{cases} \mathscr{F}(\mathbf{v}, \bar{\mathbf{w}}) = \mathbf{X}(\mathbf{v})\, \mathscr{F}(0, \bar{\mathbf{w}}), \\ \mathscr{G}_i(\mathbf{v}, \bar{\mathbf{w}}) = \mathbf{X}(\mathbf{v}) \mathscr{G}_i(0, \bar{\mathbf{w}}) \;, \\ \mathscr{P}(\mathbf{v}, \bar{\mathbf{w}}) = \mathbf{X}(\mathbf{v})\, \mathscr{P}(0, \bar{\mathbf{w}}), \end{cases}$$

where $\bar{\mathbf{w}}$ *denotes objective quantities that are, in general, different from* $\mathbf{w}$. *And the global intrinsic density, flux, and production are given by*

$$\mathscr{F}(0, \bar{\mathbf{w}}) = \sum_{\alpha=1}^{n} \mathbf{F}^\alpha(\mathbf{v}^\alpha = \mathbf{u}^\alpha, \mathbf{w}^\alpha),$$

$$\mathscr{G}_i(0, \bar{\mathbf{w}}) = \sum_{\alpha=1}^{n} \mathbf{F}_i^\alpha(\mathbf{v}^\alpha = \mathbf{u}^\alpha, \mathbf{w}^\alpha),$$

$$\mathscr{P}(0, \bar{\mathbf{w}}) = \sum_{\alpha=1}^{n} \mathbf{P}^\alpha(\mathbf{v}^\alpha = \mathbf{u}^\alpha, \mathbf{w}^\alpha)$$

The proof is given in [525].

31.2 Polyatomic-Gas Mixture Based on ET$_6$

We summarize the model of a mixture, in which each component is described by the ET$_6$ theory. The system of balance laws of a single fluid has the form (31.1) with

$$\mathbf{F}_0 = \begin{pmatrix} \rho \\ \rho v_i \\ \rho v^2 + 2\rho\varepsilon \\ \rho v^2 + 3(p+\Pi) \end{pmatrix}, \quad \mathbf{G}_i = \begin{pmatrix} 0 \\ (p+\Pi)\delta_{ij} \\ 2(p+\Pi)v_i \\ 2(p+\Pi)v_i \end{pmatrix}, \quad \mathbf{P} = \begin{pmatrix} 0 \\ 0 \\ 0 \\ P_{ll} \end{pmatrix},$$

where Π denotes the dynamic pressure [128, 131, 132, 322]. In this model, both shear viscosity and heat conductivity are neglected.

According to Definition 31.1, the system of balance laws of a mixture is given in the form (31.2) with[1]

$$\mathbf{F}^\alpha = \begin{pmatrix} \rho^\alpha \\ \rho^\alpha v_i^\alpha \\ \rho^\alpha (v^\alpha)^2 + 2\rho^\alpha\varepsilon^\alpha \\ \rho^\alpha (v^\alpha)^2 + 3(p^\alpha+\Pi^\alpha) \end{pmatrix}, \quad \mathbf{G}_\alpha^i = \begin{pmatrix} 0 \\ (p^\alpha+\Pi^\alpha)\,\delta_{ij} \\ 2\,(p^\alpha+\Pi^\alpha)\,v_i^\alpha \\ 2\,(p^\alpha+\Pi^\alpha)\,v_i^\alpha \end{pmatrix},$$

$$\mathbf{P}^\alpha = \begin{pmatrix} 0 \\ 0 \\ 0 \\ P_{ll}^\alpha \end{pmatrix}, \quad \mathbf{f}^\alpha = \begin{pmatrix} \tau^\alpha \\ m_j^\alpha \\ 2e^\alpha \\ 2\omega^\alpha \end{pmatrix}.$$

The global system is in the form (31.3) with

$$\mathcal{F} = \begin{pmatrix} \rho \\ \rho v_i \\ \rho v^2 + 2\rho\varepsilon \\ \rho v^2 + 3(p+\Pi) \end{pmatrix}, \quad \mathcal{G}_i = \begin{pmatrix} 0 \\ -t_{ij} \\ -2t_{ij}v_j + 2q_i \\ -2t_{ij}v_j + 2Q_i \end{pmatrix}, \quad \mathcal{P} = \begin{pmatrix} 0 \\ 0 \\ 0 \\ \mathcal{P}_{ll} \end{pmatrix}.$$

We notice that there appear (1) the global stress tensor $t_{ij} = -(p+\Pi)\delta_{ij} + \sigma_{\langle ij\rangle}$ where p is the total pressure, Π is the global dynamic pressure, and $\sigma_{\langle ij\rangle}$ is the global deviatoric shear stress tensor, (2) the global heat flux q_i, and (3) a new global

[1]The production terms $\hat{e}_1$ and $\hat{\omega}_1$ in [525] differ from the present ones by a factor 2. The present choice is better when we compare the results with the previous ones in the case of Eulerian gas.

quantity Q_i with the same dimension as q_i. Taking into account the expression of $\mathbf{X}$ given in (12.2) and Theorem 31.1, we obtain the following global quantities:

$$t_{ij} = -\sum_{\alpha=1}^{n} \left\{ \rho^\alpha u_i^\alpha u_j^\alpha + \left(p^\alpha + \Pi^\alpha\right) \delta_{ij} \right\},$$

$$\Pi = -\frac{1}{3} t_{ll} - p = \sum_{\alpha=1}^{n} \left(\Pi^\alpha + \frac{1}{3} \rho^\alpha \left(u^\alpha\right)^2 \right),$$

$$\sigma_{\langle ij \rangle} = t_{\langle ij \rangle} = -\sum_{\alpha=1}^{n} \rho^\alpha u_{\langle i}^\alpha u_{j\rangle}^\alpha,$$

$$q_i = \sum_{\alpha=1}^{n} \left\{ \frac{1}{2} \rho^\alpha \left(u^\alpha\right)^2 + \rho^\alpha \varepsilon^\alpha + p^\alpha + \Pi^\alpha \right\} u_i^\alpha,$$

$$Q_i = \sum_{\alpha=1}^{n} \left\{ \frac{1}{2} \rho^\alpha \left(u^\alpha\right)^2 + \frac{5}{2} \left(p^\alpha + \Pi^\alpha\right) \right\} u_i^\alpha,$$

where the total pressure p and the total internal energy ε are defined as follows:

$$p = \sum_{\alpha=1}^{n} p^\alpha, \quad \rho\varepsilon = \rho\varepsilon_I + \frac{1}{2} \sum_{\alpha=1}^{n} \frac{1}{3} \rho^\alpha \left(u^\alpha\right)^2, \quad \rho\varepsilon_I = \sum_{\alpha=1}^{n} \rho^\alpha \varepsilon^\alpha.$$

For production terms, we have

$$\mathbf{P}^\alpha \left(\mathbf{v}, \mathbf{w}^\alpha\right) = \mathbf{X}(\mathbf{v}) \hat{\mathbf{P}}^\alpha, \quad \mathbf{f}^\alpha \left(\mathbf{v}, \mathbf{w}^\alpha\right) = \mathbf{X}(\mathbf{v}) \hat{\mathbf{f}}^\alpha$$

that imply

$$\begin{aligned} &P_{ll}^\alpha = \hat{P}_{ll}^\alpha, \quad \tau^\alpha = \hat{\tau}^\alpha, \quad \mathbf{m}^\alpha = \hat{\tau}^\alpha \mathbf{v} + \hat{\mathbf{m}}^\alpha, \\ &e^\alpha = \hat{\tau}^\alpha \frac{v^2}{2} + \hat{\mathbf{m}}^\alpha \cdot \mathbf{v} + \hat{e}^\alpha, \quad \omega^\alpha = \hat{\tau}^\alpha \frac{v^2}{2} + \hat{\mathbf{m}}^\alpha \cdot \mathbf{v} + \hat{\omega}^\alpha. \end{aligned} \tag{31.6}$$

From (31.4)$_3$, we have

$$\sum_{\alpha=1}^{n} \tau^\alpha = \sum_{\alpha=1}^{n} \mathbf{m}^\alpha = \sum_{\alpha=1}^{n} e^\alpha = 0, \quad \mathscr{P}_{ll} = \sum_{\alpha=1}^{n} P_{ll}^\alpha,$$

which, from (31.6) and $(2.53)_2$, are equivalent to

$$\sum_{\alpha=1}^{n}\hat{\tau}^{\alpha}=\sum_{\alpha=1}^{n}\hat{\mathbf{m}}^{\alpha}=\sum_{\alpha=1}^{n}\hat{e}^{\alpha}=0,\quad \mathscr{P}_{ll}=\hat{\mathscr{P}}_{ll}=\sum_{\alpha=1}^{n}\hat{P}_{ll}^{\alpha}.$$

In the present case, the system (31.5) becomes

$$\begin{aligned}
&\frac{\partial\rho}{\partial t}+\frac{\partial\rho v_k}{\partial x_k}=0,\\
&\frac{\partial\rho v_i}{\partial t}+\frac{\partial(\rho v_i v_k-t_{ik})}{\partial x_k}=0,\\
&\frac{\partial(\frac{1}{2}\rho v^2+\rho\varepsilon)}{\partial t}+\frac{\partial}{\partial x_k}\left\{(\frac{1}{2}\rho v^2+\rho\varepsilon)v_k-t_{kj}v_j+q_k\right\}=0,\\
&\frac{\partial}{\partial t}\left\{\rho v^2+3(p+\Pi)\right\}+\frac{\partial}{\partial x_k}\Big\{\Big[\rho v^2+3(p+\Pi)\Big]v_k\\
&\qquad\qquad -2t_{kj}v_j+2Q_k\Big\}=\hat{\mathscr{P}}_{ll},\\
&\frac{\partial\rho^b}{\partial t}+\frac{\partial\rho^b v_k^b}{\partial x_k}=\hat{\tau}^b,\\
&\frac{\partial\rho^b v_i^b}{\partial t}+\frac{\partial}{\partial x_k}\left\{\rho^b v_i^b v_k^b+(p^b+\Pi^b)\delta_{ik})\right\}=\hat{m}_i^b+\hat{\tau}^b v_i,\\
&\frac{\partial}{\partial t}\left(\rho^b(v^b)^2+2\rho^b\varepsilon^b\right)+\frac{\partial}{\partial x_k}\left\{\left[\rho^b(v^b)^2+2\rho^b\varepsilon^b+2p^b+2\Pi^b\right]v_k^b\right\}\\
&\qquad\qquad =2\hat{e}^b+\hat{\tau}^b v^2+2\hat{\mathbf{m}}^b\cdot\mathbf{v},\\
&\frac{\partial}{\partial t}\left\{\rho^b(v^b)^2+3(p^b+\Pi^b)\right\}+\frac{\partial}{\partial x_k}\left\{\left[\rho^b(v^b)^2+5(p^b+\Pi^b)\right]v_k^b\right\}\\
&\qquad\qquad =\hat{P}_{ll}^b+2\hat{\omega}^b+\hat{\tau}^b v^2+2\hat{\mathbf{m}}^b\cdot\mathbf{v},\\
&\qquad\qquad (b=1,2,\ldots n-1).
\end{aligned}\tag{31.7}$$

31.2.1 Binary Mixture in One-Space Dimension

The expressions of the intrinsic production terms in the right-hand side of the system (31.7) were obtained in [525] by using the entropy principle. Since these are very complex in general, we consider, as a simpler case, a binary mixture of rarefied gases with no chemical reaction in the one space-dimension. In the present case, we assume

$$p_\alpha = \rho_\alpha \frac{k_B}{m_\alpha} T_\alpha, \quad \varepsilon_\alpha = \frac{D_\alpha}{2} \frac{k_B}{m_\alpha} T_\alpha.$$

The average temperature of the multi-temperature mixture (28.26) is given by

$$(\rho_1 c_{V_1} + \rho_2 c_{V_2}) T = \rho_1 c_{V_1} T_1 + \rho_2 c_{V_2} T_2. \tag{31.8}$$

The total pressure and the internal energy in equilibrium are in the same form as in the case of single-component gas:

$$p = p_1 + p_2 = \rho \frac{k_B}{m} T, \qquad \rho \varepsilon_I = \rho_1 \varepsilon_1 + \rho_2 \varepsilon_2 = \frac{D}{2} \rho \frac{k_B}{m} T, \tag{31.9}$$

provided that we introduce the average mass $m = m(c)$ and the global degrees of freedom D of the mixture in the following form:

$$\frac{1}{m(c)} = \frac{c}{m_1} + \frac{1-c}{m_2}, \qquad D = c \frac{m(c)}{m_1} D_1 + (1-c) \frac{m(c)}{m_2} D_2, \tag{31.10}$$

where c is the concentration related to the mass densities:

$$\rho_1 = \rho c, \quad \rho_2 = \rho (1-c).$$

Let us consider the case with only one-space dimension x. Then we have

$$\mathbf{v} = (v, 0, 0), \quad \mathbf{v}_\alpha = (v_\alpha, 0, 0), \quad \mathbf{u}_\alpha = (u_\alpha, 0, 0).$$

Since we have assumed that no chemical reaction occurs, we have $\tau^\alpha = 0$. From (12.46), we also assume

$$P^\alpha_{ll} = -3 \frac{\Pi_\alpha}{\bar{\tau}_\alpha},$$

where $\bar{\tau}_\alpha$ indicate the relaxation times. Then the system (31.7) now becomes

$$
\begin{aligned}
&\frac{\partial \rho}{\partial t}+\frac{\partial \rho v}{\partial x}=0,\\
&\frac{\partial \rho v}{\partial t}+\frac{\partial}{\partial x}\left(\rho v^2+p+\Pi-\sigma\right)=0,\\
&\frac{\partial}{\partial t}\left(\frac{1}{2}\rho v^2+\rho\varepsilon\right)+\frac{\partial}{\partial x}\left\{\left(\frac{1}{2}\rho v^2+\rho\varepsilon+p+\Pi-\sigma\right)v+q\right\}=0,\\
&\frac{\partial}{\partial t}\left\{\rho v^2+3(p+\Pi)\right\}+\frac{\partial}{\partial x}\left\{\left[\rho v^2+5(p+\Pi)-2\sigma\right]v+2Q\right\}\\
&\qquad\qquad=-3\Pi_1\left(\frac{1}{\bar{\tau}_1}-\frac{1}{\bar{\tau}_2}\right)-\frac{3}{\bar{\tau}_2}\left(\Pi+\frac{1}{2}\sigma\right),\\
&\frac{\partial \rho_1}{\partial t}+\frac{\partial \rho_1 v_1}{\partial x}=0,\\
&\frac{\partial \rho_1 v_1}{\partial t}+\frac{\partial}{\partial x}\left\{\rho_1 v_1^2+p_1+\Pi_1\right\}=\hat{m}_1,\\
&\frac{\partial}{\partial t}\left(\rho_1 v_1^2+2\rho_1\varepsilon_1\right)+\frac{\partial}{\partial x}\left\{\left[\rho_1 v_1^2+2\rho_1\varepsilon_1+2p_1+2\Pi_1\right]v_1\right\}=2(\hat{e}_1+\hat{m}_1 v),\\
&\frac{\partial}{\partial t}\left\{\rho_1(v_1)^2+3(p_1+\Pi_1)\right\}+\frac{\partial}{\partial x}\left\{\left[\rho_1 v_1^2+5(p_1+\Pi_1)\right]v_1\right\}=-\frac{3\Pi_1}{\bar{\tau}_1}+2(\hat{\omega}_1+\hat{m}_1 v),
\end{aligned}
\tag{31.11}
$$

where the global specific internal energy ε, the global dynamic pressure Π, the diviatoric part of the global shear stress σ, the global heat flux q, and the new quantity Q are given by

$$
\begin{aligned}
\sigma &= -\frac{2}{3}\sum_{\alpha=1}^{2}\rho_\alpha u_\alpha^2=-\frac{2}{3}\frac{\rho_1\rho}{\rho_2}u_1^2,\\
\rho\varepsilon &= \rho\varepsilon_I+\frac{1}{2}\sum_{\alpha=1}^{2}\rho_\alpha u_\alpha^2=\rho\varepsilon_I-\frac{3}{4}\sigma,\\
\Pi &= \sum_{\alpha=1}^{2}\left(\Pi_\alpha+\frac{1}{3}\rho_\alpha u_\alpha^2\right)=\Pi_1+\Pi_2-\frac{\sigma}{2},\\
q &= \sum_{\alpha=1}^{2}\left(\frac{1}{2}\rho_\alpha u_\alpha^2+\rho_\alpha\varepsilon_\alpha+p_\alpha+\Pi_\alpha\right)u_\alpha,\\
Q &= \sum_{\alpha=1}^{2}\left\{\frac{1}{2}\rho_\alpha u_\alpha^2+\frac{5}{2}\left(p_\alpha+\Pi_\alpha\right)\right\}u_\alpha,
\end{aligned}
\tag{31.12}
$$

and the total p and the intrinsic energy ε_I are given by (31.9). We observe that the global production in Eq. (31.11)$_4$ is obtained, from (31.12)$_3$, as follows:

$$\hat{\mathscr{P}}_{ll} = -3\frac{\Pi_1}{\bar{\tau}_1} - 3\frac{\Pi_2}{\bar{\tau}_2} = -3\Pi_1\left(\frac{1}{\bar{\tau}_1} - \frac{1}{\bar{\tau}_2}\right) - \frac{3}{\bar{\tau}_2}\left(\Pi + \frac{1}{2}\sigma\right).$$

It is noticeable that there appear, in the system (31.11), global viscous stress and global heat flux due to the presence of diffusion velocity even if we neglect these dissipative quantities in the field equations for each component.

The entropy production is given by

$$\Sigma = \hat{m}_1\hat{\Lambda}_1 + \hat{e}_1(2\hat{\eta}_1) + \hat{\omega}_1(2\hat{\zeta}_1),$$

where $\hat{\Lambda}_1$, $\hat{\eta}_1$, and $\hat{\zeta}_1$ are the Lagrange multipliers (components of the main field) with respect to the last 3 balance laws in (31.11). The expressions of $\hat{m}_1$, $\hat{e}_1$, and $\hat{\omega}_1$ are obtained, by imposing the conditions that Σ is positive and quadratic as usual in nonequilibrium thermodynamics, as follows:

$$\hat{m}_1 = \psi_{11}\hat{\Lambda}_1, \quad \hat{e}_1 = 2(\theta_{11}\hat{\eta}_1 + \kappa_{11}\hat{\zeta}_1), \quad \hat{\omega}_1 = 2(\kappa_{11}\hat{\eta}_1 + \phi_{11}\hat{\zeta}_1).$$

The components $\hat{\Lambda}_1$, $\hat{\eta}_1$, and $\hat{\zeta}_1$ are, in the present case, given by [525]

$$\hat{\Lambda}_1 = -\frac{u_1}{T_1}\left(1 - \frac{\Pi_1}{p_1}\right) + \frac{u_2}{T_2}\left(1 - \frac{\Pi_2}{p_2}\right),$$
$$\hat{\eta}_1 = \frac{1}{2T_1}\left(1 - \frac{\Pi_1}{p_1}\right) - \frac{1}{2T_2}\left(1 - \frac{\Pi_2}{p_2}\right),$$
$$\hat{\zeta}_1 = -\frac{D_1}{(D_1 - 3)}\frac{1}{2T_1}\frac{\Pi_1}{p_1} + \frac{D_2}{(D_2 - 3)}\frac{1}{2T_2}\frac{\Pi_2}{p_2},$$

where $\psi_{11} > 0$ and the matrix

$$\begin{pmatrix} \theta_{11} & \kappa_{11} \\ \kappa_{11} & \phi_{11} \end{pmatrix}$$

is positive definite.

We emphasize that the system (31.11) satisfies the convexity condition and the K-condition. Therefore the system is symmetric hyperbolic in terms of the main field components, and has global smooth solutions for all time, which converge to the equilibrium ones provided that the initial data are sufficiently smooth. We refer to the paper [525] for details.

Lastly we show that the associate equilibrium subsystem of (31.11) is given by

$$\begin{aligned}
&\frac{\partial \rho}{\partial t}+\frac{\partial \rho v}{\partial x}=0,\\
&\frac{\partial \rho_1}{\partial t}+\frac{\partial \rho_1 v}{\partial x}=0,\\
&\frac{\partial \rho v}{\partial t}+\frac{\partial}{\partial x}\left(\rho v^2+p\right)=0,\\
&\frac{\partial}{\partial t}\left(\frac{1}{2}\rho v^2+\rho\varepsilon_I\right)+\frac{\partial}{\partial x}\left\{\left(\frac{1}{2}\rho v^2+\rho\varepsilon_I+p\right)v\right\}=0,
\end{aligned}$$

where p and ε_I are given in (31.9) with average temperature T in (31.8) and m, D in (31.10).

Chapter 32
Relativistic Mixture of Gases and Relativistic Cucker-Smale Model

Abstract We present in this chapter a relativistic model of a mixture of Euler gases with multi-temperature. We explicitly determine the production terms resulting from the interchange of energy-momentum between the components by using the entropy principle.

We use the analogy between the homogeneous case of a mixture of gases and the thermo-mechanical Cucker-Smale (TCS) flocking model in a classical setting to derive a relativistic counterpart of the TCS model. Moreover, we employ a theory of *principal subsystem* to derive the relativistic Cucker-Smale (CS) model. Then we provide a sufficient framework leading to the exponential flocking in terms of communication weights. We also show that the relativistic CS model reduces to the classical CS model when the speed of light tends to infinity in any finite-time interval.

32.1 Introduction

The previous theory of a mixture is in the non-relativistic framework. In the relativistic case, there are important contributions by Hutter and Müller [526], by Kremer and his collaborators, in particular, in connection with the Boltzmann-Chernikov relativistic kinetic theory [527, 528], and by Mushick [529]. For a relativistic mixture of gases based on the kinetic theory, we refer to [280, 448]. Recently, Ha, Kim, and Ruggeri [163] have constructed a relativistic counterpart for a mixture of gases with multi-temperature in the same spirit of the classical approach in [158, 160]. In this chapter, we explain this theory. And we also show a possible analogy between the mixture theory and the Cucker-Smale (CS) models that permits to construct a relativistic CS model in the case when flocking agents (particles) move with velocities close to the speed of light.

We also discuss a delicate point in the construction of a relativistic counterpart for the CS model from the relativistic thermo-mechanical Cucker-Smale (TCS) model. Of course, this is not so evident because, in relativity, energy and momentum are strictly connected through the energy-momentum tensor. Thus, we cannot simply impose the constraint that the temperatures of all the components have a

T. Ruggeri, M. Sugiyama, *Classical and Relativistic Rational Extended Thermodynamics of Gases*, https://doi.org/10.1007/978-3-030-59144-1_32

constant value as in the non-relativistic case. The key idea to obtain the appropriate relativistic CS model lies in the concept of the main field and the technique of principal subsystems introduced in 2.4 and discussed, in particular, in Sect. 5.3.2.

The importance of a relativistic model for a mixture of gases is evident in many possible applications in physics and engineering: plasma physics, atomic physics, aerospace engineering, etc. For example, the analyses of charged particles inside a tokamak [530–532], and of spaceships exploring outer space [503].

32.2 Relativistic System for a Mixture of Euler Fluids

In this section, we present a thermodynamically consistent relativistic model for a mixture of Euler gases using the ingredients of the previous chapters and "*Truesdell's principle*". Let V_A^α and $T_A^{\alpha\beta}$ be the particle flux and energy-momentum tensor associated with the component A defined by

$$V_A^\alpha = \rho_A U_A^\alpha, \qquad T_A^{\alpha\beta} = \frac{e_A + p_A}{c^2} U_A^\alpha U_A^\beta - p_A g^{\alpha\beta}, \quad A = 1, \cdots, n.$$

Then we assume the following $5n$ balance laws similar to the classical equations (28.1):

$$\partial_\alpha V_A^\alpha = \tau_A, \qquad \partial_\alpha T_A^{\alpha\beta} = \mathscr{M}_A^\beta, \quad A = 1, \cdots, n, \tag{32.1}$$

where τ_A and $\mathscr{M}_A^\beta$ are production terms satisfying the zero-sum constraints:

$$\sum_{A=1}^n \tau_A = 0, \qquad \sum_{A=1}^n \mathscr{M}_A^\beta = 0, \quad \text{for any } \beta = 0, 1, 2, 3. \tag{32.2}$$

Since τ_A is mainly due to the chemical reactions, for simplicity, we may assume

$$\tau_A = 0, \quad A = 1, \cdots, n.$$

The case with τ_A different from zero can be obtained easily by extending the following procedure.

The system (32.1) can be rewritten in space-time coordinates like a single fluid with production terms due to the thermo-mechanical diffusion (5.14):

$$\begin{aligned}
&\partial_t (\rho_A \Gamma_A) + \partial_i (\rho_A \Gamma_A v_A^i) = 0,\\
&\partial_t \left(\Gamma_A^2 v_A^j \left(\rho_A + \frac{p_A + \rho_A \varepsilon_A}{c^2} \right) \right) + \partial_i \left(p_A \delta^{ij} + \Gamma_A^2 v_A^i v_A^j \left(\rho_A + \frac{p_A + \rho_A \varepsilon_A}{c^2} \right) \right) = m_A^j,\\
&\partial_t \left(p_A (\Gamma_A^2 - 1) + \rho_A \left(\Gamma_A^2 \varepsilon_A + c^2 \Gamma_A (\Gamma_A - 1) \right) \right)
\end{aligned} \tag{32.3}$$

$$+ \partial_i \left(\left(p_A \Gamma^2 + \rho_A \left(\varepsilon_A \Gamma_A^2 + c^2 \Gamma_A (\Gamma_A - 1) \right) \right) v_A^i \right) = \mathscr{E}_A,$$

where

$$\Gamma_A = \frac{1}{\sqrt{1 - \frac{v_A^2}{c^2}}}, \qquad \mathscr{E}_A = c\mathscr{M}_A^0, \qquad m_A^i = \mathscr{M}_A^i. \tag{32.4}$$

As in the classical case, we define the global quantities ρ, v^j, and ε such that the sum of the density in (32.3) is the same as the one of a single fluid (5.14):

$$\rho\Gamma = \sum_{A=1}^{n} (\rho_A \Gamma_A), \qquad \Gamma = \frac{1}{\sqrt{1 - \frac{v^2}{c^2}}},$$

$$\Gamma^2 v^j \left(\rho + \frac{p + \rho\varepsilon}{c^2} \right) = \sum_{A=1}^{n} \left(\Gamma_A^2 v_A^j \left(\rho_A + \frac{p_A + \rho_A \varepsilon_A}{c^2} \right) \right), \tag{32.5}$$

$$p(\Gamma^2 - 1) + \rho \left(\Gamma^2 \varepsilon + c^2 \Gamma (\Gamma - 1) \right) = \sum_{A=1}^{n} \left(p_A (\Gamma_A^2 - 1) + \rho_A \left(\Gamma_A^2 \varepsilon_A + c^2 \Gamma_A (\Gamma_A - 1) \right) \right).$$

We assume also that $p = \sum_{A=1}^{n} p_A$ so that the Eqs. (32.5) define, in a unique way, the total density ρ, the average velocity $\mathbf{v} \equiv (v^j)$, and the total internal energy ε. They reduce to the classical ones in the limit $c \to \infty$.

It is well known that it is difficult to define a center of mass in relativity, but the one in (32.5) seems to be very natural, since, in homogeneous solutions (independent of space coordinate), the global quantities defined in the left-hand side of (32.5) are conserved as in the classical context. The global average velocity v^j can be associated with the global four-velocity:

$$U^\alpha \equiv \Gamma \left(c, v^i \right), \qquad U^\alpha U_\alpha = c^2.$$

32.2.1 Entropy Principle

Since the main part of the left-hand side of $(32.1)_1$ (or equivalently (32.3)) is the same as the single fluid (5.14), each component satisfies the entropy law:

$$\partial_\alpha \left(\rho_A S_A U_A^\alpha \right) = \Sigma_A, \quad \leftrightarrow \quad \partial_t (\rho_A S_A \Gamma_A) + \partial_i \left(\rho_A S_A \Gamma_A v_A^i \right) = \Sigma_A.$$

Similar to (5.4), the main field for (32.3) is given by

$$\mathbf{u}'_A \equiv \frac{1}{T_A}\left(-\left(g_A + c^2\right), U_{A\beta}\right).$$

Now, we use $(2.11)_2$ to see

$$\Sigma_A = \frac{U_{A\alpha}}{T_A}\mathscr{M}^\alpha_A. \tag{32.6}$$

Thus, the entropy law of the full system is expressed as

$$\partial_\alpha h^\alpha = \Sigma, \quad \text{with} \quad h^\alpha = \sum_{A=1}^{n} \rho_A S_A U^\alpha_A, \quad \Sigma = \sum_{A=1}^{n} \Sigma_A.$$

Therefore, it follows from (32.6) that the global entropy production satisfies

$$\Sigma = \sum_{A=1}^{n} \frac{U_{A\alpha}}{T_A}\mathscr{M}^\alpha_A.$$

Again, we use $(32.2)_2$ to obtain

$$\Sigma = \sum_{b=1}^{n-1}\left(\frac{U_{b\alpha}}{T_b} - \frac{U_{n\alpha}}{T_n}\right)\mathscr{M}^\alpha_b. \tag{32.7}$$

We define the following relativistic diffusion four-vectors:

$$W^\alpha_A = U^\alpha_A - \Gamma_A U^\alpha, \tag{32.8}$$

and we have the following space-time decomposition for W^α_A:

$$W^\alpha_A = \Gamma_A\left((1-\Gamma)c, u^i_A\right) \quad \text{with} \quad u^i_A = v^i_A - \Gamma v^i. \tag{32.9}$$

We substitute U^α_A from (32.8) into (32.7) to get

$$\Sigma = \sum_{b=1}^{n-1}\left(\frac{W_{b\alpha} + \Gamma_b U_\alpha}{T_b} - \frac{W_{n\alpha} + \Gamma_n U_\alpha}{T_n}\right)\mathscr{M}^\alpha_b = \sum_{b=1}^{n-1}\left\{\frac{W_{b\alpha}}{T_b} - \frac{W_{n\alpha}}{T_n} + U_\alpha\left(\frac{\Gamma_b}{T_b} - \frac{\Gamma_n}{T_n}\right)\right\}\mathscr{M}^\alpha_b.$$

We use the space-time decomposition (32.4) and (32.9), and covariant-contravariant relation $U_i = -U^i$ to obtain

$$\begin{aligned}
\Sigma &= \sum_{b=1}^{n-1}\left\{\frac{W_{b0}}{T_b} - \frac{W_{n0}}{T_n} + U_0\left(\frac{\Gamma_b}{T_b} - \frac{\Gamma_n}{T_n}\right)\right\}\mathscr{M}_b^0 + \sum_{b=1}^{n-1}\left\{\frac{W_{bi}}{T_b} - \frac{W_{ni}}{T_n} + U_i\left(\frac{\Gamma_b}{T_b} - \frac{\Gamma_n}{T_n}\right)\right\}\mathscr{M}_b^i \\
&= \sum_{b=1}^{n-1}\left\{\frac{\Gamma_b(1-\Gamma)c}{T_b} - \frac{\Gamma_n(1-\Gamma)c}{T_n} + \Gamma c\left(\frac{\Gamma_b}{T_b} - \frac{\Gamma_n}{T_n}\right)\right\}\mathscr{M}_b^0 \\
&\quad - \sum_{b=1}^{n-1}\left\{\frac{\Gamma_b u_{bi}}{T_b} - \frac{\Gamma_n u_{ni}}{T_n} + \Gamma v_i\left(\frac{\Gamma_b}{T_b} - \frac{\Gamma_n}{T_n}\right)\right\}\mathscr{M}_b^i \\
&= \sum_{b=1}^{n-1}\left(\frac{\Gamma_b}{T_b} - \frac{\Gamma_n}{T_n}\right)\mathscr{E}_b - \sum_{b=1}^{n-1}\left\{\frac{\Gamma_b u_{bi}}{T_b} - \frac{\Gamma_n u_{ni}}{T_n} + \Gamma v_i\left(\frac{\Gamma_b}{T_b} - \frac{\Gamma_n}{T_n}\right)\right\} m_b^i.
\end{aligned}$$

We now define

$$\hat{\mathbf{m}}_b = \mathbf{m}_b \equiv (m_b^i), \qquad \hat{\mathscr{E}}_b = \mathscr{E}_b - \Gamma v_i\, \hat{m}_b^i. \tag{32.10}$$

Then, we have

$$\Sigma = \sum_{b=1}^{n-1}\left[\left(\frac{\Gamma_b}{T_b} - \frac{\Gamma_n}{T_n}\right)\hat{\mathscr{E}}_b + \left(\frac{\Gamma_n u_{ni}}{T_n} - \frac{\Gamma_b u_{bi}}{T_b}\right)\hat{m}_b^i\right].$$

By the entropy principle, Σ must be nonnegative and zero only at equilibrium. We imitate the same procedure as for the classical case (28.20) to construct a quadratic form so that Σ is positive:

$$\hat{\mathbf{m}}_b = \frac{1}{n}\sum_{c=1}^{n-1}\psi_{bc}\left(\frac{\Gamma_n \mathbf{u}_n}{T_n} - \frac{\Gamma_c \mathbf{u}_c}{T_c}\right), \qquad \hat{\mathscr{E}}_b = \frac{1}{n}\sum_{c=1}^{n-1}\theta_{bc}\left(\frac{\Gamma_c}{T_c} - \frac{\Gamma_n}{T_n}\right), \tag{32.11}$$

where the matrices (ψ_{bc}) and (θ_{bc}) are constant positive definite matrices. Finally, we substitute the relations (32.11) into (32.10) to have the explicit expression for the production terms:

$$\begin{aligned}
\mathbf{m}_b &= \frac{1}{n}\sum_{c=1}^{n-1}\psi_{bc}\left(\frac{\Gamma_n \mathbf{u}_n}{T_n} - \frac{\Gamma_c \mathbf{u}_c}{T_c}\right), \qquad b = 1, 2, \ldots, n-1, \\
\mathscr{E}_b &= \frac{1}{n}\sum_{c=1}^{n-1}\theta_{bc}\left(\frac{\Gamma_c}{T_c} - \frac{\Gamma_n}{T_n}\right) + \Gamma \mathbf{v}\cdot\frac{1}{n}\sum_{c=1}^{n-1}\psi_{bc}\left(\frac{\Gamma_n \mathbf{u}_n}{T_n} - \frac{\Gamma_c \mathbf{u}_c}{T_c}\right).
\end{aligned} \tag{32.12}$$

We notice that

$$m_b^i, \ \mathscr{E}_b, \quad b = 1, 2, \ldots, n-1 \quad \text{given by (32.12)} \quad \text{and} \quad m_n^i = -\sum_{b=1}^{n-1} m_b^i, \quad \mathscr{E}_n = -\sum_{b=1}^{n-1} \mathscr{E}_b.$$

Note that system (32.3) converges to the corresponding equations for classical mixture of Euler gases (28.1), (28.8), (28.21) in the classical limit.

32.3 Relativistic Thermo-Mechanical Cucker-Smale Model

In this section, we present a relativistic counterpart of the TCS model that we presented in Chap. 30 by considering spatially homogeneous relativistic Euler equations (32.3). Moreover, using the method of principal subsystem, we also derive a relativistic counterpart of the CS model.

32.3.1 Thermo-Mechanical Ensemble

In this subsection, we derive a particle model for a relativistic thermo-mechanical Cucker-Smale ensemble that is consistent with the second law of thermodynamics. We basically follow the same strategy as in the classical case (see Chap. 30). First, we assume that the Euler equations (32.3) lie on the pressureless regime:

$$p_A = 0, \quad A = 1, \cdots, n, \tag{32.13}$$

and are spatially homogeneous. In this case, the spatial derivatives in (32.3) vanish, and system (32.3) becomes the following ODE system:

$$\begin{aligned} &\frac{d}{dt}(\rho_A \Gamma_A) = 0, \quad A = 1, 2, \ldots, n, \quad t > 0, \\ &\frac{d}{dt}\left(\rho_A \Gamma_A^2 \mathbf{v}_A \left(1 + \frac{\varepsilon_A}{c^2}\right)\right) = \mathbf{m}_A, \\ &\frac{d}{dt}\left(\rho_A \left(\Gamma_A^2 \varepsilon_A + c^2 \Gamma_A(\Gamma_A - 1)\right)\right) = \mathscr{E}_A. \end{aligned} \tag{32.14}$$

For simplicity, we assume that the internal energy depends only on T linearly:

$$\varepsilon_A = c_{V_A} T_A,$$

and we further set the specific heat at constant volume c_{V_A} to be unity. Thus, we have

$$\varepsilon_A = T_A, \quad A = 1, \cdots, n. \tag{32.15}$$

From $(32.14)_1$, $\rho_A \Gamma_A$ remains constant along the dynamics (32.14). Hence, without loss of generality, we also assume this quantity to be unity:

$$\rho_A \Gamma_A = 1, \quad A = 1, \cdots, n. \tag{32.16}$$

Then we substitute the relations (32.15) and (32.16) to the second and third equations of (32.14) and introduce the position variable $\mathbf{x}_A$ to obtain the system of relativistic TCS model:

$$\begin{aligned} &\frac{d\mathbf{x}_A}{dt} = \mathbf{v}_A, \quad A = 1, 2, \cdots, n, \quad t > 0, \\ &\frac{d}{dt}\left(\Gamma_A \mathbf{v}_A \left(1 + \frac{T_A}{c^2}\right)\right) = \mathbf{m}_A, \\ &\frac{d}{dt}\left(\Gamma_A T_A + c^2(\Gamma_A - 1)\right) = \mathscr{E}_A. \end{aligned} \tag{32.17}$$

In the present case, the global quantities defined in (32.5) become

$$\begin{aligned} &\rho\Gamma = n, \\ &\Gamma \mathbf{v}\left(1 + \frac{T}{c^2}\right) = \sum_{A=1}^{n}\left(\Gamma_A \mathbf{v}_A \left(1 + \frac{T_A}{c^2}\right)\right) = \text{constant}, \\ &\Gamma T + c^2(\Gamma - 1) = \sum_{A=1}^{n}\left(\Gamma_A T_A + c^2(\Gamma_A - 1)\right) = \text{constant}, \end{aligned} \tag{32.18}$$

where we have defined the average temperature T such that $\varepsilon = T$. We further require the initial data to satisfy

$$\mathbf{v}(0) = 0, \quad T(0) = T_0.$$

Then relations (32.18) yield

$$\mathbf{v}(t) = 0, \quad T(t) = T_0,$$

provided that the initial data satisfy the constraints:

$$\sum_{A=1}^{n}\left(\Gamma_A(0)\mathbf{v}_A(0)\left(1+\frac{T_A(0)}{c^2}\right)\right)=0, \quad \sum_{A=1}^{n}\left(\Gamma_A(0)T_A(0)+c^2(\Gamma_A(0)-1)\right)=T_0. \tag{32.19}$$

This corresponds to the choice of the reference frame as the one with $\mathbf{v}=0$. We use the privileged frame with $\mathbf{v}=0$ and, from $(32.9)_2$, (32.10), and (32.11), we get

$$\mathbf{u}_A=\mathbf{v}_A, \quad \mathbf{m}_b=\frac{1}{n}\sum_{c=1}^{n-1}\psi_{bc}\left(\frac{\Gamma_n\mathbf{v}_n}{T_n}-\frac{\Gamma_c\mathbf{v}_c}{T_c}\right), \quad \mathscr{E}_b=\frac{1}{n}\sum_{c=1}^{n-1}\theta_{bc}\left(\frac{\Gamma_c}{T_c}-\frac{\Gamma_n}{T_n}\right).$$

We now transform $(n-1)\times(n-1)$ matrices (ψ_{bc}) and (θ_{bc}) to $n\times n$ matrices (ϕ_{AB}) and (ζ_{AB}) as in (30.3). With this change of matrices, system (32.17) becomes

$$\begin{aligned}
&\frac{d\mathbf{x}_A}{dt}=\mathbf{v}_A, \quad A=1,2,\cdots,n, \quad t>0,\\
&\frac{d}{dt}\left(\Gamma_A\mathbf{v}_A\left(1+\frac{T_A}{c^2}\right)\right)=\frac{1}{n}\sum_{B=1}^{n}\phi_{AB}\left(\frac{\mathbf{v}_B\Gamma_B}{T_B}-\frac{\mathbf{v}_A\Gamma_A}{T_A}\right),\\
&\frac{d}{dt}\left(\Gamma_A T_A+c^2(\Gamma_A-1)\right)=\frac{1}{n}\sum_{B=1}^{n}\zeta_{AB}\left(\frac{\Gamma_A}{T_A}-\frac{\Gamma_B}{T_B}\right),
\end{aligned} \tag{32.20}$$

being subject to the initial data constrained by (32.19). Therefore, this is a natural choice for the relativistic TCS model. Note that

$$\lim_{c\to\infty}\Gamma_A=1 \quad \text{and} \quad \lim_{c\to\infty}c^2(\Gamma_A-1)=\frac{v_A^2}{2}.$$

Thus, as $c\to\infty$, system (32.20) reduces to the classical TCS model (30.4).

Next, we discuss the entropy associated with the relativistic system (32.20). The Gibbs relation (8.12) under the simplifying assumptions (32.13) and (32.15):

$$p_A=0 \quad \text{and} \quad \varepsilon_A=T_A, \quad A=1,\cdots,n$$

yields

$$T_A dS_A=dT_A, \quad \text{i.e.,} \quad dS_A=d\log T_A, \quad A=1,2,\ldots n.$$

Thus, one has

$$S_A=\log T_A, \quad A=1,\cdots,n.$$

Now we introduce a global entropy S as

$$S = \sum_{A=1}^{n} S_A = \sum_{A=1}^{n} \log T_A.$$

According with the construction of the model it is easy to verify the non-decreasing property of the entropy S for the relativistic system (32.20):

$$\frac{dS}{dt} = \frac{1}{2n} \sum_{A,B=1}^{n} \phi_{AB} \left| \frac{\Gamma_B \mathbf{v}_B}{T_B} - \frac{\Gamma_A \mathbf{v}_A}{T_A} \right|^2 + \frac{1}{2n} \sum_{A,B=1}^{n} \zeta_{AB} \left(\frac{\Gamma_A}{T_A} - \frac{\Gamma_B}{T_B} \right)^2 \geqslant 0. \tag{32.21}$$

32.3.2 *Mechanical Ensemble*

In this subsection, we study a mechanical Cucker-Smale ensemble. For the non-relativistic (classical) case, this corresponds to the case where temperatures of the constituents are assumed to have a constant. For the relativistic setting under consideration, we derive the relativistic CS model from the relativistic TCS model using the theory of principal subsystem (see Sect. 2.4). More precisely, we set

$$\frac{T_A}{\Gamma_A} = T^*, \quad A = 1, \cdots, n. \tag{32.22}$$

Now, we substitute the ansatz (32.22) in the second equation of (32.20) and delete the last equation in (32.20) to get

$$\begin{aligned} &\frac{d\mathbf{x}_A}{dt} = \mathbf{v}_A, \quad A = 1, 2, \ldots, n, \quad t > 0, \\ &\frac{d}{dt}\left(\Gamma_A \mathbf{v}_A \left(1 + \frac{T^* \Gamma_A}{c^2}\right)\right) = \frac{1}{nT^*} \sum_{B=1}^{n} \phi_{AB} \left(\mathbf{v}_B - \mathbf{v}_A\right), \end{aligned} \tag{32.23}$$

being subjected to the constrained initial data:

$$\sum_{A=1}^{n} \left(\Gamma_A(0) \mathbf{v}_A(0) \left(1 + \frac{T^* \Gamma_A(0)}{c^2}\right)\right) = 0.$$

We consider subentropy associated with the principal subsystem (32.23). Under the conditions (32.13) and (32.15)–(32.16), the subentropy $\bar{h}^0_A$ in (2.35) now becomes

$$\bar{h}^0_A = S^*_A - \frac{1}{T^*}\left(T^*\Gamma^2_A + c^2(\Gamma_A - 1)\right).$$

Thus, the total subentropy $\bar{S}$ and the subentropy production $\bar{\Sigma}$ can be obtained as follows:

$$\begin{aligned}\bar{h}^0 = \bar{S} = S^* - \frac{1}{T^*}\sum_{A=1}^{n}\left(T^*\Gamma^2_A + c^2(\Gamma - 1)\right) &= \sum_{A=1}^{n}\left(\log T^*\Gamma_A - \Gamma^2_A - \frac{c^2}{T^*}(\Gamma_A - 1)\right)\\ = n\log T^* + \frac{1}{T^*}\sum_{A=1}^{n}\left(T^*\left(\log\Gamma_A - \Gamma^2_A\right) - c^2(\Gamma_A - 1)\right),\\ \bar{\Sigma} = \Sigma^* - \frac{1}{T^*}\left(\frac{1}{n}\sum_{A,B=1}^{n}\zeta_{AB}\left(\frac{1}{T^*} - \frac{1}{T^*}\right)\right) &= \frac{1}{2n(T^*)^2}\sum_{A,B=1}^{n}\phi_{AB}\,|\mathbf{v}_A - \mathbf{v}_B|^2.\end{aligned}$$

Moreover, in the paper [163], an emergent flocking estimate for the relativistic CS model was obtained, and its rigorous classical limit to the CS model in any finite-time interval when the speed of light tends to infinity was proved.

Part IX
Maxwellian Iteraction, Objectivity, and Outlook

Chapter 33
Hyperbolic Parabolic Limit, Maxwellian Iteration, and Objectivity

Abstract In this chapter, we discuss the parabolic limit of RET via the Maxwellian iteration, and we observe that the usual constitutive equations, which are nonlocal in space, are approximations of some balance laws of RET when some relaxation times are very small. An important consequence is that these equations need not satisfy the objectivity principle. To avoid misunderstanding, we should mention that the principle still continues to be valid for constitutive equations.

We also discuss the point that, under suitable assumptions, the conditions dictated by the entropy principle in the hyperbolic case guarantee the validity of the entropy principle also in the parabolic limit. Lastly we express our opinion concerning the limitation of the parabolic regularized version of RET theories.

33.1 Different Types of Constitutive Equation

We have seen that the physical laws in continuum theories are expressed by the *balance laws*, which, under regularity conditions, assume the form (1.2).

In order to have a closed system, we need *constitutive equations*. A very rough mathematical definition of constitutive equations may be considered as the equations that are necessary to close the system. That is, choosing an independent field $\mathbf{u} \in \mathbb{R}^N$, we have to give relations between the $5N$ components of the vectors $\mathbf{F}^0, \mathbf{F}^i, \mathbf{f}$ and the N components of the unknown vector $\mathbf{u}$. But of course, as we will see below, this definition has no physical meaning because the additional equations must represent the real constitutive properties of the material.

For a long time, constitutive equations have been made in an empirical way. They belong substantially to the one of the following three big classes:

- *Local constitutive equations* Examples are:
  - stress-strain relation in nonlinear elasticity: $\mathbf{t} \equiv \mathbf{t}(\mathbf{E})$ (Hooke's law in the linear case);

T. Ruggeri, M. Sugiyama, *Classical and Relativistic Rational Extended Thermodynamics of Gases*, https://doi.org/10.1007/978-3-030-59144-1_33

- The caloric and thermal equations of state in Euler fluids that express the internal energy and the pressure as functions of the mass density and the temperature:

$$\varepsilon \equiv \varepsilon_E(\rho, T), \quad p \equiv p(\rho, T).$$

- All rational extended thermodynamic (RET) theories in which we assume the local dependence of the vectors of the density, fluxes, and productions (2.1).

Introducing such constitutive equations into the balance laws, we obtain a differential system which, in general, is hyperbolic.

- *Non-local type (in space)* In the case of one-component dissipative fluid, examples are:

  - Fourier's law: $(1.11)_3$;
  - Navier-Stokes' law: $(1.11)_{1,2}$;
  - In the case of mixture of dissipative fluids with n constituents, well-known examples are the Navier-Stokes', Fourier's, and Fick's laws (28.13).

  When we introduce such constitutive equations into the balance laws, we obtain a system of differential equations where some spatial derivatives are of second order and the time derivatives are of first order. These differential systems have a parabolic structure.

- *Non-local type (in time)* Examples are as follows: visco-elastic materials or, in general, all materials in which the stress depends not only on the present deformation but also on the history of the deformation (constitutive equations with memory). Except for the case of exponential memory kernel, the mathematical structure of such systems is of integro-differential type.

33.2 Frame-Dependence of the Heat Flux

In the modern constitutive theory, all the constitutive equations must obey two universal principles that are the first two in the Axioms of RET (see Sect. 2.2): *objectivity principle* and *entropy principle*.

A long debate came out in the literature after Ingo Müller published a famous paper [533] in which he proved that the Fourier and Navier-Stokes "constitutive" equations (1.11) violate the objectivity principle. At that time Müller was convinced that his result indicates that the objectivity principle is not a valid principle. And then a huge literature appeared between supporters and non-supporters of the objectivity principle. Several authors added artificial time derivatives to try to recover the objectivity for the heat equation and for the stress. Here we record observations on the subject made independently by Bressan [534] and by Ruggeri [535]. They observed that a possible interpretation of Müller's result is that the objectivity principle is indeed universal, but the Fourier and Navier-Stokes laws are not "true" constitutive equations.

The precise and convincing answer was presented by RET. In fact we have seen in Sect. 6.3.11 that the Navier-Stokes and Fourier laws are approximations of the balance laws of 14 moment, and therefore these are not constitutive equations. It is not necessary for the Navier-Stokes and Fourier laws to satisfy the frame indifference principle.

33.2.1 Maxwellian Iteration and the Parabolic Limit

To reveal the relationship between extended and classical models, a formal iterative scheme known as the Maxwellian iteration is applied [25, 104]. In general, the first iterates are obtained from the right-hand side of balance laws by putting the "zeroth" iterates—equilibrium values—into the left-hand side. The second iterates are obtained from the right-hand side by putting the first iterates into the left-hand side, and so on. Therefore the Maxwellian iteration is substantially composed of (1) an identification of the relaxation times and (2) a formal power expansion of the solution in terms of the relaxation times: a sort of Chapman–Enskog procedure at macroscopic level.

We proved, in Sect. 6.3.11 (6.35), that:

Statement 33.1 *The Fourier and Navier-Stokes laws (1.11) are the first order approximation of the Maxwellian iteration of the RET balance laws (6.34)$_{4,5,6}$. See Sect. 6.3.11. Therefore they are not true constitutive equations and need not satisfy the objectivity principle.*

A similar situation exists in the case of mixture of fluids holding the Fick law for the mass diffusion. In fact as we have seen in Sect. 28.7 that

Statement 33.2 *The Fick law (28.49) is the first approximation in the Maxwellian iteration of momentum balance equation of each component (28.43)$_5$.*

Moreover

Statement 33.3 *The new non-local "constitutive equation" for the temperature differences (28.50) is the first approximation obtained by the Maxwellian iteration of the energy balance of each component (28.43)$_6$.*

Another simple example in the context of the mixture theory is well-known Darcy's law for porous media saying that the relative velocity between the fluid part $\mathbf{v}_F$ and the solid one $\mathbf{v}_S$ is proportional to the pressure gradient in the fluid (see, e.g. [188]):

$$\nabla p_F = -\frac{k}{\mu}\left(\mathbf{v}_S - \mathbf{v}_F\right), \tag{33.1}$$

where k and μ are, respectively, the permeability and the viscosity.

Statement 33.4 *Darcy's law (33.1) is an approximation to the balance law of the linear momentum for the fluid flowing through the porous solid being treated as a rigid body, i.e., (33.1) is a limit case of (see [536]):*

$$\rho_F \, \dot{\mathbf{v}}_F + \nabla \, p_F = -\frac{k}{\mu} \left(\mathbf{v}_S - \mathbf{v}_F \right).$$

We have seen that, using the Maxwellian iteration, we can obtain from RET—at least formally—the usual non-local constitutive equations of the classical theory. Therefore the parabolic systems of classical theories appear, from a physical point of view, to be approximations of the corresponding hyperbolic systems when some relaxation times are very small:

- Navier Stokes' and Fourier's laws as a limit case of new balance equations for shear viscous tensor, dynamic pressure, and heat flux in RET;
- Fick's law as a limit case of momentum equations of each component in a mixture with single-temperature;
- The new diffusion equation for the difference of the temperatures in mixture with multi-temperature as a limit case of the energy balance equation of each constituent.
- Darcy's law for porous material is a limit case of momentum equation.

Although the previous non-local equations are not constitutive equations but approximations of balance laws, the non-local equations have been very useful. In many applications, the relaxation times are sometimes quite small and the non-local equations are relevant in such situations. The advantage of the non-local approximation is that, in this limit, we are able to measure non-observable quantities like heat flux, viscous stress and, in particular, velocity and temperature of each component in a mixture of fluids using the classical constitutive equations.

An interesting analysis on constitutive equations can be read also in the book of Signer [537].

33.3 Maxwellian Iteration and the Entropy Principle

Clearly a major open problem in this framework is the rigorous proof of the convergence of the solutions via Maxwellian iterations. To make a little step toward this proof, first of all, we have to focus our attention on another very subtle point: is the entropy principle preserved in the Maxwellian iteration scheme? In other words: if the "full" hyperbolic theory satisfies the entropy inequality, are we sure that the corresponding parabolic limit satisfies automatically a suitable entropy inequality?

Here we give some results due to Ruggeri [105] in one space-dimension (but the results remain valid in any space-dimension):

Theorem 33.1 *If the system of balance laws of RET is endowed with a convex entropy and the processes are not far from equilibrium, the entropy principle is preserved in the Maxwellian iteration. And as a consequence, if the original hyperbolic system is entropic, the parabolic limiting system is also entropic. Instead, for processes far from equilibrium, in general, this is not true.*

Proof Let us consider, as in all RET theories, that the system is split into two blocks of M conservation equations (2.42) and of $N - M$ balance laws (2.43). And, with the notation $\mathbf{u}' \equiv (\mathbf{v}, \mathbf{w})$ (we omit the prime), it was proved in Sect. 2.5 that the equilibrium manifold in the main field components is the hyperplane $\mathbf{w} = 0$ (see (2.45)).

In RET, there are many interesting cases where the processes are not far from equilibrium. In this case, we have

$$h' = h'^0 = \frac{1}{2}\mathbf{K}(\mathbf{v})\mathbf{w}\cdot\mathbf{w} + h'_E(\mathbf{v}), \tag{33.2}$$

and therefore if we consider the case of one space-dimension, we assume the following symmetric system, which is linear in $\mathbf{w}$:

$$\mathbf{H}\,\partial_t\mathbf{v} + \mathbf{A}\,\partial_x\mathbf{v} + \mathbf{B}\,\partial_x\mathbf{w} = 0, \tag{33.3}$$

$$\mathbf{K}\,\partial_t\mathbf{w} + \mathbf{B}^T\,\partial_x\mathbf{v} + \mathbf{C}\,\partial_x\mathbf{w} = -\mathbf{L}\mathbf{w}, \tag{33.4}$$

where $\mathbf{v} \in \mathbb{R}^M$, $\mathbf{w} \in \mathbb{R}^N$, $\mathbf{H} \equiv h'_{\mathbf{vv}}$ $(M \times M) \in Sym^+$ (symmetric positive definite matrix), $\mathbf{A} \equiv k_{\mathbf{vv}}$ $(M \times M) \in Sym$ (symmetric matrices), $\mathbf{B} \equiv k_{\mathbf{vw}}$ $(M \times N)$, $\mathbf{C} \equiv k_{\mathbf{ww}}$ $(N \times N) \in Sym$ and $k = h'^1(\mathbf{v}, \mathbf{w})$. Moreover $\mathbf{L} \equiv \mathbf{L}(\mathbf{v})$ $(N \times N) \in Sym^+$ because of the residual inequality $\Sigma = -\mathbf{L}\mathbf{w}\cdot\mathbf{w} \leqslant 0$.

According with the Maxwellian iteration procedure (Sect. 33.2.1), the first iterates are obtained from the right-hand side of balance laws by putting the "zeroth" iterates—equilibrium values—on the left-hand side, i.e., putting $\mathbf{w} = 0$ on the left side of (33.4) we have

$$\bar{\mathbf{B}}^T\,\partial_x\mathbf{v} = -\mathbf{L}\mathbf{w}^{(1)},$$

where $\bar{\mathbf{B}} \equiv \mathbf{B}(\mathbf{v}, \mathbf{0})$. Then substituting this into (33.3) we obtain as the first Maxwellian iteration (taking only the principal part, i.e., we do not consider the first-order spatial derivative):

$$\mathbf{H}\,\partial_t\mathbf{v} \simeq \mathbf{D}\,\partial_{xx}\mathbf{v}, \tag{33.5}$$

where the diffusion matrix $\mathbf{D}$ is given by

$$\mathbf{D} = \bar{\mathbf{B}}\mathbf{L}^{-1}\bar{\mathbf{B}}^T \in Sym^+. \tag{33.6}$$

Consequently the entropy principle is preserved in the passage from the hyperbolic system to the parabolic limit.

In the general case, however, the entropy principle is not necessarily preserved. In this case, instead of (33.3) and (33.4), we have

$$\mathbf{H}\,\partial_t\mathbf{v} + \mathbf{G}\,\partial_t\mathbf{w} + \mathbf{A}\,\partial_x\mathbf{v} + \mathbf{B}\,\partial_x\mathbf{w} = 0,$$

$$\mathbf{G}^T\,\partial_t\mathbf{v} + \mathbf{K}\partial_t\mathbf{w} + \mathbf{B}^T\partial_x\mathbf{v} + \mathbf{C}\,\partial_x\mathbf{w} = -\mathbf{L}\mathbf{w}$$

with $\mathbf{G} \equiv h'_{\mathbf{vw}}$ $(M \times N)$ and $\mathbf{K} \equiv h'_{\mathbf{ww}}$ $(N \times N) \in Sym^+$. After the Maxwellian iteration, we obtain a parabolic system similar to (33.5) but with a diffusion matrix $\mathbf{D}$:

$$\mathbf{D} = \bar{\mathbf{B}}\mathbf{L}^{-1}\bar{\mathbf{B}}^T - \bar{\mathbf{B}}\mathbf{L}^{-1}\bar{\mathbf{G}}^T\bar{\mathbf{H}}^{-1}\bar{\mathbf{A}}. \tag{33.7}$$

This matrix, in general, does not have a definite sign. Therefore the most general system for which the entropy principle is preserved in the parabolic limit is the following one: The matrix $\mathbf{D}$ of the system given by (33.7) belongs to Sym^+. We call these special systems *entropy-principle preserving systems*.

A simple entropy-principle preserving system is the one for which h' is given by the sum of two functions, one is the function of only $\mathbf{v}$ and the other is the function of only $\mathbf{w}$. An interesting case of the processes not far from equilibrium (33.2) belongs to this special class. In fact in this case we have $\mathbf{G} \equiv 0$ and the diffusion matrix (33.7) reduces to (33.6) and becomes symmetric positive definite. This result seems to indicate that the parabolic classical theories have a limiting validity only near equilibrium according with the assumption of local equilibrium!

33.4 Regularized System and Non-subshock Formation

In the previous section, we considered the full Maxwellian iteration for all relaxation times assuming implicitly that all of them have the same order of magnitude. In this section, we consider relaxation times with different order of magnitude.

An interesting idea is the one proposed by Torrilhon and Struchtrup [75] who have constructed a sophisticated method to obtain parabolic extended systems called *regularized* systems from the hyperbolic systems of RET. Philosophy in its construction, however, seems to be similar to the Maxwellian iteration. By this method, they were able to derive, in the case of monatomic gas, a regularized 13 Grad system that is a natural parabolic extension of the NSF system of equations. As the regularized system contains first-order derivatives in time and, in some equations, second-order derivatives in space, it is of parabolic type. This fact is in sharp contrast to the hyperbolic 13 Grad system. To obtain the regularized 13 Grad system, the authors start from the RET system with 23 moments. Therefore we can regard the regularized parabolic Grad system as a sort of approximation of the

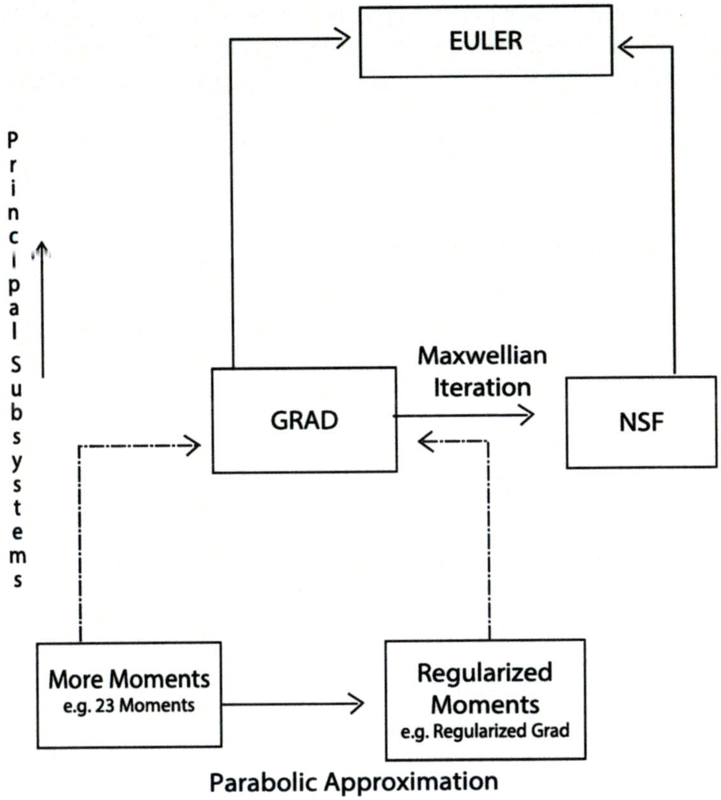

Fig. 33.1 Systems and subsystems, and the parabolic limit

hyperbolic ET_{23}. By the same procedure, the 14-moment system (6.38) for rarefied polyatomic gases was regularized by Rahimi and Struchtrup in [308] starting from RET with 36 moments.

In Fig. 33.1, the relationship between the different models is sketched: We can see that the Euler-fluid subsystem is effectively a particular case of the Navier-Stokes and Fourier system. Nevertheless the Euler-fluid subsystem is also a principal subsystem of the Grad system. And moreover the Navier-Stokes and Fourier system itself is a particular case of the Grad system when the Maxwellian iteration is used. This situation is also valid if we take the RET theory with many moments. When we adopt the regularization approximation, we have, as parabolic counterpart, the regularized moment-equations. In any case, given an N-moment hyperbolic system, all the previous systems of moments are principal subsystems according to the general *nesting structure* due to Boillat and Ruggeri [165]. Moreover we can see that all the parabolic limits can be considered as approximations of the hyperbolic system as sketched in Fig. 33.1.

The regularized extended system appears very interesting. But one of the main assertions that the parabolization of a system of equations avoids the formation of subshock in the shock structure solution, in the present authors' opinion, leads to a misunderstanding. For example, if we accept that the NSF theory is an approximation of the 14-moment hyperbolic system, the validity of the theory cannot exceed the validity of the 14-moments system. Therefore, from the Table 9.2, we notice that, for 14 moments, there appears a subshock at the critical Mach number depending on D but in the range $1.34 \leqslant M_{cr} \leqslant 1.65$. The upper and lower bounds correspond to $D \to \infty$ and $D \to 3$, respectively. Therefore we cannot expect that the NSF theory can be valid for any Mach number but we can expect at most until the critical Mach number of the 14-moment theory even though no subshock formation exists in the theory. The same is true for the regularized theory. Its validity cannot exceed the critical Mach number of the starting hyperbolic RET theory.

We have a clear explanation to the no-evidence of the subshock in experiments thanks to the mathematical properties of the nesting theories of RET. In fact, for a fixed truncation order N, the validity of the RET theory, as far as shock waves are concerned, is up to the Mach number where the shock velocity s reaches the maximum characteristic velocity evaluated in equilibrium. Beyond this limitation, from the Boillat and Ruggeri Theorem 3.1, a subshock emerges, and the model is no longer valid. We need to increase the number of truncation N so as to take more fields into the model. In this case, according to the properties of the principal subsystem, i.e., Theorem 2.3 and inequality (9.35), the maximum characteristic velocity increases. Therefore we have now a subshock formation with larger critical Mach number. In other words, we need more moments in order to let a theory be valid for larger Mach numbers. In the limit of infinite Mach number, we substantially deal with the Boltzmann equation itself to predict smooth shock structure!

33.5 Conclusion

In conclusion, RET seems to indicate in clear manner that non-local relations are not constitutive equations but approximations of balance laws. The true constitutive equations are in local form and they obey the material frame difference. The physical systems are hyperbolic in agreement with the relativity principle that any disturbance propagates with finite speed. Nevertheless, non-local equations such as the usual Fourier, Navier-Stokes, Fick, Darcy laws, and others are useful to measure non-observable quantities and they are good approximations in many practical problems. The Maxwellian iteration preserves the entropy principle at least for processes not far from equilibrium. Hyperbolic systems with dissipation (balance laws with production terms) can have global smooth solutions provided that the initial data are small.

Chapter 34
Open Problems and Outlook

Abstract We list up some open problems, and discuss briefly the perspective on RET.

34.1 Open Problems

In classical and relativistic RET, for both monatomic and polyatomic gases, there still remain interesting open problems. Some of them are listed below.

34.1.1 Open Mathematical Questions

- **Does $f_N \to f$ as $N \to \infty$?:**
  We have seen that the Boltzmann equation corresponds to a set of an infinite number of balance laws of moments. Therefore, physically we may expect that if we take into account more moments, then the solution of the system of RET should converge to the solution of the kinetic theory. In the monatomic-gas case, this expectation seems to be affirmative. As seen in the cases of sound wave and of light scattering, with the increase of the number of moment, the solutions of RET are always in better agreement with the experimental data and also with those of the kinetic theory [25]. While mathematically, from the viewpoint of the closure via MEP, a fundamental question arises: does the truncated distribution function f_N tend to the distribution function f that is the solution of the Boltzmann equation, when $N \to \infty$? As mentioned in Sect. 4.2.1, we currently do not have rigorous evidence of the existence of the limit, nor even an estimate of the error between the functions: $\| f - f_N \|$ in some norm in terms of the truncation tensor index N.
- **Subshock with the propagation velocity less than the maximum characteristic velocity in the unperturbed state**:
  We have seen, in the toy models explained in Sect. 3.6.1 and in a mixture of gases in Chap. 29, that a subshock with the propagation velocity less than the maximum

T. Ruggeri, M. Sugiyama, *Classical and Relativistic Rational Extended Thermodynamics of Gases*, https://doi.org/10.1007/978-3-030-59144-1_34

characteristic eigenvalue evaluated in front of the shock can also arise. For the toy models and the mixture, all important requirements of RET, namely, the entropy principle, convexity of the entropy, dissipative character satisfying the Shizuta-Kawashima condition are fulfilled. Instead, until now, in all RET theories with any number of moments, a shock is continuous until its propagation velocity reaches the maximum characteristic velocity in the unperturbed state. This fact seems to indicate that there still exist some missing mathematical ingredients that assure RET of this remarkable property.

- **Implications of the truncation index N and the order of expansion α**:
  In order to understand a RET theory properly, for example, to understand its applicability range, we need to know the implications of the indexes N and α deeply from both mathematical and physical viewpoints, where N is the tensorial index of truncation and α is the order of expansion of the distribution function f_N around the equilibrium distribution function $f^{(E)}$ with respect to the nonequilibrium variables. Roughly speaking, using the RET theory with larger N we can analyze nonequilibrium processes with smaller space-time scales, while using the RET theory with larger α we can describe nonequilibrium phenomena further from equilibrium. However, at present, we do not know well the interplay of N and α in RET. Then we have a problem: is there some intrinsic relationship between these indexes in RET? Maybe this question can help us also to understand deeply the problem of bounded domain in RET (see Sect. 4.6.4).

  Concerning this subject we have found a new fertile research subject in RET, that is, the analysis of highly nonequilibrium phenomena by using the RET theory with higher index α. This research field has been just explored. See also Sects. 4.5.2, 4.7, and 6.4.2.

34.1.2 Open Physical Problems

- **How many hierarchies of moments are there in RET of polyatomic rarefied gases?**:
  Following the recent paper of Pennisi and Ruggeri [296], we have considered the classical limit of the relativistic moment theory of a monatomic gas in Sect. 5.6. And we proved that there exists a unique choice for the classical moments for a prescribed index of truncation. A similar analysis was also made for a polyatomic gas in Sect. 27.2, and also in this case we found a unique choice of classical moments in the classical limit. We found an interesting fact that there are, in general, not only F- and G-hierarchies that were postulated in the previous studies but also some mixed new hierarchies (see Theorem 27.1). This question may inspire new studies for understanding well the consequences of these new hierarchies.

- **Relativistic gas having molecular rotational and vibrational modes with different relaxation times**:
The RET theory of a relativistic polyatomic gas has been constructed in Chaps. 26 and 27 where the internal modes of a molecule are treated as a unit. However, in contrast to the non-relativistic RET theory, we have at present no systematic RET theory of a relativistic polyatomic gas where the rotational mode and the vibrational mode of a molecule are treated individually. A first tentative was given in [538]
- **Origin of the dynamic pressure and the bulk viscosity**:
The main causes of the dynamic pressure seem to be the following two:

  (a) the existence of internal modes of a constituent molecule of a gas such as molecular rotation and vibration;
  (b) the existence of intermolecular interaction in a gas.

  In order to highlight these causes, it is advantageous to adopt a simplified model where the shear stress $\sigma_{\langle ij \rangle}$ and the heat flux q_i are ignored and only the dynamic pressure Π is taken into account. The cause (a) has been studied by the ET_6 and ET_7 theories of a rarefied polyatomic gas, where there is no intermolecular interaction, in Chaps. 12–14. It was discovered that the dynamic pressure comes from the energy exchange among the internal modes of a molecule. On the other hand, for a dense polyatomic gas, we must take into account both the causes (a) and (b). However, in the ET_6^{D} and ET_7^{D} theories explained in Part VI, we have taken into account only the cause (a) by assuming that the cause (a) is dominant over the cause (b) (see Remark 24.3 in Sect. 24.3.2). The detailed study of the cause (b) from the viewpoint of RET is remained as an open problem in this book. This problem may be clarified in the simplest way by studying a dense monatomic gas, where there is no internal molecular mode.
- **RET theory of dense gases with shear viscosity and heat conductivity**:
The ET_7^{D} theory is a simplified model for nonequilibrium phenomena in a dense polyatomic gas with dissipation only due to the dynamic pressure. Therefore it is evident that, as a next step of the study, we should construct a RET theory that takes into account also shear stress and heat flux.
- **RET of relativistic dense gases**:
Construction of the RET theory of relativistic dense gases is also interesting and important.
- **Comparison between RET and other nonequilibrium thermodynamic theories**:
As mentioned in Chap. 1, there are several nonequilibrium thermodynamic theories. RET has been originally constructed by the careful comparison between the theory of continuum mechanics and the kinetic theory. Therefore such a RET theory is valid only for rarefied gases. However, in the recent developments of RET of dense gases, where the kinetic theory is not quite helpful, it is necessary to make clear the connections between RET and other approaches. There are several works on such a comparison, see, for example, [71] and the recent papers

[72] and [73] in the special volume on nonequilibrium thermodynamic theories edited by Ván [539]. This topic certainly deserves a future study.

34.1.3 Applications of RET

- **Nonequilibrium shock wave with phase transition:**
  As mentioned in Sect. 3.3, it is interesting to study the problem on shock wave and phase transition. This problem was firstly studied in the case of Euler fluid [206]. The work is important also for many practical applications.
- **Applications of RET to various practical subjects in physics, chemistry, engineering sciences, biology, etc.**:
  RET is expected to make contributions to many fields like nano-technology, physical biology, aeronautical engineering, space science. We believe there are huge possibilities for RET to show its value and usefulness for such studies.
- **Applications of RET of relativistic gases with internal structure**:
  The RET theory of relativistic polyatomic gases is explained in Part VII. One of the most important points of this theory is that the bulk viscosity may become large as in the classical context. Instead, for a rarefied gas without internal structure, the bulk viscosity is zero in the classical regime and is extremely small in the relativistic regime. Many papers studying the relativistic regime and, in particular, cosmology pointed out the essential role of the bulk viscosity in an appropriate description of the early-universe expansion. Therefore, in our opinion, the theory of a relativistic gas with internal structure may afford a possible natural framework to answer this problem. Another important question is about the propagation of shock wave in these circumstances. We have seen the remarkable difference between monatomic gas and polyatomic gas with high bulk viscosity in shock wave phenomena in the classical case (see Chaps. 17 and 18). Therefore, the study of shock wave in a relativistic polyatomic gas seems to be quite interesting.

References

1. I. Müller, W. Weiss, *Entropy and Energy, A Universal Competition* (Springer, Berlin, 2005)
2. I. Müller, Entropy and energy, a universal competition. Entropy **10**, 462 (2008)
3. E. Fermi, *Thermodynamics* (Dover, New York, 1956)
4. I. Müller, *A History of Thermodynamics, The Doctrine of Energy and Entropy* (Springer, Berlin, 2007)
5. I. Müller, *Thermodynamics* (Pitman, Boston, 1985)
6. H.B. Callen, *Thermodynamics* (Wiley, New York, 1960)
7. S.R. de Groot, P. Mazur, *Non-Equilibrium Thermodynamics* (North-Holland, Amsterdam, 1963)
8. L.D. Landau, E.M. Lifshitz, *Fluid Mechanics* (Pergamon Press, London, 1958)
9. L. Saint-Raymond, Hydrodynamic limits of the Boltzmann equation, in *Lecture Notes in Mathematics*, vol. 1971 (Springer, Berlin, 2009)
10. M. Slemrod, From Boltzmann to Euler: Hilbert's 6th problem revisited. Comput. Math. Appl. **65**, 1497 (2013)
11. A.N. Gorban, I. Karlin, Hilbert's 6th problem: exact and approximate hydrodynamic manifolds for kinetic equations. Bull. Am. Math. Soc. **51**, 186 (2014)
12. T. Ruggeri, M. Sugiyama, Rational extended thermodynamics: a link between kinetic theory and continuum theory. Rend. Fis. Acc. Lincei. **31**, 33 (2020)
13. H. Spohn, in *Large Scale Dynamics of Interacting Particles*. Springer Series: Theoretical and Mathematical Physics (Springer, Berlin, 1991)
14. C.B. Morrey, On the derivation of the equations of hydrodynamics from statistical mechanics. Commun. Pure Appl. Math. **8**, 279 (1955)
15. H. Grad, Principles of the kinetic theory of gases, in *Handbuch der Physik*, vol. 12 (Springer, Heidelberg, 1958), pp. 205–294
16. O.E. Lanford, Time evolution of large classical dynamical system, in *Lecture Notes Physics*, vol. 38 (Springer, Berlin, 1975), pp. 1–111
17. S. Chapman, T.G. Cowling, *The Mathematical Theory of Non-Uniform Gases* (Cambridge University, Cambridge, 1970)
18. T. Nishida, Fluid dynamical limit of the nonlinear Boltzmann equation to the level of the incompressible Euler equation. Commun. Math. Phys. **61**, 119 (1978)
19. R. Caflisch, The fluid dynamical limit of the nonlinear Boltzmann equation. Commun. Pure Appl. Math. **33**, 651 (1980)
20. C. Eckart, The thermodynamics of irreversible processes I. The single fluids. Phys. Rev. **58**, 267 (1940)

T. Ruggeri, M. Sugiyama, *Classical and Relativistic Rational Extended Thermodynamics of Gases*, https://doi.org/10.1007/978-3-030-59144-1

21. C. Carathéodory, Untersuchungen über die Grundlagen der Thermodynamik. Math. Ann. **67**, 355 (1909)
22. L. Pogliani, M.N. Berberan-Santos, Constantin Carathéodory and the axiomatic thermodynamics. J. Math. Chem. **28**, 1–3 (2000)
23. J.B. Boyling, Caratheodory's principle and the existence of global integrating factors. Commun. Math. Phys. **10**, 52 (1968)
24. N. Pottier, *Nonequilibrium Statistical Physics* (Oxford University, Oxford, 2010)
25. I. Müller, T. Ruggeri, *Rational Extended Thermodynamics*, 2nd edn. (Springer, New York, 1998)
26. C. Cattaneo, Sulla conduzione del calore. Atti Sem. Mat. Fis. Univ. Modena **3**, 83 (1948)
27. D.D. Joseph, L. Preziosi, Heat waves. Rev. Mod. Phys. **61**, 41 (1989)
28. D.D. Joseph, L. Preziosi, Addendum to the paper Heat waves. Rev. Mod. Phys. **62**, 375 (1990)
29. B. Straughan, *Heat Waves* (Springer, New York, 2011)
30. B.D. Coleman, M. Fabrizio, D.R. Owen, On the thermodynamics of second sound in dielectric crystals. Arch. Rat. Mech. **80**, 135 (1983)
31. A. Morro, T. Ruggeri, Non equilibrium properties of solids through second sound measurements. J. Phys. C Solid State Phys. **21**, 1743 (1988)
32. T. Ruggeri, Struttura dei sistemi alle derivate parziali compatibili con un principio di entropia e termodinamica estesa. Suppl. Boll. UMI. **261**, 4 (1985)
33. T. Ruggeri, The binary mixtures of Euler fluids: a unified theory of second sound phenomena, in *Continuum Mechanics and Applications in Geophysics and the Environment*, eds. by B. Straughan, R. Greve, H. Ehrentraut, Y. Wang (Springer, Berlin, 2001), pp.79–91
34. R. Spigler, More around Cattaneo equation to describe heat transfer processes. Math. Meth. Appl. Sci. (2020). https://doi.org/10.1002/mma.6336
35. A. Sellitto, V. Zampoli, P.M. Jordan, Second-sound beyond Maxwell-Cattaneo: Nonlocal effects in hyperbolic heat transfer at the nanoscale. Int. J. Eng. Sci. **154**, 103328 (2020)
36. L. Landau, Theory of the superfluidity of Helium II. J. Physique U.S.S.R. **5**, 71 (1941)
37. V. Peshkov, Second sound in Helium II. J. Phys. U.S.S.R. **8**, 381 (1944)
38. L.A. Sidorenkov, M. K. Tey, R. Grimm, Y.-H. Hou, L. Pitaevskii, S. Stringari, Second sound and the superfluid fraction in a Fermi gas with resonant interactions. Nature **498**, 78 (2013)
39. T. Ruggeri, A. Muracchini, L. Seccia, Shock waves and second sound in a rigid heat conductor: A critical temperature for NaF and Bi. Phys. Rev. Lett. **64**, 2640 (1990)
40. T. Ruggeri, A. Muracchini, L. Seccia, Continuum approach to phonon gas and shape changes of second sound via shock wave theory. Nuovo Cimento D **16**, 15 (1994)
41. T. Ruggeri, A. Muracchini, L. Seccia, Second sound and characteristic temperature in solids. Phys. Rev. B **54**, 332 (1996)
42. F. Ancona, A. Marson, A wavefront tracking algorithm for $N \times N$ nongenuinely nonlinear conservation laws. J. Differ. Equ. **177**, 454 (2001)
43. R.A. Guyer, J A. Krumhansl, Solution of the linearized phonon Boltzmann equation. Phys. Rev. **148**, 766 (1966)
44. R.A. Guyer, J.A. Krumhansl, Thermal conductivity, second sound and phonon hydrodynamic phenomena in nonmetallic crystals. Phys. Rev. **148**, 778 (1966)
45. W. Dreyer, H. Struchtrup, Heat pulse experiments revisited. Contin. Mech. Thermodyn. **5**, 3 (1993)
46. W. Larecki, Z. Banach, Influence of nonlinearity of the phonon dispersion relation on wave velocities in the four-moment maximum entropy phonon hydrodynamics. Phys. D **266**, 65 (2014)
47. W. Larecki, Z. Banach, Consistency of the phenomenological theories of wave-type heat transport with the hydrodynamics of a phonon gas. J. Phys. A Math. Theor. **43**, 385501 (2010)
48. B.-Y. Cao, M. Grmela, Z.-Y. Guo, Y.-C. Hua, B.-D. Nie, Two temperature extension of phonon hydrodynamics. J. Non-Equilib. Thermodyn. **45**, 291 (2020)
49. I. Müller, *Zur Ausbreitungsgeschwindigkeit von Störungen in Kontinuierlichen Medien*, PhD thesis (RWTH Aachen University, Aachen, 1966)

50. W. Israel, Nonstationary irreversible thermodynamics: a causal relativistic theory. Ann. Phys. **100**, 310 (1976)
51. D. Jou, J. Casas-Vázquez, G. Lebon, *Extended Irreversible Thermodynamics*, 4th edn. (Springer, Heidelberg, 2010)
52. G. Lebon, D. Jou, J. Casas-Vázquez, *Understanding Non-equilibrium Thermodynamics* (Springer, Berlin, 2008)
53. T. Ruggeri, Symmetric-hyperbolic system of conservative equations for a viscous heat conducting fluid. Acta Mech. **47**, 167 (1983)
54. I.-S. Liu, I. Müller, Extended thermodynamics of classical and degenerate ideal gases. Arch. Rat. Mech. Anal. **83**, 285 (1983)
55. I.-S. Liu, I. Müller, T. Ruggeri, Relativistic thermodynamics of gases. Ann. Phys. **169**, 191 (1986)
56. I. Müller, T. Ruggeri, *Extended Thermodynamics*, 1st edn. (Springer, New York, 1993)
57. C. Truesdell, *Rational Thermodynamics* (McGraw-Hill, New York, 1969)
58. B.D. Coleman, W. Noll, The thermodynamics of elastic materials with heat conduction and viscosity. Arch. Ration. Mach. Anal. **13**, 167 (1963)
59. I. Müller, On the entropy inequality, Arch. Rational Mech. Anal. **26**, 118 (1967)
60. M. Šilhavý, *The Mechanics and Thermodynamics of Continuous Media* (Springer, Berlin, 1997)
61. M. Grmela, H.C. Öttinger, Dynamics and thermodynamics of complex fluids I. Development of a generic formalism. Phys. Rev. E **56**, 6620 (1997)
62. H.C. Öttinger, M. Grmela, Dynamics and thermodynamics of complex fluids. II. Illustrations of a general formalism. Phys. Rev. E **56**, 6633 (1997)
63. H.C. Öttinger, *Beyond Equilibrium Thermodynamics* (Wiley, Hoboken, 2004)
64. M. Grmela, GENERIC guide to the multiscale dynamics and thermodynamics. J. Phys. Commun. **2**, 032001 (2018)
65. I. Peshkov, M. Pavelka, E. Romenski, M. Grmela, Continuum mechanics and thermodynamics in the Hamilton and the Godunov-type formulations. Continuum Mech. Thermodyn. **30**, 1343 (2018)
66. S.K. Godunov, E.I. Romenskii, *Elements of Continuum Mechanics and Conservation Laws* (Kluwer Academic/Plenum Publishers, New York, 2003)
67. I. Müller, D. Reitebuch, W. Weiss, Extended thermodynamics—consistent in order of magnitude. Contin. Mech. Thermodyn. **15**, 113 (2003)
68. B.C. Eu, *Kinetic Theory of Nonequilibrium Ensembles, Irreversible Thermodynamics, and Generalized Hydrodynamics*. Nonrelativistic Theories, vol. 1 (Springer, Switzerland, 2016)
69. B.C. Eu, *Kinetic Theory of Nonequilibrium Ensembles, Irreversible Thermodynamics, and Generalized Hydrodynamics*. Relativistic Theories, vol. 2 (Springer, Switzerland, 2016)
70. I. Gyarmati, *Non-Equilibrium Thermodynamics, Field Theory and Variational Principles* (Springer, Berlin, 1970)
71. V.A. Cimmelli, D. Jou, T. Ruggeri, P. Ván, Entropy principle and recent results in non-equilibrium theories. Entropy **16**, 1756 (2014)
72. D. Jou, Relationships between rational extended thermodynamics and extended irreversible thermodynamics. Phil. Trans. R. Soc. A. **37820190172** (2020). https://doi.org/10.1098/rsta.2019.0172
73. H.C. Öttinger, H. Struchtrup, M. Torrilhon, Formulation of moment equations for rarefied gases within two frameworks of non-equilibrium thermodynamics: RET and GENERIC. Phil. Trans. R. Soc. A. **37820190174** (2020). https://doi.org/10.1098/rsta.2019.0174
74. I. Müller, W. Weiss, Thermodynamics of irreversible processes—past and present. Eur. Phys. J. H **37**, 139 (2012)
75. M. Torrilhon, H. Struchtrup, Regularized 13 moment equations: shock structure calculations and comparison to Burnett models. J. Fluid Mech. **513**, 171 (2004)
76. H. Struchtrup, *Macroscopic Transport Equations for Rarefied Gas Flows* (Springer, Berlin, 2005)

77. A. Bobylev, Å. Windfäll, Boltzmann equation and hydrodynamics at the Burnett level. Kinet. Rel. Models (KRM) **5-2**, 237 (2012)
78. M.N. Kogan, *Rarefied Gas Dynamics* (Springer, New York, 1969)
79. C. Cercignani, *The Boltzmann Equation and Its Applications* (Springer, New York, 1988)
80. Y. Sone, *Kinetic Theory and Fluid Dynamics* (Birkhäuser, Boston, 2002)
81. Y. Sone, *Molecular Gas Dynamics, Theory, Techniques, and Applications* (Birkhäuser, Boston, 2007)
82. I. Müller, Entropy: a subtle concept in thermodynamics, in *Entropy*, eds. by A. Greven, G. Keller, G. Warnecke (Princeton University, Princeton, 2003). Chapter 2
83. D.N. Zubarev, *Nonequilibrium Statistical Thermodynamics* (Consultants Bureau, New York, 1974)
84. H. Grad, On the kinetic theory of rarefied gases. Commun. Pure Appl. Math. **2**, 331 (1949)
85. E.T. Jaynes, Information theory and statistical mechanics. Phys. Rev. **106**, 620 (1957); E.T. Jaynes, Information theory and statistical mechanics II. Phys. Rev. **108**, 171 (1957)
86. J.N. Kapur, *Maximum Entropy Models in Science and Engineering* (Wiley, New York, 1989)
87. M.N. Kogan, On the principle of maximum entropy, in *Rarefied Gas Dynamics*, vol. I, 359–368 (Academic Press, New York, 1967)
88. W. Dreyer, Maximization of the entropy in non-equilibrium. J. Phys. A Math. Gen. **20**, 6505 (1987)
89. C.D. Levermore, Moment closure hierarchies for kinetic theories. J. Stat. Phys. **83**, 1021 (1996)
90. G. Boillat, T. Ruggeri, Moment equations in the kinetic theory of gases and wave velocities. Continuum Mech. Thermodyn. **9**, 205 (1997)
91. P. Degond, C. Ringhofer, Quantum moment hydrodynamics and the entropy principle. J. Stat. Phys. **112**, 587 (2003)
92. M. Trovato, L. Reggiani, Maximum entropy principle and hydrodynamic models in statistical mechanics. Riv. Nuovo Cimento Soc. Ital. Fis. **35**, 99 (2012)
93. W. Larecki, Z. Banach, Entropic derivation of the spectral Eddington factors. J. Quant. Spectrosc. Radiat. Transfer **11**, 22486 (2011)
94. W. Larecki, Z. Banach, One-dimensional maximum entropy radiation hydrodynamics: three-moment theory. J. Phys. A Math. Theor. **45**, 385501 (2012)
95. M. Junk, Domain of definition of Levermore's five-moment system. J. Stat. Phys. **93**, 1143 (1988)
96. M. Junk, Maximum entropy for reduced moment problems. Math. Models Methods Appl. Sci. **10**, 1001 (2000)
97. M. Junk, A. Unterreiter, Maximum entropy moment systems and Galilean invariance. Continuum Mech. Thermodyn. **14**, 563 (2002)
98. W. Dreyer, M. Kunik, Maximum entropy principle revisited. Continuum Mech. Thermodyn. **10**, 331 (1998)
99. G.M. Kremer, *An Introduction to the Boltzmann Equation and Transport Processes in Gases* (Springer, Berlin, 2010)
100. I.-S. Liu, Extended thermodynamics of fluids and virial equations of state. Arch. Rational Mech. Anal. **88**, 1 (1985)
101. G.M. Kremer, Extended thermodynamics of non-ideal gases. Physica A **144**, 156 (1987)
102. G.M. Kremer, On extended thermodynamics of ideal and real gases, in *Extended Thermodynamics Systems*, eds. by S. Sieniutycz, P. Salamon (Taylor and Francis, New York, 1992)
103. T. Arima, S. Taniguchi, T. Ruggeri, M. Sugiyama, Extended thermodynamics of dense gases. Continuum Mech. Thermodyn. **24**, 271 (2011)
104. E. Ikenberry, C. Truesdell, On the pressure and the flux of energy in a gas according to Maxwell's kinetic theory. J. Rational Mech. Anal. **5**, 1 (1956)
105. T. Ruggeri, Can constitutive relations be represented by non-local equations? Quart. Appl. Math. **70**, 597 (2012)

106. T. Arima, S. Taniguchi, T. Ruggeri, M. Sugiyama, Dispersion relation for sound in rarefied polyatomic gases based on extended thermodynamics. Continuum Mech. Thermodyn. **25**, 727 (2013)
107. T. Arima, S. Taniguchi, T. Ruggeri, M. Sugiyama, A study of linear waves based on extended thermodynamics for rarefied polyatomic gases. Acta Appl. Math. **132**, 15 (2014)
108. T. Arima, S. Taniguchi, M. Sugiyama, Light scattering in rarefied polyatomic gases based on extended thermodynamics. Proc. Symp. Ultrasonic Electronics **34**, 15 (2013)
109. S. Taniguchi, T. Arima, T. Ruggeri, M. Sugiyama, Thermodynamic theory of the shock wave structure in a rarefied polyatomic gas: Beyond the Bethe-Teller theory. Phys. Rev. E **89**, 013025 (2014)
110. S. Taniguchi, T. Arima, T. Ruggeri, M. Sugiyama, Shock wave structure in a rarefied polyatomic gas based on extended thermodynamics. Acta Appl. Math. **132**, 583 (2014)
111. E. Barbera, F. Brini, M. Sugiyama, Heat transfer problem in a van der Waals gas, Acta Appl. Math. **132**, 41 (2014)
112. T. Arima, S. Taniguchi, T. Ruggeri, M. Sugiyama, Monatomic rarefied gas as a singular limit of poyatomic gas in extended thermodynamics. Phys. Lett. A **377**, 2136 (2013)
113. C. Borgnakke, P.S. Larsen, Statistical collision model for Monte Carlo simulation of polyatomic gas mixture. J. Comput. Phys. **18**, 405 (1975)
114. J.-F. Bourgat, L. Desvillettes, P. Le Tallec, B. Perthame, Microreversible collisions for polyatomic gases. Eur. J. Mech. B/Fluids **13**, 237 (1994)
115. M. Pavić, T. Ruggeri, S. Simić, Maximum entropy principle for rarefied polyatomic gases. Physica A **392**, 1302 (2013)
116. T. Ruggeri, Maximum entropy principle closure for 14-moment system for a non-polytropic gas. Ric. Mat. (2020). https://doi.org/10.1007/s11587-020-00510-y
117. I.M. Gamba, M. Pavić-Čolić, *On the Cauchy Problem for Boltzmann Equation Modelling a Polyatomic Gas* (2020). arXiv:2005.01017 [math-ph]
118. S.-B. Yun, Ellipsoidal BGK model for polyatomic molecules near Maxwellians: a dichotomy in the dissipation estimate. J. Differ. Equ. **266**, 5566 (2019)
119. T. Arima, A. Mentrelli, T. Ruggeri, Molecular extended thermodynamics of rarefied polyatomic gases and wave velocities for increasing number of moments. Ann. Phys. **345**, 111 (2014)
120. G. Boillat, T. Ruggeri, On the evolution law of the weak discontinuities for hyperbolic quasi-linear systems. Wave Motion **1**, 149 (1979)
121. T. Ruggeri, Stability and discontinuity waves for symmetric hyperbolic systems, in *Non-Linear Wave Motion* ed. by A. Jeffrey (Longman Press, New York, 1989), pp.148–161
122. A. Muracchini, T. Ruggeri, L. Seccia, Dispersion relation in the high frequency limit and non-linear wave stability for hyperbolic dissipative systems. Wave Motion **15**, 143 (1992)
123. Z. Banach, W. Larecki, T. Ruggeri, Dispersion relation in the limit of high frequency for a hyperbolic system with multiple eigenvalues. Wave Motion **51**, 955 (2014)
124. G. Boillat, T. Ruggeri, On the shock structure problem for hyperbolic system of balance laws and convex entropy. Continuum Mech. Thermodyn. **10**, 285 (1998)
125. T. Arima, T. Ruggeri, M. Sugiyama, S. Taniguchi, Monatomic gas as a singular limit of polyatomic gas in molecular extended thermodynamics with many moments. Ann. Phys. **372**, 83 (2016)
126. M.S. Cramer, Numerical estimates for the bulk viscosity of ideal gases. Phys. Fluids **24**, 066102 (2012)
127. S. Taniguchi, T. Arima, T. Ruggeri, M. Sugiyama, Effect of dynamic pressure on the shock wave structure in a rarefied polyatomic gas. Phys. Fluids **26**, 016103 (2014)
128. T. Arima, T. Ruggeri, M. Sugiyama, S. Taniguchi, On the six-field model of fluids based on extended thermodynamics. Meccanica **49**, 2181 (2014)
129. J. Meixner, Absorption und dispersion des schalles in gasen mit chemisch reagierenden und anregbaren komponenten. I. Teil. Ann. Phys. **43**, 470 (1943)
130. J. Meixner, Allgemeine theorie der schallabsorption in gasen und flussigkeiten unter beruck-sichtigung der transporterscheinungen. Acoustica **2**, 101 (1952)

131. T. Ruggeri, Non-linear maximum entropy principle for a polyatomic gas subject to the dynamic pressure. Bull. Inst. Math. Acad. Sin. **11**, 1 (2016)
132. T. Arima, S. Taniguchi, T. Ruggeri, M. Sugiyama, Nonlinear extended thermodynamics of real gases with 6 fields. Int. J. Non-Linear Mech. **72**, 6 (2015)
133. S. Taniguchi, T. Arima, T. Ruggeri, M. Sugiyama, Overshoot of the nonequilibrium temperature in the shock wave structure of a rarefied polyatomic gas subject to the dynamic pressure. Int. J. Non-Linear Mech. **79**, 66 (2015)
134. M. Bisi, T. Ruggeri, G. Spiga, Dynamical pressure in a polyatomic gas: interplay between kinetic theory and extended thermodynamic. Kinet. Relat. Mod. **11**, 71 (2017)
135. T. Arima, T. Ruggeri, M. Sugiyama, Rational extended thermodynamics of a rarefied polyatomic gas with molecular relaxation processes. Phys. Rev. E **96**, 042143 (2017)
136. T. Arima, T. Ruggeri, M. Sugiyama, Extended thermodynamics of rarefied polyatomic gases: 15-field theory incorporating relaxation processes of molecular rotation and vibration. Entropy **20**, 301 (2018)
137. W.G. Vincenti, C.H. Kruger, Jr., *Introduction to Physical Gas Dynamics* (Wiley, New York, 1965)
138. Ya.B. Zel'dovich, Yu.P. Raizer, *Physics of Shock Waves and High-Temperature Hydrodynamic Phenomena* (Dover, New York, 2002)
139. E.F. Smiley, E.H. Winkler, Z.I. Slawsky, Measurement of the vibrational relaxation effect in CO_2 by means of shock tube interferograms. J. Chem. Phys. **20**, 923 (1952)
140. E.F. Smiley, E.H. Winkler, Shock-tube Measurements of vibrational relaxation. J. Chem. Phys. **22**, 2018 (1954)
141. W.C. Griffith, W. Bleakney, Shock waves in gases, Am. J. Phys. **22**, 597 (1954)
142. W. Griffith, D. Brickl, V. Blackman, Structure of shock waves in polyatomic gases. Phys. Rev. **102**, 1209 (1956)
143. N.H. Johannesen, H.K. Zienkiewicz, P.A. Blythe, J.H. Gerrard, Experimental and theoretical analysis of vibrational relaxation regions in carbon dioxide. J. Fluid Mech. **13**, 213 (1962)
144. W.C. Griffith, A. Kenny, On fully-dispersed shock waves in carbon dioxide. J. Fluid Mech. **3**, 286 (1957)
145. H.A. Bethe, E. Teller, *Deviations from Thermal Equilibrium in Shock Waves* (Engineering Research Institute/University of Michigan, New York)
146. D. Gilbarg, D. Paolucci, The structure of shock waves in the continuum theory of fluids. J. Rat. Mech. Anal. **2**, 617 (1953)
147. S. Kosuge, K. Aoki, T. Goto, Shock wave structure in polyatomic gases: numerical analysis using a model Boltzmann equation. AIP Conf. Proc. **1786**, 180004 (2016)
148. S. Kosuge, K. Aoki, Shock-wave structure for a polyatomic gas with large bulk viscosity. Phys. Rev. Fluids **3**, 023401 (2018)
149. S. Kosuge, H.-W. Kuo, K. Aoki, A kinetic model for a polyatomic gas with temperature-dependent specific heats and its application to shock-wave structure. J. Stat. Phys. **177**, 209 (2019)
150. T. Arima, T. Ruggeri, M. Sugiyama, Duality principle from rarefied to dense gas and extended thermodynamics with six fields. Phys. Rev. Fluids **2**, 013401 (2017)
151. T. Arima, T. Ruggeri, M. Sugiyama, Rational extended thermodynamics of dense polyatomic gases incorporating molecular rotation and vibration. Phil. Trans. R. Soc. A **378**, 20190176 (2020)
152. C. Eckart, The thermodynamics of irreversible processes III: Relativistic theory of the simple fluid. Phys. Rev. **58**, 919 (1940)
153. S. Pennisi, T. Ruggeri, Relativistic extended thermodynamics of rarefied polyatomic gas. Ann. Phys. **377**, 414 (2017)
154. M.C. Carrisi, S. Pennisi, T. Ruggeri, Monatomic limit of relativistic extended thermodynamics of polyatomic gas. Continuum Mech. Thermodyn. **31**, 401 (2019)
155. E. Barbera, I. Müller, M. Sugiyama, On the temperature of a rarefied gas in non-equilibrium. Meccanica **34**, 103 (1999)

156. J. Au, I. Müller, T. Ruggeri, Temperature jumps at the boundary of a rarefied gas. Cont. Mech. Thermodyn. **12**, 19 (2000)
157. J. Casas-Vázquez, D. Jou, Temperature in non-equilibrium states: a review of open problems and current proposals. Rep. Prog. Phys. **66**, 1937 (2003)
158. T. Ruggeri, S. Simić, On the hyperbolic system of a mixture of eulerian fluids: a comparison between single and multi-temperature models. Math. Meth. Appl. Sci. **30**, 827 (2007)
159. H. Gouin, T. Ruggeri, Identification of an average temperature and a dynamical pressure in a multitemperature mixture of fluids. Phys. Rev. E **78**, 016303 (2008)
160. T. Ruggeri, S. Simić, Average temperature and Maxwellian iteration in multitemperature mixtures of fluids. Phys. Rev. E **80**, 026317 (2009)
161. S.-Y. Ha, T. Ruggeri, Emergent dynamics of a thermodynamically consistent particle model. Arch. Ration Mech. Anal. **223**, 1397 (2017)
162. F. Cucker, S. Smale, Emergent behavior in flocks. IEEE Trans. Automat. Control **52**, 852 (2007)
163. S.-Y. Ha, J. Kim, T. Ruggeri, From the relativistic mixture of gases to the relativistic Cucker-Smale flocking. Arch. Rational Mech. Anal. **235**, 1661–1706 (2020)
164. T. Ruggeri, A. Strumia, Main field and convex covariant density for quasi-linear hyperbolic systems: Relativistic fluid dynamics. Ann. Inst. H. Poincaré Sect. A **34**, 65 (1981)
165. G. Boillat, T. Ruggeri, Hyperbolic principal subsystems: entropy convexity and subcharacteristic conditions. Arch. Rational Mech. Anal. **137**, 305 (1997)
166. G. Boillat, Sur l'existence et la recherche d'équations de conservation supplémentaires pour les systémes hyperboliques. C. R. Acad. Sci. Paris A **278**, 909 (1974)
167. T. Ruggeri, Galilean invariance and entropy principle for systems of balance laws. The structure of the extended thermodynamics. Continuum Mech. Thermodyn. **1**, 3 (1989)
168. Y. Shizuta, S. Kawashima, Systems of equations of hyperbolic-parabolic type with applications to the discrete Boltzmann equation. Hokkaido Math. J. **14**, 249 (1985)
169. B. Hanouzet, R. Natalini, Global existence of smooth solutions for partially dissipative hyperbolic systems with a convex entropy. Arch. Rational Mech. Anal. **169**, 89 (2003)
170. W-A. Yong, Entropy and global existence for hyperbolic balance laws. Arch. Rational Mech. Anal. **172**, 247 (2004)
171. S. Bianchini, B. Hanouzet, R. Natalini, Asymptotic behavior of smooth solutions for partially dissipative hyperbolic systems with a convex entropy. Commun. Pure Appl. Math. **60**, 1559 (2007)
172. T. Ruggeri, D. Serre, Stability of constant equilibrium state for dissipative balance laws system with a convex entropy. Quart. Appl. Math. **62**, 163 (2004)
173. J. Lou, T. Ruggeri, Acceleration waves and weak Shizuta-Kawashima condition. Suppl. Rend. Circ. Mat. Palermo **78**, 187 (2006)
174. T. Ruggeri, Global existence of smooth solutions and stability of the constant state for dissipative hyperbolic systems with applications to extended thermodynamics, in *Trends and Applications of Mathematics to Mechanics STAMM 2002* (Springer, Berlin, 2005)
175. T. Ruggeri, Entropy principle and relativistic extended thermodynamics: global existence of smooth solutions and stability of equilibrium state. Il Nuovo Cimento B **119**, 809 (2004); *Lecture notes of the International Conference in honour of Y. Choquet-Bruhat: analysis, Manifolds and Geometric Structures in Physics*, ed. by G. Ferrarese, T. Ruggeri (2004)
176. T. Ruggeri, Extended relativistic thermodynamics, in *General Relativity and the Einstein Equations*, ed. by Y. Choquet Bruhat (Oxford University, Oxford, 2009), pp. 334–340
177. K.O. Friedrichs, P.D. Lax, Systems of conservation equation with a convex extension. Proc. Nat. Acad. Sci. USA **68**, 1686 (1971)
178. T. Ruggeri, Convexity and symmetrization in relativistic theories. Privileged time-like congruence and entropy. Continuum Mech. Thermodyn. **2**, 163 (1990)
179. S.K. Godunov, An interesting class of quasilinear systems. Sov. Math. **2**, 947 (1961)
180. G. Boillat, in *Recent Mathematical Methods in Nonlinear Wave Propagation*. CIME Course, Lecture Notes in Mathematics, vol. 1640, ed. by T. Ruggeri (Springer, Berlin, 1995), pp. 103–152

181. T. Ruggeri, Godunov Symmetric systems and rational extended thermodynamics, in *Continuum Mechanics, Applied Mathematics and Scientific Computing: Godunov's Legacy*, ed. by G.V. Demidenko, E. Romenski, E. Toro, M. Dumbser (Springer, Cham, 2020), 321. ISBN 978-3-030-38869-0
182. I.-S. Liu, Method of Lagrange multipliers for exploitation of the entropy principle. Arch. Rational Mech. Anal. **46**, 131 (1972)
183. G. Boillat, Involutions des systèmes conservatifs. C. R. Acad. Sci. Paris **307**, 891 (1988)
184. C.M. Dafermos, Quasilinear hyperbolic systems with involutions. Arch. Rat. Mech. Analys. **94**, 373 (1986)
185. S. Gavrilyuk, H. Gouin, Symmetric form of governing equations for capillary fluids, in *Monographs and Surveys in Pure and Applied Mathematics*, vol. 106, ed. by G. Iooss, O. Gueŝ, A. Nouri (Chapman & Hall/CRC, London, 2000), pp. 306–312
186. H. Gouin, T. Ruggeri, Symmetric form for the hyperbolic-parabolic system of fourth-gradient fluid model, Ric. Mat. **66**, 491 (2017)
187. H. Gouin, Symmetric forms for hyperbolic-parabolic systems of multi-gradient fluids. ZAMM (2019). https://doi.org/10.1002/zamm.201800188
188. T. Ruggeri, Maximum of entropy density in equilibrium and minimax principle for an hyperbolic system of balance laws, in *Contributions to Continuum Theories, Anniversary Volume for Krzysztof Wilmanski*, ed. by B. Albers (2000), pp.207–214. WIAS-Report No.18
189. S. Kawashima, Large-time behavior of solutions to hyperbolic-parabolic systems of conservation laws and applications. Proc. R. Soc. Edimburgh **106A**, 169 (1987)
190. A.E. Fischer, J.E. Marsden, The Einstein evolution equations as a first-order quasi-linear symmetric hyperbolic system. Commun. Math. Phys. **28**, 1 (1972)
191. A. Majda, *Compressible Fluid Flow and Systems of Conservation Laws in Several Space Variables* (Springer, New York, 1984)
192. C.M. Dafermos, *Hyperbolic Conservation Laws in Continuum Physics* 3rd edn. Grundlehren der Mathematischen Wissenschaften, vol. 325 (Springer, Berlin, 2010)
193. T. Nishida, *Nonlinear Hyperbolic Equations and Related Topics in Fluid Dynamics* (Département de Mathematique, Université de Paris-Sud/Mathématiques d'Orsay, No. 78-02, Orsay, 1978)
194. L. Hsiao, T.-P. Liu, Convergence to non linear diffusion waves for solutions of a system of hyperbolic conservation laws with damping. Commun. Math. Physics **143**, 599 (1992)
195. Y. Zeng, Gas dynamics in thermal nonequilibrium and general hyperbolic systems with relaxation. Arch. Ration. Mech. Anal. **150**, 255 (1999)
196. C.M. Dafermos, Periodic BV solutions of hyperbolic balance laws with dissipative source. J. Math. Anal. Appl. **428**, 405 (2015)
197. S. Pennisi, T. Ruggeri, A new method to exploit the entropy principle and galilean invariance in the macroscopic approach of extended thermodynamics. Ric. Mat. **55**, 159 (2006)
198. G. Boillat, *La Propagation des Ondes* (Gauthier-Villars, Paris, 1965)
199. G. Boillat, Chocs caractéeristiques, C. R. Acad Sci. Paris A **274**, 1018 (1972)
200. V.D. Sharma, *Quasilinear Hyperbolic Systems, Compressible Flows, and Waves*. Chapman & Hall/CRC Monographs and Surveys in Pure and Applied Mathematics, vol. 142. (CRC Press, Boca Raton, 2010)
201. P.O.K. Krehl, *History of Shock Waves, Explosions and Impact: A Chronological and Biographical Reference* (Springer, Heidelberg, 2009)
202. D.G. Morris, An investigation of the shock-induced transformation of graphite to diamond. J. Appl. Phys. **51**, 2059 (1980)
203. J. Vink, Supernova remnants: the X-ray perspective. Astron Astrophys Rev. **20**, 49 (2012)
204. C.G. Chaussy, F. Eisenberger, B. Forssmann, Extracorporeal shockwave lithotripsy (ESWL): a chronology. J. Endourology **21**, 1249 (2007)
205. T. Ruggeri, S. Taniguchi, Shock waves in hyperbolic systems of non-equilibrium thermodynamics, in *Applied Wave Mathematics II*, ed. by A. Berezovski, T. Soomere (Springer, Cham, 2019), pp.167–186

206. N. Zhao, A. Mentrelli, T. Ruggeri, M. Sugiyama, Admissible shock waves and shock-induced phase transitions in a van der Waals fluid. Phys. Fluids **23**, 086101 (2011)
207. P.D. Lax, *Hyperbolic Systems of Conservation Laws and the Mathematical Theory of Shock Waves*. CBMS-NSF, Regional Conference Series in Applied Mathematics, vol. 11 (SIAM, New York, 1973)
208. T.-P. Liu, The entropy condition and the admissibility of shocks. J. Math. Anal. Appl. **53**, 78 (1976)
209. T.-P. Liu, Admissible solutions of hyperbolic conservation laws. Mem. Am. Math. Soc. **240**, 12 (1981)
210. T.-P. Liu, T. Ruggeri, Entropy production and admissibility of shocks. Acta Math. Appl. Sinica, English Series. **19**, 1 (2003)
211. S. Taniguchi, A. Mentrelli, T. Ruggeri, M. Sugiyama, N. Zhao, Prediction and simulation of compressive shocks with lower perturbed density for increasing shock strength in real gases. Phys. Rev. E **82**, 036324 (2010)
212. N. Zhao, M. Sugiyama, T. Ruggeri, Phase transition induced by a shock wave in hard-sphere and hard-disc systems. J. Chem. Phys. **129**, 054506 (2008)
213. S. Taniguchi, A. Mentrelli, N. Zhao, T. Ruggeri, M. Sugiyama, Shock-induced phase transition in systems of hard spheres with internal degrees of freedom. Phys. Rev. E **81**, 066307 (2010)
214. Y. Zheng, N. Zhao, T. Ruggeri, M. Sugiyama, S. Taniguchi, Non-polytropic effect on shock-induced phase transitions in a hard-sphere system. Phys. Lett. A **374**, 3315 (2010)
215. S. Taniguchi, N. Zhao, M. Sugiyama, Shock-induced phase transitions from gas phase to solid phase. J. Phys. Soc. Japan **80**, 083401 (2011)
216. S. Taniguchi, M. Sugiyama, Shock-induced phase transitions in systems of hard spheres with attractive interactions. Acta Appl. Math. **122**, 473 (2012)
217. G. Boillat, T. Ruggeri, Reflection and transmission of discontinuity waves through a shock wave, General theory including also the case of characteristic shocks. Proc. Roy. Soc. Edinb. **83A**, 17 (1979)
218. T. Ruggeri, Interaction between a discontinuity wave and a shock wave: critical time for the fastest transmitted wave, example of the polytropic fluid. Appl. Anal. **11**, 103 (1980)
219. M. Pandey, V.D. Sharma, Interaction of a characteristic shock with a weak discontinuity in a non-ideal gas. Wave Motion **44**, 346 (2007)
220. A. Mentrelli, T. Ruggeri, M. Sugiyama, N. Zhao, Interaction between a shock and an acceleration wave in a perfect gas for increasing shock strength. Wave Motion **45**, 498 (2008)
221. H. Alsmeyer, Density profiles in argon and nitrogen shock waves measured by the absorption of an electron beam. J. Fluid Mech. **74**, 497 (1976)
222. R. Becker, Stoßwelle und Detonation, Z. f. Physik **8**, 321 (1922)
223. H.W. Mott-Smith, The solution of the Boltzmann equation for a shock wave. Phys. Rev. **82**, 885 (1951)
224. K. Zoller, Zur Struktur des Verdichtungsstoßes. Zeitschr. f. Physik **130**, 1 (1951)
225. J.D. Foch, On higher order hydrodynamic theories of shock structure. Acta Physica Austriaca Suppl. **10**, 123 (1973)
226. G.A. Bird, Aspects of the structure of strong shock waves. Phys. Fluids **13**, 1172 (1970)
227. T. Ruggeri, Breakdown of shock-wave-structure solutions. Phys. Rev. E **47**, 4135 (1993)
228. W. Weiss, Continuous shock structure in extended thermodynamics. Phys. Rev. E **52**, R5760 (1995)
229. T. Ruggeri, Non existence of shock structure solutions for hyperbolic dissipative systems including characteristic shocks. Appl. Anal. **57**, 23 (1995)
230. D. Gilbarg, The existence and limit behavior of the one-dimensional shock layer. Am. J. Math. **73**, 256 (1951)
231. K. Zumbrun, W-A. Yong, Existence of relaxation shock profiles for hyperbolic conservation laws. SIAM J. Appl. Math. **60**, 1565 (2000)
232. S. Simić, Shock structure in continuum models of gas dynamics: stability and bifurcation analysis. Nonlinearity **22**, 1337 (2009)

233. S. Simić, Shock structure in the mixture of gases: stability and bifurcation of Equilibria, Chapter 8, in *Thermodynamics—Kinetics of Dynamic Systems*, ed. by J.C. Moreno Pirajãn (InTech, Rijeka, 2011)
234. P. Lax, Hyperbolic systems of conservation laws II. Commun. Pure Appl. Math. **10**, 537 (1957)
235. J. Smoller, *Shock Waves and Reaction-Diffusion Equations* (Springer, Berlin, 1994)
236. D. Serre, *Systems of Conservation Laws 1: Hyperbolicity, Entropies, Shock Waves* (Cambridge University, Cambridge, 1999)
237. E.F. Toro, *Riemann Solvers and Numerical Methods for Fluid Dynamics* (Springer, Berlin, 2009)
238. M. Dumbser, W. Boscheri, M. Semplice, G. Russo, Central weighted ENO schemes for hyperbolic conservation laws on fixed and moving unstructured meshes. SIAM J. Sci. Comput. **39**, A2564 (2017)
239. C. Klingenberg, Relaxation-projection schemes, the ultimate approximate Riemann solvers, in *Continuum Mechanics, Applied Mathematics and Scientific Computing: Godunov's Legacy*, ed. by G.V. Demidenko, E. Romenski, E. Toro, M. Dumbser (Springer, Cham, 2020), p. 237. ISBN 978-3-030-38869-0
240. T.-P. Liu, Linear and nonlinear large-time behavior of solutions of general systems of hyperbolic conservation laws. Commun. Pure Appl. Math. **30**, 767 (1977)
241. T.-P. Liu, Large-time behavior of solutions of initial and initial-boundary value problems of a general system of hyperbolic conservation laws. Commun. Math. Phys. **55**, 163 (1977)
242. A. Mentrelli, T. Ruggeri, Asymptotic behavior of Riemann and Riemann with Structure problems for a 2×2 hyperbolic dissipative system. Suppl. Rend. Circ. Mat. Palermo II **78**, 201 (2006)
243. F. Conforto, A. Mentrelli, T. Ruggeri, Shock structure and multiple sub-shocks in binary mixtures of Eulerian fluids. Ric. Mat. **66**, 221 (2017)
244. V. Artale, F. Conforto, G. Martalò, A. Ricciardello, Shock structure and multiple sub-shocks in Grad 10-moment binary mixtures of monatomic gases. Ric. Mat. **68**, 485 (2019)
245. M. Bisi, F. Conforto, G. Martalò, Sub-shock formation in Grad 10-moment equations for a binary gas mixture. Continuum Mech. Thermodyn. **28**, 1295 (2016)
246. S. Taniguchi, T. Ruggeri, On the sub-shock formation in extended thermodynamics. Int. J. Non-Linear Mech. **99**, 69 (2018)
247. F. Brini, T. Ruggeri, On the Riemann problem in extended thermodynamics, in *Proceedings of the 10th International Conference on Hyperbolic Problems (HYP2004); Osaka, September 13–17, 2004*, vol. I (Yokohama Publication, New York, 2006), p. 319
248. F. Brini, T. Ruggeri, On the Riemann problem with structure in extended thermodynamics, in *Supplementary Rendiconti del Circolo Matematico di Palermo "Non Linear Hyperbolic Fields and Waves, A tribute to Guy Boillat", Serie II*, vol. 78 (2006), p. 21
249. T.-P. Liu, Nonlinear hyperbolic-dissipative partial differential equations, in *Recent Mathematical Methods in Nonlinear Wave Propagation, Lecture Notes in Mathematics*, vol. 1640, ed. by T. Ruggeri (Springer, Berlin, 1996), pp. 103–136
250. F. Brini, T. Ruggeri, Second-order approximation of extended thermodynamics of a monatomic gas and hyperbolicity region. Continuum Mech. Thermodyn. **32**, 23 (2020)
251. W. Weiss, *Zur Hierarchie der Erweiterten Thermodynamik* (Dissertation, TU Berlin, 1990)
252. G. Boillat, T. Ruggeri, Maximum wave velocity in the moments system of a relativistic gas. Continuum Mech. Thermodyn. **11**, 107 (1999)
253. G. Boillat, T. Ruggeri, Relativistic gas: moment equations and maximum wave velocity. J. Math. Phys. **40**, 6399 (1999)
254. F. Brini, T. Ruggeri, Maximum velocity for wave propagation in a relativistic rarefied gas. Cont. Mech. Thermodyn. **11**, 331 (1999)
255. J. McDonald, M. Torrilhon, Affordable robust moment closures for CFD based on the maximum-entropy hierarchy. J. Comput. Phys. **251**, 500 (2013)

256. F. Brini, T. Ruggeri, Entropy principle for the moment systems of degree α associated to the Boltzmann equation. Critical derivatives and non controllable boundary data. Continuum Mech. Thermodyn. **14**, 165 (2002)
257. M.R.A. Abdelmalik, E.H. van Brummelen, Moment closure approximations of the Boltzmann equation based on φ divergences. J. Stat. Phys. **164**, 77 (2016)
258. H.L. Hamburger, Hermitian transformations of deficiency-index (1, 1), Jacobian matrices, and undetermined moment problems. Am. J. Math. **66**, 489 (1944)
259. J. McDonald, C.P.T. Groth, Towards realizable hyperbolic moment closures for viscous heat conducting gas flows based on a maximum-entropy distribution. Continuum Mech. Thermodyn. **25**, 573 (2013)
260. F. Laurent, Characterization of the moment space corresponding to the Levermore basis. Comptes Rendus Math. **358**, 97 (2020)
261. F. Brini, Hyperbolicity region in extended thermodynamics with 14 moments. Continuum Mech. Thermodyn. **13**, 1 (2001)
262. T. Ruggeri, M. Trovato, Hyperbolicity in extended thermodynamics of Fermi and Bose Gases. Continuum Mech. Thermodyn. **16**, 551 (2004)
263. Z. Cai, Y. Fan, R. Li, On hyperbolicity of 13-moment system. Kinetic and related models **7**, 415 (2014)
264. Z. Cai, Y. Fan, R. Li, Globally hyperbolic regularization of Grad's moment system. Commun. Pure Appl. Math. **67**, 464 (2014)
265. I. Müller, T. Ruggeri, Stationary heat conduction in radially symmetric situations—an application of extended thermodynamics. J. Non-Newtonian Fluid Mech. **119**, 139 (2004)
266. H. Struchtrup, W. Weiss, Maximum of the local entropy production becomes minimal in stationary processes. Phys. Rev. Lett. **80**, 5048 (1998)
267. E. Barbera, I. Müller, D. Reitebuch, N. Zhao, Determination of boundary conditions in extended thermodynamics via fluctuation theory. Continuum Mech. Thermodyn. **16**, 411 (2004)
268. T. Ruggeri, J. Lou, Heat conduction in multi-temperature mixtures of fluids: the role of the average temperature. Physics Lett. A **373**, 3052 (2009)
269. H. Struchtrup, Heat transfer in the transition regime: solution of boundary value problems for Grad's moment equations via kinetic schemes. Phys. Rev. E **65**, 041204 (2002)
270. H. Struchtrup, W. Weiss, Temperature jump and velocity slip in the moment method. Continuum Mech. Thermodyn. **12**, 1 (2000)
271. N. Zhao, M. Sugiyama, Analysis of heat conduction in a rarefied gas at rest with a temperature jump at the boundary by consistent-order extended thermodynamics. Continuum Mech. Thermodyn. **18**, 367 (2007)
272. S. Taniguchi, A. Iwasaki, M. Sugiyama, Relationship between Maxwell boundary condition and two kinds of stochastic thermal wall. J. Phys. Soc. Jpn. **77**, 124004 (2008)
273. A. Mentrelli, T. Ruggeri, Shock structure in extended thermodynamics with second order maximum entropy principle closure. Contin. Mech. Thermodyn. **33**, 125 (2021).
274. T. Ruggeri, Shock Waves in Hyperbolic Dissipative Systems in "Nonlinear waves and dissipative effects", in *Proceedings of the Euromech Colloquium 270; Reggio Calabria, 25–28 September 1990*, ed. by D. Fusco, A. Jeffrey. Pitman Research Notes in Mathematical Series, vol. 227 (Longman Scientific & Technical, New York, 1991), pp. 256–264
275. I. Müller, Extended thermodynamics: a theory of symmetric hyperbolic field equations. Entropy **10**, 477 (2008)
276. T. Ruggeri, The entropy principle from continuum mechanics to hyperbolic systems of balance laws: the modern theory of extended thermodynamics. Entropy **10**, 319 (2008)
277. A.M. Anile, *Relativistic Fluids and Magneto-Fluids* (Cambridge University, Cambridge, 1989)
278. A.M. Anile, V. Romano, Non parabolic band transport in semiconductors: closure of the moment equations. Continuum Mech. Thermodyn. **11**, 307 (1999)

279. V.D. Camiola, G. Mascali, V. Romano, *Charge Transport in Low Dimensional Semiconductor Structures The Maximum Entropy Approach*. Series: The European Consortium for Mathematics in Industry (Springer, Berlin, 2020)
280. C. Cercignani, G.M. Kremer, *The Relativistic Boltzmann Equation: Theory and Applications* (Birkhäuser, Basel, 2002)
281. L. Rezzolla, O. Zanotti, *Relativistic Hydrodynamics* (Oxford University, Oxford, 2013)
282. N.A. Chernikov, Microscopic foundation of relativistic hydrodynamics. Acta Phys. Polonica **27**, 465 (1965)
283. J.L. Synge, The Relativistic Gas (North Holland, Amsterdam, 1957)
284. T. Ruggeri, Q. Xiao, H. Zhao, Nonlinear hyperbolic waves in relativistic gases of massive particles with Synge energy. Arch. Rational Mech. Anal. **239**, 1061 (2021)
285. T. Ruggeri, Q. Xiao, Classical and ultrarelativistic limits of the Riemann problem for the relativistic Euler fluid with Synge energy. Ric. Mat. (2020). https://doi.org/10.1007/s11587-020-00502-y
286. C. Rogers, T. Ruggeri, On invariance in 1+1-dimensional isentropic relativistic gasdynamics. Wave Motion **94**, 102527 (2020)
287. C. Rogers, T. Ruggeri, W.K. Schief, On relativistic gasdynamics, in *Invariance Under a Class of Reciprocal-type Transformations and Integrable Heisenberg Spin Connections*. Proceedings A Royal Society (2020)
288. H. Freistühler, Relativistic barotropic fluids: a Godunov-Boillat formulation for their dynamics and a discussion of two special classes. Arch. Rational Mech. Anal. **232**, 473 (2019)
289. J. Smoller, B. Temple, Global solutions of the relativistic Euler equations. Commun. Math. Phys. **156**, 67 (1993)
290. H. Ott, Lorentz Transformation der Wařmeundder Temperatur. Zeitschrift für Physik **175**, 70 (1963)
291. H. Arzeliés, Relativistic transformation of temperature and some other thermodynamical quantities (in French). N. Cim. **35**, 792 (1965)
292. C. Fariás, V.A. Pinto, P.S. Moya, What is the temperature of a moving body? Sci. Rep. **7**, 17657 (2017)
293. T. Ruggeri, Relativistic extended thermodynamics: general assumptions and mathematical procedure. *Corso CIME Noto (Giugno 1987)*. Lecture Notes in Mathematics 1385, ed. by A. Anile, Y. Choquet-Bruhat (Springer, Berlin, 1987), pp. 269–277
294. Y. Choquet Bruhat, *General Relativity and the Einstein equations* (Oxford University, Oxford, 2009)
295. G.M. Kremer, Extended thermodynamics of ideal gases with 14 fields. Ann. I. H. P., Phys. Theor. T. **45**, 419 (1986)
296. S. Pennisi, T. Ruggeri, Classical limit of relativistic moments associated with Boltzmann-Chernikov equation: optimal choice of moments in classical theory. J. Stat. Phys. **179** 231–246 (2020)
297. W. Dreyer, W. Weiss, The classical limit of relativistic extended thermodynamics. Ann. Inst. H. Poincaré, Sect. A **45**, 401 (1986)
298. H. Engholm Jr., G.M. Kremer, Thermodynamics of a diatomic gas with rotational and vibrational degrees of freedom. Int. J. Eng. Sci. **32**, 1241 (1994)
299. G.M. Kremer, Extended thermodynamics and statistical mechanics of a polyatomic ideal gas. J. Non-Equil. Therm. **14**, 363 (1989)
300. G.M. Kremer, Extended thermodynamics of molecular ideal gases. Continuum Mech. Thermodyn. **1**, 21 (1989)
301. I.-S. Liu, G.M. Kremer, Hyperbolic system of field equations for viscous fluids. Mat. Aplic. Comput. **9**, 123 (1990)
302. I.-S. Liu, J.A. Salvador, Hyperbolic system for viscous fluids and simulation of shock tube flows. Continuum Mech. Thermodyn. **2**, 179 (1990)
303. M.C. Carrisi, M.A. Mele, S. Pennisi, On some remarkable properties of an extended thermodynamics model for dense gases and macromolecular fluids. Proc. R. Soc. A **466**, 1645 (2010)

304. F. Mallinger, Generalization of the Grad theory to polyatomic gases. INRIA-Res. Rep. **RR-3581**, 1 (1998)
305. F. Brini, T. Ruggeri, Hyperbolicity of first and second order extended thermodynamics theory of polyatomic rarefied gases. Int. J. Non-Linear Mech. **124**, 103517 (2020)
306. T. Arima, T. Ruggeri, M. Sugiyama, S. Taniguchi, Recent results on nonlinear extended thermodynamics of real gases with six fields Part I: general theory. Ric. Mat. **65**, 263 (2016)
307. H. Struchtrup, The BGK model for an ideal gas with an internal degree of freedom. Transp. Theory Stat. Phys. **28**, 369 (1999)
308. B. Rahimi, H. Struchtrup, Capturing non-equilibrium phenomena in rarefied polyatomic gases: a high-order macroscopic model. Phys. Fluids **26**, 052001 (2014)
309. P.L. Bhatnagar, E.P. Gross, M. Krook, A model for collision processes in gases. I. Small amplitude processes in charged and neutral one-component systems. Phys. Rev. **94**, 511 (1954)
310. C. Baranger, Y. Dauvois, G. Marois, J. Mathè, J. Mathiaud, L. Mieussens, A BGK model for high temperature rarefied gas flows. Eur. J. Mech. B/Fluids **80**, 1 (2020)
311. A.B. Bhatia, *Ultrasonic Absorption: An Introduction to the Theory of Sound Absorption and Dispersion in Gases, Liquids, and Solids* (Dover, New York, 1985)
312. W.P. Mason, in *Physical Acoustics: Principles and Methods*, vol. II-A (Academic, London, 1965)
313. Y. Stupochenko, S. Losev, A. Osipov, *Relaxation in Shock Waves* (Springer, Berlin, 1967)
314. E.V. Kustova, On the simplified state-to-state transport coefficients. Chem. Phys. **270**, 177 (2001)
315. S. Brull, J. Schneider, On the ellipsoidal statistical model for polyatomic gases. Continuum Mech. Thermodyn. **20**, 489 (2009)
316. T. Arima, A. Mentrelli, T. Ruggeri, Extended thermodynamics of rarefied polyatomic gases and characteristic velocities. Rend. Lincei Mat. Appl. **25**, 275 (2014)
317. V.M. Zhdanov, The kinetic theory of a polyatomic gas. Sov. Phys. JETP **26**, 1187 (1968)
318. F.J. McCormack, Kinetic equations for polyatomic gases: the 17-moment approximation. Phys. Fluids **11**, 2533 (1968)
319. L. Desvillettes, R. Monaco, F. Salvarani, A kinetic model allowing to obtain the energy law of polytropic gases in the presence of chemical reactions. Eur. J. Mech. B. Fluids **24**, 219 (2005)
320. M.C. Carrisi, S. Pennisi, An 18 moments model for dense gases: entropy and Galilean relativity principles without expansions. Entropy **17**, 214 (2015)
321. T. Arima, T. Ruggeri, M. Sugiyama, S. Taniguchi, Molecular extended thermodynamics: comparison between rarefied polyatomic and monatomic gas closures. Ric. Mat. **66**, 1 (2017)
322. T. Arima, S. Taniguchi, T. Ruggeri, M. Sugiyama, Extended thermodynamics of real gases with dynamic pressure: an extension of Meixner's theory. Phys. Lett. A **376**, 2799 (2012)
323. C. Currò, N. Manganaro, Differential constraints and exact solutions for the ET6 model. Ric. Mat. **68**, 179 (2019)
324. B.D. Coleman, M.E. Gurtin, Thermodynamics with internal state variables. J. Chem. Phys. **47**, 597 (1967)
325. G.A. Maugin, W. Muschik, Thermodynamics with internal variables. Part I. General concepts. J. Non-Equilib. Thermodyn. **19**, 217 (1994)
326. G.A. Maugin, W. Muschik, Thermodynamics with internal variables. Part II. Appl. J. Non-Equilib. Thermodyn. **19**, 250 (1994)
327. P. Ván, D. Berezovski, J. Engelbrecht, Internal variables and dynamic degrees of freedom. J. Non-Equilib. Thermodyn. **33**, 235 (2008)
328. M.A. Leontovich, *Introduction to Thermodynamics. Statistical Physics* (Nauka, Moscow, 1983)
329. J. Casas-Vázquez, D. Jou, Nonequilibrium temperature versus local-equilibrium temperature. Phys. Rev. E **49**, 1040 (1994)
330. R.G. Gordon, W. Klemperer, J.I. Steinfeld, Vibrational and rotational relaxation. Annu. Rev. Phys. Chem. **19**, 215 (1968)

331. P. Secchi, Existence theorems for compressible viscous fluid having zero shear viscosity. Rend. Sem. Padova **70**, 73 (1983)
332. V. Shelukhin, Vanishing shear viscosity in a free-boundary problem for the equations of compressible fluids. J. Diff. Equ. **167**, 73 (2000)
333. H. Frid, V. Shelukhin, Vanishing shear viscosity in the equations of compressible fluids for the flows with the cylinder symmetry. SIAM J. Math. Anal. **31**, 1144 (2000)
334. L.D. Landau, E.M. Lifshitz, *Statistical Physics* (Pergamon, Oxford, 1980)
335. L. Tisza, Supersonic absorption and Stokes' viscosity relation. Phys. Rev. **61**, 531 (1942)
336. T. Ruggeri, M. Sugiyama, Recent developments in extended thermodynamics of dense and rarefied polyatomic gases. Acta Appl. Math. **132**, 527 (2014)
337. E. Nagnibeda, E. Kustova, *Non-Equilibrium Reacting Gas Flow* (Springer, Berlin, 2009)
338. C. Currò, N. Manganaro, Nonlinear wave interactions for a model of extended thermodynamics with six fields. Ric. Mat. **68**, 131 (2019)
339. K.F. Herzfeld, F.O. Rice, Dispersion and absorption of high frequency sound waves. Phys. Rev. **31**, 691 (1928)
340. L.I. Mandelstam, M.A. Leontovich, On the theory of sound absorption in fluids. Zh. Eksp. Teor. Fiz. **7**, 438 (1937)
341. K.F. Herzfeld, T. A. Litovitz, *Absorption and Dispersion of Ultrasonic Waves* (Academic Press, New York, 1959)
342. T. Arima, M. Sugiyama, Dispersion relation of a rarefied polyatomic gas with molecular relaxation processes based on extended thermodynamics. AIP Conf. Proc. **2132**, 130007 (2019)
343. T.G. Winter, G.L. Hill, High-temperature ultrasonic measurements of rotational relaxation in hydrogen, deuterium, nitrogen, and oxygen. J. Acoust. Soc. Am. **42**, 848 (1967)
344. E.J. Rhodes, Jr., The velocity of sound in hydrogen when rotational degrees of freedom fail to be excited. Phys. Rev. **70**, 932 (1946)
345. C. Sluijter, H. Knaap, J. Beenakker, Determination of rotational relaxation times of hydrogen isotopes by sound absorption measurements at low temperatures. I. Physica **30**, 745 (1964)
346. E.S. Stewart, J.L. Stewart, Rotational dispersion in the velocity, attenuation, and reflection of ultrasonic waves in hydrogen and deuterium. J. Acoust. Soc. Am. **24**, 194 (1952)
347. E.M. Lifshitz, L.P. Pitaevskii, *Statistical Physics, Part 1*, 3rd edn. (Pergamon Press, Oxford, 1980)
348. E.M. Lifshitz, L.P. Pitaevskii, *Statistical Physics, Part 2* (Pergamon Press, Oxford, 1980)
349. A.A. Radzig, B.M. Smirnov, *Reference Data on Atoms, Molecules, and Ions* (Springer, Berlin, 1985)
350. C.G. Sluijter, R.M. Jonkman, Sound absorption measurements in para hydrogen at 170°K. Physica **30**, 1670 (1964)
351. H.J.M. Hanley, R.D. McCarty, H. Interman, The viscosity and thermal conductivity of dilute gaseous hydrogen from 15 to 5000 K. J. Res. Nat. Bur. Stand. Sect. A **74**, 331 (1970)
352. M.J. Assael, S. Mixafendi, W.A. Wakeham, The viscosity of normal deuterium in the limit of zero density. J. Phys. Chem. Ref. Data **16**, 189 (1987)
353. S.C. Saxena, W.K. Saxena, Thermal conductivity data for hydrogen and deuterium in the range 100–1100 degrees. C. J. Phys. A **3**, 309 (1970)
354. B.C. Eu, Y.G. Ohr, Generalized hydrodynamics, bulk viscosity, and sound wave absorption and dispersion in dilute rigid molecular gases. Phys. Fluids **13**, 744 (2001)
355. G. Emanuel, Bulk viscosity of a dilute polyatomic gas. Phys. Fluids. A **2**, 2252 (1990)
356. W.E. Meador, G.A. Miner, L.W. Townsend, Bulk viscosity as a relaxation parameter: fact or fiction? Phys. Fluids. A **8**, 258 (1996)
357. G. Emanuel, Bulk viscosity as a relaxation parameter: fact or fiction? [Phys. Fluids **8**, 258 (1996)]. Phys. Fluids A **8**, 1984 (1996)
358. G. Emanuel, Bulk viscosity in the Navier–Stokes equations. Int. J. Engng Sci. **36**, 1313 (1998)
359. R.E. Graves, B.M. Argrow, Bulk viscosity: Past to present. J. Thermophys. Heat Transfer **13**, 337 (1999)

360. H.J. Bauer, Phenomenological theory of the relaxation phenomena in gases, in *Physical Acoustics II Part A*, ed. by W.P. Mason (Academic Press, New York, 1965), pp. 47–131
361. G.M. Kremer, O.V. Kunova, E.V. Kustova, G.P. Oblapenko, The influence of vibrational state-resolved transport coefficients on the wave propagation in diatomic gases. Physica A **490**, 92 (2018)
362. L.D. Landau, E.M. Lifshitz, *Quantum Mechanics, Non-Relativistic Theory* (Pergamon, Oxford, 1977)
363. F.D. Shields, Measurements of thermal relaxation in CO_2 extended to 300°C. J. Acoust. Soc. Am. **31**, 248 (1959)
364. F.D. Shields, Sound absorption in the halogen gases. J. Acoust. Soc. Am. **32**, 180 (1960)
365. JSME Data Book, *Thermophysical Properties of Fluids* (Japan Society of Mechanical Engineers, Tokyo, 1983)
366. H.O. Kneser, Zur Dispersionstheorie des Schalles. Ann. Physik **403**, 761 (1931)
367. D.G. Bougin, Kinetics of absorption of ultrasonic waves. Nature **122**, 133 (1928)
368. J.J. Markham, T.R. Beyer, R.B. Lindsay, Absorption of sound in fluids. Rev. Modern Phys. **23**, 353 (1951)
369. G.A. Bird, *Molecular Gas Dynamics and the Direct Simulation of Gas Flows* (Clarendon Press, Oxford, 1994)
370. F.J. McCormack, Kinetic moment equations for a gas of polyatomic molecules with many internal degrees of freedom. Phys. Fluids **13**, 1446 (1970)
371. W.M. Haynes, D.R. Lide (eds), *CRC Handbook of Chemistry and Physics*, 91st edn. (CRC Press, Boca Raton, 2010)
372. J.D. Au, M. Torrilhon, W. Weiss, The shock tube study in extended thermodynamics. Phys Fluids **13**, 2423 (2001)
373. S. Taniguchi, T. Arima, T. Ruggeri, M. Sugiyama, Shock wave structure in rarefied polyatomic gases with large relaxation time for the dynamic pressure. J. Phys. Conf. Ser. **1035**, 012009 (2018)
374. R. Nagaoka, S. Taniguchi, T. Ruggeri, Similarity solution of strong spherical shock waves in a rarefied polyatomic gas based on extended thermodynamics. AIP Conf. Proc. **2153**, 020014 (2019)
375. P. Andries, P. Le Tallec, J.-P. Perlat, B. Perthame, The Gaussian-BGK model of Boltzmann equation with small Prandtl number. Eur. J. Mech. B/Fluids **19**, 813 (2000)
376. E. Kustova, M. Mekhonoshina, A. Kosareva, Relaxation processes in carbon dioxide. Phys. Fluids **31**, 046104 (2019)
377. I.V. Alekseev, A.A. Kosareva, E.V. Kustova, E.A. Nagnibeda, Shock waves in carbon dioxide: Simulations using different kinetic-theory models. AIP Conference Proceedings **2132**, 060005 (2019)
378. M. Sugiyama, A study of steady flow of a polyatomic gas by extended thermodynamics with six fields. J. Phys. Conf. Ser. **1035**, 012008 (2018)
379. H.W. Liepmann, A. Loshko, *Elements of Gasdynamics* (Wiley, New York, 1957)
380. L.P. Kadanoff, P.C. Martin, Hydrodynamic equations and correlation functions. Ann. Phys. **24**, 419 (1963)
381. R.D. Mountain, Spectral distribution of scattered light in a simple fluid. Rev. Mod. Phys. **38**, 205 (1966)
382. R.D. Mountain, Density fluctuations in fluids having an internal degree of freedom. J. Res. Natl. Bur. Stand. **72**, 95 (1968)
383. S. Yip, M. Nelkin, Application of a kinetic model to time-dependent density correlations in fluids. Phys. Rev. **135**, 1241 (1964)
384. A. Sugawara, S. Yip, Kinetic model analysis of light scattering by Molecular Gases. Phys. Fluids **10**, 1911 (1967)
385. R.C. Desai, L.R. Kapral, Translational hydrodynamics and light scattering from molecular fluids. Phys. Rev. A **6**, 2377 (1972)
386. C.D. Boley, R.C. Desai, G. Tenti, Kinetic models and Brillouin scattering in a molecular gas. Can. J. Phys. **50**, 2158 (1972)

387. G. Tenti, C.D. Boley, R.C. Desai, On the kinetic model description of Rayleigh-Brillouin scattering from molecular gases. Can. J. Phys. **52**, 285 (1974)
388. W. Marques Jr., Light scattering and sound propagation in polyatomic gases with classical degrees of freedom. Continuum Mech. Thermodyn. **16**, 517 (2004)
389. A.S. Fernandes, W. Marques Jr., Kinetic model analysis of time-dependent problems in polyatomic gases. Physica A **373**, 97 (2007)
390. W. Weiss, I. Müller, Light scattering and extended thermodynamics. Continuum Mech. Thermodyn. **7**, 123 (1995)
391. T.J. Greytak, G.B. Benedek, Spectrum of light scattered from thermal fluctuations in gases. Phys. Rev. Lett. **17**, 179 (1966)
392. T.J. Greytak, Ph.D. thesis, Massachusetts Institute of Technology (1967)
393. Q.H. Lao, P.E. Schoen, B. Chu, Rayleigh-Brillouin scattering of gases with internal relaxation. J. Chem. Phys. **64**, 3547 (1976)
394. E.H. Hara, A.D. May, H.P.F. Knaap, Rayleigh–Brillouin scattering in compressed H_2, D_2, and HD. Can. J. Phys. **49**, 420 (1971)
395. E.H. Hara, A.D. May, Rayleigh–Brillouin scattering in compressed H_2, D_2, and HD II. Can. J. Phys. **50**, 1227 (1972)
396. M. Hubert, A.D. May, The Rayleigh–Brillouin spectrum of normal and parahydrogen: a test of model solutions of the Wang Chang Uhlenbeck equation. Can. J. Phys. **53**, 343 (1975)
397. Z. Gu, M.O. Vieitez, E.J. van Duijn, W. Ubachs, A Rayleigh-Brillouin scattering spectrometer for ultraviolet wavelengths. Rev. Sci. Instrum. **83**, 053112 (2012)
398. Z. Gu, W. Ubachs, W. van de Water, Rayleigh–Brillouin scattering of carbon dioxide. Opt. Lett. **39**, 3301 (2014)
399. X. Pan, M.N. Shneider, R.B. Miles, Coherent Rayleigh-Brillouin scattering in molecular gases. Phys. Rev. A **69**, 033814 (2004)
400. X. Pan, M.N. Shneider, R.B. Miles, Power spectrum of coherent Rayleigh-Brillouin scattering in carbon dioxide. Phys. Rev. A **71**, 045801 (2005)
401. M.O. Vieitez, E.J. van Duijn, W. Ubachs, B. Witschas, A. Meijer, A.S. de Wijn, N.J. Dam, W. van de Water, Coherent and spontaneous Rayleigh-Brillouin scattering in atomic and molecular gases and gas mixtures. Phys. Rev. A **82**, 043836 (2010)
402. T. Arima, E. Barbera, F. Brini, M. Sugiyama, The role of the dynamic pressure in stationary heat conduction of a rarefied polyatomic gas. Phys. Lett. A **378**, 2695 (2014)
403. E. Barbera, F. Brini, On stationary heat conduction in 3D symmetric domains: an application of extended thermodynamics. Acta Mech. **215**, 241 (2010)
404. E. Barbera, F. Brini, G. Valenti, Some non-linear effects of stationary heat conduction in 3D domains through extended thermodynamics. EPL **98**, 54004 (2012)
405. E. Barbera, F. Brini, Heat transfer in gas mixtures: Advantages of an extended thermodynamics approach. Phys. Lett. A **375**, 827 (2011)
406. E. Barbera, F. Brini, Heat transfer in a binary gas mixture between two parallel plates: an application of linear extended thermodynamics. Acta Mech. **220**, 87 (2011)
407. E. Barbera, F. Brini, Heat transfer in multi-component gas mixtures described by extended thermodynamics. Meccanica **47**, 655 (2012)
408. C. Truesdell, ZAMM **33**, 345 (1953)
409. E. Barbera, I. Müller, Heat conduction in a non-inertial frame, in *Rational Continua, Classical and New*. Podio-Guidugli, Brocato (Springer, Milano 2002), pp. 1–10
410. W.P. Teagan, G.S. Springer, Heat-transfer and density-distribution measurements between parallel plates in the transition regime. Phys. Fluids **11**, 497 (1968)
411. H.B. Callen, T.A. Welton, Irreversibility and generalized noise. Phys. Rev. **83**, 34 (1951)
412. M.S. Green, Brownian motion in a gas of noninteracting molecules. J. Chem. Phys. **19**, 1036 (1951)
413. R. Kubo, The fluctuation-dissipation theorem and Brownian motion, 1–16, in *1965 Tokyo Summer Lectures in Theoretical Physics*, ed. by R. Kubo (Part I. Many-Body Theory) (Shokabo and Benjamin, New York, 1966)
414. R.F. Fox, Gaussian stochastic processes in physics. Phys. Reports **48**, 179 (1978)

415. M. Kac, J. Logan, *Fluctuation phenomena, studies in statistical mechanics* Chap. 1, vol. VII, ed. by E.W. Montroll, J.L. Lebowitz (North-Holland, Amsterdam, 1979)
416. J.M. Ortiz de Zarate, J.V. Sengers, *Hydrodynamic Fluctuations* (Elsevier, Amsterdam, 2006)
417. G. Karniadakis, A. Beskok, N. Aluru, *Microflows and Nanoflows: Fundamentals and Simulation* (Springer, New York, 2005)
418. W. Kang, U. Landman, Universality crossover of the pinch-off shape profiles of collapsing liquid nanobridges in vacuum and gaseous environments. Phys. Rev. Lett. **98**, 064504 (2007)
419. R.D. Astumian, P. Hänggi, Brownian motors. Phys. Today **55**, 33 (2002)
420. G. Oster, Darwin's motors. Nature (London) **417**, 25 (2002)
421. A.L. Garcia, M. Malek-Mansour, G.C. Lie, E. Clementi, Numerical integration of the fluctuating hydrodynamic equations. J. Stat. Phys. **47**, 209 (1987)
422. A.L. Garcia, C. Penland, Fluctuating hydrodynamics and principal oscillation pattern analysis. J. Stat. Phys. **64**, 1121 (1991)
423. N. Sharma, N.A. Patankar, Direct numerical simulation of the Brownian motion of particles by using fluctuating hydrodynamic equations. J. Comput. Phys. **201**, 466 (2004)
424. G. De Fabritiis, R. Delgado-Buscalioni, P.V. Coveney, Multiscale modeling of liquids with molecular specificity. Phys. Rev. Lett. **97**, 134501 (2006)
425. G. De Fabritiis, M. Serrano, R. Delgado-Buscalioni, P.V. Coveney, Fluctuating hydrodynamic modeling of fluids at the nanoscale. Phys. Rev. E **75**, 026307 (2007)
426. J.B. Bell, A.L. Garcia, S.A. Williams, Numerical methods for the stochastic Landau-Lifshitz Navier–Stokes equations. Phys. Rev. E **76**, 016708 (2007)
427. J.B. Bell, A.L. Garcia, S.A. Williams, Computational fluctuating fluid dynamics. ESAIM: M2AN **44**, 1085 (2010)
428. R. Schmitz, Fluctuations in nonequilibrium fluids. Phys. Reports **171**, 1 (1988)
429. P. Español, Stochastic differential equations for non-linear hydrodynamics. Physica A **248**, 77 (1998)
430. A. Ikoma, T. Arima, S. Taniguchi, N. Zhao, M. Sugiyama, Fluctuating hydrodynamics for a rarefied gas based on extended thermodynamics. Phys. Lett. A **375**, 2601 (2011)
431. T. Arima, A. Ikoma, S. Taniguchi, M. Sugiyama, N. Zhao, Fluctuating hydrodynamics based on extended thermodynamics. Note Mat. **32**, 227 (2012)
432. H.S. Green, *The Molecular Theory of Fluids* (Dover, New York, 1969)
433. J.P. Boon, S. Yip, *Molecular Hydrodynamics* (Dover, New York, 1991)
434. J.-P. Hansen, I.R. McDonald, *Theory of Simple Liquids* 3rd edn. (Academic Press, London, 2006)
435. D.J. Evans, G. Morriss, *Statistical Mechanics of Nonequilibrium Liquids* (Cambridge University, Cambridge, 2008)
436. M.C. Carrisi, S. Pennisi, T. Ruggeri, M. Sugiyama, Extended thermodynamics of dense gases in the presence of dynamic pressure. Ric. Mat. **64**, 403 (2015)
437. N. Hirai, H. Eyring, Bulk viscosity of liquids. J. Appl. Phys. **29**, 810 (1958)
438. A. Onuki, Dynamic equations and bulk viscosity near the gas-liquid critical point. Phys. Rev. E **55**, 403 (1997)
439. A. Onuki, Dynamic van der Waals theory. Phys. Rev. E **75**, 036304 (2007)
440. T. Arima, Characteristic velocity and dispersion relation of linear waves in extended thermodynamics of a van der Waals gas. Ric. Mat. (2020). https://doi.org/10.1007/s11587-020-00518-4
441. T. Arima, M. Sugiyama, Nonequilibrium pressure and temperatures in extended thermodynamics of gases with six fields. Ric. Mat. **68**, 211 (2019)
442. T. Arima, Six-field extended thermodynamics models representing molecular energy exchange in a dense polyatomic gas. J. Phys. Conf. Ser. **1035**, 012002 (2018)
443. R. Bass, J. Lamb, Ultrasonic relaxation of the vibrational specific heat of carbon dioxide, sulphur hexafluoride, nitrous oxide, cyclopropane and methyl chloride in the liquid state. Proc. R. Soc. A **247**, 168 (1958)
444. D.Y. Peng, D.B. Robinson, A rigorous method for predicting the critical properties of multicomponent systems from an equation of state. AIChE J. **23**, 137 (1977)

445. J.M. Prausnitz, R.N. Lichtenthaler, E.G. de Azevedo, *Molecular Thermodynamics of Fluid-Phase Equilibria* (Prentice Hall PTR, Upper Saddle River, 1998)
446. R.S. Chatwell, J. Vrabec, Bulk viscosity of liquid noble gases. J. Chem. Phys. **152**, 094503 (2020)
447. S. Pennisi, T. Ruggeri, Relativistic Eulerian rarefied gas with internal structure. J. Math. Phys. **59**, 043102 (2018)
448. S.R. de Groot, W.A. Van Leeuwen, C.G. Van Weert, *Relativistic Kinetic Theory: Principles and Applications* (North-Holland, Amsterdam, 1980)
449. S. Weinberg, Entropy generation and the survival of protogalaxies in an expanding universe. Astrophys. J. **168**, 175 (1971)
450. M.C. Carrisi, S. Pennisi, T. Ruggeri, Integrability properties for relativistic extended thermodynamics of polyatomic gas. Ric. Mat. **68**, 57 (2019)
451. M.C. Carrisi, S. Pennisi, T. Ruggeri, Production terms in relativistic extended thermodynamics of gas with internal structure via a new BGK model. Ann. Phys. **405**, 298 (2019)
452. S. Pennisi, T. Ruggeri, A new BGK model for relativistic kinetic theory of monatomic and polyatomic gases. J. Phys. Conf. Ser. **1035**, 012005 (2018)
453. C. Marle, Modèle cinétique pour l'ètablissement des lois de la conduction de la chaleur et de la viscositè en thèorie de la relativitè. C.R. Acad. Sci., Ser. I (Mathematique) **260**, 6539 (1965)
454. J.L. Anderson, H.R. Witting, A relativistic relaxational time model for the Boltzmann equation. Physica **74**, 466 (1974)
455. M. Takamoto, S. Inutsuka, The relativistic kinetic dispersion relation: comparison of the relativistic Bhatnagar-Gross-Krook model and Grad's 14-moment expansion. Physica A **389**, 4580 (2010)
456. S. Pennisi, T. Ruggeri, Ultra-relativistic limit of extended thermodynamics of rarefied polyatomic gas, in *LE MATEMATICHE*, vol. LXXIV(II) (2019), p. 369
457. J.M. Burgers, *Flow Equations for Composite Gases* (Academic Press, New York, 1969)
458. D. Kannappan, T.K. Bose, Transport properties of a two-temperature argon plasma. Phys. Fluids **20**, 1668 (1977)
459. D. Kannappan, T.K. Bose, Transport properties of a two-temperature helium plasma. Phys. Fluids **23**, 1473 (1980)
460. T.K. Bose, R.V. Seeniraj, Two-temperature noble gas plasmas. I. Thermodynamics and transport coefficients. J. Indian Inst. Sci. **64**, 181 (1983)
461. A. Sellitto, V.A. Cimmelli, D. Jou, Influence of electron and phonon temperature on the efficiency of thermoelectric conversion. Int. J. Heat Mass Transfer **80**, 344 (2015)
462. K. Xu, X. He, C. Cai, Multiple temperature kinetic model and gas-kinetic method for hypersonic non-equilibrium flow computations. J. Comput. Phys. **227**, 6779 (2008)
463. I. Müller, A thermodynamic theory of mixtures of fluids. Arch. Ration. Mech. Anal. **28**, 1 (1968)
464. K. Hutter, *Continuum Methods of Physical Modeling* (Springer, New York, 2004)
465. K.R. Rajagopal, L. Tao, *Mechanics of Mixtures* (World Scientific, Singapore, 1995)
466. K. Wilmanski, *Continuum Thermodynamics—Part I: Foundations* (World Scientific, Singapore, 2008)
467. R.J. Atkin, R.E. Craine, Continuum theories of mixtures: basic theory and historical development. Quart. J. Mech. Appl. Math. **29**, 209 (1976)
468. T. Ruggeri, Multi-temperature mixture of fluids. Theoret. Appl. Mech. **36**, 207 (2009)
469. T. Ruggeri, Some recent results on multi-temperature mixture of fluids, in *Continuous Media with Microstructure*, ed. by A. Bettina (Springer, Berlin, 2010), pp. 39–57
470. S. Simić, M. Pavić-Čolić, D. Madjarević, Non-equilibrium mixtures of gases: modelling and computation. Riv. Mat. Univ. Parma **6**(1), 135214 (2015). https://open.uns.ac.rs/handle/123456789/6216
471. T.K. Bose, *High-Temperature Gas Dynamics* (Springer, Berlin, 2003)

472. C. Eckart, The thermodynamics of irreversible processes II. Fluid Mixtures. Phys. Rev. **58**, 269 (1940)
473. L. Onsager, Reciprocal Relations in Irreversible Processes I. Phys. Rev. **37**, 405 (1931); Reciprocal Relations in Irreversible Processes II. Phys. Rev. **38**, 2265 (1931)
474. M. Pavić-Čolić, Multi-velocity and multi-temperature model of the mixture of polyatomic gases issuing from kinetic theory. Phys. Lett A **383**, 2829 (2019)
475. T. Ruggeri, S. Simić, Nonlinear wave propagation in binary mixtures of Euler fluids. Continuum Mech. Thermodyn. **16**, 125 (2004)
476. T. Ruggeri, Global existence, stability and non linear wave propagation in binary mixtures, in *Proceedings of the International Meeting in honour of the Salvatore Rionero 70th Birthday, Napoli 2003*, ed. by P. Fergola, F. Capone, M. Gentile, G. Guerriero (World Scientific, Singapore, 2004), pp. 205–214
477. T. Ruggeri, Some recent mathematical results in mixtures theory of Euler fluids, in *Proceedings of WASCOM 2003*, ed. by R. Monaco, S. Pennisi, S. Rionero, T. Ruggeri (World Scientific, Singapore, 2004), pp. 441–454
478. T. Ruggeri, S. Simić, Mixture of gases with multi-temperature: identification of a macroscopic average temperature, in *Proceedings Mathematical Physics Models and Engineering Sciences* (Liguori Editore, Napoli, 2008), p. 455
479. T. Ruggeri, S. Simić, Mixture of gases with multi-temperature: Maxwellian iteration, in *Asymptotic Methods in Non Linear Wave Phenomena*, ed. by T. Ruggeri, M. Sammartino (World Scientific, Singapore, 2007), p. 186
480. T. Ruggeri, S. Simić, Non-equilibrium temperatures in the mixture of gases via Maxwellian iteration. Ric. Mat. **66**, 293–312 (2017)
481. V. Giovangigli, M. Massot, Entropic structure of multicomponent reactive flows with partial equilibrium reduced chemistry. Math. Meth. Appl. Sci. **27**, 739 (2004)
482. F. Conforto, R. Monaco, F. Schürrer, I. Ziegler, Steady detonation waves via the Boltzmann equation for a reacting mixture. J. Phys. A **36**, 5381 (2003)
483. M. Groppi, G. Spiga, Kinetic theory of a chemically reacting gas with inelastic transitions. TTSP **30**, 305 (2001)
484. F. Conforto, A. Jannelli, R. Monaco, T. Ruggeri, On the Riemann problem for a system of balance laws modelling a reactive gas mixture. Physica A **373**, 67 (2007)
485. D. Madjarević, S. Simić, Shock structure in helium-argon mixture—a comparison of hyperbolic multi-temperature model with experiment. EPL (Europhys. Lett.) **102**, 44002 (2013)
486. L.N. Harnet, E. Muntz, Experimental investigation of normal shock wave velocity distribution functions in mixtures of argon and helium. Phys. Fluids **10**, 565 (1972)
487. K. Abe, H. Oguchi, *An Analysis of Shock Waves in Binary Gas Mixtures with Special Regard to Temperature Overshoot, Report No. 511.* (Institute of Space and Aeronautical Science/University of Tokyo, Tokyo, 1974)
488. G.A. Bird, The structure of normal shock waves in a binary gas mixture. J. Fluid Mech. **31**, 657 (1968)
489. D. Madjarević, T. Ruggeri, S. Simić, Shock structure and temperature overshoot in macroscopic multi-temperature model of mixtures. Phys. Fluids **26**, 106102 (2014)
490. A. Raines, Study of a shock wave structure in gas mixtures on the basis of the boltzmann equation. Eur. J. Mech. B. Fluids **21**, 599 (2002)
491. R. Monaco, Shock-wave propagation in gas mixtures by means of a discrete velocity model of the Boltzmann equation. Acta Mech. **55**, 239 (1985)
492. M. Groppi, K. Aoki, G. Spiga, V. Tritsch, Shock structure analysis in chemically reacting gas mixtures by a relaxation-time kinetic model. Phys. Fluids **20**, 117103 (2008)
493. M. Hantke, C. Matern, G. Warnecke, Analytical results for the Riemann problem for a weakly hyperbolic two-phase flow model of a dispersed phase in a carrier fluid, in *Continuum Mechanics, Applied Mathematics and Scientific Computing: Godunov's Legacy*, ed. by G.V. Demidenko, E. Romenski, E. Toro, M. Dumbser (Springer, Cham, 2020), p. 169. ISBN 978-3-030-38869-0

494. M. Hantke, F. Thein, A numerical method for two phase flows with phase transition including phase creation, in *Continuum Mechanics, Applied Mathematics and Scientific Computing: Godunov's Legacy*, ed. by G.V. Demidenko, E. Romenski, E. Toro, M. Dumbser (Springer, Cham, 2020), p. 177. ISBN 978-3-030-38869-0
495. P. Degond, S. Motsch, Large-scale dynamics of the persistent Turing Walker model of fish behavior. J. Stat. Phys. **131**, 989 (2008)
496. Y. Kuramoto, International symposium on mathematical problems in mathematical physics. Lecture Notes Theor. Phys. **30**, 420 (1975)
497. J. Toner, Y. Tu, Flocks, herds, and schools: a quantitative theory of flocking. Phys. Rev. **58**, 4828 (1998)
498. C.M. Topaz, A.L. Bertozzi, Swarming patterns in a two-dimensional kinematic model for biological groups. SIAM J. Appl. Math. **65**, 152 (2004)
499. T. Vicsek, A. Czirók, E. Ben-Jacob, I. Cohen, O. Schochet, Novel type of phase transition in a system of self-driven particles. Phys. Rev. Lett. **75**, 1226 (1995)
500. A.T. Winfree, Biological rhythms and the behavior of populations of coupled oscillators. J. Theor. Biol. **16**, 15 (1967)
501. N.E. Leonard, D.A. Paley, F. Lekien, R. Sepulchre, D.M. Fratantoni, R.E. Davis, Collective motion, sensor networks and ocean sampling. Proc. IEEE **95**, 48 (2007)
502. D.A. Paley, N.E. Leonard, R. Sepulchre, D. Grunbaum, J.K. Parrish, Oscillator models and collective motion. IEEE Control Syst. **27**, 89 (2007)
503. L. Perea, P. Elosegui, G. Gómez, Extension of the Cucker-Smale control law to space flight formation. J. Guidance Control Dyn. **32**, 526 (2009)
504. J.A. Carrillo, M. Fornasier, J. Rosado, G. Toscani, Asymptotic flocking dynamics for the kinetic Cucker-Smale model. SIAM J. Math. Anal. **42**, 218 (2010)
505. J. Cho, S.-Y. Ha, F. Huang, C. Jin, D. Ko, Emergence of bi-cluster flocking for the Cucker-Smale model. Math. Models Methods Appl. Sci. **26**(6), 1191–1218 (2016)
506. F. Cucker, J.-G. Dong, Avoiding collisions in flocks. IEEE Trans. Autom. Control **55**, 1238 (2010)
507. M. Fornasier, J. Haskovec, G. Toscani, Fluid dynamic description of flocking via Povzner-Boltzmann equation. Physica D **240**, 21 (2011)
508. S.-Y. Ha, J.-G. Liu, A simple proof of Cucker-Smale flocking dynamics and mean field limit. Commun. Math. Sci. **7**, 297 (2009)
509. S.-Y. Ha, E. Tadmor, From particle to kinetic and hydrodynamic description of flocking. Kinet. Rel. Models **1**, 415 (2008)
510. Z. Li, X. Xue, Cucker-Smale flocking under rooted leadership with fixed and switching topologies. SIAM J. Appl. Math. **70**, 3156 (2010)
511. S. Motsch, E. Tadmor, A new model for self-organized dynamics and its flocking behavior. J. Stat. Phys. **144**, 923 (2011)
512. J. Shen, Cucker-Smale flocking under hierarchical leadership. SIAM J. Appl. Math. **68**, 694 (2007)
513. D. Bhaya, Light matters: Phototaxis and signal transduction in unicellular cyanobacteria. Mol. Microbiol. **53**, 745754 (2004)
514. M. Burriesci, D. Bhaya, Tracking phototactic responses and modeling motility of Synechocystis sp. Strain PCC6803. J. Photochem. Photobio. **91**, 7786 (2008)
515. S.-Y. Ha, D. Levy, Particle, kinetic and fluid models for phototaxis. Discrete Contin. Dyn. Syst. B **12**, 77 (2009)
516. D. Levy, T. Requeijo, Modeling group dynamics of phototaxis: from Particle systems to PDEs. Disc. Cont. Dyn. Sys. B **9**, 108128 (2008)
517. D. Levy, T. Requeijo, Stochastic models for phototaxis. Bull. Math. Bio. **70**, 16841706 (2008)
518. S.-Y. Ha, J. Kim, T. Ruggeri, Emergent behaviors of thermodynamic Cucker-Smale particles. SIAM J. Math. Anal. **50**, 3092 (2018)
519. S.-Y. Ha, J. Kim, C. Min, T. Ruggeri, X. Zhang, Uniform stability and mean-field limit of thermodynamic Cucker-Smale model. Quart. Appl. Math. **77**, 131 (2019)

520. S.-Y. Ha, J. Kim, C. Min, T. Ruggeri, X. Zhang, A global existence of classical solution to the hydrodynamic Cucker-Smale model in presence of temperature field. Anal. Appl. **16**, 757 (2018)
521. Y.-P. Choi, S.-Y. Ha, J. Jung, J. Kim, Global dynamics of the thermomechanical Cucker-Smale ensemble immersed in incompressible viscous fluids. Nonlinearity **32**, 1597 (2019)
522. Y.-P. Choi, S.-Y. Ha, J. Jung, J. Kim, On the coupling of kinetic thermomechanical Cucker-Smale equation and compressible viscous fluid system. J. Math. Fluid Mech. **22**, 4 (2020)
523. M.-J. Kang, S.-Y. Ha, J. Kim, W.J. Shim, Hydrodynamic limit of the kinetic thermomechanical Cucker-Smale model in a strong local alignment regime. Commun. Pure Appl. Anal. **19**(3), 1233–1256 (2020)
524. S.-Y. Ha, H. Park, T. Ruggeri, W. Shim, Emergent behaviors of thermodynamic Kuramoto ensemble on a regular ring lattice. J. Stat. Phys. **181**, 917 (2020)
525. T. Arima, T. Ruggeri, M. Sugiyama, S. Taniguchi, Galilean invariance and entropy principle for a system of balance laws of mixture type. Rend. Lincei Mat. Appl. **28**, 495 (2017)
526. K. Hutter, I. Müller, On mixtures of relativistic fluids. Helv. Phys. Acta **48**, 675 (1975)
527. V. Moratto, G.M. Kremer, Mixtures of relativistic gases in gravitational fields: Combined Chapman-Enskog and Grad method and the Onsager relations. Phys. Rev. E **91**, 052139 (2015)
528. G.M. Kremer, W. Marques Jr., Grad's moment method for relativistic gas mixtures of Maxwellian particles. Phys. Fluids **25**, 017102 (2013)
529. W. Muschik, Entropy identity inducing non-equilibrium thermodynamics of relativistic multi-component systems and their Newtonian limits (2018). arXiv:1806.01648v1
530. C.M. Ko, Q. Li, Relativistic Vlasov-Uehling-Uhlenbeck model for heavy-ion collisions. Phys. Rev. C **37**, 2270 (1988)
531. M. Nielsen, C. Providência, J. da Providência, Collective modes in hot and dense nuclear matter. Phys. Rev. C **47**, 200 (1993)
532. K. Saito, T. Maruyama, K. Soutome, Collective modes in hot and dense matter. Phys. Rev. C **40**, 407 (1989)
533. I. Müller, On the frame dependence of stress and heat flux. Arch. Rational Mech. Anal. **45**, 241 (1972)
534. A. Bressan, On relativistic heat conduction in the stationary and nonstationary cases, the objectivity principle and piezoelasticity. Lett. Nuovo Cimento **33**, 108 (1982)
535. T. Ruggeri, Generators of hyperbolic heat equation in nonlinear thermoelasticity. Rend. Sem. Mat. Padova, **68**, 79 (1982)
536. K.R. Rajagopal, On a hierarchy of approximate models for flows of incompressible fluids through porous solids. Math. Models Methods Appl. Sci. **17**, 215 (2007)
537. D.A. Signer, Stability of non-linear constitutive formulations for viscoelastic fluids, in *Springer Briefs in Applied Sciences and Technology* (2014), pp. 9–88
538. M.C. Carrisi, S. Pennisi, A 16 moments model in relativistic extended thermodynamics of rarefied polyatoinc gas. Ric. Mat. (2019). https://doi.org/10.1007/s11587-019-00468-6
539. P. Ván, Nonequilibrium thermodynamics: emergent and fundamental. Phil. Trans. R. Soc. A. **37820200066** (2020). https://doi.org/10.1098/rsta.2020.0066

Author Index

T. Ruggeri, M. Sugiyama, *Classical and Relativistic Rational Extended Thermodynamics of Gases*, https://doi.org/10.1007/978-3-030-59144-1

X

Y

Z

Subject Index

T. Ruggeri, M. Sugiyama, *Classical and Relativistic Rational Extended Thermodynamics of Gases*, https://doi.org/10.1007/978-3-030-59144-1